SMART GRIDS
Infrastructure, Technology, and Solutions

Electric Power and Energy Engineering

Series Editor
John D. McDonald

Smart Grids: Infrastructure, Technology, and Solutions, *Stuart Borlase*

SMART GRIDS
Infrastructure, Technology, and Solutions

Edited by
STUART BORLASE

CRC Press
Taylor & Francis Group
Boca Raton London New York

CRC Press is an imprint of the
Taylor & Francis Group, an **informa** business

Cover art: The editor would like to thank Rick Giammaria (courtesy of Pepco Holdings Inc) for providing the substation control room photo and Alstom Grid for providing the EMS control room photo for the book cover artwork.

CRC Press
Taylor & Francis Group
6000 Broken Sound Parkway NW, Suite 300
Boca Raton, FL 33487-2742

© 2013 by Taylor & Francis Group, LLC
CRC Press is an imprint of Taylor & Francis Group, an Informa business

No claim to original U.S. Government works

Printed in the United States of America on acid-free paper
Version Date: 20120706

International Standard Book Number: 978-1-4398-2905-9 (Hardback)

This book contains information obtained from authentic and highly regarded sources. Reasonable efforts have been made to publish reliable data and information, but the author and publisher cannot assume responsibility for the validity of all materials or the consequences of their use. The authors and publishers have attempted to trace the copyright holders of all material reproduced in this publication and apologize to copyright holders if permission to publish in this form has not been obtained. If any copyright material has not been acknowledged please write and let us know so we may rectify in any future reprint.

Except as permitted under U.S. Copyright Law, no part of this book may be reprinted, reproduced, transmitted, or utilized in any form by any electronic, mechanical, or other means, now known or hereafter invented, including photocopying, microfilming, and recording, or in any information storage or retrieval system, without written permission from the publishers.

For permission to photocopy or use material electronically from this work, please access www.copyright.com (http://www.copyright.com/) or contact the Copyright Clearance Center, Inc. (CCC), 222 Rosewood Drive, Danvers, MA 01923, 978-750-8400. CCC is a not-for-profit organization that provides licenses and registration for a variety of users. For organizations that have been granted a photocopy license by the CCC, a separate system of payment has been arranged.

Trademark Notice: Product or corporate names may be trademarks or registered trademarks, and are used only for identification and explanation without intent to infringe.

Library of Congress Cataloging-in-Publication Data

Smart grids : infrastructure, technology, and solutions / editor, Stuart Borlase.
     p. cm. -- (Electric power and energy engineering)
   Summary: "This book discusses the smart grid, currently the hottest topic in the electric utility market. It details in-development and existing technologies composing the smart grid solution framework used in electrical infrastructure. International contributors share their perspective to explore the role of the utility industry and standards in smart grid development. They detail the initiatives and organizations driving efforts. Detailing recent successful real-world implementations, this reference examines critical factors for successful smart grid deployment, from utility, regulatory, and consumer perspectives. It then explores the technologies and business drivers that will propel next-generation smart grid technology"-- Provided by publisher.
   Includes bibliographical references and index.
   ISBN 978-1-4398-2905-9 (hardback)
   1. Smart power grids. I. Borlase, Stuart.

TK3105.S56 2012
621.31--dc23   2012018249

Visit the Taylor & Francis Web site at
http://www.taylorandfrancis.com

and the CRC Press Web site at
http://www.crcpress.com

Contents

Foreword vii
Preface... ix
Acknowledgments... xi
Editor xiii
Contributors xv

Chapter 1 Overview of the Electric Utility Industry 1

Chapter 2 What Is Smart Grid, Why Now?... 15

Chapter 3 Smart Grid Technologies.. 61

Chapter 4 Smart Grid Barriers and Critical Success Factors 497

Chapter 5 Global Smart Grid Initiatives 531

Chapter 6 Smart Grid: Where Do We Go from Here?... 549

Index... 561

Foreword

I am proud to serve as Chairman, President, and Chief Executive Officer of Pepco Holdings, Inc. (PHI), a utility holding company with power delivery companies that have served southern New Jersey, the Delmarva Peninsula, and the Washington, District of Columbia, area for more than 100 years. While we have served the nation's capital since the Cleveland administration, the technologies of the late 1800s are not that much different from what we see in today's electric system. Copper cable, wooden poles, transformers, and meters were as integral to the system then as they are now.

Today, the industry is changing that picture and focusing on building a smarter grid that leverages a wide range of digitally based advanced technologies that will automate many electric system functions. This emerging "smart grid" has been described as the convergence of electric system and information technologies. In other words, Thomas Edison meets Bill Gates: the electric system that has served us so well since the nineteenth century will be integrated with digital technologies to provide utility customers the enhanced information, services, and reliability that are so critical in the twenty-first century.

The foundation of the smart grid starts with an Advanced Metering Infrastructure or AMI. Through the installation of smart meters and supporting technologies, customers will have access to enriched energy data; renewable generation and electric vehicles will be more easily integrated into the electric grid; and innovative rate structures and programs will be implemented that enable customers to assume greater control of their energy consumption.

In this new, smarter world, customers will be able to view their energy usage via mobile devices, the Internet, or special home monitors, and the utility will do the same through a meter data management system. And, instead of one total consumption figure once a month, customer hourly data will be transmitted wirelessly multiple times a day, or on demand if needed. The meter will also serve as a grid sensor, which can help to detect power fluctuations and outages and be used to remotely connect or disconnect service.

While a smart grid necessarily entails smart meters, it involves far more pervasive technologies across the entire grid—a supporting communications infrastructure, and an array of intelligent devices on distribution and transmission systems and within substations, which are able to isolate electrical problems, make automatic power adjustments, and speed overall restoration. Ultimately, a smart grid creates an advanced analytical platform for better operational awareness and real-time optimization of the distribution and transmission network. All this new technology, in short, will create a "smarter" electric grid.

Back in 2006, in response to the critical energy and environmental challenges facing the nation and the world, PHI began to promote a vision of an alternative energy future. We referred to this vision as our "Blueprint for the Future"—a partnership between the utility and its customers that leverages technology and information to drive energy efficiency and increase reliability. We envisioned a world where customers would be empowered to manage their energy use and costs, governments could reach energy reduction goals, and all would benefit from fewer carbon emissions and a cleaner environment. At the time, one could not look at the nation's energy companies, state regulatory and legislative bodies, and federal agencies and see strong alignment or coalescence of thought on the "smart grid" topic.

Now, as I write this in 2012, it is inspiring to look back and see how much has changed in our national smart grid discussion. Through the efforts of advocates and early adopters—and particularly the leadership of President Obama's administration—the smart grid has been transformed from a theoretical notion to a practical reality that is increasingly recognized as an essential step into the future.

It is important to recognize that although technology is critical to the smart grid, in isolation it is not transformational. To fully achieve the benefits of the smart grid requires a strong partnership between utilities, government, regulators, and the public—and new energy behaviors. Smart meters do not reduce carbon emissions, people do by utilizing enhanced energy information, responding to price signals, and leveraging financial incentives that will help us collectively achieve the goals of energy independence and emissions reduction. The concept of transformation is essential to the smart grid, and it is a theme that I cite often. Historically, the relationship between the utility and the customer has been transparent. People are used to the daily "miracle" of electricity—of flipping a switch to create light or pushing a button to activate a TV, computer, or other electronic device that helps our society function. Going forward, customers will have a wide variety of tools to manage their energy use, which help to reduce energy costs and meet the environmental and energy challenges before us.

The energy industry is at the center of this transformation—optimizing the integration of existing and future energy supplies, including the connection of renewables; leveraging demand side resources such as dynamic pricing and direct load control of customer appliances; and offering customers access to enhanced energy information and tools to lower energy use to help keep supply and demand in balance. Exciting opportunities also are forming in the electric transportation market, where utilities are being called upon to develop charging infrastructures for expected growth in plug-in electric vehicles. In sum, fertile ground is being tilled for growing new business partnerships, products, and services.

Of course, with change comes challenge. When the smart grid is fully deployed and customers take advantage of its capabilities, electricity sales growth will be significantly affected. Statistics show that even now, in many regions, energy consumption has declined in absolute terms. Reduced energy use is a good thing, but regulators need to recognize that the cost of delivering power is relatively fixed, regardless of consumption levels. Regulatory mechanisms need to be developed that recognize this reality and that recognize the importance of financially healthy utilities.

As you will discover in Chapter 1, the utility industry is arguably the most regulated in existence. Understanding the nature of this utility–regulator relationship is essential to a sufficient understanding of the industry. And a sufficient understanding of the industry is essential for all stakeholders seeking to advance smart grid adoption.

Many in the industry define the smart grid by the myriad evolving technologies that we are evaluating, designing, and implementing. I would suggest to you that the smart grid transformation can best be advanced if we define it less by the technology and more by the benefits that the technology brings:

Lower energy use
Better reliability
Faster restoration
New jobs
New markets

The smart grid is *the* best approach to meeting our industry's challenges and customer expectations—the shortest distance between our current state and our desired future. As we enter the second decade of the new century, we have a better collective understanding of what this means and of what can be. I am as optimistic as ever about the benefits to be realized through our collective efforts.

Joe Rigby
Chairman, President, and Chief Executive Officer
Pepco Holdings, Inc.
Washington, District of Columbia

Preface

A Global Perspective on How the Smart Grid Integrates 21st Century Technology with the 20th Century Power Grid

Never before has there been so much attention on the electric utility industry and such widespread use of the phrase "smart grid." From where we are today, the smart grid has promised the glory of electric grid transformation, utility heroes, incredible technology breakthroughs, dreams of consumer enamorment, a cleaner environment, and job creation. Utilities and vendors alike have been clamoring to be the first in smart grid, many ideas of smart grid have been presented, and there has been week after week of some kind of smart grid meeting or conference. There have certainly been some changes in the industry since the marketing hype around intelligent grid and smart grid first started, and there have been some serious efforts at making sense of it all through collaboration and demonstrations worldwide. However, fundamental questions still remain: what exactly is smart grid? how do we get there? and the favorite is smart grid real?

While smart grid is an elusive term with many different interpretations, there seems to be a gathering of critical mass in the thinking of the electric utility industry around the needs of smart grid and the reform in regulation, technologies, and applications. It is without a doubt that the smart grid initiative has gained much interest and backing worldwide, driven in part by regulatory initiatives in advanced metering and from funding opportunities, not only in the United States, but also in other regions of the world. Smart grid has led to the focus on more efficient and environmentally friendly electric energy solutions and has helped heighten the awareness and commitment from consumers in energy reduction. Changes in the utility industry have historically been very slow. Yes, some technologies touted as smart grid have been around for many years, but I think that in most cases with these existing technologies there has not been a concerted drive to move away from pilots and demonstration projects to more scalable and widespread deployments with sustainable benefits. I define smart grid in terms of not what it is, but what it achieves and the benefits it brings to the utility, consumer, society, and environment. The smart grid should focus on advancements in technologies and synergies in the integration of IT and operations applications that will bring more far-reaching benefits than imagined today and bring about tremendous change to the industry. Are we at this level of smart grid today—of course not. Will we get there—who knows at this point in time. Smart grid seems to be in a dormant phase, spurred to this point by an initial focus on deploying smart meters and system-wide AMI through major regulatory and funding initiatives, currently held in check by global market and financial crises, and hesitant to progress based on a mixed review of past successes and failures.

This book discusses the current buzzword in the electric utility market, "smart grid," and answers the following questions: what is smart grid? why is the concept of a smart grid receiving so much attention? and what are utilities, vendors, and regulators doing about smart grid? This book is a blend of views from numerous authors, as intended from the outset. It describes the impetus for change in the electric utility industry and discusses the business drivers, benefits, and market outlook of the smart grid initiative. It identifies the technical framework of enabling technologies and smart solutions and describes the role of technology developments and coordinated standards in smart grid, including various initiatives and organizations helping to drive the smart grid effort. This book presents both current technologies and forward-looking ideas on new technologies, and discusses barriers and critical factors for a successful smart grid from a utility, regulatory, and consumer perspective. This book concludes with a summary of recent smart grid initiatives around the world and an outlook of the drivers and technologies for the next generation smart grid. While this book is focused more on the smart grid market in the United States, I included as much of a global perspective as possible.

The smart grid effort still has a long way to go. It will continually evolve, driven by economies and regulations across the world. Who knows the final state of the effort, but it looks like it has sufficient momentum to bring significant advances in technology and changes in the industry. There will be winners and there will be losers not just in vendor technologies and their smart grid solutions, but also in the utilities and their smart grid deployments. Consumers will be among the winners and losers, too. I think the focus in the near term should be to secure buy-in from regulators, policy makers, and consumers for the need of smart grid in terms of the value it brings to all parties. Smart grid to consumers should be more than just incentives or savings on their electric bills—smart grid should enable consumer choice rather than dictate consumer choice. I think the momentum in smart grid will be revealed in the next few years after the hype and funding have faded; smart grid demonstrations have been proven or disproven; and the myriad smart grid meetings, conferences, expositions, and trade shows have diminished enough that the utility industry has time to sit down and think more about what smart grid is today, what it should be in the future, and how to get there.

This book brings together the knowledge and views of a diverse array of experts and leaders in their respective fields. While I have contributed personally to several sections of this book, I am indebted to the efforts and perseverance of a long list of over 80 people who contributed material. It has been a challenge to solicit material from contributors during the throes of smart grid initiatives around the world.

Note that the author contributions in this book do not necessarily reflect the views and opinions of the affiliations of the contributing authors.

Stuart Borlase can be contacted at stuart.borlase@gmail.com

Acknowledgments

The editor would like to thank Rick Giammaria (courtesy of Pepco Holdings, Inc.) for providing the substation control room photo and Alstom Grid for providing the EMS control room photo for the book cover artwork.

Editor

Stuart Borlase works as a Business Development Manager for the Siemens Transmission division in Cary, North Carolina. Stuart has overall responsibility for the strategy and development of the Siemens U.S. substation turnkey projects business with a key focus on providing unique solutions that meet customers' needs in the ever-changing transmission market.

Stuart has more than 20 years of technical and business experience in the electric utility T&D industry. His experience includes leadership positions in sales, business development, marketing, product development, engineering, and consulting. He previously worked as Business Development Director responsible for strategic direction, leadership, and growth in the development and deployment of GE Energy's global smart grid solutions.

Stuart received his Master of Engineering and Doctor of Engineering degrees in Electrical Engineering from Texas A&M University. He is a Senior Member of the IEEE and a registered Professional Engineer.

Contributors

Witold P. Bik
S&C Electric Company
Alameda, California

Stuart Borlase
Siemens Energy, Inc.
Raleigh, North Carolina

Steven Bossart
National Energy Technology Laboratory
United States Department of Energy
Morgantown, West Virginia

Thomas Bradley
Department of Mechanical Engineering
Colorado State University
Fort Collins, Colorado

Stephen Byrum
GE Energy—Digital Energy
Rockville, Virginia

Mary Carpine-Bell
GE Energy—Digital Energy
Atlanta, Georgia

David P. Chassin
Pacific Northwest National Laboratory
Battelle Memorial Institute
Richland, Washington

Yousu Chen
Pacific Northwest National Laboratory
Battelle Memorial Institute
Seattle, Washington

John Chowdhury
Center for Utility Innovation
Irving, Texas

Catherine Dalton
Beckwith Electric Company, Inc.
Duluth, Georgia

Keith Dodrill
National Energy Technology Laboratory
United States Department of Energy
Morgantown, West Virginia

Michael G. Ennis
S&C Electric Company
Chicago, Illinois

Johan Enslin
Energy Production and Infrastructure Center
William States Lee College of Engineering
University of North Carolina at Charlotte
Charlotte, North Carolina

Jiyuan Fan
GE Energy—Digital Energy
Atlanta, Georgia

Xiaoming Feng
ABB Corporate Research Center
Raleigh, North Carolina

Harry Forbes
ARC Advisory Group
Dedham, Massachusetts

Jay Giri
Alstom Grid
Redmond, Washington

Erich Gunther
EnerNex
Knoxville, Tennessee

James P. Hanley
GE Energy—Digital Energy
Atlanta, Georgia

Tim Heidel
Advanced Research Projects Agency— Energy
United States Department of Energy
Washington, District of Columbia

Gerald T. Heydt
School of Electrical, Computer and Energy Engineering
Arizona State University
Tempe, Arizona

Miriam Horn
Environmental Defense Fund
New York, New York

Gale Horst
Electric Power Research Institute
Knoxville, Tennessee

Régis Hourdouillie
Ericsson
Paris, France

Zhenyu (Henry) Huang
Pacific Northwest National Laboratory
Battelle Memorial Institute
Richland, Washington

Carroll Ivester
Consultant
Jefferson, Georgia

Marco C. Janssen
UTInnovation B.V
Duiven, the Netherlands

Henry Jones
SmartSynch, Inc.
Jackson, Mississippi

Mladen Kezunovic
Department of Electrical and Computer Engineering
Texas A&M University
College Station, Texas

Chris King
eMeter, a Siemens business
San Mateo, California

Neil Kirby
Alstom Grid
Philadelphia, Pennsylvania

Soorya Kuloor
GRIDiant Corporation
Durham, North Carolina

Rajat Majumder
ABB Corporate Research Center
Raleigh, North Carolina

Art Maria
AT&T, Inc.
Redmond, Washington

Paul E. Marken
GE Energy—Digital Energy
Columbia City, Indiana

Christopher McCarthy
S&C Electric Company
Chicago, Illinois

John McDonald
GE Energy—Digital Energy
Atlanta, Georgia

Bob McFetridge
Beckwith Electric Company, Inc.
Middlesex, North Carolina

Mehrdad Mesbah
Alstom Grid
Massy, France

Joe Miller
Horizon Energy Group
Bloomington, Illinois

Marita Mirzatuny
Environmental Defense Fund
Austin, Texas

Rita Mix
AT&T, Inc.
Atlanta, Georgia

Salman Mohagheghi
Colorado School of Mines
Golden, Colorado

Rui Menezes de Moraes
Operador Nacional do Sistema Elétrico
Rio de Janeiro, Brazil

Contributors

Thomas Morris
Electrical and Computer Engineering
Mississippi State University
Mississippi State, Mississippi

Mirrasoul J. Mousavi
ABB Corporate Research Center
Raleigh, North Carolina

Lauren Navarro
Environmental Defense Fund
Sacramento, California

Charles W. Newton
Newton-Evans Research Company, Inc.
Ellicott City, Maryland

Reynaldo Nuqui
ABB Corporate Research Center
Raleigh, North Carolina

Mica Odom
Environmental Defense Fund
Austin, Texas

Jiuping Pan
ABB Corporate Research Center
Raleigh, North Carolina

Manu Parashar
Alstom Grid
Redmond, Washington

Michael Pesin
Seattle City Light
Seattle, Washington

Steve Pullins
Horizon Energy Group
Maryville, Tennessee

Casey Quinn
NSG Engineering Solutions, LLC
Fort Collins, Colorado

Steven Radice
Ventyx, an ABB company
Atlanta, Georgia

V.R. Ramanan
ABB Corporate Research Center
Raleigh, North Carolina

Bruce A. Renz
Renz Consulting, LLC
Columbus, Ohio

Dietmar Retzmann
Siemens AG
Erlangen, Germany

Greg Robinson
Xtensible Solutions
Cape Canaveral, Florida

Julio Romero Agüero
Quanta Technology
Raleigh, North Carolina

Walter Sattinger
Swissgrid AG
Laufenburg, Switzerland

James Stoupis
ABB Corporate Research Center
Raleigh, North Carolina

Tim Taylor
Ventyx, an ABB company
Raleigh, North Carolina

Matthew Thomson
Sterling Infosystems
Marietta, Georgia

Jean-Charles Tournier
ABB Corporate Research Center
Raleigh, North Carolina

Steve Turner
Beckwith Electric Company Inc.
Largo, Florida

David M. Velazquez
Pepco Holdings, Inc.
Washington, District of Columbia

Aleksandar Vukojevic
GE Energy—Digital Energy
Atlanta, Georgia

Matt Wakefield
Electric Power Research Institute
Knoxville, Tennessee

Paul Wilson
GE Energy—Digital Energy
Santa Barbara, California

Bartosz Wojszczyk
GE Energy—Digital Energy
Atlanta, Georgia

Eric Woychik
Itron, Inc.
Oakland, California

Alex Zheng
Silver Spring Networks
Redwood City, California

Daniel Zimmerle
Department of Mechanical Engineering
Colorado State University
Fort Collins, Colorado

1 Overview of the Electric Utility Industry

Stuart Borlase, Tim Heidel, and Charles W. Newton

CONTENTS

1.1 United States: Historical Perspective 2
    1.1.1 Electrification and Regulation 3
    1.1.2 Northeast Blackout of 1965 4
    1.1.3 Energy Crisis of 1973–1974 5
    1.1.4 Deregulation 5
    1.1.5 Western Energy Crisis of 2000–2001 6
    1.1.6 Northeast Blackout of 2003 9
1.2 Other World Regions 9
    1.2.1 Western and Eastern Europe 10
    1.2.2 Latin America 10
    1.2.3 Middle East and Africa 10
    1.2.4 Asia–Pacific Region 11
1.3 Utility Regulatory Systems 11
Acknowledgments 13
Reference 13

When electricity was first made available in the late nineteenth century, it was through central stations serving a group of nearby customers. Generation and distribution was localized and long-haul transmission was not yet in the picture. Demand for the service was high with a larger and larger distribution network being covered. Systems once isolated from one another were becoming interconnected. Out of this emerged the basic operating structure of the grid still in place today:

1. Large power plants generate electricity* and transmit it at high-voltage levels
2. To interconnected transmission lines that transmit electricity over long distances
3. To distribution substations where a transformer steps down the voltage†
4. To deliver it over relatively short distances to a network of smaller, local transformers, which step the voltage down further to levels safe and appropriate for the homes or businesses it serves

---

* Most large power plants function in a similar fashion: using an energy source to drive a rotating turbine attached to a generator. These turbines can be driven by water, wind steam, or hot gases. Steam requires nuclear fission or the burning of a fossil fuel like coal, while hot gases require the burning of natural gas or oil. A combined cycle plant uses both hot gases and steam—they typically burn natural gas in a gas turbine and use the excess heat to create steam to power a steam turbine.
† Transmission lines carry *alternating current* (AC) electricity at voltages ranging from 110,000 V (110 kV) to 1,200,000 V (1.2 MV), which are eventually stepped down to 110/220 V for residential use. When electricity is transmitted at higher voltage levels, less of it is lost along the way; *line loss* is currently about 7% in the United States. *Direct current* or DC power may be more suitable for transmitting power over long distances if the reduced energy loss offsets the required investment in stations at each end of the line to convert it back to AC.

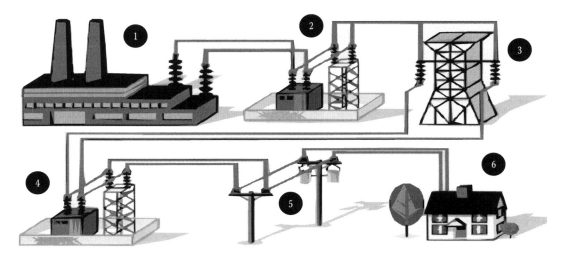

FIGURE 1.1 Electric utility interconnection overview. (Courtesy of the Advisory Board of the Utility Executive Course, University of Idaho, Moscow, ID.)

These elements are illustrated in Figure 1.1 including (1) a power plant, (2) a transmission substation, (3) a transmission line, (4) a distribution substation, (5) a distribution line/transformer, and (6) an end user.

While the basic operating structure of the grid has largely remained the same over the decades, the practices used to plan and operate the grid and the regulatory structures that govern the industry have evolved substantially since that time. The history of the industry, in particular the United States, is essentially a timeline of regulatory responses to a relatively small number of key events.

While electric power is now available to approximately 4.8 billion people around the world, more than 1.8 billion people are left "in the dark" with no, or very limited, access to electricity. Developing nations continue to lag in the provision of electricity to their citizenry. Globally, more than 1.6 billion electricity meters are installed at end-use locations (houses, apartments, commercial establishments, and industrial sites and factories), measuring usage information that provides the global electric utility industry with revenues of more than one trillion dollars annually.

This chapter aims to provide context for the more focused, technical chapters that follow. A full understanding of the new challenges and opportunities the industry will face over the coming decades requires an understanding of the factors that have shaped the utility industry's history. The regulatory structures that exist today will also fundamentally shape the development of the smart grid. The enormity and complexity of the electricity delivery network, coupled with its social, economic, regulatory, and political operating environments, directly impact the understanding, acceptance, and ultimate promotion of the smart grid.

1.1 UNITED STATES: HISTORICAL PERSPECTIVE

The electric utility industry in the United States today is highly fragmented, operating under a variety of different industry and regulatory structures. Much of the heterogeneity is a result of the history of the industry and the strong influence of ever-evolving regulatory structures. There are over 3100 investor-owned utilities, municipals, cooperatives, and federal and state agencies that deliver electric power in the United States today. These entities collectively deliver electric power across 50 states, 3 interconnections, and 8 distinct "reliability regions" and own over 160,000 miles of high-voltage transmission lines, 60,000 transmission and distribution substations, and millions of miles of distribution networks. This vast and intricate network keeps the lights on and systems running for over 142 million residential, commercial, industrial, and governmental customers.[1]

1.1.1 Electrification and Regulation

When electricity was first made available in the late nineteenth century, it was provided by relatively small central stations serving a group of nearby customers. Generation and distribution were initially highly localized. However, demand for electric service grew quickly leading to the development of larger and larger distribution networks. Rapid technology improvements also enabled improvements in both electric power generation and transport. Systems once isolated from one another eventually became *interconnected*. Eventually, the interconnection of localized systems led to substantial industry consolidation by the end of the 1920s.

The interconnection of once isolated systems brought both benefits and risks. The biggest benefit was that generation could be shared among distribution networks. Since power plants have significant economies of scale, this allowed electricity to become cheaper to produce. Reliability was also improved as the failure of a local generator could be offset by another generator farther away—without customers even knowing that there had been a problem. Such was, and is the case, the vast majority of the time. However, the fact that localized distribution networks were now interdependent exposed utilities to the risk of disruptive events miles away. Eventually, the interconnection of localized systems led to substantial industry consolidation by the end of the 1920s.

This consolidation resulted in a handful of holding companies controlling more than 80% of the U.S. electric power market. While utilities had been state regulated since as early as 1907, the state public utility commissions (PUCs)* had limited or no control over the actions of interstate holding companies. These holding companies were often highly leveraged[†] and financial failures were not uncommon. In addition, some holding companies were being operated essentially as "pyramid schemes," in which resources were transferred from utilities at the bottom to the parent company at the top—to the benefit of a small number of large investors at the expense of ratepayers and smaller investors. For a service as vital as electricity to the economy, this was an unsustainable situation. It was eventually addressed in the 1930s during a wave of legislative reform that followed the stock market crash of 1929.

The *Public Utility Holding Company Act of 1935* (PUHCA), in particular, had an enormous impact on the structure of the industry. In sum, this legislation

1. Broke up the large holding companies that dominated the industry
2. Gave the Federal Power Commission (predecessor of today's Federal Energy Regulatory Commission or FERC)[‡] power over activities that crossed state jurisdictional boundaries, such as electric transmission and wholesale power pricing
3. Gave the Securities and Exchange Commission (SEC) the power to regulate holding companies in a way that state PUCs never could

---

* PUC is a general term for a state regulatory agency. State regulatory agencies can go by a variety of names.
† Excessively reliant on debt to fund their activities.
‡ The FERC is an independent regulatory agency within the Department of Energy. According to its most recent strategic plan, its top priorities remain interstate/national matters: (1) promote the development of a strong energy infrastructure, (2) support competitive markets, and (3) prevent market manipulation. At present, FERC is composed of up to five commissioners appointed by the President for 5 year terms, with one appointed by the President to be the Chair. No more than three commissioners can belong to the same political party and there is no Presidential or Congressional review of the FERC's decisions.

In direct response to the *cross-subsidization** that took place in the pyramid schemes of the 1920s, PUHCA required new cost accounting complexities that remain today in nearly all utility holding companies. PUHCA regulation is also the reason why the parent companies of most investor-owned utilities† are typically based in the United States with holdings concentrated within the industry (i.e., not industrial conglomerates) and why most mergers and acquisitions take place between geographically contiguous entities. Simply put, policymakers preferred the electric industry to be run by local electric companies, not by outside speculators, and this legislation helped accomplish that goal.

It was a noble plan, though not without its flaws. For one, as the world around it changed, the ability of utilities to adapt was greatly limited by the PUHCA. For example, utilities were constrained in their ability to reduce operating risk by diversifying their activities. PUHCA was also a "deal-breaker" for many acquisition opportunities; nonenergy businesses would essentially have had to overhaul their business model (i.e., divest nonenergy businesses) and subject themselves to higher levels of regulatory scrutiny in order to "buy in" to the industry.

1.1.2 Northeast Blackout of 1965

From 1935 to 1965, the utility industry was stable and relatively uneventful. Transmission interconnection had become so pervasive that isolated power systems in the continental United States were essentially nonexistent. Oversight of these interdependencies was in place via the North American Power Systems Interconnection Committee (NAPSIC)—which had been formed by the industry in the early 1960s to help ensure effective governance of the nation's transmission system—and by *regional reliability councils*.‡

However, on November 9, 1965, a confluence of events—a minor power surge, an improperly configured system protection component, and extremely cold weather pushing the electric system near peak capacity—triggered a cascading blackout that affected 25 million people in parts of New York, New Jersey, New England, and Ontario. A review of what happened and why it happened revealed that effective governance of the nation's transmission system had not been ensured—specifically, that interconnection pervasiveness was not accompanied by the appropriate level of interconnection planning and operations. In other words, though a utility's service reliability was heavily dependent on the reliability of its neighbor utilities, this did not prevent independent operating standards and procedures, system protection schemes, and restoration practices from evolving. In response to constituent outcry about the blackout, more formalized oversight was legislated through the *Electric Reliability Act of 1967*.

As part of this Act, external scrutiny of the industry increased. The North American Electric Reliability Council (NERC) was formed on June 1, 1968, as a successor to the NAPSIC. Its charter was to promote electric reliability, adequacy, and security by driving utilities to common policies and procedures. Also, out of the *Electric Reliability Act of 1967* came the impetus for large-scale energy management systems (EMSs) that utilities use to efficiently and reliably remotely monitor and control their transmission networks and the development of SCADA systems to remotely monitor and control distribution networks.

---

* Funding one entity with the assets and resources of another.
† Investor-owned utilities serve the largest number of customers in the United States. In addition to investor-owned utilities, there is another classification of utilities called publicly owned utilities. Publicly owned utilities are often referred to as "municipals" (municipality-owned) or "cooperatives" (customer-owned), the latter typically serving rural areas.
‡ Regional reliability councils remain in place today, covering the continental United States and much of Canada. Examples of reliability councils include the Northeast Power Coordinating Council (NPCC), the Electric Reliability Council of Texas (ERCOT), and the Western Electricity Coordinating Council (WECC).

Overview of the Electric Utility Industry

1.1.3 Energy Crisis of 1973–1974

The Arab oil embargo of 1973 and 1974 drove the U.S. economy into recession and prompted unprecedented interest in conservation and renewable energy. For the first time since average retail price data have been tracked, the *real** cost to the consumer for electricity increased. In *nominal*[†] terms, electric bills essentially doubled from 1973 to the end of the decade. In response, many electric utilities shifted their marketing focus from consumption to conservation—promoting investment in home insulation, higher-efficiency heating and air conditioning equipment, and other energy efficiency measures through financial assistance programs to residential and business customers. The federal government also attempted to promote more-efficient generation technologies and to encourage new players to enter into the generation market through the Public Utility Regulatory Policies Act (PURPA).

Put forth as part of the National Energy Act of 1978, PURPA created incentives for nonutilities (e.g., chemical refineries, paper mills) to produce power, and required utilities to buy that power. In order to create enough of an incentive for these nonutilities to make the necessary upfront investment, certain risks were transferred from the nonutility to the utility (and, therefore, ultimately to its customers). This was done through purchased power contracts, which were often long term in nature. When oil prices fell during the 1980s, these *cogeneration* contracts proved to be a significant drag on utility earnings—and on the energy efficiency PURPA sought to promote. Ultimately, PURPA was used by many utilities in their arguments that less regulation, not more, was needed to drive efficiencies in the electric industry.

1.1.4 Deregulation

The first major attempt at deregulation of the electric power industry was the *Energy Policy Act of 1992*, which sought to drive efficiency in the industry through wholesale[‡] competition. As airline deregulation had driven down prices in the 1980s, it was believed that the price of electricity to the end user would go down if the price of generation to the electric delivery company was determined by a free market. Many economists argued that electricity was not a natural monopoly, but rather the *delivery* of electricity was; *generation* of electricity was not. If power plants could be exposed to competition, it was believed, then the most efficient generation operations would prevail and prices would drop below those set by state regulators.

Policymakers and regulators recognized the potential for new electricity markets to be *gamed*—rules manipulated and loopholes exploited, to the benefit of a few at the expense of the many. It was understood that control over transmission assets—the high-voltage lines that link power plants (generation) and customers (distribution)—could be used to stifle competition. In anticipation of this, FERC was given the ability to mandate utilities to provide access to the transmission grid, preventing them from keeping competition out of their market by denying the entry of outside power to the transmission "highway." Policymakers also understood the value of information related to transmission, and created *standards of conduct* designed to ensure that all players in the marketplace had access to the same information at the same time[§] and to keep information from finding its way from the regulated side of utilities to the deregulated side—an information flow that could create a significant competitive advantage for a utility's generation business.

---

* Net of inflation.
[†] Inclusive of inflation.
[‡] The wholesale market is where bulk power is bought and sold by grid operators based on immediate or long-term system load levels, while the retail market—which was deregulated in certain states later in the 1990s—is where electric supply choices can be made by the end user.
[§] This is done through OASIS—open access same-time information system.

FERC relied heavily on *independent system operators* (ISOs) and *regional transmission operators* (RTOs)* to help ensure a functioning marketplace for wholesale electricity. ISOs and RTOs were given responsibility for managing transmission assets that in most cases are owned by one utility but essential to multiple utilities. ISO and RTOs were established across a wide number of states and regions in the late 1990s in the United States including California, Texas, New York, New England, the Mid-Atlantic, and the Midwest, as depicted in the map on the following page. If it had been in FERC's power to do so, it would have mandated—in the interest of marketplace efficiencies—that all transmission assets be governed and operated by an independent agency such as an ISO or RTO. However, FERC did not—and still does not—have this authority. Not all state PUCs or utilities believed that their interests would be best served by abdicating transmission asset responsibility to an independent agency. As a result, RTOs and ISOs help oversee only about two-thirds of the nation's electricity consumption (Figure 1.2).

The results of restructuring have been mixed. In the PJM market and in Texas, for example, deregulation has been considered a very real success. In California, as described in the following, it was, at least at first, a very vivid disaster. The difference between success and failure in these situations has often boiled down to specific details of market design.

1.1.5 Western Energy Crisis of 2000–2001

> In the final analysis, it doesn't matter what you crazy people in California do, because I've got smart guys who can always figure out how to make money.—Enron CEO Ken Lay to the Chairman of the California Power Authority (2000)
>
> I inherited the energy deregulation scheme which put us all at the mercy of the big energy producers. We got no help from the Federal government. In fact, when I was fighting Enron and the other energy companies, these same companies were sitting down with Vice President Cheney to draft a national energy strategy.—California Governor Gray Davis (2003)

On September 23, 1996, deregulation of the electric market was passed into law in California by unanimous vote of the state legislature. This legislation required that investor-owned utilities (i.e., Pacific Gas and Electric in the North, Southern California Edison in the South, and San Diego Gas and Electric) divest their generation business. Power plants were sold off to independent power producers (e.g., Enron, Mirant, Reliant, Williams, Dynegy, AES), who would then sell this energy to the regulated utilities responsible for power delivery to residential and business customers.

Of great concern to the utilities were *stranded assets*—capital that they had previously invested and which, under the new rules, they would be unable to recover. In return for asset recovery, the utilities agreed to retail price caps. Though the price the utilities would be paying to purchase energy would change with the market, the amount the utilities could pass on to the customer was fixed. It can take several decades for bad legislation to become apparent. In California, it took less than 5 years.

The spot market for electricity began operating in April of 1998. Caps were removed from wholesale prices in May of 2000, while caps remained on retail prices. Energy prices began to rise in May of 2000. Rolling blackouts first started in June 2000 and lasted through May 2001, including 2 days in mid-March when 1.5 million customers were affected. A State of Emergency was declared in January 2001, with the state of California having to step in for the utilities (which were essentially insolvent due to rising wholesale prices and retail price caps) to buy power at market rates and financed through significant levels of long-term debt. Pacific Gas & Electric filed for bankruptcy in April 2001. Southern California Edison nearly did the same. In aggregate, the two utilities took on

---

* U.S. ISOs and RTOs include California ISO (CAISO), Electric Reliability Council of Texas (ERCOT), Midwest ISO (MISO), New York ISO (NY ISO), New England RTO, PJM Interconnection (PJM), and Southwest Power Pool (SPP). Some of these overlap with regional reliability councils.

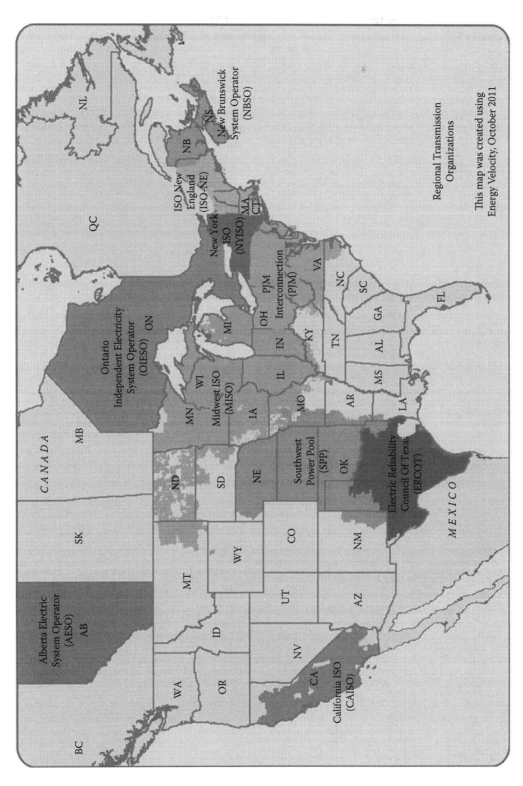

FIGURE 1.2 Regional Transmission Organizations. (From FERC, Washington, DC.)

an additional $20B in debt and saw their credit ratings* downgraded to the level of junk bonds. The State of Emergency was not lifted until November 13, 2003.

As it was happening, there was no consensus on the key factor driving the Western Energy Crisis. In retrospect, it was a combination of the following:

- *Weather:* It was hot and dry. The worst Pacific Northwest hydroelectric year in history drove down supply and unusually hot weather over much of the West drove up demand—with drought-fueled fires knocking out key transmission lines along the way.
- *Capacity:* From 1993 to 1999, California's peak load demand had grown by over 15% while growth in capacity was virtually nonexistent. In addition, the ability to easily exchange power back and forth throughout the region was constrained by transmission line capacity.
- *Flawed market design:* Utilities reduce their exposure to energy supply fluctuations through a number of strategies, most notably long-term, fixed-cost (aka *hedged*) power contracts. During the summer of 2000, only 50% of the energy purchased by California utilities was hedged compared to 85%–90% by utilities in the PJM market. Market rules forced California utilities to be excessively reliant on the inherently riskier *spot market* (i.e., "that day's price") to meet demand.
- *Corporate malfeasance:* The flaws in the deregulated marketplace were being manipulated, most notably by Enron. One of the most common techniques involved the exploitation of supply constraints to drive up prices. Wholesale energy companies' business models often focus on peak demand days—and the ability to meet that demand using *peaking units*†—as a key driver of profitability. Enron's business model involved *creating* peak demand days by shutting off power plants for unplanned maintenance and then selling their remaining capacity into the market at exorbitant rates to meet the needs of a captive market. This is just one example of the many schemes that Enron employed to game the market.
- *Failed oversight:* It is evident now that California and FERC were operating under an inconsistent set of assumptions. An implicit assumption was made by the California legislature that FERC would play a role in keeping out-of-state interests from manipulating the market. An implicit assumption was made by FERC that the wholesale markets they were advocating could and would be designed in a manner that did not require extensive oversight to prevent manipulation.

Though labeled a state or regional crisis, the impact of this series of events was felt well beyond California and the West, extending throughout the industry as the market responded to the dramatic levels of uncertainty in what had historically been a relatively stable industry. Any number of statistics could underscore this point, but here is one particularly striking one that captures the state of the industry in the first years of the new century: Over the 3 year period from 2000 to 2002, there were 65 upgrades compared to 342 downgrades of electric utility credit ratings.

The California crisis also substantially slowed the momentum that had emerged for markets in the late 1990s. In the years that followed, some regions publically considered reverting to the traditional model of industry regulation (though no regions actually switched) and no new ISOs/RTOs were formed. FERC has repeatedly reaffirmed its support for wholesale market competition over the past decade and those regions with organized markets continue to evolve their market designs. However, today only about two-thirds of the nation's electricity consumption occurs in regions with organized wholesale markets.

---

* Third-party assessments of a company's ability to repay its debts.
† Peaking power plants that can be brought on-line and off-line quickly, as opposed to base load power plants that require much more time to "turn on and off."

It is unclear if or when further deregulation will occur in the U.S. electric power industry. Today, the electric power industry in the United States remains in a state of partial deregulation. While many utilities continue to operate in open wholesale and retail markets, the Enron debacle is still fresh in enough people's minds to dampen any enthusiasm for expanding deregulation further. Though the Western Energy Crisis in the United States had everything to do with wholesale markets and little to do with retail markets, the distinction is not clear in the public's mind—and the impact on existing or proposed regulatory reform efforts has been significant. The industry remains in a state of partial consolidation, with merger and acquisition activity well below the pace projected by many in recent years.

1.1.6 NORTHEAST BLACKOUT OF 2003

Only a couple of years after the California crisis, August 14, 2003, saw a massive power outage that affected 50 million people in Michigan, Indiana, Ohio, Pennsylvania, Maryland, New York, Vermont, Connecticut, and Ontario. As with the Northeast Blackout of 1965, the initial cause was a fairly innocuous one that eventually triggered a system imbalance that cascaded across neighboring utilities. The investigation eventually identified the root cause to be a handful of high-voltage transmission lines—which, by the laws of physics, sag (literally) as load increases—coming into contact with overgrown trees in Ohio and going off-line. As with the Northeast Blackout of 1965, failures in other parts of the system protection process allowed this outage to spread wider—specifically, a problem that caused alarms on First Energy's EMS to go unnoticed.

In addition, as with the Northeast Blackout of 1965, the governmental response to this issue has been to legislate greater oversight of the industry. As part of the *Energy Policy Act of 2005*, FERC was authorized to designate a national Electric Reliability Organization (ERO). On July 20, 2006, FERC certified NERC as the ERO for the United States. With this designation, NERC's *guidelines* for system operation and reliability became *standards*. This Act gave NERC the power to exact financial penalties for entities operating out of compliance with the standards.

As should be readily apparent, the history of the electric power industry is one in which a relatively small number of disruptive events (major blackouts, market manipulation) resulted in significant governmental responses to those events. Because legislation rarely is written in a manner that allows it to adapt to changes in the marketplace, the impact of governmental intervention on the structure of the industry is felt for years to come and not always in the manner in which it was intended.

1.2 OTHER WORLD REGIONS*

Globally, electric power is now available to approximately 4.8 billion people. However, more than 1.8 billion people are left "in the dark" with no, or very limited, access to electricity. Developing nations continue to lag in the provision of electricity to their citizenry. Globally, more than 1.6 billion electricity meters are installed at end-use locations (houses, apartments, commercial establishments, and industrial sites and factories), measuring usage information that provides the global electric utility industry with revenues of more than 1 trillion dollars annually.

There are about 43,000 medium and large electric power generating facilities outside of North America providing electricity to people in more than 200 countries. These power stations transmit electricity through a network of about 58,000 transmission substations, with more than 120,000 large and very large power transformers installed. There are approximately 190,000 distribution substations outside of North America and about 25,000 industrially operated substations.

---

* *Source:* Newton-Evans Research Company internal estimates.

1.2.1 Western and Eastern Europe

Western European nations have a total of about 19,300 generating facilities providing electricity to over 400 million residents via a strongly interconnected (mainland international) transmission network. The HV/EHV network includes more than 14,000 transmission substations, some 44,000 primary distribution substations, and thousands more of secondary distribution substations. Nearly 60% of the western European power generation capacity can be found in just three countries (Germany, France, and the United Kingdom).

Central and eastern European nations have an installed power generation base of more than 2700 large and medium plants, with a capacity of more than 425 GW. Most residents of central and eastern Europe have access to electricity. There are more than 12,000 transmission substations and 32,000 distribution substations in the combined central-eastern European region.

Some of the world's largest electric power utilities are found in western Europe, where state-run or quasi-state-owned utilities dominate in some countries (EDF in France, EDP in Portugal, ENEL in Italy), while some nations have 5–20 major electric utilities (United Kingdom, Denmark, the Netherlands, Spain, and others). A few countries (e.g., Germany and Switzerland) have scores or hundreds of small municipal or rural area utilities with a few large to very large urban utilities.

In Eastern Europe, a number of countries continue to operate state-controlled electric power companies. In the forefront of these is Russia's UES, generating and transmitting electricity to about 60 mid-size to quite large distribution utilities in the country's larger cities. Russia accounts for just over one-half of the total generating capacity for the entire central and eastern European region; Ukraine is second and Poland third in generating capacity and in populations served with electricity.

1.2.2 Latin America

Two countries (Brazil and Mexico) dominate the Central and South American regions in terms of population (300 million out of a region-wide 565 million inhabitants), electricity production (about 55% of the total); and in the investment in existing T&D infrastructure. Argentina and Venezuela are next in terms of the status of electricity infrastructure development.

The entire region provides about 250 GW of electric production capacity, has nearly 5,000 transmission substations and 18,000 distribution substations already in operation. There are still millions of residents in the region without access to electricity, but each year brings some progress with new areas being served by power utilities and by micro-grid developments based on renewable energy sources.

Latin American countries are home to about 3600 large and medium power generation facilities, most of which are hydropower facilities (other than Mexico). Some of the world's largest hydropower facilities are found in South America, including the world-class Itaipu Binacional hydropower facility, just behind China's Three Gorges in terms of its production capacity (12,600 MW).

1.2.3 Middle East and Africa

The Middle Eastern countries of the Mashreq and Maghreb regions provide more than 200 GW of mostly gas (and oil)-fired electric power capacity to more than 350 million users out of a total of about 400 million residents. More than one-half of the region's inhabitants reside in three countries (Egypt, Turkey, and Iran). There are more than 4200 transmission substations delivering power to about 100 million end use electric power sites.

The African nations currently rely on coal-fired plants for the majority of electricity generation, but coal is expected to be overtaken by gas-fired plants by 2020. Nonetheless, coal consumption continues to increase, with new plants being largely combined cycle gas fueled facilities.

Sub-Saharan African nations have about 750 million inhabitants, but only 90 GW of electricity production capacity. More than one-half of the existing generating capacity is in South Africa. There are an estimated 450 million or more people in sub-Saharan Africa without direct access to a reliable electric power supply. As renewable energy production methods develop and their costs decrease, African countries will be able to adopt them more rapidly than at present.

1.2.4 Asia–Pacific Region

This vast region includes the two most populous, rapidly developing nations in the world, India and China. Across the expanse of the Asia–Pacific region, there are more than 14,000 large power generation facilities in operation. China and India both have more than 2150 of the large (mostly coal-fired) power plants in the region.

South Asia as a subregion includes 1.6 billion people, with less than 200 GW of electricity production capacity, of which India holds the major share of people (1.1 billion) and electricity production capacity (160 GW). The country also has most of the substations in the region (about 17,500 out of about 22,000 in total). Pakistan and Bangladesh are other large countries of South Asia neighboring India, together having 315 million residents, but only about 32 GW of capacity.

Other Asian and Pacific countries have more than 2.1 billion inhabitants, of which 1.3 billion live in China. China accounts for about one-half of this regions electricity production capacity and one-half of the power delivery infrastructure. Japan is second in terms of electricity production and delivery infrastructure, though Indonesia is a more populous country (235 million Indonesians and 127 million Japanese). South Korea is third in electricity production and delivery, with 49 million residents.

The entire Asia–Pacific and South Asian regions represent more than one-half of the world's population and have invested greatly to account for about 30% of the world's electricity production capacity.

Non-OECD (Organization for Economic Co-operation and Development) countries in Asia will be making impressive gains in the use of renewable energy production, but the reliance on coal-fired plants, primarily in China, is still expected to double by 2020.

1.3 UTILITY REGULATORY SYSTEMS

The nature of the electric industry cannot be understood without understanding the nature of how the industry is regulated. The electric industry is, arguably, the most externally controlled industry in the United States and the majority of nations around the world. The impact of this regulation on how and why utilities do what they do cannot be overstated.

Regulatory oversight of electric utilities is necessary because they are natural monopolies. The term *natural monopoly* applies to industries where the best outcome, in terms of the societal interest, will be one and only one provider of that good or service in a given market. Society does not benefit, the reasoning goes, when overlapping subway systems, water mains, or electric delivery networks are attempted.

The most common natural monopoly occurs in a market where the cost of entry is exceedingly high—such as a "poles and wires" company or any other entity for which significant capital resources are required to "open up shop." Investors will not fund any venture without some reasonable assurance that they will be able to earn a return on their investment. If multiple entities are allowed to build competing distribution networks, then no such assurance exists that any of these entities will be able to earn a return on the capital they have invested. Rational investors anticipate this and therefore they will not put capital toward stringing wires on poles unless they are guaranteed to be the sole provider of electricity to that market.

With monopoly, however, comes market power—specifically the power to set profit-maximizing prices with no concern for a competitive pricing. No rational public policymaker will agree to such a sole-provider arrangement without being able to control prices.

So, a deal is struck: in order for a society to benefit from the provision of a vital service, (1) public policymakers grant an exclusive franchise to an entity to provide electricity to homes and businesses and (2) the entity must agree to a customer service obligation and consent to pricing controls and third-party oversight.

Electric power utilities are always subject to some form of regulation or oversight. This can be at the national, regional, or local level. For example, in the United States, investor-owned utilities are regulated by FERC and state PUCs, while municipal and cooperative utilities are regulated by local communities and/or boards of directors made up of their members. Many countries throughout the world have national regulatory bodies including OFGEM in the United Kingdom, Commission de Regulation de l'Energie in France, CRE in Mexico, and so on. In many Middle Eastern and African countries, the regulatory function is provided by the ministry of energy.

The design of regulatory systems has a strong impact on incentives for shareholder-owned utilities. The nature of these incentives will also have important impacts on the pace of smart grid development throughout the world. For example, because a regulated utility serves 100% of their designated market, customer growth is driven not by product-based or price-based competition but by the underlying growth in the market as a whole. Further, because prices are fixed by regulatory tariffs, utilities are severely limited in their ability to drive revenue increases through pricing strategies. In addition, because prices are set by regulatory tariffs, utilities often give regulatory relationship management the same or greater emphasis than customer relationship management. In fact, some will argue that the regulator is the customer.

Profitability is determined largely by an administratively determined regulated rate of return on a utility's asset base, so great emphasis is placed by the utilities on investing capital in a prudent manner and on the preparation and defense of rate cases. To grossly oversimplify the rate-making process, utilities and regulators reach agreement on

1. What assets are essential to service delivery (i.e., the *rate base*)
2. What an appropriate rate of return on those assets should be

With these two variables in place, an allowable annual return (i.e., net income as a percentage of assets) can be calculated. Rates for different classes of customers are then set, which are projected to result in that level of return. This does not eliminate all variability in a utility's earnings, but it does create a far more predictable environment than that in which a typical nonregulated entity operates.

Profitability is determined in the following way:

- Utilities can make very large capital investments and take on relatively high levels of debt with a much lower degree of uncertainty than an unregulated company. The business case for such an investment is driven primarily by the regulatory recoverability of the investment rather than—at unregulated companies—by the anticipated impact of the investment on revenue or expense.
- The distinction between capital costs (spending that ends up on the company's books as an asset, such as the labor and equipment required to put a new transformer in service) and operating costs (spending that does not end up on the company's books as an asset, such as an administrator's salary) is nontrivial. Expense that can be charged to an asset should eventually earn the utility a regulated rate of return.

With profitability capped by a regulated rate of return on the asset base, utilities have at times had a disincentive to drive down spending. For example, a utility with a regulated rate of return of 10% generates significant operational efficiencies that allow it to earn a rate of return of 12%, only to have to "return" those excess earnings to the ratepayers during the next rate case. This phenomenon has contributed to a number of recent trends, including utilities going many years between rate cases, utilities proposing rate caps to regulators in return for other concessions, and utilities and regulators establishing *performance-based rates* in which rate of return is driven by factors other than cost of service (e.g., service levels).

ACKNOWLEDGMENTS

The editor would like to thank the University of Idaho's Utility Executive Course and the UEC Advisory Board for the material used in this chapter to describe the history of the U.S. electric utility industry.

REFERENCE

1. U.S. Energy Information Administration (EIA), http://www.eia.gov/

2 What Is Smart Grid, Why Now?

*Stuart Borlase, Steven Bossart, Keith Dodrill, Tim Heidel,
Miriam Horn, John McDonald, Marita Mirzatuny,
Lauren Navarro, Charles W. Newton, Mica Odom,
Matt Wakefield, Bartosz Wojszczyk, and Eric Woychik*

CONTENTS

2.1	Smart Grid or Smarter Grid?	16
2.2	Smart Grid Drivers	18
2.3	Benefits: More Than a Business Case	20
	2.3.1 Utility Benefits	21
	2.3.2 Consumer Benefits	22
	2.3.2.1 Conservation Effect of Real-Time Information and Pricing	22
	2.3.2.2 Cutting Peak Demand and Expanding Demand Response	22
	2.3.3 Environmental Benefits	25
	2.3.4 Enabling High Penetration of Clean, Renewable Energy	26
	2.3.5 Integrating Electric Vehicles to the Grid	27
	2.3.6 Benefit Synergies	28
2.4	U.S. Electric Utility Industry Challenges	28
	2.4.1 Generation and Energy Resource Mix Changes	29
	2.4.1.1 Coal	29
	2.4.1.2 Natural Gas	29
	2.4.1.3 Nuclear	29
	2.4.1.4 Oil	30
	2.4.1.5 Renewable Generation	30
	2.4.1.6 Energy Storage	31
	2.4.1.7 Consumer Demand Management	31
	2.4.2 Transmission Expansion	31
	2.4.3 New Demands	31
	2.4.4 New Technical Opportunities	32
	2.4.5 Regulatory Challenges	32
2.5	Federal Smart Grid Influences in the United States	32
	2.5.1 Energy Independence and Security Act of 2007, Title XIII	32
	2.5.2 American Recovery and Reinvestment Act of 2009	33
	2.5.3 The U.S. Department of Energy	34
	2.5.3.1 Demonstration and Investment Grants for Smart Grid Projects	34
	2.5.3.2 Smart Grid Task Force	35
	2.5.3.3 Electricity Advisory Committee	35
	2.5.4 The National Institute of Standards and Technology	36
	2.5.4.1 Smart Grid Interoperability Panel	37
	2.5.4.2 Smart Grid Advisory Committee	39
	2.5.5 The Federal Energy Regulatory Commission	39

2.6　International Treaties and Nongovernment Organizations 40
　　2.6.1　Treaties and Negotiations 40
　　2.6.2　National and Regional Action and Precedent 42
　　　　　2.6.2.1　Asia/Pacific 42
　　　　　2.6.2.2　Europe 43
　　　　　2.6.2.3　Africa/Latin America 43
　　2.6.3　Nongovernmental Organizations 44
2.7　Smart Grid Industry Initiatives 44
　　2.7.1　EPRI IntelliGrid™ Methodology 44
　　2.7.2　EPRI's Smart Grid Demonstration Initiative 45
　　2.7.3　Smart Energy Alliance 47
　　2.7.4　GridWise Alliance 47
　　2.7.5　GridWise Architecture Council 47
　　2.7.6　International Electrotechnical Commission Technical Committee 57 47
　　2.7.7　IEC Strategic Group (SG 3) on Smart Grid 49
　　2.7.8　U.K. Low-Carbon Transition Plan 49
　　2.7.9　Electricity Networks Strategy Group 49
　　2.7.10　OFGEM Low Carbon Networks Fund 49
　　2.7.11　European Renewable Energy Strategy 49
　　2.7.12　Centre for Sustainable Electricity and Distributed Generation 50
　　2.7.13　Smart Grid Information Clearinghouse 50
2.8　Smart Grid Market Outlook 50
　　2.8.1　Market Drivers 50
　　2.8.2　Market Potential 52
　　2.8.3　Smart Grid and IT Expenditure Forecasts 53
References 55
Bibliography 59

2.1　SMART GRID OR SMARTER GRID?

The smart grid concept has experienced major hype in the past few years. Many consider "smart grid" a cliché. It seems as if almost everything is dubbed "smart" in the electric system, even down to a "smart bolt" in a transmission tower. Smart grid has gained worldwide recognition and the "smart" catchphrase has carried over into other industries, adding significantly to the confusion. What was initially considered "intelligent grid" (which may have initially implied too much), smart grid is now interpreted in a hundred and one different ways. The many interpretations of "smart grid" depend on the perspective of those speaking about it—utilities, vendors, consultants, academics, or consumers. However, it is very apparent that smart grid reflects a significant change in the way people think about the generation, delivery, and use of electric energy. As a result, the smart grid has come to represent an essential change in the way we address the energy demand, security, and the environmental challenges we face.

While the grid is a marvel in engineering design and may be one of mankind's greatest achievements, it has yet to be transformed into a modern grid. A truly modern smart grid would include sustainable concepts that leverage proven, cleaner, cost-effective technologies available today or under development. Whether one believes in the smart grid or not, most in the utility industry agree that smart grid efforts place major emphasis on improvements based on new technologies for the electricity grid and for consumers. The smart grid effort still has a long way to go, and will continue to evolve, driven by the economies and regulations across the globe. Who knows what the final state of these efforts will produce, but it looks to have sufficient momentum to bring significant advances in technology and in the industry as a whole. There will be winners and there will be losers, not just among technologies and vendors, but also with utilities and their smart grid deployments.

Consumers will also be among the winners and losers, as seen recently with challenges in early technology choices and consumer acceptance of electricity pricing.

Some argue that the electric grid is really not that "dumb." The current electric grid is operated by complex software programs and automation routines and protected by microprocessor-based relays. Though true in some parts of the world, the evolution of electric infrastructure has been slow, hampered by economics, demographics, regulation, and a host of other factors. What is similar across the globe are the challenges of operating electric systems to deliver power reliably and cost effectively to consumers. People compare the electric infrastructure to the telecommunications industry using a well-known analogy: "Graham Bell would not recognize cell phones if he were alive today, but Thomas Edison would still recognize the electric utility system if he were alive today." There are many reasons for this difference, but one stands out, the massive electric infrastructure cannot be replaced as quickly and as cost effectively as the telecommunications infrastructure. Electricity infrastructure is simply more capital intensive. A second reason is that developments in telecommunications processing and IT devices have far exceeded those in the electric industry. Also, the environment that drives the need for advanced communications has evolved much faster for the communications industry than the electric industry. One can argue further that comparing electricity to consumer-driven personal communications is not the same, since consumers are more willing to pay for additional services on a cell phone than to pay for electricity as a commodity. This seems to indicate where the smart grid should be heading—focus on the consumer, but not just to try reducing consumer electricity bills, but also to provide other services. Still in some countries and locations, the focus is more rudimentary, to supply more reliable service to customers, or simply to make electricity available to more people. The electric industry can realize major capital and operating benefits through use of advanced smart grid technologies. So while we may not have such a "dumb" grid, our thinking should be more along the lines of a "smarter grid"—not about doing things a lot differently than our current practice today, but rather about doing them smart through developing more advanced capabilities. This suggests we share communications infrastructures, fill in product gaps, and leverage existing technologies to a much greater extent while driving toward a higher level of integration that will achieve synergies across enterprise applications. Smart grid initiatives can be harnessed to drive the development of even more advanced technologies to accelerate the pace we have seen in recent decades. These technologies can be more pervasive and build upon much-needed platforms for enterprise-wide solutions that deliver far-reaching benefits for utilities and customers.

What constitutes a smart grid? Numerous utilities have staked claims to be first with a smart grid. While these claims may include smart technologies, most smart grid solutions today focus on responses to metering requirements and use a pilot program to showcase a technology under the smart grid banner. In the broader sense, smart grid is not an off-the-shelf product nor is it something you install and turn on the next day. Rather, it is an integrated set of technologies that net incremental savings in capital expenditures, operation and maintenance costs, and customer and societal benefits.

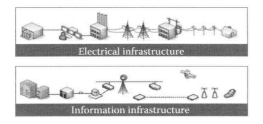

This smart grid is the integration of electrical and communications infrastructures with advanced process automation and information technologies within the existing electrical network. Smart grid represents a complete change in the way utilities, politicians, customers, and other industry participants think about electricity delivery and its related services. This new thought process will likely lead to different smart grid technologies and solutions, but will benefit from greater integration of utility engineering and operations, and new business models.

The set of solutions that will provide these benefits is vast. At the same time, significant smart grid benefits may not be easily achieved by all utilities and their consumers since new solutions are required only if needed and if operational strategies dictate. Global, regional, and national economics and growth will serve as the cornerstones for investments in smart grid infrastructure and in greater use of integrated communications and information technologies. Drivers will include national and state government policy directives and incentives to enable energy futures and development of smart infrastructure. When the stimulus funding and country initiatives have come and gone, the ensuing years will indicate how much of today's fervor and hype have influenced investment, innovation, and the achievable benefits of a "smarter grid."

More broadly, the concept of smart grid has many definitions and interpretations, which depend on the specific country and region, and the various industry stakeholders' drivers and desirable outcomes and benefits. A preferred view of smart grid may not be what it is, but what it does and how it benefits utilities, consumers, the environment, and the economy. The European Technology Platform (comprising European stakeholders and the surrounding research community) defines "a Smart Grid [as] an electricity network that can intelligently integrate the actions of all users connected to it—generators, consumers and those that do both, in order to efficiently deliver sustainable, economic and secure electricity supply" [1]. In North America, two of the more dominant smart grid definitions stem from the Department of Energy (DOE) and the Electric Power Research Institute (EPRI), respectively. These are as follows:

- U.S. DOE: "Grid 2030 envisions a fully automated power delivery network that monitors and controls every customer and node, ensuring two-way flow of information and electricity between the power plant and the appliance, and all points in between" [2].
- EPRI: "The term 'Smart Grid' refers to a modernization of the electricity delivery system so it monitors, protects, and automatically optimizes the operation of its interconnected elements—from the central and distributed generator through the high-voltage network and distribution system, to industrial users and building automation systems, to energy storage installations and to end-use consumers and their thermostats, electric vehicles, appliances, and other household devices" [3].

Beyond specific, stakeholder-driven definitions, smart grid should refer to the entire power grid from generation, through transmission and distribution (T&D) infrastructure all the way down to a wide array of electricity consumers. Smart grid is essentially aimed to modernize the twentieth century grid for a twenty-first century society. A well-thought-out smart grid initiative builds on the existing infrastructure, provides a greater level of integration at the enterprise level, and has a long-term focus. It is not a one-time solution, but a change in how utilities look at a set of technologies that can enable both strategic and operational processes. Smart grid is the means to leverage benefits across applications and remove the major barriers of silos of organizational thinking. Smart grid pilot projects driven by regulatory pressure that focus on the impact of new meters on consumers will evolve to technology-rich, system-wide smart grid deployments that demonstrate well-proven and quantified benefits. A key component to effectively enable full value of smart grid realization is technology with the functionalities and capabilities to achieve cohesive end-to-end integrated, scalable, and interoperable solutions.

2.2 SMART GRID DRIVERS

The smart grid is essential to enable a future that is prosperous and sustainable. All stakeholders must be aligned around a common vision to fully modernizing today's grid. Throughout the twentieth century, the electric power delivery infrastructure has served many countries well to provide adequate, affordable energy to homes, businesses, and factories. Once a state-of-the-art system, the electricity grid brought a level of prosperity unmatched by any other technology in the world. But a twenty-first century economy cannot be built on a twentieth century electric grid. There is an urgent need for major improvements in the world's power delivery system and in the technology areas needed to make these improvements possible.

A number of converging factors will drive the energy industry to modernize the electric grid. These factors can be combined in five major groups, as follows:

Policy and Legislative Drivers

- Electric market rules that create comparability and monetize benefits
- Electricity pricing and access to enable smart grid options
- State regulations to allow smart grid deferral of capital and operating costs
- Compatible Federal and state policies to enable full integration of smart grid benefits

Economic Competitiveness

- Creating new businesses and new business models and adding "green" jobs
- Technology regionalization
- Alleviate the challenge of a drain of technical resources in an aging workforce

Energy Reliability and Security

- Improve reliability through decreased outage duration and frequency
- Reduce labor costs, such as manual meter reading and field maintenance, etc.
- Reduce nonlabor costs, such as the use of field service vehicles, insurance, damage, etc.
- Reduce T&D system delivery losses through improved system planning and asset management
- Protect revenues with improved billing accuracy, prevention and detection of theft and fraud
- Provide new sources of revenue with consumer programs, such as energy management
- Defer capital expenditures as a result of increased grid efficiencies and reduced generation requirements
- Fulfill national security objectives
- Improve wholesale market efficiency

Customer Empowerment

- Respond to consumer demand for sustainable energy resources
- Respond to customers increasing demand for uninterruptible power
- Empower customers so that they have more control over their own energy usage with minimal compromise in their lifestyle
- Facilitate performance-based rate behavior

Environmental Sustainability

- Response to governmental mandates
- Support the addition of renewable and distributed generation (DG) to the grid

Many of these drivers are country and region specific and differ according to unique governmental, economic, societal, and technical characteristics. For developed countries, issues such as grid loss reduction, system performance and asset utilization improvement, integration of renewable energy sources, active demand response and energy efficiency are the main reasons in adopting smart grid. Many developed countries experience system reliability degradation resulting from aging grid infrastructure. Inadequate access to "strong" T&D grid infrastructure limits the potential benefits of integration of renewable energy generation.

The recent emergence of workshops, seminars, and conferences related to smart grid is one key indicator of the momentum of the industry in answering this urgent need. The world is now witnessing a mass movement of smart grid deployments to enable many of the smart grid characteristics needed to meet the future needs of T&D grids. With such, it is imperative to maintain the momentum around a clear systems vision to ensure the proper transformation is realized. Significant deployments of technologies such as advanced metering infrastructure (AMI) have emerged all across the industry. This is ahead of many of the enablers that will truly reap the benefits of the technology. Refitting or reconfiguring may be required costing consumers and utilities. We must continue to move in a coordinated effort in the evolution of the smart grid to capture the benefits and avoid unnecessary cost.

2.3 BENEFITS: MORE THAN A BUSINESS CASE

The smart grid provides enterprise-wide solutions that deliver far-reaching benefits for both utilities and their end customers. Utilities that adopt smart grid technologies can reap significant benefits in reduced capital and operating costs, improved power quality, increased customer satisfaction, and a positive environmental impact. With these capabilities come questions: What is the potential of the smart grid? Is there one set of technologies that can enable both strategic and operational processes? How do the technologies fit together? How do you leverage benefits across applications?

Smart grid delivery should not be based on only enabling solutions but on integrated solutions that address business and operating concerns and deliver meaningful, measurable, and sustainable benefits to the utility, the consumer, the economy, and the environment (Figure 2.1).

Various components come into play when considering the impact of smart grid technologies. Utilities and customers can benefit in several ways. Rate increases are inevitable, but smart grids

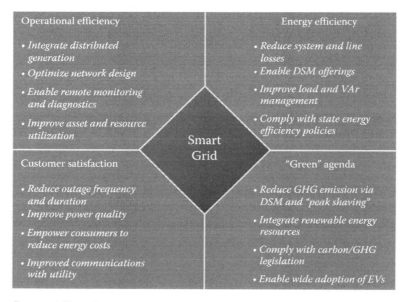

FIGURE 2.1 Smart grid benefits.

can offer the prospect of increased utility earnings, together with reduced rate increases (plus improved quality of service). Viewing smart grid programs in the context of, for example, a "green" program for customer choice or a cost reduction program to moderate customer rate increases can help define utility drivers and shape the smart grid roadmap. A smart grid program should have a robust business case where numerous groups in the utility have discussed and agreed upon the expected benefits and costs of smart grid candidate technologies and a realistic implementation plan. In some cases, the benefits are modestly incremental, but a smart grid plan should minimize the lag in realized benefits that typically occur after a step change in technology. A smart grid deployment is also intended to allow smoother and lower cost migrations to new technologies and avoid the need to incur "forklift" costs. A good smart grid plan should move away from the "pilot" mentality and depend on wisely implemented field trials or "phased deployments" that provide the much-needed feedback of cost, benefit, and customer acceptance that can be used to update and verify the business case.

Of utmost importance in implementing smart grid solutions are the tangible, quantifiable, and meaningful results:

- Improving the utility's power reliability, operational performance, and overall productivity
- Delivering increases in energy efficiencies and decreases in carbon emissions
- Empowering consumers to manage their energy usage and save money without compromising their lifestyle
- Optimizing renewable energy integration and enabling broader penetration

2.3.1 Utility Benefits

Improving grid reliability and operational efficiency is possible using more intelligence in the delivery network to monitor power flow in realtime and improve voltage control to optimize delivery efficiency and eliminate waste and oversupply. This will reduce overall energy consumption and related emissions while conserving finite resources and lowering the overall cost of electricity. Software applications—including smart appliances, home automation systems, etc.—that manage load and demand distribution help to empower consumers to manage their energy usage and save money without compromising their lifestyle—encouraging consumers to become smart consumers in smart homes, by giving them access to time-of-use rates and real-time pricing signals that will help them to save on electricity bills and cut their power usage during peak hours. This also helps to improve overall system delivery efficiency and reduce the number of power plants and transmission lines that will need to be built.

In 2008, the United States had electricity distribution losses adding up to 271 billion kilowatt hours [4], more than 6% of total net generation. Xcel Energy estimates that the smart grid can reduce those losses by 30%, utilizing optimal power factor performance and system balancing [5]. U.S. DOE estimates that conservation voltage reduction and advanced voltage control can reduce greenhouse gas (GHG) emissions from electricity by 2% nationally in 2030 [6].

The rapid deployment of smart grid technologies across the country reflects the multiple operational and reliability benefits utilities expect to realize, including savings on operation and maintenance costs and the avoidance of costly outages. Operational and energy efficiency benefits are highly valuable, but will not always—by themselves—justify the ratepayer expense. Boston Consulting Group (BCG) estimates that just 60% of the cost of smart grid deployment can be justified through the utility business case alone. Making smart meters, for example, a "winning proposition," per BCG, will require that 20%–30% of a utility's customers use the new technology to reduce their overall consumption or peak demand by 15%–20%. "Falling short of that threshold," says Pattabi Seshadri, a consultant at BCG's energy practice, "will likely prevent the utility from delivering the necessary return on investment" [7].

2.3.2 CONSUMER BENEFITS

Environmental, health, and other social benefits of the smart grid can contribute real value to these calculations, if the grid is designed to capture them. Capturing those social benefits is especially important because it is customers, ultimately, who are financing this new grid. Under the current regulatory structure, investor-owned utilities propose investments, regulators approve those investments—the rate of return the utility will earn on them—and consumers (ratepayers) foot the bill. As witnessed in the United States in Indiana, Maryland, and elsewhere, regulators around the country are requiring utilities to demonstrate that they will deliver long-term benefits to consumers commensurate with the public's investments. Designing to maximize those benefits will in turn benefit utilities. As J.D. Power and Associates found in a recent consumer survey: "Utility providers that develop smart systems with customer satisfaction in mind may be able to get things right the first time, ultimately saving in long-term development and implementation costs" [8].

Fortunately, a well-designed smart grid can deliver significant additional benefits, which can repay that investment many times over. Consumers will benefit from reduced bills and much greater control: the ability to use electricity when it is cheapest and to produce and sell power and other services into the grid when demand and prices are high. Entrepreneurs and their employees will benefit from new opportunities to provide energy services—from storage at substations to behind-the-meter "energy apps." Communities will enjoy greater energy security, as they rely increasingly on distributed energy resources in their own backyards. The most valuable benefit could be the opportunity to radically reduce the hidden costs of electricity to the environment and public health.

The smart grid will enable significant reductions in both overall energy consumption [9] and peak use of electricity by giving customers real-time information and pricing, facilitating much broader use of demand response, providing the necessary information to support "continuous commissioning" in the built environment, increasing the capacity of existing transmission lines, and reducing T&D line losses.

2.3.2.1 Conservation Effect of Real-Time Information and Pricing

Numerous studies have found that giving customers real-time energy usage information cuts consumption by 5%–15%. Adding pricing incentives and automated home energy management tools, such as programmable thermostats and smart appliances linked to home area networks, can double those savings [10,11].

A June 2010 report from the American Council for an Energy-Efficient Economy (ACEEE) found that U.S. consumers could cut their household electricity use as much as 12% and save $35 billion or more over the next 20 years if U.S. utilities go beyond AMI deployment to include a wide range of energy-use feedback tools that engage consumers in using less energy. ACEEE based its findings on a review of 57 different residential sector feedback programs between 1974 and 2010, concluding that "to realize potential feedback-induced savings, advanced meters must be used in conjunction with in-home (or on-line) displays and well-designed programs that successfully inform, engage, empower, and motivate people" [12].

The Pacific Northwest National Laboratory (PNNL) and the Brattle Group have found that conservation tends to be strongest when feedback is based on actual usage data, provided on a frequent basis over a year or more, involves goal setting and choice with specific behavioral recommendations, and involves normative or historical comparisons [6].

2.3.2.2 Cutting Peak Demand and Expanding Demand Response

Existing demand response programs, focused on large industrial users, can currently deliver 37 GW nationwide [13]. Without new programs, that capacity will grow little over the coming decade, to just 38 GW by 2019, saving just 4% compared to a scenario with no demand response programs at all [13]. A smart grid will almost quadruple those savings, according to modeling done for Federal Energy Regulatory Commission (FERC): large-scale deployment of AMI, enabling technologies, and dynamic pricing will enable peak reductions of 138 GW by 2020 [13].

A smart grid greatly expands the potential participants in demand response programs by making it possible to send the necessary signals, including dynamic prices, to residences and small- and medium-sized businesses. A Battelle-PNNL pilot, for instance, using predefined customer preferences and fast, autonomous controls on clothes dryers and water heaters to respond to ancillary service signals on very short timescales, achieved peak residential demand reductions of 16%, and average demand reductions of 9%–10% for extended periods of time [14]. In Oklahoma Gas and Electric's pilot, customers with smart thermostats achieved peak demand reductions of 57% [15].

Dynamic pricing is particularly valuable for cutting peak power demand. Analyzing a range of experiments, Brattle's Ahmad Faruqui found that time-of-use rates cut peak demand by 3%–6% and critical peak pricing (CPP) cut peak demand by 13%–20%. When accompanied with enabling technologies, CPP cut peak demand by 27%–44% [10].

A number of studies have shown that customers respond to, and appreciate, time-of-use rates. PowerCentsDC—an American Recovery and Reinvestment Act of 2009 (ARRA)-funded pilot in the nation's capital—ran from July 2008 through October 2009. This voluntary program chose 900 customers at random, providing each with a smart meter and smart thermostat and assigning them to one of the three pricing plans. One of those plans, a Critical Peak Rebate, rewarded customers for reducing their use below baseline during critical peaks. It cut peak use by 13%, with low-income customers achieving savings in line with others' results. Nearly three-quarters of the customers who participated were satisfied with the program and 93% preferred the dynamic rates over the utility's standard rates [16]. A September 2010 meta-study for the Edison Foundation Institute for Electric Efficiency found similar results in its assessment of recent dynamic pricing programs at Connecticut Light and Power, Baltimore Gas and Electric, and Pacific Gas and Electric. They not only found that low-income customers did shift load in response to dynamic pricing but also found that because they began with a flatter load, they saved money even when they did not shift load [17].

The economic benefits of these peak reductions are broadly shared, even by consumers who do not shift their consumption. Shifting just 5% of peak demand reduces prices substantially for everyone [18], both because the most expensive peak power plants do not get turned on and because new peakers need not be built [18].

Peak shaving delivers huge environmental and health benefits that 138 GW of peak reductions forecast by FERC is equivalent to the output of 1300 peaking power plants [13]. Many of these plants—often inefficient natural gas turbines—are located in or near major population centers, where their smog-forming emissions harm public health. As with the coal fleet, the National Academy of Sciences (NAS) study found that just 10% of natural gas-fired power plants contribute a majority (65%) of the air pollution damages from all of the 498 plants they studied. Replacing those plants with smart-grid enabled efficiency and demand response would significantly reduce public health impacts as well as GHG emissions, cutting 100–200 million tons of CO_2 per year—5%–10% of total GHG pollution from the U.S. power sector in 2007.

A concerted effort to make full use of demand response opportunities in regions now served by the dirtiest coal-fired power plants could also multiply benefits for human health by altering the economic calculus for those plants [19]. The entry of low-cost demand side resources into the PJM market, for example, has put downward pressure on the capacity revenues earned by marginal power plants for being on standby to meet demand spikes. This downward price pressure contributed to the decision to retire two old, marginal coal plants in Philadelphia, and is putting financial pressure on other high-polluting, marginal coal, oil, and natural gas–fired units in the region; it may well cut more pollution than the direct effects of avoided demand [20].

The biggest environmental gains of demand response will come from the combined effects of these shifts on the overall generation portfolio: providing demand side balancing for renewables in place of fossil-fueled backup generation, avoiding the need for new peaker plants, and hastening retirement of old dirty coal. Whether load shifting will also directly reduce emissions will depend on the current resource mix: since the emissions from one source of electricity are effectively traded for those of another, the environmental result will depend on the emissions profile of that second source. For example, carbon emissions will go down when the use shifts from inefficient, simple cycle natural gas–fired plants that serve peak loads to efficient, combined cycle plants that serve intermediate loads [6]. One analysis of 12 North American Electric Reliability Corporation subregions showed that most regions would shift to natural gas and reduce carbon emissions, but a few would shift to coal and increase carbon emissions [6]. As clean energy makes up a larger portion of base load generation, shifting away from peak power will have an increasingly positive impact [6]. Applying an algorithm with CO_2 reductions as its primary objective and adding energy storage will make possible still greater reductions in CO_2 [6].

A 2003 Synapse model of demand response in New England indicates that a systemwide analysis will also be necessary to capture critical health benefits. It found that if demand response was used for more efficient unit commitment, reduced operation of oil- and gas-fired steam units, and increased operation of combined-cycle units in New England, it would significantly reduce NO_x, SO_2, and CO_2 in summer months. Those benefits would not be realized, however, if it simply shifted load to on-site diesel- or natural gas–fueled internal combustion (IC) engines [21].

As the smart grid improves the ability to measure real-time environmental impacts of dispatch decisions, it will facilitate prioritization of cleaner alternatives. Because power plant dispatch presents thousands of options for rearranging the generation mix [6], Charles River Associates (CRA) and others have been developing sophisticated modeling tools to precisely measure the actual carbon impact of electricity use in realtime, or "marginal carbon intensity" (MCI). CRA's analyses indicate that the real-time and locational variability of carbon emissions is as great as the variability of electricity prices: both depend on which marginal generators are brought on line or displaced as the system is redispatched to accommodate changes in load and transmission congestion [22]. PJM Interconnection—which administers the competitive wholesale market serving 51 million people in Delaware, Illinois, Indiana, Kentucky, Maryland, Michigan, New Jersey, North Carolina, Ohio, Pennsylvania, Tennessee, Virginia, West Virginia, and the District of Columbia—has begun applying a similar analysis to estimate CO_2 reductions from demand response, energy efficiency measures, and increases in carbon-free generation [23].

PJM is among the leaders in incorporating demand response into wholesale markets: more than 9000 MW of demand-side capacity resources participated in its 2010 capacity market, equivalent to 120 grid-scale, gas-fired combustion turbines or 18 medium size coal-fired power plants. Roughly a quarter of this, 2444 MW, participated as an economic resource, responding solely to market price signals to provide service to the grid. The remainder was emergency capacity, which jumps into service at the direction of the grid operator. In terms of actual energy delivered, PJM received 94,000 MWh from all demand side resources in 2010, 60% on price signal alone. Though some of this DR may have come from on-site generators, most came from avoided energy consumption, translating to about 77,000 t of avoided carbon. In short, PJM's demand response market rules enabled about 6% of the region's total peak load to be served by demand side resources, up from less than 2% 4 years ago but still less than half the 15% potential DR Brattle found in this region.

Smart grid–enabled monitoring of chillers, control systems, and other equipment in large (>100,000 ft^2) commercial buildings can detect suboptimal performance and prescribe operational improvements or maintenance, thus achieving overall electricity savings of 9% [24]. Applied in 20% of such buildings nationwide, the annual energy savings would be 8.8 billion kWh, avoiding 5 million metric tons of CO_2 emissions [25]. A smart grid will also provide detailed consumption data: utilizing that data for improved diagnostics in residential and commercial buildings will allow for accurate targeting of efficiency investments in HVAC, lighting, and other systems, translating to a 3% reduction in U.S. CO_2 emissions from the electricity sector in 2030 [6].

2.3.3 ENVIRONMENTAL BENEFITS

Smart grids will enable broader deployment and optimal inclusion of cleaner, greener energy technologies into the grid from localized and distributed resources, including rooftop solar, combined heat and power plants and DG, thereby reducing dependence on coal and foreign oil and promoting a sustainable energy future. Electric and plug-in hybrid electric vehicle (EV) integration will bring another distributed resource to market, but one at scale—with supporting rates and billing mechanisms that can help flatten the load profile and reduce the need for additional peaking power plants and transmission lines potentially reducing the carbon footprint and fostering energy security and independence.

Electricity generation and use in the United States is one of the biggest sources of pollution on the planet, accounting for more than one-fifth of the world's CO_2 emissions [26]. The U.S. power plants also draw a huge fraction of the nation's freshwater supply. Nearly 40% of all domestic water withdrawals in the United States are used for cooling thermoelectric power plants. Depending on the cooling system, that water may be returned to the source at a higher temperature and with diminished quality, or evaporate and be lost for good [27]. In the Interior West, for example, where power plants rely primarily on recirculating cooling systems, approximately 56% of the water is lost to evaporation [28]. Conventional power plants in Arizona, Colorado, New Mexico, Nevada, and Utah consumed an estimated 292 million gal of water a day (MGD) in 2005—approximately equal to the water consumed by Denver, Phoenix, and Albuquerque combined. By 2030, water use for power production in the Rocky Mountain/Desert Southwest region is projected to grow by 200 MGD—that water would otherwise be available to meet the needs of almost 2.5 million people [28]. In Texas, power plants consume as much water as 3 million people, each using 140 gal per person per day [29]. With climate change already impacting water resources—reducing, for instance, snowpack in the West, a major source of freshwater [28]—and with U.S. energy demand projected to grow 1.7% per year through 2030, these stresses will only grow.

Monetizing these environmental impacts gives a clearer sense of the real price we currently pay for conventional electricity generation and use. A report from the NAS on "Unpriced Consequences of Energy Use and Production" estimates that in 2005 alone, environmental externalities from U.S. electricity production cost $120 billion. That figure in fact underestimates the true costs, the report notes, because it does not include the costs of climate change or damage to ecosystems [19]. Half of that $120 billion comes from aggregate damages from sulfur dioxide (SO_2), nitrogen oxides (NO_x), and particulate matter (PM) from production of coal-fired electricity at 406 plants, for an average of $1.56 million per plant. Natural gas plants tend to be less polluting due to their cleaner fuel and smaller size, but are not without cost—averaging $1.49 million in annual damages per plant [19].

These are not theoretical costs but real costs—for water, health care, and premature deaths—borne directly by citizens. In Utah, for instance, burning coal to provide electricity for its residents and for neighboring states produces health and water impacts of up to $2.1 billion dollars per year [30]. These costs include hospital visits from respiratory injuries and asthma and the use of 24 billion gallons of water annually, adding as much as $45 per MWh to the cost of fossil fuel generation [30]. Those harmful externalities, in other words, effectively double the true cost of that electricity.

Like the NAS numbers, the Utah figure does not include costs from GHG emissions. Nationally, the costs of climate change impacts related to real estate loss due to sea level rise, damages from more extreme hurricanes, increased energy costs to keep comfortable in a warmer world, and water supply impacts are forecasted to exceed $270 billion by 2025 [31]. U.C. Berkeley researchers David Roland-Holst and Fredrich Kahrl found that, if no action is taken to avert the worst effects of global warming, California alone will face damages of "tens of billions per year in direct costs, even higher indirect costs, and expose trillions of dollars of assets to collateral risk." Costs in the water, energy, tourism and recreation, agriculture, forestry, and fisheries sectors will be as high as $23 billion annually, with another $24 billion annually in public health costs [32].

Air pollution impacts are not evenly distributed: the NAS study notes that just 10% of coal-fired power plants account for 43% of all damages. For those dirtiest plants, the damages cost a stunning 12 c/kWh [19], five times greater than the price the plants pay for coal today [33]. The distribution is even more extreme for natural gas—the top 10% of the most polluting facilities produce 65% of air pollution related damages [19]. Developing smart grid–enabled alternatives to those plants will be particularly valuable.

The smart grid has the potential to radically reduce costly damage to the environment and public health—while increasing energy independence and security and creating new industries and jobs—by enabling

1. Increased reliance on clean, renewable energy, including DG
2. Vastly improved efficiency of electricity production, transportation, and use, including the ability to shift demand to lower impact times and supply resources
3. Decarbonization of the transport sector
4. Reduced water impacts—wind, solar photovoltaics (PVs), and demand side resources use very little or no water to generate [28]

Though the benefits will grow as innovation flourishes, early studies and pilots suggest how extensive smart grid environmental benefits can be.

2.3.4 Enabling High Penetration of Clean, Renewable Energy

Global investment in clean energy surged 30% in 2010 to $243 billion, according to a Pew report based on Bloomberg New Energy Finance data [34]. A total of 17,000 MW of new solar capacity were added globally, and 40,000 MW of new wind capacity, bringing total renewable energy installations to 388,000 MW worldwide, exceeding the 377,000 MW of global nuclear energy capacity for the first time. China led the world with a 39% increase in investment to a record $54.4 billion, consolidating its position as the world's largest manufacturer of wind turbines and solar modules, and overtaking the United States in terms of installed renewable energy capacity.

China has set a goal of installing 20,000 MW of solar energy by 2020, and the European Union of generating 20% of its power from renewable sources. In the United States, 30 states have renewable portfolio standards (RPS), requiring utilities to buy an increasing amount of their electricity from renewable sources; in spring 2011, California increased its RPS to 33%, and Governor Jerry Brown announced a goal of 12,000 MW of DG statewide.

A smart grid capable of handling the challenges presented by renewables—which are intermittent, dependent on weather influences that cannot be dispatched and are difficult to forecast [14]—will support increasing renewable generation. In fact, to maintain grid stability above 15% or 20% renewable penetration, operators will require real-time system information to manage that generation, balance intermittent supply with flexible demand, handle multidirectional power flows, and maintain power quality, as the International Energy Agency (IEA) notes in its April 2011 Smart Grid Technology Roadmap [35]. A KEMA study similarly finds that accommodating the 33% RPS in California will require "major alteration to system operations," such as increased storage to balance volatility [36]. Ben Kortlang of KPCB, a Silicon Valley venture capital firm, puts it most simply: "without [a smart grid], most of the other green technology won't work" [37].

China and Europe have consequently matched their leadership in renewables deployment with leadership in deployment of smart grid technologies. Ireland's transmission system operator, EirGrid, is deploying smart grid technologies—including high-temperature, low-sag conductors and dynamic line rating special protection schemes—to manage the high proportion of wind energy on its system. Operation of the system is being improved through state-of-the-art modeling and decision-support tools that provide real-time system stability analysis, wind farm dispatch capability, improved wind forecasting, and contingency analysis. Such smart grid approaches are expected to facilitate real-time penetrations of wind up to 75% by 2020 (EirGrid, 2010). In Spain, Red Eléctrica has established a Control Centre of Renewable Energies (CECRE), the first control center in the world for all wind farms over 10 MW.

The IEA forecasts that by 2050, solar power could deliver between 20% and 25% of global electricity—reducing CO_2 emissions by almost 6 billion tons per year [38].

A smart grid will make possible these high levels of penetration by

- Balancing intermittency with demand side resources and storage [6] rather than backup fossil fuel generation [6,14]
- Enhancing, forecasting, and monitoring to better predict the impacts of intermittent generation [6,14]
- Managing reverse power flows from DG, using voltage regulators, batteries, and short-circuit protections that adapt "on-the-fly" to protect the grid and the safety of workers [6]
- Using advanced communication and control technologies to reduce the need for additional ancillary services, fossil fuel generation, improved voltage control, and short-circuit protection [6]

Michael Jung and Peter Yeung of Silver Spring Networks found that demand response and smart-grid networked energy storage can facilitate an additional 10% of renewable generation in the overall generation mix by 2030, expanding the market and delivering 0.3 Gt of CO_2 GHG reductions [11]. EPRI's March 2011 report, "Estimating the Costs and Benefits of the Smart Grid," [39] has slightly higher numbers, estimating an additional 13% of renewable generation enabled by the smart grid.

2.3.5 Integrating Electric Vehicles to the Grid

The move to the electrification of vehicles worldwide continues to gather momentum with the rollout of the Chevy Volt, Nissan Leaf, and China's BYD; the Tesla IPO and partnership with Toyota, and the two dozen other car companies that have announced plans to sell EVs within the next several years. With help from stimulus and venture funding, including $2.4 billion in federal grants, the United States is predicted to dominate EV sales in 2015 [40]. A recent report from Greentech Media predicts 3.8 million electric cars on the road worldwide by 2016, with 1.5 million in the United States, 1.5 million in Europe, and 760,000 in Asia [41].

Michigan-based auto industry analyst Alan Baum predicts that by model year 2015, the new car market will have 108 electric-drive models, including 18 plug-in hybrids (PHEV), 32 EVs, and 6 fuel-cell electric cars, and that such advanced technology vehicles could account for 5% of annual sales in the United States [42].

The need to integrate these EVs into the grid underlines "the ultimate importance of a smart grid," says Goldman Sachs analyst Daniela Costa [43]. Vehicle to Grid (V2G) technologies, though not yet proven, may also enable the use of plug-in car batteries as distributed storage, capable of supplying energy and ancillary services to the grid. From 2015 to 2020, Zpryme predicts global V2G vehicle unit sales will grow from 103,900 to 1.06 million, with the market value for vehicles, infrastructure, and technology growing to almost $44 billion. Regulators are taking note. Delaware, for example, passed a law in 2009 requiring utilities to compensate electric car owners for power sent back to the grid at the same rate they pay to charge the battery [40].

The environmental benefits to be gained from vehicle electrification are significant. Transportation makes up 27% of the nation's GHG emissions: 62% of those emissions come from cars and light trucks [44]. U.S. DOE estimates that a switch to EVs would improve energy consumption per Vehicle Miles Traveled by 30%, reduce CO_2 emissions by 30%, cut imports of foreign oil by 52% [6], and lower the risk of environmental calamities like the Deepwater Horizon explosion in the Gulf of Mexico. Electrification of the vehicle fleet will also improve urban air quality and public health: A 2007 NRDC/EPRI study on EVs concluded that PHEVs would reduce ozone, PM from mobile sources, and deposition rates for acids, nutrients, and mercury. (Their modeling used a 2030 scenario in which electricity demand was met with present-day coal-fired technology with current regulations and controls: in that scenario, direct PM emissions from power plants would increase by 10% from added coal-fired generation [45].)

Maximizing the environmental benefits of EVs—and preventing disruptions on the distribution grid from spikes in load—will require smart grid technologies to manage cars as "smart loads," charging when power is clean and cheap. As a "flexible load," EVs will significantly improve the capacity factor and economics of intermittent generation, especially wind, reducing the need to curtail wind at hours of low demand. Their management will require control signals from the ISO and from local utilities for such services as peak load shedding or relief on targeted parts of the distribution system, with "intelligent mediation" between driver needs and grid operator needs, and distribution automation technologies to further protect against local problems from large EV load draws [11,40]. By directing charging away from peak times, the smart grid will reduce the need to build additional transmission and generation to power EVs, and contribute 3% in CO_2 reductions above those achievable through unmanaged charging [6]. Continuing advances in technology, such as the Nissan Leaf's fast-charge option that can recharge a car to 80% capacity in just 30 min, will give customers still more flexibility in managing when they charge [46].

All of these improvements add up. A preliminary analysis by Environmental Defense Fund of the leading studies to date finds that a well-designed smart grid can enable 30% reductions in carbon and health-damaging pollutants from the electric sector and 25% from the transport sector by 2030 [47].

2.3.6 BENEFIT SYNERGIES

Increasing reliance on distributed and demand side resources, reducing line losses, and increasing capacity of existing transmission lines through the use of dynamic thermal rating [48] and wide area control technology—all could reduce the need for new transmission and generation units [6], saving money and avoiding impacts on land and wildlife. The California Public Utility Commission recognized that value in its June 2010 decision on smart grid deployment plans: "The Smart Grid can decrease the need for other infrastructure investments and these benefits should be taken into account when planning infrastructure" [49].

Such analyses elsewhere in the country have resulted in the deferral of several transmission lines. Synapse, for instance, has provided expert testimony on electric power transmission issues on behalf of consumer advocates and environmental groups in Pennsylvania, Virginia, and Maine. The key issue in all three cases was "how recent increases in demand response and energy efficiency affect utility and RTO forecasts of the need for new transmission over the next decade." In Virginia, they demonstrated that—factoring in efficiency and demand response resources under development in PJM's easternmost states—an AEP/APS proposed 765 kV line would not be needed within the 10 year planning period. PJM sensitivity studies confirmed Synapse's estimates, and the transmission line application was withdrawn [50].

A well-designed smart grid will help electricity customers meet their need for affordable, adaptable, and efficient power. It will equip communities to protect public health, conserve water, and promote energy self-sufficiency and local economic development. And it will maximize the diversity of clean, low-carbon energy production, reducing the overall environmental footprint of the largest and most polluting industry in the world.

2.4 U.S. ELECTRIC UTILITY INDUSTRY CHALLENGES

The U.S. electric utility industry is likely to face a variety of new challenges over the next several decades including substantial changes to the generation mix and the need for new transmission to enable enhanced utilization of existing capacity and to enable the integration of new renewable

What Is Smart Grid, Why Now?

generation resources. The industry will also encounter a wide variety of technical opportunities over the next several decades as smart grid visions start to become reality. Some of these technical opportunities will require innovation on the part of both utilities and their regulators.

Some of the challenges that will be faced by the utility industry have already started to emerge, including the following:

- Annual electricity use in the typical home and average retail price per kilowatt hour continue to trend up year over year.
- The transformer fleet is aging, and the load on each transformer is continuing to rise; when frequency and severity of loss are taken into consideration, electric utilities face the highest risk [51].
- T&D losses amount to almost 6% of net generation [52].
- There has been a 3% increase per year in outage duration and a 4% increase per year in outage frequency over the past 5 years [53].
- Power outages and power quality disruptions cost U.S. businesses $150–$200 billion per year [54].
- There has been a well-documented decline in U.S. energy R&D spending and the underinvestment by both the public and private sector has become a focus of policy debate [55].
- By 2030, given the aforementioned points and the fact that electricity consumption is expected to increase at least 30% by 2020 [52], almost $1.5 trillion of investment is projected to be needed—$298 billion in transmission, $582 billion in distribution, and $85 billion in AMI and demand response [56].

This section introduces many of the new challenges that are likely to have a strong impact on the evolution of the utility industry over the next several decades.

2.4.1 GENERATION AND ENERGY RESOURCE MIX CHANGES

The nation's portfolio of generation and energy resources is undergoing significant change as the impact of various market and legislative forces is felt. Each option presents a different value proposition.

2.4.1.1 Coal

Coal provides nearly 50% of the United States's electricity generation at a relatively low average cost of $27 per MWh [57]. However, its share of electricity generation has been in decline for most of the past decade, while international demand has helped drive its cost steadily upward. While the Energy Information Administration forecasts coal to still produce 45% of the nation's electricity output in 2025, "cap and trade" legislation that constrains carbon emissions will effectively serve as a tax on coal that will necessarily drive producers to revisit and reallocate their fuel portfolio.

2.4.1.2 Natural Gas

Natural gas fuels nearly 22% of the nation's electricity, up from 15% in 1999, making it the second largest generation source. While these plants are far cleaner to operate than coal plants and far easier to build than either coal or nuclear plants, natural gas has proven to be a particularly volatile commodity, increasing fourfold in price between 1999 and 2008. In 2008, the average cost to produce electricity from natural gas was $84 per MWh, up by $20 per MWh in only 1 year. In subsequent years (2009–mid-2011), the cost to produce electricity with natural gas fell sharply.

2.4.1.3 Nuclear

Despite the fact that no new nuclear plants have been built in the United States in decades, nuclear plants still provide 20% of the nation's electricity at a price point (average of $17 per MWh) below coal and well below natural gas. In fact, of the $17 per MWh price, more than 70% goes to nonfuel operating and maintenance (O&M) expenses which helps to illustrate the plant operation challenges inherent with this power source. Nuclear power benefits from (1) a fuel source (uranium) that is relatively stable in price and (2) an increasing recognition by environmentalists that its minimal carbon footprint offsets the low probability risk of catastrophic failure. Nevertheless, plant construction is exceedingly slow due to regulatory, licensing, and siting issues.

2.4.1.4 Oil

In 2008, oil accounted for less of the nation's electricity than wind (only 1%). This represents a 70% reduction in only 3 years. This was driven in large part by dramatic spikes in the cost of oil, which increased the price point from $60 per MWh in 2004 to $180 per MWh in 2008.

2.4.1.5 Renewable Generation

Ever-heightening concern about the impact of power plant emissions on the environment and the climate has led to increased interest in renewable generation sources. Renewable energy accounted for 10% of the nation's electricity generation mix in 2010, with hydro making up the vast majority of this generation followed by wind, biomass, geothermal, and solar. However, every state in the United States has a statewide renewable electricity [58] goal and 29 states and the District of Columbia have renewable electricity standard mandates, known as RPS. RPS-type mechanisms have also been adopted in several other countries, including Western Europe (Britain, Italy, Poland, Sweden, and Belgium), [59] and in Latin America (Chile and Brazil). These mandates vary, but most are stated as a percentage of renewable energy in the generation portfolio by a specified date. In the United States, for example,

- Wisconsin 10% by 2015
- Virginia 15% by 2025
- Colorado 20% by 2020
- New York 25% by 2013
- California 33% by 2020
- Maine 40% by 2017

These policies, combined with the potential for federal carbon-constraining legislation and rising fuel commodity prices, have spurred significant investment in renewable generation. This can be seen in the trend in generation capacity brought online by fuel type, where wind's share of new capacity has grown from less than 10% in 2004 and 2005 to more than 40% by 2008.

A wide variety of "renewable" resources, each with their own challenges, can be used to generate electric power, including

Hydro: Hydro produces electricity at the lowest price point (approximately $10 per MWh) among the major generation sources. In addition, unlike solar or wind power, hydro is "dispatchable," while wind and solar output vary dramatically based on weather conditions. Hydro generation can be "ramped up" or "ramped down" depending on system requirements, though overall capacity is greatly influenced by precipitation amounts and resulting water levels. The largest hydro plant in the world is the Three Gorges Dam in China, which cost $30 billion to build with capacity in excess of 20 GW.

Wind: Wind energy is growing more rapidly in importance than any other generation source, and by mid-2011 wind power installations in the United States stood at 35,000 MW of wind energy in service. Globally, more than 50% of all wind generation is found in Europe, where wind accounts for much higher proportions of national "energy portfolios" than in the United States; Denmark generates nearly 20% of its power from wind, the largest proportion in the world. The largest wind energy complex ("wind farm") in the world is in Roscoe, Texas, with capacity of nearly 800 MW. It cost more than $1 billion to build with a footprint of nearly 100,000 acres in western Texas.

Solar: Solar energy is growing in importance, but remains more distributed (i.e., "micro") in usage than other renewable sources. Solar Energy Generating Systems (SEGS) in the Mojave Desert is the largest solar energy complex in the world, with capacity of more than 350 MW. Built in the 1980s, it utilizes nearly 1,000,000 mirrors over 1,600 acres in California. While SEGS is an example of a technology called concentrating solar power (CSP), there is a trend toward PV arrays for larger-scale generation. The Olmedilla Photovoltaic Park in Spain is the largest PV array in the world with a capacity of 60 MW, though much larger PV arrays are under consideration.

Geothermal: Geothermal energy is produced from the heat stored in the earth, with large-scale plants located near boundaries between tectonic plates. The largest geothermal generation complex in the world is located 70 miles north of San Francisco, with a 1.5 GW capacity. As with hydro, wind, and solar, the upfront costs of geothermal energy—primarily, the cost of drilling a geothermal well typically around 2 miles in depth—far outweigh the ongoing operating costs. Unlike hydro, wind, and solar, geothermal energy sources will be depleted by the power plant over time.

Biomass: Biomass energy is produced through methods more similar to fossil fuel plants, namely the burning of fuel. With biomass, however, the fuel is renewable—typically wood and waste. The most common wood source is waste product from the paper manufacturing industry. The most common "trash" sources are municipal or manufacturing waste and gas from landfills. Biomass plants are typically smaller in size than any of the other renewable sources noted. The largest biomass plant in the world is currently in Wales installed in 2010, with a capacity of 350 MW at a cost of around $650 million.

2.4.1.6 Energy Storage

There are a number of potential electricity storage technologies being researched and developed in 2011. Among these are pumped storage, compressed air, multiple battery chemistry formulations, flywheels, superconducting magnetic energy storage, and others. More information can be found at the Electricity Storage Association website (www.electricitystorage.org).

Battery storage technology can have a significant impact on the proportion of wind and solar energy in a generation portfolio. As this storage becomes increasingly cost-effective and scalable, wind and solar energy essentially will be to some extent "dispatchable" and load management will be greatly facilitated.

2.4.1.7 Consumer Demand Management

Environmental advocates have long maintained that comparing the relative merits of coal, nuclear, and natural gas (90% of the nation's generation portfolio) alone is fundamentally flawed because it does not include the demand reduction option. Conservation, if viewed as an energy source, can be a suitable and equivalent alternative to a new power plant. The ability to fully leverage this option, however, depends in large part on two factors: (1) technology that better enables customers to manage and control their usage and (2) rates that send price signals to customers while removing the financial disincentives for utilities to drive demand reduction.

2.4.2 Transmission Expansion

The changes in the generation mix described above will likely require substantial new transmission growth over the coming decades. Transmission network expansion, especially projects that connect renewable generation to densely populated regions of the country, will help the nation utilize its existing generation fleet more fully while providing stimulus for further investment in additional renewable capacity.

2.4.3 New Demands

EVs or expanded demand response resources could not only impact the generation mix but could also require other changes to how the power system is planned and operated.

2.4.4 NEW TECHNICAL OPPORTUNITIES

Technical challenges the industry will face include the following:

- Managing an increasing number of operating contingencies that differ from "system as design" expectations (e.g., in response to wind and solar variability)
- Facilitating the introduction of intermittent renewable and distributed energy resources with limited controllability and dispatchability
- Mitigating power quality issues (voltage and frequency variations) that cannot be readily addressed by conventional solutions
- Integrating highly distributed, advanced control and operations logic into system operations
- Developing sufficiently fast response capabilities for quickly developing disturbances
- Operating systems reliably despite increasing volatility of generation and demand patterns, given increasing wholesale market demand elasticity
- Increasing the adaptability of advanced protection schemes to rapidly changing operational behavior (due to the intermittent nature of renewable and DG resources)

2.4.5 REGULATORY CHALLENGES

To meet operations challenges, the industry is looking toward new technology while still relying on much that is a century old. However, the expectations of the end user have changed dramatically. Increasingly, utilities are attempting to build regulatory support—with mixed results—for smart grid investments. Utilities may find that these operations challenges cannot be met through new technology unless accompanied by increased investment in core technology. Investment, particularly in transmission infrastructure, has been far outpaced by load growth—significantly so in certain parts of the country—due, in large part, to difficulties in getting projects of this magnitude planned, approved, permitted, and funded.

The industry is returning to its reliance on rate cases to secure the level of revenue necessary to maintain a vital component of the national infrastructure, but is doing so without the same level of regulatory support it enjoyed prior to deregulation. While rate case frequency has increased, the average awarded return on equity for shareholder-owned electric utilities in the United States has declined steadily. This reflects, in part, the industry's mixed success in rebuilding the regulatory relationships damaged by deregulation initiatives that either failed to generate the expected results or were outright disasters. Rebuilding trust will be essential, whether seeking approval for new technology or simply reaching reasonable outcomes on rate cases. Going forward, the trend is for utilities to submit rate cases far more frequently to regulators than in the recent past. These regulatory discussions are also increasingly turning to matters of technology that could provide enhanced service to customers, including the ability to manage their usage more proactively.

2.5 FEDERAL SMART GRID INFLUENCES IN THE UNITED STATES

2.5.1 ENERGY INDEPENDENCE AND SECURITY ACT OF 2007, TITLE XIII

In December 2007, Congress passed, and the President approved, Title XIII of the Energy Independence and Security Act of 2007 (EISA). EISA provided the legislative support for the U.S. DOE's smart grid activities and reinforced its role in leading and coordinating national grid modernization efforts. Title XIII of the EISA is titled smart grid and established smart grid policy creating several initiatives and studies. Key provisions of Title XIII are as follows [60]:

- Section 1301 of Title XIII of the EISA established the policy of the United States to support modernization of the nation's electricity T&D system to maintain a reliable and security electricity infrastructure.

- Section 1302 established a requirement for the DOE to report to Congress every 2 years on the status of smart grid deployments and any regulatory or government barriers to smart grid.
- Section 1303 establishes within the DOE the Smart Grid Advisory Committee and the Federal Smart Grid Task Force.
- Section 1304 authorizes the DOE to develop a "Smart Grid Regional Demonstration Initiative."
- Section 1305 directs the National Institute of Standards and Technology (NIST), with DOE and others, to develop a Smart Grid Interoperability Framework.
- Section 1306 authorizes the DOE to develop a "Federal Matching Fund for Smart Grid Investment Costs."
- Section 1307 requires each state and utilities to consider investment in smart grid technologies and systems when investing in grid maintenance, expansion, and upgrades.
- Section 1308 requires the DOE to conduct a study on the effect that private wire laws has on combined heat and power development.
- Section 1309 requires the DOE to conduct a study on security attributes of smart grid systems.

Characteristics of a smart grid as described by Title XIII of the EISA include the following:

1. Increased use of digital information and controls technology to improve reliability, security, and efficiency of the electric grid
2. Dynamic optimization of grid operations and resources, with full cybersecurity
3. Deployment and integration of distributed resources and generation, including renewable resources
4. Development and incorporation of demand response, demand side resources, and energy efficiency resources
5. Deployment of "smart" technologies (real-time, automated, and interactive technologies that optimize the physical operation of appliances and consumer devices) for metering, communications concerning grid operations and status, and distribution automation
6. Integration of "smart" appliances and consumer devices
7. Deployment and integration of advanced electricity storage and peak-shaving technologies, including plug-in electric and hybrid EVs, and thermal storage air conditioning
8. Provision to consumers of timely information and control options
9. Development of standards for communication and interoperability of appliances and equipment connected to the electric grid, including the infrastructure serving the grid
10. Identification and lowering of unreasonable or unnecessary barriers to adoption of smart grid technologies, practices, and services

2.5.2 AMERICAN RECOVERY AND REINVESTMENT ACT OF 2009

The American Recovery and Reinvestment Act of 2009—abbreviated ARRA, and commonly referred to as the Stimulus or the Recovery Act—is an economic stimulus package enacted by the 111th U.S. Congress in February 2009 and signed into law on February 17, 2009, by President Barack Obama. To respond to the late 2000s recession, the primary objective for ARRA was job preservation and creation. Secondary objectives were to provide temporary relief programs for those most impacted by the recession and invest in infrastructure, education, health, and "green" energy. The approximate total cost of the economic stimulus package was estimated to be $787 billion at the time of passage. The Act included direct spending in infrastructure, education, health, and energy; federal tax incentives; expansion of unemployment benefits; and other social welfare provisions [61].

Some of the ARRA funding was issued by each state, which evaluated proposals and issued grants; some of the funding was directed by the DOE, and some of the ARRA funding was issued in the form of loan guarantees. More than 300 recipients received ARRA funding specifically for smart grid related projects and programs, including the following [62]:

- Ninety-nine for Smart Grid Investment Grants (SGIGs) with a total obligation of $3.48 billion
- Thirty-two for smart grid regional and energy storage demonstration projects with a total obligation of $620 million
- Fifty-two for workforce development programs with a total obligation of $100 million
- Six for interconnection transmission planning with a total obligation of $80 million
- Forty-nine for state assistance for electricity policies with a total obligation of $48.62 million
- Fifty for enhancing state energy assurance with a total obligation of $43.5 million
- Forty-three for enhancing local government energy assurance with a total obligation of $8.02 million
- One for interoperability standards and framework with a total obligation of $12 million

A wide range of activity and funding levels exists within these categories. The numerous projects funded by the ARRA stimulus program are now well underway. The results from these projects and programs will go a long way toward stimulating further private investment into smart grid development among American utilities.

2.5.3 THE U.S. DEPARTMENT OF ENERGY

The U.S. DOE is a cabinet-level department of the U. S. government responsible for research and development of energy technology—from electric cars to sustainable energy projects—such as clean-coal production, energy conservation, nuclear cleanup from the Cold War era, energy data collection, and selling government-made power to the public [63]. The DOE sponsors more basic and applied scientific research than any other U.S. federal agency; most of this is funded through its system of U. S. Department of Energy National Laboratories. The agency is administered by the U. S. Secretary of Energy.

2.5.3.1 Demonstration and Investment Grants for Smart Grid Projects

The ARRA provided the largest single energy grid modernization investment in U.S. history, funding a broad range of technologies that will spur the nation's transition to a smarter, stronger, more efficient and reliable electric system. The end result will promote energy-saving choices for consumers, increase efficiency, and foster the growth of renewable energy sources like wind and solar.

Ninety-nine private companies, utilities, manufacturers, cities, and other partners received the SGIG awards. The $3.48 billion in SGIG awards are part of the ARRA, matched by industry funding for a total public–private investment worth over $8 billion. Along with the SGIG projects, the DOE awarded $620 million for 32 Smart Grid Demonstration Projects (SGDP) around the country, matched with about $1 billion of industry investment to demonstrate advanced smart grid technologies and integrated systems. These demonstration projects, which include large-scale energy storage, smart meters, distribution and transmission system monitoring devices, and a range of other smart technologies, will act as models for deploying integrated smart grid systems on a broader scale. Applicants stated that the smart grid funded projects will create tens of thousands of jobs, and consumers will benefit from these investments in a stronger, more reliable grid and serve as a catalyst for a full deployment of the smart grid in the United States. The smart grid projects will provide invaluable data on the benefits and cost-effectiveness of the smart grid, including energy and cost savings. An analysis by the EPRI estimates that implementing smart grid technologies could reduce electricity use by more than 4% by 2030 [64]. The demonstration projects will also help verify the technological and business viability of new smart technologies and show how fully integrated smart grid systems can be readily adapted and implemented around the country. Applicants say this investment will create thousands of new job opportunities that will include manufacturing workers, engineers, electricians, equipment installers, IT system designers, cybersecurity specialists, and business and power system analysts.

The SGDP funding awards are divided into two main smart grid areas of application. In the first area of SGDP, 16 awards totaling $435.2 million will support fully integrated, regional smart grid demonstrations in 21 states, representing over 50 utilities and electricity organizations with a combined customer base of almost 100 million consumers. The projects include streamlined communications technologies that will allow different parts of the grid to exchange data in realtime, sensing and control devices that help grid operators monitor and control the flow of electricity to avoid disruptions and outages, smart meters and in-home systems that empower consumers to reduce their energy use and save money, energy storage options, and onsite and renewable energy sources that can be integrated onto the electrical grid. In the second area of SGDP, an additional 16 awards for a total of $184.8 million will help fund utility-scale energy storage projects that will enhance the reliability and efficiency of the grid, while reducing the need for new electricity plants. Improved energy storage technologies will allow for expanded integration of renewable energy resources, such as wind and PV systems, and will improve frequency regulation and peak energy management. The selected projects include advanced battery systems (including flow batteries), flywheels, and compressed air energy systems (CAES).

In addition to the SGIGs and SGDPs, the DOE awarded ARRA funds to the NIST to develop an Interoperability Framework ($12M); transmission analysis and planning ($80M), state regulator assistance ($43.5M), state and local planning for smart grid resiliency ($51.5M), and workforce development ($100M).

Federal Technical Project Officers will manage the 99 grants awarded under the SGIG and the 32 cooperative agreements awarded under the SGDP. Metrics and benefits analysis will be conducted on both the SGIG and SGDP projects. Consumer behavior studies will be part of selected SGIG projects. Results and other information concerning the smart grid ARRA projects will be posted at www.smartgrid.gov. A biannual report by the DOE to Congress provides updates on the status of smart grid efforts and technologies implemented through the smart grid ARRA projects.

2.5.3.2 Smart Grid Task Force

The Federal Smart Grid Task Force was established under Title XIII of the EISA and includes representatives from multiple federal agencies. The lead organization for the task force is the DOE's Office of Electricity Delivery and Energy Reliability. The DOE Office of Energy Efficiency and Renewable Energy and the DOE National Energy Technology Laboratory are also members of the task force. Other representatives on the task force include the FERC, the Department of Commerce, the Environmental Protection Agency (EPA), the Department of Homeland Security, the Department of Agriculture, the Department of Defense, the Department of State, the Federal Communications Commission (FCC), and the National Aeronautics and Space Administration (NASA). The mission of the Smart Grid Task Force is to ensure awareness, coordination, and integration of the diverse activities related to smart grid technologies, practices, and services. The Smart Grid Task Force collaborates with the DOE's Electricity Advisory Committee (EAC) and implements administration policies articulated by the DOE National Science and Technology Council (NSTC) subcommittee on smart grid while coordinating federal research, development, and demonstration; international activities; and outreach and education efforts [65].

2.5.3.3 Electricity Advisory Committee

The EAC, which reports to the assistant secretary for Electricity Delivery and Energy Reliability of the DOE, was established to enhance leadership in electricity delivery modernization and provide senior level counsel to the DOE on ways in which the United States can overcome the many barriers to moving forward, including the deployment of smart grid technologies, research and development of energy storage technologies, renewable energy resource system integration, and new transmission infrastructure [66]. The mission of the EAC is to assist the DOE in implementing the Energy Policy Act of 2005, the EISA, and modernizing the nation's electricity delivery infrastructure. The EAC was established under the Federal Advisory Committee Act.

The EAC provides advice and recommendations to the DOE on several broad subjects including the following:

- Ensuring electricity supply adequacy, including plans for new base load generation capacity needed to meet demand growth, likely retirements of existing capacity, energy efficiency programs, and development of new renewable-based capacity
- DOE's electricity-related research and development programs
- Modernization of the United States' electricity infrastructure (generation, transmission, and distribution), regionally and nationally, including implementation of smart grid technology and energy storage technology
- Enhancing the resilience of the nation's energy supply infrastructure to cope with natural disasters, equipment failures, human errors, and man-made disruptions

2.5.4 THE NATIONAL INSTITUTE OF STANDARDS AND TECHNOLOGY

NIST, an agency of the U.S. Department of Commerce, was founded in 1901 as the nation's first federal physical science research laboratory. NIST works with the industry to develop and apply technology, measurements, and standards. As outlined in the EISA, NIST has been given "primary responsibility to coordinate development of a framework that includes protocols and model standards for information management to achieve interoperability of smart grid devices and systems" [67].

The NIST smart grid roadmap is the phase one of a three-phase plan to establish standards, priorities, and a framework to achieve smart grid interoperability. The second phase of the plan is the Smart Grid Interoperability Panel (SGIP). The SGIP is an ongoing, public–private organization that provides an open process through which stakeholders can participate in coordinating, harmonizing, and accelerating smart grid standards development. The third phase of the plan is the establishment of a framework for testing conformity with smart grid standards and certifying the compliance of smart grid devices and systems.

To help guide the industry, NIST defines interoperability as follows:

> The capability of two or more networks, systems, devices, applications, or components to exchange and readily use information - securely, effectively, and with little or no inconvenience to the user. The Smart Grid will be a system of interoperable systems. That is, different systems will be able to exchange meaningful, actionable information. The systems will share a common meaning of the exchanged information, and this information will elicit agreed-upon types of response. The reliability, fidelity, and security of information exchanges between and among Smart Grid systems must achieve requisite performance levels.

The EPRI was selected by NIST to facilitate the development of the smart grid interoperability roadmap. NIST designated the National Electrical Manufacturers Association (NEMA) to sit on the standards roadmap task force. In addition to NIST, NEMA will work closely with other organizations, such as EPRI, to ensure uniform standards that will simplify new product design, engineering, manufacturing, and installation. The NIST roadmap contains several important items that shape the work of the SGIP:

- A conceptual reference model to present a shared view of smart grid's complex system of systems and to facilitate design of smart grid architecture (see Figure 2.2)

What Is Smart Grid, Why Now?

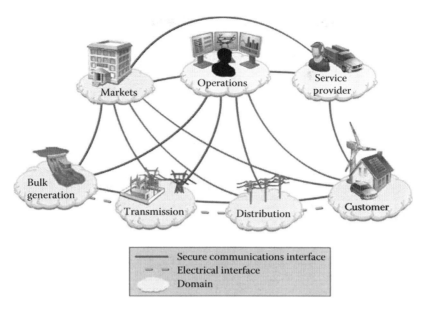

FIGURE 2.2 NIST smart grid conceptual reference model. (From National Institute of Standards and Technology (NIST), U.S. Department of Commerce, NIST framework and roadmap for smart grid interoperability standards, Relase 2.0, NIST Special Publication 1108R2, February 2012.)

- An initial set of 75 smart grid standards for implementation to address issues identified by NIST and focused on priorities identified in the FERC smart grid policy, specifically demand response and consumer energy efficiency, wide area situational awareness, energy storage, electric transportation, AMI, distribution grid management, network communications, and cybersecurity
- Priorities for developing additional standards and making revisions to existing standards, with supporting action plans to resolve major gaps affecting interoperability and security of smart grid components
- Initial steps toward a smart grid cybersecurity strategy to assess risks and to identify requirements to address those risks

2.5.4.1 Smart Grid Interoperability Panel

Initiated by NIST and established in November 2009, the SGIP is dedicated to the development of a framework for interoperability of smart grid devices and systems. According to the SGIP charter

> The Smart Grid Interoperability Panel is a membership-based organization created by an Administrator under a contract from NIST to provide an open process for stakeholders to participate in providing input and cooperating with NIST in the ongoing coordination, acceleration and harmonization of standards development for the Smart Grid. The SGIP also reviews use cases, identifies requirements and architectural reference models, coordinates and accelerates Smart Grid testing and certification, and proposes action plans for achieving these goals. The SGIP does not write standards, but serves as a forum to coordinate the development of standards and specifications by many standards development organizations.

Thus, the SGIP not only identifies and addresses standardization priorities, but also plays a leadership role in facilitating and developing an information architecture, a cybersecurity strategy, and a framework for testing and certification. It focuses on analysis and coordination of efforts in helping NIST fulfill its responsibilities under EISA. The NIST roadmap is the starting point for this activity. The SGIP structure depicted in Figure 2.3 enables the SGIP to accomplish its complex and

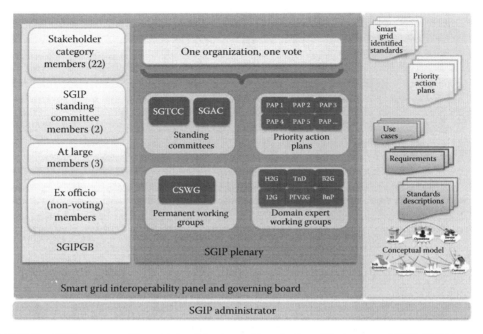

FIGURE 2.3 SGIP structure. (From National Institute of Standards and Technology (NIST), U.S. Department of Commerce, NIST framework and roadmap for smart grid interoperability standards, Relase 2.0, NIST Special Publication 1108R2, February 2012.)

urgent work. The SGIP membership is led by three core teams—NIST, plenary officers, and a governing board—and it is fully supported by an administrator. The governing board maintains a broad community-based perspective by having a breadth of experience, knowledge, and involvement. It also holds consensus as a core value, ensuring that all legitimate views and proposals are considered. Key responsibilities include approving and prioritizing work programs, facilitating dialogue with standards development organizations, and arranging for necessary resources for the SGIP.

The NIST membership is large and diverse by design, as it is free and open to all who share the smart grid vision. To date, it consists of some 1700 individuals from 590 member organizations (90%—United States, 5%—Canada, 5%—other international) representing 22 stakeholder categories. Furthermore, the membership is organized into the following standing committees, working groups and teams, and is now supported by a Program Management Office:

1. Standing Committees and Working Groups
    a. Architecture (SGAC)
    b. Cybersecurity (CSWG)
    c. Test and Certification (SGTCC)
2. Domain Expert Working Groups
    a. Transmission and Distribution (T&D)
    b. Industry to Grid (I2G)
    c. Building to Grid (B2G)
    d. Home to Grid (H2G)
    e. Vehicle to Grid (V2G)
    f. Business and Policy (BnP)
3. Priority Action Plan (PAP) Teams
    a. Meter Upgradeability Standard
    b. Role of IP in the smart grid

What Is Smart Grid, Why Now? 39

    c. Wireless Communications for the smart grid
    d. Common Price Communication Model
    e. Common Scheduling Mechanism
    f. Standard Meter Date Profiles
    g. Common Semantic Model for Meter Data Tables
    h. Electric Storage Interconnection Guidelines
    i. Common Information Model (CIM) for Distribution Grid Management
    j. Standard DR and DER Signals
    k. Standard Energy Usage Information
    l. Common Object Models for Electric Transportation
    m. IEC 61850 Objects/DNP3 Mapping
    n. Time Synchronization, IEC 61850 Objects/IEEE C37.118 Harmonization
    o. T&D Power Systems Model Mapping
    p. Harmonize PLC Standards for Appliance Communications in the Home
    q. Wind Plant Communications

With the work being undertaken by NIST and the SGIP, combined with a new focus on telecommunications and cybersecurity initiatives, the road ahead looks friendlier for continued public investments and an increased level of private funding for the mid-size and smaller market participants coupled with direct investments being planned by the major utilities.

2.5.4.2 Smart Grid Advisory Committee

In September 2010, NIST named 15 individuals from U.S. industry, academia, and trade and professional organizations to serve on its Smart Grid Federal Advisory Committee (SGFAC). The NIST Smart Grid Advisory Committee provides input to NIST on the smart grid interoperability standards, priorities, and gaps and on the overall direction, status, and health of the smart grid implementation by the smart grid industry, including identification of issues and needs. Input to NIST will be used to help guide the activities of the SGIP activities and also to assist NIST in directing smart grid-related research and standards activities. The duties of the Committee are solely advisory in nature [69].

2.5.5 THE FEDERAL ENERGY REGULATORY COMMISSION

In response to an energy crisis, the U.S. Congress passed the DOE Organization Act in 1977, which consolidated various energy-related agencies into the DOE. Congress insisted that a separate independent regulatory body be retained in the DOE, called the FERC, preserving its independent status "within" the Department. FERC has jurisdiction over interstate electricity sales, wholesale electric rates, hydroelectric licensing, natural gas pricing, and oil pipeline rates. FERC also reviews and authorizes liquefied natural gas (LNG) terminals, interstate natural gas pipelines, and non-federal hydropower projects [70]. The U.S. Energy Policy Act of 2005 expanded FERC's authority to impose mandatory reliability standards on the bulk transmission system and to impose penalties on entities that manipulate the electricity and natural gas markets.

    To help support the modernization of the nation's electric system consistent with Title XIII of the EISA, FERC is focusing on issues associated with a smarter grid. FERC adopted a smart grid policy in July 2009. The new policy adopts as a priority the early development by industry of smart grid standards to

- Ensure the cybersecurity of the grid
- Provide two-way communications among regional market operators, utilities, service providers, and consumers
- Ensure that power system operators have equipment that allows them to operate reliably by monitoring their own systems as well as neighboring systems that affect them

- Coordinate the integration into the power system of emerging technologies such as renewable resources, demand response resources, electricity storage facilities, and electric transportation systems

The policy also provides for early adopters of smart grid technologies to recover smart grid costs if they demonstrate that those costs serve to protect cybersecurity and reliability of the electric system, and have the ability to be upgraded, among other requirements. Importantly, the policy statement also explains that by adopting these standards for smart grid technologies, FERC will not interfere with any state's ability to adopt whatever advanced metering or demand response program it chooses. In adopting this policy, FERC continues to abide by the Federal Power Act's jurisdictional boundaries between federal and state regulation of rates, terms and conditions of transmission service, and sales of electricity [71]. Most of FERC's priorities were expected; however, FERC added a controversial aspect: It strayed from standard practice and approved an interim rate policy for early adopters. Normally, utilities are only allowed to raise their rates to recover expenses involving technologies that have been proven to be cost-effective. State regulators plan to let power providers who pioneer the technology pass their costs on to their customers, before national standards are approved and before analysts have determined the most cost-effective technologies. Under FERC's interim rate policy, for a power provider to be allowed to recover its costs from ratepayers, it will have to meet four requirements: It must advance the goals of the EISA, demonstrate that the reliability and security of the bulk-power system will not be adversely affected, minimize stranded costs, and agree to provide feedback to the DOE that is useful to the development of the smart grid and its interoperability standards [72].

FERC's interest and responsibilities in the smart grid area derive from its authority over the rates, terms and conditions of transmission, and wholesale sales in interstate commerce, its responsibility for approving and enforcing mandatory reliability standards for the bulk-power system in the United States, and a recently enacted law requiring the Commission to adopt interoperability standards and protocols necessary to ensure smart grid functionality and interoperability in the interstate transmission of electric power and in regional and wholesale electricity markets [73].

2.6 INTERNATIONAL TREATIES AND NONGOVERNMENT ORGANIZATIONS

2.6.1 Treaties and Negotiations

The impetus for smart grid development around the world ranges from the need to develop basic utility infrastructure where none exists to concerns regarding climate change and environmental sustainability. The IEA views smart grids as "the essential key to reliable electricity networks that can integrate a broader portfolio of responses to sustainability criteria and take account of evolving energy approaches among both power producers and power consumers" [74], and has identified key drivers, including incentives and legislation. Its "2011 Technology Roadmap on Smart Grids" outlines a path to halve the carbon footprint of global electricity consumption: from 40% of global CO_2 emissions today to 21% by 2050, an annual reduction of over 20 Gt of CO_2, with smart grid technologies essential to achieving these emissions reductions [75]. That modernization and decarbonization of the electric sector will require a global investment of about $13 trillion over the next 20 years [56].

While no global treaties center specifically and entirely on smart grid, it is increasingly a focus of multinational negotiations, partnerships, and international NGOs.

The 2009 Copenhagen Accord, supported by 120 developed and developing countries, saw the 10 largest emitters pledge concrete targets and actions [76], creating incentives for the development of smart grid globally as a means to reduce GHGs. China pledged to cut its carbon emissions per unit of economic growth by 40%–45% by 2020 from 2005 levels, and to begin domestic carbon trading programs during its 12th Five-Year Plan period (2011–2015) to meet that target.

The Major Economies Forum on Energy and Climate (MEF), representing the 17 largest economies of the world, also announced at Copenhagen a new Global Partnership centered on a "suite of plans which span ten climate-related technologies that together address more than 80 percent of the energy sector carbon dioxide emissions reduction potential identified by the IEA." One of the 10 is a smart grid plan, led by Italy and South Korea, which calls for "establishing a platform to enable cross-country and cross-regional development and coordination of smart grid technology standards" [77]. Outlined in this plan are (1) a description of smart grid technologies, which are "expected to enable improved power reliability, safety and security, energy efficiency, environmental impact, and financial performance... [as well as] grid integration for renewable power generation by facilitating load management and storage technologies integration," (2) the climate mitigation potential of smart grid as a "a key enabler for other CO_2-reduction technologies and solutions," and (3) potential actions to "accelerate the development and deployment of smart grid technology" [78]. These actions include

> supporting innovation by working with current research initiatives to integrate and align development efforts across the globe, accelerating deployment, enabling cross-country and cross-regional smart grids standards development and coordination, and facilitating information sharing by developing and managing a central global repository with past and ongoing smart grids R&D, pilot, and full-scale deployment efforts by different entities [77]

Following on the MEF plan, in July 2010 the United States, South Korea, Italy, Japan, and 12 other nations launched the International Smart Grid Action Network (ISGAN), a framework for high-level voluntary coordination among governments, the IEA, and the International Organization for Standardization to speed global development of smart grids. Supported by more than 15 national-level governments, ISGAN focuses on aspects of smart grid where governments have regulatory authority, expertise, convening power, or other leverage, which include policy, standards and regulations, finance and business models, technology and systems development, user and consumer engagement, and workforce skills and knowledge.

The World Economic Forum (WEF) is also engaged in transnational smart grid coordination. It has established an international public–private portfolio of 10 large-scale integrated SGDPs across different regulatory regimes and implementing environments in order to clarify what can be learned from existing pilot projects. WEF's goals include defining success criteria, establishing a value case for future pilots and creating a roadmap for smart grid rollout. WEF will also develop recommendations to inform investment and policymaking, catalyze partnerships across the smart grid value chain to share risk, and illustrate the value proposition to investors and governments [79].

In July 2010, the U.S. Secretary of Energy Steven Chu hosted the first Clean Energy Ministerial in Washington DC, aimed at accelerating global cleantech investment. Participating countries, including the United Kingdom, China, Italy, Norway, South Korea, and 19 others, account for 70% of global GHG emissions and 80% of global GDP. Their energy ministers committed hundreds of millions of dollars to 11 efficiency and renewable energy initiatives aimed at avoiding the need to build 500 mid-size power plants during the next 20 years, including efforts focused on EV deployment and smart grid. Russia and the United States announced a "company-to-company pilot project" and utilities formed partnerships focused on smart grid technology application to optimize T&D and advance innovation.

The European Technology Platform Initiative, launched in 2005, has provided another forum "to promote the positive development of numerous European electricity networks corresponding with the European Union's 2020 goals for renewable energy and energy efficiency" [80]. Bilateral partnerships include that between Japan and Spain, who have partnered with Mitsubishi and Hitachi to advance smart infrastructure to support EVs in southern Spain. The Japan and Spain Innovation Program will "not only provide smart grid access, but also global business expertise from the Japanese companies, energy management and ICT systems" [80].

Other relevant international negotiations and collaborations include (1) the International Renewable Energy Agency (IRENA), which seeks to promote the adoption of all forms of renewable energy worldwide that includes among its signatories 48 African states, 37 European, 34 Asian, 15 American, and 9 Australia and Oceania States [81], (2) the International Conference on Integration of Renewable and Distributed Energy Resources, which addresses "technical, market, and regulatory issues as a prerequisite to prepare massive DER integration into power grids," and (3) the International Electric Vehicle Symposium uniting the World Electric Vehicle Association, the European Association for Battery, Hybrid and Fuel Cell Electric Vehicles (AVERE), the Electric Drive Transportation Association (EDTA), and the Electric Vehicle Association of Asia Pacific (EVAAP).

2.6.2 NATIONAL AND REGIONAL ACTION AND PRECEDENT

The high-speed transition to smarter grids in some nations and regions—accelerated by "green stimulus" packages from the 11 major world economies totaling $173.6 billion [82]—has set the stage for further negotiations toward global standards, protocols, and treaties.

2.6.2.1 Asia/Pacific

China is currently outpacing the rest of the world in developing smart grid infrastructure to meet domestic needs and secure global leadership in engineering and manufacturing next-generation grid technologies. In 2009, the Chinese government spent more on building a smarter grid than on power generation, according to the China Electricity Council [83]. In 2010 China's State Grid Corporation, which controls the grid that covers 80% of mainland China, committed to build out its "strong and smart" grid by 2020. By spring 2011, the company had moved into its "comprehensive construction phase," rolling out 50 million smart meters as well as multiple smart substations, automated distribution systems, intelligent communities, and comprehensive pilot projects. Having already deployed the most extensive EV charging network in the world, the 12th Five Year Plan mapped plans for more than 2000 new recharging and switching stations and more than 200,000 recharging poles, creating a network to accommodate a predicted 500,000 EVs by 2015. Already managing more than 28 GW of wind power in its service area, State Grid is designing its smart infrastructure to support 100 GW of wind and 5 GW of solar PV by 2015. It has also led on formulating standards, including 15 standards for smart substations, the world's first. By 2020, the completion of its "Strong and Smart Grid" will, State Grid predicts, reduce CO_2 emissions by 1.65 billion tons, reducing carbon intensity by 8.8% compared with 2005, 20% of the overall target [84].

South Korea is moving nearly as quickly. In 2009, it created the Korea Smart Grid Institute and in 2010 announced four projects in Illinois. A $25 million smart buildings project in Chicago led by Korea Telecom and LG will enable DR aggregation and sale into the wholesale market (PJM); the University of Chicago and Korean National Research Institute will collaborate on research and development of cybersecurity and network resilience; IIT and the Korea Electrotechnology Research Institute will work together on distribution automation, DG, storage integration, and building energy management to tie microgrids into the main grid; and the Korean Electrical Engineering and Science Research Institute will help deepen IIT's smart grid training and bring senior level executives to learn best practices and skills. According to Choi Kyunghwan, South Korea's Minister of Knowledge Economy, "our primary goal is to expand to a worldwide smart grid. If we have a first success story [in Illinois], it will be easier to expand" [85]. Within South Korea, more than 150 enterprises in telecom, appliance manufacture, and transport are participating in a $100 million demonstration project on the island of Jeju linking 6000 homes equipped with smart appliances and EVs with renewable generation.

Toshiba's May 2011 $2.3 billion purchase of Landis & Gyr, which controls 30% of the global smart meter market, positions Japan as a strong competitor to Korea in the export market [86], Japan's New Energy and Industrial Technology Development Organization (NEDO) has gained a foothold in the U.S. market with pilots in New Mexico and a $37 million project with six companies, including Hitachi and Sharp, in Hawaii, focused on applications of smart grid for island states and nations.

In its study of global pilots for the WEF, Accenture notes the particular promise of island nations for quick smart grid adoption, with dense urban areas conducive to EV use and streamlined regulatory environments [87]. Their dependency on external sources of energy and high vulnerability to the effects of global warming provide further incentive to diversify the energy mix toward microgeneration and renewables.

The WEF report particularly spotlighted Singapore as an ideal place to launch an EV network due to its (1) contained urban area; (2) top–down policy environment; (3) highly efficient and reliable grid and sophisticated IT sector; (4) key vulnerabilities to climate change include coastal land loss, increased flooding, impact on water resources, and spread of disease; (5) energy security concerns, given its lack of indigenous oil, gas, hydro, and geothermal resources; and impracticality of nuclear energy, given its population density and size; (6) belief in a liberalized market to provide a platform for innovation, but with market failures corrected through standards and regulations; and (7) an ambition to develop leading-edge expertise that can be exported to larger markets [88].

In Australia, the state of Victoria became one of the first in the world to mandate full deployment of smart meters, while the national government, with partners in the power and water sectors, committed $100 million to develop a Smart City demonstration project and support installation of Australia's first commercial-scale smart grid [89]. New Zealand's Transpower has allocated $10 million in initial funding for smart grid investment [80].

2.6.2.2 Europe

Europe's early adoption of a cap on carbon emissions, deep financial and policy support for renewable generation (including a binding goal of 20% renewable energy by 2020) [90], and new carbon regulations for new passenger cars [91], which will accelerate the move to PEVs [75], has impelled it to lead in the smart grid space as well.

The Smart Grids European Technology Platform for Electricity Networks of the Future began its work in 2005. The European Community recently mandated that all 27 member states give citizens information on their power use, and the European Union set a target to deploy smart meters for more than 80% of customers by 2020.

Individual nations are also driving progress. Italy, Denmark, and Ireland are all global leaders in scale adoption of AMI and demand side balancing of high levels of renewables [92]. A large-scale North Sea Grid connecting nine countries was announced in December 2009 [90]. In France, the electricity distribution operator ERDF—the largest vertically integrated electric utility in the world—deployed 300,000 smart meters and 7000 low-voltage transformers in a pilot project based on an advanced communications protocol named Linky. If deemed a success, ERDF will replace all 35 million meters with Linky smart meters by 2016, at a cost of 4–6 billion Euros [75].

In July 2009, the United Kingdom published the Low Carbon Transition Plan (LCTP), including a commitment that every home would be fitted with smart meters by the end of 2020, requiring installation of 47 million meters. Implementation is being led jointly by the Department of Energy and Climate Change (DECC) and the Office of Gas and Electricity Markets (Ofgem), with strong support from the British Electrotechnical and Allied Manufacturers Association [93]. It has been described by DECC as "the biggest energy industry change since the changeover to North Sea Gas" [94].

Business model changes have not kept pace with technology advances. Bob Heile, Chairman of IEEE's Working Group on Wireless Personal Area Networks, believes the European Union still has significant work to do in developing new pricing models to encourage off-peak usage. As currently envisioned, load shifting will require customers to proactively adjust their power usage based on information they receive independently of metering infrastructure via the Internet [95].

2.6.2.3 Africa/Latin America

In Africa, the primary activity to date has centered on the Desertec and Super Smart Grid initiatives, aimed at linking Europe to the enormous solar and wind resource in the North African deserts [96]. South Africa has begun deploying smart technologies to address reliability issues, and BPL is deploying smart grid technologies and broadband over power line communications technology on a section of the distribution network for Ghana's national utility [97]. New initiatives may emerge from the first Middle East and North Africa (MENA) conference on smart grids, to be held in Cairo in December 2011.

Latin America is also in an exploratory stage. Brazil's utility association (APTEL) is working with the government on narrowband power line carrier trials with a social and educational focus. Several utilities are also managing smart grid pilots, including Ampla, a power distributor in Rio de Janeiro State owned by the Spanish utility Endesa, which is deploying smart meters and secure networks to reduce losses from illegal connections. AES Eletropaulo, a distributor in São Paulo State, has developed a smart grid business plan using the existing architecture developed by the IntelliGrid Consortium, an initiative of the California-based EPRI [75].

Other nations not directly investing in smart grid are adding renewable energy, EVs, and DG at a pace that will ultimately require more advanced grid management. David Wheeler, a senior fellow at the Center for Global Development notes that since 1990, developing countries have accounted for 55% of the global increase in clean energy development. A groundbreaking agreement by a group of island states, led by Grenada, to increase energy efficiency, lower fossil fuel consumption, and adopt hard targets for lowering tons of GHG emissions [98] underscores the need for new utility infrastructure. Green Energy Corp, a Colorado-based company, has developed what it calls the "first universal plan to provide developing nations with renewable power" and begun piloting their "Global Energy Model, or GEM" in Haiti, aiming to develop a model for Ethiopia, Sudan, Pakistan, and the Congo [99].

2.6.3 Nongovernmental Organizations

To date, only a few international NGOs have engaged in advocacy around smart grid implementation, regulation, and market structure. The Smart Green Grid Initiative (SGGI)—an industry consortium—was an official delegate in Copenhagen [100]. Environmental Defense Fund, headquartered in the United States but active in China, India, Korea, Southeast Asia, Mexico, the Caribbean, and Brazil, as well as in ongoing international climate negotiations, is engaged in smart grid deployments and advocacy in a number of states and regions, including the Pecan Street Project in Austin, Texas. The GridWise Alliance has a mission to "facilitate effective collaboration among all stakeholders, and promote, educate, and advocate for the adoption of innovative smart grid solutions that will achieve economic and environmental benefits for customers, communities and shareholders" [101]. The GridWise Alliance was instrumental in helping Japan initiate "a close working relationship [with] the United States to develop and deploy the smart grid in each nation" [102].

To facilitate further international smart grid dialogue and collaboration, the Global Smart Grid Federation was established in April of 2010 to "support rapid implementation of Smart Grid technologies, foster the international exchange of ideas and best practices on energy issues, and create avenues for dialogue and cooperation between the public and private sectors in countries around the world" [103].

2.7 SMART GRID INDUSTRY INITIATIVES

There are numerous smart grid activities worldwide that are driven by alliances, working groups, committees, and independent organizations. Any efforts undertaken to build intelligence into the electrical network must be tightly coupled with regulatory body positions and initiatives, and governmental policy. These organizations often provide the incentive, justification, and mandates for utilities to make the initial investments in smart grid. Some smart grid initiatives are summarized in the following.

2.7.1 EPRI IntelliGrid™ Methodology

The IntelliGrid methodology came from a multiyear study by EPRI in the 2001–2004 period known as the "Integrated Energy and Communication Systems Architecture" (IECSA). The results of this work are publicly available [104] and have been used internationally by utilities to develop smart grid roadmaps for their enterprise systems and grid infrastructures and support ongoing development of communications architectures that will enable interoperability between products and systems. The methodology was approved by the IEC in 2008 (IEC/PAS 62559) as a publicly available specification for developing requirements for energy systems.

The EPRI approach to architecting smart grid standards uses a combination of utility operations experience with communications and information technology knowledge. This integrated approach ensures a "no regrets" path for smart grid technology standards development that can save utilities millions of dollars in research and development and helps to accelerate the adoption of those standards. IntelliGrid provides the methodology, tools, and recommendations for standards and technologies when implementing smart grid systems, such as advanced metering, distribution automation, demand response, and wide area measurements. Examples of such accomplishments in standards development include EPRI's contributions to ANSI C12 standards for metering and the EPRI CCAPI (Control Center Application Program Interface) project for development of the IEC 61970/61968 standards for CIM for back-end IT integration and the IEC 61850 standard for substation automation. The IntelliGrid program at EPRI is a robust research program continuing to perform strategic and innovative research related to applying standards, communications technology, and advanced system integration to enable advanced smart grid applications for operational improvements with a strong emphasis on Informational and Operational Technology (IT & OT) Convergence.

2.7.2 EPRI's Smart Grid Demonstration Initiative

In an effort to advance the industry to achieve the benefits of integrating DER, the EPRI is collaborating directly with 18 utilities and indirectly with numerous others in a 6 year international smart grid demonstration initiative integrating DER (DG, storage, renewable generation, and demand response) into a "virtual power plant" (Figure 2.4). Of the 18 utilities participating in the initiative, 11 of them will make the significant investment to be a demonstration host-site and sharing research activities and results to advance the industry. All of the collaborating utilities, whether a host-site or

FIGURE 2.4 EPRI smart grid demonstration framework. (Copyright 2009 Electric Power Research Institute, Inc. All rights reserved.)

not, are directly participating and contributing to the research and gaining the benefits of first-hand knowledge of technology assessments, architecture design, cost–benefit analysis, best practices, and lessons learned. To achieve the goal of advancing the industry, key results from this initiative are being made publicly available on EPRI's Smart Grid Resource Center [105] and the funding utilities should be commended for their contribution to the industry.

The design of EPRI's large-scale demonstration initiative includes implementation of a common, scientific framework and structure so that the results of the effort are reported on and measured in a common methodology as well as enabling the coordination of similar research from project to project more efficiently gaining the benefits of collaboration—the whole is greater than the sum of its parts. To achieve this goal, EPRI's IntelliGrid methodology is being applied as a consistent scientific process to define requirements, assess technologies, develop the architecture to minimize the risk of technology obsolescence, and perform cost–benefit analysis.

In order to challenge the project teams to scope projects that will have the most impact on advancing the industry in identifying approaches for systemwide interoperability and integration of DER, EPRI has identified six critical elements or criteria for smart grid demonstration host-sites:

1. Integration of multiple types of DER: Projects must integrate at least two types out of the different types of DER into a common platform including DG, renewable generation, storage and demand response.
2. Connect retail customers to wholesale and/or grid conditions: Communications are one of the most significant technology advancements to enable a smart grid and provide the ability to link supply with demand. Projects must expose DER devices and/or consumers to some type of market or grid reliability signal to directly control, or incent a change in, the operation of DER accordingly.
3. Deployment of critical integration technologies and standards: Projects must apply existing and emerging technologies and standards for DER integration, in particular use of NIST identified standards and specifications are recommended.
4. Integration into system planning and operations: Projects must integrate DER into utility system planning and operations. Tools and techniques for integration are identified to accommodate DER on an equal footing as supply-side resources in utility operations and planning processes.
5. Compatibility with the initiative goals and approach: Commit to applying the IntelliGrid methodology and sharing of information to advance the industry.
6. Leverage additional funding sources as applicable: The utility must commit financial resources to ensure success and one way to do that as well as strengthen the collaborative is to leverage resources from government, academia, vendors, research organizations, and other utilities.

The project deployments will continue through 2014 and as of late 2011 has 23 international utility members including Ameren Services Company, American Electric Power Service Corporation, CenterPoint, Central Hudson Gas & Electric, Consolidated Edison, Duke Energy, Electricité de France, Entergy, Ergon Energy, ESB Networks, Exelon (ComEd & PECO), FirstEnergy, Hawaiian Electric Company, Hydro-Québec, Kansas City Power & Light, Public Services Company of New Mexico, Sacramento Municipal Utility District, Southern California Edison, Southern Company, Southwest Power Pool Inc., Salt River Project, Tennessee Valley Authority, Wisconsin Public Service, and a Resident Research from Tokyo Electric Power company representing the United States, Canada, Ireland, France, Australia, and Japan.

The EPRI Smart Grid Resource Center (http://www.smartgrid.epri.com) is the primary means of sharing public information with one of the significant contributions being the Use Case Repository (http://smartgrid.epri.com/Repository/Repository.aspx). The use case repository documents functional requirements of industry smart grid projects. The industry benefits from the contributions of many other generous utilities that have gone through the use case process to determine their own smart grid system functional requirements and architectural design. In addition to sharing information on the smart grid resource center, EPRI is active in participating and coordinating smart grid related events and conferences.

2.7.3 SMART ENERGY ALLIANCE

The Smart Energy Alliance (SEA) is a collaboration effort of Capgemini, Cisco Systems, GE Energy, Hewlett-Packard, Intel, and Oracle to help utilities transform their T&D operations. The next generation of distribution systems will bring the worlds of IT, communications, and energy systems closer together than ever before. Using a flexible, modular framework, the SEA's open collaboration delivers solutions that leverage deep industry expertise, technology leadership and complimentary capabilities to help customers drive greater productivity and profitability [106]. Since 2007, SEA member companies have been working with utilities and other stakeholders to help the industry move from a "smart energy" roadmap to reality.

2.7.4 GRIDWISE ALLIANCE

The GridWise Alliance, founded in 2003 and organized by the U.S. DOE, is a coalition of stakeholders along the energy supply chain from utilities to large tech companies to academia to venture capitalists to emerging tech companies who are advocating for a smarter grid for the public good. This variety of stakeholders gives the Alliance a diversity of perspectives that enables interactive dialogue between members. Being a consensus-based organization, the assortment of opinions produces deliberate and highly reflected upon resolutions to key issues [107]. The GridWise Alliance is not a research and development effort, but is more of a vehicle to coordinate activities to leverage technologies into developing an integrated power system capable of spanning all components and applications from generator to customer appliances. The GridWise Alliance worked with the U.S. DOE's Office of Electricity Delivery and Energy Reliability on a major, national conference, "GridWeek," as part of its accelerated efforts to develop a smart grid in the United States.

2.7.5 GRIDWISE ARCHITECTURE COUNCIL

Working closely with the GridWise Alliance, is the GridWise Architecture Council (GWAC). The GWAC was formed by the U.S. DOE to promote and enable interoperability among the many entities that interact with the nation's electric power system. The GWAC members are a balanced team representing the many constituencies of the electricity supply chain and users. The GWAC maintains a broad perspective of the GridWise vision and provides industry guidance and tools that make it an available and practical resource for the various implementations of smart grid technology [108]. After forming in May 2004, GWAC focused its initial work on establishing interoperability principles that have been recognized by many electric industry stakeholders and to which many are subscribing. Currently, GWAC is working with stakeholders to formulate a technical framework for electric system interoperability and integration issues. As such, the GWAC is developing checklists that stakeholders can use to recognize, advance, and implement interoperability in their projects. The GWAC also is building community awareness of the intrinsic value of flexibility and usability offered by the GridWise interoperability principles as well as promoting an industry culture of continual improvement in the level of interoperability.

2.7.6 INTERNATIONAL ELECTROTECHNICAL COMMISSION TECHNICAL COMMITTEE 57

The International Electrotechnical Commission (IEC) Technical Committee (TC) 57 [109] develops and maintains international standards for power system control equipment and systems including energy management systems (EMSs), supervisory control and data acquisition (SCADA), distribution automation, teleprotection, and associated information exchange for real-time and non-real-time information used in the planning, operation, and maintenance of power systems.

- WG 3—Telecontrol protocols
- WG 10—Power system IED communication and associated data models (IEC 61850)
- WG 13—Energy management system application program interface (EMS—API, CIM, IEC 61970)
- WG 14—System interfaces for distribution management (SIDM, CIM, IEC 61968)
- WG 15—Data and communication security
- WG 16—Deregulated energy market communications
- WG 17—Communications systems for distributed energy resources (DER)
- WG 18—Hydroelectric power plants—communication for monitoring and control
- WG 19—Interoperability within TC 57 in the long term
- WG 20—Planning of (single-sideband) power line carrier systems (IEC 60495) and planning of (single-sideband) power line carrier systems (IEC 60663)

User groups promote the use of TC 57 standards and provide valuable feedback for the continuous improvement of the standards. User groups related to IEC TC57 are shown in Figure 2.5.

The CIM Users Group is under the administrative umbrella of the UCA International Users Group (UCAIug), a not-for-profit corporation. The UCAIug as well as its member groups (CIMug, OpenAMI, and IEC 61850) draws its membership from utility user and supplier companies. The mission of the CIMug is to manage and to communicate issues concerning the CIM model and to serve as the primary means for developing the CIM model consensus and consistency across the industry.

There is also a U.S. National Committee Technical Advisory Group (TAG) for the IEC. The TAG is responsible for the development of U.S. positions on working documents and proposals distributed by IEC and the development of U.S. proposals for international consideration.

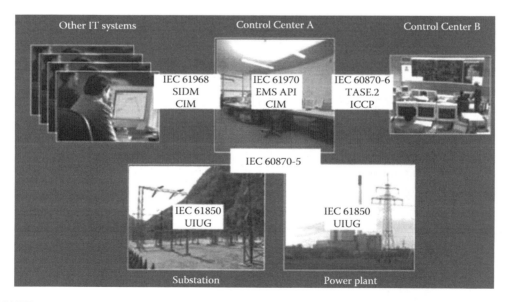

FIGURE 2.5 IEC TC 57 user groups (From International Electrotechnical Commission (IEC), Electrotechnical Committee (TC) 57.)

2.7.7 IEC Strategic Group (SG 3) on Smart Grid

The IEC set up a Strategic Group (SG 3) on smart grid in 2009 with the primary responsibility for the development of a framework that includes protocols and model standards to achieve interoperability of smart grid devices and systems. IEC SG 3 provides advice on fast-moving ideas and technologies likely to form the basis for new international standards or IEC TCs in the area of smart grid technologies [111]. SG 3 has developed the framework and provides strategic guidance to all TCs involved in smart grid work and has developed the smart grid roadmap that covers standards for interoperability, transmission, distribution, metering, connecting consumers, and cybersecurity. The IEC through SG 3 is working in close collaboration with smart grid projects around the globe, including U.S. NIST. IEC Standards are recognized as being crucial in the development of smart grids worldwide.

2.7.8 U.K. Low-Carbon Transition Plan

In July 2009, the U.K. Government published "The UK Low Carbon Transition Plan—National Strategy for Climate and Energy." This plan outlines the broad set of policy measures, targets, and principles that will allow the United Kingdom to deliver its five-point plan to tackle climate change. It also provides the framework against which the role of smart grid can be identified and offers a benchmark against which the smart grid vision must be tailored.

2.7.9 Electricity Networks Strategy Group

The Electricity Networks Strategy Group (ENSG) provides a high-level forum that brings together key stakeholders in electricity networks to support the U.K. Government and Ofgem in meeting the long-term energy challenges of tackling climate change and ensuring secure, clean, and affordable energy. The group is jointly chaired by the U.K. DECC and Ofgem, and its broad aim is to identify and coordinate work to help address key strategic issues that affect the transition of electricity networks to a low-carbon future. The ENSG smart grid working group has been tasked by DECC and Ofgem to produce a high-level smart grid vision and smart grid "routemap" (roadmap) [112]. The ENSG endorses this smart grid routemap as a high-level description of the way in which a U.K. smart grid could be delivered to contribute to the realization of government carbon targets and end-customer benefits. The ENSG believes that it is critical to deliver a range of well-targeted pilot projects between 2010 and 2015 in the expectation that many of them will prove to be technically and economically successful, and therefore available for U.K.-wide application from 2015 onward.

2.7.10 OFGEM Low Carbon Networks Fund

Ofgem in the United Kingdom has been actively engaged, as joint chair of the ENSG's smart grid working group, in the delivery of the smart grid roadmap in the United Kingdom. Ofgem established a £500 million Low-Carbon Networks Fund (LCNF), which is an important part of the 5th Electricity Distribution Price Control Review for the U.K. utility market.

2.7.11 European Renewable Energy Strategy

The European Renewable Energy Directive sets an obligatory target of 20% renewable energy in final energy consumption as well as a 10% target in transport for 2020. Against the background of the EU's ambition to move toward a reduction of 80%–95% of GHG emissions in a 2050 perspective, it is clear that a further strong growth in renewables will be needed beyond the 2020 targets [113]. The Renewable Energy Strategy sets out how everyone has a role to play in promoting renewable energy. Smart grid plans in Europe are strongly influenced by the renewable energy directive.

2.7.12 CENTRE FOR SUSTAINABLE ELECTRICITY AND DISTRIBUTED GENERATION

The Centre for Sustainable Electricity and Distributed Generation is a collaborative venture building on extensive ongoing research at Imperial College London, the University of Cardiff and Strathclyde University, and funded by the Department for Business, Enterprise and Regulatory Reform (BERR). The Centre undertakes a range of research projects to bridge the gap between academic research and the needs of industry to work toward meeting the 2020 targets on renewable energy for the United Kingdom. Its focus is on providing fundamental research aimed at achieving cost-effective integration of renewable generation and DG into operation and development of the U.K. electricity system. The centre investigates the technical and economic performance of transmission and active distribution networks and as well as the devices and systems that are connected to them with a view to develop and evaluate new concepts and solutions using software simulation or hardware testing as appropriate [114]. The activities of the Centre also contribute to the realization of the DTI/Ofgem DGCG objectives through coordination with its Technical Steering Group program, the DTI Electricity System Technical Issues Steering Group, and DTI/Ofgem/IEE Technical Architecture project.

2.7.13 SMART GRID INFORMATION CLEARINGHOUSE

The Smart Grid Information Clearinghouse (SGIC) [115] is a cooperative effort among Virginia Tech, the IEEE Power and Energy Society, and EnerNex Corporation to design, populate, manage, and maintain a public website containing information on smart grid. Contents in the SGIC include demonstration projects, use cases, standards, legislation, policy and regulation, lessons learned, best practices, and research and development activities. It is envisioned that the SGIC will be the essential gateway that connects the smart grid community to the relevant sources of national and international smart grid information. The SGIC is intended to be the first stop for stakeholders seeking information on smart grid.

2.8 SMART GRID MARKET OUTLOOK

Smart grid products—computers, electronics, and advanced materials for electric power transmission and delivery—are gradually transforming the world's electric power systems. Each year smart infrastructure equipment and systems take a larger percentage of the grid infrastructure market, which struggled to reach $85 billion worldwide in 2009 and continued to show only halting growth in 2010 to an estimated $89–$90 billion, spurred on by electrification projects in the developing regions. Although barriers remain that can delay the changeover to a more intelligent grid, global trends assure that this market will expand for the next decade and beyond, though at a slower rate that earlier forecasts had suggested.

2.8.1 MARKET DRIVERS

One of the keys to smart grid investment will be the recovery of the industrial sector of electricity end-users. While making up only a very small percentage of total customers, the industrial sector accounts for nearly 20% of all electricity consumption in North America (see Figure 2.6). The percentage of industrial electricity use varies across the world, but in some developing nations, industrial usage accounts for as much as 55% of total consumption.

The outlook for total consumption in the United States had been one of moderate growth for the 2011 and 2012 years, as depicted in Figure 2.7. However, the November 2011 release of the Energy Outlook by the U.S. DOE calls for a *decrease* in total electricity consumption in 2012 of about 0.6%. Longer-term outlooks continue to indicate modest to moderate growth in consumption in the developed nations with strong growth likely continuing in developing nations.

What Is Smart Grid, Why Now? 51

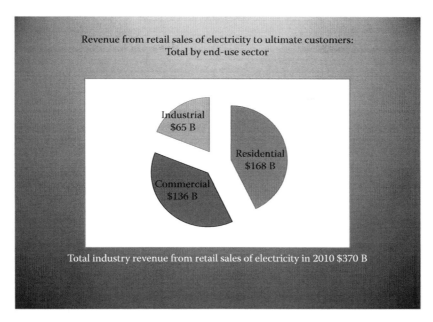

FIGURE 2.6 Primary electricity end-user segments in the United States. (Courtesy of Newton-Evans Research—based on DOE EIA source information.)

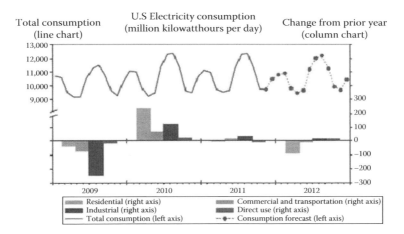

FIGURE 2.7 (See color insert.) U.S. outlook for electricity consumption in 2011–2012. (From U.S. Energy Information Administration (EIA), Short-term Energy Outlook, November 2011.)

The downturn in industrial and commercial electric power usage over the 2008–2011 years is actually greater than illustrated in Figure 2.8, because retail electricity prices are actually rising, as reflected in the residential usage curve. Even with increased prices, the revenues from commercial and industrial sales of electricity in 2011 remain at or below the 2008 level.

As reported in the April 2011 edition of Newton-Evans' CAPEX for smart grid and T&D report series, the positive news in the first quarter of 2011 concerning continuation of the rebound from the global recession of 2008–2009 had outdistanced the negative factors that could influence the economic picture for electric power industry investments in 2011 and 2012. Planned upticks in CAPEX budgeting for more of the included smart grid investment categories among many of the utility officials that were surveyed for that study remained in effect for months after the cautious

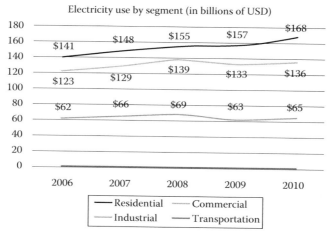

FIGURE 2.8 (See color insert.) U.S. electricity use by segment 2006–2010. (Courtesy of Newton-Evans Research—based on DOE EIA source information.)

outlooks released by such stalwart and respected organizations such as the World Bank, the Carnegie Endowment, and the Conference Board. However, no forecast produced through early March 2011 included the effects of such devastating natural disasters as the three-pronged calamity that has struck Japan, the world's third largest economy. By the autumn of 2011, financial and economic alarm bells were ringing, especially in the Eurozone countries, with Greece and Italy having substantial difficulties in righting their respective economic ships of state.

Current concerns expressed by the major economic forecasting services organizations and financial nongovernmental organizations such as the World Bank and IMF are centered on the lack of financial reforms being enacted as the economy heals; ongoing concerns with the Euro Debt situation affecting several western European countries; currency tensions among some countries; counter-cyclical government economic measures, especially in the United States; and the fragility of the banking industry recovery in some countries. Further inaction at the federal level in the United States has resulted in a failure to enact a comprehensive national energy policy, and in turn that inaction is holding back smart grid investments as much as the economy.

2.8.2 Market Potential

Many utilities will start on their path to developing a smarter grid from the control center outward, with upgraded SCADA and energy management technology. Implementations of OMS and FDIR/FLISR will likely follow. Other utilities will work inward from customer premises-based smart meters to field area networks, then to distribution automation. Regulators and Commercial and Industrial (C&I) customers would like utilities to begin with additional demand response initiatives and increased reliability measures.

During the first quarter of 2011, Newton-Evans Research Company published the findings from its fourth annual study of capital investment trends made by the world's electric power utilities that are related to smart grid development and infrastructure upgrades. The consensus view among major T&D equipment manufacturers is one of positive growth year-over-year for 2010 and onward compared with 2009's dismal results. Currently, it appears that any increases may not be sufficient to reach the values of industry shipments (including imports) obtained in 2008. T&D equipment and systems sales for the United States amounted to about $18 billion in 2009, with ancillary supplies, towers, poles, cables, lines, and T&D services amounting to an additional $18–$21 billion. Worldwide, the amount spent on T&D equipment hovered around $85 billion in 2009, and

What Is Smart Grid, Why Now? 53

expectations are for T&D equipment and systems expenditures to rebound to $90 billion or more this year, with GDP growth looking very strong in some large developing markets (China, India, Brazil, GCC) compared with the expected 1%–3% growth anticipated in much of the West.

As noted earlier in this section, the U.S. DOE's highly regarded Energy Information Administration has released its preliminary Annual Energy Outlook for 2012. The new report forecasts 2011 electricity utilization at levels comparable to 2010, suggesting either (or both) a gain in energy efficiency or a continuing low-growth economic outlook. In addition, the department has forecast an increase in the nation's electricity generating capacity amounting to 13 GW for 2010, with more than 60% of this net addition coming from renewable sources, principally wind energy.

Commercial use of electricity may increase by 2.1% compared with less than 1% increase for residential customers, according to the Energy Information Administration. The DOE also foresees a further 2% decrease in industrial electricity consumption and a slight decrease in transportation sector consumption.

These findings support those utilities reporting decreases in CAPEX budgets who have indicated such decreases are being caused first and foremost by the economic outlook for 2011, a more important factor for their decisions than are regulatory mandates.

The 2011 Newton-Evans survey also requested utility officials to provide the reasons for their increases in CAPEX plans, looking into the rationale for change in year-over-year budget plans. In summary, smart grid initiatives were cited as being more important factors than either regulatory mandates or government stimulus programs. These initiatives are going forward despite a less-than-rosy economic outlook at this time, spurred on in part by government stimulus, by regulatory directives, and by good business practices. Indications are strong for a continuing slow economic recovery and further slight improvements in outlook for GDP growth in the West, better utility access to capital markets and for smart grid initiatives overall. For developing nations, the consensus view holds that growth will continue at good rates over the 2012–2013 periods, but will be affected negatively by the slowdown in capital investment and consumer spending among Western nations.

Figures 2.9 and 2.10 summarize the estimated 2010 and 2014 market sizes globally for major smart grid–related investment categories. Other analysts and researchers have opined that the dollar amount of smart grid investments may be somewhat larger than the Newton-Evans' estimates. It really comes down to what is included in the smart grid "basket" of systems, equipment, products, and services.

2.8.3 SMART GRID AND IT EXPENDITURE FORECASTS

In examining the current outlook for smart grid–related expenditures, we have made a serious attempt to avoid double counting potential revenues from all of the components of information systems spending and the emerging smart grid sector of utility investment.

While the utility enterprise-wide IT portions of the chart in Figure 2.9 include all major components of IT (hardware, software, services, and staffing), the "pure" smart grid components tend to be primarily in hardware, in our view. Significant overlap with both administrative and operational IT supporting infrastructure is a vital component for all smart grid programs underway at this time.

Between "traditional IT" and the evolving smart grid components of IT and automation for electric power transmission and distribution, nearly $25 billion will likely be invested this year by the world's electric utilities. Nearly one-third of all information technology investments made by electric utilities were at least partially "smart grid" related.

By 2014, as shown in the chart in Figure 2.10, the total value of the various pie segments is expected to increase substantially, with all of IT and "smart grid" spending combined possibly exceeding $40 billion.

What some observers may include in a broader portrait of smart grid is the total value of T&D capital spending, which is already approaching $100 billion globally, and will likely top $120

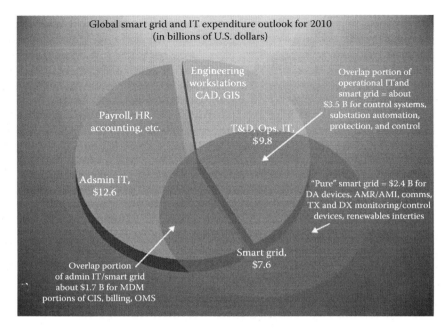

FIGURE 2.9 Global smart grid and IT expenditure outlook for 2010. (Copyright 2012 Newton-Evans Research Company. All rights reserved.)

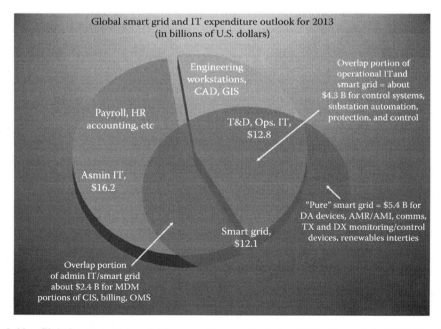

FIGURE 2.10 Global smart grid and IT expenditure outlook for 2014. (Copyright 2012 Newton-Evans Research Company. All rights reserved.)

billion by 2014. The majority of new procurements of infrastructure equipment will continue to be made with an eye to including as much "smart" content as is available and practical from the manufacturers and integrators. What we are limiting our definition to here is "edge" investment, the components of the twenty-first century digital transport and delivery systems being added on or incorporated into the building blocks (power transformers, lines, switchgears, etc.) of electric power transmission and delivery.

There is a case to be made to include at least some of the automation investments made by power generation facilities and for the energy resource planning software and systems used by utilities in generation planning as part of the development of a smart grid.

REFERENCES

1. http://www.smartgrids.eu
2. GRID 2030; A National Vision for Electricity's Second 100 Years, United States Department of Energy, Office of Electric Transmission and Distribution, 2003.
3. Report to NIST on the Smart Grids Interoperability Standards Roadmap, EPRI, 2009.
4. http://tonto.eia.doe.gov/cfapps/ipdbproject/IEDIndex3.cfm?tid=2&pid=2&aid=9
5. Xcel Energy; Smart grid: A white paper, http://bannercenterforenergy.com/pfd%20files/smart%20Grid%20Revelant%20Docs/11-SmartGridWhitePaper[1].pdf, 5, February, 2008.
6. US Department of Energy; The smart grid, an estimation of energy and CO2 benefits, http://www.pnl.gov/main/publications/external/technical_reports/PNNL-19112.pdf, January 2010.
7. Boston Consulting Group; Press release: Smart meters hold great promise for energy and cost savings, but utilities need to improve customer education to reap the rewards, Says the Boston Consulting Group, http://www.bcg.com/media/PressReleaseDetails.aspx?id=tcm:12-48247; May 19, 2010.
8. JD Power and Associates; Although awareness of smart grid technology substantially boosts residential customer satisfactions with electric utility providers, Awareness to be low, tends http://businesscenter.jdpower.com/news/pressrelease.aspx?ID=2010114; July 1, 2010
9. Electric Power Research Institute; Methodological approach for estimating the benefits and costs of smart grid demonstration projects, http://my.epri.com/portal/server.pt?Abstract_id=000000000001020342, January 2010; 2-15 and 2-25 (Table 2-2).
10. Faruqui, A. and Sanem, S.; Household response to dynamic pricing of electricity—A survey of the experimental evidence; January 10, 2009, 2.
11. Jung, M. and Peter Y.; Connecting smart grid and climate change, Silver Spring Networks; http://www.silverspringnet.com/pdfs/SSN_WP_ConnectingSmartGrid-1109.pdf
12. Ehrhardt-Martinez, K., Donnelly, K.A., and Laitner, J.A.; Advanced metering initiatives and residential feedback programs: A meta-review for household electricity-saving opportunities, ACEEE: June 2010 http://www.aceee.org/e105.htm
13. Federal Energy Regulatory Commission; A national assessment of demand response potential, http://www.ferc.gov/legal/staff-reports/06-09-demand-response.pdf, June 2009.
14. Battelle Energy Technology; Investigating smart grid solutions to integrate renewable sources of energy in to the electric transmission grid, http://www.battelle.org/electricity/vadari_davis.pdf, 2009, 8.
15. http://www.silverspringnet.com/newsevents/pr-020211.html
16. PowerCentsDC; PowerCents DC program final report, http://www.powercentsdc.org/ESC%2010-09-08%20PCDC%20Final%20Report%20-%20FINAL.pdf, 2010.
17. The Edison Foundation Institute for Electric Efficiency; The impact of dynamic pricing on low income customers, http://www.electric-efficiency.com/reports/index.htm
18. Peter Fox-Penner; *Smart Power: Climate Change, the Smart Grid, and the Future of Electric Utilities*; Island Press, Washington, DC, 2010, 44.
19. National Academies of Sciences Committee on Health, Environmental, and Other External Costs and Benefits of Energy Production and Consumption; National Research Council et al; The hidden costs of energy: Unpriced consequences of energy production and use, http://www.nap.edu/catalog.php?record_id=12794
20. Colin, M. and Brownstein, M.; EDF Analysis Based on Data From PJM, 2010.
21. Keith et al.; *Modeling Demand Response and Air Emissions in New England; Synapse Energy Economics*, Cambridge, MA, 2, 6, 2003.

22. Rudkevich, A.; *Locational Carbon Footprint and Renewable Portfolio Standards*, Charles River Associates.
23. PJM; PJM Reports new carbon dioxide emissions data, http://ftp.pjm.com/~/media/about-pjm/newsroom/2010-releases/20100325-pjm-reports-new-carbon-dioxide-emissions-data.ashx
24. Mills, E.; *The Cost-Effectiveness of Commercial-Buildings Commissioning: A Meta-Analysis of Energy and Non-Energy Impacts in Existing Buildings and New Construction in the United States*, LBNL, Berkeley, CA; December 2004.
25. Electric Power Research Institute; The green grid energy savings and carbon emissions reductions enabled by a smart grid; June 2008.
26. Pew Center on Global Climate Change; Electricity emissions in the United States, http://www.pewclimate.org/technology/overview/electricity, May 2009.
27. United States Department of Energy; Energy demands on water resources: Report to congress on the interdependence of energy and water, http://www.sandia.gov/energy-water/docs/121-RptToCongress-EWwEIAcomments-FINAL.pdf; 9: National energy technology laboratory, water and energy, Addressing the critical link between the nation's water resources and reliable and secure energy, http://www.netl.doe.gov/technologies/coalpower/ewr/pubs/water_brochure.pdf, 5, October 2006: National energy technology laboratory; IEP—water energy interface; http://www.netl.doe.gov/technologies/coalpower/ewr/water/power-gen.html
28. Environmental Defense Fund and Western Resource Advocates; Protecting the lifeline of the west: How climate and clean energy policies can safeguard water, http://www.westernresourceadvocates.org/water/lifeline/lifeline.pdf
29. Environmental Defense Fund and The University of Texas Austin; Energy water nexus in Texas, http://www.edf.org/sites/default/files/Energy_Water_Nexus_in_Texas.pdf, p. 26, April 2009.
30. Synapse Energy Economics, Inc; *Co-Benefits of Energy Efficiency and Renewable Energy in Utah*; Feb 2010, 4.
31. Ackerman, S. et al.; The cost of climate change: What we will pay if climate change continues unchecked, http://www.nrdc.org/globalWarming/cost/contents.asp, May 2008.
32. Holst, D. and Fredrich K.; California climate risk and response, UC Berkeley Center for Energy, Resources, and Economic Sustainability; http://www.nextten.org/pdf/report-CCRR/California_Chmate_Risk_and_Response.pdf, Paper 8102801, November 2008.
33. Estimated 2010 Cost of Coal at 2.22 cents per KWH based on: US Energy Information Administration; Electric power monthly, http://www.eia.doe.gov/cneaf/electricity/epm/epm_sum.html, accessed July 2010; and US Energy Administration; Average operating heat rate for selected energy sources, http://www.eia.doe.gov/cneaf/electricity/epa/epat5p3.html, accessed July 2010.
34. http://www.pewenvironment.org/uploadedFiles/PEG/Newsroom/Press_Release/China%20Leads%20All%20Other%20G-20%20Members%20in%20Clean%20Energy%20Investments.pdf
35. http://www.iea.org/Papers/2011/SmartGrids_roadmap.pdf
36. KEMA; Press release: Kema sees opportunities for storage applications in high-peneration renewables, http://www.kema.com/news/pressroom/press-releases/2010/KEMA-sees-opportunities-for-storage-applications-in-high-penetration-renewables.aspx, June 2010: KEMA; Research Evaluation of Wind Generation, Solar Generation, and Storage Impact on the California Grid; http://www.energy.ca.gov/2010publications/CEC-500-2010-010/CEC-500-2010-010.PDF, June 2010.
37. Valley Wireless; Valley wireless and smart grid: Future of the electricity and role of IT, http://valleywireless.us/2.html, NuEnergen, Getting smarter—Smart grid on the rise, http://www.nuenergen.com/2010/01/getting-smarter-smart-grid-on-the-rise/, January 2010.
38. International Energy Agency; Press release: IEA sees great potential for solar providing up to a quarter of world electricity by 2050, http://www.iea.org/press/pressdetail.asp?PRESS_REL_ID=301, May 2010.
39. http://my.epri.com/portal/server.pt?space=CommunityPage&cached=true&parentname=ObjMgr&parentid=2&control=SetCommunity&CommunityID=404&RaiseDocID=000000000001022519&RaiseDocType=Abstract_id
40. Zpryme; Smart grid insights 2010, http://www.zpryme.com/SmartGridInsights/2010_V2G_Report_Zpryme_Smart_Grid_Insights_ZigBee_Alliance_Sponsor.pdf, 20, 2010.
41. Greentech Media; http://www.grist.org/green-cars/charging-ahead-report-predicts-3.8-million-electrics-on-road-by-2016
42. Green Auto Blog. http://green.autoblog.com/2010/09/20/auto-industry-analyst-predicts-well-have-more-than-100-hybrid-a/
43. ABB, Siemens, Schneider gain from smart grids demand. Bloomberg. 16 Sept 2010 http://www.energystorageforum.com/2010/09/abb-siemens-schneider-to-gain-from-smart-grids-demand/

44. US EPA; Inventory of U.S. Greenhouse gas emissions and sinks: 1990—2008, March 2010.
45. Electric Power Research Institute; Environmental assessment of plug-in hybrid electric vehicles; http://mydocs.epri.com/docs/CorporateDocuments/SectorPages/Portfolio/PDM/PHEV-ExecSum-vol1.pdf, 2007.
46. Kuo, I.; Nissan to take orders for its Leaf electric car starting tomorrow. Reuters. http://www.reuters.com/article/2010/08/31/idUS286000615420100831, 30 Aug 2010.
47. Studies assessed include the PNNL and Silver Spring Studies as well as: Becker, Thomas A., Ikhlaq Sidhu, Burghardt Tenderich; Annual Energy Outlook 2010: Electric Vehicles in the United States: A New Model with Forecasts to 2030, US Energy Information Administration: http://cet.berkeley.edu/dl/CET_Technical%20Brief_EconomicModel2030_f.pdf, v: Austin Energy; 2010 Annual Report of System Information.
48. Kyeon H., Boddeti, M., Sarma, N.D.R., Dumas, J., Adams, J., and Soon-Kin, C.; High wire act: ERCOT balances transmission flows for Texas-size savings using its dynamic thermal ratings application, *IEEE Power and Energy Magazine*, 8(1), 37–45, Jan/Feb 2010.
49. California Public Utilities Commission; Decision adopting requirements for smart grid deployment plans pursuant to senate bill 17, http://docs.cpuc.ca.gov/PUBLISHED/FINAL_DECISION/119902.htm, 123, June 2010.
50. Synapse Energy Economics Inc; Reports and news from synapse energy economis, http://www.synapse-energy.com/Newsletter/2010-03-Newsletter.shtml
51. William H. Bartley, P.E.; Keeping the lights on: An action plan for America's aging utility transformers by The Hartford Steam Boiler Inspection and Insurance Company, http://www.hsb.com/Thelocomotive
52. U.S. Energy Information Administration (EIA), http://www.eia.gov/
53. Integrated resource planning for renewables options by steye pullins, DistribuTECH/TransTECH 2008.
54. Petersen, J. Active power—A solid investment opportunity and a valuable object lesson for investors, 2010, http://www.altenergystocks.com/archives/energy_storage/flywheel/
55. Norberg-Bohm, V., Belfer Center for Science and International Affairs, John F. Kennedy School of Government, Harvard University; The role of government in energy technology innovation: Insights for government policy in the energy sector, October 2002, http://www.innovations.harvard.edu/download-doc.html?id=5603
56. Transforming America's Power Industry: The Investment Challenge 2010–2030, report by Brattle Group to the EEI, November 2009.
57. Megawatt hours. While megawatts are a measure of power, megawatt hours are a measure of the energy produced from that power.
58. Power produced from sustainable resources, such as solar, wind, water, and biomass.
59. http://en.wikipedia.org/wiki/Renewable_portfolio_standard#cite_note-1
60. http://www.gpo.gov/fdsys/pkg/PLAW-110publ140/pdf/PLAW-110publ140.pdf
61. Public Law 111 - 5 - American Recovery and Reinvestment Act of 2009. http://www.gpo.gov/fdsys/pkg/PLAW-111publ5/content-detail.html
62. Electric Light & Power; http://www.elp.com/index/display/article-display/8929430514/articles/utility-automation-engineering-td/volume-16/issue-11/features/arra-paves-smart-grid-path-with-cash.html
63. http://www.whorunsgov.com/institutions/energy
64. EPRI; *Assessment of Achievable Potential from Energy Efficiency and Demand Response Programs in the U.S. (2010–2030)*, Palo Alto, CA, January 2009.
65. http://energy.gov/oe/technology-development/smart-grid
66. http://www.oe.energy.gov/eac.htm
67. http://www.nist.gov/smartgrid/nistandsmartgrid.cfm
68. NIST Special Publication 1108R2; NIST framework and roadmap for smart grid interoperability standards, Release 2.0, National Institute of Standards and Technology (NIST), U.S. Department of Commerce, February 2012.
69. http://www.nist.gov/smartgrid/upload/Overview-and-Executive-Summary.pdf
70. http://www.ferc.gov/about/ferc-does/ferc101.pdf
71. http://tdworld.com/smart_grid_automation/ferc-smart-grid-policy-0709/
72. http://insideclimatenews.org/news/20090717/ferc-adopts-smart-grid-policy-rules-raising-rates
73. http://www.ferc.gov/industries/electric/indus-act/smart-grid.asp
74. Goethe, S.;*Smarter Grids...And More of Them*. IEA. July 2009. http://www.iea.org/impagr/cip/pdf/Issue60SGoethe.pdf
75. Elzinga, D. et al.; *Technology Roadmap: Smart Grids*. International Energy Agency. http://www.iea.org/papers/2011/smartgrids_roadmap.pdf

76. International Climate Policy Post-Copenhagen. Communication From The Commission To The Council; The European Parliament, The European Economic And Social Committee And The Committee Of The Regions. 2010. http://forestindustries.eu/content/post-copenhagen-point-view-eu
77. Pellerin, C.; Nations work together despite slow progress on copenhagen accord, America.gov. January 2010. http://www.america.gov/st/energy-english/2010/January/20100126081441lcnirellep0.6686975.html
78. Technology Action Plan Executive Summary; Major economies forum on climate and energy, December 2009. http://www.majoreconomiesforum.org/images/stories/documents/MEF%20Exec%20Summary%2014Dec2009.pdf
79. World Economic Forum.; Accelerating smart grid investments project overview,
80. Lesser, S. Top ten smart grid projects globally. Clean Techies Blog. April 2011. http://blog.cleantechies.com/2011/04/06/top-ten-smart-grid-projects-globally/
81. Vision and Mission of IRENA.; Preparatory commission for international renewable energy agency, http://www.irena.org/ourMission/vision.aspx?mnu=mis
82. Zpryme Research & Consulting, LLC.: Smart grid insights: V2G. July 2010. http://www.zpryme.com/SmartGridInsights/2010_V2G_Report_Zpryme_Smart_Grid_Insights_ZigBee_Alliance_Sponsor.pdf
83. Reitenbach, G.: Which country's grid is the smartest? POWER magazine. January 2010. http://www.powermag.com/environmental/Which-Countrys-Grid-Is-the-Smartest_2345_p3.html
84. State Grid: China's Smart Grid in Comprehensive Construction. State Grid Energy Research Institute. March 2011. http://www.sgeri.sgcc.com.cn/english/Center/News/97624.shtml
85. Wernau, J.; South Korea launches energy savings project in Chicago. Chicago Tribune. July 2010. http://articles.chicagotribune.com/2010-07-21/business/ct-biz-0722-korean-tech-20100721_1_buildings-aon-center-south-korea
86. Osawa, J.; Toshiba to buy swiss power meter maker. *Wall Street Journal*. May 2011. http://online.wsj.com/article/SB10001424052748704904604576332501074832920.html
87. Accelerating Smart Grid Investments. World Economic Forum. 2009 https://members.weforum.org/pdf/SlimCity/SmartGrid2009.pdf
88. Electrifying Singapore: Drivers and Roadblocks. Green Leap Forward. May 2009. http://greenleapforward.com/2009/01/23/electrifying-singapore-drivers-and-roadblocks/
89. Australia to develop its first commercial smart grid. International Business Times. October 2010. http://www.ibtimes.com/articles/70351/20101010/australia-to-develop-its-first-commercial-smart-grid.htm
90. GLG Expert Contributor. Review of global smart grid initiatives. Gerson Lehrman Group. February 2010. http://www.glgroup.com/News/Review-of-Global-Smart-Grid-Initiatives-46466.html
91. Europe: EU car CO2 regulation published.; Automotive world. June 2009. http://www.automotiveworld.com/news/environment/76912-europe-eu-car-co2-regulation-published
92. Frost and Sullivan; Europe the front-runner at copenhagen in the smart grid space, asserts Frost and Sullivan. Transmission & Distribution World. December 2009. http://tdworld.com/smart_grid_automation/europe-smart-grid-copenhagen-1209/
93. BEAMA Calls for Government to Plug 'Real Gaps' in Strategy. BEAMA. March 2010. http://www.beama.org.uk/en/news/index.cfm/ECC-report
94. Coaster, M.; Smart metering implementation programme: Consumer protection. Office of Gas and Electricity Markets. July 2010. http://www.ofgem.gov.uk/eserve/sm/Documentation/Documents1/Smart%20metering%20-%20Consumer%20Protection.pdf
95. Heile, B.; Surveying the international AMI landscape, *IEEE's 802.15 Working Group on Wireless Personal Area Networks*, P2030 Work Group Task Force 3. http://www.elp.com/index/display/article-display/3987256252/articles/utility-automation-engineering-td/volume-15/Issue_7/Features/Surveying_the_International_AMI_Landscape.html, June 2010.
96. Super Smart Grid. 2010. http://www.supersmartgrid.net/
97. Jyothi S.; BPLG and BPL Africa deploy smart grid solutions in ghana for VRA green technology world, January 2009. http://green.tmcnet.com/topics/green/articles/48256-bplg-bpl-africa-deploy-smart-grid-solutions-ghana.htm
98. Andreassen, J.; Small islands' huge move toward renewable energy, global carbon market 'pioneering model' for world. Environmental Defense Fund. December 17, 2010. http://www.edf.org/pressrelease.cfm?contentID=11520
99. Smart grids in Developing Countries. 3 Jan 2011. http://www.globalsmartgridfederation.org/news_20110103_smart.html

100. GE; Whirlpool and others launch smart green rid initiative aimed at including smart grid in copenhagen meetings. PR Newswire. October 2010. http://www.prnewswire.com/news-releases/ge-whirlpool-and-others-launch-smart-green-grid-initiative-aimed-at-including-smart-grid-in-copenhagen-meetings-64999317.html
101. Joining Forces to Realize a Smart Grid. GridWise Alliance. 2011. http://www.gridwise.org/gridwisealli_about.asp
102. GridWise Alliance signs MOU with Japan smart community alliance, GridWise Alliance. April 2010. http://tdworld.com/smart_grid_automation/gridwise-alliance-japan-mou-0410
103. Global Smart Grid Federation. 2011. http://www.globalsmartgridfederation.org/about.html
104. http://intelligrid.epri.com
105. http://www.gridwise.org
106. http://www.gridwiseac.org
107. http://tc57.iec.ch/index-tc57.html
108. International Electrotechnical Commission (IEC) Technical Committee (TC) 57.
109. http://www.iec.ch/smartgrid/development/
110. http://www.ensg.gov.uk
111. http://ec.europa.eu/energy/renewables/consultations/20120207_renewable_energy_strategy_en.htm
112. http://www.sedg.ac.uk/
113. http://www.sgiclearinghouse.org/
114. US Energy Information Administration (EIA), Short-term energy outlook, November 2011.

BIBLIOGRAPHY

Sources for Smart Grid Market Outlook section.

Newton-Evans Research Company, Global CAPEX and O&M expenditure outlook for electric power transmission and distribution investments, 2010–2011 funding outlook for smart grid development.

Newton-Evans Research Company, The worldwide smart grid market in mid-2011, A reality check and five year outlook through 2015.

Nomura Global Economics, 2011 global economic outlook.

U.S. Department of Energy, Energy Information Administration, Annual energy outlook for 2011 and 2012, updated November 2011.

U.S. Federal Reserve Outlook, November 2011.

International Monetary Fund, September 2011, World economic outlook.

3 Smart Grid Technologies

CONTENTS

- 3.1 Technology Drivers 67
  - 3.1.1 Transformation of the Grid 67
  - 3.1.2 Characteristics of a Smart Grid 70
  - 3.1.3 Smart Grid Technology Framework 73
- 3.2 Smart Energy Resources 79
  - 3.2.1 Renewable Generation 79
    - 3.2.1.1 Regulatory and Market Forces... 79
    - 3.2.1.2 Centralized and Distributed Generation 81
    - 3.2.1.3 Technologies 82
    - 3.2.1.4 Renewable Energy Needs in a Smart Grid 86
  - 3.2.2 Energy Storage 86
    - 3.2.2.1 Regulatory and Market Forces Driving Energy Storage and Smart Grid Impact 86
    - 3.2.2.2 Centralized and Distributed Energy Storage 89
    - 3.2.2.3 Technologies 91
  - 3.2.3 Electric Vehicles 96
    - 3.2.3.1 Regulatory and Market Forces Driving Electric Vehicles and Smart Grid Impact 96
    - 3.2.3.2 Technologies 97
    - 3.2.3.3 Vehicle to Grid 99
  - 3.2.4 Microgrids 103
    - 3.2.4.1 Microgrid Definition 103
    - 3.2.4.2 Microgrid Drivers 104
    - 3.2.4.3 Microgrid Benefits 105
    - 3.2.4.4 Challenges to the Development of Microgrids 106
    - 3.2.4.5 Microgrid Pilot Projects 107
    - 3.2.4.6 Types of Microgrid 108
    - 3.2.4.7 Building Blocks of a Microgrid 109
  - 3.2.5 Energy Resources Integration Challenges, Solutions, and Benefits 112
    - 3.2.5.1 Integration Standards 112
    - 3.2.5.2 Renewable Generation Integration Impacts 113
    - 3.2.5.3 Electric Vehicle Impacts on Electric Grid Systems 120
- 3.3 Smart Substations 126
  - 3.3.1 Protection, Monitoring, and Control Devices (IEDs) 127
  - 3.3.2 Sensors 127
  - 3.3.3 SCADA 128
    - 3.3.3.1 Master Stations 128
    - 3.3.3.2 Remote Terminal Unit 130
  - 3.3.4 Substation Technology Advances 131

	3.3.5	Platform for Smart Feeder Applications 134
	3.3.6	Interoperability and IEC 61850 136
		3.3.6.1 Process Level 141
		3.3.6.2 Bay Level 141
		3.3.6.3 Station Level 141
		3.3.6.4 IEC 61850 Benefits 142
	3.3.7	IEC 61850-Based Substation Design 142
		3.3.7.1 Paradigm Shift in Substation Design 144
		3.3.7.2 IEC 61850 Substation Hierarchy 146
		3.3.7.3 IEC 61850 Substation Architectures 148
		3.3.7.4 Station-Bus-Based Architecture 148
		3.3.7.5 Station and Process Bus Architecture 150
	3.3.8	Role of Substations in Smart Grid 152
		3.3.8.1 Engineering and Design 153
		3.3.8.2 Information Infrastructure 154
		3.3.8.3 Operation and Maintenance 154
		3.3.8.4 Enterprise Integration 155
		3.3.8.5 Testing and Commissioning 155
3.4	Transmission Systems 155
	3.4.1	Energy Management Systems 156
		3.4.1.1 History of Energy Management Systems 156
		3.4.1.2 Current EMS Technology 156
		3.4.1.3 Advances in Energy Management Systems for the Smart Grid 157
		3.4.1.4 Control System Cybersecurity Considerations 162
	3.4.2	FACTS and HVDC 169
		3.4.2.1 Power System Developments 170
		3.4.2.2 Flexible AC Transmission Systems 171
		3.4.2.3 High-Voltage Direct Current 181
	3.4.3	Wide Area Monitoring, Protection and Control 188
		3.4.3.1 Overview 188
		3.4.3.2 Drivers and Benefits of WAMPAC 191
		3.4.3.3 WAMPAC Needs in a Smart Grid 192
		3.4.3.4 Major WAMPAC Activities 194
		3.4.3.5 Role of WAMPAC in a Smart Grid 203
	3.4.4	Role of Transmission Systems in Smart Grid 210
3.5	Distribution Systems 212
	3.5.1	Distribution Management Systems 212
		3.5.1.1 Distribution SCADA 212
		3.5.1.2 Trends in Distribution SCADA and Control 214
		3.5.1.3 Current Distribution Management Systems 216
		3.5.1.4 Advanced Distribution Management Systems 218
	3.5.2	Volt/VAr Control 221
		3.5.2.1 Inefficiency of the Power Delivery System 221
		3.5.2.2 Voltage Fluctuations on the Distribution System 222
		3.5.2.3 Effect of Voltage on Customer Load 223
		3.5.2.4 Drivers, Objectives, and Benefits of Voltage and VAr Control 224
		3.5.2.5 Volt/VAr Control Equipment inside the Substation 226
		3.5.2.6 Volt/VAr Control Equipment on Distribution Feeders 231
		3.5.2.7 Volt/VAr Control Implementation 234
		3.5.2.8 Volt/VAr Optimization 239

Smart Grid Technologies

	3.5.3	Fault Detection, Isolation, and Service Restoration 245
		3.5.3.1 Faults on Distribution Systems 246
		3.5.3.2 Drivers, Objectives, and Benefits of FDIR 247
		3.5.3.3 FDIR Equipment...247
		3.5.3.4 FDIR Implementation 253
		3.5.3.5 Reliability Needs in a Smarter Grid 260
	3.5.4	Outage Management 261
	3.5.5	High-Efficiency Distribution Transformers 264
3.6	Communications Systems...266
	3.6.1	Communications: A Key Enabler of the Smart Grid....................266
	3.6.2	Communications Requirements for the Smart Grid 267
		3.6.2.1 AMI Communications 267
		3.6.2.2 Communications for Smart Grid Operations............268
		3.6.2.3 Home Area Network 273
	3.6.3	Wireless Network Solutions for Smart Grid 274
		3.6.3.1 Cellular..274
		3.6.3.2 RF Mesh..281
	3.6.4	Communication Standards and Protocols 282
		3.6.4.1 IEC 61850..283
		3.6.4.2 DNP3 and IEC 60870-5 283
		3.6.4.3 IEEE C37.118 285
		3.6.4.4 IEC 61968-9 and MultiSpeak...............................285
		3.6.4.5 ANSI C12.19, ANSI C12.18, ANSI C12.21, and ANSI C12.22 285
		3.6.4.6 High-Reliability Protocols 286
		3.6.4.7 Time Synchronization Protocols...........................286
	3.6.5	Communications Challenges in the Smart Grid............................287
		3.6.5.1 Harnessing Technology Complexity 287
		3.6.5.2 Legacy Integration, Migration, and Technology Life Cycle....................287
		3.6.5.3 Communications Service Planning and Evolution Trends 288
		3.6.5.4 Cybersecurity for Wireless Networks 290
		3.6.5.5 Management and Organization Challenges 294
	3.6.6	Communications in the Smart Grid: An Integrated Roadmap.......294
3.7	Monitoring and Diagnostics 296
	3.7.1	Architectures 297
		3.7.1.1 Tier 1: Local Level..298
		3.7.1.2 Tier 2: Station/Feeder Level 298
		3.7.1.3 Tier 3: Centralized Control Room Level 299
	3.7.2	Wireless Sensor Networks 301
	3.7.3	Diagnostics 301
	3.7.4	Future Trends...304
3.8	Geospatial Technologies..304
	3.8.1	Technology Roadmap 304
		3.8.1.1 Age of Paper...305
		3.8.1.2 Emergence of Digital Maps 306
		3.8.1.3 From Maps to Geospatial Information Systems 307
		3.8.1.4 Across the Enterprise..307
		3.8.1.5 Developing World 309
	3.8.2	Changing Grid 309
	3.8.3	Geospatial Smart Grid...310
		3.8.3.1 Core Spatial Functionality 310

		3.8.3.2	Planning and Designing the Grid	312
		3.8.3.3	Operating and Maintaining the Grid	315
		3.8.3.4	Mobile Geospatial Technologies	316
		3.8.3.5	Engaging the Consumer	320
	3.8.4	Smart Grid Impact on Geospatial Technology		320
		3.8.4.1	Coping with Scale	321
		3.8.4.2	Moving to Realtime	321
		3.8.4.3	Supporting Distributed Users	322
		3.8.4.4	Usability	322
		3.8.4.5	Visualization	323
		3.8.4.6	Standards	323
		3.8.4.7	Data Quality	324
		3.8.4.8	More Open: Sensors and Other Data Sources	324
		3.8.4.9	More Closed: Security	325
		3.8.4.10	More Closed: Privacy	325
	3.8.5	Future Directions		326
		3.8.5.1	Architecture	326
		3.8.5.2	Cloud	326
		3.8.5.3	Place for Neo-Geo	327
3.9	Asset Management			327
	3.9.1	Drivers		328
		3.9.1.1	Safety	328
		3.9.1.2	Reliability	328
		3.9.1.3	Financial	329
		3.9.1.4	Regulatory	329
	3.9.2	Optimizing Asset Utilization		329
	3.9.3	Asset Management Implementation		331
	3.9.4	Where Smart Grid Meets Business: The Electric Utility Perspective on Asset Management		332
		3.9.4.1	Asset Condition Monitoring	332
		3.9.4.2	More Effective Management of the Workforce	335
		3.9.4.3	Examples of Utility Asset Management Applications	343
	3.9.5	Where Smart Grid Meets Consumer Reality: The Consumer Perspective on Asset Management		345
		3.9.5.1	On-Site Generation	345
		3.9.5.2	Managing Energy Demand and Consumption	346
	3.9.6	Centralized, Data-Driven Asset Management		347
		3.9.6.1	Data Collection	347
		3.9.6.2	Integration and Analysis	348
		3.9.6.3	Decision Making	348
		3.9.6.4	Work Execution	349
	3.9.7	Geospatial Integration for Asset Management		349
	3.9.8	Advanced Asset Management for the Smart Grid		351
3.10	Smart Meters and Advanced Metering Infrastructure			356
	3.10.1	Evolution of the Electric Meter		356
	3.10.2	Evolution of Meter Reading		359
	3.10.3	AMI Drivers and Benefits		361
	3.10.4	AMI Protocols, Standards, and Initiatives		362
		3.10.4.1	ANSI C.12.18 and C.12.19	362

Smart Grid Technologies

    3.10.4.2 IEC 61968-9 Common Information Model..363
    3.10.4.3 IEC 62056 DLMS-COSEM Standard 363
    3.10.4.4 North American Electric Reliability Corporation:
          Critical Infrastructure Protection Security Requirements 363
    3.10.4.5 National Institute of Standards and Technology.....................................363
    3.10.4.6 Smart Energy Profile 364
    3.10.4.7 Common Information Model 364
    3.10.4.8 802.16e 364
  3.10.5 AMI Security 364
    3.10.5.1 Strategy 364
    3.10.5.2 AMI Security Requirements..364
    3.10.5.3 AMI Security Threats 366
    3.10.5.4 Applying Security Specification to AMI 366
  3.10.6 AMI Needs in the Smart Grid..366
    3.10.6.1 Meter Data Reads...367
    3.10.6.2 Internal Device Management..367
    3.10.6.3 Remote Configuration 367
    3.10.6.4 Firmware Upgrades..367
    3.10.6.5 Time Synchronization...367
    3.10.6.6 Local Connectivity 367
    3.10.6.7 Testing and Diagnostics 367
    3.10.6.8 Other Functions..367
    3.10.6.9 Supporting the Customer Interface..367
    3.10.6.10 Integration with Utility Enterprise Applications 368
3.11 Consumer Demand Management 369
  3.11.1 Demand Management Mechanisms 370
  3.11.2 Consumer Load Patterns and Behavior 371
  3.11.3 Conserved versus Deferred Energy 374
  3.11.4 Supply Side of the Equation 376
  3.11.5 Consumer Side of the Equation 377
  3.11.6 Utility–Customer Interaction 378
  3.11.7 Value of Demand Management 379
  3.11.8 Demand Management Enablers in the Smart Grid...382
    3.11.8.1 Beyond Peak Shifting 383
    3.11.8.2 Utility Demand Response Management 385
    3.11.8.3 Empowering Consumers 393
3.12 Convergence of Technologies and Enterprise Level Integration 397
  3.12.1 Synergies of Integrated Applications...398
  3.12.2 Examples of Converging Technologies 400
    3.12.2.1 Integrating Distribution Operation Applications 400
    3.12.2.2 Integration of AMI into the Distribution Operations Environment 404
    3.12.2.3 Multiple Smart Grid Functions Benefit Outage Management 405
    3.12.2.4 Integrating Workforce, Asset, and Network Management Systems........406
  3.12.3 Enterprise Integration 409
  3.12.4 Data Integration versus Application Integration.. 410
  3.12.5 Enterprise Service Bus 410
  3.12.6 Service-Oriented Architecture 411
  3.12.7 Enterprise Information Management.. 411
    3.12.7.1 Establishing an EIM Framework 412

		3.12.7.2 Role of an Enterprise Semantic Model 414
		3.12.7.3 ESM Architecture 416
		3.12.7.4 ESM Information Sources 418
		3.12.7.5 Developing and Implementing an EIM Master Plan 419
		3.12.7.6 EIM Benefits 419
		3.12.7.7 Integrating OT and IT Systems... 420

3.13 High-Performance Computing for Advanced Smart Grid Applications 425
    3.13.1 Computational Challenges in a Smart Grid... 426
        3.13.1.1 Data Complexity 427
        3.13.1.2 Modeling Complexity 429
        3.13.1.3 Computational Complexity 430
    3.13.2 Existing Functions Improved by HPC.. 433
        3.13.2.1 Parallelized State Estimation 433
        3.13.2.2 Parallel Contingency Analysis 433
    3.13.3 New Functions Enabled by HPC 437
        3.13.3.1 Dynamic State Estimation 438
        3.13.3.2 Real-Time Path Rating 440
    3.13.4 HPC in the Smart Grid 443
3.14 Cybersecurity.. 444
    3.14.1 Defining Security 444
        3.14.1.1 Confidentiality 444
        3.14.1.2 Integrity 445
        3.14.1.3 Availability 445
        3.14.1.4 Control 446
        3.14.1.5 Authenticity 446
        3.14.1.6 Usability 446
    3.14.2 Communications Model.. 447
    3.14.3 Security Functions 448
        3.14.3.1 Layered Security Model.. 448
        3.14.3.2 Authentication 449
        3.14.3.3 Authorization 450
        3.14.3.4 Auditing 450
        3.14.3.5 Key Management 451
        3.14.3.6 Message Integrity 451
        3.14.3.7 Network Integrity 452
        3.14.3.8 System Integrity 452
    3.14.4 Security Threats.. 452
        3.14.4.1 People 453
        3.14.4.2 Process 454
        3.14.4.3 Technology 454
    3.14.5 Cybersecurity in the Smart Grid 458
        3.14.5.1 Authentication and Authorization Services 458
        3.14.5.2 Certificate Services 459
        3.14.5.3 Network Security Services 459
Glossary 460
3.15 Smart Grid Standardization Work 461
    3.15.1 Introduction to Standards and Technology... 461

3.15.2 Standards Development Organizations 463
        3.15.2.1 Selected SDO Groups Focused on Smart Grid Standards 464
3.15.3 Alliances.. 465
        3.15.3.1 Example of an Alliance on Smart Grid Standards.............................. 465
3.15.4 User Groups.. 465
3.15.5 Smart Grid Standards Assessment 466
3.15.6 Smart Grid Gap Identification and Decomposition.. 468
        3.15.6.1 Generation 469
        3.15.6.2 Transmission... 469
        3.15.6.3 Distribution... 470
        3.15.6.4 AMI Communications Technologies 470
        3.15.6.5 Consumer 471
        3.15.6.6 Enterprise Integration... 471
        3.15.6.7 NERC CIP Standards 471
3.15.7 Beyond Standardization... 472
3.15.8 Key Issues... 473
        3.15.8.1 Deployment of Technologies Still Under Development 473
        3.15.8.2 Lack of Market Power For Smaller Utilities 474
        3.15.8.3 Interoperability Weak Spots... 474
        3.15.8.4 Enterprise Application Integration.. 474
        3.15.8.5 Lack of Standard Distribution LANs, Especially Wireless Mesh and BPL 475
        3.15.8.6 Too Many HAN Standards... 476
        3.15.8.7 Gateway Definition Between Utility and Premise 477
        3.15.8.8 Common Information Model (CIM) 477
        3.15.8.9 Legacy Transmission and Distribution Automation.......................... 477
        3.15.8.10 Poor Business Cases in Isolation, High Initial Investment When Integrated 478
        3.15.8.11 Merging Organizations 478
        3.15.8.12 Applying Holistic Security 479
3.15.9 Best Practices 479
3.15.10 Legislation and Regulations 480
3.15.11 Advancing Smart Grid Standards 480
Glossary 483
References.. 487
Bibliography 494

3.1 TECHNOLOGY DRIVERS

Stuart Borlase, Steven Bossart, Keith Dodrill, Joe Miller,
Steve Pullins, Bruce A. Renz, and Bartosz Wojszczyk

3.1.1 Transformation of the Grid

Current transmission and distribution grids were not designed with smart grid in mind. They were designed for the cost-effective, rapid electrification of developing economies. The requirements of smart grid are quite different, and, therefore, the reengineering of the current grid is imminent. This engineering work will take many forms including enhancements and extensions to the existing grid, inspection and maintenance activities, preparation for distributed generation and storage, and the development and deployment of an extensive two-way communications system.

The "heavy metal" electric delivery system of transmission lines, distribution feeders, switches, breakers, and transformers will remain the core of the utility transmission and distribution infrastructure. Many refer to this as the "dumb" part of the grid. While some changes in the inherent design of these components can be made, for example, the use of amorphous metal in transformers to reduce losses, the "smarts" in the T&D system are typically related to advances in the monitoring, control, and protection of the "dumb" equipment. Substations therefore play an essential role as the operational interface to the T&D equipment in the field. Advances in technology over the years and the introduction of microprocessor-based monitoring, control, protection, and data acquisition devices have made a marked improvement in the operation and maintenance of the transmission and distribution network. However, changes in the way the T&D system is utilized and operated in a smarter grid will create significant challenges.

Since the invention of electric power technology and the establishment of centralized generation facilities, the greatest changes in the utility industry have been driven not by innovation but by system failures and regulatory/government reactions to those failures. Smart grid technologies have the potential to be the first true "game changing" technology since alternating current supplanted direct current in the late 1800s. As an example, the design of today's power system took advantage of the economies of scale through the establishment of large centralized generation stations. Supply and demand are continuously balanced by dispatching the appropriate level of generation to satisfy load. This operating model schedules the dispatch of generation to meet the day ahead forecast load. This supply dispatch model is the predominant method of balancing supply and demand today. One vision to optimize the end-to-end system would entail not just supply dispatch but also a complementary dispatch of demand resources as well. Currently, generation is matched to supply consumer load plus a reserve margin, and often expensive generating plants are used to satisfy peak demand or supply reserve energy in the case of contingencies. Today, much is being done in the areas of consumer demand management to leverage the use of demand resources that are located behind-the-meter. Customer-owned generation, storage, and controllable loads that can be curtailed or turned on or off upon request can change the load as seen by the utility. Enabled by the smart grid, consumer demand management and this new dispatch model could significantly improve the optimization of the electric system, create new markets, and establish new consumer participation opportunities. The capability to dispatch consumer demand and alternative energy resources in the grid could also place downward pressure on the wholesale price of electricity.

From the transmission perspective, increased amounts of power exchanges and trading will add more stress to the grid. The smart grid challenge will be to reduce grid congestion, ensure grid stability and security, and optimize the use of transmission assets and low-cost generation sources. In order to keep generation, transmission, and consumption in balance, the grids must become more flexible and more effectively controlled. The transmission system will require more advanced technologies such as FACTS and HVDC to help with power flow control and ensure stability.

Substations in a smart grid will move beyond basic protection and traditional automation schemes to bring complexity around distributed functional and communications architectures, more advanced local analytics, and the management of vast amounts of data. There will be a migration of intelligence from the traditional centralized functions and decisions at the energy management and distribution management system (DMS) level to the substations and feeders in order to enhance responsiveness of the T&D system. System operation applications will become more advanced in

Smart Grid Technologies

being able to coordinate the distributed intelligence in the substations and feeders in the field to ensure system-wide reliability, efficiency, and security.

As supply constraints continue, there will be more focus on the distribution network for cost reduction and capacity relief. Monitoring and control requirements for the distribution system will increase, and the integrated smart grid architecture will benefit from data exchange between smarter distribution field devices and enterprise applications. The emergence of widespread distributed generation and consumer demand response (DR) programs also introduces considerable impact to utility operations. The smart grid will see an increase in utility and consumer-owned resources on the distribution system. Utility customers will be able to generate electricity to the grid or consume electricity from the grid based on determined rules and schedules. This means that consumers will no longer be pure consumers but sellers or buyers of energy, switching back and forth from time to time. This will require that the grid operates with two-way power flows and monitors and controls the generation and consumption points on the distribution network. The distributed generation will be from disparate and mostly intermittent sources and subject to great uncertainty. Real-time pricing and consumer demand management will require advanced analytics and forecasting of the electricity consumption of individual consumers. Figure 3.1 illustrates some expected transformations of the grid.

Smart grid technologies will generate a tremendous amount of real-time and operational data with the increase in sensors and the need for more information on the operation of the system.

Decentralized information technology is changing the rules in the electric utility industry. The impact of this change on the utility industry is analogous, in part, to what digitized information has

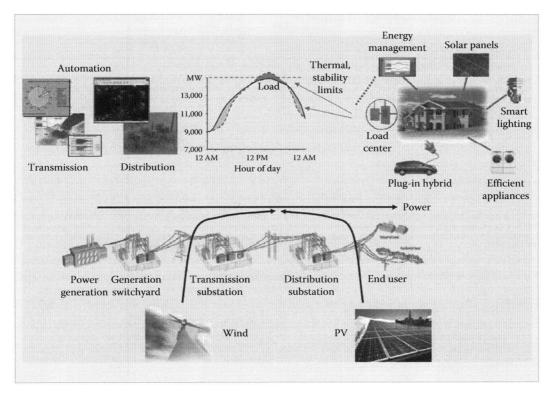

FIGURE 3.1 Transformation of the grid. (From Fan, J. and Borlase, S., Advanced distribution management systems for smart grids, *IEEE Power and Engineering*, Copyright March/April 2009 IEEE.)

done or is doing to the publishing and entertainment industries. While the exact circumstances of the utility industry are unique and nontrivial differences exist, the fundamentals remain—a disruptive technology creates operational opportunities and challenges, resulting in winners and losers within the industry incumbents. To this point, much of the "business case" for smart grid technology is driven by reduced usage of traditional energy sources, which necessarily means that not every industry stakeholder will be aligned on this issue. Companies that are driven primarily by generation evaluate smart grid opportunities differently than companies that are driven primarily by power delivery—for reasons that are or should become obvious.

3.1.2 Characteristics of a Smart Grid

The "smart grid" and similarly denominated programs have been proposed as an effort to integrate the three critical developments in the future grid: expansion of the grid infrastructure to accommodate renewable resources and microgrids; penetration of information technology to implement full digital control in generation, transmission, and distribution systems; and development of new applications. Further, the smart grid programs respond to the political, public, and scientific community requests to deploy high percent of low-CO_2-emitting renewable energy resources. Although there are a number of incarnations of these programs, perhaps the most widely used is from the U.S. DOE (Department of Energy). In June 2008, the U.S. DOE's National Energy Technology Laboratory (NETL) convened a diverse stakeholder meeting that established seven principal characteristics that define the functions of smart grid. The stakeholders converged around the following seven principal characteristics of smart grid.

DOE's NETL has defined the main tenets of the smart grid having the following seven goals [2]:

1. Enabling active participation by consumers in DR
2. Accommodating all generation and storage options
3. Enabling new products, services, and markets
4. Providing power quality (PQ) for twenty-first century needs
5. Optimizing assets and operating efficiently
6. Self-healing from power disturbance events
7. Operating resiliently against physical and cyber attack

First, it will enable active participation by consumers. The active participation of consumers in electricity markets will bring tangible benefits to both the grid and the environment. The smart grid will give consumers information, control, and options that allow them to engage in new "electricity markets." Grid operators will treat willing consumers as resources in the day-to-day operation of the grid. Well-informed consumers will have the ability to modify consumption based on balancing their demands and resources with the electric system's capability to meet those demands. Programs such as DR will offer consumers more options to participate in their energy usage and costs. The ability to reduce or shift peak demand allows utilities to minimize capital expenditures and operating expenses while also providing substantial environmental benefits by reducing line losses and minimizing the operation of inefficient peaking power plants. In addition, emerging products like the plug-in hybrid electric (PHEVs) and all electric vehicles (EVs) will result in substantially improved load factors while also providing significant environmental benefits.

Second, it will accommodate all generation and storage options. It will seamlessly integrate all types and sizes of electrical generation and storage systems using simplified interconnection processes and universal interoperability standards to support a "plug-and-play" level of convenience. Large central power plants including environmentally friendly sources, such as wind and solar farms and advanced nuclear plants, will continue to play a major role even as large numbers of smaller distributed energy resources (DERs), including plug-in EVs, are deployed. Various capacities from small to large will be interconnected at essentially all voltage levels and will include DERs such as photovoltaic, wind, advanced batteries, plug-in hybrid vehicles, and fuel cells. It will be easier and more profitable for commercial users to install their own generation such as highly efficient combined heat and power installations and electric storage facilities.

Third, it will enable new products, services, and markets. The smart grid will link buyers and sellers together—from the consumer to the regional transmission organization (RTO)—and all those in between. It will facilitate the creation of new electricity markets ranging from the home energy management system (EMS) at the consumers' premises to the technologies that allow consumers and third parties to bid their energy resources into the electricity market. Consumer response to price increases felt through real-time pricing will mitigate demand and energy usage, driving lower-cost solutions and spurring new technology development. New, clean, energy-related products will also be offered as market options. The smart grid will support consistent market operation across regions. It will enable more market participation through increased transmission paths, aggregated DR initiatives, and the placement of energy resources including storage within a more reliable distribution system located closer to the consumer.

Fourth, the smart grid will provide PQ for the twenty-first century society and our increasing digital loads. It will monitor, diagnose, and respond to PQ deficiencies, leading to a dramatic reduction in the business losses currently experienced by consumers due to insufficient PQ. New PQ standards will balance load sensitivity with delivered PQ.

The smart grid will supply varying grades of PQ at different pricing levels. Additionally, PQ events that originate in the transmission and distribution elements of the electrical power system will be minimized, and irregularities caused by certain consumer loads will be buffered to prevent impacting the electrical system and other consumers.

Fifth, it will optimize asset utilization and operate efficiently. Operationally, the smart grid will improve load factors, lower system losses, and dramatically improve outage management performance. The availability of additional grid intelligence will give planners and engineers the knowledge to build what is needed when it is needed, extend the life of assets, repair equipment before it fails unexpectedly, and more effectively manage the work force that maintains the grid. Operational, maintenance, and capital costs will be reduced, thereby keeping downward pressure on electricity prices.

Sixth, it will anticipate and respond to system disturbances (self-heal). It will heal itself by performing continuous self-assessments to detect and analyze issues, take corrective action to mitigate them, and, if needed, rapidly restore grid components or network sections. It will also handle problems too large or too fast-moving for human intervention. Acting as the grid's "immune system," self-healing will help maintain grid reliability, security, affordability, PQ, and efficiency. The self-healing grid will minimize disruption of service by employing modern technologies that can acquire data, execute decision-support algorithms, avert or limit interruptions, dynamically control the flow of power, and restore service quickly. Probabilistic risk assessments based on real-time measurements will identify the equipment, power plants, and lines most likely to fail. Real-time contingency analyses will determine overall grid health, trigger early warnings of trends that could result in grid failure, and identify the need for immediate investigation and action. Communications with local and remote devices will help analyze faults, low voltage, poor PQ, overloads, and other undesirable system conditions. Then appropriate control actions will be taken, automatically or manually as the need determines, based on these analyses.

Seventh and finally, the smart grid will operate resiliently against attack and natural disaster. The smart grid will incorporate a system-wide solution that reduces physical and cyber vulnerabilities

TABLE 3.1
DOE Seven Characteristics of a Smart Grid

Today's Grid	Principal Characteristic	Smart Grid
Consumers do not interact with the grid and are not widely informed and educated on their role in reducing energy demand and costs	Enables consumer participation	Full-price information available, choose from many plans, prices, and options to buy and sell
Dominated by central generation, very limited distributed generation and storage	Accommodates all generation and storage options	Many "plug-and-play" DERs complement central generation
Limited wholesale markets, not well integrated	Enables new markets	Mature, well-integrated wholesale markets, growth of new electricity markets
Focus on outages rather than PQ	Meets PQ needs	PQ a priority with a variety of quality and price options according to needs
Limited grid intelligence is integrated with asset management processes	Optimizes assets and operates efficiently	Deep integration of grid intelligence with asset management applications
Focus on protection of assets following fault	Self-heals	Prevents disruptions, minimizes impact, and restores rapidly
Vulnerable to terrorists and natural disasters	Resists attack	Deters, detects, mitigates, and restores rapidly and efficiently

and enables a rapid recovery from disruptions. Its resilience will deter would-be attackers, even those who are determined and well equipped. Its decentralized operating model and self-healing features will also make it less vulnerable to natural disasters than today's grid. Security protocols will contain elements of deterrence, detection, response, and mitigation to minimize impact on the grid and the economy. A less susceptible and more resilient grid will make it a more difficult target for malicious acts.

These seven characteristics represent the unique yet interdependent features that define the smart grid. Table 3.1 summarizes these seven points and contrasts today's grid with the vision for the smart grid.

These seven points have come to define the smart grid for many, although there are variants to the list that emphasize such points as encouraging renewable resources deployed in the transmission, subtransmission, and distribution system; emphasis on the use of sensors and sensory signals for direct automatic control; accelerating automation particularly in the distribution system; and intelligently (optimally) managing multiobjective issues in power system operation and design. The seven cited DOE elements may be viewed more generically as making the grid as follows:

- *Intelligent*: capable of sensing system overloads and rerouting power to prevent or minimize a potential outage; of working autonomously when conditions require resolution faster than humans can respond and cooperatively in aligning the goals of utilities, consumers, and regulators
- *Efficient*: capable of meeting increased consumer demand without adding infrastructure
- *Quality focused*: capable of delivering the PQ necessary (free of sags, spikes, disturbances, and interruptions) to power our increasingly digital economy and the data centers, computers, and electronics necessary to make it run
- *Accommodating*: accepting energy from virtually all fuel source including solar and wind as easily and transparently as coal and natural gas; capable of integrating any and all better

ideas and technologies (e.g., energy storage technologies) as they are market-proven and ready to come online
- *Resilient*: increasingly resistant to attacks and natural disasters as it becomes more decentralized and reinforced with smart grid security protocols
- *Motivating*: enabling real-time communication between the consumer and utility so consumers can tailor their energy consumption based on individual preferences, like price and/or environmental concerns
- *Green*: slowing the advance of global climate change and offering a genuine path toward significant environmental improvement
- *Opportunistic*: creating new opportunities and markets by means of its ability to capitalize on plug-and-play innovation wherever and whenever appropriate

3.1.3 Smart Grid Technology Framework

Beyond a specific, stakeholder-driven definition, smart grid should refer to the entire power grid from generation, through transmission and distribution infrastructure all the way down to a wide array of electricity consumers. The concept of a smart grid embraces all the monitoring, control, and data acquisition functions across the T&D and low-voltage networks with the need for more advanced integration and meaningful information exchange between the utility and the electricity network and between the utility and customers. The smart grid will therefore be an enabler of system-wide solutions in the areas of network operations, asset management, distributed generation management, advanced metering, enterprise data access, public and private transport, etc. (Figure 3.2).

The smart grid is a framework for solutions. It is both revolutionary and evolutionary in nature because it can significantly change and improve the way we operate the electrical system today, while providing for ongoing enhancements in the future. It represents technology solutions that optimize the value chain, allowing us to drive more performance out of the infrastructure we have

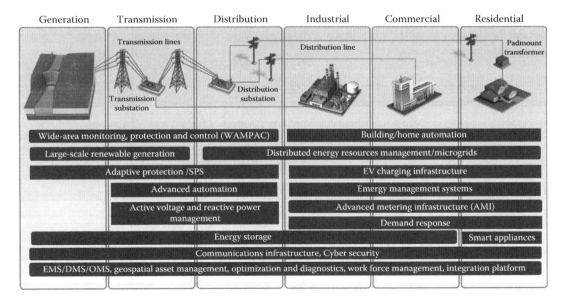

FIGURE 3.2 Smart grid technologies span the entire electric grid. (© Copyright 2012 GE Energy. All rights reserved.)

and to better plan for the infrastructure we will be adding. It requires collaboration among a growing number of interested and invested parties, in order to achieve significant, system-level change. The smart grid will embrace more renewable energy, public and private transport, buildings, industrial complexes, houses, increase grid efficiency, and transfer real-time energy information directly to the consumer—empowering them to make smarter energy choices.

From a high-level system perspective, the smart grid can be considered to contain the following major components:

- Smart sensing and metering technologies that provide faster and more accurate response for consumer options such as remote monitoring, time-of-use (TOU) pricing, and demand-side management (DSM)
- An integrated, standards-based, two-way communications infrastructure that provides an open architecture for real-time information and control to every endpoint on the grid
- Advanced control methods that monitor critical components, enabling rapid diagnosis and precise responses appropriate to any event in a "self-healing" manner
- A software system architecture with improved interfaces, decision support, analytics, and advanced visualization that enhances human decision making, effectively transforming grid operators and managers into knowledge workers

Interoperability between the different smart grid components is paramount. A framework can be used that defines the components at three levels, the electricity infrastructure level, the smart infrastructure level, and the smart grid solution level (Figure 3.3). At each of these levels, different applications exist that need to interoperate among themselves (horizontally) and with the levels above or below (vertically).

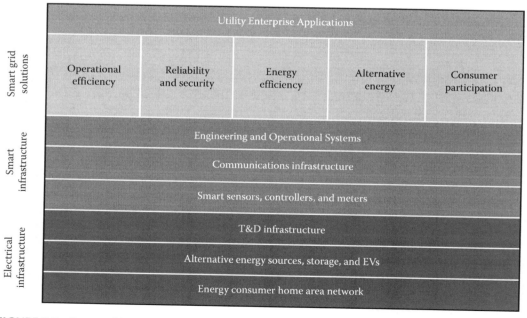

FIGURE 3.3 Smart grid technology framework.

The smart grid will provide a scalable, integrated architecture that delivers not only increased reliability and capital and O&M savings to the utility but also cost savings and value-added services to customers (AMR/advanced metering infrastructure [AMI]/ADI). A well-designed smart grid implementation can benefit from more than just AMI. Numerous technologies now touted under the smart grid banner are currently implemented to various degrees in utilities. The smart grid initiative uses these building blocks to drive toward a more integrated and long-term infrastructure than is intended to realize incremental benefits in operational efficiency and data integration while leveraging open standards. For example, building on the benefits of an AMI with extensive communication coverage across the distribution system helps to improve outage management and enables IVVC (integrated volt/VAr control). In addition, a high-bandwidth communications network provides opportunities for enhanced customer service solutions, such as Internet access, through a home area network (HAN), and a more attractive return on investment. New smart grid–driven technologies, such as advanced analytics and visualization, will continue to offer incremental benefits and strengthen a renewed interest in the consumer interface, AMI, DSM, and other customer-centric technologies, such as PHEVs.

Many industry reports define a wide range of smart grid technologies. These technologies can be broadly captured under the following areas:

- *Low carbon*: for example, large-scale renewable generation, DERs, EVs, carbon capture and sequestration (CCS)
- *Grid performance*: for example, advanced distribution and substation automation (self-healing), wide area adaptive protection schemes (special protection schemes), wide area monitoring and control systems (PMU-based situational awareness), asset performance optimization and conditioning (CBM), dynamic rating, advanced power electronics (e.g., FACTS, intelligent inverters, etc.), high-temperature superconducting (HTS), and many others
- *Grid-enhanced applications*: for example, DMSs; EMSs; outage management systems (OMS); DR; advanced applications to enable active voltage and reactive power management (IVVC, CVVC); advanced analytics to support operational, nonoperational, and BI decision making; DER management; microgrid and virtual power plant (VPP); work force management; geospatial asset management (GIS); KPI dashboards and advanced visualization; and many others
- *Customer*: for example, AMI, home/building automation (HAN), EMSs and display portals, EV charging stations, smart appliances, and many others
- Cybersecurity and data privacy
- Communications and integration infrastructure

Within the smart grid technology landscape, a broad range of hardware, software, application, and communications technologies are at various levels of maturity. In some cases, the technology is well developed (proven performance over time); however, in many areas, the technologies are still at an early stage of maturity and have yet to be deployed at scale deployments.

Many stakeholders determine smart grid technology selection and rollout based on the following factors

1. Business risk
2. Technical risk
3. Technical functionality and capability, availability and maturity

1. *Business risk:* Innovative leaders or fast followers in the rollout of their smart technology:
   a. *Innovative leaders (invest to lead)* set a course that enables the utility to achieve an industry technology leadership position across all business units based on investment in new, and very often, time unproven performance technologies.
   b. *Fast followers (deploy when justified)* provide direction to all business units to deploy smart technology only when it can be economically justified with defined level of maturity.
2. *Technical risk:* Technology deployed in any particular location and grid system will vary, based on a number of factors such as technology maturity, complexity required to integrate with the existing/legacy systems and technologies, existing network performance, financial analysis, customer preference and acceptance, and state and/or federal regulatory influence. This suggests a smart grid design around the criteria as follows:
   a. *Core architecture* consisting of mature technologies with time-proven performance and higher/most certain delivered overall benefits
   b. *Interoperable and scalable architecture* to enable integration with existing/legacy systems, maximize future flexibility, and minimize risk of technology obsolescence
   c. *Secure and open standard architecture*
   d. *New technologies* added incrementally as they mature and/or are cost justified
3. *Technical functionalities and capability availability and maturity:* Smart grid technology selection needs to be defined through the wide range of functionalities and capabilities. Effective selection of available and mature functionalities and capabilities should support the objectives as follows:
   a. *Business requirements*: deliver well-defined and quantifiable smart grid benefits to all stakeholders
   b. *Architecture and integration*: integrate them as one cohesive end-to-end scalable and interoperable solution
   c. *Performance*: proof of performance in support of full-value realization

Table 3.2 provides a summary of key functionalities and capabilities that can be considered for a wide range of smart grid technologies. The presented functionalities and capabilities can be grouped into four categories: infrastructure, metering, grid, and home/building.

Within the smart grid technology landscape, a broad range of hardware, software, application and communications technologies are at various levels of maturity and deployment. In some cases, the technology is well developed (proven performance over time); however, in many areas, the technologies are still at an early stage of maturity and have yet to be demonstrated at scale deployments.

A high-level review of the smart grid technology functionalities and capabilities landscape suggests representative maturity levels and development trends as shown in Table 3.3. This assessment is based on the scale/level of deployed technologies in existing smart grid projects across the globe.

The development of new capabilities and enabling technologies will be critical to fulfilling the grand promise of the smart grid. Smart grid investments should be directed toward holistic grid solutions that will differentiate utility smart grid initiatives. Smart grid, however, is more than simply new technology. Smart grid will have a significant impact on a utility's processes. Perhaps more importantly, it is also about the new information made available by these technologies and the new customer–utility relationships that will emerge. Enabling technologies such as smart devices, communications and information infrastructures, and operational software are instrumental in the development and delivery of smart grid solutions. Each utility customer will

TABLE 3.2
Smart Grid Technology Functionalities and Capabilities

No.	Functionalities and Capabilities	Description
Infrastructure		
1	Communication and security	Underlying communications to support real-time operational and nonoperational smart technology performance
2	Embedded EVs, large-scale renewable generation, DERs	Integration of high penetration of EVs, large-scale renewable generation, and DERs can lead to situations in which the distribution network evolves from a "passive" (local/limited automation, monitoring, and control) system to one that actively (global/integrated, self-monitoring, semiautomated) responds to the various dynamics of the electric grid. This poses a challenge for the design, operation, and management of the power grid as the network no longer behaves as it once did. Consequently, the planning and operation of new systems must be approached somewhat differently with a greater amount of attention paid to global system challenges. In addition, integration of large-scale renewable energy resources presents a challenge with dispatchability and controllability of these resources. Energy storage systems can offer a substantial contribution to alleviate such potential problems by decoupling the production and delivery of energy
Metering		
1	Remote consumer price signals	Function that provides TOU pricing information
2	Granular energy consumption data/information	Function with the ability to collect, store, and report customer energy consumption data/information for any required time intervals or near realtime
3	Identify outage location, extent remotely	Metering function capable of sending signal when meter goes out and identifying themselves after power restoration
4	Remote connection, disconnection, reconnection	Function capable of remotely controlling "on" and "off" smart asset
5	Remote configuration	Function capable of being remotely configured for functionality changes and firmware and software updates
6	Optimize retailer cash flow	Ability for a retail energy service provider to manage its revenues through more effective cash collection and debt management
Grid		
1	Embedded sensing, automation, protection, and control	Wide area system monitoring and advanced system analytics: a real-time, PMU-based grid monitoring system combined with advanced analytics consisting of intelligent fault and outage detection. PMU-based state estimation enabling real-time dynamic and static system stability analysis, risk and margin evaluation, power system optimization, special protection schemes arming, etc., therefore providing planners/system operators and engineering with capabilities to effectively predict possible severe grid disturbances leading to major power system outages and blackouts Wide area adaptive protection, control, and automation: protection philosophy that permits and seeks to make adjustments in various protection functions automatically in order to make them more attuned to the prevailing power system conditions. The purpose of adaptive protection includes mitigate wide area disturbances, improve power system transmission capacity, improve power system reliability, change of operational criteria from *The power system should withstand the most severe credible contingency*—(n-1 or n-2 criterion) to *The power system should withstand the most severe credible contingency followed by protective remedial actions from the wide area protection/emergency control system*

(continued)

TABLE 3.2 (continued)
Smart Grid Technology Functionalities and Capabilities

No.	Functionalities and Capabilities	Description
2	Advanced system operation	The modern grid relies on fully or semiautomated grid operation with a certain level of human intervention provided from the control centers. Advanced system operation tools are comprised of dynamic security assessment and wide area monitoring system (WAMS) and control capabilities
3	Advanced system management	Advanced asset management enables two-key smart grid capabilities • Optimum equipment performance leading to effective asset utilization. This can be accomplished by implementing real-time, dynamic rating applications at grid level. This allows for planed transfer capabilities and grid assets above the manufacturer's "nameplate" ratings • Maintenance efficiency of network components, attained by implementing condition- and performance-based maintenance
4	Advanced system planning	Smart grid system planning considering real-time system impacts from large integration of renewable energy resources, high penetration of distributed generation, and chargeable and dischargeable EVs
5	Intentional islanding (microgrids) and aggregated load and generation management (VPP)	Intentional islanding and/or grid-parallel operation of electric subsystem. Allows for optimum, multiple load/generation balancing to enable reliable and cost-effective operation

Home/building

No.	Functionalities and Capabilities	Description
1	Aggregated DR	Aggregation of demand to reduce peak load and help balance the system more efficiently
2	EMS	Ability to control in-home appliances, distributed generation, and EVs to provide an optimum energy consumption

TABLE 3.3
Smart Grid Technology Landscape

	Functionalities and Capabilities	Maturity Level	Development Trend
1	Communication and security	Developing	Fast
2	Embedded EVs, large-scale renewable generation, DERs	Developing	Fast
3	Metering	Mature	Fast
4	Embedded sensing automation protection and control	Developing	Fast
5	Advanced system operation	Developing	Moderate
6	Advanced system management	Mature	Fast
7	Advanced system planning	Developing	Moderate
8	Intentional islanding (microgrids) and aggregated load and generation management (VPP)	Developing	Moderate
9	Home/building	Developing	Fast

begin the smart grid journey based upon past actions and investments, present needs, and future expectations.

3.2 SMART ENERGY RESOURCES

Thomas Bradley, Johan Enslin, Régis Hourdouillie,
Casey Quinn, Julio Romero Aguero, Aleksandar Vukojevic,
Bartosz Wojszczyk, Alex Zheng, and Daniel Zimmerle

3.2.1 RENEWABLE GENERATION

3.2.1.1 Regulatory and Market Forces

Many countries across the world, including the United States, developed regulation to enable integration of more renewable energy into the overall generation portfolio mix. These include renewable energy portfolio standards (RPS) attached with interconnection initiatives like renewable tax credits and feed-in tariffs. Some of these requirements for renewable energy are so aggressive that utilities are concerned about the grid performance and system operational impacts of intermittent nature of renewable energy generation (e.g., wind and solar).

In 2002, California established its RPS program, with the goal of increasing the percentage of renewable energy in the state's electricity mix to 20% by 2017. On November 17, 2008, Governor Arnold Schwarzenegger signed Executive Order S-14-08 requiring that California utilities reach the 33% renewable goal by 2020. Achievement of a 33% by 2020 RPS would reduce generation from nonrenewable resources by 11% in 2020. This is currently the most aggressive RPS proposed by any of the U.S. states. Other state governments have similar, although at lower penetration levels, but also aggressive RPS allocations [1].

As electric utilities prepare to meet their state's RPS, for example, 33% by 2020 in California, and to comply with Global Warming Solutions Act of 2006 (AB 32), it becomes evident that U.S. utilities must adapt its engineering practices and planning and operations in order to maintain the high levels of service, reliability, and security. The state initiatives require integration of significantly higher levels of renewable energy, such as wind and solar, which exhibit intermittent generation patterns. Due to the geographic location of renewable resources, the majority of the expected new renewable generation additions will be connected via one or two utility's transmission systems. This presents unique challenges to these utilities as the level of planned intermittent renewable generation in relation to their installed system capacity reaches unprecedented and disproportionate levels as compared to other utilities in the state.

Entities in the United States, such as CEC, NERC, CAISO, NYSERDA, SPP, CPUC, etc., have initiated and funded several studies on the integration of large levels of renewable energy, and most of these studies concluded that with 10%–15% intermitted renewable energy penetration levels, traditional planning and operational practices will be sufficient. However, once a utility exceeds the 20% penetration levels of renewable resources, it may require a change in engineering, planning, and operational practices with the development of a smarter grid. These studies support continuing transmission and renewable integration planning studies and recommend that smart grid demonstration project installations should be conducted by the different power utilities.

The United States, and especially California, has a different set of electric system characteristics than in Europe, but there is no experience or research in Europe that would lead us to think that it is technically impossible to achieve 20%–30% intermitted penetration levels at most U.S. utilities. Long transmission distances between generation resources and load centers characterize the network in the United States and especially in the WECC region. There are areas now in Europe that are highly penetrated with intermittent renewable, especially wind generation, at levels of around 30%–40%.

Large-scale wind and solar generation will affect the physical operation of the grid. The areas of focus include frequency regulation, load profile following, and broader power balancing. The variability of wind and solar regimes across resource areas, the lack of correlation between wind and solar generation volatility and load volatility, and the size and location of the wind plants relative to the system in most U.S. states suggest that impacts on regulation and load profile requirement resource smoothing will be large at above 20% penetration levels [1].

The European experience taught us that there are consequences of integrating these levels of wind resources on network stability that have to be addressed as wind resources reach substantial levels of penetration. A list of the major issue categories follows:

- New and in-depth focus on system planning. Steady-state and dynamic considerations are crucial.
- Accurate resource and load forecasting becomes highly valuable and important.
- Voltage support. Managing reactive power compensation is critical to grid stability. This also includes dynamic reactive power requirements of intermittent resources.
- Evolving operating and power balancing requirements. Sensitivity to existing generator ramp rates to balance large-scale wind and solar generation, providing regulation and minimizing start–stop operations for load-following generators.
- Increased requirements on ancillary services. Faster ramp rates and a larger percentage of regulation services will be required, which can be supplied by responsive storage facilities.
- Equipment selection. Variable-speed generation (VSG) turbines and advanced solar inverters have the added advantage of independent regulation of active and reactive power. This technology is essential for large-scale renewable generation.
- Strong interconnections. Several large energy pump-storage plants are available in Switzerland that are used for balancing power. Larger regional control areas make this possible.

Technical renewable integration issues should not delay efforts to reach the renewable integration goals. However, focus has increased on planning and research to understand the needs of the system, for example, research on energy storage options.

Studies and actual operating experience indicate that it is easier to integrate wind and solar energy into a power system where other generators are available to provide balancing power and precise load-following capabilities. The greater the number of wind turbines and solar farms operating in a given area, the less their aggregate production is variable. High penetration of intermittent resources (greater than 20% of generation meeting load) affects the network in the following ways [1]:

- Thermal and contingency analysis
- Short circuit
- Transient and voltage stability
- Electromagnetic transients
- Protection coordination
- Power leveling and energy balancing
- Power quality

The largest barrier to renewable integration in the United States is sufficient transmission facilities and associated cost-allocation in the region to access the renewable resources and connecting these resources to load centers. Other key barriers include environmental pressure and technical interconnection issues such as forecasting, dispatchability, low-capacity factors, and intermittency impacts on the regulation services of renewable resources.

In the United States, the sources of the major renewable resources are remote from the load centers in California and the Midwest states. This results in the need for addition of new major transmission facilities across the country. Wind and solar renewable energy resources normally have capacity factors between 20% and 35%, compared to higher than 90% with traditional nuclear and coal generation. These low-capacity factors place an even higher burden on an already scarce transmission capacity. Identification, permitting, cost-allocation, approval, coordination with other stakeholders, engineering, and construction of these new transmission facilities are major barriers, costly, and time consuming.

Although energy production using renewable resources is pollution free, wind and solar plants need to be balanced with fast ramping regulation services like peaker generator or hydro generation plants. Existing regulation generation is too slow and is polluting much more during ramping regulation service. The increased requirements in regulation services counteract the emission savings from these renewable resources. Currently the frequency regulation requirement at the CAISO is around 1% of peak load dispatch, or about 350 MW. This is currently mainly supplied by peaker generating plants and results in higher emission levels. It has been calculated that around 2% regulation would be required for integrating 20% wind and solar resources by 2010 and 4% to integrate 33% renewables by 2020 [1].

With the integration of wind and solar generation, the output of fossil fuel plant needs to be adjusted frequently, to cope with fluctuations in output. Some power stations will be operated below their maximum output to facilitate this, and extra system balancing reserves will be needed. Efficiency may be reduced as a result with an adverse effect on the emissions. At high penetrations (above 20%), wind and solar energy may need to be "spilled" or curtailed because the grid cannot always utilize the excess energy.

3.2.1.2 Centralized and Distributed Generation

In the early beginnings of the electric industry, power generation used to be comprised of a series of small generators installed at large customer facilities, towns, and cities. As the demand for reliability electricity supply increased and the industry developed, the need for larger generators and interconnected power systems grew as well. Large-scale centralized generation dominated the power industry for decades until growing environmental and socioeconomic concerns and rising interest in power system efficiency improvement favored the construction of smaller-scale generation facilities (particularly those of renewable nature) closer to customer loads over the construction of large power plants and long transmission lines. This trend, prompted, for instance, by the Public Utilities Regulatory Policy Act (PURPA) of 1978 or the Energy Policy Act of 1992, has led to the emergence of the distributed energy resource (DER) concept, which includes distributed generation (DG), distributed storage (DS), and other customer energy resources implemented through programs such as demand response, load management, etc.

DERs are distribution-level energy sources that have smaller generating capacities than utility-scale generation resources. Examples include reciprocating diesel engines, natural gas–powered microturbines, large batteries, small to utility-scale renewable generation (photovoltaic [PV], wind, etc.) and fuel cells. DG usually refers to generation only (not storage) energy resources at the distribution level. There are many potential configurations for DERs, from basic backup functions all the way up to a full microgrid.

Providing backup has been the most basic and prevalent application of DERs. Backup generators are usually small diesel generators designated as support for specific loads. Under this configuration, the grid has primary responsibility for providing power; the backup generator only kicks in when the grid has been compromised. Figure 3.4 shows privately owned and utility-owned backup generators powering the grid during an outage. The problem with these configurations is that they lead to what is called *low asset utilization*, since the backup generators do not run unless the grid is unavailable. Because they have relatively low asset utilization rates, the cost of delivered energy over the lifetime of backup generators tends to be very high. Those high costs drive

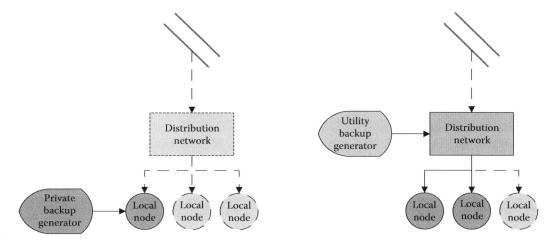

FIGURE 3.4 Privately and utility-owned backup generator configurations.

private backup generator customers to opt for smaller backup generators that are generally not large enough to pick up the entire load. When an outage occurs, most of the load must be dropped with only critical loads, such as emergency lighting, remaining active. Furthermore, these critical loads are often on a separate circuit, meaning that even if the backup generators were large enough, they would not be able to power regular loads. Utilities follow similar logic, putting backup generators only on circuits where critical operations, such as hospitals or high-tech businesses, are located. In many situations, it would be helpful to the system to have the DER operating much of the time. But without embedded intelligence in these resources, they cannot be effectively integrated into the rest of the system.

The last two decades have seen the resurgence of grid-connected DG, either independent power producers (IPPs) or utility-owned DG has started to dot distribution grids around the world. This DG application has the objective of supplying service to the grid in a continuous fashion, that is, in the same way as conventional centralized generators. The main difference of this approach is the location (close to the loads), installed capacity (smaller size), and type or lack of ancillary services (e.g., voltage regulation and frequency regulation) that the DG provides. Furthermore, it requires interconnection with the distribution system using synchronous, induction, and electronically coupled generators. This can represent a significant challenge since distribution systems have historically designed to be operated in a radial fashion, without any special considerations for DG, and it may lead to impacts that could affect the operation of both, distribution systems and DG, particularly for intermittent DG such as solar PV and wind. Smart grid technologies can play a significant role in facilitating the integration of DG and mitigating impacts on the distribution grid.

3.2.1.3 Technologies

There are several renewable sources of electric energy (generically called renewables). The main difference between renewables and other conventional energy sources is that renewables provide energy that is considered to be cleaner with respect to pollution. Other distinguishing difference is that renewable energy sources do not deplete natural resources in the process of creating of power. The third difference is that renewables are scalable to the appropriate size anywhere from single-house applications all the way up to large-scale renewables, which can supply power to thousands of homes. Some of the most common renewable energy resources are introduced in the next sections.

3.2.1.3.1 Solar PV

Solar PV generation has experienced a tremendous growth in recent years due to growing demand for renewable energy sources. PV represents a method of generating electric power in solar panels that are exposed to the light. Power generated is based on the conversion of the energy of the sun rays. Solar panels consist of solar cells that contain PV material, which exhibits PV effect. Solar cell that is exposed to light transfers electrons between different bands inside the material. This in turn results in potential difference between two electrodes, which caused direct current (DC) to flow.

There are several main PV applications, such as solar farms, building, auxiliary power supply in transportation devices, stand-alone devices, and satellites. Utilities around the world started incorporating solar farms into their generation portfolios mostly during the last decade. In order to incorporate solar farm into utility power system, alternating current (AC)/DC converter is needed as well as the corresponding relay protection. The main issue with PVs is intermittency. Since PV is unreliable power source that cannot always be counted on, several efforts have been undertaken to increase the reliability of PVs. One of the most successful ones was adding the battery storage where electric energy is stored during the off-peak hours or curtailment period, and then reused when PVs are not available.

In recent years, PV has been used in smaller-size applications, because of high inefficiency of solar cells. However, major advances in the design, materials, and manufacturing have made PV industry one of the fastest growing energy sources. Today, solar PV represents less than 0.5% of total global power generation capacity.

3.2.1.3.2 Solar Thermal

Solar thermal energy (STE) is a technology that converts solar energy into thermal energy (heat).

There are three types of collector levels that are based on the temperature levels: low, medium, and high. In practice, low-temperature collectors are placed flat to heat swimming pools or space heating, medium-temperature collectors are flat plates used for heating of water or air, and high-temperature collectors are used for electric power production.

Heat represents the measure of the thermal energy that particular object contains, and three main factors, specific heat, mass, and temperature, define this value. In essence, heat gain is accumulated from the sun rays hitting the surface of the object. Then, heat is transferred by either conduction or convection. Insulated thermal storage enables STE to produce electricity during the days that have no sunlight. The main downside to STE plants is the efficiency, which is a little over 30% at best for solar dish/stirling engine technology, while other technologies are far behind.

3.2.1.3.3 Wind

Wind power is obtained by converting the energy of the wind by wind turbines into electricity. Even though wind energy dates back from early centuries when it was being used to propel the ships, today's applications are more geared toward utilities and supply of the power to larger regions. The main drivers of success of any wind farm are the average wind speed in the area and close proximity to the transmission power system.

Wind energy is highly desirable renewable energy source because it is clean technology that produces no greenhouse gas emissions. The main downside of wind power is its intermittency and visual impact that it creates on the environment. During the normal operation, all of the power of the wind farm must be utilized when it is available. If it is not used, the wind farm is either curtailed or power generated can be used to charge the battery energy storage (BES) if one is associated with wind farm.

Wind power is higher at higher speed of wind, but since the speed of the wind constantly changes, power comes and goes in short intervals. Inconsistency in power output is the main reason why wind farms cannot be used in utility's base-load generation portfolio. Capacity factor of a wind power turbine ranges anywhere from 20% to 40%.

3.2.1.3.4 Biomass and Biogas

Who would have thought that one day technology will be able to produce the electricity from the fuel made out of living and dead biological material? Dead trees, wood chips, plant or animal matter used for production of fibers, chemical, or heat all refer to biomass. Technologies associated with biomass conversion to electrical energy include releasing energy in form of heat or electricity or conversion to a different form such as combustible biogas or liquid biofuel. The downside of biomass as a fuel is increased air pollution. The biomass industry has recently experienced an upswing, and the electricity in the United States produced by biomass plants is around 1.4% of the total U.S. electricity supply. Another form of biogas can be produced from algae. Algae produce oil that can be converted for industrial use and also produce a biomass that is converted to a synthetic natural gas, which can be used to generate electricity.

3.2.1.3.5 Geothermal

Geothermal power is extracted from the earth through natural processes. There are several technologies in use today, such as binary cycle power plants, flash steam power plants, and dry steam power plants. The main issue with geothermal power is low thermal efficiency of geothermal plants, even though capacity factor can be quite high (up to 96%).

Geothermal plants can be different in size. Geothermal power is reliable and cost effective (no fuel), but initial capital costs associated with deep drilling as well as earth exploration are main deterring factors from higher penetration of geothermal resources.

3.2.1.3.6 Wave Power

There are two types of ocean power that can be harnessed: wave power and tidal power. Wave power is associated with the energy produced by ocean waves that are on the surface and converting that energy for the generation of electricity. Today, wave farms have been installed in Europe. Currently, this type of renewable does not have significant penetration, because it is highly unreliable, and it requires large wave energy converter to be deployed. The first such farm in the United States is expected to be a wave park in Reedsport, Oregon. The PowerBuoy technology that will be used for this project will have modular, ocean-going buoys, and rising and falling of the waves will cause buoy to move, creating mechanical energy that will be later converted to electric energy and transmitted offshore through the underwater transmission line.

3.2.1.3.7 Hydro

Hydropower plants use the energy of the moving water as a main source for producing electricity. Water fall and gravitational force of this falling water hit the blades on the rotor, which causes rotor to turn, thus producing electricity. Most of the time, hydropower plants are built in places where there is not an abundance of water, but the water is very fast moving (like mountainous areas), and in the valleys where there is an abundance of water, but the water is moving slowly.

3.2.1.3.8 Fuel Cells

Fuel cells are an electrochemical cell that converts the source fuel into an electric energy. Reaction within the cell between a fuel and oxidant with presence of electrolyte generates electricity. Reactants flow into the cell, and reaction produces the flow out of it, while electrolyte remains in it. Fuel cells must be replenished from outside.

3.2.1.3.9 Tidal Power

Tidal power converts the energy of tides into electricity. The most common tidal power technologies are tidal stream generators and tidal barrages. Tidal stream generators rotate underwater and produce electricity using the kinetic energy of tidal streams. Tidal barrage uses a dam located across a tidal estuary to produce electricity using the potential energy of water. Water flows into the barrage

Smart Grid Technologies

during high tide and then it is released during low tide while moving a set of turbines. New technologies such as dynamic tidal power are being discussed and evaluated; this technology is intended to take advantage of a combination of the kinetic and potential energy of tides.

3.2.1.3.10 Combined Heat and Power

Base-load or combined heat and power (CHP) operation modes give the DER primary responsibility for supporting the load. CHP takes its name from the fact that heat from the DER can be used to supply heat locally, increasing the overall efficiency of the system. The DER operates nearly continuously, but the load is still connected to the grid in most cases, especially if the load is too large to be supported by the DER alone. A grid outage does not significantly change the operating mode of the DER, although it may be required to drop part of the load if they are unable to support it in its entirety. Figure 3.5 shows a base-load/CHP DER unit supporting a load during normal operation and during a grid outage, respectively.

This configuration offers much higher asset utilization rates than the backup generator configuration because the DER is essentially running at all times. This improves the economics of the DER purchase compared to the backup generation. However, it may not improve the economics of the operation as a whole. That is because the constant use of a base-load or CHP system means its energy costs can run almost as high as the equipment. In contrast, backup generation has low energy costs because it is seldom used.

For example, if power from a backup generator costs $0.50/kWh to produce 5% of your energy use, and grid electricity costs $0.10/kWh for the other 95%, then the blended energy cost is $0.12/kWh. In contrast, if power from a base-load generator costs $0.15/kWh and comprises 60% of your energy use, and grid electricity costs $0.10/kWh and comprises the other 40%, then the blended rate is $0.13/kWh. Therefore, the cost of production from the base-load or CHP DER application matters more than the cost of the backup generation, because it represents a larger fraction of the total energy costs.

It is clear that for a base-load or CHP DER application to be economical, the generator must be carefully chosen and matched with a load in a way that delivers energy at a lower cost than the grid. One way to accomplish this is to use the excess heat from the DER to warm a nearby industry or residential area. This delivers value through heat as well as electricity. Despite an aggressive build-out of CHP and base-load applications in the 1970s, there are many viable candidates in the United States and a great deal more in developing countries.

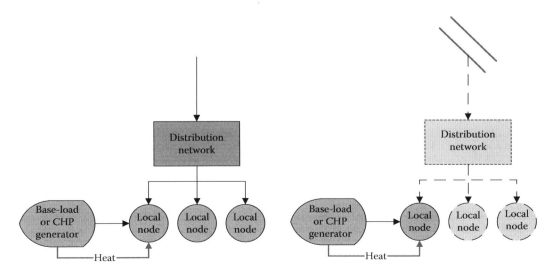

FIGURE 3.5 Base-load or CHP generator supporting local node during normal operation and outage.

3.2.1.4 Renewable Energy Needs in a Smart Grid

To integrate high penetration levels like 33% intermittent renewable resources by 2020 in California, several planning and operational solutions should be followed. There is no silver bullet but requires a combined effort on three major levels that can be used in a smart grid strategy [1]

- Generation mix to utilize different complementary resources
- Advanced smart grid transmission facilities, including fast responsive energy storage, FACTS, HVDC, WAMPAC, etc.
- Smart grid applications on distribution networks including distribution automation, fast demand response, including distributed resources (DRs) on the distribution feeders, distributed energy storage, controlled charging of plug-in hybrid electric vehicles (PHEV), demand-side management (DSM), etc.

The purpose of increased transmission planning is to identify complete and preferred transmission plans and facilities to integrate these high levels of renewables. The clear goal would be to develop a staged transmission expansion plan, facilities, and storage options to integrate this potential level renewable penetration levels.

Most of the models for these advanced wind and solar facilities have not been fully developed yet and need to be validated. The generator models for wind and solar generation technologies need to be upgraded and validated to include short-circuit models, dynamic variance models like clouding and short-term wind fluctuations.

The European experience with high levels of intermittent resources up to 80% penetration levels does not transfer fully due to the difference in U.S. grid design and load density. The integration of renewable energy at this scale will have significant impact, especially if the addition of energy storage devices (central and distributed) and FACTS devices utilized to counterbalance the influence of the intermittent generation sources. Utilities and ISOs in the United States should conduct RD&D projects and commence studies to fulfill its obligation to accurately and reliably forecast the impacts on future system integrated resource planning. Due to the long lead time for some of the proposed technology solutions, it is recommended that utilities engage these challenges sooner versus later. If technical challenges manifest, a timely solution cannot be implemented if studies, demonstration installations, and field tests still have to be conducted. Additionally, utilities should study all conceivable options that may severely affect transmission system integrity and stability. Otherwise, utilities may experience unintended consequences due to unforeseen technical issues resulting from high penetrations of new renewable energy sources.

3.2.2 ENERGY STORAGE

Energy storage in general is a very old concept, even though it was not recognized as such. For instance, solar energy has been transformed and stored in the form of fossil fuels that are used today in a large number of applications. Energy storage concepts have not been widely applied to power systems until recently due mainly to technological and economic limitations given the large volumes of energy that typically are of interest in the power industry. Some exceptions are pumped hydro and uninterruptible power supply (UPS) systems. However, energy storage concepts have been commonly applied to other areas of electrical engineering such as electronics and communications, where the amounts of energy to be stored are easier to manage.

3.2.2.1 Regulatory and Market Forces Driving Energy Storage and Smart Grid Impact

Grid energy storage or the ability to store energy within the power delivery grid can arguably be regarded as the "holy grail" of the power industry, and it is expected to play a key role in facilitating

Smart Grid Technologies

the integration of DRs and plug-in electric vehicles (PEVs)* and fully enabling the capabilities, higher efficiency, and operational flexibilities of the smart grid. The main challenge with electric energy is that it must be used as soon as it is generated, or if not, it must be converted into other forms of energy. During the times when their assistance is not required, storage systems accumulate energy. Later on, stored energy is dispatched into the power system for certain periods of time, thus decreasing the demand for generation and assisting the system when needed.

The ability of storing energy in an economic, reliable, and safe way would greatly facilitate the operation of power systems. Unfortunately, high costs and technology limitations have constrained the large-scale application of storage systems. Historically, pumped hydro has been the most common application of energy storage technologies on power system level applications. Nevertheless, the last two decades have seen the emergence and practical applications of new technologies such as battery systems, flywheels, etc., prompted by the increasing interest and need to integrate intermittent resources and PEVs, growing demand for high reliability, for instance, via implementation of microgrids, and the need for finding alternative technologies to provide ancillary services and system capacity deferral among others. There is growing interest worldwide in this area, and regulatory mechanisms and incentives are being proposed and debated, such as the U.S. Congress Storage Act of 2009 (S. 1091), which called for amendment to the Internal Revenue Code to

- Allow a 20% energy tax credit for investment in energy storage property directly connected to the electrical grid (i.e., state systems of generators, transmission lines, and distribution facilities) and designed to receive, store, and convert energy to electricity and deliver such electricity for sale
- Make such property eligible for new, clean, renewable energy bond financing
- Allow a 30% energy tax credit for investment in energy storage property used at the site of energy storage
- Allow a 30% nonbusiness energy property tax credit for the installation of energy storage equipment in a principal residence

There are several main applications where energy storage systems can be used. Some of those include frequency regulation, spinning reserve, peak shaving/load shifting, and renewable integration.

3.2.2.1.1 Frequency Regulation

In practice, there always exists a mismatch between generation and load in a power system. This mismatch results in frequency variations. System operators are always trying to match the generation to the load, so that the frequency can be as close as possible to 60 Hz. Variability of the frequency is further increased by the addition of renewables, such as solar and wind. Any power system is required to maintain the frequency within the desired limits. Any large variations from 60 Hz will cause unwanted system instability and can bring the whole system down. As noted earlier, system operators are trying to balance the generation and load by varying the output of proper generating units based on the system frequency. This type of regulation is called frequency regulation. In addition to having the whole system being able to supply power for the desired load, utility operators always have extra amount of generation that is known as spinning reserve. This spinning reserve has to be enough to provide enough power for frequency regulation purposes as well as be enough to support the tripping of the largest generating unit in the system to prevent the power interruptions. The amount of regulation capacity is most based on historical records and might vary on several factors such as time of the day, time of the year, etc.

---

* PEV—Plug-in electric vehicle, typically meant to include the entire family of grid-rechargeable vehicles, including plug-in hybrids (PHEVs), battery electric vehicles (BEVs or EVs), and extended range electric vehicles (EREVs).

One basic difference between the regulated and deregulated markets is that deregulated markets have a market for ancillary services such as frequency regulation. In this market, reserve capacities of every generating unit can be bid and market price is paid for capacity reserved for the regulation as well as actual provided energy. The system works as follows: control system for particular balancing authority sets the outputs of each generation asset. The system computes the difference between the power output and load demand (adjusted with frequency error bias) called area control error or ACE. From this signal, another signal called automatic generation control or AGC is extracted and sent to regulation service provider. This provided in turn adjusts its power output based on the AGC signal that was received. Frequency increase requires provider to supply additional power to the grid, which is equivalent to energy storage system discharging the energy to the system. On the opposite side, frequency decrease requires provider to remove power from the grid, which is equivalent to power system charging the energy system.

In the past, thermal generators or hydro facilities have been used to provide frequency regulation due to their fast response, which is needed for effective regulation. However, this was not the most optimum way for economic dispatch because of the losses and increased wear and tear on the generating sources. In addition, these are base-load generating plants, so output had to be reduced in order to provide frequency regulation capacity, which in turn caused higher-cost generating units to be online in order to support the load. Energy storage that provides frequency regulation allows for better optimization of generation assets. In addition, every MW of renewable resources added to the system will require between 3% and 10% increase in regulation service.

3.2.2.1.2 Spinning Reserve

As mentioned earlier, the total generation in a region that belongs to one utility system is equal to the load demand plus some spinning reserve. The amount of spinning reserve is equal or larger to the highest power-producing unit connected to the system plus some margin. The reason for this is the need for immediate power if the largest unit trips off-line. Knowing that it takes certain amount of time to start any generating unit, having energy storage systems provides additional benefit, because those systems can be immediately deployed. In reality, during the high-load periods, majority of thermal and hydro units are dispatched and run at its maximum and cannot be utilized as spinning reserve. So, in order to have spinning reserve, additional units are needed. Note that during light- or medium-load conditions, these generating units have output less than maximum, with the difference being designated as spinning reserve. Committing generating resources for spinning reserves is mandatory, but it results in increased operating costs and decreased efficiency. Energy storage systems help in reduction of spinning reserves provided by thermal and hydro generating units and allow dispatchers to set operating points at maximum levels during the economic dispatch. Similar to frequency regulation market, in deregulated markets, there exists a spinning reserve service market, where generation owners bid to provide this service. The only downside to energy storage systems is that they provide output only for a limited amount of time and not infinite. After the energy storage system has started providing energy to the utility system, additional generating units must be deployed before the output of energy storage systems runs out in order to avoid service interruptions.

3.2.2.1.3 Peak Shaving/Load Shifting

Load demand is always changing, and utilities employ different techniques in order to predict daily load curves. Major inputs into load estimation are temperature, load demand during the last 7–10 days, and historical data. Based on the estimated load curves, economic dispatch is created to identify generating units that will be supplying the needed power along with spinning reserves and uncertainty in load estimation. Every generating unit has the operating kWh cost associated with it, and economic dispatch is based on these costs. Units with lowest operating cost are used for base loading and run most of the time. For example, nuclear, hydro, and modern coal plants are almost exclusively used for base load. Note here that these units also have the highest capital cost during the construction. In order to cover the peak load demand, utility must bring on-line its higher operating cost generating units. For example, plants that have combustion turbines (CTs) might only be utilized few hours during the whole year to cover the peak load. In order to level demand and move energy usage towards the off-peak hours, energy has to be stored first. This can be done during the time with low demand, because the cost of generation is low. This energy can be supplied from energy storage systems to the grid during the peak times.

3.2.2.1.4 Renewable Integration

Energy storage, power electronics and communications have a key role to play to mitigate the intermittency and ramping requirements of large-scale renewable energy penetration of wind and solar energy. Since its inception, wind and solar technologies have made major breakthroughs and became more reliable. Utilities are constantly incorporating these two renewable resources into their generation portfolios. However, the biggest issue associated with wind and solar power is their unpredictability and variability of the output. In addition, these technologies also require regulation. Solar and wind energy productions are not dispatchable and result typically in high levels of power and associated voltage fluctuations. Common problems in remote wind production areas include low capacity factors for all the wind farms, impacts of line contingencies on wind farm operations, curtailment of wind farm outputs during high production times, and high ramp rate requirements [1].

In most urban regions, PV flat-plate collectors are predominately used for solar generation and can produce power production fluctuations with a sudden (seconds time-scale) loss of complete power output. With partial PV array clouding, large power fluctuations can also result at the output of the PV solar farm with large power quality impacts on distribution networks. It is clear that these power variations on large-scale penetration levels can produce several power quality and power balancing problems. Cloud cover and morning fog require fast ramping and fast power balancing on the interconnected feeder. Furthermore, several other solar production facilities are normally planned in close proximity on the same electrical distribution feeder that can result in high levels of voltage fluctuations and even flicker. Reactive power and voltage profile management on these feeders are common problems in areas where high penetration levels are experienced.

Energy storage systems can be used for smoothing the power out of renewable sources. This can be accomplished by limiting the rate of change of the output of a renewable resource. Energy storage systems can either add or remove power from the system as needed in order to smooth the power output of a renewable resource. One of the most promising solutions to mitigate these integration issues is by implementing a hybrid fast-acting energy storage and STATCOM in a smart grid solution. Several fast-reacting energy storage solutions are currently available on the market. For mitigating the mentioned wind and solar integration problems, the energy storage device needs to be fast acting and a storage capability of typically 15 min–4 h and a STATCOM that is larger than the battery power requirements to have adequate dynamic reactive power capabilities. Figure 3.6 shows a STATCOM—BESS application for mitigating the wind farm related integration issues [2]. The main components and technical characteristics of this smart energy storage solution are described as follows:

- 8 MW/4 h battery
- 20 MVAr inverters for BESS and STATCOM
- Integrated control and HMI (human-machine interface) of STATCOM and BESS system
- Substation communications interface for integrating the BESS solution into a distribution automation and ISO market participation environment

3.2.2.2 Centralized and Distributed Energy Storage

Energy storage applications can be centralized or distributed. The selection of the type of solution and technology to be used in an application is a function of the type of problem to be addressed and a series of technical and economic considerations such as ratings, size and weight, capital costs, life efficiency, and per-cycle cost. Figure 3.7 shows a summary of the installed grid-connected energy storage technologies worldwide.

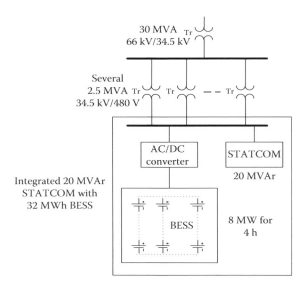

FIGURE 3.6 Basic Schematic of STATCOM-BESS application. (From Enslin, J., Dynamic reactive power and energy storage for integrating intermittent renewable energy, Invited Panel Session, Paper PESGM2010-000912, *IEEE PES General Meeting*, Minneapolis, MN, July 25–29, 2010.)

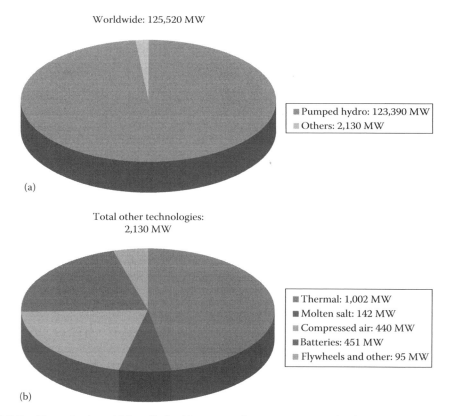

FIGURE 3.7 (**See color insert.**) Installed grid-connected energy storage technologies worldwide as of April of 2010. (From Current Energy Storage Project Examples, California Energy Storage Alliance (CESA), http://www.storage alliance.org/presentations/CESA_OIR_Storage_Project_Examples.pdf)

Smart Grid Technologies

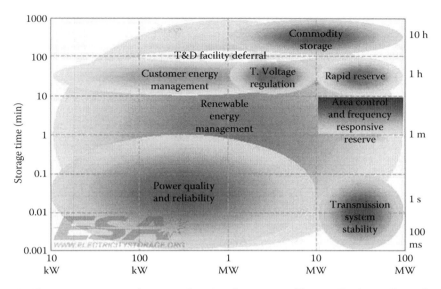

FIGURE 3.8 Energy storage requirements for electric power utility applications. (Data from Sandia Report 2002-1314; Electricity Storage Association (ESA), Utility Support, http://www.electricitystorage.org/technology/technology_applications/utility_support/)

Centralized energy storage applications consist of large MW-size facilities usually connected to transmission system level voltages; these applications are typically used for providing ancillary services during short periods of time (e.g., seconds or minutes) and for intermittent renewable generation integration. DS consists of kW and smaller MW-size facilities connected to distribution system level voltages, either at distribution substations, feeders, or customer facilities; this includes applications such as community energy storage (CES) and vehicle-to-grid (V2G). CES is a concept that is increasingly being studied with applications ranging 25–75 kWh and devices similar to pad-mounted distribution transformers. Distributed energy storage in general is typically used for intermittent renewable generation integration, distribution reliability improvement, and capacity deferral; therefore they are required to have larger storage times (e.g., minutes or hours), as shown in Figure 3.8. This application is also increasingly being considered for integration of PEVs. The electricity storage association (ESA) provides a very comprehensive description of the recommended applications and advantages and disadvantages of each technology, which are summarized in Table 3.4 and Figure 3.8 and discussed in the next sections.

The coordinated implementation of smart grid technologies such as distributed energy storage, communications, control, power electronics, and power system technologies allows the seamless integration of intermittent DG and adds further capabilities to it including controllability (i.e., dispatchability) and firmness. These capabilities can be used for capacity planning applications (e.g., capacity deferral), increased operational flexibility during outages (intentional islanding), and reliability improvement. Furthermore, DS in the smart grid context may be used to mitigate impacts caused by both, DG (especially PV) and PEVs; this idea is described in general terms in Figures 3.9 and 3.10.

3.2.2.3 Technologies

Energy storage methods can be divided into several groups: chemical, electrical, electrochemical, mechanical, thermal, and biological. Some of the most common examples of energy storage systems connected to utility power grid include the following:

- BES
- Superconducting magnetic energy storage (SMES)

TABLE 3.4
Energy Storage Technology Comparisons

Storage Technologies	Main Advantages (Relative)	Disadvantages (Relative)	Power Application	Energy Application
Pumped storage	High capacity, low cost	Special site requirement		●
CAES	High capacity, low cost	Special site requirement, need gas fuel		●
Flow batteries: PSB VRB ZnBr	High capacity, independent power and energy ratings	Low energy density	◐	●
Metal-air	Very high energy density	Electric charging is difficult		●
NaS	High power and energy densities, high efficiency	Production cost, safety concerns (addressed in design)	●	●
Li-ion	High power and energy densities, high efficiency	High production cost, requires special charging circuit	●	○
Ni-Cd	High power and energy densities, efficiency		●	◐
Other advanced batteries	High power and energy densities, high efficiency	High production cost	●	○
Lead-acid	Low capital cost	Limited cycle life when deeply discharged	●	○
Flywheels	High power	Low energy density	●	○
SMES, DSMES	High power	Low energy density, high production cost	●	
E.C. Capacitors	Long cycle life, high efficiency	Low energy density	●	◐

Source: Electricity Storage Association (ESA), Technology Comparison, http://www.electricitystorage.org/technology/storage_technologies/technology_comparison

- Flywheel energy storage (FES)
- Compressed air energy storage (CAES)
- Ultracapacitors
- Pumped hydro

These energy storage systems will be separately investigated later on in this chapter.

Smart Grid Technologies

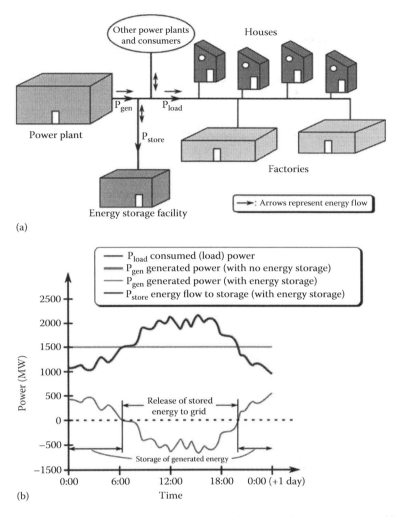

FIGURE 3.9 (**See color insert for b only.**) Conceptual description of grid energy storage. (a) Network power flows and (b) Energy storage and release cycles. (From Wikipedia, Grid Energy Storage, http://en.wikipedia.org/wiki/Grid_energy_storage)

3.2.2.3.1 Battery Energy Storage

BES is mostly used for load leveling, peak shaving, and frequency regulation. Today, there are two types of batteries based on their chemical structure. One type is called power battery, and these batteries are capable of delivering fast charge/discharge. These batteries are mainly used for frequency regulation. Another type of battery has a slow charge/discharge times, and those batteries are used for load leveling and peak shaving.

3.2.2.3.2 Superconducting Magnetic Energy Storage

SMES stores energy in the magnetic field that is created due to the flow of DC in a superconducting coil. The coil has been cooled cryogenically to below its superconducting critical temperature. SMES consists of three parts: bidirectional AC/DC inverter system, superconducting coil, and cryogenically cooled refrigerator. DC charges the superconducting coil, and when the coil is charged, it stores magnetic energy until it is released. This energy is released by discharging the coil. Bidirectional inverter is used to convert AC to DC power and vice versa during the coil

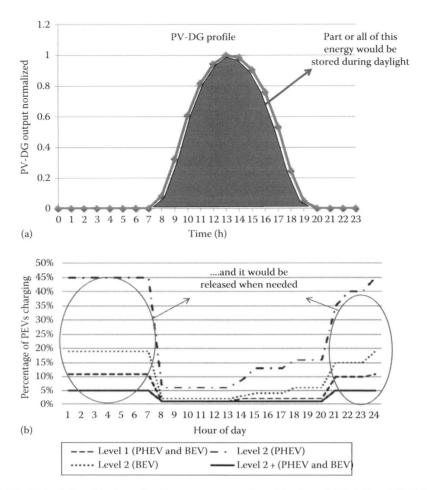

FIGURE 3.10 Potential application of grid energy storage for mitigation of PV-DG and PEV impacts. (a) PV-DG energy stored and (b) PHEV and BEV energy released. (From Agüero, J.R., Steady state impacts and benefits of solar photovoltaic distributed generation (PV-DG) on power distribution systems, *CEATI 2010 Distribution Planning Workshop*, Toronto, Canada, June 2010.)

charging/discharging cycles. The cost of SMES is high today because of its superconducting wires and refrigeration energy use, and its main use is for reducing the loading during the peak times.

The main technical challenges associated with SMES are large size, mechanical support due to high forces, superconducting cable manufacturing, infrastructure required for installation, low levels of critical current when superconducting properties of materials break down, levels of critical magnetic field, and health effects due to exposure to large magnetic fields.

3.2.2.3.3 Flywheel Energy Storage

FES operates on the principle of rotational energy—flywheel is being accelerated to a very high speed, and maintaining the rotation at such speed stores energy. When energy is demanded from the system, flywheel rotational speed is reduced. In order to reduce friction during the rotation, vacuum chamber is used to place the rotor. Rotor is connected to electric motor or generator. FES is not affected by the change of temperature, and stored energy is easily calculated, but the main danger is the explosion of flywheel when tensile strength of a flywheel is exceeded.

Smart Grid Technologies

3.2.2.3.4 Compressed Air Energy Storage

Energy generated at one point in time (off-peak) can be stored and later on used during different periods of time (peak). CAES represents one viable option. There are three types of air storage: adiabatic, diabatic, and isothermic. Adiabatic storage retains the heat that is produced by compression and later returns the heat to the air when the air is expanded to generate power. Diabatic storage dissipates some portion of heat as waste. In order for air to be used after it is removed from storage, it must be heated again prior to expansion in the turbine to power the generating unit. Isothermal storage operates under the same temperature conditions by utilizing the heat exchanger. These exchangers account for some losses.

Most CAES systems currently in operation do not utilize the compressed air to directly generate electricity [8]. Rather, the compressed air is fed into simple-cycle CTs, reducing the compression work in the standard recuperated Brayton cycle. In this mode, the CAES system serves to precompress combustion air during off-peak periods, improving the output of the CT during on-peak periods.

3.2.2.3.5 Ultracapacitors

Ultracapacitors or supercapacitors are the sources of DC energy. In order to be able to be connected to the power grid, a bidirectional AC/DC inverter is needed. Because of fast charge/discharge rates, ultracapacitors are used only during the short interruptions and voltage sags.

Unlike batteries where energy is stored chemically, ultracapacitors store this energy electrostatically. Ultracapacitors consist of two electrodes called collector plates, which are suspended with an electrolyte. Dielectric separator is placed between the collector plates in order to prevent the charges from moving from one electrode to another. Applied potential difference between the two collector plates causes negative ions in electrolyte to be attracted to the positive collector plate and positive ions in electrolyte to be collected on negative collector plate.

Ultracapacitors have several advantages and disadvantages comparing to batteries. Some of disadvantages include lower amount of energy stored per unit of weight, more complex control and switching equipment, high self-discharge, additional voltage balancing, safety issues, while some of advantages include long life, low cost per cycle, good reversibility, high rate of charge/discharge, high efficiency, and high output power.

3.2.2.3.6 Pumped Hydro

Pumped hydro storage method stores energy in the form of water, which is pumped from a reservoir on a lower elevation to a reservoir on a higher elevation. This is done during the off-peak hour when the cost of production of electricity necessary to run the pumps is lower. During the high-demand period, this water is released through the turbines. Pumped hydro is the highest-capacity storage system currently available. It is used for load flattening, frequency control, and reserve generation. However, the cost of building pumped hydro storage is very high.

3.2.2.3.7 Thermal Energy Storage

Thermal energy storage consists of a series of technologies that store thermal energy in reservoirs (e.g., using molten salt of ice) when electricity production is cheap (e.g., during off-peak, when most of the electricity is produced by using efficient and relatively inexpensive "base" units) and release it for heating or cooling purposes when electricity production is expensive (e.g., during peak, when electricity is produced by using costly "peaking" units), which equates to electricity production savings and/or T&D capacity deferral due to load shaving.

Recent developments in thermal storage have investigated conversion of stored heat directly into electricity, using Brayton or Rankine cycles [9]. Work on these systems has been catalyzed by thermal storage systems utilized for concentrating solar power, where excess heat captured during the day is stored for power generation in the evening. Round-trip efficiency of electrical-thermal storage remains problematic, with typical verified efficiencies below 30%. As a result, much attention is currently focused on increasing the temperature of thermal storage to greater than 500°C, utilizing phase-change materials to reduce system size and augmenting thermal storage material to improve thermal conductivity within the storage tanks.

3.2.3 Electric Vehicles

3.2.3.1 Regulatory and Market Forces Driving Electric Vehicles and Smart Grid Impact

With the implementation of smart grid technologies and the associated improvements in the reliability, sustainability, security, and economics of the electric grid comes the opportunity to include vehicles as an active participant in the smart grid. Although electrification of segments of the transportation energy sector does not require any technological or systemic advancements of the electric grid over what is presently available, the large scale of the transportation energy sector will provide long-term challenges to the legacy systems of the electric grid along with considerable opportunities for improved power, energy, and economic management in a smart grid system.

Electric transit (including electric trains and catenary trolleybuses) has a long history of integration with the electric grid. Electric transit has traditionally always operated at large, centralized scales, "tethered" to the grid. These technologies require a more-or-less continuous provision of electricity during operation of the vehicle. The introduction of high-density energy storage has introduced a watershed change in electric transportation: distributed, small vehicles operating in an untethered mode. PEVs are defined as vehicles that can store and use electricity from the electric grid. There are many types of PEVs under development by automakers. Electric vehicles (EVs) are vehicles whose only source of motive energy is stored in batteries. By definition, all EVs are "PEVs," charging their storage system—typically batteries—from the grid (Figure 3.11).

Hybrid electric vehicles (HEVs) combine conventional engines and electric drive trains to provide motive power from either internal combustion fuel or energy stored in batteries. In conventional HEVs, all motive energy is supplied by the internal combustion engine (ICE), with the battery providing limited energy buffering during operation. These vehicles do not interact with the grid. In contrast, PHEVs are PEVs that have the capability to charge their batteries from the electric grid, thereby allowing either liquid fuel or the grid electricity to be the ultimate energy source for the vehicle [10]. Plug-in fuel cell vehicles (PFCVs) are fuel cell, EV hybrids where the electrical energy storage system for the vehicle can be charged from the electric grid. As with PHEVs, either the fuel cell reactants or grid electricity can be the ultimate source of motive energy [11,12].

Relative to a conventional internal combustion vehicle or conventional HEV baseline, there are numerous benefits that come with the electrification of transportation energy through PEVs [10]:

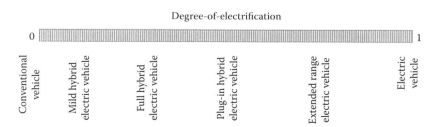

FIGURE 3.11 Degree of Vehicle Electrification. (Discussion of the Benefits and Impacts of Plug-In Hybrid and Battery Electric Vehicles, Electric Power Research Institute, MIT Energy Initiative paper, Draft 6, April 2010, http://web.mit.edu/mitei/docs/reports/duvall-hybrid-electric.pdf)

- Reduced petroleum consumption
- Lower life-cycle greenhouse gas and criteria pollutant emissions
- Lower fueling costs
- Lower life-cycle cost of ownership

Because of these potential benefits, there is a steady and growing interest in the development of PEVs.

Numerous traditional and entrepreneurial automakers have research, development, and limited production plug-in vehicle programs. Mitsubishi, General Motors, Nissan, and others have launched production plug-in vehicle programs. The rate of introduction of PEVs into the world vehicle fleet is expected to increase due to pressures from regulators such as Environmental Protection Agency (United States), California Air Resources Board, and others. The increasing commercial and private investment in PEVs will drive a corresponding investment in electrical infrastructure servicing PEVs. This investment in infrastructure will include public and in-home electric charger installations, which will incorporate passive or active forms of communication to facilitate the integration of large fleets of PEVs onto the electric grid.

The following sections will examine the potential impact of PEVs on the existing grid, describe methods of using smart grid technologies alleviate foreseen problems, and investigate potential opportunities enhance the performance of the electric grid using PEVs.

3.2.3.2 Technologies

3.2.3.2.1 Battery Electric Vehicles

A battery electric vehicle (BEV) is a type of EV that uses rechargeable battery packs to store electrical energy and an electric motor (DC or AC depending on the technology) for propulsion. Intrinsically it is a PEV since the battery packs are charged via the electric vehicle supply equipment (EVSE), that is, by "plugging-in" the BEV. The North American standard for electrical connectors for EVs is the SAE J1772, which is being maintained by the Society of Automotive Engineers (SAE) [13]. The standard defines two charging levels AC level 1 (120 V, 16 A, single-phase) and AC level 2 (208–240 V, up to 80 A, single-phase). Furthermore, additional work is being conducted on standardizing level 3 (300–600 V, up to 400 A, DC). A variety of technologies are being used for manufacturing the battery pack, including lead acid, lithium ion, nickel metal hydride, etc. The technical requirements of the batteries are different than those of conventional vehicles and include higher ampere-hour capacity, power-to-weight ratio, energy-to-weight ratio, and energy density. Since BEVs do not have combustion motor, their operation fully depends on charging from the electric grid. Therefore, uncontrolled charging cycles of BEVs for large market penetration levels may cause significant impacts on power distribution systems. Commercial examples of this type of vehicle are the Nissan Leaf, Mitsubishi MiEV, and the Tesla Roadster. The main criticism about BEVs is the reduced driving range (between 100 and 200 mi before recharging) when compared with conventional vehicles (>300 mi) [14].

3.2.3.2.2 Hybrid Electric Vehicles

An HEV is a type of EV that uses a combination of a conventional ICE and an electric motor for propulsion. HEVs use different technologies to improve efficiency and reduce emissions; such technologies include using regenerative breaking, using the ICE to generate electricity to recharge batteries or power the electric motor, and using the electric motor during most of the time and reserving the ICE for propulsion only when needed. Commercial examples of this type of vehicle include the Toyota Prius and the Honda Insight. HEVs are not PEVs, since they can operate autonomously without need of recharging batteries using the power grid. Therefore, no impact on the power grid is expected from proliferation of this type of EV. HEVs are, as of 2011, the highest selling EVs in the market. US HEV sales for the 2009–2010 period exceeded 250,000 units, as shown in Figure 3.12.

98 Smart Grids: Infrastructure, Technology, and Solutions

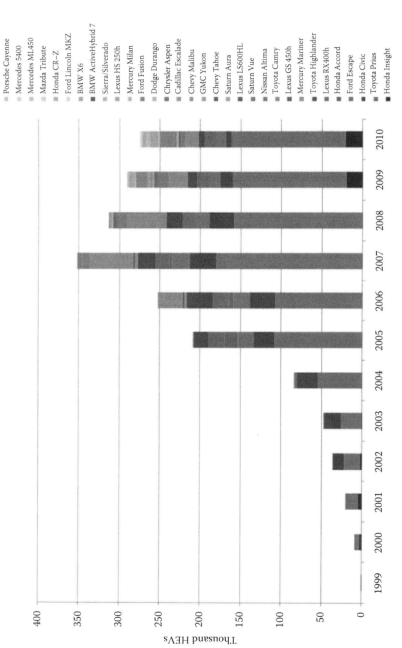

FIGURE 3.12 (See color insert.) U.S. HEV sales from 1999 to 2010. (From U.S. DOE alternative fuel vehicles (AFVs) and hybrid electric vehicles (HEVs), http://www.afdc.energy.gov/afdc/data/vehicles.html#afv_hev)

Smart Grid Technologies

3.2.3.2.3 Plug-in Hybrid Electric Vehicles

A PHEV is a type of EV that has an ICE and an electric motor (like an HEV) and a high-capacity battery pack that can be recharged by plugging-in the car to the electric power grid (like a BEV). There are two basic PHEV configurations [16]:

- Series PHEVs or extended range electric vehicles (EREVs): only the electric motor turns the wheels; the ICE is only used to generate electricity. Series PHEVs can run solely on electricity until the battery needs to be recharged. The ICE will then generate the electricity needed to power the electric motor. For shorter trips, these vehicles might use no gasoline.
- Parallel or blended PHEVs: Both the engine and electric motor are mechanically connected to the wheels, and both propel the vehicle under most driving conditions. Electric-only operation usually occurs only at low speeds.

The main advantage of PHEVs with respect to BEV is longer driving range. With respect to conventional vehicles, the advantage is reduced fossil fuel consumption and greenhouse gas emissions. However, the price of a PHEV is higher than that of conventional vehicles. Commercial examples of PHEVs are the Chevrolet Volt and the Fisher Karma.

3.2.3.3 Vehicle to Grid

3.2.3.3.1 Utilization of EVs for Grid Support

The first generation of PEVs is expected to be more costly than traditional vehicles. Early estimates place this premium at $10,000 or more. Due to this expected additional cost, research has been conducted to determine if PEVs can provide additional services to help offset the added expense of a PEV. Studies have shown that vehicles sit unused, on average, for more than 90% of the day [17]. Using this fact, researchers have conducted studies on the ability of PEVs to provide grid support services to provide a source of revenue for the vehicle owner. If this revenue helped offset the initial cost of the plug-in vehicle, it could increase the incentive for consumers to purchase PEVs. The primary means for monetizing the capabilities of PEVs are proposed participation in a deregulated ancillary service market. Studies, to date, have determined that frequency regulation is the component of the ancillary service market most compatible with plug-in vehicle capabilities and will provide the largest financial incentive to vehicle owners [18–20].

There are two primary types of power interactions possible between the vehicle and the electric grid. Grid-to-vehicle charging (G2V) consists of the electric grid providing energy to the plug-in vehicle through a charge port. G2V is the traditional method for charging the batteries of EVs and PHEV. A V2G capable vehicle has the additional ability to provide energy back to the electric grid. V2G provides the potential for the grid system operator to call on the vehicle as a distributed energy and power resource.

In order for PEVs to achieve wide-spread near-term penetration in the ancillary service market, the two primary stakeholders in the plug-in vehicle ancillary service transaction must be satisfied: the grid system operator and the vehicle owner. The grid system operator demands industry standard availability and reliability for regulation services. The vehicle owner demands a robust return on their investment in the additional hardware required to perform the service.

Since PEVs are not stationary but instead have stochastic driving patterns, these resources possess unique availability and reliability profiles in comparison to conventional technologies providing ancillary services. In addition to this, the power rating of an individual plug-in vehicle is significantly less than the power capacity of conventional generation systems that utilities normally contract for ancillary services. These key aspects of PEVs create unique challenges for their integration and acceptance into conventional power regulation markets to provide ancillary services.

The connection between the grid system operator and the PEV to provide grid support services can be classified as one of two types that have been proposed to date: a direct, deterministic

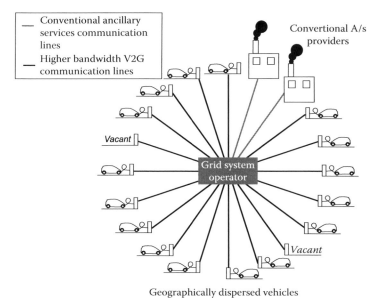

FIGURE 3.13 Example V2G network showing geographically dispersed communications connections under the direct, deterministic architecture. (From Quinn, C. et al., *J. Power Sources*, 195(5), 1500, 2010.)

architecture and an aggregative architecture. The direct, deterministic architecture, shown conceptually in Figure 3.13, assumes that there exists a direct line of communication between the grid system operator and the plug-in vehicle so that each vehicle can be treated as a deterministic resource to be commanded by the grid system operator. Under the direct, deterministic architecture, the vehicle is allowed to bid and perform services while it is at the charging station. When the vehicle leaves the charging station, the contracted payment for the previous full hours is made, and the contract is ended. The direct, deterministic architecture is conceptually simple, but it has recognized problems in terms of near-term feasibility and long-term scalability.

First, there exists no near-term information infrastructure to enable the required line of communication. The direct, deterministic architecture cannot use the conventional control signals that are currently used for ancillary service contracting and control because the small, geographically distributed nature of PEVs is incompatible with the existing contracting frameworks. For example, the peak power capabilities of individual vehicles (1.8 kW [11], 17 kW [22]) are below the 1 MW threshold that is required of many ancillary service hourly contracts [23].

In the longer-term, the grid system operator might be required to centrally monitor and control all of the PEVs subscribed in the power control region—a potentially overwhelming communications and control task [24]. As these millions of vehicles engage and disengage from the grid, the grid system operator would need to constantly update the contract status, connection status, available power, vehicle state of charge, and driver requirements quantify the power it can deterministically command. This information would need to be fed into the operator's market system to determine contract sizes and clearing prices.

The aggregative architecture is shown conceptually in Figure 3.14. In the aggregative architecture, an intermediary is inserted between the vehicles performing ancillary services and the grid system operator. This aggregator receives ancillary service requests from the grid system operator and issues power commands to contracted vehicles that are both available and willing to perform the required services. Under the aggregative architecture, the aggregator can bid to perform ancillary services at any time, while the individual vehicles can engage and disengage from the aggregator as they arrive at and leave from charging stations. This allows the aggregator to bid into the ancillary

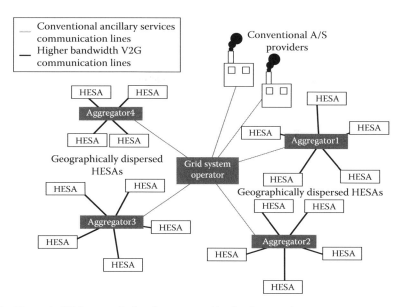

FIGURE 3.14 Example V2G network showing geographically dispersed communications connections under the aggregative architecture. (From Quinn, C. et al., *J. Power Sources*, 195(5), 1500, 2010.)

service market using existing contract mechanisms and compensate the vehicles under its control for time that they are available to perform ancillary services. As such, this aggregative architecture attempts to address the two primary problems with the direct, deterministic architecture.

First, the larger scale of the aggregated power resources commanded by the aggregator, and the improved reliability of aggregated resources connected in parallel allows the grid system operator to treat the aggregator like a conventional ancillary service provider. This allows the aggregator to utilize the same communications infrastructure for contracting and command signals that conventional ancillary service providers use, thus eliminating the concern of additional communications workload placed on the grid system operator.

In the longer term, the aggregation of PEVs will allow them to be integrated more readily into the existing ancillary service command and contracting framework, since the grid system operator needs only directly communicate with the aggregators. The communications network between the aggregator and the vehicles is of a more manageable scale than communications network required under the direct, deterministic architecture. The aggregative architecture is therefore more extensible than the direct, deterministic architecture as it allows for the number of vehicles under contracts to expand by increasing the number of aggregators, increasing the size of aggregators, or both. Since many distribution utilities are installing "advanced metering" systems, allowing two-way communication with individual consumers, these utilities could potentially enter the ancillary service market by providing such aggregation services using their metering communications networks.

From the perspective of the grid system operator, the aggregative architecture represents a more feasible and extensible architecture for implementing PEVs as ancillary service providers. For the system operator, the aggregative architecture is an improvement relative to the direct, deterministic architecture because it allows PEVs to make use of the current market-based, command and control architectures for ancillary services. Aggregators can control their reliability and contractible power to meet industry standards by controlling the size of their aggregated plug-in vehicle fleet, thereby providing the grid system operator with a buffer against the stochastic availability of individual vehicles. This allows the aggregator to maintain reliability equivalent to conventional ancillary service providers including conventional power plants. Because the payments from the grid system operator for ancillary services are equal for both architectures, the direct, deterministic architecture offers no apparent advantages from the perspective of the grid system operator.

From the perspective of the vehicle owner, the direct, deterministic architecture is preferred relative to the aggregative architecture. The initial allowable investment for the aggregative architecture is approximately 40% of the initial allowable investment for the direct, deterministic architecture [21]. The substantially higher initial investments allowed by the direct, deterministic architecture suggest that the average vehicle owner will prefer the direct, deterministic architecture.

These divergent preferences of the vehicle owners and the system operator highlight a fundamental problem that must be overcome before PEVs can be successfully implemented into the ancillary service market. The differing requirements of the stakeholders make only the aggregative architecture acceptable to both parties. The direct, deterministic architecture is unacceptably complex, unreliable, and unscalable to utilities and grid system operators. The aggregative architecture more than halves the revenue that can be accrued by the vehicle owners but still allows for a positive revenue stream. Only the aggregative architecture is mutually acceptable to all stakeholders and can provide a more feasible pathway for realization of a near-term utilization of PEVs for ancillary service provision.

3.2.3.3.2 Utilization of EVs for Energy Buffering

There exists a daily load cycle for the U.S. electric grid. In general, the grid is relatively unloaded during the night and reaches peak loading during the afternoon hours in most U.S. climates. Balancing authorities dispatch power plants to match the power generation to the time-varying load. Types of generation resource are dispatched differently to meet different portions of the load. Nuclear and large thermal plants are typically dedicated to relatively invariant "base-load" power. Dispatchable generation with fast response rates (e.g., CTs), hydropower, and energy storage can be dispatched to meet predicted and actual load fluctuations. By combining generation types, the control authority meets the time-varying load with a time-varying power generation, while meeting constraints imposed by environmental requirements, emission caps, transmission limitations, power markets, generator maintenance, unplanned outages, and more.

Even at relatively low market penetrations, plug-in vehicles will represent a large new load for the electric grid, requiring the generation of more electrical energy. In one set of scenarios analyzed by NREL researchers, a 50% plug-in market penetration corresponded to a 4.6% increase in grid load during peak hours of the day [25]. When vehicle charging and discharging can be controlled, other studies have found that as many as 84% of all U.S. cars, trucks, and SUVs (198 million vehicles) could be serviced using the present generation and transmission capacity of the U.S. electrical grid [26]. Controlling the electrical demand of PEVs will determine the infrastructure, environmental and economic impacts of these vehicles. Smart grid technologies can provide the control, incentives, and information to enable the successful transition to PEVs, but these technologies must reconcile the requirements of the electricity infrastructure with the expectations and economic requirements of the vehicle owner.

The simplest and most effective means for controlling the energy consumption of PEVs is direct utility control of charging times. Under this scenario, the utility would only allow consumers to charge during off-peak hours. By filling the nightly valley in electrical load, PEVs would reduce the hourly variability of the load profile. This has the effect of improving the capacity factor of base-load power plants, reducing total emissions, reducing costs, and eliminating the load growth due to plug-in vehicle market penetration. From a utility perspective, having direct control of the vehicle charging is ideal. From a consumer perspective, the willingness of vehicle owners to tolerate utility control of charging times depends on the type of plug-in vehicle that is being considered. For BEVs, the charger is the only source of energy for the vehicle, and being limited to charging during off-peak periods would significantly limit the usability of the vehicle and reduce its consumer acceptability. For PHEVs, the vehicle can operate with normal performance and reduced fuel economy when charging is not available. The degree to which consumers would tolerate increased fueling costs due to utility control of charging is under debate.

A more acceptable means for using smart grid technologies to control the energy consumption of plug in vehicles is by providing incentives for off-peak charging through a time-of-use (TOU) rate. A TOU rate is an electricity rate structure where the cost of electricity varies with time. Smart grid technologies such as advanced metering and consumer information feedback are necessary conditions for implementation of TOU tariffs. TOU rates are generally designed to represent the fact that electricity is more expensive during the day (when the grid is highly loaded) and less expensive during the night (when the grid is lightly loaded), so as to incentivize the conservation of electricity during the day. Special TOU rate structures have been designed for EV use so as to encourage EV owners to charge their vehicles at night, thereby conserving electricity during hours of peak demand. These legacy EV TOU rate structures have also been made available to PHEV owners. In theory, TOU rates should be able to be designed so as to provide an economic incentive for plug-in vehicle owners to charge their vehicles at night. In practice, the TOU rate can provide robust economic incentives for EV owners to charge their vehicle during off-peak periods because electricity is the only fuel cost for EVs. When TOU rates are applied to low all-electric range PHEVs, they can only provide partial compensation for the increase in vehicle fuel consumption that is caused by delaying charging until off-peak periods. For high all-electric range PHEVs, TOU rates are very effective at incenting off-peak charging of PHEVs. In summary, achieving the goals of controlling the energy consumption of many PEVs cannot be achieved solely by incenting off-peak charging through TOU rates [27].

These results do not necessarily suggest that an increase in peak load is inevitable with the introduction of PEVs. Instead of the smart grid being used to enable consumer controls, punitive pricing structures, and price volatility, smart grid must be used to engage the consumer in understanding how they can improve the sustainability and economy of the vehicle/grid systems. Consumer education and real-time information exchange between the utility and consumers will be a critical component of controlling the energy consumption rate and timing of plug-in vehicles.

3.2.4 MICROGRIDS

Microgrid has become a concept much talked about within the smart grid evolution. The microgrid market is in its infancy, and it is difficult to clearly estimate its size and potential. Pike Research—a smart energy practice specializing in new energy technologies—has tried to give an estimate of the potential for the microgrid market. They foresee the number of microgrid to increase from about 100 today to 2000 by 2015. Local generation could increase from 422 MW in 2010 to 3 GW in 2015. The following chapters describe what a microgrid may bring to end consumers and utilities, what the drivers and the challenges are, and what the requirements may be for microgrid automation.

3.2.4.1 Microgrid Definition

A microgrid is an integrated energy system consisting of interconnected loads and DESs that can operate connected to the grid or in an intentional island mode. The objective is to ensure better energy reliability, security, and efficiency. Some solutions have been typically available for improving energy reliability and efficiency in industrial plants, in commercial buildings, for military or university campuses. A new breed of microgrid is now becoming a reality for utilities wishing to integrate local generation or implementing grid relief solutions in areas that are poorly served by the transmission grid. Microgrids may be a quick alternative to the building or reinforcement of transmission lines. Main challenges for utilities are to guarantee grid reliability, stability, and security and also to optimize energy efficiency.

Scale and location of the microgrid are important factors. Microgrids should be constructed at the low-voltage (LV) or medium-voltage (MV) level. The key defining characteristics of a microgrid are as follows:

- Provides sufficient and continuous energy to a significant portion of the internal demand
- Has its own internal control and optimization strategy

- Can be islanded and reconnected with minimal service disruption
- Can be used as a flexible controlled entity to provide services/optimization for the grid or the energy market
- Applicable to various voltage levels (usually 1–20 kV)
- Has storage capacity

So as not to confuse with other grid components, a microgrid is defined as NOT

- One microturbine in a commercial building (this only DG)
- A group of individual generation sources that are not coordinated but run optimally for a narrowly defined load
- A load or group of loads that cannot be easily separated from the grid or controlled (facility/building management)

A microgrid's capacity to self-manage and define operational strategies concerning islanding mode and self-reorganization makes it the ultimate "smart grid" offering.

3.2.4.2 Microgrid Drivers

Several drivers are pushing towards strong and quick deployment of microgrids:

- *Environmental incentives*: Owners and operators of microgrid DG capacities should benefit in most countries from governmental incentives to help renewable implementation. This makes microgrids a new and efficient way to develop renewable energies and attain goals set by many countries in this area (e.g., the European Union 2020 plan).
- *Cost-effective access to electricity*: Microgrids could constitute a relatively cheap and efficient step toward rural electrification. For many emerging countries, low rural population density and high electrical infrastructure prices represent too big a hurdle to completely electrify a territory. Microgrids could be a more gradual solution to solve this issue.
- *Reliability*: In areas where grid saturation is a problem and causes black- or brown-outs regularly, microgrids could offer a solution to alleviate pressure without heavy investment in large-scale power plants and high-voltage power lines. This, in turn, would give the end consumer a more reliable electrical supply.
- *Security*: The islanding capacity is also one way to improve grid resilience in case of unforeseen difficulties, an important factor for several sensitive end consumers such as military bases, hospitals, or server farms.
- *Energy efficiency*: DG has the benefit of reducing losses from long distance electrical transmission. Losses currently vary from 7% (in most advanced countries) to 25% and more in several emerging countries.
- *Renewable energy implementation*: Microgrids could be a strong accelerator towards renewable energy implementation. Many countries have set ambitious goals in this matter—the 2020 plan for European countries, for example—microgrids constitute one way to achieve those goals. It is, in general, a way to accelerate the development of smart grids with an easy integration.
- *Progress in energy storage technologies*: Energy storage is a vital part of the microgrid system. The storage market should be multiplied by 7 by 2015 and should then represent a market of about $2.5 billion [28]. Also development of electrical vehicles can be seen as a plug-and-play storage capacity and for this reason, can have a large impact on microgrid development and research in the coming years.

Smart Grid Technologies

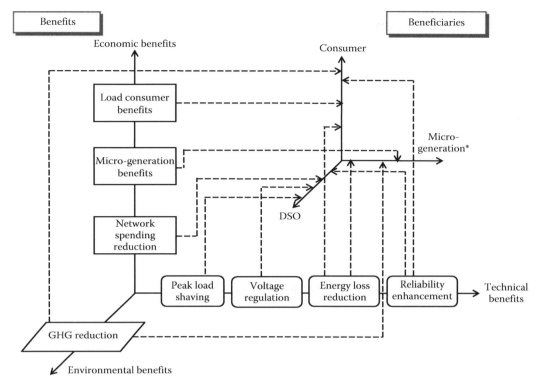

FIGURE 3.15 Microgrid benefits. *If owners of the micro-generation capability are neither DSO nor consumer. (EU More Microgrids project highlights, December 2009, www.microgrids.eu)

3.2.4.3 Microgrid Benefits

It is clear that microgrid may bear many benefits especially when renewable energy is generated and consumed locally. Microgrids should reduce electrical losses, increase grid stability and security, and as a whole, reduce spending for both consumers and distribution system operators (DSOs). Indeed, microgrids may benefit at the same time DSOs, end consumers, and the microgeneration operators (this group may of course be the DSO or an association of end consumers). The ensuing section details briefly these various benefits (Figure 3.15).

3.2.4.3.1 Economic Benefits

- *Load consumer benefits*: Microgrid automation systems can encompass relatively complex price setting mechanisms. It is possible to imagine systems in which dynamic pricing software calculates in realtime the cheapest source of energy: main grid electricity or local generation sources (e.g., rooftop PV panels or wind farm integrated to the microgrid).
- *Microgeneration benefits*: Many countries have introduced incentives to accelerate the implementation of renewable energies. Such schemes usually include a subsidized price for the owner of a renewable energy generation system (PV, wind, small hydro, biomass) to sell back to the electric company the electricity produced at higher than market price. This can also be considered as a microgrid benefit.
- *Network spending reduction*: As mentioned earlier, in areas where the existing infrastructure is under high demand or where there is no existing electrical infrastructure (e.g., rural areas in developing countries), microgrid implementation could represent a much cheaper alternative to transmission infrastructure costs. Network spending is in that respect reduced or at the very least postponed.

3.2.4.3.2 Environmental Benefits

Greenhouse gas reduction: Microgrids may rely heavily on local renewable energy sources. Furthermore, DG drastically reduces electrical losses incurred on HV transport lines (losses that can be translated in tons of CO_2 emission reduction).

3.2.4.3.3 Technical Benefits
- *Peak load shaving*: Dynamic pricing coupled with the availability of local generation may be a powerful tool to shave or shift loads. It has been shown that a dynamic shaving can lessen peak load demand by 10% and general consumption by up to 15%.
- *Reliability enhancement*: Thanks to their potential high-quality local automation capacities, microgrids should improve general grid stability and electricity reliability.
- *Voltage regulation*: The possibility to get energy from the grid or locally may help to improve voltage quality of electricity provided that the right automation solutions are in place.
- *Energy loss reduction*: As explained earlier, local generation reduces the need to transmit electricity on long distance, thus reducing the energy losses.

Obviously, identification of microgrid benefits is a multiobjective and multiparty coordination task, which will strongly depend on business structure and models. However, it seems clear that microgrids have a lot to offer to all players of the electrical grid.

3.2.4.4 Challenges to the Development of Microgrids

There exists a certain number of factors that can slow down the development and deployment of microgrids. These limits can be divided into two groups: technical challenges and legal challenges.

The first technical challenge concerns the balance management between load and generation. To improve grid efficiency, it is important that the DG be scheduled, that the demand be controlled, and that clear systems and securities be implemented to know when to consume the local generation or when to store it. All these imply also that the grid be able to forecast energy production and make "smart" decisions based on the forecasts. The microgrid must also be able to evaluate very precisely its reserves so as to take good decisions at multiple horizons: "a week, a day, 15 min, a few seconds."

Protection and safety represent other technical issues. Pilot programs are designed at the moment to see how a grid reacts when it is switched from a normal mode of operation to an islanded mode (and vice versa). Research has not yet tried to evaluate precisely the consequences of unplanned outages and the reorganization needed to be done in the microgrid. A second point concerns "black start," the process of restoring a power station to operation without relying on external energy sources. The final challenge concerns two-way electricity flows. One idea concerning microgrids is that if one was to regroup multiple microgrids, they could feed each other with energy. For this system to function, there needs to be two-way electricity flows at the transformer. In most current cases, this is not a possibility due to security rules and equipment capability.

The third technical challenge concerns everything to do with the microgrid and main grid interconnection. There needs to be real-time monitoring of power flow, precise measurement tools to calculate and monitor key electrical characteristics (voltages, flows, angles), Volt VAr analysis, a controlled frequency and harmonics in islanded mode, and of course the reconnection procedures after an islanded period. For the moment, pilot programs do not emphasize any specific challenges concerning these various points, but as complexity and size increase with real-case larger microgrids, difficulties may arise, and so it is important to keep all these issues in mind.

Smart Grid Technologies

The final technical challenge concerns the information and communication parameters linked to the electrical aspects of the microgrid. It is important to design a data management system capable of handling all the data generated by a complete microgrid but at the same time has no redundant operations and is economical in proportion to the size of the microgrid.

One last idea concerning microgrids would be to create a dynamic peer-to-peer market to transfer electricity (and do the money settlement) inside the microgrid between the various generators/consumers of electricity. This last point bears various technical issues, and many more tests have to be conducted before an operational peer-to-peer microgrid is created.

Legal and regulatory challenges are numerous: who operates the microgrid and what are the operator's responsibilities? Can the microgrid participants sell their electricity to other participants, to other microgrids, to the transmission system operator, etc.? These challenges also need to be addressed by pilot projects.

3.2.4.5 Microgrid Pilot Projects

Currently, operational microgrids are all pilot programs, generally small scale and with no proven return on investment. However, these various projects should be seen as the first steps to prove the business model and the technical feasibility on a large scale.

Currently, three areas in the world have started implementing several microgrid programs: Europe, Japan, and North America.

Europe subsidized a major research program called More Microgrids (budget of €8.5 million) during the 2006–2010 period. Partly funded by the European Union, partly funded by the private sector, the project implemented eight microgrids in various locations in Europe, both North and South. Most microgrids are small-scale low-voltage pilot projects to research on technical issues and feasibility. Two of these programs are operated in laboratories. However, one project stands out by it size: Bornholm Island in Denmark. Bornholm is a Danish Island with 28,000 inhabitants. Electricity is generated locally through local sources in low and medium voltage (oil, coal, and wind). It is also linked to an underwater high-voltage cable from Sweden. After an accident, this underwater cable was cut, and the grid became islanded, becoming in effect a microgrid for several months. This is, up to now, the only "microgrid" of this size and voltage levels in the world. The next step in European research is to develop projects of larger scale to identify and solve size-related issues. In 2010, the European Commission launched a call for demonstration program within the Framework Program 7 (Energy 2010—7.1.1—large-scale demonstration of smart distribution networks with DG and active customer participation). Two ambitious projects called Ecogrid and GRID4EU are starting at the end of 2011. One example of subproject of GRID4EU is the NiceGrid project: specification and deployment of a medium-voltage microgrid in a new area of Nice (France) with strong concentration of PV generation.

In *Japan*, the main research institute in charge of microgrid research is NEDO (the New Energy and Industrial Technology Development Organization). Japan is the world leader in pure numbers of microgrids, but again, most of them are very small, and they focus mostly on the integration of renewable energies. Private research is led by Mitsubishi Electric Corporation (MELCO) which has already developed several specific microgrid products (inverters and management system).

North America probably has the most advanced research when it comes to microgrids. Canada, through the CANMET Energy and Technology Research Center, has several pilot microgrid programs, specially focusing on DER integration standards and codes and also net-metering.

In the United States, the Consortium for Electric Reliability Technology Solutions (CERTS) and Power Systems Engineering Research Center (PSERC) are two main research institutes. With the bailout grants given out by the Obama administration, total budget for smart grid research increased to approximately $8.1 billion in 2009. About 10%–15% ($800 million) should go toward various microgrid research projects. Research in the United States is mostly developed for institutional projects (military or university campuses): out of the 455 MW of current microgrid capacity given by Pike Research, 320 MW is from campus microgeneration.

The *rest of the world* is not as advanced in microgrid research. However, other countries in Asia and in the Middle East have started investing in this type of technology. Several projects, some of which highly ambitious such as the Masdar Smart City in Abu Dhabi, have been launched (at least at the specification stage). Microgrids have also got a lot of interest over the past years in China.

As a whole, microgrid programs are at a very early stage, and each project is maximum of a few MW. Europe and Asia are mostly developing community or industrial/commercial microgrids while America is focusing more on institutional and campus projects.

Case Study 1: San Diego Gas and Electric Microgrid Project

San Diego Gas and Electric microgrid project, with support from the Department of Energy (DOE) and the California Energy Commission (CEC), has begun and one of the largest-scale microgrid demonstration projects in the United States. The project, which is to be implemented in the desert city of Borrego, will combine distribution-side technologies with consumer-side technologies, improve system reliability, and reduce peak load by more than 15%. These technologies include DER resources, advanced energy storage, residential solar panels, and demand response resources. The project has received about $10 million in federal and state funding and projects a total project cost of about $15 million over 3 years. The project will incorporate an advanced microgrid controller that will integrate and optimize the utility-side DER with consumer-side resources such as demand response. The microgrid controller will also coordinate with other utility systems such as the distribution management system (DMS), outage management system (OMS), and advanced metering infrastructure (AMI). This kind of integrated microgrid controller program could potentially be expanded to the rest of the utility's distribution system to improve DER management across the utility's service area.

3.2.4.6 Types of Microgrid

The largest number of identified type of microgrid are, in this order, institutional microgrids (hospitals, university, or military campuses), followed by commercial/industrial (factories, server farms, commercial malls, business towers), and finally community grids (multiple houses or apartment buildings, some commercial buildings). The latest is very small today, but a huge increase is expected when regulatory and business barriers are lifted.

One possible customer segmentation is the following:

Blue ocean: This segment consists of areas not yet connected to the country's main grid. In this case, there is no preexisting infrastructure. One of the main drivers in this case is the possibility to supply good-quality electricity without spending in building transmission lines.

Network relief: Areas where the main grid is saturated and hence has problems on voltage stability and peak demand are the key markets for this segment. Microgrids will increase stability and defer expensive investments in large-scale infrastructure.

Energy security: This particular segment deals with all institutions where it is strategically important to get stable and good-quality electricity without any interruption. Hospitals, military campuses, refineries, and the like can potentially be islanded for long periods of time in case of main grid outage. Within this category, it is possible to define subcategories depending on the criticality: typically hospitals need a higher level of service (a few seconds of interruption can cost a life) than industries.

Energy efficiency: This particular segment's main motivations are environmental concerns and profits made by the sale of renewable energies. In this particular case, a microgrid is only one solution since the islanding characteristic that defines a microgrid is optional. This segment may include university campuses, office buildings, small communities, etc.

From a *technical point of view*, different categories of microgrids can also be defined along different level of voltages. Depending on the microgrid architecture, level of voltages impacted, and components included (substations, feeders, etc.), automation processes and telecommunication tools may differ greatly. Figure 3.16 shows four possible categories.

Smart Grid Technologies

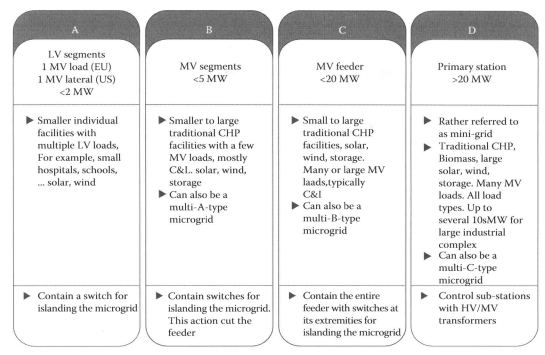

FIGURE 3.16 Microgrid categories. (US DOE/CEC Microgrids Research Assessment, Navigant Consulting Inc., May 2006.)

3.2.4.7 Building Blocks of a Microgrid

A typical microgrid will contain physical systems, control systems, and interfaces with the other systems at the utility.

3.2.4.7.1 Physical Systems

A microgrid is composed largely of off-the-shelf physical components. None of the physical components discussed here are specific to a microgrid application. In fact, most of these same physical systems are used in other ways by the utilities. It is the combination of the physical systems under an advanced control scheme that creates a microgrid application. Nonetheless, it is worth understanding the basic building blocks of the system. These include the following:

- *Sensors*: Sensors, and more generally information input, are required to determine whether criteria for islanding or reconnecting have been met. Sensors are the eyes and ears of the microgrid.
- *Switches*: Intelligent switches are a vital part of the microgrid because they allow quick reconfiguration of the components in the microgrid. Switches allow the microgrid to electrically disconnect or reconnect with the grid, section-off areas of the circuit, or bring various DER components on or off-line.
- *Power electronics*: Power electronics allow for DC-to-AC or AC-to-DC conversions, as well as voltage changes for DER components. This allows DER such as energy storage or microturbine generators to be grid-connected despite nongrid-conforming generating modes.
- *Energy storage*: Energy storage helps smooth rapid changes due to external events or characteristics of DER in the microgrid. For example, in the event of a blackout, energy storage

can help to support the system for a few minutes while generators start up. In order to perform these functions, the storage must be sufficiency large and able to respond quickly.
- *Generators*: Generators can take a variety of forms but are most commonly diesel- or natural gas–based combustion engines. These generators are necessary even when renewable alternatives are available because they provide consistent energy that can be relied on with relatively high certainty.
- *Protection equipment*: Protection equipment is always necessary, regardless of whether or not the DER is configured in a microgrid. Nonetheless, there are special precautions that must be taken in a microgrid because of the added level of complexity involved with becoming a separate electrical entity. Specifically, the major issues that must be accounted for are protection when disconnecting and reconnecting with the grid, and ensuring that there is an appropriate level of fault detection and protection in each part of the islanded microgrid system. Control systems for protection equipment will also have to be modified to fit the operating paradigm of the microgrid.
- *Metering*: Advanced metering must be in place at the substation and preferably in the residential neighborhood so that conditions and power flow can be monitored in realtime.

3.2.4.7.2 Control Systems

Taking the example of the microgrid category D (multimedium voltage feeder microgrid), three functional levels of automation can be identified:

- Local controllers necessary to control individual components of the microgrid: load controllers, energy storage controllers, and microgeneration source controllers. These local controllers response to orders sent by the microgrid controller and react to real-time conditions (detection of a fault, etc.) in order to guarantee the reliability of the different components. Protection relays and smart switches are also part of this level of automation.
- Local microgrid controllers provide real-time monitoring and control functions for all the components within their control boundaries. Their main objective is to ensure power reliability and quality. It may bear some basic local optimization functions based on economics of local generation and storage.
- The master microgrid management system allows optimizing the overall microgrid/collection of microgrids under its control. It defines set points for all the loads and generation in order to optimize the efficiency of the microgrid depending on electricity prices and local generation cost. It also provides forecast and real-time estimation of loads and generation to the DMS. Finally, the master microgrid management system is responsible for the disconnection/reconnection to the main grid. It may also interact directly with the market or with curtailment service providers (Figure 3.17).

3.2.4.7.3 Interfaces with Other Systems

The microgrid controller relies on other systems to deliver information to it, as well as execute some of its requests. In some cases, the utility can use its systems to deliver commands to the microgrid controller. Though it can operate independently, the microgrid controller is of most use to the grid when it understands the external pressures on the system and reacts accordingly. Interfaces with the DMS/OMS and the AMI system are particularly important.

3.2.4.7.4 Microgrid General Functions

The architecture earlier is functional. It means that it may have to be adapted depending on the size of the microgrid and the regulatory constraints. In any case, it is mandatory that all functions are fulfilled somewhere in the microgrid:

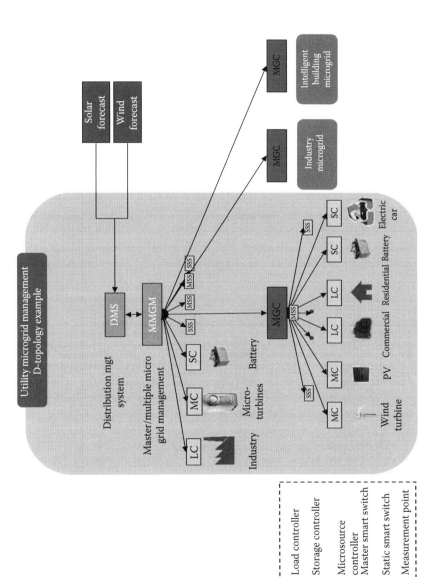

FIGURE 3.17 Possible microgrid automation architecture.

- Manage exchange with the main grid (disconnection, reconnection, market functions, emergency control if required).
- Operate reliably for microgrid consumers in islanded mode and in connected mode.
- Optimize asset utilization (DG, load, transformers) within the microgrid while in islanded mode or in connected mode.

3.2.5 Energy Resources Integration Challenges, Solutions, and Benefits

Large-scale distributed renewable generation is planned to form part of the renewable energy portfolio. The additional increased use of DERs, including storage and PHEVs, results in bidirectional power flows, protection issues on utility distribution systems that were not designed to accommodate active generation and storage at the distribution level. The technologies and operational concepts to properly integrate (DRs) into the existing distribution feeders need to be addressed with smart grid solutions to avoid negative impacts on system reliability and safety.

Electric distribution system is sensitive to DR based on several factors. The most important ones are size, location type, and also a characteristic of the distribution system. Even though DR can be connected anywhere on the system (substation, primary feeder, low-voltage or secondary, customer), its location has the biggest impact on the distribution system. The farther the DR's PCC is, the weaker the system it must be connected to. There are several impacts on distribution power system based on the DR connection. Some of those include short-circuit current levels, location of DR point of interconnection (POI) (substation, at primary lines, at secondary lines), system losses, reactive power flow, impact on lateral fusing, reverse power flow, islanding, and voltage and frequency control.

Every DR system consists of the following major parts:

- *Prime movers*—this represents the primary source of power. There are several prime movers available today, such as reciprocating engines, CTs, microturbines, wind turbines, PV systems, fuel cells, and storage technologies.
- *Power converter*—this represents the way that power is converted from one entity to another. Synchronous generators, induction generators, double-fed asynchronous generators, inverters, and static power converters are examples of power converters.
- *Transformer, switches, relays, and communications devices*—these devices enable the protection of the DR from the distribution system and vice versa.

There are several types of DR interconnection systems. They can be divided into several groups:

- *Inverter-based systems*—these systems are used in batteries, fuel cells, PV, microturbine, and wind turbine applications. Some systems such as batteries, fuel cells, and PV generate DC power, and inverter is bidirectional DC/AC converter. Microturbines generate AC power with high frequency that is later on converted to DC and then to AC with 60 Hz frequency.
- Systems that run parallel to the distribution system and interconnection system require synchronization with the common bus. These systems are used in load peak shaving, emergency power supply, prime power, and cogeneration.

3.2.5.1 Integration Standards

Most small and large-scale distributed renewable generation resources are currently governed by the IEEE 1547 [28] set of standards that include references to UL1741 for interconnecting on low-voltage networks. These standards were developed toward the end of the 1990s when DG, especially distributed PV and wind generation, was at very low penetration levels. IEEE 1547 describes the interconnection issues of DG resources in terms of voltage limits, anti-islanding, power factor, and reactive power production mainly from a safety and utility operability point of view.

There are, however, concerns on some of the practical impacts of the IEEE 1547 standard on distribution feeder design, operation, and safety. These include reactive power injection, voltage regulation, voltage ride-through, and power quality of high levels of inverters penetrating the distribution network without any coordinated control. Currently there are several IEEE standard working groups working on different application notes and setting the requirements for a future update on IEEE 1547.

Larger wind generation facilities above 10 MW are now required to have low-voltage ride-through (LVRT) capability to increase system reliability. New generation interconnection requirements have been adopted by FERC as part of the FERC Order 661, docket RM05-4-0000 NOPR, mainly for large wind and solar power facilities, larger than 20 MW. These provisions are updated and adopted as Appendix G to the LGIA [30]. FERC requires now also renewable energy plants to be able to provide sufficient dynamic voltage support and reactive power if the utility's system impact study shows that it is needed to maintain system reliability. This implies that wind generators should have dynamic reactive capability for the entire power factor range and that dynamic reactive power capability must be required in every instance.

Currently there is also an industry-wide initiative on the smart grid inoperability panel [31]. This initiative is coordinated by NIST and EPRI. The main purpose of this panel is to develop interconnection and communications requirements for DRs, including PV, energy storage, and demand response.

Some of the standard communication and protocol profiles and standards for PV generation and storage systems are using DNP3 and IEC-61850. The purpose of defining a standard communication profile is to make it easier to interconnect DRs with increased security levels.

3.2.5.2 Renewable Generation Integration Impacts

Studies and actual operating experience indicate that it is easier to integrate PV solar and wind energy into a power system where other generators are available to provide balancing power, regulation, and precise load-following capabilities. The greater the number of intermittent renewable generation is operating in a given area, the less their aggregate production is variable.

A summary of the main renewable generation integration is provided followed by a more in-depth description.

Typical T&D system–related problems include [32] the following:

- PV and wind capacity factors in the range of 15%–30%.
- No dispatch capability of PV solar and wind farms without storage.
- Ultrafast ramping requirements (400–1000 MW/min).
- Most existing PV inverters do not provide reactive power and voltage support capability.
- Existing PV inverters do not have LVRT capability.
- Most PV plants are noncompliant with FERC—large generator interconnection procedure (LGIP).
- IEEE-1574 provides contradicting guidance with LVRT and nonislanding requirements.
- Reactive power management and coordination within feeders are not designed with high PV and wind production in mind.
- Power quality, especially voltage fluctuations, temporary over voltage (TOV), flicker, and harmonics may be out of IEEE-519 and other standards.
- Lack of coordination control of existing reactive power support.

High penetration of intermittent resources (greater than 20% of generation meeting load) affects the network in the following ways:

- *Power flow and reactive power*: Interconnected transmission and distribution lines must not be overloaded. Reactive power should be generated throughout the network, not only at the interconnection point and should be compensated locally through the feeders. Due to PV and wind power variations and required ramp rates larger than 1 MW/s, fast-acting reactive power sources should be employed throughout the feeders and network.

- *Short circuit*: Impact of additional generation sources to the short-circuit current ratings of existing electrical equipment on the network should be determined. PV inverters normally do not contribute short-circuit duty to the feeder networks.
- *Transient stability*: Dynamic behavior of the system during contingencies, sudden load changes, and disturbance clouds can affect stability and power quality. Voltage and angular stability during these system disturbances and production variance are very important. In most cases, fast-acting reactive-power compensation equipment, including SVCs and distributed STATCOMs, is required for improving the transient stability and power quality of the network. PV array clouding in larger PV plants may require energy storage facilities to provide smoothing for the PV plant output.
- *Electromagnetic transients*: Ensure these fast operational switching transients have a detailed representation of the connected equipment, capacitor banks, their controls and protections, the converters, and DC links. Due to PV power fluctuations, these network equipments may switch much more than originally intended.
- *Protection and islanding*: Investigate how unintentional islanding and reverse power flow may have a large impact on existing protection schemes, philosophy, and settings. Large levels of PV production will reverse power flows during certain times of the day, and protection circuits need to be able to protect the distribution feeders under these conditions. Problems were reported with PV inverter nonislanding circuitry in regions with high PV power production.
- *Power leveling and energy balancing*: Due to the fluctuating and uncontrollable nature of wind power as well as the uncorrelated generation from PV power and load, PV power generation has to be balanced with other very fast controllable generation sources. These include gas, hydro, or renewable power-generating sources, as well as fast-acting energy storage, to smooth out fluctuating power from wind generators and increase the overall reliability and efficiency of the system. The costs associated with capital, operations, maintenance, and generator stop-start cycles have to be taken into account.
- *Power quality*: Fluctuations in the PV and wind power production and the strength of the T&D network at interconnection points have direct consequences to the power quality. As a result, large voltage fluctuations may result in voltage variations outside the regulation limits, as well as violations on flicker and other power quality standards.
- *Other DER facilities*: Several other DER technologies are currently being integrated on the distribution feeders as part of smart grid initiatives including PEVs, CHP generation, and distributed energy storage. The coordination of these DER devices is crucial to determine the combined impacts on the distribution feeders and networks.

3.2.5.2.1 Intermittency

In most urban regions, PV flat-plate collectors are predominately used for solar generation and can produce power production fluctuations with a sudden (seconds time-scale) loss of complete power output. PV generation penetration within residential and commercial feeders approaches 4–8 MW per feeder. With partial PV array clouding, large power fluctuations result at the output of the PV solar farm with large power quality impacts on distribution networks [32].

During cloudy and foggy days, large power fluctuations are measured on the feeders with high penetration levels and can produce several voltage quality, protection, uncoordinated reactive power demand, and power balancing problems. Cloud cover and morning fog require fast ramping and fast power balancing. Furthermore, several other solar production facilities are normally planned in close proximity on the same electrical distribution feeder that can result in high levels of voltage fluctuations and even flicker on the feeder. Reactive power and voltage profile management on these feeders are common problems in areas where high penetration levels are experienced.

Feeder automation and smart grid communications are therefore crucial to solve these intermittency problems.

3.2.5.2.2 Short-Circuit Levels

Short-circuit current levels vary greatly with respect to impedance of the feeder and length of the conductor. Addition of DR affects the values of short-circuit currents, thus inadvertently affecting the relay settings. One measure that is of interest is the ratio of rated output current of DR with respect to the available current at POI. For DRs on feeder primary voltage levels, if this ratio is ≥1%, then DR will have noticeable impact on voltage regulation, power quality, and voltage flicker. If the DR is on secondary or low-voltage levels, ratio of <1% can have major impact on secondary voltage.

3.2.5.2.3 DR with POI at Substation

Substation represents the strongest point of the distribution system. Voltage levels can range anywhere from 12 to 34.5 kV, and transformers typically have capacities in 12–20 MVA. Placing DR in the substation represents less of a challenge for the distribution system since DR acts as another power source. The only additional requirement is the modification of protection and control schemes that will account for the addition of DR. However, if capacity of DR is 15%–20% of the substation load, then additional issues arise such as voltage regulation, equipment ratings, fault levels, and protective relaying. If capacity of the DR is close to the substation load, then issues will arise with voltage regulation on LTC. Current transformer at LTC will not measure the difference between the current in the substation and current supplied by DR. Since this difference is low, LTC will interpret as if the total load is light and will not boost the voltage appropriately, thus causing the low voltage at the end of the line. If capacity of DR is larger than substation load, then it will export power into transmission system, thus creating additional protection and control issues.

3.2.5.2.4 DR with POI at Primary and Secondary Lines

Distribution system has higher impedance on primary feeder lines, so DR placed anywhere on these lines will have more influence on the system than comparable DR placed in the substation. To begin with, most of the distribution systems were designed for one-way power flow: from the transformer to the end customer. DR placed on the feeder can cause reverse power flows, and it requires additional protection and/or control equipment. Generally, security and safety of all protective devices may be compromised if DR causes fault levels to change by more than 5%. The main factor that affects the effect of DR on the system is the strength of the distribution system. The closer the distance to the other strong power source, the stronger the system is.

There are a couple of ratios called stiffness ratios of stiffness factors that predict the impact of DR on the system:

$$\text{Primary\_stiffness\_ratio} = \frac{\text{available\_distribution\_system\_fault\_current\_at\_POI}}{\text{DR\_steady\_state\_full\_load\_output\_current}}$$

$$\text{Secondary\_stiffness\_ratio} = \frac{\text{fault\_current\_of\_power\_system}}{\text{fault\_current\_of\_DR}}$$

As a general rule, stiffness ratios of more than 100 are less likely to create voltage problems. One additional stiffness ratio is defined in IEEE P1547-D8 standard as

$$\text{Stiffness\_ratio} = \frac{\text{system\_fault\_current\_including\_DR}}{\text{DR\_fault\_current}}$$

This stiffness ratio is used when trying to evaluate the impact of DR on system fault levels.

3.2.5.2.5 System Losses and Reactive Power Flow

Reducing system losses represents one of the main challenges of power utility's T&D system today. Utilizing DRs reduces system losses if DR is properly sized and placed. In order to obtain the maximum loss reduction in a radial distribution circuit with single DR, the DR has to be placed at a position where the output current of DR is equal to half of the load demand. The reason for this is that the distance that power has to travel from sources to loads is minimum, which in turn, minimizes losses. However, if DR is too large, then it can actually cause losses to increase.

3.2.5.2.6 Equipment Loading, Maintenance, and Life Cycle

In the same way that low to moderate penetration levels of DG (either conventional or intermittent) reduce equipment loading, moderate to high penetration levels or a condition that leads to reverse power flow may increase equipment loading up to a point where this can become a concern from an equipment rating perspective and lead to equipment overload. Similarly, the interaction among intermittent DG (PV and wind) and voltage control and regulation equipment such as load tap changers (LTC), line voltage regulators, and voltage-controlled capacitor banks may lead to frequent operation of these equipment (frequent tap changes and status changes). This, in turn, increases maintenance requirements, and, ultimately, if it is not properly addressed, it may impact equipment life cycle. The smart grid plays a key role in this regard, since the ability of continuously monitoring equipment and additional controllability that can be achieved, for instance, via phasor measurement units (PMUs) and DS, allows the system operator to avoid this type of conditions. Furthermore, DS and dynamic Volt VAr control and compensation using smart technologies such as inverters and flexible AC distribution systems (FACDS) allow mitigating impacts due to intermittent DG and significant impacts on additional voltage control and regulation equipment.

3.2.5.2.7 Impacts on Protection Systems

A key impact of the integration of DR in power distribution grids is that on protection systems, which have been traditionally designed to be operated in a radial fashion. The integration of DR may lead to reverse power flows through feeder sections and substations; therefore, this is a situation that the distribution grid, in general, has not been designed, built, and is not prepared for.

It has been a long standing practice of utilities to protect laterals with fuses. Utilities generally use two philosophies for protection coordination, fuse clearing and fuse saving, and in some case, a combination of both, fuse clearing where fault currents are high and fuse saving where fault currents are moderate to low. For the case of fuse saving, relays upstream trip before the fuse blows, and once the fault is cleared, the breaker recloses. This action has to be fast because the breaker has to trip before fuse starts to melt and gets damaged. Depending on the severity of the fault, these schemes sometimes cannot operate correctly. DR causes fuse saving schemes to be even more complex, because of the increased fault currents. In addition, DR increases the fault current level through the fuse, but not necessarily through the breaker. Furthermore, the addition of DR causes issues with fuse-to-fuse coordination. Choosing correct fuse sizes, relay settings, and DR tripping settings can alleviate this problem.

Additional impacts on protection systems are modification of the "reach" of protective devices such as circuit reclosers and relays due to the feeder load offset effect of DR, particularly for the case of large DG, and potential overvoltage issues during unintentional islanding conditions, which are a function of the DR's interconnection transformer configuration. This situation can be particularly severe when the configuration of the medium-voltage side of the interconnection transformer is delta.

A critical component of protective devices on distribution networks is overcurrent relays. These relays have instantaneous and time-delayed settings, which cause the distribution breakers to trip if fault current levels have been exceeded. In addition, on 34.5 kV long distribution lines, sometimes distance relays that are overcurrent relay supervised are used because it might be hard to distinguish

Smart Grid Technologies

between the high load currents and low fault currents. The commonality between all these relays is that they are designed and built for one-way flow. However, reverse power flow can cause protection devices to misoperate.

Smart grid technologies can play an important role in mitigating these impacts, for instance, by using adaptive protection systems, which allow the settings of protective devices to adapt to the varying system conditions, either feeder loading and configuration or DG output. Most important is to recognize the need for distribution protection system to evolve; this is expected to become more important as the penetration level of DG and PEVs and other smart grid technologies increases. As the complexity of operating the smart distribution system increases, the need for replacing conventional protective devices, specifically fuses, will grow as well. It is likely that the distribution grid of the future will be similar to modern transmission systems, from a protection system standpoint.

3.2.5.2.8 Intentional and Unintentional Islanding

Islanding happens when part of the utility system has been isolated by operation of one or more protective devices, and DR that is installed in that isolated part of the system continues to supply power to the customers in that area. This is very dangerous situation because of several reasons:

- DR might not be able to maintain proper system parameters such as voltage and frequency and can damage the customer equipment.
- The islanded area might be out of phase, so utility system might not be able to reconnect the islanded area.
- Security issues associated with utility workers working on downed lines that are back-fed from DR.
- Improper grounding can lead to high voltages during the islanding.

DRs that can self-excite are capable of islanding, while non-self-exciting DRs can island only if certain conditions have been met.

There are a couple of techniques that are used to prevent islanding: frequency regulation and voltage regulation. During the normal operation, frequency and voltage are fluctuating within certain ranges. For frequency, the settings are set to anywhere from 0.5 to 1.0 Hz from nominal frequency of 60 Hz. Allowed voltage variations are from 120 ± 6 V. Thus, having frequency, undervoltage and overvoltage protection can prevent islanding. The reason why relays trip very fast is because DR units can rarely match the power demand in the area, which in turn results in change in frequency and voltage that is detected by relays.

Additional issue is reconnection after the fault. When fault on the feeder that has DR occurs, breakers trip, and depending on reclosing sequence, they can reclose up to three times before getting locked out. The reclosing sequence normally has three reclosing shots, one of which is instantaneous (of course, there is a delay of several cycles because that represents the time that it takes breaker's mechanism to reclose). IEEE Std. 1547 recommends DR to trip before any breaker reclosing occurs. After the DR trips off-line, for safety reasons, it is not advisable to have control logic programmed such that DR reconnects to the system immediately after the normal power supply has been established. After voltage and frequency have been restored to its normal limits, DR shall be allowed to be reconnected.

There are, however, some situations when the load on the island is balanced with DR. In that case, several techniques such as voltage shift and frequency shift are used to detect islanding. This protection should operate within few seconds after islanding has occurred.

3.2.5.2.9 Voltage Regulation and Control

Currently, electric distribution systems are capable of handling one-way power flow—from the substation downstream to the customers. In such system, voltages are highest at the substation, and they are the lowest at the end of the line. However, this assumes that there are no distributed energy sources on distribution line. Depending on the size of DRs online, and its placement on the feeder, it is possible to have the voltage at the end of the line to be higher than the voltage at the substation.

Voltage regulation in distribution power networks is specified in ANSI C84.1 standards. In essence, nominal voltage of 120 V is desired, while expected deviations are ±5% or equivalently 114–126 V. It describes the process and equipment that is needed in order to keep the voltage within the limits provided herein. According to IEEE Std. 1547, "The DR shall not actively regulate voltage at PCC. The DR shall not cause the Area EPS service voltage at other local EPSs to go outside the requirements of ANSI C84.1 standards."

Having distribution energy sources significantly increases voltage regulation and relay protection. Voltage on distribution networks is controlled by voltage regulators and capacitor banks. Voltage drop depends on the wire size, type of conductor, length of the feeder, loads on the feeder, and power factor. DR can affect the voltage on distribution feeder in couple of ways:

- If DR is injected into power system, then it will reduce the amount of current needed from the substation for all the loads, thus automatically reducing the voltage drop.
- If DR supplies or absorbs reactive power to the system, it will affect the voltage drop on the whole feeder. If DR supplied reactive power, the voltage drop will be reduced, and if DR absorbs reactive power, it will increase voltage drop.

There are several operating problems associated with inclusion of DRs into distribution networks:

- *Low voltage*—many feeders utilize voltage regulator (VR) with LDC compensation. If DR is placed downstream from the VR, and it represents the significant part of the load downstream (i.e., DR can supply vast majority of the load downstream), then VR will lower its settings, causing the voltage to drop. If DR does not inject sufficient reactive power into EPS, low-voltage condition will persist.
- *High voltage*—if the distribution system is near upper limit of 126 V, injecting real and reactive power by DR can push this voltage over the limit. At the same time, DR can have a setting for high voltage out of range lower than its tripping point, so DR will stay online. Solving this problem will require either the DR to reduce its output or voltage increase that will trip DR.
- *Voltage unbalance*—small-scale DR devices are single phase most of the time. Thus, injecting the power will have effect only on one phase, and voltage difference can change between the phases, thus creating high unbalance. This unbalance can exist even if the voltages are within ANSI C84.1 range. To alleviate this problem, it might be smart to connect DR to highest loaded phase and transfer single-phase load from the highest loaded phase to two other phases.
- *Excessive operations*—DR devices (e.g., wind or solar) can be very unpredictable, and their output can be intermittent. The output of DR can change rapidly, and this can cause voltage regulating devices to operate excessively. Majority of these devices have daily maximum limit number of operations. The solution here is changing the time-delay settings on voltage regulating devices to provide better coordination with DR.
- *Improper regulation during reverse power flow conditions*—if there is a feeder with VR that has LDC, and there is a DR located downstream of VR, it is possible that DR can be quite large so that not only it can supply power to all load in the area, but it might be also capable of supplying the load upstream from it. VRs should detect then reverse power flow. However, VR now assumes that the source now is stronger than substation (which is not the case). VR will try to raise its tap to the limit in one direction or the other. Tap change produces now voltage change that is opposite from what control algorithm expects. In order to find the solution for this problem, so all VR with LDC have to be in a specific mode in reactive bidirectional mode to operate with DRs in reverse power flow.

- *Improper regulation during alternate feed configurations*—this problem is the one where a feeder gets a portion of the load from another feeder through the tie switch. Having DRs on the system complicates things in such a way that any of the five problems from earlier is possible.

3.2.5.2.10 Frequency Control

Small-scale DR itself cannot exert frequency control, which is generally reserved for large synchronous generators. This is not the case of large-scale (MW-size) DR, which depending on the size and regulatory framework may be allowed to provide ancillary services; this is the case, for instance, of large-scale DS and DG. Potentially the wide-area controllability that can be achieved via smart grid technologies can allow the implementation of the "virtual power plant" concept, which consists of the aggregated and coordinated dispatch, and operation of a large number of DR (either small-scale, medium-scale, or utility-scale) may allow providing this type of ancillary service. Similarly, the implementation of the microgrid concept requires the availability of DR with frequency control capability; this can be accomplished by means of DG, the combination of intermittent DG and DS, or using DS alone.

3.2.5.2.11 Dispatchability and Control

IEEE Std. 1547 states that each DR unit of 250 kVA or more or DR aggregate of 250 kVA or more at since PCC shall have provisions for monitoring its connection status, real and reactive power output, and voltage at the point of DR connection. Monitoring the exchange of information and control for DR systems should support interoperability between DR devices and area EPS. Use of standard commands and protocols and data definitions enables this interoperability. In addition, this reduces costs for data translators, manual configuration, and special devices. DR can be dispatched as a unit for energy export as needed, according to a certain schedule, during peak periods, shut down for maintenance, used for ancillary services such as load regulation, energy losses, spinning reserve, voltage regulation, and reactive power supply.

3.2.5.2.12 Power Quality

Important potential impacts of DR integration are voltage rise, voltage fluctuation, flicker, voltage unbalance, voltage sags and swells, and increased total harmonic distortion (THD). All these impacts may affect the overall distribution grid power quality. Voltage rise and voltage fluctuation are a natural consequence of the interconnection of DR on the power distribution grid, as previously discussed, their magnitude is a function of the grid's stiffness factor and DR output. Furthermore, extreme intermittency due to cloud cover may lead to rapid voltage fluctuations; this has motivated some utilities to require evaluating potential impacts on flicker as a requisite for authorizing DR interconnection. Voltage unbalance can be accentuated by large penetration levels of single-phase DR, particularly if different technologies and capacities are used, and they are uncoordinatedly connected to different phases of the power distribution grid. Voltage sags and swell can be the consequence fault current contributions and sudden connection and disconnection of utility-scale DG. Increased harmonic distortion may be caused by large proliferation of electronically coupled DG; here, it is worth noting that despite the fact that individual inverters may comply with standard requirements pertaining to harmonic injection, it is the interaction and cumulative effect of harmonics produced by a large number of inverters that could have a negative effect on feeders' THD levels. As previously indicated, smart grid technologies and intelligent control of DR inverters can help alleviate issues related with voltage rise, voltage fluctuation, and intermittency. Other issues such as voltage sags and swells due to larger fault currents may be mitigated using, for instance, superconducting fault current limiters. Finally, issues related with increased voltage unbalance and THD should be addressed in the planning stage of the smart grid, where maximum penetration levels and location of DR must be carefully evaluated. Another potential and more complex solution is the coordinated dispatch of these technologies via the virtual power plant concept.

3.2.5.3 Electric Vehicle Impacts on Electric Grid Systems

PEVs are seen as having the potential to improve multiple facets of the transportation sector. However, for PEVs to have a significant positive impact on the transportation sector, a substantial fraction of the existing vehicle fleet must be converted to PEVs. Any significant conversion of this type will impose a large demand on the electric sector if not properly administered. Therefore, to realize transportation improvements on a grand scale without creating concurrent electrical problems, changes in the electric and transportation sectors must be collaborative and occur concurrently.

The charging of PEVs is the most important interaction between electrified transportation and the electric grid, and is the area in which smart grid technologies can provide tools to assimilate the two sectors. Plug-in vehicle charging is divided into two main categories: "smart" charging and unconstrained charging. Unconstrained charging is the simplest form of plug-in vehicle charging and allows the vehicle owner to plug in at any time of the day without any limitations [33]. Constrained charging is defined as any charging strategy in which the electricity provider and vehicle are able to cooperatively implement charging strategies with an aim to limit plug-in vehicle charging loads so as to maximize the economic efficiency of vehicle charging. The first generation of PEVs will likely charge without input or restriction from the utility. Due to the initial low volume of vehicles, this will likely have a low impact on the electric grid [34,35]. However, most research, to date, has shown that as PEVs penetrate the market, unconstrained charging will need to be replaced with some level of constrained or "smart" charging to reduce the possibility of exacerbating peak electric demands [33,36]. Studies have shown that "smart" charging can potentially permit replacement of at least 50% of the traditional vehicle fleet with PEVs without the need to increase generation or transmission capacity. Larger penetrations also present opportunities for the electric sector to regulate the system more effectively, resulting in more uniform daily load profiles, better capital utilization, and reduced operational costs [33,36].

The electric utility sector has also expressed concern regarding expected increased loads on residential transformers and other electric grid components. Studies have shown that the acceptance of HEVs such as the Toyota Prius has typically occurred nonuniformly throughout geographic areas, with high concentrations in certain areas and little-to-no adoption in others. The adoption of PEVs is expected to follow a similar pattern [37].

Increased loading on residential transformers poses a problem for the electricity provider as most residential transformers are already approaching their recommended use capacities. In addition, although "smart" charging of PEVs will help the electric sector reduce peak demands, "smart" charging may force transformers—especially residential transformers—to be fully utilized for the majority of the day. Increased use will reduce the amount of equipment rest and cooling time, which could shorten the operational life of the transformers and other electric grid equipment [38]. These studies agree, however, that these pressures will not result in decreases in reliability or functionality of distributions systems. They will merely require changes in distribution system maintenance schedules.

The most prevalent strategies currently being pursued to implement smart charging are as follows:

- *Financial (TOU pricing, critical peak pricing, real-time pricing)*—Charging different rates at different times of day to incentivize users to change their behavior
- *Direct (delayed charging, demand response)*—Curtailment of charging activities, enabled by smart charging chips or charger-side intelligence in a demand-response type program
- *Information based (home area network, smart meters and displays)*—Giving users information and signals to help them make informed decisions about the cost and impact of charging on the grid [33,35,39,40]

Due to the variation in the energy sources used throughout the electric sector, some charging strategies may prove more advantageous and effective than others. All of the "smart" charging

Smart Grid Technologies

strategies require some level of communication between the PEV, vehicle owner, and the electricity provider or grid system operator. For direct and financial smart-charging strategies, the plug-in vehicle or owner must be able to receive and process pricing and/or power control signals sent by the electricity provider [36]. More advanced charging strategies, especially market-oriented or two-way power flow strategies, require reliable, two-way communication between the plug-in vehicle and the electricity provider or the grid system operator [20,36]. Two-way communication is required because the electricity provider or grid system operator needs to know the state of charge (SOC) of all the PEVs connected in order to forecast the charging load for the valley-filling algorithm and the availability of PEVs to provide V2G frequency control. Research has shown that the communication task can be achieved by integrating broadband over PowerLine and HomePlug™, Zigbee™, or cellular communications technologies into a stationary charger or into the PHEV's power electronics [40].

Regardless of the type of smart-charging strategy utilized, the required charging infrastructure and strategies will impose constraints on the electric grid. The largest impact smart charging will have on the electric grid is associated with the communications requirements needed between PEVs and owners, and the electricity provider or grid system operator. The simplest method (in terms of communication) for the electric sector to control charging behavior is to implement TOU rates. TOU rates can be relayed to PEV owners through rate plans that only change based on time of day and year and require the installation of an electric meter capable of metering energy transfer in realtime for billing purposes. However, it is yet to be determined if TOU rates are strong enough motivators to affect the charging habits of the majority of plug-in vehicle owners. The next level of complexity available for the electric sector is the use of real-time data communication. Control could be based upon one-way communication: For example, vehicles could charge only when real-time rates drop below a set threshold. Several proposed control strategies (e.g., V2G) would require two-way communication. However, for a large number of PEVs, real-time data transfer has been seen as an overwhelming task [19].

To help guide the development of the charging infrastructure, which is required for PEVs, the SAE J1772 standard has been developed. The standard requires plug-in vehicle power transfer connections to be able to operate on single phase 120 or 240 V and also support communications capabilities. The power transfer equipment can either be a separate component or can be integrated into the power electronic equipment and electric motor. In order for PEVs to be capable of V2G, either an inverter must be added to the vehicle's power electronics or equipment capable of utilizing the on-board charger as both an inverter and a rectifier would need to be used [39]. Although various power levels of charging have been proposed, level 1 charging (110 V, 15 A) is currently the most common. Level 2 and level 3 rapid chargers have increased power ratings, but the installation of level 2 and level 3 chargers can be a slow and costly process, especially for residential installations [41,42] (Table 3.5).

It is clear that some level of smart-charging infrastructure will be needed as PEVs begin to penetrate the transportation market. Smart grid technologies provide a variety of charging methods that can help ensure PEV customer satisfaction while maintaining a balance between plug-in vehicle charging demand and the electric grid's resources. However, "smart" charging of PEVs will require a large investment in electric grid and communications infrastructure and will significantly increase the workload of the electric sector.

3.2.5.3.1 Equipment Loading, Maintenance, and Life Cycle

Arguably the most significant impact of PEVs charging on the power grid is increase in equipment loading, specifically on distribution transformers and lines. Here, it is worth noting that the severity of this impact is a function of the charging scenarios, charging strategy (uncontrolled or controlled charging), market penetration level, and distribution feeder characteristics (existing loading, voltage level, load profile, etc.). In order to determine the impact of PEV charging on the grid, it is

Table 3.5
SAE Charging Configurations and Ratings Terminology

AC		DC	

AC Level 1	• PEV includes on-board charger
	• 120 V, 1.4 kW @ 12 A
	• 120 V, 1.4 kW @ 12 A
	• Estimated charge time:
	  • PHEV: 7 hrs (SOC* – 0% to full)
	  • BEV: 17 hrs (SOC – 20% to full)
		*DC Level 1	• EVSE includes an off-board charger
			• 200-450 VDC, up to 36 kW (80 A)
			• Estimated charge time (20 kW off-board charger):
			  • PHEV: 22 min (SOC* – 0% to 80%)
			  • BEV: 1.2 hrs (SOC – 20% to 100%)

AC Level 2	• PEV includes on-board charger
	• 240 V, up to 19.2 kW (80 A)
	• Estimated charge time for 3.3 kW on-board charger:
	  • PEV: 3 hrs (SOC* – 0% to full)
	  • BEV: 7 hrs (SOC – 20% to full)
	• Estimated charge time for 7 kW on board charger:
	  • PEV: 1.5 hrs (SOC* – 0% to full)
	  • BEC: 3.5 hrs (SOC – 20% to full)
	• Estimated charge time for 20 kW on-board charger:
	  • PEV: 22 min (SOC* – 0% to full)
	  • BEC: 1.2 hrs (SOC – 20% to full)
		*DC Level 2	• EVSE includes an off-board charger
			• 200-450 VDC, up to 90 kW (200 A)
			• Estimated charge time (45 kW off-board charger):
			  • PHEV: 10 min (SOC* – 0% to 80%)
			  • BEV: 20 min (SOC – 20% to 80%)

*AC Level 3 (TBD)	• > 20 kW, single-phase and three-phase	*DC Level 3 (TBD)	• EVSE includes an off-board charger • 200-600 VDC (proposed) up to 240 kW (400 A) • Estimated charge time (45 kW off-board charger): • BEV (only): <10 min (SOC – 0% to 80%)

Source: http://www.sae.org/smartgrid/chargingspeeds.pdf
SOC, state of charge; EVSE, electric vehicle supply equipment
* Not finalized

necessary to conduct preliminary studies to determine (a) charging scenarios, like the one shown in Figure 3.18, which indicates the expected level 1 and level 2 charging profiles of PEVs (PHEVs and BEVs), that is, the time of day when charging is expected to occur and the likely charging demands in percentage of PEVs and (b) market penetration levels, which indicate the amount of PEVs that are expected to be charged in a geographic area as a function of time. Studies and common sense indicate that residential PEV charging is expected to occur during the late afternoons and early evenings, when commuters get back home. Unfortunately, in many cases, this coincides with peak feeder loading conditions, which have a direct impact on increasing distribution transformer and line loadings.

Once charging and market penetration scenarios are determined, it is necessary to conduct power flow analyses under a series of varying loading conditions to determine equipment loading. These simulations consist of superimposing PEV loads on expected customer or distribution transformer loads and running power flow analyses to determine feeder electrical variables (voltages, currents, etc.). The complexity of these analyses will vary depending on the accuracy sought, and they may include conducting statistical analyses to model the uncertainty about charging and market penetration scenarios. These analyses must be conducted for uncontrolled charging scenarios, to determine "worst case" impacts, and under controlled charging scenarios that are designed

Smart Grid Technologies

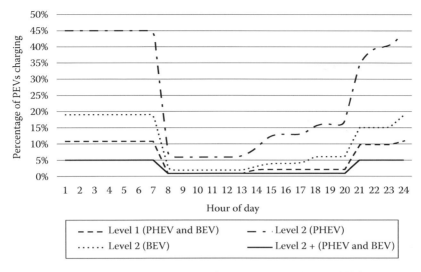

FIGURE 3.18 Example of an expected PEV charging scenario (projected 2020). (From Xu, L. et al., A framework for assessing the impact of plug-in electric vehicle to distribution systems, *2011 IEEE PSCE*, Phoenix, AZ, March 2011.)

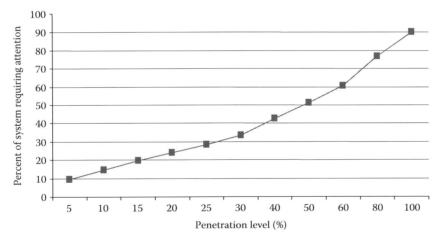

FIGURE 3.19 Example of percent of distribution system impacted versus PEV market penetration (uncontrolled charging). (From Dow, L. et al., A novel approach for evaluating the impact of electric vehicles on the power distribution system, *2010 IEEE PES General Meeting*, Minneapolis, MN, July 2010.)

to mitigate expected impacts. Controlled scenarios aim at modifying PEV charging profiles by providing incentives or penalties via TOU rates or exerting charging load control or management to displace charging to off-peak hours.

The literature indicates that under uncontrolled scenarios, transformer overloads are expected to occur even at low penetration levels; this is shown in Figure 3.19. Despite the fact that, at first sight, smart grid technologies such as controlled charging appear to be a mitigation measure for equipment loading impacts, it has the disadvantage of shifting charging to off-peak hours, for example, during early morning. This ultimately leads to (a) increasing load coincidence and creating new peaks that may also overload distribution transformers and lines, especially for large market penetration levels; this is shown in Figure 3.20 and (b) "flattening" distribution

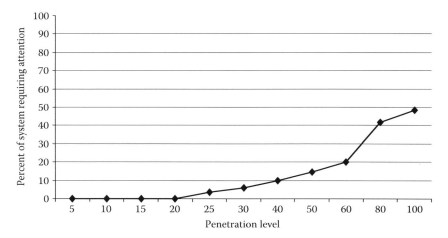

FIGURE 3.20 Example of percent of distribution system impacted versus PEV market penetration (controlled charging). (From Dow, L. et al., A novel approach for evaluating the impact of electric vehicles on the power distribution system, *2010 IEEE PES General Meeting*, Minneapolis, MN, July 2010.)

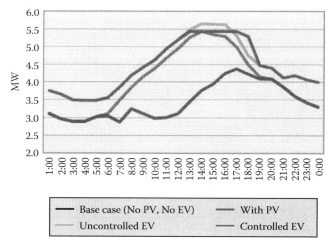

FIGURE 3.21 (**See color insert.**) Example of feeder load under PV and EV penetration scenarios. (From Agüero, J.R., *IEEE Power Energy Mag.*, September/October 2011, 82–93.)

transformer load profiles, that is, increasing their load factors. Obviously the former is undesired, and even though the latter seems attractive, it may have a negative impact on equipment maintenance and life cycle, since off-peak loading conditions allow distribution transformers to cool down. Therefore, incentives and load control or management strategies must be carefully designed and applied to avoid creating further impacts. Other solutions to equipment overload are conventional approaches such as capacity increase (transformer upgrade, line reconductoring, etc.). Furthermore, the coordinated control and dispatch of local DER, such as DG and DS and the implementation of demand response, is a promising alternative for solving these issues, as shown in Figure 3.21. Finally, a combination of all the aforementioned approaches (conventional and smart grid technologies) is recommended. As indicated previously, the smart grid will play a critical role in enabling these solutions.

Smart Grid Technologies

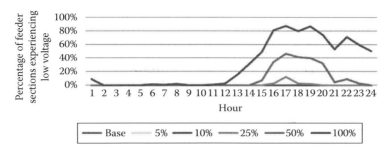

FIGURE 3.22 (**See color insert.**) Example of percentage of feeder sections experiencing low voltage for various PEV penetration levels. (From Agüero, J.R. and Dow, L., Impact studies of electric vehicles, Quanta Technology, Raleigh, NC, 2011.)

3.2.5.3.2 Voltage Regulation and Feeder Losses

The additional currents flowing through distribution transformers and lines due to moderate-to-high penetration scenarios of PEV may lead to an increase on voltage drop along distribution feeders that can cause low-voltage violations, particularly on areas located far from distribution substations. An example of this is shown in Figure 3.22 for various PEV penetration levels. This issue can be addressed by installing additional line voltage regulators and switched capacitor banks, as well as by the coordinated dispatch and control of local DER, such as DG and DS, and the implementation of DR and load control/management. PEV's charging loads are expected to have a power factor close to unity, thanks to power factor correction (PFC) systems. However, as the penetration level increases, higher charging loads imply higher currents and distribution line and load transformer losses. Therefore, PEV proliferation is expected to increase distribution system losses. Again, the combined implementation of conventional and smart grid solutions via the additional communications and control capabilities enabled by the smart grid is expected to be the more successful approach for ensuring adequate voltage regulation and minimizing the impact of PEVs on distribution losses. This also highlights the need of multiobjective optimization approaches for a coordinated utilization of all available resources.

3.2.5.3.3 Power Quality

As indicated in previous sections, increased harmonic distortion may be caused by large proliferation of inverter-based equipment, including PEV charging facilities; here, it is worth noting that despite the fact that individual inverters may comply with standard requirements pertaining to harmonic injection, it is the interaction and cumulative effect of harmonics produced by a large number of inverters (including PEV and electronically coupled DG inverters) that could have a negative effect on feeders' THD levels. This is an area that requires attention and further research, since it is expected to become more important as the deployment of these technologies grows. As previously indicated, issues related with increase on THD should be addressed in the planning stage of the smart grid, where maximum penetration levels and location of DG and PEVs must be carefully evaluated.

3.2.5.3.4 Others

Almost since the first sales of hybrid vehicles, there has been considerable interest in using the vehicles as auxiliary power supplies—backup generators or supplemental power systems. In some geographical areas, there remains a substantial risk of power failure due to natural disasters such as storms or floods. Owners of PHEVs in these areas could tap into their vehicles' electrical systems for backup power in the event of power failure. Indeed, several informal projects have utilized electric vehicles for this purpose, connecting directly to the traction battery [47] or operating solely off the vehicle's 12 V convenience power [48].

These efforts have been hampered by lack of support from vehicle manufacturers and lack of suitable inverters capable of both operation off grid and supporting EV battery voltages. This application remains a hobbyist niche but could rapidly become a de facto standard if many vehicles are equipped with inverters to support V2G operations, and vehicle manufacturers see the backup-power market as a potential added feature in their product offering. Serious safety issues must also be addressed, including electrical safety with both DC and AC circuits and the buildup of emissions if the vehicle is unintentionally operated in enclosed spaces.

Limitations on the size of household electrical services will also impact the introduction of EVs—particularly the selection of charging solutions. Many newer houses in the United States are equipped with 100 A electrical services, while older homes may have smaller services, and larger homes may have 200 A services or larger. Regardless of the absolute service size, in most cases, the installed service was properly sized for the anticipated loads in the household. Similarly, multiunit developments also size electrical services to meet electrical codes with limited spare capacity.

Although electrical codes remain relatively conservative, allowing for increased demand, introduction of a new, large electricity demand will likely violate those codes and possibly overload the electrical service. Further, electrical codes do not generally allow the introduction of additional circuits on the understanding that those circuits will not be utilized simultaneously with existing household loads. That is, although vehicle connections could be electronically limited to nighttime charging, when other household loads are low, there are currently few mechanisms in electrical codes to allow for such expansion.

The layout of household electrical services also presents issues. While newer homes frequently have the incoming electrical service in the garage area, in many homes, the electrical service entrance is located far from the garage—a location that has traditionally experienced far lower loads than other parts of the house. The expense of modifying the incoming electrical panel, and pulling new circuits to the garage areas, will likely retard the adoption of level 2 and level 3 charging. Vehicles charged at level 1 will typically require continuous electrical connections all night to reach a full state-of-charge. Therefore, if only level 1 charging is widely implemented, many of the most promising control mechanisms (controlled charging, V2G, etc.) offered by integrating EVs into the electrical grid will be inaccessible.

Finally, it should be noted that vehicle manufacturers currently have little incentive to modify vehicles to support grid support functions. Indeed, many proposed solutions, including V2G, controlled charging, and backup power applications, are likely to negatively impact battery life and/or decrease customer satisfaction—primary goals of the vehicle manufacturers.

Ultimately, integration of PEVs into both the transportation and electricity sectors is a system problem, requiring system solutions. Viable solutions will need to balance competing goals of vehicle owners, grid operators, and vehicle manufacturers, as well as address issues as diverse as electrical code compliance and dispersed communication.

3.3 SMART SUBSTATIONS

Stuart Borlase, Marco C. Janssen, and Michael Pesin

An electrical substation is a focal point of an electricity generation, transmission, and distribution system where voltage is transformed from high to low or reverse using transformers. Electric power flows through several substations between generating plants and consumer and usually is changed in voltage in several steps. There are different kinds of substations such as transmission substations, distribution substations, collector substations, and switching substation. The general functions of a substation include the following:

- Voltage transformation
- Connection point for transmission and distribution power lines

Smart Grid Technologies

- Switchyard for electrical transmission and/or distribution system configuration
- Monitoring point for control center
- Protection of power lines and apparatus
- Communication with other substations and regional control center

Substations and feeders are the source of critical real-time data for efficient and safe operation of the utility network. Real-time data, also called operational data, are instantaneous values of power system analog and status points such as volts, amps, MW, MVAR, circuit breaker status, and switch position. These data are time critical and are used to protect, monitor, and control the power system field equipment. There is also a wealth of operational (non-real-time) data available from the field devices. Nonoperational data consist of files and waveforms such as event summaries, oscillographic event reports, or sequential event records, in addition to supervisory control and data acquisition (SCADA)-like points (e.g., status and analog points) that have a logical state or a numerical value. Nonoperational data are not needed by the SCADA dispatchers to monitor and control the power system, but the data can help make operation and management of system assets more efficient and reliable.

3.3.1 Protection, Monitoring, and Control Devices (IEDs)

Intelligent electronic devices (IEDs) are microprocessor-based devices with the capability to exchange data and control signals with another device (IED, electronic meter, controller, SCADA, etc.) over a communications link. IEDs perform protection, monitoring, control, and data acquisition functions in generating stations, substations, and along feeders and are critical to the operations of the electric network.

IEDs are widely used in substations for different purposes. In some cases, they are separately used to achieve individual functions, such as differential protection, distance protection, overcurrent protection, metering, and monitoring. There are also multifunctional IEDs that can perform several protection, monitoring, control, and user interfacing functions on one hardware platform.

IEDs are a key component of substation integration and automation technology. Substation integration involves integrating protection, control, and data acquisition functions into a minimal number of platforms to reduce capital and operating costs, reduce panel and control room space, and eliminate redundant equipment and databases. Automation involves the deployment of substation and feeder operating functions and applications ranging from SCADA and alarm processing to integrated volt/VAr control (IVVC) in order to optimize the management of capital assets and enhance operation and maintenance (O&M) efficiencies with minimal human intervention.

The main advantages of multifunctional IEDs are that they are fully IEC 61850 compatible and compact in size and that they combine various functions in one design, allowing for a reduction in size of the overall systems and an increase in efficiency and improvement in robustness and providing extensible solutions based on mainstream communications technology.

IED technology can help utilities improve reliability, gain operational efficiencies, and enable asset management programs including predictive maintenance, life extensions, and improved planning.

3.3.2 Sensors

The main functionality of sensors is to collect data from power equipment at the substation yard such as transformers, circuit breakers, and power lines. With the introduction of digital and optical technologies in combination with communication, new sensors are becoming available to acquire different types of asset-related information. Original copper-wired analog apparatus can now be replaced by optical apparatus with fiber-based sensors for monitoring and metering. The most prominent advantages of such sensors are higher accuracy, no saturation, reduced size and weight, safe and environment friendly (avoid oil or SF_6), higher performance, wide dynamic range, high bandwidth, and low maintenance. The main advantages of optical sensors are the wide frequency bandwidth, wide dynamic range, and high accuracy. Furthermore, these new sensors allow monitoring and control to be implemented with two important application features:

a. Single sensor may serve different types of IEDs.
b. Single sensor may serve a large number of IEDs via process bus.

Those sensors also need accurate time synchronization of the inputs and the samples being placed on the process bus.

3.3.3 SCADA

SCADA refers to a system or a combination of systems that collects data from various sensors at a plant or in other remote locations and then sends these data to a central computer system, which then manages and controls the data and remotely controls devices in the field.

SCADA is a term that is used broadly to portray control and management solutions in a wide range of industries. The electric power industry has a specific set of requirements that applied to SCADA systems.

The primary purpose of an electric utility SCADA system is to acquire real-time data from the field devices located at the power plants, transmission and distribution substations, distribution feeders, etc., provide control of the field equipment, and present the information to the operating personnel. Realtime to the monitoring and control of substations and feeders is typically in the range of 1–5 s.

SCADA systems are globally accepted as a means of real-time monitoring and control of electric power systems, particularly generation and transmission systems. RTUs (remote terminal units) are used to collect analog and status telemetry data from field devices, as well as communicate control commands to the field devices. Installed at a centralized location, such as the utility control center, are front-end data acquisition equipment, SCADA software, operator graphical user interface (GUI), engineering applications that act on the data, historian software, and other components.

Recent trends in SCADA include providing increased situational awareness through improved GUIs and presentation of data and information, intelligent alarm processing, the utilization of thin clients and web-based clients, improved integration with other engineering and business systems, and enhanced security features.

Typically, control and data acquisition equipment compose a system with at least one master station, one or more RTUs, and a communications system. The electric utility master station is usually located at an energy control center (ECC), and RTUs are installed at the power plants, transmission and distribution substations, distribution feeder equipment, etc.

3.3.3.1 Master Stations

The master station is a computer system responsible for communicating with the field equipment and includes a human machine interface (HMI) in the control room or elsewhere. In smaller SCADA systems, the master station may be composed of a single PC. In larger SCADA systems, the master station may include multiple redundant servers, distributed software applications, and disaster recovery sites.

A large electric utility master station or energy management system (EMS) typically has the following:

- One or more data acquisition servers (DAS) or front-end processors (FEP) that interface with the field devices via the communications system
- Real-time data server(s) that contains real-time database(s) (RTDB)
- Historical server(s) that maintains historical database
- Application server(s) that runs various EMS applications
- Operator workstations with an HMI

Smart Grid Technologies

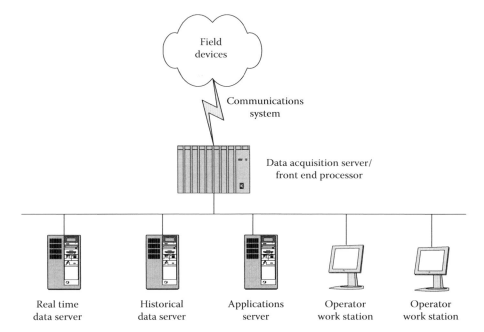

FIGURE 3.23 Typical modern EMS architecture (simplified).

In most modern EMSs, hardware components are connected via one or more local area networks (LANs). Many systems have a secure interface to the corporate networks to make EMS data available to the corporate users (Figure 3.23).

There are several different types of the modern master stations. In general, master stations can be divided into five different categories based on their functionality. However, in some cases the functions can cross over from one type of system to another:

1. SCADA master station
2. SCADA master station with automatic generation control (AGC)
3. EMS
4. Distribution management system (DMS)
5. Distribution automation (DA) master

SCADA master station primary functions:

- Data acquisition
- Remote control
- User interface
- Areas of responsibility
- Historical data analysis
- Report writer

SCADA/AGC system primary functions (in addition to SCADA master station):

- AGC
- Economic dispatch (ED)/hydroallocator
- Interchange transaction scheduling

EMS primary functions (in addition to SCADA/AGC system):

- Network configuration/topology processor
- State estimation
- Contingency analysis
- Three phase balanced operator power flow
- Optimal power flow
- Dispatcher training simulator

DMS primary functions:

- Interface to automated mapping/facilities management (AM/FM) or geographic information system (GIS)
- Interface to customer information system (CIS)
- Interface to outage management
- Three phase unbalanced operator power flow
- Map series graphics

DA system primary functions:

- Two-way distribution communications
- Fault identification/fault isolation/service restoration
- Voltage reduction
- Load management
- Power factor control
- Short-term load forecasting

All types of master stations are interfaced with the field devices. Historically in electric utilities, these devices were RTUs. In recent years, with the proliferation of IEDs, many of these devices are taking over the RTU functionality.

3.3.3.2 Remote Terminal Unit

The RTU is a microprocessor-based device that interfaces with a SCADA system by transmitting telemetry data to the master station and changing the state of connected devices based on control messages received from the master station or (in some modern systems) commands generated by the RTU itself. The RTU provides data to the master station and enables the master station to issue controls to the field equipment. Typical RTUs have physical hardware inputs to interface with field equipment and one or more communication ports (Figure 3.24).

Different RTUs process data in different ways, but in general there are several internal software modules that are common among most RTUs:

- Central RTDB that interfaces with all other software modules.
- Physical I/O application—acquires data from the RTU hardware components that interface with physical I/O.
- Data collection application (DCA)—acquires data from the devices with data communications capabilities via communication port(s). For example, IEDs.
- Data processing application (DPA)—presents data to the master station or HMI.
- Some RTUs also have data translation applications (DTA) that manipulate data before they are presented to the master station or support stand-alone functionality at the RTU level (Figure 3.25).

Smart Grid Technologies

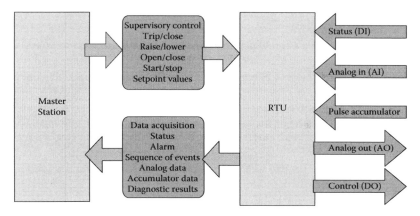

FIGURE 3.24 SCADA system data flow architecture.

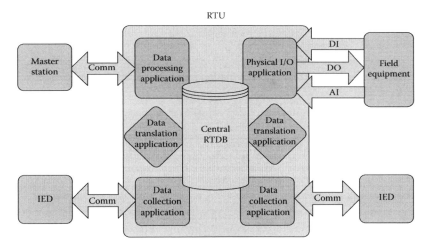

FIGURE 3.25 RTU software architecture.

3.3.4 Substation Technology Advances

Early generations of SCADA systems typically employed one RTU at every substation. With this architecture, all cables from the field equipment had to be terminated at the RTU. RTUs have typically offered limited expansion capacities. For analog inputs, the RTU required the use of transducers to convert higher level voltages and currents from CT and PT outputs into the milliamp and volt level. Most RTUs had a single communication port and were only capable of communicating with one master station. The communication between an RTU and its master station was typically achieved via proprietary bit-oriented communication protocols. As technology advanced, RTUs became smaller and more flexible. This allowed for a distributed architecture approach, with one smaller RTU for one or several pieces of substation equipment. This resulted in lower installation costs with reduced cabling requirements. This architecture also offered better expansion capabilities (just add more small RTUs). In addition, the new generation of RTUs was capable of accepting higher level AC analog inputs. This eliminated the need for intermediate transducers and allowed direct wiring of CTs and PTs into the RTU. This also enabled RTUs to have additional functionality, such as digital fault recording (DFR) and power quality (PQ) monitoring.

There were also advances in communications capabilities, with additional ports available to communicate with IEDs. However, the most significant improvement was the introduction of an open communications protocol. The older SCADA systems used proprietary protocols to communicate between the master station and the RTUs. Availability of an open and standard (for the most part) utility communications protocol allowed utilities to choose vendor-independent equipment for the SCADA systems. The de facto standard protocol for electric utilities SCADA systems in North America became DNP3.0. Another open communications protocol used by utilities is MODBUS. The MODBUS protocol came from the industrial manufacturing environment. The latest communication standard adopted by utilities is IEC 61850. IEC61850 is a very powerful and flexible network-based, object-oriented communication standard that allows for utilities to move their next-generation substations that are flexible and expandable; allows for the implementation of multivendor solutions; and in addition to the communication, also facilitates a standardized engineering approach allowing for optimization of utility engineering and maintenance processes.

Another technology that aided SCADA systems was network data communications. The SCADA architecture based on serial communications protocols put certain limitations on system capabilities. With a serial SCADA protocol architecture,

- There is a static master/slave data path that limits the device connectivity
- Serial SCADA protocols do not allow multiple protocols on a single channel
- There are issues with exchanging new sources of data, such as oscillography files, PQ data, etc.
- Configuration management has to be done via a dedicated "maintenance port"

The network-based architecture offers a number of advantages:

- *There is significant improvement in speed and connectivity*: An Ethernet-based LAN greatly increases the available communications bandwidth. The network layer protocol provides a direct link to devices from anywhere on the network.
- *Availability of logical channels*: Network protocols support multiple logical channels across multiple devices.
- *Ability to use new sources of data*: Each IED can provide another protocol port number for file or auxiliary data transfer without disturbing other processes (e.g., SCADA) and without additional hardware.
- *Improved configuration management*: Configuration and maintenance can be done over the network from a central location.

The network-based architecture in many cases also offers a better response time, ability to access important data, and reduced configuration and system management time. Take, for example, SCADA systems that have been around for many years. These were simple remote monitoring and control systems exchanging data over low-speed communications links, mostly hardwired. In recent years, with the proliferation of microprocessor-based IEDs, it became possible to have information extracted directly from these IEDs either by an RTU or by other substation control system components. This is achieved by using the IED communications capabilities, allowing it to communicate with the RTU, data concentrator, or directly with the master station. As more IEDs were installed at the substations, it became possible to integrate some of the protection, control, and data acquisition functionality. A lot of the information previously extracted by the RTUs now became available from the IEDs. However, it may not be practical to have the master station communicate directly with the numerous IEDs in all the substations. To enable this data flow, a new breed of devices called substation servers is utilized. A substation server communicates with all the IEDs at the substation, collects all information from the IEDs, and then communicates back to the central master station. Because the IEDs at the substation use many different communications protocols, the substation

Smart Grid Technologies

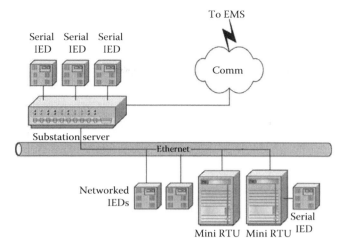

FIGURE 3.26 Server-based substation control system architecture.

server has to be capable of communicating via these protocols, as well as the master station's communications protocol. A substation server allows the SCADA system to access data from most substation IEDs, which were only accessible locally before.

With the substation server-based SCADA architecture (Figure 3.26), all IEDs (including RTUs) are polled by the substation server. The IEDs and RTUs with network connections are polled over the substation LAN. The IEDs with only serial connection capabilities are polled serially via the substation server's serial RS232 or RS485 ports (integrated or distributed). In addition to making additional IED data available, the substation server significantly improves overall SCADA system communication performance. With the substation server-based architecture, the master station has to communicate directly with only the substation server instead of multiple RTUs and IEDs at the substation. Also, a substation server's communications capability is typically superior to that of an IED. This, and the reduced number of devices directly connected to the master station, contributes to a significantly improved communications performance in a polled environment.

Data available in the substation can be divided into two types: operational or real-time data and nonoperational data. Operational data are real-time data required for operating utility systems and performing EMS software applications such as AGC. These data are stored by EMS applications and available as historical data. Nonoperational data are historical, real-time, and file type data used for analysis, maintenance, planning, and other utility applications.

Modern IEDs, such as protection relays and meters, have a tremendous amount of information. Some of these devices have thousands of data points available. In addition, many IEDs generate file type data such as DFR or PQ files. A typical master station is not designed to process this amount of data and this type of data. However, a lot of this information can be extremely valuable to the different users within the utility, as well as, in some cases, the utility's customers. To take advantage of these data, an extraction mechanism independent from the master station needs to be implemented.

Operational data and nonoperational data have independent data collection mechanisms. Therefore, two separate logical data paths should also exist to transfer these data (Figure 3.27). One logical data path connects the substation with the EMS (operational data). A second data path transfers nonoperational data from the substation to various utility information technology (IT) systems. With all IEDs connected to the substation data concentrator, and sufficient communications infrastructure in place, it also becomes possible to have a remote maintenance connection to most of the IEDs. This functionality is referred to as either "remote access" or "pass-through." Remote access or pass-through is the ability to have a virtual connection to remote devices via a secure network.

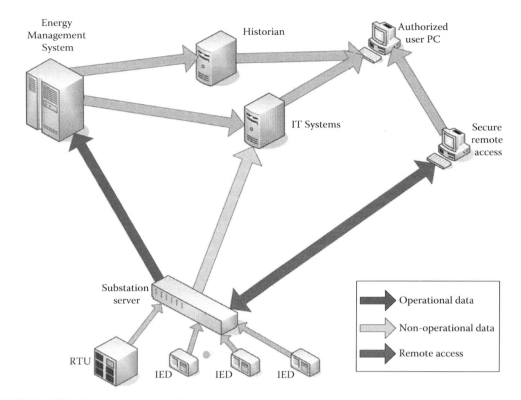

FIGURE 3.27 (**See color insert**). Substation data flow.

This functionality significantly helps with troubleshooting and maintenance of remote equipment. In many cases, it can eliminate the need for technical personnel to drive to a remote location. It also makes real-time information from individual devices at different locations available at the same computer screen that makes the troubleshooting process more efficient.

Figure 3.28 shows the conceptual migration path from basic SCADA functionality through integration and automation to a full smart grid substation solution.

An advanced substation integration architecture (Figure 3.29) offers increased functionality by taking full advantage of the network-based system architecture, thus allowing more users to access important information from all components connected to the network. However, it also introduces additional security risks into the control system. To mitigate these risks, special care must be taken when designing the network, with special emphasis on the network security and the implementation of user authentication, authorization, and accounting. It is very important that a substation communication and physical access security policy is developed and enforced.

3.3.5 Platform for Smart Feeder Applications

Monitoring, control, and data acquisition of the electricity network will extend further down to the distribution pole-top transformer and perhaps even to individual customers, either through the substation communications network, by means of a separate feeder communications network, or tied into the advanced metering infrastructure (AMI). More granular field data will help increase operational efficiency and provide more data for other smart grid applications, such as outage management. Higher speed and increased bandwidth communications for data acquisition and control will be needed. Fault detection, isolation, and service restoration (FDIR) on the distribution system

Smart Grid Technologies

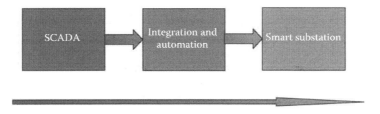

SCADA	Integration and automation	Smart substation
		Enabler of system-wide interfaces and applications
		Substation monitoring and security
	Enterprise access to non-operational data	Enterprise access to non-operational data
EMS/DMS interface	EMS/DMS interface	EMS/DMS interface
	Local HMI	Local HMI
	Substation and feeder automation	Substation and feeder automation
	IED integration	IED integration
		Process bus
Power equipment and sensors (transformers, breakers, reclosers, CTs, PTs, etc.)	Power equipment and sensors (transformers, breakers, reclosers, CTs, PTs, etc.)	Power equipment and sensors (transformers, breakers, reclosers, CTs, PTs, etc.)

FIGURE 3.28 Substation smart grid migration.

will require a higher level of optimization and will need to include optimization for closed-loop, parallel circuit, and radial configurations. Multilevel feeder reconfiguration, multiobjective restoration strategies, and forward-looking network loading validation will be additional features with FDIR. IVVC will include operational and asset improvements, such as identifying failed capacitor banks and tracking capacitor bank, tap changer, and regulator operation to provide sufficient statistics for opportunities to optimize capacitor bank and regulator placement in the network. Regional IVVC objectives may include operational or cost-based optimization. This will all require advanced smart substation and feeder solutions and a broader perspective on how integration of substation and feeder data and T&D automation can benefit the smart grid.

> Realizing the promises and benefits of a smarter grid—from improved reliability, to increased efficiency, to the integration of more renewable power – will require a smarter distribution grid, with advanced computing power and two-way communications that operate at the speed of our 21st Century digital society. The problem – only approximately 10% of the 48,000 distribution substations on today's grid in the U.S. are digitized. Upgrading these substations to meet today's energy challenges will require time, resources and money [1].

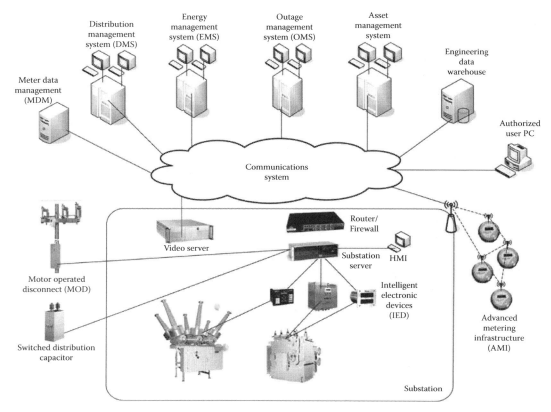

FIGURE 3.29 Smart substations in the smart grid architecture. (© Copyright 2012 Michael Pesin. All rights reserved.)

3.3.6 Interoperability and IEC 61850

IEC 61850 is a vendor-neutral, open systems standard for utility communications, significantly improving functionality while yielding substantial customer savings. The standard specifies protocol-independent and standardized information models for various application domains in combination with abstract communications services, a standardized mapping to communications protocols, a supporting engineering process, and testing definitions. This standard allows standardized communication between IEDs located within electric utility facilities, such as power plants, substations, and feeders but also outside of these facilities such as wind farms, electric vehicles, storage systems, and meters. The standard also includes requirements for database configuration, object definition, file processing, and IED self-description methods. These requirements will make adding devices to a utility automation system as simple as adding new devices to a computer using "plug and play" capabilities. With IEC 61850, utilities will benefit from cost reductions in system design, substation wiring, redundant equipment, IED integration, configuration, testing, and commissioning. Additional cost savings will also be gained in training, MIS operations, and system maintenance.

IEC 61850 has been identified by the National Institute of Standards and Technology (NIST) as a cornerstone technology for field device communications and general device object data modeling. IEC 61850™ Part 6 defines the configuration language for systems based on the standard. Peer-to-peer communication mechanisms such as the Generic Object Oriented System Event (GOOSE) will minimize wiring between IEDs. The use of peer-to-peer communication in combination with the use of sampled values (SVs) from sensors will minimize the use of copper wiring throughout the substation, leading to significant benefits in cost savings, more compact substation designs, and advanced and more flexible automation systems, to name a few. With high-speed Ethernet, the IEC 61850-based communications system will be able to manage all of the data available at the process level as well as at the station level.

The IEC 61850 standard was originally designed to be a substation communications solution and was not designed to be used over the slower communications links typically used in DA. However, as wide area and wireless technologies (such as WiMAX) advance, IEC 61850 communications to devices in the distribution grid will become possible. It is therefore possible that IEC 61850 will eventually be used in all aspects of the utility enterprise. At this time, an IEC WG is in the process of defining new logical nodes (LNs) for distributed resources—including photovoltaic, fuel cells, reciprocating engines, and combined heat and power.

With the introduction of serial communication and digital systems, the way we look at secondary systems is fundamentally changing. Not only are these systems still meant to control, protect, and monitor the primary system but we expect these systems to provide more information related to a realm of new functions. Examples of new functions include the monitoring of the behavior, the aging, and the dynamic capacity of the system. Many of the new functions introduced in substations are related to changing operating philosophies, the rise of distributed generation, and the introduction of renewable energy. For protection, new protection philosophies are being introduced focused more on the dynamic adaption of protection functions to the actual network topology, wide area protection and monitoring, the introduction of synchrophasors, and many more.

This tendency is not new. Ever since the introduction of the first substation automation systems and digital protection, we have been searching for ways to make better use of the technologies at hand. After many experiments and discussions, this has led to the development of IEC 61850, originally called "Communication networks and systems in substations." It has now evolved into a worldwide standard called "Communication networks and systems for power utility automation," providing solutions for many different domains within the power industry.

The concepts and solutions provided by IEC 61850 are based on three cornerstones:

- *Interoperability*: The ability of IED from one or several manufacturers to exchange information and use that information for their own functions.
- *Free configuration*: The standard shall support different philosophies and allow a free allocation of functions, for example, it will work equally well for centralized (RTU based) or decentralized (substation control system based) configurations.
- *Long-term stability*: The standard shall be future proof, that is, it must be able to follow the progress in communications technology as well as evolving system requirements.

This is achieved by defining a level of abstraction that allows for the development of basically any solution using any configuration that is interoperable and stable in the long run. The standard defines different logical interfaces within a substation that can be used by functions in that substation to exchange information between them. This is shown in Figure 3.30.

IEC 61850 does not predefine or prescribe communications architectures. The interfaces shown in Figure 3.30 are logical interfaces. IEC 61850 allows in principle any mapping of these interfaces on communications networks. A typical example could be to map interfaces 1, 3, and 6 on what we call a station bus. This bus is a communications network focused on the functions at bay and station level. We also could map interfaces 4 and 5 on a process bus, a communications network focused on the process and bay level of a substation. The process bus may in such a case be restricted to one bay, while the station bus might connect functions located throughout the substation. However, it may be possible as well to map interface 4 on a point-to-point link connecting a process-related sensor to the bay protection.

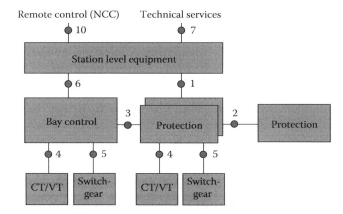

FIGURE 3.30 Interfaces within a substation automation system. (© Copyright 2012 Marco Janssen. All rights reserved.)

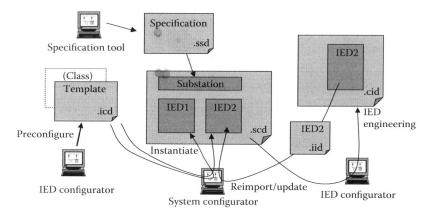

FIGURE 3.31 Engineering approach in IEC 61850. (© Copyright 2012 Marco Janssen. All rights reserved.)

IEC 61850 is, in principle, restricted to digital communications interfaces. However, IEC 61850 specifies more than the communications interfaces. It includes domain-specific information models. In case of substation, a suite of substation functions have been modeled, providing a virtual representation of the substation equipment. The standard, however, also includes the specification of a configuration language. This language defines a suite of standardized, XML-based files that can be used to define in a standardized way the specification of the system, the configuration of the system, and the configuration of the individual IEDs within a system. The files are defined such that they can be used to exchange configuration information between tools from different manufacturers of substation automation equipment. This is shown in Figure 3.31.

The definitions in IEC 61850 are based on a layered approach. In this approach the domain-specific information models, abstract communications services, and the actual communications protocol are defined independently. This basic concept is shown in Figure 3.32.

IEC 61850 is divided in parts, and in parts 7-3 and 7-4xx, the information model of the substation equipment is specified. These information models include models for primary devices such as circuit breakers and instrument transformers such as CTs and VTs. They also include the models for secondary functions such as protection, control, measurement, metering, monitoring, and synchrophasors.

Smart Grid Technologies 139

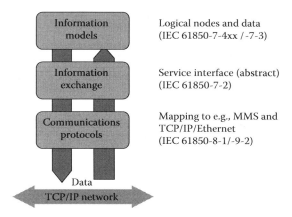

FIGURE 3.32 Concept of the separation of application and communications in IEC 61850. (© Copyright 2012 Marco Janssen. All rights reserved.)

In order to have access to the information contained in the information models, the standard defines protocol-independent, abstract communications services. These are described in part 7-2, such that the information models are coupled with communications services suited to the functionality making use of the models. This definition is independent from any communications protocol and is called the abstract communications service interface (ACSI). The major information exchange models defined in IEC 61850-7-2 are the following:

- Read and write data
- Control
- Reporting
- GOOSE
- SV transmission

The first three models are based on a client/server relation. The server is the device that contains the information while the client is accessing the information. Read and write services are used to access data or data attributes. These services are typically used to read and change configuration attributes. Control model and services are somehow a specialization of a write service. The typical use is to operate disconnector, earthing switches, and circuit breakers. The reporting model is used for event-driven information exchange. The information is spontaneously transmitted when the value of the data changed.

The last two models are based on a publisher/subscriber concept. In IEC 61850 for this the term peer-to-peer communication is introduced to stress that publisher subscriber/communication involves mainly horizontal communication among peers. These communications models are used for the exchange of time-critical information. The device, being the source of the information, is publishing the information. Any other device that needs the information can receive it. These models are using multicast communication (the information is not directed to one single receiver).

The GOOSE concept is a model to transmit event information in a fast way to multiple devices. Instead of using a confirmed communications service, the information exchange is repeated regularly. Application of GOOSE services are the exchange of position information from switches for the purpose of interlocking or the transmission of a digital trip signal for protection-related functions.

The model for the transmission of SVs is used when a waveform needs to be transmitted using digital communication. In the source device, the waveform is sampled with a fixed sampling frequency. Each sample is tagged with a counter representing the sampling time and transmitted over the communications network. The model assumes synchronized sampling, that is, different devices are sampling the waveform at exactly the same time. The counter is used to correlate samples from different sources. That approach creates no requirements regarding variations of the transmission time.

While IEC 61850-8-x specifies the mapping of all models from 7-2 with the exception of the transmission of SVs, IEC 61850-9-x is restricted to the mapping of the transmission of SV model. While IEC 61850-9-2 is mapping the complete model, IEC 61850-9-1 is restricted to a small subset using a point-to-point link providing little flexibility. Both mappings are using Ethernet as communications protocol.

Of course, in order to create real implementations, we need communications protocols. These protocols are defined in parts 8-x and 9-x. In these parts is explained how real communications protocols are used to transmit the information in the models specified in IEC 61850-7-3 and -7-4xx using the abstract communications services of IEC 61850-7-2. In the terminology of IEC 61850, this is called "specific communication service mapping" (SCSM).

Through this approach, an evolution in communications technologies is supported since the application and its information models and the information exchange models are decoupled from the protocol used, allowing for upgrading the communications technology without affecting the applications.

The core element of the information model is the LN. An LN is defined as the smallest reusable piece of a function. It, as such, can be considered as a container for function-related data. LNs contain data, and these data and the associated data attributes represent the information contained in the, part of the, function. The name of an LN class is standardized and comprises always four characters. Basically, we can differentiate between two kinds of LNs:

- LNs representing information of the primary equipment (e.g., circuit breaker—XCBR or current transformer—TCTR). These LNs implement the interface between the switchgear and the substation automation system.
- LNs representing the secondary equipment including all substation automation functions. Examples are protection functions, for example, distance protection—PDIS or the measurement unit—MMXU.

The standard contains a comprehensive set of LNs allowing to model many, if not all, substation functions. In case a function does not exist in the standard extension rules for LNs, data and data attributes have been defined allowing for structured and standardized extensions of the standard information models.

The mappings currently defined in IEC 61850 (part 8-x and 9-x) are using the same communications protocols. They differentiate between the client/server services and the publisher/subscriber services. While the client/server services are using the full seven-layer communication stack using MMS and TCP/IP, the publisher/subscriber services are mapped on a reduced stack, basically directly accessing the Ethernet link layer.

For the transmission of the SVs, IEC 61850-9-2 is using the following communications protocols:

- *Presentation layer*: ASN.1 using basic encoding rules (BER) [ISO/IEC 8824-1 and ISO/IEC 8825]
- *Data link layer*: Priority tagging/VLAN and CSMA/CD [IEEE 802.1Q and ISO/IEC 8802-3]
- *Physical layer*: Fiber optic transmission system 100-FX recommended [ISO/IEC 8802-3]

Ethernet is basically a nondeterministic communications solution. However, with the use of switched Ethernet and priority tagging, a deterministic behavior can be achieved. Using full duplex switches, collisions are avoided. Tagging the transmission of SVs—which requires, due to the cyclic behavior, a constant bandwidth—with a higher priority than the nondeterministic traffic used, for example, reporting of events, ensures that the SVs always get through.

The model for the transmission of SVs as specified in IEC 61850-7-2 is rather flexible. The configuration of the message being transmitted is done using an SV control block. Configuration options include the reference to the dataset that defines the information contained in one message, the number of individual samples that are packed within one message, and the sampling rate.

While the flexibility makes the concept future proof, it adds configuration complexity. That is why the UCA users group has prepared the "Implementation guideline for digital interface to instrument transformers using IEC 61850-9-2." This implementation guideline is an agreement of the vendors participating in the UCA users group, how the first implementations of digital interfaces to instrument transformers will be. Basically, the implementation guideline is defining the following items:

- A dataset comprising the voltage and current information for the three phases and for neutral. That dataset corresponds to the concept of a merging unit (MU) as defined in IEC 60044-8.
- Two SV control blocks: a first one for a sample rate of 80 samples per period, where for each set of samples an individual message is sent and a second one for 256 samples per period, where 8 consecutive set of samples are transmitted in one message.
- The use of scaled integer values to represent the information including the specification of the scale factors for current and for voltage.

3.3.6.1 Process Level

Process level technology is a maturing technology. Designed primarily to interface with nonconventional CTs and VTs, a process level communication will also include "transitional" hardware that will interface with existing copper CTs and VTs. The benefits of the process near implementation of the IEC 61850-based technology include elimination of copper, the elimination of CT saturation, and avoidance of CT open circuits, which are a serious safety hazard.

With this solution, new designs become possible, where electronic transformers are used instead of conventional transformers in the switchyard. The voltage and current signals are captured at the primary side, converted to the optic signals by an MU, and transferred to the protection and control devices via optical fibers. This can lower the requirement of transformer insulation and reduce the conducted and radiated interference suffered in the analog signal transmitted through legacy wiring. Intelligent control units are used as an intermediate link to circuit breaker controls. The intelligent control unit also converts analog signals from primary devices (such as circuit breaker and switches) into digital signals and sends it to the protection and control devices via process bus. At the same time, the tripping and reclosing commands issued by protection and control devices will be converted into analog signals to control the primary equipment. Large amount of copper wiring between IEDs and primary devices in conventional substations are replaced by optical fibers.

3.3.6.2 Bay Level

All the IEDs in the control house fully support IEC 61850. Synchronous phasor measurements are realized by phasor measurement units (PMUs) or in protection IEDs. PMUs are used for wide area power system monitoring and control, improving state estimation and archiving more reliable system performance. GOOSE messaging and SV network over the process bus are used. The interoperation between IEDs is realized by GOOSE messages sent over the network. The Ethernet switch is used to process the message priority to realize the GOOSE exchange scheme between relays.

3.3.6.3 Station Level

At the station level, an MMS-based communications network is used. This also provides the communications link between SCADA, control centers, and IEDs located at the bay level.

3.3.6.4 IEC 61850 Benefits

High-speed peer-to-peer communications between IEDs connected to the substation LAN based on exchange of GOOSE messages can successfully be used to replace hardwiring for different protection and control applications. Sampled analog values communicated from MUs to different protection devices connected to the communications network replace the copper wiring between the instrument transformers in the substation yard and the IEDs. IEC 61850 is a communications standard that allows the development of new approaches for the design and refurbishment of substations. A new range of protection and control applications results in significant benefits compared to conventional hardwired solutions. It supports interoperability between devices from different manufacturers in the substation, which is required in order to improve the efficiency of microprocessor-based relays applications and implement new distributed functions.

Process-bus-based applications offer some important advantages over conventional hardwired analog circuits. The first very important one is the significant reduction in the cost of the system due to the fact that multiple copper cables are replaced with a small number of fiber optic cables. Using a process bus also results in the practical elimination of CT saturation because of the elimination of the current leads resistance. Process-bus-based solutions also improve the safety of the substation by eliminating one of the main safety-related problems—an open current circuit condition. Since the only current circuit is between the secondary of a current transformer and the input of the MU is located right next to it, the probability for an open current circuit condition is very small. It becomes nonexistent if optical current sensors are used. The process bus improves the flexibility of the protection, monitoring, and control systems. Since current circuits cannot be easily switched due to open circuit concerns, the application of bus differential protection, as well as some backup protection schemes, becomes more complicated. This is not an issue with process bus, because any changes will only require modifications in the subscription of the protection IEDs receiving the sampled analog values over IEC 61850 9-2.

IEC 61850-based substation systems provide some significant advantages over conventional protection and control systems used to perform the same functions in the substations:

- Reduced wiring, installation, maintenance, and commissioning costs
- Optimization possibilities in the design of the high voltage system in a substation
- Improved interoperability due to the use of standard high-speed communications between devices of different manufacturers over a standard communications interface
- Easy adaptation to changing configurations in the substation
- Practical elimination of CT saturation and open circuits
- Easier implementation of complex schemes and solutions as well as easier integration of new applications and IEDs by using GOOSE messages and SVs that are multicasted on the communications network and that the applications and IEDs can simply subscribe to

It has been shown that the greatest benefits of using IEC 61850 may not be found in initial deployment, but it will be IEC 61850's additional flexibility later in the substation life cycle that shows the greatest benefits. Table 3.6 shows the factors. Of the three factors for which IEC 61850 is believed to show a clear benefit, only the *configuration* benefits could be realized on the first installation by a utility.

The result is a significant improvement in configuration time as well as a reduction in the errors introduced by having to configure both the IED and server, as in a traditional approach. An expected 75% reduction in labor costs when configuring a substation represents a significant savings. For a more complex device that would normally take a day to configure, the savings could be even higher, perhaps approaching 90% (Figure 3.33).

3.3.7 IEC 61850-BASED SUBSTATION DESIGN

In a smart grid environment, availability of and access to information is key. Standards like IEC 61850 allow the definition of the available information and access to that information in a

Smart Grid Technologies

TABLE 3.6
Anticipated IEC 61850 Benefits

Description	Network	Legacy	Impact
Equipment purchase	$	$	–
Installation	$	$	0
Configuration	$$$	$	+
Equipment migration	$$$	$	+
Application additions	$$$	$	+

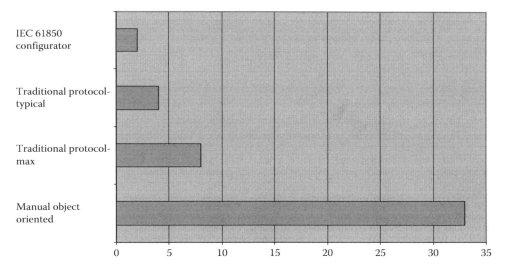

FIGURE 3.33 Approximate time (min) to configure an IEC 61850 client to communicate with a 200-point IED.

standardized way. IEC 61850 Communication Networks and Systems for Utility Automation is a standard for communications that creates an environment that will allow significant changes in the way the power system is protected and operated. In addition these concepts can also be used outside of the substation, allowing the implementation of wide area protection using standardized communications.

The IEC 61850 standard Communication Networks and Systems for Utility Automation allows the introduction of new designs for various functions, including protection inside and outside substations. The levels of functional integration and flexibility of communications-based solutions bring significant advantages in costs at various levels of the power system. This integration affects not only the design of the substation but almost every component and/or system in it such as protection, monitoring, and control by replacing the hardwired interfaces with communication links. Furthermore, the design of the high voltage installations and networks can be reconsidered regarding the number and the location of switchgear components necessary to perform the primary function of a substation in a high voltage network. The use of high-speed peer-to-peer communications using GOOSE messages and SVs from MUs allows for the introduction of distributed and wide area applications. In addition, the use of optical LANs leads in the direction of copperless substations.

3.3.7.1 Paradigm Shift in Substation Design

For many years, the current generation substation designs have been based on that functionality, and over time, we have developed several typical designs for the primary and secondary systems used in these substations. Examples of such typical schemes for the primary equipment are shown in Figure 3.34 and include the breaker and a half scheme, the double busbar scheme, the single busbar scheme, and the ring bus scheme. These schemes have been described and defined in many documents including Cigré Technical Brochure 069 General guidelines for the design of outdoor AC substations using fact controllers.

For the secondary equipment (protection, control, measurement, and monitoring), typical schemes have also been in use, but here we have seen more development in new concepts and philosophies. Typical concepts for secondary equipment include redundant protection for transmission

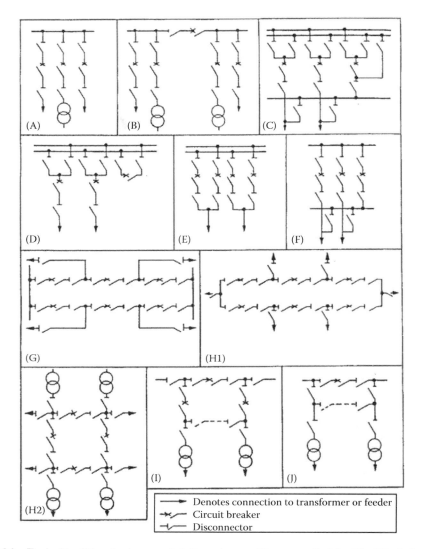

FIGURE 3.34 Typical traditional primary substation schemes. (From page 8 of the Cigré Technical Brochure 069, General guidelines for the design of outdoor AC substations using FACTS controllers. Coyright Cigré.) (Labels A–J refer to different substation primary plant topologies as they are used in the Cigré report that this comes from.)

Smart Grid Technologies

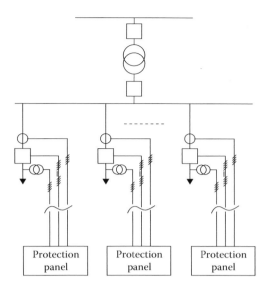

FIGURE 3.35 Typical conventional substation design. (From Apostolov, A. and Janssen, M., IEC 61850 Impact on Substation Design, paper number 0633. © Copyright 2008 IEEE.)

system using different operating principles and manufacturers and separate systems for control, measurements, monitoring, data acquisition, operation, etc. At distribution, integrated protection and control at feeder level is a common solution. In general, it can be said that the concepts used for the secondary systems have been based on the primary designs and the way the utility wants to control, protect, and monitor these systems.

In general the existing or conventional substations are designed using standard design procedures for high voltage switchgear in combination with copper cables for all interfaces between primary and secondary equipment.

Several different types of circuits are used in the substation:

- Analog (current and voltage)
- Binary—protection and control signals
- Power supply—DC or AC

A typical conventional substation design is shown in Figure 3.35.

Depending on the size of the substation, the location of the switchgear components, and the complexity of the protection and control system, there very often are a huge number of cables with different lengths and sizes that need to be designed, installed, commissioned, tested, and maintained.

A typical conventional substation has multiple instrument transformers and circuit breakers associated with the protection, control, and monitoring and other devices being connected from the switchyard to a control house or building with the individual equipment panels.

These cables are cut to a specific length and bundled, which makes any required future modification very labor intensive. This is especially true in the process of refurbishing old substations where the cable insulation is starting to fail.

The large amount of copper cables and the distances that they need to cover to provide the interface between the different devices expose them to the impact of electromagnetic transients and the possibility for damages as a result of equipment failure or other events.

The design of a conventional substation needs to take into consideration the resistance of the cables in the process of selecting instrument transformers and protection equipment, as well as their connection to the instrument transformers and between themselves. The issues of CT saturation are of special importance to the operation of protection relays under maximum fault conditions. Also ferroresonance in voltage transformers has to be considered in relation to the correct operation of the protection and control systems.

Failures in the cables in the substation may lead to misoperation of protection or other devices and can represent a safety issue. In addition open CT circuits, especially when it occurs while the primary winding is energized, can cause severe safety issues as the induced secondary e.m.f. can be high enough to present a danger to people's life and equipment insulation.

The earlier discussion is definitely not a complete list of all the issues that need to be taken into consideration in the design of a conventional substation. It provides some examples that will help better understand the impact of IEC 61850 in the substation.

In order to take full advantage of any new technology, it necessary to understand what it provides. The next part of this chapter gives a short summary of some of the key concepts of the standard that have the most significant impact on the substation design.

3.3.7.2 IEC 61850 Substation Hierarchy

In a smart grid environment, availability of and access to information is key. Standards like IEC 61850 allow the definition of the available information and access to that information in a standardized way.

The IEC 61850 standard Communication Networks and Systems for Utility Automation allows the introduction of new designs for various functions, including protection inside and outside substations. The levels of functional integration and flexibility of communications-based solutions bring significant advantages in costs at various levels of the power system. This integration affects not only the design of the substation but almost every component and/or system in it such as protection, monitoring, and control by replacing the hardwired interfaces with communications links. Furthermore, the design of the high voltage installations and networks can be reconsidered regarding the number and the location of switchgear components necessary to perform the primary function of a substation in a high voltage network. The use of high-speed peer-to-peer communications using GOOSE messages and SVs from MUs allows for the introduction of distributed and wide area applications. In addition, the use of optical LANs leads in the direction of copperless substations.

The development of different solutions in the substation protection and control system is possible only when there is good understanding of both the problem domain and the IEC 61850 standard. The modeling approach of IEC 61850 supports different solutions from centralized to distributed functions. The latter is one of the key elements of the standard that allows for utilities to rethink and optimize their substation designs.

A function in an IEC 61850-based integrated protection and control system can be local to a specific primary device (distribution feeder, transformer, etc.) or distributed and based on communications between two or more IEDs over the substation LAN.

Considering the requirements for the reliability, availability, and maintainability of functions, it is clear that in conventional systems numerous primary and backup devices need to be installed and wired to the substation. The equipment as well as the equipment that they interface with must then be tested and maintained.

The interface requirements of many of these devices differ. As a result specific multicore instrument transformers were developed that allow for accurate metering of the energy or other system parameters on the one hand and provide a high dynamic range used by, for example, protection devices.

With the introduction of IEC 61850, different interfaces have been defined that can be used by substation applications using dedicated or shared physical connections—the communications links between the physical devices. The allocation of functions between different physical devices defines the requirements for the physical interfaces and, in some cases, may be implemented in more than one physical LAN or by applying multiple virtual network on a physical infrastructure.

Smart Grid Technologies

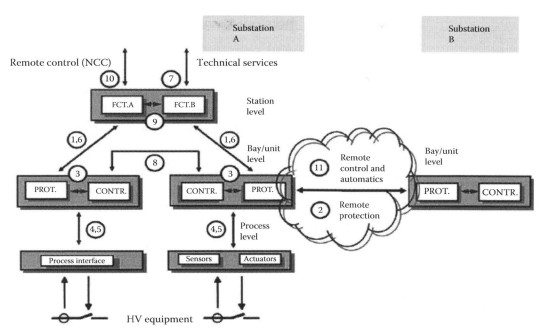

FIGURE 3.36 Logical interfaces in IEC 61850. (From IEC TR 61850–1, Copyright IEC.)

The functions in the substation can be distributed between IEDs on the same or on different levels of the substation functional hierarchy—station, bay, or process as shown in Figure 3.36.

A significant improvement in functionality and reduction of the cost of integrated substation protection and control systems can be achieved based on the IEC 61850-based communications as described in the following.

One example where a major change in substation is expected is at the process level of the substation. The use of nonconventional and/or conventional instrument transformers with digital interface based on IEC 61850-9-2 or the implementation guideline IEC 61850-9-2 LE results in improvements and can help eliminate issues related to the conflicting requirements of protection and metering IEDs as well as alleviate some of the safety risks associated with current and voltage transformers.

The interface of the instrument transformers (both conventional and nonconventional) with different types of substation protection, control, monitoring, and recording equipment as defined in IEC 61850 is through a device called an MU. The definition of an MU in IEC 61850 is as follows:

> Merging unit: interface unit that accepts multiple analog CT/VT and binary inputs and produces multiple time synchronized serial unidirectional multi-drop digital point to point outputs to provide data communication via the logical interfaces 4 and 5.

MUs can have the following functionality:

- Signal processing of all sensors—conventional or nonconventional
- Synchronization of all measurements—three currents and three voltages
- Analog interface—high- and low-level signals
- Digital interface—IEC 60044-8 or IEC 61850-9-2

It is important to be able to interface with both conventional and nonconventional sensors in order to allow the implementation of the system in existing or new substations.

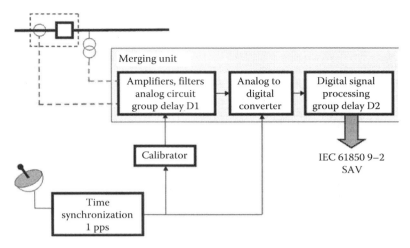

FIGURE 3.37 Concept of the MU. (From Apostolov, A. and Janssen, M., IEC 61850 Impact on Substation Design, paper number 0633. © Copyright 2008 IEEE.)

The MU has similar elements as can be seen from Figure 3.37 as a typical analog input module of a conventional protection or multifunctional IED. The difference is that in this case the substation LAN performs as the digital data bus between the input module and the protection or functions in the device. They are located in different devices, just representing the typical IEC 61850 distributed functionality.

Depending on the specific requirements of the substation, different communications architectures can be chosen as described hereafter.

IEC 61850 is being implemented gradually by starting with adaptation of existing IEDs to support the new communications standard over the station bus and at the same time introducing some first process-bus-based solutions.

3.3.7.3 IEC 61850 Substation Architectures

IEC 61850 is being implemented gradually by starting with adaptation of existing IEDs to support the new communications standard over the station bus and at the same time introducing some first process-bus-based solutions.

3.3.7.4 Station-Bus-Based Architecture

The functional hierarchy of station-bus-based architectures is shown in Figure 3.38. It represents a partial implementation of IEC 61850 in combination with conventional techniques and designs and brings some of the benefits that the IEC 61850 standard offers.

The current and voltage inputs of the IEDs (protection, control, monitoring, or recording) at the bottom of the functional hierarchy are conventional and wired to the secondary side of the substation instrument transformers using copper cables.

The aforementioned architecture however does offer significant advantages compared to conventional hardwired systems. It allows for the design and implementation of different protection schemes that in a conventional system require significant number of cross wired binary inputs and outputs. This is especially important in large substations with multiple distribution feeders connected to the same medium voltage bus where the number of available relay inputs and outputs in the protection IEDs might be the limiting factor in a protection scheme application. Some examples of such schemes are a distribution bus protection based on the overcurrent blocking principle, breaker failure protection, trip acceleration schemes, or a sympathetic trip protection.

Smart Grid Technologies 149

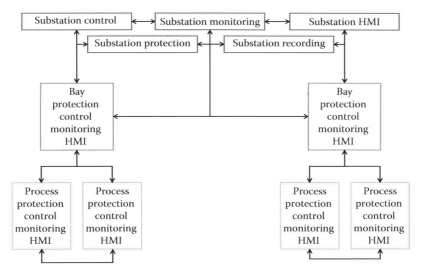

FIGURE 3.38 Station bus functional architecture. (© Copyright 2012 Marco Janssen. All rights reserved.)

The relay that detects the feeder fault sends a GOOSE message over the station bus to all other relays connected to the distribution bus, indicating that it has issued a trip signal to clear the fault. This can be considered as a blocking signal for all other relays on the bus. The only requirement for the scheme implementation is that the relays connected to feeders on the same distribution bus have to subscribe to receive the GOOSE messages from all other IEDs connected to the same distribution bus.

The reliability of GOOSE-based schemes is achieved through the repetition of the messages with increased time intervals until a user-defined time is reached. The latest state is then repeated until a new change of state results in sending of a new GOOSE message. This is shown in Figure 3.39.

The repetition mechanism does not only limit the risk that the signal is going to be missed by a subscribing relay. It also provides means for the continuous monitoring of the virtual wiring between the different relays participating in a distributed protection application. Any problem in a device or in the communications will immediately, within the limits of the maximum repetition time interval, be detected and an alarm will be generated and/or an action will be initiated to resolve the problem. This is not possible in conventional hardwired schemes where problems in the wiring or in relay inputs and outputs can only be detected through scheduled maintenance.

One of the key requirements for the application of distributed functions using GOOSE messages is that the total scheme operating time is similar to or better than the time of a hardwired conventional scheme. If the different factors that determine the operating time of a critical protection

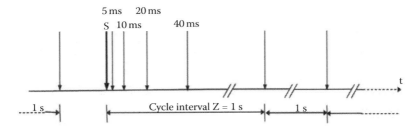

FIGURE 3.39 GOOSE message repetition mechanism. (From IEC TR 61850, Copyright IEC.)

scheme such as breaker failure protection are analyzed, it is clear that it requires a relay to initiate the breaker failure protection through a relay output wired into an input. The relay output typically has an operating time of 3–4 ms and it is not unusual that the input may include some filtering in order to prevent an undesired initiation of this critical function.

As a result, in a conventional scheme, the time over the simple hardwired interface, being the transmission time between the two functions, will be between 0.5 and 0.75 cycles—longer than the required 0.25 cycles defined for critical protection applications in IEC 61850-based systems.

Another significant advantage of the GOOSE-based solutions is the improved flexibility of the protection and control schemes. Making changes to conventional wiring is very labor intensive and time consuming, while changes of the "virtual wiring" provided by IEC 61850 peer-to-peer communications require only changes in the system configuration using the substation configuration language (SCL)-based engineering tools.

3.3.7.5 Station and Process Bus Architecture

Full advantage of all the features available in the new communications standard can be taken if both the station and process bus are used. Figure 3.40 shows the functional hierarchy of such a system.

IEC 61850 communications-based distributed applications involve several different devices connected to a substation LAN. MUs will process the sensor inputs, generate the SVs for the three phase and neutral currents and voltages, format a communications message, and multicast it on the substation LAN so that it can be received and used by all the IEDs that need it to perform their functions. This "one to many" principle similar to that used to distribute the GOOSE messages provides significant advantages as it not only eliminates current and voltage transformer wiring but it also supports the addition of new ideas and/or applications using the SVs in a later stage as these can simply subscribe to receive the same sample stream.

Another device, the IO unit (IOU) will process the status inputs, generate status data, format a communications message, and multicast it on the substation LAN using GOOSE messages.

All multifunctional IEDs will receive the SVs messages as well as the binary status messages. The ones that have subscribed to these data then process the data, make a decision, and operate by sending another GOOSE message to trip the breaker or perform any other required action.

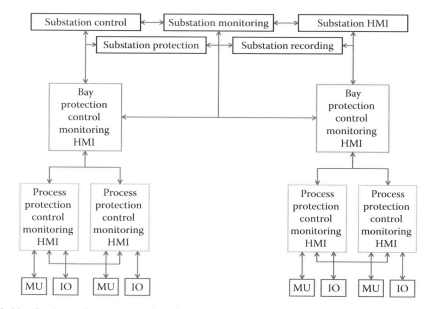

FIGURE 3.40 Station and process bus functional architecture. (© Copyright 2012 Marco Janssen. All rights reserved.)

Smart Grid Technologies 151

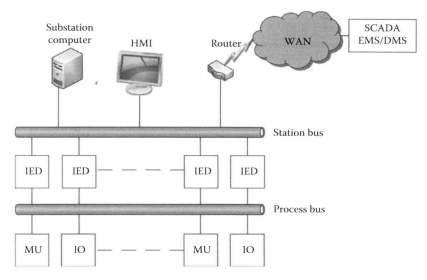

FIGURE 3.41 Communications architecture for process and station bus. (© Copyright 2012 Marco Janssen. All rights reserved.)

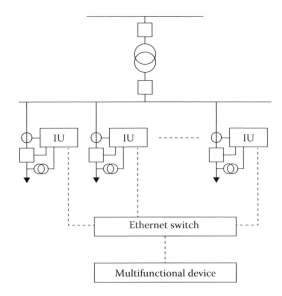

FIGURE 3.42 Alternative substation design. (From Apostolov, A. and Janssen, M., IEC 61850 Impact on Substation Design, paper number 0633. © Copyright 2008 IEEE.)

Figure 3.41 shows the simplified communications architecture of the complete implementation of IEC 61850. The number of switches for both the process and substation buses can be more than one depending on the size of the substation and the requirements for reliability, availability, and maintainability.

Figure 3.42 is an illustration of how the substation design changes when the full implementation of IEC 61850 takes place. All copper cables used for analog and binary signals exchange between devices are replaced by communications messages over fiber. If the DC circuits between the substation battery and the IEDs or breakers are put aside, the "copperless" substation is a fact. We then can even go a step further and combine all the functions necessary for multiple feeders into one multifunctional device, thus eliminating a significant amount of individual IEDs. Of course the opposite is also possible. Since all the information is available on a communication bus, we can choose to implement relatively simple or even single function devices that share their information on the network, thus creating a distributed function.

The next possible step when using station and process bus is the optimization of the switchgear. In order for the protection, control, and monitoring functions in a substation to operate correctly, several instrument transformers are placed throughout the high-voltage installation. However with the capability to send voltage and current measurements as SVs over a LAN, it is possible to eliminate some of these instrument transformers. One example is the voltage measurements needed by distance protections. Traditionally voltage transformers are installed in each outgoing feeder. However if voltage transformers are installed on the busbar, the voltage measurements can be transmitted over the LAN to each function requiring these measurements. These concepts are not new and have already been applied in conventional substations. In conventional substations, however, it requires large amounts of (long) cables and several auxiliary relays limiting or even eliminating the benefit of having less voltage transformers.

Process-bus-based applications offer important advantages over conventional hardwired analog circuits. The first very important one is the significant reduction in the cost of the system due to the fact that multiple copper cables are replaced with a small number of fiber optic cables.

Using a process bus also results in the practical elimination of CT saturation of conventional CTs because of the elimination of the current leads resistance. As the impedance of the MU current inputs is very small, this results in the significant reduction in the possibility for CT saturation and all associated with its protection issues. If nonconventional instrument transformers can be used in combination with the MUs and process bus, the issue of CT saturation will be eliminated completely as these nonconventional CTs do not use inductive circuits to transduce the current.

Process-bus-based solutions also improve the safety of the substation by eliminating one of the main safety-related problems—an open current circuit condition. Since the only current circuit is between the secondary of a current transformer and the input of the MU is located right next to it, the probability for an open current circuit condition is very small. It becomes nonexistent if optical current sensors are used.

Last, but not least, the process bus improves the flexibility of the protection, monitoring, and control systems. Since current circuits cannot be easily switched due to open circuit concerns, the application of bus differential protection, as well as some backup protection schemes, becomes more complicated. This is not an issue with process bus because any changes will only require modifications in the subscription of the protection IEDs receiving the sampled analog values over IEC 61850 9-2.

3.3.8 ROLE OF SUBSTATIONS IN SMART GRID

Substations in a smart grid will move beyond basic protection and traditional automation schemes to bring complexity around distributed functional and communications architectures, more advanced local analytics, and data management. There will be a migration of intelligence from the traditional centralized functions and decisions at the energy management and DMS level to the substations to enhance reliability, security, and responsiveness of the T&D system. The enterprise system applications will become more advanced in being able to coordinate the distributed intelligence in the substation and feeders in the field to ensure control area and system-wide coordination and efficiency.

The integration of a relatively large scale of new generation and active load technologies into the electric grid introduces real-time system control and operational challenges around reliability and security of the power supply. These challenges, if not addressed properly, will result in degradation of service, diminished asset service life, and unexpected grid failures, which will impact the financial performance of the utility's business operations and public relationship image. If these challenges are met effectively, optimal solutions can be realized by the utility to maximize return on investments in advanced technologies. To meet these needs, a number of challenges must be addressed:

- Very high numbers of operating contingencies different from "system as design" expectations
- High penetration of intermittent renewable and distributed energy resources, with their (current) characteristic of limited controllability and dispatchability
- PQ issues (voltage and frequency variation) that cannot be readily addressed by conventional solutions
- Highly distributed, advanced control and operations logic
- Slow response during quickly developing disturbances
- Volatility of generation and demand patterns and wholesale market demand elasticity
- Adaptability of advanced protection schemes to rapidly changing operational behavior due to the intermittent nature of renewable and DG resources

In addition, with wide deployment of smart grid, there will be an abundance of new operational and nonoperational devices and technologies connected to the wide area grid. The wide range of devices will include smart meters; advanced monitoring, protection, control, and automation; EV chargers; dispatchable and nondispatchable DG resources; energy storage; etc. Effective and real-time management and support of these devices will introduce enormous challenges for grid operations and maintenance. To effectively address all these challenges, it is necessary to engineer, design, and operate the electric grid with an overarching solution in mind, enabling overall system stability and integrity. A smart grid solution, from field devices to the utility's control room, utilizing intelligent sensors and monitoring, advanced grid analytical and operational and nonoperational applications, comparative analysis and visualization, will enable wide area and real-time operational anomaly detection and system "health" predictability. These will allow for improved decision-making capabilities, PQ, and reliability. An integrated approach will also help to improve situational awareness, marginal stress evaluation, and congestion management and recommend corrective action to effectively manage high penetration of new alternative generation resources and maximize overall grid stability

Some expected smart substation transformations are summarized later.

3.3.8.1 Engineering and Design

Future substation designs will be driven by current and new well-developed technologies and standards, as well as some new methodologies which are different from the existing philosophy. The design requirements for the next-generation substations will be based on the total cost of ownership and shall be aimed at either cost reduction while maintaining the same technical performance or performance improvement while assuring a positive cost benefit ratio. Based on these considerations, smart substation design may take the form of (a) retrofitting existing substations with a major replacement of the legacy equipment with minimal disruption to the continuity of the services, (b) deploying brand-new substation designs using the latest off-the-shelf technologies, or (c) greenfield substation design that takes energy market participation, profit optimization, and system operation risk reduction into combined consideration.

Designing the next-generation substations will require an excellent understanding of primary and secondary equipment in the substation, but also the role of the substation in the grid, the region, and the customers connected to it. Signals for monitoring and control will migrate from analog to digital, and the availability of new types of sensors, such as nonconventional current and voltage instrument transformers, will require shifting the engineering and design process from a T&D network focus to also include the substation information and communications architecture. This will require a better understanding of communications networks, data storage, and data exchange needs in the substation. As with other communications networks used in other process or time-critical industries, redundancy, security, and bandwidth are an essential part of the design process. Smart substations will require protocols specific to the needs of electric utilities while ensuring interconnectivity and interoperability of the protection, monitoring, control, and data acquisition devices. One approach to overcoming these challenges is to modify the engineering and design documentation process so that it includes detailed communication schematics and logic charts depicting this virtualized circuitry and data communications pathways.

3.3.8.2 Information Infrastructure

Advances in processing technology have been a major enabler of smarter substations with the cost-effective digitization of protection, monitoring, and control devices in the substation. Digitization of substation devices has also enabled the increase in control and automation functionality and, with it, the proliferation of real-time operational and nonoperational data available in the substation. The availability of the large amounts of data has driven the need for higher speed communications within the substation as well as between the substation and feeder devices and upstream from the substation to SCADA systems and other enterprise applications, such as outage management and asset management. The key is to filter and process these data so that meaningful information from the T&D system can be made available on a timely basis to appropriate users of the data, such as operations, planning, asset maintenance, and other utility enterprise applications.

Central to the smart grid concept is design and deployment of a two-way communications system linking the central office to the substations, intelligent network devices, and ultimately to the customer meter. This communications system is of paramount importance and serves as the nervous system of the smart grid. This communications system will use a variety of technologies ranging from wireless, RF, and broadband over power line (BPL) most likely all within the same utility. The management of this communications network will be new and challenging to many utilities and will require new engineering and asset management applications. Enhanced security will be required for field communications, application interfaces, and user access. An advanced EMS and DMS will need to include data security servers to ensure secure communications with field devices and secure data exchange with other applications. The use of IP-based communications protocols will allow utilities to take advantage of commercially available and open-standard solutions for securing network and interface communications.

IEC 61850 will greatly improve the way we communicate between devices. For the first time, vendors and utilities have agreed upon an international communications standard. This will allow an unprecedented level of interoperability between devices of multiple vendors in a seamless fashion. IEC 61850 supports both client/server communications as well as peer-to-peer communications. The IEC process bus will allow for communication to the next generation of smart sensors. The self-description feature of IEC 61850 will greatly reduce configurations costs, and the interoperable engineering process will allow for the reuse of solutions across multiple platforms. Also because of a single standard for all devices training, engineerings and commissioning costs can be greatly reduced.

3.3.8.3 Operation and Maintenance

The challenge of operations and maintenance in advance substations with smart devices is usually one of acceptance by personnel. This is a critical part of the change management process. Increased amounts of data from smart substations will increase the amount of information available to system operators to improve control of the T&D network and respond to system events. Advanced data integration and automation applications in the substation will be able to provide a faster response to changing network conditions and events and therefore reduce the burden on system operators, especially during multiple or major system events. For example, after a fault on a distribution feeder, instead of presenting the system operator with a lockout alarm, accompanied by associated low volts, fault passage indications, battery alarms and so on, leaving it up to the operator to drill down, diagnose, and work out a restoration strategy, the applications will instead notify the operator that a fault has occurred and analysis and restoration is in progress in that area. The system will then analyze the scope of the fault using the information available; tracing the current network model; identifying current relevant safety documents, operational restrictions, and sensitive customers; and locating the fault using data from the field. The master system automatically runs load flow studies identifying current loading, available capacities, and possible weaknesses, using this information to develop a restoration strategy. The system then attempts an isolation of the fault and maximum restoration of customers with safe load transfers, potentially involving multilevel feeder reconfiguration to prevent cascading overloads to adjacent circuits. Once the reconfiguration is complete, the system can alert the operator to the outcome and even automatically dispatch the most appropriate crew to the identified faulted section.

3.3.8.4 Enterprise Integration

Enterprise integration is an essential component of the smart grid architecture. To increase the value of an integrated smart grid solution, the smart substation will need to interface and share data with numerous other applications. For example, building on the benefits of an AMI with extensive communication coverage across the distribution system and obtaining operational data from the customer point of delivery (such as voltage, power factor, loss of supply, etc.) help to improve outage management and IVVC implementation locally at the substation level. More data available from substations will also allow more accurate modeling and real-time analysis of the distribution system and will enable optimization algorithms to run, reducing peak load and deferring investment in transmission and distribution assets. By collecting and analyzing nonoperational data, such as key asset performance information, sophisticated computer-based models can be used to assess current performance and predict possible failures of substation equipment. This process combined with other operational systems, such as mobile workforce management, will significantly change the maintenance regime for the T&D system.

3.3.8.5 Testing and Commissioning

The challenge of commissioning a next-generation substation is that traditional test procedures cannot adequately test the virtual circuitry. The best way to overcome this challenge is to use a system test methodology, where functions are tested end to end as part of the virtual system. This allows performance and behavior of the control system to be objectively measured and validated. Significant changes will also be seen in the area of substation interaction and automation database management and the reduction of configuration costs. There is currently work under way to harmonize the EPRI CIM model and enterprise service bus IEC 61968 standards with the substation IEC 61850 protocol standards. Bringing these standards together will greatly reduce the costs of configuring and maintaining a master station through plug and play compatibility and database self-description.

3.4 TRANSMISSION SYSTEMS

Transmission systems are the bulk power delivery systems of electric utilities; they carry millions of megawatts of energy each day. There will be an increased focus on the transmission level of the smart grid as transmission systems become more complex and more interconnected and serve as the power delivery system for more renewable energy sources. There are several monitoring and control technologies that ensure efficient operation, safety, and reliability at the transmission level of the grid and are key to any smart grid deployment. While some smart grid technologies for transmission systems may take proactive action to automatically control the network and have a localized effect at their point of connection, other technologies are transmission systems in their own right and may deliver energy from one location in a precisely controlled manner to the load center. These intelligent systems are able to offer dynamic control of not only power flow but many other aspects of a stable network, including voltage, reactive power, frequency, etc.

3.4.1 Energy Management Systems

Jay Giri and Thomas Morris

3.4.1.1 History of Energy Management Systems

Operating the electric grid at close to normal frequency, without causing any unexpected disconnections of load or generation, is known as maintaining electrical integrity or "normal synchronous operation." The first centralized control centers designed to maintain the integrity of the electric grid were implemented in the 1950s.

Control centers use a software and hardware system called an energy management system (EMS). Based on a centralized command and control paradigm, the EMS has evolved over the past six decades into much larger and more complex systems through computer automation. But the newer systems have the same simple mission as the original system: "Keep power available on at all times."

An EMS monitors and manages flows in the higher-voltage transmission network. A distribution management system (DMS) monitors and manages flows in the lower-voltage distribution network.

> *Real-time monitoring of grid conditions.* The first EMS application placed in control centers across the country was known as the supervisory control and data acquisition (SCADA) system. SCADA allows electric system operators to visually monitor grid conditions from a central location and to take control and remedial actions remotely via the SCADA system if adverse conditions are detected. The initial SCADA systems were hardwired analog systems.
>
> *Maintaining system frequency.* The next function implemented at control centers was load frequency control (LFC). The objective of LFC is to automatically maintain system frequency as load changes by changing generation output accordingly. In the early implementations of LFC, the control center operator visually monitored the system frequency measurement and periodically sent incremental change signals to generators via analog-wired connections or by placing phone calls to generating plant operators to keep generation output close to system load demand. Later, as analog systems transitioned to digital, LFC became the first automated application to help the control center operator keep power available at all times.
>
> *Sharing electricity with neighbors.* The next progression in system monitoring and control was interconnecting one power utility with neighboring utilities to increase overall grid reliability by allowing power sharing during emergencies and to exchange cheaper power during normal operations.
>
> *Modern control centers.* Figure 3.43 shows the suite of real-time and off-line functions that comprise a modern control center.

3.4.1.2 Current EMS Technology

There are over 3000 electric service territories in the United States responsible for managing their portion of the electric grid. Most of the very high voltage transmission substations in the United States have sensors and meters that monitor real-time operating conditions and have the means to remotely operate transmission equipment, such as circuit breakers and transformer tap changers. Less than 25% of the distribution substations have any remote monitoring and control capability, and the final supply to the end user typically has no technology at all. However, this is changing

Smart Grid Technologies

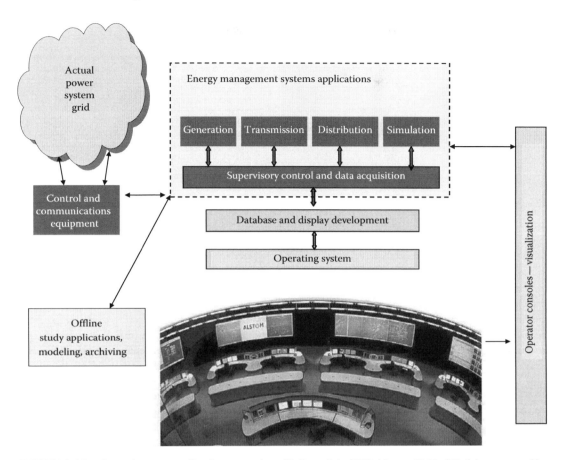

FIGURE 3.43 Control center applications overview. (© Copyright 2012 Alstom Grid. All rights reserved.)

with technology evolution and the reduction in monitoring and control device costs. Smart grid has been driving increased implementations of intelligent residential meters and other technologies and applications that will help drive more visibility of the T&D network through SCADA.

Figure 3.44 shows a typical modern-day EMS control center environment with the different display screens the operator uses at the console to monitor and control grid conditions. The control center consists of many such operator consoles, as well as large wallboards or digital displays that provide a bird's eye view of the entire system. The operators' responsibilities are to monitor data on their consoles, coordinate with other operators within their control center, coordinate with plant operators, and periodically exchange information with neighboring system EMS operators. The majority of the time, the grid is relatively quiescent with no adverse conditions. But when a disturbance suddenly occurs, the operators each need to perform their specific individual tasks and need to coordinate with other operators in the control center in order to use their collective expertise to identify specific actions that may need to be taken to mitigate the impact of the disturbance.

3.4.1.3 Advances in Energy Management Systems for the Smart Grid

3.4.1.3.1 Grid Operator Visualization Advances

Timely visualization of real-time grid conditions is essential for successful grid operations.

In the aftermath of the 1965 blackout of the northeast United States and Canada, the findings from the blackout report included the following: "control centers should be equipped with display

FIGURE 3.44 EMS control center operator console. (© Copyright 2012 Alstom Grid. All rights reserved.)

and recording equipment which provide operators with as clear a picture of system conditions as possible." Since then many more blackouts have occurred, small and large, around the world, and in almost all cases, improvements in visibility of grid conditions were identified as one of the primary recommendations.

On August 14, 2003, the largest blackout in the history of the North American power grid occurred. Subsequently, numerous experts from across the industry were brought together to create a blackout investigation team. A primary objective of this team was to perform in-depth postevent analyses to identify the root causes and, more importantly, to make recommendations on what could be done to prevent future occurrences of such events. The report (the United States–Canada, 2004) identified four root causes: inadequate system understanding, *inadequate situational awareness* (SA), inadequate tree trimming, and inadequate reliability coordinator diagnostic support. This report gave a sudden new prominence to the term "situation awareness" or "situational awareness."

There are several definitions of SA. Very simply, SA means to be constantly aware of the health of changing power system grid conditions. Other definitions include "Being cognizant of the current power system state and the potential imminent impact on grid operations" and "The perception of the elements in the environment within a volume of time and space, the comprehension of their meaning, and projection of their status in the near future" [1].

An essential aspect of SA for grid operations is being able to extract and concisely present the information contained in the vast amount of ever-changing grid conditions. An advanced visualization framework (AVF) is necessary to be able to present real-time conditions in a timely, prompt manner. AVF needs to provide the ability to efficiently navigate and drill down, to discover additional information, such as the specific location of the problem. More importantly, AVF needs to provide the ability to identify and implement corrective actions in order to mitigate any risks to successful grid operations. Operators do not just want to only know that there is a problem now or that a problem is looming in the immediate horizon. They also want to know how to fix the problem.

A number of visualization products have been developed over the past decade. These include Powerworld [2], RTDMS [3], Space-Time Insight, and so on.

The following are examples of currently available ALSTOM technology.

A frequently cited human limitation has been described as Miller's magical number seven, plus or minus two [4]. Miller's observation was that humans have a limited capacity for the number of items or "chunks" of information that they can maintain in their working memory. Therefore, as increasing volumes of data are streamed into the control center, one must keep in mind that there is a limit on how much of these data are actually useful to the operator. As per Miller, the operator can typically handle only five to nine such "chunks" of information. The limitation with the traditional display technologies has been that they approach the problem by "rolling up" (aggregating) the data and then allowing the operator to "drill down" for details. The result is a time-consuming and cognitively expensive process [5].

Smart Grid Technologies 159

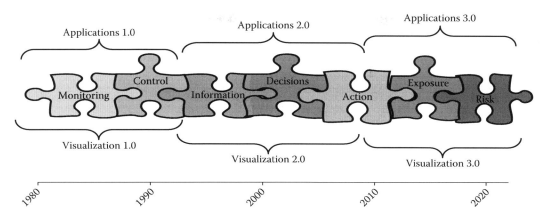

FIGURE 3.45 Evolution of EMS SA capabilities. (© Copyright 2012 Alstom Grid. All rights reserved.)

The challenge is to translate the ever-increasing deluge of measurement data being brought into the control center into useful, bite-size, digestible "chunks" of information. This has been the primary objective of all the recent SA developments for grid operations: "converting vast volumes of grid data into useful information and showing it on a display screen."

As the saying goes, "a picture is worth a thousand words." More importantly, the *correct picture is worth a million words!* What this means is that providing the grid operator with a concise depiction of voluminous data is meaningful, whereas providing a depiction of voluminous data that *needs immediate operator attention* is immensely more meaningful! This is the objective of an advanced, intelligent SA, to provide timely information that may need prompt action, for current system conditions.

The way to develop the "correct" picture is to organize the visualization presentation around *operator goals*; that is, what is the task result the operator is seeking? Use-cases need to be developed to document the specific actions an operator takes in order to reach a specific goal. These use-cases can then be used to develop efficient navigation capabilities to quickly go from receipt of an alert to analyzing the "correct picture" and determining the appropriate course of action.

These are the requirements upon which today's advanced visualization and SA capabilities have been developed. SA capabilities continue to be developed and enhanced to help improve grid operations. Figure 3.45 shows how SA has evolved with analytical tools over the past few decades and what is foreseen for the immediate future.

Generation 1 visualization and applications were focused on monitoring and control; these capabilities were developed in the 1980s and 1990s. Generation 2 focused on creating information from data to facilitate decision making in order to take corrective action; these capabilities were developed in the last two decades and are operational in many control centers around the world. Generation 3 is foreseen to focus on developing real-time measures to determine exposure and associated risk related to ensuring integrity of the grid. This generation will likely be focused on stochastic analytics, as well as heuristic and intelligent systems, for the development of advanced risk management and mitigation applications and visualization capabilities. These developments will be aided by ongoing technology advances such as subsecond, synchronous measurements, coupled with fast-acting, subsecond controllers.

3.4.1.3.2 Decision Support Systems
Most control center operator decisions today are essentially *reactive*. Current information, as well as some recent history, is used to reactively make an assessment of the current state and its vulnerability. Operators then extrapolate from current conditions and postulate future conditions based on personal experience and planned forecast schedules.

The next step is to help operators make decisions that are *preventive*. Once there is confidence in the ability to make reactive decisions, operators will need to rely on "what-if" analytical tools to be able to make decisions that will prevent adverse conditions if a specific contingency or disturbance were to occur. The focus therefore shifts from "problem analysis" (reactive) to "decision making" (preventive).

The industry trend next foresees *predictive* decision making, and in the future, decisions will be *proactive*. These types of decision-making process are the foundation of a decision support system (DSS) that will be essential to handle operation of smarter grids with increasing complexity and more diverse generation and load types. The DSS will use more accurate forecast information and more advanced analytical tools to be able to confidently predict system conditions and use what-if scenarios to be able to take action now in order to preclude possible problematic scenarios in the future. The components of DSS include the following:

- AVF
- Geospatial views of the grid
- Dynamic dashboards generated on demand
- Holistic views combining data from multiple diverse sources
- Use-case analysis to enhance ergonomics
- Advanced, fast, alert systems
- Root-cause analysis to quickly identify sources of problems
- Diagnostic tools that recommend corrective actions
- Look-ahead analysis to predict imminent system conditions

Figure 3.46 is an overview of a look-ahead analytical tool to help the operator make preventive decisions in order to obviate potential problems. The current system state is used to calculate projected future system states based on load forecasts, generation schedules, etc., to determine whether conditions in the future are safe. As the figure depicts, if the projections indicate a problem is imminent, the operator could then determine and implement an action, in advance, to ensure that the problem is avoided.

Figure 3.47 is one example of a DSS implemented by ALSTOM grid. It consists of a central DSS server and database. The DSS server provides information to a map board (for wide area visualization) and to operator workstations. A power system simulator is used as a look-ahead engine to forecast immediate future conditions. This look-ahead data, together with traditional EMS SCADA and state estimator data, are shown at the operator workstations to facilitate and improve timely, preventive decision making.

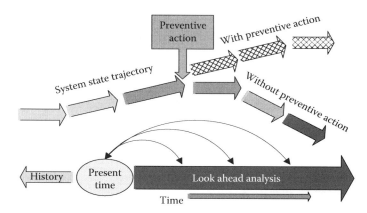

FIGURE 3.46 Look-ahead analysis for preventive control. (© Copyright 2012 Alstom Grid. All rights reserved.)

Smart Grid Technologies

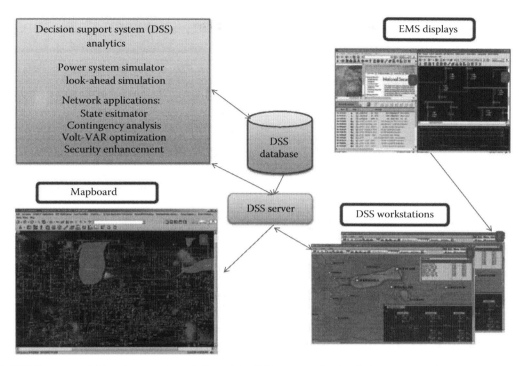

FIGURE 3.47 DSS implementation. (© Copyright 2012 Alstom Grid. All rights reserved.)

3.4.1.3.3 Control Centers of the Future

Automation of the grid will evolve toward more decentralized, intelligent, and localized control. This is the vision of smart grid at the transmission level. Evolution toward a "smarter" transmission system grid is imminent and will take many forms of predictive and corrective actions: from avoiding system congestion while maximizing efficiency and minimizing supply costs to reacting quickly to system faults while maintaining power to as many customers as possible. These are goals not only at the transmission level but also at the distribution level of the electric grid.

The future will likely see more generation sources closer to the load centers. Residential subdivisions could have their own local fuel cells supplying power to 20 or 30 households; this will result in the creation of local microgrids that will attempt to optimize benefits for that local area. This would reduce dependence on the transmission grid to transfer power from remote locations to populated load centers. As renewable energy costs become more competitive, there will be growth in generation sources such as wind power, solar cells, and possibly geothermal, tidal, and ocean power. Customers will be able to monitor the current price of electricity and decide whether to turn on the dishwasher or not using a "smart metering" scheme. This will flatten the utility's load demand profile and make generation dispatch more predictable.

In addition to more local generation, use of renewable energy and increased customer control, new types of measurements will be deployed aggressively worldwide. Already, globally synchronized measurements taken in the subsecond range, such as the phasor measurement units (PMUs) described earlier, are being used in control centers to facilitate earlier and faster detection of problems and to make it easier to assess conditions across the grid. Novel control center applications will be developed to use this new type of synchronized measurement technology to further improve the ability to maintain the integrity of the power system. These applications will also be able to identify disturbances, unplanned events, and stability problems at a much faster rate.

Continual development of control center applications and tools will play a critical role in driving smart grid advances in the transmission arena: wide area measurements and control, congestion alleviation, increased power delivery efficiency and reliability, and system-wide stability and security.

3.4.1.4 Control System Cybersecurity Considerations

3.4.1.4.1 Introduction

The North American Electric Reliability Corporation (NERC) Critical Infrastructure Protection (CIP) Standards 002–009 [6] require utilities and other responsible entities to place critical cyber assets within an electronic security perimeter. The electronic security perimeters must be subjected to vulnerability analyses, use access control technologies, and include systems to monitor and log the electronic security perimeter access. The Federal Energy Regulatory Commission (FERC) requires responsible entities involved in bulk electricity transmission to adhere to the NERC CIP standards. No such regulation exists for the electric distribution systems in the United States. Electronic perimeter security minimizes the threat of illicit network penetrations; however, persons with electronic access to control systems within the electronic security perimeter still remain a threat. Such persons include hackers who have penetrated the electronic security perimeter via external network connections, disgruntled insiders, and hackers who may penetrate wireless interconnection points within the electronic security perimeter.

SCADA systems remotely monitor and control grid physical assets. SCADA systems are used in power transmission and distribution systems for SA and control. Present-day SCADA systems are commonly connected to corporate intranets which may have connections to the Internet. SCADA communications protocols such as MODBUS, DNP3, and Allen Bradley's Ethernet Industrial Protocol lack authentication features to prove the origin or age of network traffic. This lack of authentication capability leads to the potential for network penetrators and disgruntled insiders to inject false data and false command packets into a SCADA system either through direct creation of such packets or replay attacks.

Modern power systems are being upgraded with the addition of PMUs and phasor data concentrators (PDCs) that facilitate wide area transmission system SA. The IEEE C37.118 protocol carries phasor measurements between PMUs and PDCs to historians and to EMSs. As with MODBUS and DNP3, the IEEE C37.118 protocol does not include a cryptographic digital signature. As such, a hacker or disgruntled insider may potentially inject false synchrophasor data into a transmission control system network without detection. Furthermore, the IEEE C37.118 protocol includes command frames used to configure PMUs and PDCs. False command frames may also be injected in a manner to similar to that used to inject false data frames.

IEC 61850 is one of the new protocol stacks development to increase interoperability among protection and control devices (IEDs—intelligent electronic devices) in the substation. However, the IEC 61850 protocol does not directly include cybersecurity features, though a separate IEC recommendation [7], IEC 62351, guides users on how to secure an IEC 61850 network installation. IEC 61850 offers features such as a standardized XML-based substation configuration language (SCL) for describing and configuring substation protection and control devices. IEC 61850 also offers standardized data-naming conventions for power system components. Such standardization greatly simplifies power system management and configuration, though it is also an enabler for hackers since it can minimize a hacker's learning curve. It is imperative that IEC 61850 installations adhere to the IEC 62351 recommendations.

3.4.1.4.2 Network Penetration Threats

There are three primary threats to process control systems: sensor measurement injection, command injection, and denial of service (DOS).

Sensor measurement injection attacks inject false sensor measurement data into a control system. Since control systems rely on feedback control loops before making control decisions, protecting the integrity of the sensor measurements is critical. Sensor measurement injection can be used by attackers to cause control algorithms to make misinformed decisions.

Command injection attacks inject false control commands into a control system. Control injection can be classified into two categories. First, human operators oversee control systems and occasionally intercede with supervisory control actions, such as opening a breaker. Hackers may attempt to inject false supervisory control actions into a control system network. Second, remote terminal units (RTUs) and IEDs protect, monitor, and control grid assets. The protection and control algorithms take the form of ladder logic, C code, and registers that perform calculations and hold key control parameters such as high and low limits, comparison, and gating control actions. Hackers can use command injection attacks to overwrite ladder logic, C code, and remote terminal register settings.

DOS attacks attempt to disrupt the communications link between the remote terminal and master terminal or human machine interface. Disrupting the communications link between master terminal or human machine interface and the remote terminal affects the feedback control loop and makes process control impossible. DOS attacks take many forms. A common DOS attack attempts to overwhelm hardware or software so that it is no longer responsive.

3.4.1.4.3 Isolating the Control System Network

Control systems should be isolated from corporate networks or LANs to minimize the potential of illicit penetration via wired networks. Corporate networks are used by most employees of a company. Corporate networks often include connections to the WWW (Internet), some via wireless LAN connections using IEEE 802.11 protocols. These portable nodes, such as laptop computers which come and go from the corporate network, generally allow the use of e-mail and typically see frequent use of USB disk drives. All of these characteristics lead to cybersecurity vulnerabilities and the need to isolate the control system network from the corporate network.

Connections to the WWW are a common point for external network penetration. Hackers commonly use port scanning tools to scan for TCP and UDP services. Contemporary port scanning equipment software such as NMAP [8] can target specific IP address ranges, find TCP and UDP services, identify service demon version numbers, and identify operating system name and version numbers. Armed with such information, hackers can use look-up tables available on the Internet to find exploits targeted at specific versions of specific network services running on specific operating system platforms. These exploits often allow hackers to bypass network defenses and penetrate the corporate network.

Wireless LANs on corporate networks are also a significant weak link. The IEEE 802.11 standards include multiple security substandards of which a predominant group has been cracked and is subject to penetration attacks [9]. The Wireless Equivalent Privacy (WEP) standard is vulnerable to exploit in less than 60 s. The TKIP portion of the Wireless Protected Access standard has also been cracked. These vulnerabilities allow an attacker in close proximity to a corporate network to penetrate the corporate network for further port scanning, eavesdropping, and network traffic injection.

Portable nodes such as laptop computers commonly travel between many networks. For instance, a corporate user may use his or her laptop at home, at the local coffee shop, or in the airport and then later connect the laptop to the corporate Internet. External networks such as the home, coffee shop, and airport networks often have less robust cybersecurity profiles and provide a convenient platform for injecting malware such as key loggers and root kits onto corporate laptops via viruses and worms. When the laptop returns to the corporate network infected with a root kit or key logger, it may offer a backdoor for hackers to then penetrate the corporate network for further port scanning, eavesdropping, and network traffic injection.

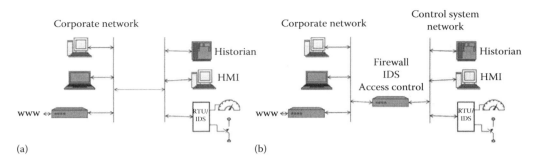

FIGURE 3.48 (a) Insecure versus (b) isolated control system.

Corporate employees almost always have e-mail access. E-mail is a very common platform for infecting computers in a corporate network. Hackers use spam e-mail to spread viruses which contain root kits and key loggers which may then be used to offer a backdoor to penetrate the corporate network for further port scanning, eavesdropping, and network traffic injection.

Another malware injection vector is through thumb drives. In April of 2010, the Industrial Control Systems Cyber Emergency Response Team (ICS-CERT) released an alert warning control system operators of the threat of USB drives with autorun features that can be used to inject malware. The advisory recommends control system operators disable CD-ROM autorun capability, establish strict policies for the use of USB drives on control system networks, and train users on the treatment of these drives.

The aforementioned penetration threats, connections to the WWW (or Internet), usage of wireless LANs, and the use of laptop computers, e-mails, and USB drives lead to the need for isolation of the control system network from the corporate network. Figure 3.48 shows two control system network architectures: an insecure architecture and an architecture secured via isolation.

Figure 3.48a shows seemingly separate corporate and control system networks. Often in such network arrangements, the corporate network and control system network will be separated via routers and often the two networks will be on separate virtual networks. However, if there is no mechanism in place to stop unauthorized network traffic from entering the control system network from the corporate network, penetrators can harm the control system via data or control injection attacks or via DOS attacks.

Figure 3.48b shows a control system network isolated from a corporate network. The diagram labels the box between the networks as a firewall, IDS or intrusion detection system (IDS), and access control system. The firewall can be used to limit access between the two networks. NAT (network address translation) firewalls hide the internal IP address of nodes on the control system network from nodes on the corporate network. This protects the control system nodes from port scanning attacks. Further, the firewalls can be used to scan the contents of network packets for signatures of known attacks. Firewalls can also be used as gateway devices which limit traffic to only certain applications on specific TCP and UDP ports.

Access control may reside in the firewall or may reside on a separate server within the control system network. Access control schemes limit network access to authorized individuals and systems. Access control schemes vary in strength. A simple access control scheme is the use of user ID and passwords. NERC CIP 007-3, Cyber Security—Systems Security Management, requires that when passwords must be used, passwords must be at least six characters long and include a mixture of letters, numbers, and special characters. The use of passwords for access control should be avoided wherever possible. Passwords systems are subject to dictionary attacks and other brute force attacks. Also password files are vulnerable to exploits, including password files in control systems [10].

A more robust access control system may use a public key infrastructure (PKI) to provide all control system users and systems with a certificate. PKI systems assign individual public/private key pairs to each user or system in a network. Certificates signed by a certification authority are used to communicate a user's or system's public key to other users or systems. When a user or system accesses a network device, a challenge response protocol can be used to allow the connecting user or system to authenticate identification by proving the user or system possesses the private key associated with the public key in their certificate. Systems within a PKI-protected network may also confirm certificate validity via an inquiry to the network certificate authority. Certificates may be revoked by a certificate authority and PKI certificates also have expiration dates. PKI provides a good means for adhering to NERC CIP requirements that require access control and encourage the use of individual user accounts with individual roles. Role-based access control allows each user to be assigned privileges, which match his or her work needs, without providing excess privileges that may allow a user to inadvertently or intentionally harm a system.

IDSs are used to monitor network activity for patterns related to cybersecurity threats. IDS systems may reside within a firewall or external to the firewall. Often, multiple types of IDSs are used on a single network. There are two basic types of IDS: signature-based IDS and statistical IDS.

A signature-based IDS scans network packets for signatures of known attacks. If a packet matches a known signature, the packet is flagged and an alert is generated. The alert may be audible, e-mail, or just written to a file for later review. Signature-based IDSs are generally deterministic, meaning that they will always detect an attack that matches a known signature. Because signature-based IDS monitors for exact pattern matches, they can be bypassed by small changes to previously known attacks. Also, signature-based IDSs cannot detect completely new attacks since by definition no signature will exist to match. Signature-based IDS systems are also relatively fast which can be important in real-time system applications. Some work has been done to develop signature-based IDS patterns for control systems using SNORT® (an open source network intrusion prevention and detection system [IDS/IPS]) for the MODBUS and DNP3 protocols [11]. Additionally, Oman and Phillips have used signature-based methods to detect SCADA cyber intrusions [12].

A statistical IDS estimates the probability that a network transaction or a group of network transactions are part of a cyber attack. The general idea of statistical IDSs is to attempt to detect intrusions that do not match a previously known intrusion signature yet are still different enough from normal traffic to warrant review. Most statistical IDSs are anomaly detectors that use data mining classifiers, such as neural networks or Bayesian networks, to classify network transactions as anomalous or normal. Many statistical IDS methodologies exist in both the research and practical domains. No statistical IDS is deterministic, all are probabilistic, meaning that all have less than 100% accuracy and all sometimes classify normal traffic as abnormal (a false-positive) and sometimes classify abnormal traffic as normal (a false negative). Control systems monitor and control critical physical processes, and therefore one of the most important cybersecurity criteria is availability. The control system must remain available for control and monitoring at all times, and a corollary to this is that control system cybersecurity solutions must do no harm to the control system. As such, IDS inaccuracies are problematic and lead to the need for statistical IDS alerts to always be sent to a human for validation before intrusion mitigation actions are taken.

Statistical IDSs are being developed specifically for use with control systems and the smart grid. These involve development of IDS inputs or (aka features) specific to control system applications and network protocols. The introduction of control system and smart-grid-specific IDS features will lead to more accurate IDSs.

3.4.1.4.4 Smart Grid Control System Cybersecurity Considerations
NERC CIP 005 requires utilities and other bulk energy system constituents to create an electronic security perimeter around critical cyber assets. Figure 3.49 is a diagram representing the primary assets found in a typical bulk electric transmission system after the addition of PMUs and PDCs. The assets are grouped into four major blocks: a control center, a PDC, and two transmission

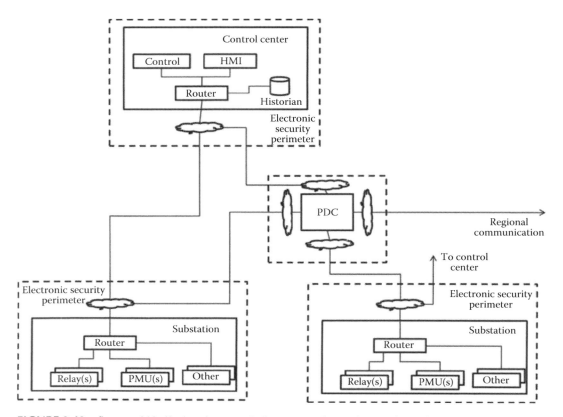

FIGURE 3.49 Smart grid bulk electric transmission system electronic security perimeters.

substations. Each block is electronically isolated by an electronic security perimeter. The electronic security perimeters are denoted as dashed lines around each isolated block. NERC CIP does not specify the methods for creating the electronic security perimeter. As such the methods vary widely and are therefore drawn as security clouds in Figure 3.49.

The security cloud should address the three basic cybersecurity core principles: confidentiality, integrity, and availability. Of these three cybersecurity core principles, it is generally agreed that they should be ranked by importance for the smart grid and for control systems as availability, integrity, and then confidentiality.

3.4.1.4.4.1 Availability The smart grid is considered critical infrastructure. Loss of SA over the bulk electric transmission systems may lead to incorrect control actions. Furthermore, loss of the ability to make control actions may lead to system damage or failure including ultimately blackouts. Such can lead to economic harm for the local or regional economy. There are two primary components to ensuring control system availability: IDSs and system design.

Loss of availability can come from DOS attack and command injection attacks that attempt to directly take control of the control system. Control injection attacks can be detected with IDSs and prevented with authentication techniques that are covered under the integrity discussion. DOS attacks attempt to deny network service by flooding a network with information at a rate faster than it can be processed. IDSs can detect and mitigate many DOS attacks.

Smart grid control systems should include IDS sensors to monitor network transactions at all entry points to the control system network or at points guaranteed to capture traffic from all entry points to the control system network. Entry points include local area network (LAN) drops, dial-up modems, wireless terminals, and connections to trusted neighbors such as regional operators and independent system operators, as well as connections to the corporate LAN.

System design also affects control system availability. First, many attack vectors can be stopped by eliminating unneeded network services. NERC CIP 007 requires bulk electric responsible entities to disable all network ports and services not used for normal or emergency operation. For instance, TCP and UDP each use port multiplexing to support many transport layer services. In total, TCP and UDP can support 64 K ports. The Internet Assigned Numbers Authority (IANA) assigns port numbers to frequently used services, such Telnet, SSH, and SMTP, etc. Many control system protocols have IANA-reserved port numbers, for example, MODBUS TCP servers listen on port 502, Allen Bradley EtherIP uses TCP port 44818, and UDP port 2222. DNP3 over TCP uses port 20000. Any unused port should not have a listening server running on any cyber system connected to the control system network. IDSs should monitor for activity on all TCP and UDP ports.

Smart grid control systems should be designed to allow each user to have a unique account ID and password. This requirement supports traceability and role-based access control. Traceability means that actions taken on the control system can be traced to an individual user. Role-based access control means that each user can be assigned roles and associated privileges (levels of authority). For example, a dispatcher may be allowed to open a breaker, while a less privileged user may not be able to open the same breaker. Legacy control system equipment may not support separate usernames. In this case, NERC CIP 007 requires entities to limit password knowledge to those individuals with a need to know. The security clouds shown in Figure 3.49 include access control features that limit access to an entire electronic security perimeter. These access control features can be certificate based, can support separate user ID and passwords for all users, and can support role-based access control.

The final design-related element to the availability principle may be obvious but is worth mentioning. All cybersecurity solutions should of course do no harm to the control system. The algorithms used to model control systems and make control decisions often have data age requirements. For instance, some EMS algorithms require data to be less than 2–4 ms old to support control decisions. "Bump-in-the-wire" (additional hardware and software in the communication link) for cybersecurity solutions, such as that shown in Figure 3.49, add latency to traffic delivery. The additional latency must never cause the system to become nonfunctional or uncontrollable. Furthermore, many proposed IDS systems include automated mitigation actions. These actions should only be taken when the IDS system is deterministic. Statistical IDS systems are probabilistic and therefore always have a probability of misclassifying network traffic. In such cases, the IDS may recommend mitigation actions, but a human should be kept in the control loop to validate mitigation recommendations.

3.4.1.4.4.2 Integrity The integrity cybersecurity principle is intended to protect network traffic from unauthorized modification. The most common method for insuring network traffic integrity is authenticating network traffic through the use of digital signature algorithms (DSAs). The MODBUS, DNP3, Allen Bradley EtherIP, IEEE C37.118, and IEC 61850 standards do not include features to authenticate network traffic. Authentication is left to the responsibility of a high layer protocol.

The security clouds in the network architecture shown in Figure 3.49 can digitally sign network traffic. Bump-in-the-wire solutions exist that can capture network traffic as it leaves an electronic security perimeter and append it with a digital signature. The security cloud in a receiving electronic security perimeter can validate the digital signature before forwarding the traffic to cyber systems inside the electronic security perimeter. The digital signatures can be based on multiple algorithms. FIPS 186 (the NIST Federal Information Processing Standards Publications for the Digital Signature Standard) specifies the NIST-recommended DSA. DSA uses public key cryptography techniques to sign network traffic. This method is often considered too slow and resource intensive for control systems; however, if cryptographic processors are used in the place of the security clouds in Figure 3.49, it is likely that DSA signing and validation can meet required latency targets for smart grid applications. The elliptic curve digital signature algorithm (ECDSA) is an alternative approach for network traffic authentication. The ECDSA is considered faster, uses smaller keys, and therefore has less storage than DSA. ECDSA is patented by CERTICOM, RSA, the U.S. National Security Agency, and Hewlett Packard. This may slow the adoption of ECDSA. A third alternative for authentication is the hashed message authentication algorithm (HMAC). HMAC is the least resource-intensive DSA of the three discussed here.

A key consideration when using digital signatures in control systems is the length of the signature and the latency added to the network traffic as a result of adding and validating the signature. This is especially pertinent for low data rate systems such as SCADA systems that often have data rates of 1,200–19,200 Bd. The Pacific Northwest National Laboratory [13] recently released a study that measures round-trip response times for DNP3 frames signed with various length HMAC authenticators. Response times varied according to the length of the authenticator, according to the data rate of the communications link and according to the type of bump-in-the-wire cyber system used to authenticate the DNP3 frames. The worst case was 1996 ms latency for 1200 Bd systems using industrial PCs to create and validate the 12 B HMAC signatures. The best case was 210 ms latency for 19,200 Bd system using industrial PCs to create and validate the 12 B HMAC signatures. Systems, such as YASIR [14], have been developed to minimize latency involved in adding digital signatures to MODBUS and DNP3 network frames.

Another important consideration when planning to use digital signatures is availability of hardware resources. Many RTUs are equipped with cryptographic resources and storage capabilities that may conflict with the needs of an adequate DSA. This presents three possibilities for control systems that support digital signatures for authentication. First, system designers may choose to use existing hardware and add a bump-in-the-wire cybersecurity solution (to sign and validate network traffic). Second, system designers may choose to upgrade existing remote or master terminal hardware to integrate the required cryptographic and storage resources. Third, system designers may choose to use a lightweight DSA that can execute on existing master terminal and remote terminal platforms.

3.4.1.4.4.3 Confidentiality The confidentiality cybersecurity principle intends to protect network traffic from unauthorized eavesdropping. The most common method for insuring network traffic confidentiality is through the use of encryption.

The need for confidentiality of control system network traffic can be a controversial topic, with many control system engineers arguing against the need for control system network traffic confidentiality. The need for confidentiality will vary for each installation. However, it must be stressed that hackers use eavesdropping to collect information about systems before executing attacks. Confidentiality minimizes the potential attacker's capabilities in this intelligence gathering stage.

The security clouds in the network architecture shown in Figure 3.49 can encrypt and decrypt network traffic. Bump-in-the-wire solutions exist that can capture network traffic as it leaves an electronic security perimeter and encrypt the network traffic. The security cloud in a receiving electronic security perimeter can decrypt network traffic before forwarding the traffic to cyber systems inside the electronic security perimeter. Encryption algorithm choice can be based on multiple algorithms.

A key consideration when using encryption in control systems is the latency added to the network traffic as a result of encryption and decryption of the traffic. This is especially pertinent for low data rate systems such as SCADA systems, which often have data rates of 1,200–19,200 Bd. Symmetric block ciphers such as AES (Advanced Encryption Standard), DES (Data Encryption Standard), or 3DES seem best suited for use in control systems. These ciphers are generally quite fast and all three have many open source implementations in software and hardware. All three of these ciphers can be used as a stream cipher (output feedback, cipher feedback, or counter mode) to speed up the encryption decryption process and thereby reduce latency.

Another important consideration when planning to use encryption is availability of hardware resources. Many RTUs are equipped with cryptographic resources and storage capabilities which may conflict with the needs of an adequate encryption algorithm. This presents three possibilities for control systems, which support encryption for confidentiality. First, system designers may choose to use existing hardware and add a bump-in-the-wire cybersecurity solution (to sign and validate network traffic). Second, system designers may choose to upgrade existing remote or master terminal hardware to integrate the required cryptographic and storage resources. Third, system designers may choose to use a lightweight encryption algorithm which can execute on existing master terminal and remote terminal platforms.

3.4.1.4.5 Conclusion

Control systems that implement feedback control loops across networked communications links must protect the availability and integrity of control and sensor measurement data. Three primary threats to the control systems found in utility control systems are sensor measurement injection, control injection, and denial of service (DOS) attacks. Current installations of synchrophasor systems have a primary application of wide area visibility. As these systems evolve from a wide area visibility role to a wide area control (WAC) role, their cyber critical asset classification will also evolve from non-critical to critical. As that evolution completes, synchrophasor system cybersecurity protections will need to be upgraded to protect system availability, integrity, and confidentiality.

3.4.2 FACTS AND HVDC

Stuart Borlase, Neil Kirby, Paul Marken, Jiuping Pan, and Dietmar Retzmann,

One basic function of electricity networks is that the amount of power produced at any given moment must match the amount of power consumed. In the middle of this balancing act is the infrastructure that must carry the power from its place of production to its point of use, that is, the transmission network. In a typical AC power system, the transmission network performs this service through transmission lines, transformers, circuit breakers, and other common equipment. The flow of electricity through the transmission system follows the basic laws of physics. For a given voltage and line impedance, one can calculate the amount of current that will flow. This current flow may be more (overloaded) or less (underutilized) than desired by the transmission operator. A transmission device that is able to change the electrical system response to a given condition is obviously a useful element in creating a smarter grid. While adding this equipment alone does not constitute having a "smart grid," measurement devices and software that calculates optimum situations are only helpful to the extent that something can be done about the situation. The ability to control the flow of real and/or reactive power, the voltage and the frequency, and other aspects of the transmission system can be key elements in optimizing the grid. Those devices that can assert control over the real or reactive power flow in a specific line or node or even region of a network are the following:

- Synchronous condensers
- FACTSs (flexible AC transmission systems) devices
- HVDC (high-voltage direct current)

These devices have the ability to implement aspects of smart control under normal, steady-state operating conditions, as well as under transient or fault events, and depending on their speed of response, may be able to automatically prevent or speed up the recovery from fault situations.

In particular, the group of high-voltage power electronics devices, such as FACTS and HVDC, provide features that avoid problems in heavily loaded power systems; they increase the transmission capacity and system stability very efficiently and assist in preventing cascading disturbances.

As load increases and changes, some system elements are going to become loaded up to their thermal limits, and wide area power trading with fast varying load patterns will contribute to increasing congestion [1,2]. In addition to this, the dramatic global climate developments call for changes in the way electricity is supplied. Environmental constraints, such as loss minimization and CO_2 reduction, will play an increasingly important role. Consequently, network planners must deal with conflicting requirement between reliability of supply, environmental sustainability, as well as economic efficiency [3,4]. The power grid of the future must be secure, cost effective, and environmentally compatible. The combination of these three tasks can be tackled with the help of ideas, intelligent solutions, as well as innovative technologies, such as HVDC and FACTS, which have the potential to cope with the new challenges. By means of power electronics, they provide a versatile range of features that are necessary to avoid many operational problems in the power systems; they increase the transmission capacity and system stability very efficiently and help prevent cascading disturbances. Features of a future smart grid such as this can be outlined as follows: flexible, accessible, reliable, and economic. Smart grid will help achieve a sustainable development.

The developing load and generation patterns of existing power systems will lead to bottlenecks and reliability problems. Therefore, the strategies for the development of large power systems go clearly in the direction of smart grid, consisting of AC/DC interconnections and point-to-point bulk power transmission "highways" (super grid solutions). FACTS technology is also an important part of this strategy, and hybrid systems offer significant advantages in terms of technology, economics, and system security.

3.4.2.1 Power System Developments

The development of electric power supply began more than 100 years ago. Residential areas and neighboring establishments were at first supplied with DC via short lines. At the end of the nineteenth century, AC transmission was introduced, using higher voltages to transmit power from remote power stations to the consumers.

In Europe, 400 kV became the highest AC voltage level, in Far East countries mostly 550 kV, and in America 550 and 765 kV. The 1150 kV voltage level was anticipated in some countries in the past, and some test lines have already been built. Figure 3.50 depicts these developments and prospects.

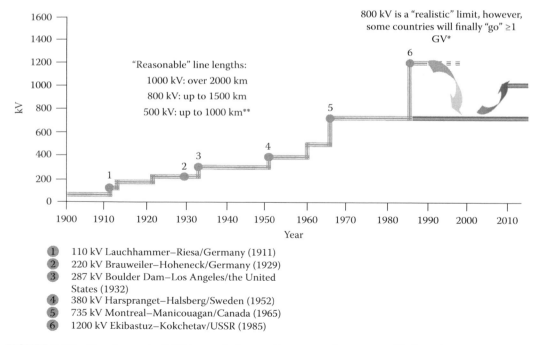

FIGURE 3.50 Development of AC transmission—milestones and prospects. (© Copyright 2012 Siemens. All rights reserved.) *China (1000 kV pilot project launched) and India (1200 kV in actual planning) are currently implementing bulk power UHV AC backbone; **Brazil: North–South interconnector.

Examples of large synchronous AC interconnections are systems in North America, Brazil, China, and India, as well as in Europe (installed capacity 631 GW, formerly known as UCTE—now CE, Continental Europe) and Russia (IPS/UPS—315 GW). IPS/UPS and CE are planned to be interconnected in the future.

It is an unfortunate consequence of increasing size of interconnected systems that the advantages of larger size diminish for both technical and economical reasons, since the energy has to be transmitted over extremely long distances through the interconnected synchronous AC systems. These limitations are related to problems with low-frequency inter-area oscillations [5–7], voltage quality, and load flow. This is, for example, the case in the CE (former UCTE) system, where the 400 kV voltage level is in fact too low for large cross border and inter-area power exchange.

FACTS technology, based on power electronics, was developed in the 1960s to improve the performance of weak AC systems, to make long-distance AC transmission feasible, and to help solve technical problems within the interconnected power systems.

FACTS systems are used both in a parallel connection (SVC [static VAr compensator], STATCOM [static synchronous compensator]), in a series connection (FSC, TCSC/TPSC, S^3C), or as a combination of both (UPFC, CSC) to control load flow and to improve dynamic conditions. These will be described in the following sections.

In the second half of the last century, high power HVDC transmission technology was introduced, offering new dimensions for long-distance transmission. This development started with the transmission of power in a range of less than 100 MW and was continuously increased. The state of the art for many years settled at 500 kV rating, as illustrated in Figure 3.50, and there are many examples of links with transmission ratings of 3 GW over large distances with only one bipolar DC line around the world today. More recent development has achieved transmission ratings of 6 GW and more over even larger distances with only one bipolar DC transmission system. Further projects with similar or even higher ratings in China, India, and other countries are going to follow.

Table 3.7 summarizes the impact of FACTS and HVDC on load flow, stability, and voltage quality when using different devices. Evaluation is based on a large number of studies and experiences from projects. For comparison, mechanically switched devices (MSC/R) are included in the table.

FACTS and HVDC applications will play an important role in the future development of smart power systems. This will result in efficient, low-loss AC/DC hybrid grids which will ensure better controllability of the power flow and, in doing so, do their part in preventing "domino effects" in case of disturbances and blackouts. By means of these DC and AC ultrahigh-power transmission technologies, the "smart grid," consisting of a number of highly flexible "microgrids," will turn into a "super grid" with bulk power energy highways, fully suitable for a secure and sustainable access to huge renewable energy resources such as hydro, solar, and wind. The state-of-the-art AC and DC technologies and solutions for smart and super grids are explained in the following sections.

In addition to these relatively complex systems using power electronics, there are other lower cost features which may be incorporated into the equipment installed in future power networks. These offer varying extents of functionality to add to the overall intelligence of the smart grid of the future, such as monitoring of transformers and switchgear which provide real-time analysis of transformer oil and other status information. Maintenance management systems can monitor and analyze this information and determine increased wear-and-tear rates, predict failure modes, and identify the need for preemptive maintenance before the next scheduled maintenance activity.

3.4.2.2 Flexible AC Transmission Systems

Reactive power compensation has been regarded as a fundamental consideration in achieving efficient electric energy delivery system. Reactive compensation may be categorized into series compensation, shunt compensation, and combined compensation, representing the intentional insertion

TABLE 3.7
FACTS and HVDC: Overview of Functions

Principle	Devices	Scheme	Impact on System Performance		
			Load Flow	Stability	Voltage Quality
Variation of the line impedance Series compensation	FSC (fixed series compensation)		●	●●●	●
	TPSC (thyristor protected series compensation)		●	●●●	●
	TCSC (thyristor controlled series compensation)		●●	●●●	●
Voltage control Shunt compensation	MSC/R (mechanically switched capacitor/reactor)		○	●	●●
	SVC (static VAr compensator)		○	●●	●●●
	STATCOM (static synchronous compensator)		○	●●	●●●
Load-flow control	HVDC – B2B, LDT		●●●	●●●	●●●
	HVDC VSC				
	UPFC (unified power flow controller)		●●	●●●	●●●

Source: ©Copyright 2012 Siemens Energy, Inc. All right reserved.
Influence (based on studies and practical experience): ○, No or low; ●, small; ●●, medium; ●●● strong.

of reactive power-producing devices in series and/or in parallel in the power circuit, either capacitive or inductive. Further flexibility can be achieved with dynamically controllable compensation to provide the required amount of corrective reactive power precisely and promptly. A family of such controllable compensation devices based on power electronics technology is often referred to as FACTS devices.

3.4.2.2.1 FACTS Developments

Since the 1960s, FACTSs have been evolving to a mature technology with high power ratings [8]. The technology, proven in various applications, became a first-rate, highly reliable one. Figure 3.51 shows the basic configurations of FACTS devices.

In Figure 3.52, the impact of series compensation on power transmission and system stability is illustrated, and Figure 3.53 depicts the increase in voltage quality by means of shunt compensation with SVC (or STATCOM).

3.4.2.2.2 Series Compensation

The conventional or fixed series compensation (FSC) is a well-established technology and has been in commercial use since the early 1960s. The basic concept of series-capacitor compensation is to reduce the overall inductive reactance of power lines by connecting series capacitors in series with the line conductors. As shown in Figure 3.54, the series-capacitor compensation equipment comprises series-capacitor banks, located in the line terminals or in the middle of the line, and overvoltage protection circuit for the capacitor bank. A photograph of a series compensation installation is shown in Figure 3.55.

Smart Grid Technologies

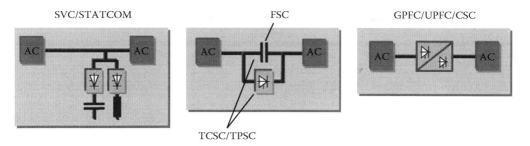

- SVC—static VAr compensator
- STATCOM—static synchronous compensator, with VSC
- FSC—fixed series compensation
- TCSC—thyristor controlled series compensation
- TPSC—thyristor protected series compensation
- GPFC—grid power flow controller (FACTS-B2B)
- UPFC—unified power flow controller (with VSC)
- CSC—convertible synchronous compensator (with VSC)

FIGURE 3.51 Transmission solutions with FACTS.

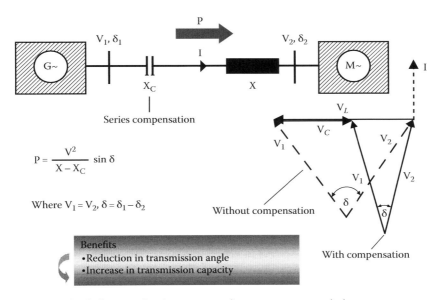

FIGURE 3.52 FACTS—influence of series compensation on power transmission.

Incorporating series capacitors in suitable power lines can improve both power system steady-state performance and dynamic characteristics. Series compensation has traditionally been used associated with long-distance transmission lines and with improving transient stability. In a transmission system, the maximum active power transferable over a certain power line is inversely proportional to the series inductive reactance of the line. Thus, by compensating the series inductive reactance to a certain degree, typically between 25% and 70%, using series capacitors, an electrically shorter line is realized and higher active power transfer and improved system performance can be achieved. In recent years, series capacitors are also applied on shorter transmission lines to

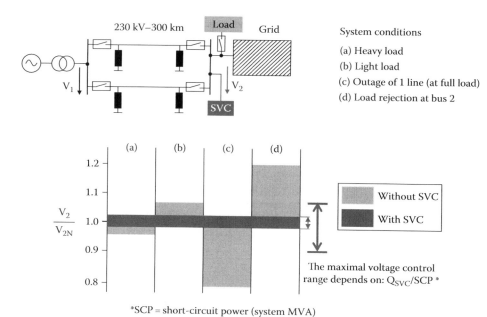

FIGURE 3.53 FACTS—improvement in voltage profile with SVC. (© Copyright 2012 Siemens. All rights reserved.)

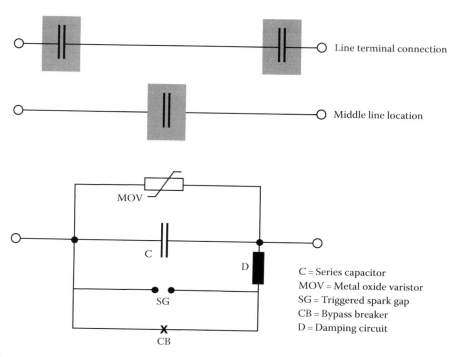

FIGURE 3.54 Common FSC locations and main circuit diagram. (Courtesy of Siemens.)

FIGURE 3.55 Photograph of a series compensation installation. (© Copyright 2012 Siemens. All rights reserved.)

improve voltage stability. In general, the main benefits of applying series compensation in transmission systems include the following

- Enhanced system dynamic stability
- Desirable load division among parallel lines
- Improved voltage regulation and reactive power balance
- Reduced network power losses

The thyristor-controlled series compensation (TCSC) is an extension of conventional series compensation technology, providing further flexibility of series compensation in transmission applications (Figure 3.56).

3.4.2.2.3 Shunt Compensation

An SVC is a regulated source of leading or lagging reactive power. By varying its reactive power output in response to the demand of an automatic voltage regulator, an SVC can maintain virtually constant voltage at the point in the network to which it is connected. An SVC is comprised of standard inductive and capacitive branches controlled by thyristor valves connected in shunt to the transmission network via a step-up transformer. Thyristor control gives the SVC the characteristic of a variable shunt susceptance. Figure 3.57 shows three common SVC configurations for reactive power compensation in electric power systems. The first configuration consists of a thyristor-switched reactor (TSR) and a thyristor-switched capacitor (TSC). Since no reactor phase control is used, no filters are needed. The second one consists of a thyristor-controlled reactor (TCR), a TSC, and harmonic filters (FC). The third one consists of a TCR, mechanically switched shunt capacitors (MSC), as well as FC.

For example, with the TCR/TSC configuration, flexible and continuous reactive power compensation can be obtained by appropriate switching of TSCs and accurate controlling of TCR, from the full inductive rating of the TCR to the full capacitive rating of the TSCs and the FC.

SVC technology has been in commercial use since the early 1970s (with over 1000 systems in service), initially developed for the steel industry to address the problem of voltage flicker with arc furnaces. The SVC is now a mature technology that is widely used for transmission applications,

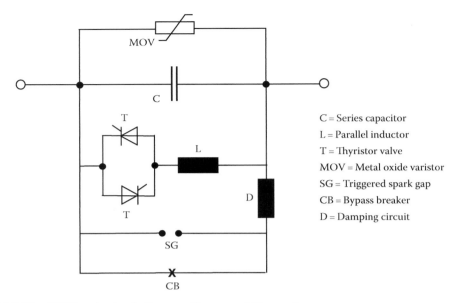

FIGURE 3.56 TCSC main circuit diagram. (Courtesy of Siemens.)

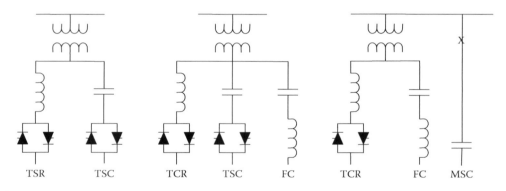

FIGURE 3.57 Common SVC configurations. (Courtesy of Siemens.)

providing voltage support in response to system disturbances and balancing the reactive power demand of large and fluctuating industrial loads. The installation can be in the midpoint of transmission interconnections or in load areas. In general, the main benefits of applying SVC technology in power transmission systems include the following:

- Improved system voltage profiles
- Reduced network power losses
- Stabilized voltage of weak systems or load areas
- Increased network power delivery capability
- Mitigated active power oscillations

An SVC installation is shown in Figure 3.58, which is part of the Lévis De-icer Substation in Québèc, Canada. This system performs regular reactive power compensation in normal operation but is also capable of reconfiguration into a DC source, to generate DC current to remove ice buildup on transmission lines.

Smart Grid Technologies

FIGURE 3.58 Photograph of an SVC installation. (© Copyright 2012 Alstom Grid. All rights reserved.)

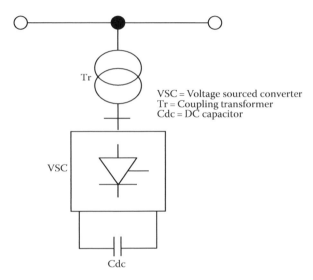

FIGURE 3.59 STATCOM main circuit diagram.

The STATCOM technology is based on power electronic concept of voltage-sourced conversion (Figure 3.59). The shunt-connected voltage-sourced converter (VSC) is comprised of solid-state switching components with turn-off capability with antiparallel diodes. Performance of the STATCOM is analogous to that of a synchronous machine generating balanced three-phase sinusoidal voltages at the fundamental frequency with controllable amplitude and phase angle. The device, however, has no inertia and does not contribute to the short circuit capacity.

The STATCOM consists of a VSC operating as an inverter with a capacitor as the DC energy source. It is controlled to regulate the voltage in much the same way as an SVC. A coupling transformer is used to connect to the transmission voltage level. In this application, only the voltage magnitude is controlled, not phase angle. By controlling the converter output voltage relative to the system voltage, reactive power magnitude and direction can be regulated. If the VSC AC output voltage is lower than the system voltage, reactive power is absorbed. If the VSC AC output voltage is higher than the system voltage, reactive power is produced.

The functions performed by STATCOM in a transmission network are quite the same as an SVC such as steady-state and dynamic voltage support and regulation, improved synchronous stability

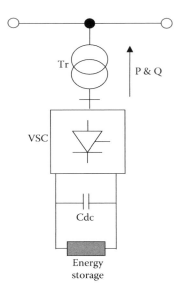

FIGURE 3.60 STATCOM with energy storage.

and transfer capability, and improved power system damping. In addition, STATCOM is also installed for power quality applications. These include the following:

- Improved dynamic load balancing
- Improved flicker control
- Faster response for load compensation

STATCOM with energy storage is an enhancement of STATCOM consisting of series-connected batteries as shown in Figure 3.60. Energy storage enables the STATCOM to generate and consume active power for a certain period of time. One typical application of STATCOM with energy storage is for integrating renewable energy source such as wind farm or solar farm that has a strongly fluctuating power production. The load balancing function with energy storage delivers active power at a scheduled power level and reactive consumption/production within operational limits, according to the power and voltage setting orders from the system operator.

These devices will form an increasingly important component of the future smart grid as a result of the increasing use of variable generation sources such as wind and solar, as the stored energy may be used to fill in the nongenerating periods from these diverse renewable sources. The capacity of the storage system will clearly need to be suitably rated to substitute the energy normally provided by the renewable source, but for short periods this is a viable solution.

3.4.2.2.4 Combined and Other Devices

More sophisticated systems to control power flow in transmission lines may be formed by combining series and shunt devices. The STATCOM described previously is a shunt-connected voltage sourced device that can regulate voltage at the point of connection through control of reactive power flow by injecting reactive current. Another device called the static synchronous series compensator (SSSC), which is similar to the STATCOM except it is series-connected, controls the magnitude and phase of an injected voltage independent of the current in the line.

In the unified power flow controller (UPFC) configuration, a STATCOM and SSSC are combined on a transmission line as shown in Figure 3.61, and they can regulate both real and reactive power in a line, allowing for rapid voltage support and power flow control. These devices require

Smart Grid Technologies

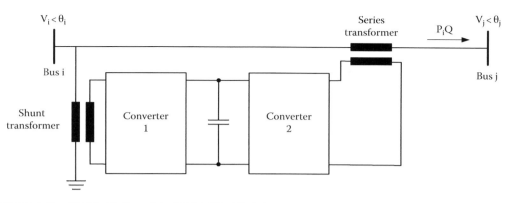

FIGURE 3.61 UPFC. (© Copyright 1999 ABB. All rights reserved.)

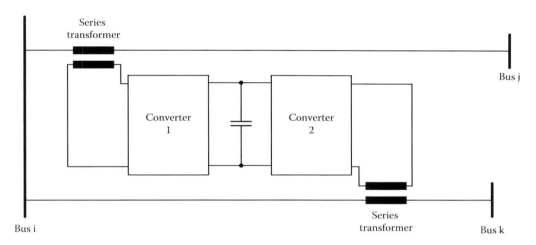

FIGURE 3.62 IPFC. (© Copyright 1999 ABB. All rights reserved.)

two converters in a back-to-back (B2B) configuration and may use the same DC capacitor in much the same way as a HVDC link.

The interline power flow controller (IPFC) is another configuration of the combined VSCs, except that the two converters are inserted on different transmission lines. The IPFC consists of two SSSC converters as shown in Figure 3.62. In this configuration, the IPFC is able to control both real and reactive power in both lines i–j and i–k by exchanging power through the DC link between them.

The SSSC, UPFC, and IPFC are applications of VSC converters, and presently these systems are not in common use; those that are in operation have been constructed as development projects [9].

3.4.2.2.5 Variable Frequency Transformer

A relatively new transmission device is the variable frequency transformer or VFT. A VFT is considered by many to be a "smart" device as it has the ability to control the amount of power flowing through it. Similar to an HVDC system, the VFT can interconnect asynchronous grids with the key difference being that the VFT provides a true AC connection. The first asynchronous AC transmission using a VFT appeared in 2003 at Hydro-Québec's Langlois Substation.

The VFT absorbs reactive power since it is an induction machine. It is normally applied with shunt banks to supply reactive power per the application's needs. As a true AC connection, the VFT allows reactive power to flow from one side to the other. As in any AC circuit, reactive power flow

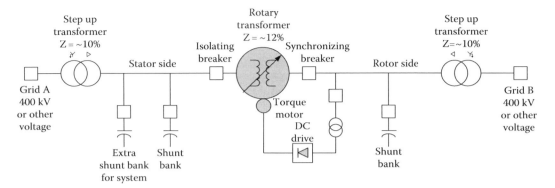

FIGURE 3.63 One-line diagram of a variable frequency transformer. (© Copyright 2012 GE Energy. All rights reserved.)

is a function of the system voltages and the series impedance. Figure 3.63 is a simplified one-line diagram of a VFT interconnection.

While many designs are theoretically possible, the present technology consists of one or multiple parallel 100 MW, 60–60 Hz, machines. There are currently five machines in commercial operation. In addition to providing flexibility in moving a controlled amount of real power between two points, which need not be synchronized, adding a VFT has also demonstrated improvements to power system dynamic performance and generator damping.

The ability of the VFT to control the flow of power through it offers network operators a transmission device that can be dispatched similar to generating assets. This can assist with operation of the power grid in a more optimized manner. The power flowing through any given transmission path may be higher or lower than operators prefer. Changing the system to adjust the flow of power through one point may have unintended consequences through another transmission path

3.4.2.2.6 Synchronous Condenser

While the synchronous condenser is not a new, high-tech device invented to contribute to the modern smart grid, it is worthy to consider that this may be the original volt/VAr controller. Once commonly found in both industrial and utility applications, the number of synchronous condensers in operation has been on the decline. Simply put, the synchronous condenser is a motor without a load connected to its shaft. Or viewed in a way more familiar to the utility industry, a condenser is similar to a generator without a prime mover. The field is under- or overexcited to absorb or produce reactive power. The machine will absorb a small amount of real power to overcome losses. When equipped with a modern generator field exciter, the speed of response is reasonably fast.

Although slower than a STATCOM and more costly than SVC, the synchronous condenser demonstrates a number of advantages over electronic solutions, including significant overload capability, short circuit level, and real rotating inertia. It is relatively compatible with harmonic issues and can even act as a harmonic sink. As more renewable sources of energy such as wind and solar have displaced traditional thermal machines, some grids have experienced a decline in rotating inertia. In other applications, synchronous condensers are required on the receiving end of large thyristor-based HVDC systems to ensure proper inverter operation. This has prompted a renewed interest in the application of synchronous condensers as part of the overall smart grid solution.

The synchronous condenser's usefulness in the smart grid is not unlike modern FACTS devices. Controlling voltage through injecting or absorbing reactive power at key points in the transmission system can allow more precise control of power flow and allow optimized transmission grid operation.

3.4.2.3 High-Voltage Direct Current

HVDC transmission is a well-established method of using controlling power flow within or between networks through power electronics systems. Originally the power flow control device was based on mercury-arc technology, though these systems have now been almost all decommissioned. Modern HVDC systems have been based on the use of thyristors as the controlled device (referred to as line commutated converter [LCC], current source converter [CSC], or conventional HVDC) for over 40 years, and more recently in the last 10 years or so, the use of the transistor (referred to as VSC) has been increasing.

The HVDC control system is designed to automatically respond to stimulus events from many sources, including the following:

- Operator input
- Routine changes in AC network conditions
- Routine network switching events
- Disturbance caused by faults within the DC system, the AC network

The intelligence incorporated into the control and protection system can be made to provide fully automated responses to all of these scenarios, such that the situation is detected and the response carried out without the need for human intervention. In this way the HVDC system can be considered as an essential component of the smart grid.

HVDC transmission systems offer many benefits over their AC counterparts, including the following:

- Power flow through the link can be precisely controlled in both magnitude and direction, either through operator action or through automated response.
- Voltage and frequency in the two AC networks can be controlled independently of each other, again either through operator action or through automated response.
- The HVDC link can be used to assist one (or even both) of the AC networks in responding to disturbances (e.g., power swing damping, by modulation of the transmitted power). This is normally fully automated since the operator is unable to respond in this timescale.

Additionally, the use of an HVDC link rather than an AC interconnection provides the following:

- Improved system stability margins due to the ability to rapidly change power transfer
- No increase of the short circuit level of the system
- No transfer of faults across the interconnected systems

In the evolution of HVDC, different applications were developed, as shown schematically in Figure 3.64.

Figure 3.65a shows the results of a simulation study based on two AC networks (A and B), which are interconnected and synchronized by a line rated at 500 MW. A short circuit fault occurs in network B at about 0.3 s; it can be seen that after about 7 s, the angular displacement of the rotors of selected generators in network A, relative to a reference generator, is increasing, that is, they cannot regain synchronism and the system is unstable. Exactly the same fault is applied in Figure 3.65b, but in this case an HVDC B2B link has been introduced between the two networks, that is, the link is now effectively asynchronous. It can be seen that within about 4 s, the rotor angle swings have been damped and stability maintained; the power flow through the link is virtually unchanged once the fault had been cleared. This is just one example of the way that networks which incorporate the controllability of HVDC can be made more intelligent, self-healing, and an integral, essential part of the smart grid of the future.

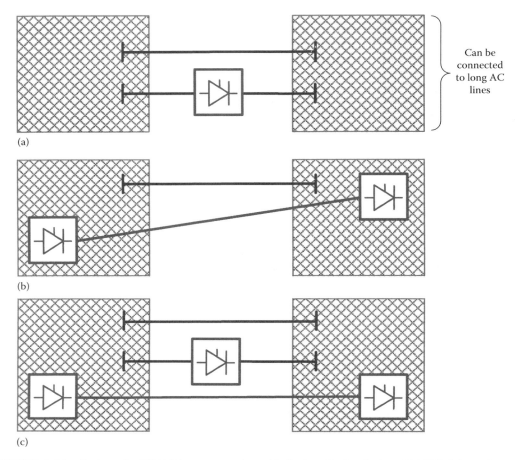

FIGURE 3.64 Options for HVDC interconnections. (a) Back-to-Back solution, (b) HVDC long-distance transmission, and (c) integration of HVDC into the AC system (hybrid solution).

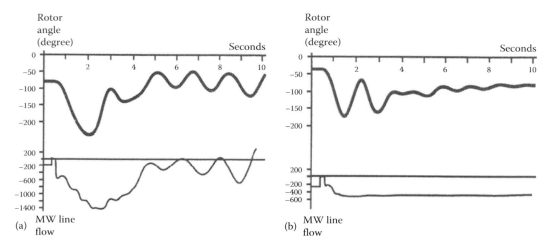

FIGURE 3.65 Post-fault response of two AC networks interconnected with (a) AC link; and (b) HVDC B2B link.

Smart Grid Technologies

Smart grid intelligence requires both functionality in individual equipment or subsystems and communications between these subsystems to allow other components of the network or hierarchy to see what is going on and, in turn, allow them to make other intelligent decisions and take controlling actions.

HVDC is one of the most intelligent subsystems within a network, since it is able to carry out precise control of power flow based on internal and external information, since it is customary for HVDC systems to pass information at a very detailed level to remote centers for monitoring, protection, and control at other locations. By coordinating the action of the HVDC control system and other control systems in the network (generators, switching, and transformer substations, FACTS devices, etc.), it is possible to build up a complete control hierarchy for the network made up of these discrete and dispersed subsystems. Intelligent systems such as this are obviously capable of responding to events much faster than the human operator, and the rapid response to most faults or other events is critical to allow the system to recover and restabilize quickly.

For example, the power flow on the VSC-HVDC systems can be optimally scheduled based on system economics and security requirements. It is also feasible to dispatch VSC-HVDC systems in real-time power grid operations. Such increased power flow control flexibility allows the system operators to utilize more economic and less pollutant generation resources and implement effective congestion management strategies.

3.4.2.3.1 HVDC Developments

In general, for transmission distances above 600 km, DC transmission is more economical than AC transmission ($\geq$1000 MW). Power transmission of up to 600–800 MW over distances of about 300 km has already been achieved with submarine cables, and cable transmission lengths of up to approximately 1000 km are at the planning stage. HVDC is now a mature and reliable technology (Figure 3.66).

The first commercial applications were cable transmissions for AC cable transmission over more than 80–120 km is technically not feasible due to reactive power limitations. Then, long-distance HVDC transmissions with overhead lines were built as they are more economical than transmission with AC lines [10]. To interconnect systems operating at different frequencies, B2B schemes were applied. B2B converters can also be connected to long AC lines (Figure 3.64a). A further application of HVDC transmission that is highly important for the future is its integration into the complex interconnected AC system (Figure 3.64c). The reasons for these hybrid solutions are basically lower transmission costs as well as the possibility of bypassing heavily loaded AC systems. Further information on the application of HVDC to handle large-scale transmission to overcome the difficulties encountered by the conventional AC networks may be found in Barker et al. [11] and MacLeod et al. [12].

The power ranges of VSC-HVDC have been improved rapidly in recent years. In the upper range, the technology now reaches 1200 MVA for symmetric monopole schemes with cables, which can be increased to 2400 MVA for bipole schemes with overhead lines.

Typical configurations of HVDC are depicted in Figure 3.67. HVDC VSC is the preferred technology for connecting islanded grids, such as offshore wind farms, to the power system [13]. This technology provides the "black-start" feature by means of self-commutated VSCs [14]. VSCs do not need any "driving" system voltage; they can build up a three-phase AC voltage via the DC voltage at the cable end, supplied from the converter at the main grid.

VSC-HVDC technology is now emerging as a flexible and economical alternative for future transmission grid expansion. In particular, embedded VSC-HVDC applications, together with the wide area monitoring system, in meshed AC grids could significantly improve overall system performance, enabling smart operation of transmission grids with improved security and efficiency. VSC-HVDC transmission also offers a superior solution for many challenging technical issues associated with integration of large-scale renewable energy sources such as offshore wind power.

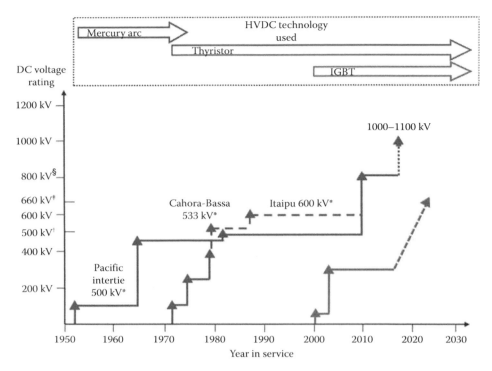

FIGURE 3.66 (**See color insert.**) Evolution of HVDC voltage rating and technology. * Multiple bridges per pole; †500 kV becomes *de facto* standard for single 12-pulse bridge per pole; ‡660 kV used as "standard" in China; §800 kV used as "standard" in China and India. (© Copyright 2012 Siemens. All rights reserved.)

- HVDC "conventional" with 500 kV (HV)/660 kV (EHV)—up to 4 GW
- HVDC "bulk" with 800 kV (UHV)—from 5 GW up to 7.5 GW
- HVDC VSC (voltage-sourced converter)
- HVDC can be combined with FACTS
- V control included

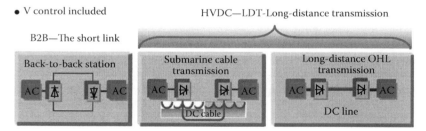

FIGURE 3.67 HVDC configurations and technologies.

3.4.2.3.2 Thyristor-Based "Conventional" HVDC

Conventional HVDC, also known as line commutated converter (LLC) systems, are based on thyristor switching technology.

Figure 3.68 shows a simplified circuit diagram of main components that make up the power circuit of a typical conventional HVDC system: these are the thyristor valves, the converter

Smart Grid Technologies

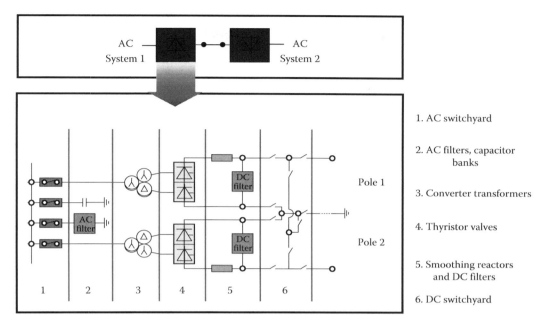

FIGURE 3.68 Conventional HVDC single-line diagram. (© Copyright 2012 Siemens. All rights reserved.)

transformers, and the AC FC. The most common configuration of conventional HVDC is using the 12-pulse bridge arrangement, which offers the best compromise of least cost and least harmonics. The AC FC perform the dual roles of (a) maintaining reactive power balance with the AC network and (b) preventing harmonics generated by the HVDC from reaching the AC system. A photograph of a conventional HVDC installation is shown in Figure 3.69.

Figure 3.70 illustrates a typical HVDC thyristor valve hall for one end of an HVDC scheme. There are two different configurations of HVDC, a point-to-point system as shown in Figure 3.68,

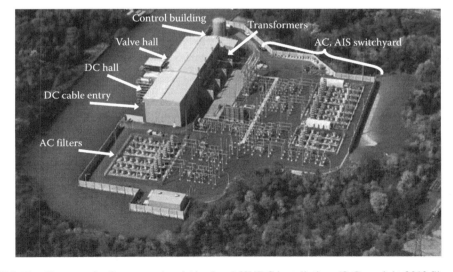

FIGURE 3.69 Photograph of a conventional (thyristor) HVDC installation. (© Copyright 2012 Siemens. All rights reserved.)

FIGURE 3.70 Conventional HVDC thyristor valve hall. (© Copyright 2012 Siemens. All rights reserved.)

where the DC connection is an overhead line or insulated cable, and a B2B system where the DC connection has zero length. The point-to-point system is used for the economical transmission of power over long distances. The B2B system can be used to isolate systems that are normally asynchronous to prevent the spread of cascading faults and to increase the stability limit on an AC line.

3.4.2.3.3 VSC-Based HVDC

VSC-HVDC is a transmission technology based on VSCs and insulated gate bipolar transistors (IGBT). The converter operates with high frequency pulse width modulation (PWM) and thus has the capability to rapidly control both active and reactive power, independently of each other.

In particular, VSC-HVDC systems are attractive solutions for transmitting power underground and under water over long distances. With extruded DC cables, power ratings from a few tens of megawatts up to more than 1000 MW are available.

Figure 3.71 shows a simplified circuit diagram of main components that make up the power circuit of a typical VSC-HVDC system: these are the IGBT converter valves, converter reactors, DC capacitors, AC FC, DC cables, and transformers.

The first VSC-HVDC schemes were based on a two level topology, where the output voltage is switched between two voltage levels; however, the most common valve configuration being currently implemented is the modular multilevel converter (MMC), due to improvements in operating efficiency. Each phase has two valves, one between the positive potential and the phase outlet and one between the outlet and the negative potential. Thus, a three-phase converter has six valves, three-phase current reactors, and a set of DC capacitors. The phase reactor permits continuous and independent control of active and reactive power. It provides low-pass filtering of the PWM pattern to give the desired fundamental frequency voltage. The converter generates harmonics related to the switching frequency. The harmonic currents are blocked by the phase reactor, and the harmonic content on the AC bus voltage is reduced by AC filters. The fundamental frequency voltage across the reactor defines the power flow (both active and reactive) between the AC and DC sides. AC filters typically contain two or three grounded or ungrounded tuned filter branches. Depending on filter performance requirements, that is, permissible voltage distortion and others, the filter configuration

Smart Grid Technologies

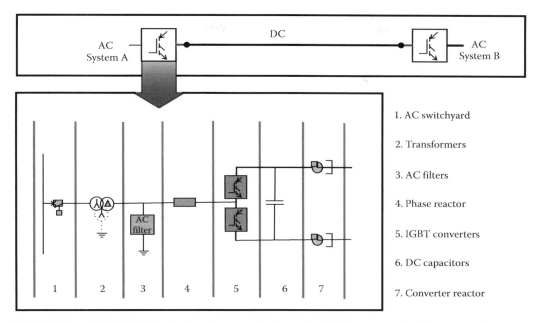

FIGURE 3.71 VSC-HVDC single-line diagram. (© Copyright 2012 Siemens. All rights reserved.)

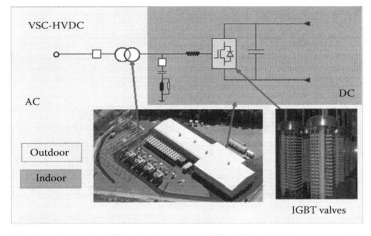

FIGURE 3.72 HVDC station with VSCs. (© Copyright 2009 ABB. All rights reserved.)

may vary between schemes. The transformer is an ordinary single- or three-phase power transformer, with tap changer. The secondary voltage and the filter bus voltage will be controlled with the tap changer to achieve the maximum active and reactive power from the converter. Figure 3.72 shows a VSC-HVDC converter station.

One attractive feature of VSC-HVDC systems is that the power direction is changed by changing the direction of the current and not by changing the polarity of the DC voltage. This makes it easier to build a VSC-HVDC system with multiple terminals. These terminals can be connected to different points in the same AC network or to different AC networks. The resulting multiterminal VSC-HVDC systems can be radial, ring, or meshed topologies.

VSC-HVDC is ideal for embedded applications in meshed AC grids. Its inherent features include flexible control of power flow and the ability to provide dynamic voltage support to the surrounding AC networks. Together with advanced control strategies, these can greatly enhance smart transmission operations with improved steady-state and dynamic performance of the grid.

Fast control of active and reactive power of VSC-HVDC systems can improve power grid dynamic performance under disturbances. For example, if a severe disturbance threatens system transient stability, fast power runback and even instant power reversal control functions can be used to help maintain synchronized power grid operation.

3.4.3 Wide Area Monitoring, Protection and Control

Jay Giri, Zhenyu (Henry) Huang, Rajat Majumder, Rui Menezes de Moraes, Reynaldo Nuqui, Manu Parashar, Walter Sattinger, and Jean-Charles Tournier

3.4.3.1 Overview

Time-synchronized measurements across widely dispersed locations in an electric power grid are a key and differentiating feature of a wide area monitoring, protection and control (WAMPAC) system. WAMPAC systems are based on the synchronized sampling of power system currents and voltage signals across the power grid using a common timing signal derived from GPS* (Figure 3.73) The sampled signals are converted into phasors—vector representations of the grid's voltage and current measurements at fundamental frequency—that are synchronized and compared across the electrically connected power system using an accurate GPS time reference. Bus voltage and current phasors define the state of an electric power grid in realtime.

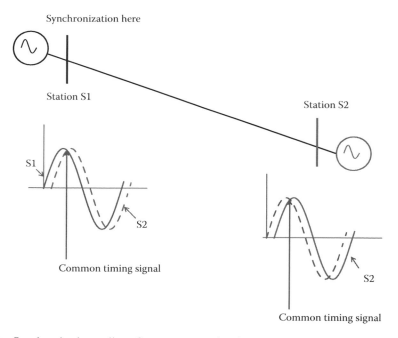

FIGURE 3.73 Synchronized sampling of power system signals.

---

* The Global Positioning System (GPS) is a satellite-based navigation and time signal system made up of a network of satellites placed into orbit by the U.S. Department of Defense. GPS was originally intended for military applications, but in the 1980s, the government made the system available for civilian use.

Smart Grid Technologies

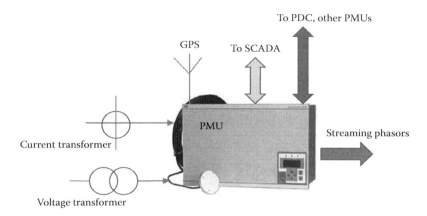

FIGURE 3.74 Phasor measurement unit (PMU).

3.4.3.1.1 Phasor Measurement Unit

The PMU, also known as a synchrophasor, is the basic building block of a WAMPAC system. The PMU samples the power system signals from voltage and current sensors and converts them into phasors. Phasors are complex number representations of the sampled signals commonly used in the design of, and inputs to, control and protection systems for bulk power transmission grids. The phasors are time tagged from a timing pulse derived from the GPS and then streamed into the wide area communications network as fast as one phasor per cycle of the power system frequency (Figure 3.74). Currently, the IEEE synchrophasor standard 37.118 defines the format by which the phasor data are transmitted from the PMU. The phasor angle information is referenced with the GPS timing pulse; for it to have physical significance, it has to be compared to (subtracted) other phasor angle measurements from the same system. Phasor angle differences provide useful information concerning system stress or modes of oscillatory disturbances in the power system. The PMU provided the critical synchronized time-lapsed information that enabled a clear understanding of the events leading to the northeast blackout of 2003 in the United States.

PMU technology has advanced significantly since Dr. Arun Phadke and his team developed the first PMU in Virginia Tech in 1988. Modern-day PMUs have become more accurate and capable of measuring a larger set of phasors in a substation. Most PMUs have binary output modules for transmitting binary signals, such as trip signals to open a circuit breaker. Some vendors have PMUs integrated within protection relays or digital fault recorders, with timing signals taken from IRIG-B* time sources instead of GPS antennae. A typical PMU connection in a transmission substation is shown in Figure 3.75. The PMU is considered one of the most promising if not the most important measurement device in modern transmission systems.

3.4.3.1.2 Time Synchronization

Time synchronization is the core of WAMPAC-based applications. WAMPAC applications rely on a precise time stamp transmitted with each PMU measurement to monitor, control, and protect the electrical network. In general, time synchronization requirements range from nanoseconds to microseconds. Time synchronization can be achieved by multiple means. All methods are based on the distribution of a common source clock signal across the network either by satellite, via the communications network (e.g., using the IEEE 1588† protocol), or using dedicated synchronization

---

* The Inter-Range Instrumentation Group (IRIG) time protocol is widely used by electric utilities to precisely communicate time to power system devices, such as breakers, relays, and meters from a clock source. IRIG-B is one of the IRIG standards in a serial communication format.
† 1588 is an IEEE protocol standard for precision clock synchronization for networked measurement and control systems.

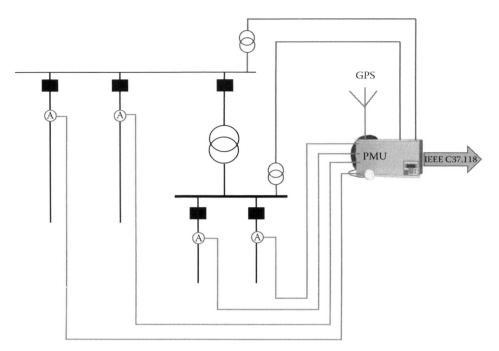

FIGURE 3.75 Typical PMU connection in a transmission substation.

networks (e.g., IRIG-B). The crucial need for a highly reliable and available time synchronization system implies the systemic use of a high-quality clock with accuracies expressed in PPM (parts per million) in order to maintain high accuracy, even in the case of a temporary loss of the synchronization signal.

3.4.3.1.3 Phasor Data Concentrator

A PDC collects phasor data from multiple PMUs or other PDCs, aligns the data by time tag to create a synchronized dataset, and then passes the data on to applications processors. For applications that process PMU data from across the grid, it is vital that the measurements are time aligned based on their original time tag to create a system-wide, synchronized snapshot of grid conditions. To accommodate the varying latencies in data delivery from individual PMUs, and to take into account delayed data packets over the communications system, PDCs typically buffer the input data streams and include a certain "wait time" before outputting the aggregated data stream. A PDC also performs data quality checks, validates the integrity or completeness of the data, and flags all missing or problematic data.

PMUs utilize various data formats (IEEE 1344, IEEE C37.118, BPA Stream, etc.*), data rates, and communications protocols (e.g., TCP, UDP, etc.) for streaming data to the PDC. On the input side, the PDC must support these different formats; additionally, it must be able to down-sample (or up-sample) the input streams to a standard reporting rate and process the various datasets into a common format output stream. There may also be multiple users of the data. Hence the PDC should be able to distribute received data to multiple users simultaneously, each of which may have different data requirements that are application specific.

The functions of a PDC can vary depending on its role or its location between the source PMUs and the higher-level applications. Broadly speaking, there are three levels of PDCs (Figure 3.76).

---

* These are examples of protocol standards used to exchange phasor data.

Smart Grid Technologies

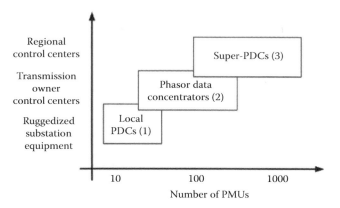

FIGURE 3.76 Levels of PDCs: (1) local or substation level, (2) transmission owner control centers, and (3) regional control center level (ISOs, RTOs).

1. *Local or substation PDC.* A local PDC is generally located at the substation for managing the collection and communication from multiple PMUs within the substation or neighboring substations and sending this time-synchronized aggregated dataset to higher-level concentrators at the control center. Since the local PDC is close to the PMU source, it is typically configured for minimal latency. It is also commonly utilized for local substation control operations. Local PDCs may include a short-term data storage system to protect against communications network failures. A local PDC is generally a hardware device that requires limited maintenance and that can operate independently if it loses communications with the rest of the communications network.
2. *Control center PDC.* This PDC operates within a control center environment and aggregates data from one utility's PMUs and substation PDCs, as well as neighboring utility PDCs. They are capable of simultaneously sending multiple output streams to different applications, such as visualization, alarms, storage, and EMS applications, each of which has its own specific data rate requirements. Control center PDC architectures are typically redundant in order to handle expected future loads and to satisfy high-availability needs of a production system regardless of PMU vendor and device type. PDCs need to be adaptable to accommodate new protocols and output formats, as well as interfaces with new applications.
3. *Super-PDC.* A Super-PDC operates on a larger, regional scale and is responsible for collecting and correlating phasor measurements from hundreds of PMUs and multiple substations and/or control center PDCs; it may also be responsible for facilitating PMU data exchange between utilities. In addition to supporting applications such as wide area monitoring system (WAMS) and visualization, and EMS and SCADA applications, it is capable of archiving a vast amount of data (typically, several Terabytes per day). Super-PDCs are therefore typically enterprise-level software systems running on clustered server hardware to accommodate scalability to meet the growing PMU deployment and utility needs.

3.4.3.2 Drivers and Benefits of WAMPAC

Microprocessor-based computer relaying, information technology, and advances in communications are changing the landscape of transmission systems monitoring. WAMPAC systems are driven by the need for alternative solutions for managing transmission reliability and security via improved SA. Electric transmission grids interconnect bulk power systems that are spread across geographical regions. As such, electric transmission grids have evolved to be very reliable and secure systems—the cost of failure is great. Now, smart grid initiatives will impose new reliability and economic requirements that will further impact how transmission systems will be monitored, protected, and controlled in the future.

Smart transmission grids are expected to be self-healing, that is, when experiencing disturbances, component failures, or cyber attacks, the grid is expected to recover. New and unconventional generating sources from renewable energy introduce operational challenges. With the increase in renewable energy sources, smart grids need to provide the most efficient transmission corridors for delivering energy to major load centers. Power systems are being operated closer to their thermal and stability limits. As a result, transmission operators need to increase their SA of the grid. The onset and early indications of disturbances and contingencies need to be visible to the operator in a timely fashion. SCADA and EMS systems need more advanced WAMPAC applications for the transmission grid to meet these challenges.

3.4.3.3 WAMPAC Needs in a Smart Grid

It is noteworthy to underline the key functional characteristics of smart grids specific to transmission systems to understand the needs for WAMPAC. First, a smart grid should be self-healing from power disturbance events. Events often cause failure or isolation of transmission lines and generation sources that could potentially lead to grid collapse. A smart grid should effectively manage a large number of renewable sources from wind, solar, and storage, including electric vehicles, and maintain the same level of power quality. There is also a great expectation that smart grids will be more efficient in transmitting electricity. With a large number of intermittent renewable generation sources, the reliable and efficient transmission of electricity is not a trivial task (Figure 3.77).

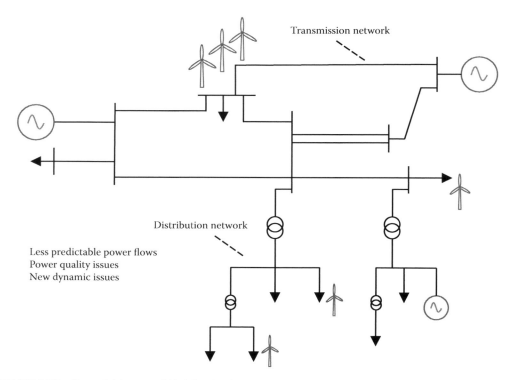

FIGURE 3.77 Potential impact of high levels of renewable energy sources in smart grids.

Smart Grid Technologies

3.4.3.3.1 Maintaining Reliability, Stability, and Security against Large Disturbances

Smart transmission grids should exhibit self-healing capabilities from power disturbance events—they are expected to endure disturbances and outages with zero or minimal impact to the grid's ability to supply and distribute power. While grids have been designed to survive large events, the integration of renewable generation will probably push these designs to their limits. Renewable generation will be integrated into both the bulk transmission level and in the distribution level. Power system events often cause failure or isolation of transmission or distribution lines and generators that could potentially lead to grid collapse. The ensuing dynamics often stresses generators and loads to the point where they disconnect from the grid, which more often than not, further stresses the remaining grid components. A typical dynamic system response to grid disturbances is in the form of power oscillations between the generating sources, which if left uncontrolled, can persist and lead to system instability. While local control methodologies address most of these grid disturbances, control based on wide area information promises to be a more effective solution. For example, WAMPAC can enable adjustment of the excitation set points of generation sources on a WAC basis in order to more effectively dampen persistent power oscillations. Large disturbances also have been known to produce voltage and current oscillations that often result in unwanted and false operation of protection systems. A need exists to communicate system events to these grid protection units to help them differentiate between disturbances requiring action to those that should be temporarily ignored. Such information transfer must be fast and handle data from several distant locations in the grid.

3.4.3.3.2 Management of Large Numbers of Intermittent Generation

Smart grids will need to support a higher penetration of intermittent generation and storage. Managing such large numbers and varieties of generation sources could exceed the processing limits of existing EMSs or DMSs. Large numbers of intermittent generators will result in highly transient power flows that can push the limit of transmission lines beyond their current carrying capabilities and cause transmission grid congestion. Smart grid operations will benefit from applications that can coordinate the management of renewable generators to mitigate these intermittent flows. This coordination will need to occur in much faster time frame than what can be realized in current grid management systems.

Smart transmission grids require an advanced level of monitoring both at the control center and substation level. Several factors drive these requirements. Increased connectivity with neighboring systems will make existing systems more sensitive to neighboring disturbances. A higher number of low inertia generators on the grid, such as renewable energy, offer less damping to the spread of system disturbances, and therefore faster propagation of disturbances can be expected. The spread of disturbances across utility boundaries will pose a challenge to system operators. Expanded visibility for tracking disturbances outside traditional control center boundaries is required. Transmission system monitoring will benefit from more field data available at a higher sampling frequency to quickly and accurately measure disturbances and dynamics so that corrective or mitigating control actions can be taken.

3.4.3.3.3 Maintaining Power Quality

A potential issue in smart grid is the degradation of power quality as large numbers of intermittent generation and power electronics loads become integrated into the transmission and distribution system. It will be challenging to maintain nominal frequency and the quality of power with highly variable generation in the grid. Frequency and voltage quality issues could be resolved with improved regulation from the grid's active power sources. Improved frequency regulation could come from using energy storage. A key requirement to addressing these power quality issues is the ability to monitor, store, and communicate data for processing and analysis in the control center so that operator actions may be initiated. However as previously mentioned, it is highly unlikely that existing grid management systems will be capable of carrying the extra volume of data transfer and most of the current measuring and monitoring systems also do not use a sampling rate high enough to capture some of these power quality issues. What is required is a monitoring system that supports the required signal sampling rate and communications bandwidth to transfer such data.

3.4.3.3.4 Increasing Transmission Efficiency

Smart grids are expected to increase utilization of existing grid assets, such as lines and transformers. In this way, the grid can be more efficient in delivering power from the source to the loads. One approach to increase transmission utilization is to dispatch generators to maximize power flow through the grid without exceeding system thermal and stability limits. Maximizing efficiency across the grid requires sufficient system-wide measurements to support system optimization applications. The efficient utilization of grid assets is also limited by the requirement for operating margins to account for potential grid instabilities or generation outages. It is anticipated that renewable generation and plug-in electrical vehicles will further complicate the estimation of these limits and that operators will probably establish higher margins for operational security. Countermeasures against instabilities can be instituted to provide security against these instabilities. More advanced decision support tools are clearly needed to ensure increased efficiency in a smart grid.

3.4.3.4 Major WAMPAC Activities

3.4.3.4.1 United States

Most groundbreaking research and initial WAMPAC applications worldwide started in the United States. Currently, smart grid initiatives from the federal government have made funds available to support the large-scale deployment of PMUs in the United States. Prior to the smart grid initiatives, there existed a working group to drive the deployment of WAMPAC in the United States: the North American Synchrophasor Working Group (NASPI). NASPI is a collaborative effort between the U.S. Department of Energy (DOE), the North American Electric Reliability Corporation (NERC), and North American electric utilities, vendors, consultants, federal and private researchers, and academics. NASPI's mission is to improve power system reliability and visibility by creating a robust, widely available and secure synchronized data measurement infrastructure for the interconnected North American electric power system with associated analysis and monitoring tools for better planning and operation and improved reliability. The NASPI architecture is referred to as the NASPI network or NASPInet—see Figure 3.78. A key effort in NASPI is the development of the phasor gateway and the super-phasor data concentrator (Super-PDC). The NASPInet consists of phasor gateways exchanging data via a phasor data bus. The overall goal of the NASPInet effort is to develop an "industry grade," secure, standardized, distributed, and expandable data communications infrastructure to support dissemination of utility synchrophasor measurements for applications across North America.

3.4.3.4.1.1 Phasor Gateway The phasor gateway is the primary interface between a utility, or another authorized party, and the data bus for synchrophasor data exchanges via NASPInet. The phasor gateway manages the connected devices on the entity's side, manages quality of service, administers cybersecurity and access rights, performs necessary data conversions, and interfaces the utility's PMU network with the data bus. The main functions of the phasor gateway include the following:

- Serve as the sole access point to the data bus for interorganizational synchrophasor traffic via a publisher-subscriber-based data exchange mechanism.
- Facilitate and administer registration of user PMUs, PDCs, and phasor signals. This is done through a name and directory service (NDS) system-wide registry. All real-time data

Smart Grid Technologies 195

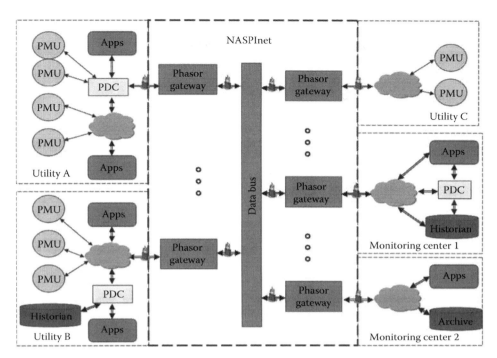

FIGURE 3.78 NASPInet conceptual architecture (phasor gateways and data bus). (Quanta Technology LLC, Phasor Gateway Technical Specification for North American Synchro–Phasor Initiative Network (NASPInet), May 29, 2009., http://www.naspi.org.)

streaming sources need to be registered through the owner's phasor gateway before their data can be published to NASPInet. This includes information such as physical location of the device, device type, device identifier, signal description, signal quality, and ownership according to the phasor gateway owner and NASPInet naming conventions. Only upon successful registration of the data source with the NDS can the phasor gateway publish data.
- Facilitate and administer the subscription and publishing of phasor data. The publish/subscribe mechanism consists of three parts: device/signal registration by publishers, subscription setup between publisher and subscriber that is initiated by subscribers, and quality of service and data security of the subscribed data. The owner of the phasor gateway that publishes the data to NASPInet maintains full control of its data distribution regarding who could subscribe to its data and which data could be subscribed to on a per-subscriber and per-signal basis. Nonsubscribers are therefore prevented from receiving the published data without a valid subscription. Subscribers are ensured that data will only come from publishers that they subscribe to.
- Administer and disseminate cybersecurity and access rights. The phasor gateway should provide system administrator functions to configure, operate, diagnose, and control the phasor gateway access rights to ensure appropriate access to, and usage of, the data on a per-user and per-signal basis, including who can add, edit, and remove users, and control each user's access rights. The security must meet corresponding NERC (North American Reliability Corporation), CIP (U.S. CIP program), FIPS (U.S. Federal Information Processing Standard), and other relevant cybersecurity standards and guidelines to safeguard reliable operation and data exchange.
- Manage traffic priority through the phasor gateway according to data service classes. It is well understood that different applications have different data requirements in terms of

latency, data rates, availability, etc. Five different classifications of applications are identified based on these requirements: Class A, feedback and control; Class B, open loop control (e.g., state estimation); Class C, visualization; Class D, postevent analysis; and Class F, R&D. The phasor gateway must support data delivery based on the priority traffic levels, that is, higher priority data are always processed and delivered before lower priority data.

- Monitor data integrity. This includes the ability to monitor both data that are forwarded to and received from the data bus for error and conformance with the data service class specifications to ensure that all transported data meet quality-of-service requirements. The types of statistics provided by the phasor gateway are the number of missing packets and missing packet rate, number of packets with data integrity checks, data stream interruptions, data stream delays, and changes in input data configuration. The phasor gateway should also have the ability to notify the administrator when there are excessive data errors or the data do not conform to the data service class specification.
- Provide logging of data transmission, access controls, and cybersecurity for analysis of all anomalies. The phasor gateway should log all user activities (e.g., access requests), system administration activities (e.g., data source registration), data subscription-related activities, quality-of-service alerts, cybersecurity alerts, application errors, etc. Therefore, any anomaly can be traced and analyzed to determine whether it is the result of NASPInet's own degradation or failures, or intentional/unintentional intrusion by unauthorized entities (hackers, intruders, unauthorized equipment connection, unauthorized user logins, etc.).
- Provide APIs for interfacing with a user's systems and applications to access data bus data and services.

3.4.3.4.1.2 Super-PDC The term "super-phasor data concentrator" or "Super-PDC" was first coined within the context of the Eastern Interconnection Phasor Project (EIPP), which was a U.S. DOE-led initiative started in 2002 to deliver immediate value of synchrophasor information within the U.S. Eastern Interconnection. The initial focus of the project involved networking existing PMU installations across the entire eastern interconnection and streaming these data to a centralized site for data concentration and archival. To support this EIPP endeavor, Tennessee Valley Authority (TVA) made a substantial investment in developing this "centralized" PDC (termed as the "Super-PDC") for the entire Eastern Interconnection that (1) was capable of gathering data from multiple PDCs and PMUs deployed across several utilities and ISOs, (2) supported a variety of phasor data transmission protocols (e.g., BPA PDCStream, IEEE C37.118, IEEE 1344, OPC, VirginaTech FNET) to ensure that all PMU capable devices within the interconnection could be integrated, (3) included a comprehensive database mechanism to manage the metadata associated with the phasor measurements, and (4) was capable of archiving huge amounts of this measurement data with fast historical data retrieval mechanisms.

The Super-PDC architecture that was developed by TVA is shown in Figure 3.79. It includes the real-time data acquisition module for parsing the data packets from various devices and protocols, the interface to TVA's proprietary DatAWare database, which maintains a 30 day rolling archive, data preprocessing module responsible for synchronization and encapsulation of these time-aligned data into a single stream, and finally the real-time broadcast module for streaming these data in realtime to applications. The Super-PDC at TVA is currently receiving data from approximately 120 PMUs across the Eastern Interconnection (the largest collection of PMU data within North America). It archives the data in a historian with no data compression, collecting approximately 36 GB per day (1 TB per month).

In 2008 the North American Electricity Reliability Corporation (NERC) contracted TVA to architect the second generation to TVA's "centralized" Super-PDC architecture, where multiple

Smart Grid Technologies

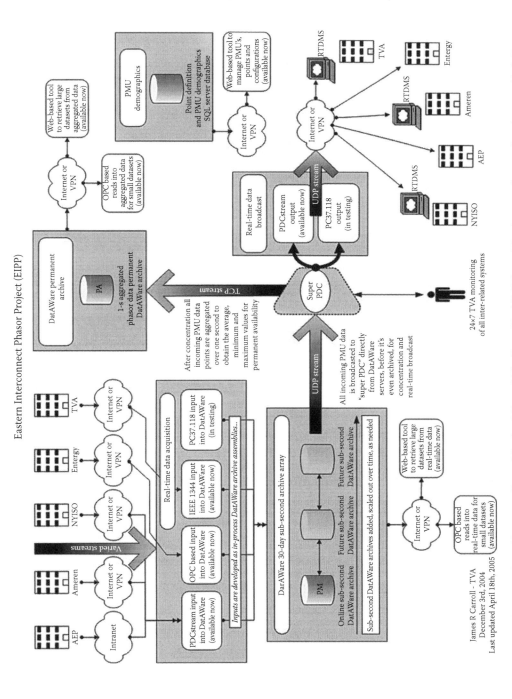

FIGURE 3.79 NASPI Super-PDC architecture. (From Myrda, P.T. and Koellner, K., NASPInet—The internet for synchrophasors, *43rd Hawaii International Conference on System Sciences (HICSS)*, Kauai, HI, January 5–8, 2010.)

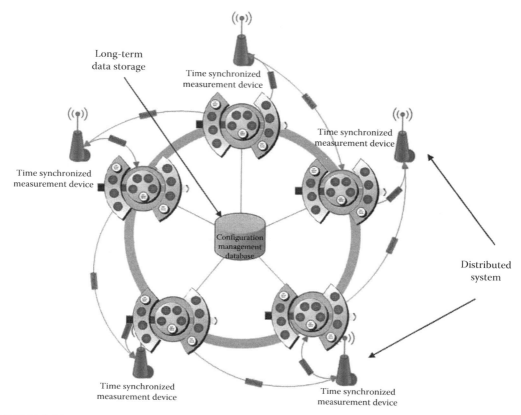

FIGURE 3.80 Generation II Super-PDC system (also known as the NERC phasor concentration system). (From Myrda, P.T. and Koellner, K., NASPInet—The internet for synchrophasors, *43rd Hawaii International Conference on System Sciences (HICSS)*, Kauai, HI, January 5–8, 2010.)

regional Super-PDCs could work together collaboratively to create a "distributed" system of data collection and concentration nodes that are centrally managed and configured, with minimal amount of information exchanged between nodes to ensure high availability. In this way, computationally intensive tasks such as data archival with I/O speed limitations could be dispersed across distributed resources. Additionally, such a distributed approach also eliminates the concern for a single point of failure associated with the earlier centralized approach (Figure 3.80).

In late 2009, TVA released the Super-PDC source code to open source development as the openPDC and formally posted the openPDC Version 1.0 source code in January 2010. The openPDC is an enhancement of the original TVA Super-PDC that has been modified for greater performance and scalability. In April 2010, TVA and NERC positioned the Grid Protection Alliance (GPA) to provide ongoing administration of the openPDC code base. GPA is a not-for-profit corporation that has been formed to support the electric utility industry.

Several U.S. utilities, under their 2010 Smart Grid Investment Grants (SGIG), have already undertaken pilot projects to implement and demonstrate various aspects of NASPInet. It is envisioned that NASPInet, once fully deployed, would support hundreds of phasor gateways and thousands of PMUs, each typically sampling data at 30 times per second.

3.4.3.4.2 Europe

Concepts of and experience with wide area monitoring systems (WAMS) in Europe date back to the years 1980–1990, when EdF, the French transmission system operator (TSO) of that time,

Smart Grid Technologies   199

developed a comprehensive plan based on phasor measurements. However, due to the fact that all the required telecommunication from the substation to the central control system and back was based on very expensive satellite channels, the system has never been put into operation. The further development of PMU technology based on accurate GPS time synchronization as well as the development of low-cost and reliable terrestrial communications channels has facilitated a restart of the phasor technology only some 10 years later. In the meantime, accurate synchronized off-line transient recorders have been developed and used for dynamic model calibration as well as for complex events analysis within the highly meshed CE system. One of the main driving factors for using more accurate measurement equipment was the increase of system dynamic challenges due to the increase of system size caused by the connection of power systems from the eastern European system to the western European system in the early 1990s. This need is intensified today by the more and more extensive use of the transmission system infrastructure due to increased market activities as well as increased power flow distances caused by renewable (wind) infeed far from the main energy consumers.

Power system equipment manufacturers have recognized the needs of system operators and have developed devices able to measure voltage and current phasors that are subsequently computed online in central PDCs. On the level of this centralized communication and data computation, together with a corresponding visualization platform, a large number of corresponding applications have been developed. Due to the nature of a relatively new technology, the ongoing WAMS activities can be divided into two categories.

3.4.3.4.2.1 Universities/Research & Development and Demonstration Projects For this kind of application, the data acquisition is performed using the public Internet connection and the data servers are located inside university or manufacturer labs. The PMUs used are mainly installed on the low voltage outlets in the buildings. Due to this fact, the related analyses are restricted to frequency and voltage phase angle as system input measurands. A few projects financed by the European Commission, such as ICOEUR, have already delivered valuable results.

3.4.3.4.2.2 Industrial Applications In the industrial applications, data acquisition is performed with the help of private TSO communications channels, where the PDC is embedded in the TSO IT environment and the output of the WAMS is already used within the operation or planning departments of the TSOs. In contrast to university-driven R&D projects, the TSO measurements are performed on the high voltage level using dedicated CT and VT measurements. Consequently, exact and high-resolution active power and reactive power measurements are also available.

Based on different technologies and corresponding software and hardware suppliers, the Continental Europe (CE) power system is monitored by receiving WAMS measurements from various transmission substations in each country. For the analysis of all major and minor events with a system-wide impact within the last years, those devices have delivered an important contribution for the related postmortem dynamic system analysis. The same measurements are continuously used for monitoring of the dynamic system performance as well as for the calibration of system dynamic models.

Some European TSOs have already integrated the PMU and corresponding PDC information within their SCADA systems. The corresponding main applications are the following:

- Voltage phase angle difference monitoring
- Line thermal monitoring
- Voltage stability monitoring (online P–V curves)
- Online monitoring of system damping (online modal analysis with online parameter estimation
- Intelligent alarming if predefined critical levels are exceeded
- Online monitoring of system loading

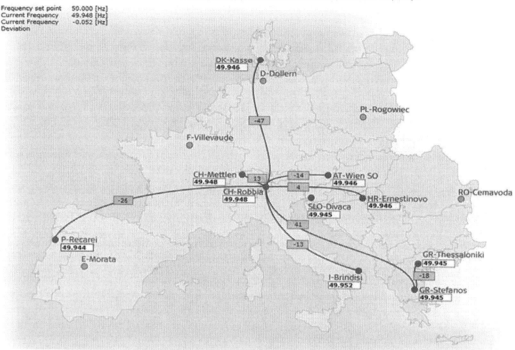

FIGURE 3.81 (**See color insert.**) Swissgrid web page showing current European PDCs links (January 2012). (© Copyright 2012 Swissgrid AG. All rights reserved.)

In order to increase system observability* beyond their own system observation area, a few European TSOs have already meshed their PDCs by exchanging PMU data online. One of these applications is a web page application setup by Swissgrid, see Figure 3.81.

Within the CE power system, more than 100 WAMS devices are currently in operation, continuously delivering high-quality measurements for system operation and system planning. Accurate time-stamped measurements have shown that they are a valuable component to ensure secure system operation. The related tools for data postprocessing have also demonstrated their maturity. However, WAMS integration with traditional SCADA systems has only reached the initial stages. In addition, the effort for enhancement of these links in combination with future implementation of dynamic security assessment (DSA), VSA (voltage security assessment), and wide area protection (WAP) systems have to be increased with the active participation of all partners (universities, manufacturers, TSOs, consultants).

3.4.3.4.3 Brazil

Brazil spans a large part of the South American continent. The distance of the far ends of the Brazilian territory (from north to south, and from east to west) is about 3900 km. Today, the

---

* Observability related to electric power systems is a necessary condition for state estimation. State estimation provides estimation of all measured and non-measured electrical quantities of the power system. Its output is used for online operation and management of the power system, such as load flow analysis.

Smart Grid Technologies

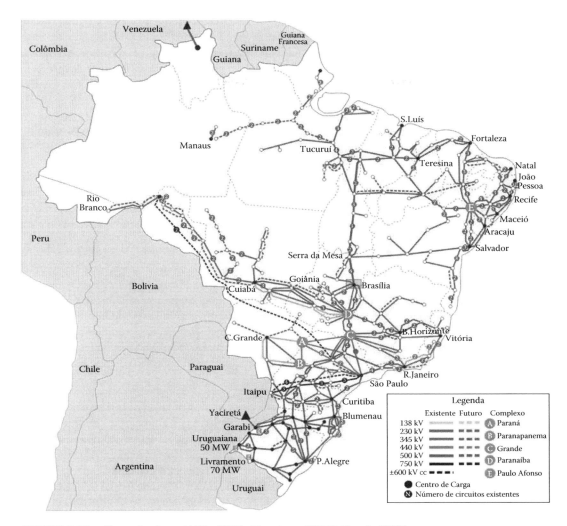

FIGURE 3.82 (**See color insert.**) The BIPS. (Courtesy of ONS, Brazil, 2010.)

Brazilian Interconnected Power System (BIPS) covers almost 70% of the Brazilian territory with a large transmission network that includes over 90,000 km of 230, 345, 440, 500, and 765 kV transmission lines, one 600 kV HVDC transmission line, approximately 400 substations and more than 170 power plants. Figure 3.82 shows the Brazilian main transmission grid.

The country's main generation source is hydroelectric. In the past, more than 80% of the total installed capacity and near 90% of the total energy production came from hydro plants. The hydro generation plants are located along 12 major hydrographic basins all over the Brazilian territory and many of them are not close to the major load centers in the southeast and south region. Some of the largest hydroelectric plants are the furthest from the load centers, resulting in bulk power transfers over long distances. Rainfall and the resulting inflow patterns are distinct among regions and may vary significantly over the year for each region, as well as between dry and wet years. In this scenario, one of the main operational tasks in the Brazilian power system is to allow the economical gains through interregional power transfers, taking advantage of the seasonal rainfall and water flow differences in each of its geoelectric regions. This is realized through optimization of the available hydro resources, mixed with complementary thermal energy. The result of this process has a direct impact on the overall operating cost of the system. As in all systems of this proportion, disturbances due to significant generation and load unbalances may cause excessive frequency variations, voltage collapse situations, and even the islanding of certain parts of the network, with loss of important load centers. Studies of the system dynamic behavior have shown inter-area low-frequency electromechanical oscillations in the range of 0.3–0.8 Hz. These oscillations are usually well damped but could, in some disturbances, spread with severe consequences. To avoid such situations, conventional (not synchronized measurement) system integrity protection schemes (SIPS) were deployed to perform predefined actions. Load shedding or generator tripping are some planned actions for expected system contingencies, such as losing one or more circuits of a major transmission path. The economic and reliable operation of the Brazilian power system must also accommodate the needs of a deregulated electricity market established since 1998, which increased the number of players in the electricity market. The main operational challenge of the Brazilian power system thus is how to achieve optimal hydro resource utilization while ensuring reliable system operation within the constraints of a long transmission system and market operation regulations.

The interest in transmission grid synchronized measurements in Brazil emerged in the 1990s due to the difficulty to assess the system dynamic performance during wide area disturbances. The PMU received the attention of the Brazilian Electric Studies Committee, a member of the Group to Coordinate the Interconnected Operation (GCOI) of the Brazilian power system. The feasibility of PMU application in the Brazilian power system was subject of preliminary studies done by this committee, with utilities' and manufacturers' participation.

With the electricity sector restructuring that happened in Brazil by the end of the 1990s, BIPS operation was transferred to a recently instituted Independent System Operator, the ONS. On March 11, 1999, only 2 months after ONS started operating the BIPS, Brazil faced a huge blackout. This blackout affected mainly the southeastern region, which accounts for the largest load in Brazil. The March 11 event analysis highlighted the need for better tools aiming at long-lasting dynamic behavior recording. Following the blackout recommendations, ONS started a project in 2000 to deploy a WAMS on the BIPS to record its dynamic performance, and in 2003, the first commercial PMU product was available in Brazil. In 2004 the regulatory environment in Brazil changed and ANEEL (the Brazilian regulatory office) decided not to allow ONS to own transmission assets. After working with ANEEL to reformulate the project strategy from the early centralized approach to a decentralized one, a resolution was passed in 2005 establishing the framework, under which the responsibilities and tasks for ONS and utilities in implementing the WAMS project were clearly defined. For ONS, its main responsibilities and tasks are as follows: (a) define and specify the WAMS architecture and equipment; (b) specify, acquire, and install the ONS PDCs; (c) define PMU placement on BIPS; (d) coordinate certification tests on PMU models to guarantee the system's integration and WAMS global performance; and (e) define the WAMS deployment schedule and coordinate the PMU installation by utilities. For utilities, their responsibilities and tasks are as follows: (a) purchase, install, operate, and maintain the PMU placed in their substations and (b) supply the communications links, complying with technical requirements, specifications, and schedules coordinated by ONS.

The deployment plan for Brazilian WAMS consists of three main components:

A phased deployment plan: ONS has adopted a phased deployment plan to address most of the challenges of this project. On the application side, ONS will focus on first deploying a sufficient number of PMUs at selected locations to facilitate the system dynamics recording for envisioned off-line applications, such as postmortem analysis, system model validation, and performance assessment. The number of PMUs will be gradually increased for real-time system operation support, such as state estimator improvement, until a full observability of BIPS's higher voltage level (345 kV and above) by phasor measurement is reached. Additional PMU installations for WAC and protection applications will be considered only at a later stage of the system deployment, as practical experience is gained with

Smart Grid Technologies

this technology. This phased deployment plan allows utilities and ONS to limit the initial capital investment, minimize the risks associated with many uncertainties of the project, and gradually gain experience on the system before making a full-scale deployment.

Top-down system design approach: ONS has adopted a top-down system design approach aimed to avoid potential future problems in its phased deployment plan. This top-down approach allows ONS to take not only the requirement of the current applications but also the need of future applications into account in the WAMS design. The system architecture is designed to be highly flexible and scalable to allow for easy system expansion later. It also allows the system design to take into account the availability of current, off-the-shelf products, maturity of technologies, as well as current communication support from the BIPS. In addition to system design, this approach enables ONS to provide unified design specifications for PMUs and any other system component that will be installed and operated at a utility's substation, such as substation phasor data concentrators (SPDC). These specifications will be used by all utilities involved in this project in their procurement process.

PMU/PDC certification test process: To ensure global performance of the Brazilian WAMS, ONS has included a PMU and PDC certification test process as an integral part of its deployment plan. The PMU certification test process included first developing a PMU test methodology and the test guidelines and then conducting the PMU certification test to ensure that all PMUs to be acquired by utilities will meet the same standards and system requirements. ONS is also envisioning the need for PDCs testing and certification. The WAMS architecture design includes the use of substation PDCs to aggregate and process the data from PMUs at the substation and then forward the data to PDCs installed at ONS control centers. Substation PDCs must therefore be verified to be interoperable with all PMU models and also with the PDCs installed at ONS' control centers. With a phased deployment plan, one of the main system design objectives of the Brazilian WAMS was the system flexibility and scalability for easy system expansion. For PDCs at ONS control centers, they will only support a small number of phasor measurements initially but they will be easily expanded to support hundreds of PMUs at the project final stage. ONS is investigating testing tools/methods that will allow it to verify whether main PDCs can meet the earlier requirements.

Another important WAMS initiative in Brazil came from the Santa Catarina Federal University (UFSC). The initiative started in 2001 as a research project carried out jointly by UFSC and a Brazilian industry partner. In 2003, the project received financial support from the Brazilian Government, which allowed the deployment of a prototype phasor measurement system. This first Brazilian system measures the distribution low voltage in nine university laboratories communicating with a PDC at UFSC over the Internet. This system recorded the BIPS dynamic performance during the latest major power system disturbances. Currently, another project from UFSC installed PMUs on three 500 kV substations in the South of Brazil.

WAMS deployment is understood as an important step to allow the Brazilian transmission system to evolve to a true smart grid. There is a common agreement that synchronized measurements will be part of the next generation of SCADA and EMSs. Without a better measurement system, it will be very difficult to develop more advanced EMS applications.

3.4.3.5 Role of WAMPAC in a Smart Grid

Smart grids will rely on various utility systems interoperating with each other. The level of interaction will be governed, among other requirements, by the ability of these systems to communicate and exchange meaningful data at sufficient time intervals. PMU data are time tagged and therefore any system interoperating with WAMPAC must include time synchronization. For example, WAMPAC systems can interoperate with network management systems to enable improved disturbance visualization into the control center. On the other hand, network management systems can supply other system information not modeled by WAMPAC systems to improve WAMPAC performance.

WAMPAC is the all encompassing term for WAM, WAP, and WAC applications of PMUs. Therefore, the integration of WAM, WAP, and WAC (WAMPAC) is in the context of the integration of the applications utilizing the phasor data originating from the PMUs and collected and disseminated by the PDCs and supporting architectures (e.g., NASPI in the United States).

3.4.3.5.1 Maintaining Reliability, Stability, and Security against Large Disturbances

3.4.3.5.1.1 Wide Area Monitoring Since power grid conditions are constantly changing, the overall health status of the grid is also constantly changing. It is the responsibility of grid operations to continually monitor real-time conditions to assess the current state of the system, to determine if corrective actions are required, and to identify and implement corrective actions if warranted. Synchrophasors and WAMPAC technology are smart-grid enabling technologies that offer great promise in terms of providing the industry with new SA tools to quickly assess the current grid conditions. Specifically, PMUs are capable of directly measuring the system state (i.e., voltage and current phasors) very accurately and at the high subsecond resolution, which is well suited for observing the dynamic behavior of the power grid and characterizing its stability. Of equal importance is the time-alignment property of these measurements that allows for comparison of phase angles from widely disparate locations to assess grid stress over a wide area. Measurement-based techniques that leverage these characteristics of WAMS technologies will complement existing EMS capabilities. Figure 3.83 illustrates how a PMU-based WAMS and network model-based EMS hybrid solution

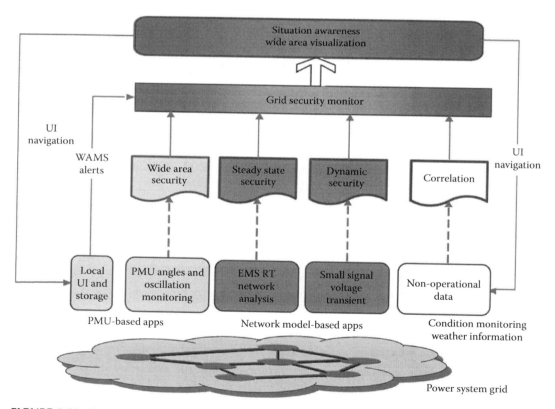

FIGURE 3.83 Integration of WAMS and EMS for enhanced grid security assessment. (© Copyright 2012 Alstom Grid. All rights reserved.)

can provide a more comprehensive grid security assessment. While measurement-based techniques may be applied to quickly and accurately assess grid conditions over a wide area basis, the model-based EMS applications offer the required context in terms of establishing dynamic security limits and suggesting corrective actions to mitigate potentially harmful conditions.

The phase angle separation information that is provided by WAMS is a good measure of grid instability and may signify potential voltage or oscillatory stability problems in the system. Similarly, rapid changes in phase angles that can be quickly detected by the high-resolution measurements can indicate sudden weakening of transmission capacity due line outages. Additionally, it is also possible to assess the current damping levels of both local and inter-area oscillations directly from the measurements, provide locational information on where the oscillations are most prominent, and alarm the operator should poor damping conditions occur (see Figure 3.84). Other measurement-based wide area security assessments include the use of localized frequency measurements from synchrophasors and observable time delays within this subsecond PMU data, along with any additional real-time EMS SCADA and transmission network topology information, to quickly identify the specific location of the origin of the disturbance, to detect and manage electrical islanding conditions, and then to monitor system restoration following grid separation.

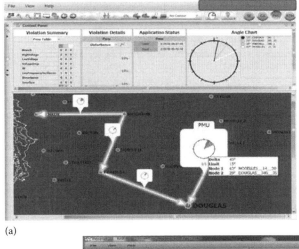

(a)

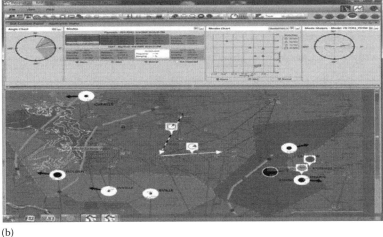

(b)

FIGURE 3.84 (a) Phase angle difference monitoring and (b) local and inter-area oscillations monitoring. (© Copyright 2012 Alstom Grid. All rights reserved.)

WAMS technologies also benefit steady-state network analysis applications. PMUs directly measure the system state. Using additional real-time measurements in the grid improves the EMS state estimator application, which helps in increasing the grid reliability and performance. A much needed *predictive* element of system operations is needed in smart grid to help the decision-making process of the control center operator. Once the operators have made an assessment of the current state and its vulnerability, operators will need to rely on "what-if" analytical tools to be able to make decisions that will prevent adverse conditions if a specific contingency or disturbance were to occur and make recommendations on corrective actions. Thus the focus shifts from "problem analysis" (reactive) to "decision making" (proactive/preventive). WAMPAC will play a significant role in future DSSs that will use more accurate forecast information and more advanced analytical tools to be able to confidently predict system conditions and analyze "what-if" scenarios in the transmission grid.

3.4.3.5.1.2 Wide Area Protection Protection in smart transmission grids will become more challenging due to entry of renewable generation and distributed energy resources. The underlying protection systems in traditional grids are largely designed based on conventional generator responses to short circuits. The pattern of fault currents flow from generators to the short circuit points is estimated with a high degree of certainty. Renewable and distributed energy resources will distort these responses, resulting in increased risk of failure to detect short circuits or increased risk of false operation of protection systems in the absence of a fault. A large number amount of distributed generation on the grid can also make the system behave dynamically different after fault clearing, thus posing greater risk of system disturbances. These system disturbances include power swings and oscillations that could propagate throughout the system. Fault detection and isolation schemes in a smarter grid will have to be revised to take into account the impact of distributed generators responding to short circuits. WAP will be able to processes multiple local and remote measurements and implement wider area protection schemes to contain or prevent the spread of disturbances in an interconnected power system. WAP could be used to isolate unstable areas of smart grids to prevent the disturbance from cascading into the other regions and also identify the separation boundaries in the grid to create islands that survive major disturbances. WAP systems are designed to protect the system when control actions fail to address the disturbance. Protection actions include system separation, controlled islanding, generator tripping, and any other actions designed to contain a large-scale disturbance from precipitating into a system collapse.

3.4.3.5.1.3 Wide Area Control Existing transmission grids are being pushed to their limits with tremendous growth of energy demand worldwide. The energy infrastructure must also take into account environmental constraints and energy-efficiency requirements while maintaining the grid stability. Substantial amounts of renewable energy and the use of HVDC and FACTS devices in transmission systems result in more complexity in grid controllability. In order to utilize these assets to improve the overall grid stability, a system-wide approach is essential. One of the major precursors of having system-wide control utilizing signals from remote locations is a very reliable communication infrastructure with a high bandwidth. PMUs are the building blocks of a wide area control system. Employing PMUs in a wide area control system that includes monitoring and control of HVDC, FACTS or power systems stabilizers can help to improve transfer capability and to counter disturbances, such as power oscillations as shown in Figure 3.85. Such remote power grid information could come from a wide area monitoring system (WAMS). WAMS/WACS applications range from monitoring (such as state estimation and voltage security monitoring) to wide area control such as the damping of power oscillations. It is envisioned that future smart transmission grid operation could be highly improved by WAMS/WACS. For example, events often cause failure or isolation of transmission lines and generators that could potentially lead to grid collapse. WAC can be used for transferring blocking or overriding signals to protection and control systems to allow grids to ride through disturbances. For example, it is well known that during voltage collapse events transformer tap changer operation to restore voltages to normal levels aggravates the voltage

Smart Grid Technologies

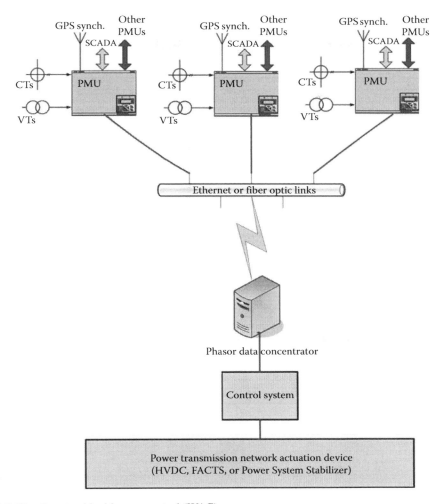

FIGURE 3.85 Smart grid wide area control (WAC).

collapse. WAC can send blocking signals to these transformers to inhibit their tap changer operations during voltage collapse.

3.4.3.5.1.4 Wide Area Stability Smart grid deployment results in both generation and loads being more dynamic and stochastic, which would make the grid be more vulnerable to adverse oscillations. Electromechanical oscillations, also known as small signal stability* problems, are one major threat to the stability and reliability of transmission grids. A poorly damped oscillation mode can become unstable, producing large-amplitude oscillations, leading to system breakup and large-scale blackout. Existing transmission capacity in most countries is derated in order to provide a margin of safety for reliable operations. There have been several incidents of system-wide low-frequency oscillations. Of them, the most notable is the August 10, 1996, U.S. western system breakup involving undamped system-wide oscillations. Figure 3.86 shows the measurement of power transfer from the Pacific Northwest to California for the August 10, 1996, event in the United States. The system deteriorated over time since the first line was tripped at 15:42:03. About 6 min later, undamped oscillations occurred and the system broke up into several islands.

---

* Small signal stability is the ability of the power system to maintain synchronism when subjected to small disturbances.

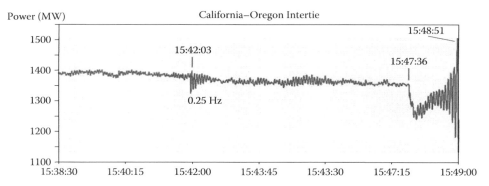

FIGURE 3.86 Undamped oscillations leading to the August 10, 1996, U.S. western system islanding event.

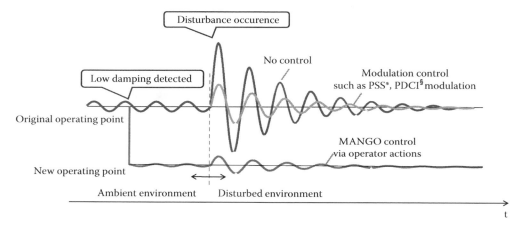

FIGURE 3.87 (See color insert.) MANGO versus modulation stability control. (* Power system stabilization; § Pacific DC Intertie damping).

The first step to address this concern is to develop real-time monitoring of low-frequency oscillations. Significant efforts have been devoted to monitoring system oscillatory behaviors from measurements in the past 20 years. The deployment of advanced sensors such as PMUs provides high-precision time-synchronized data needed for detecting oscillation modes. A category of measurement-based modal analysis techniques, also known as ModeMeter, uses real-time phasor measurements to estimate system oscillation modes and their damping. There is yet a need for new methods to bring modal information from a monitoring tool to actionable steps. The methods should be able to correlate low damping with grid operating conditions in a real-time manner, so that operators can respond by adjusting operating conditions when low damping is observed.

Modal Analysis for Grid Operations (MANGO) is a U.S. effort funded by the U.S. DOE to address the problem of adequately detecting transmission grid power oscillations and to establish a procedure to aid grid operation decision making for mitigating inter-area oscillations [15]. Compared to alternative modulation-based methods, MANGO aims to improve damping through adjustment of operating points, whereas the modulation-based methods do not change the grid operating points. Figure 3.87 illustrates the difference of these two types of damping improvement

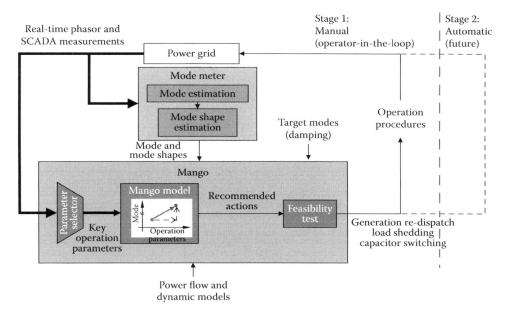

FIGURE 3.88 Proposed MANGO framework.

methods. Modulation control retains the operating point but improves damping through automatic feedback control. Figure 3.88 illustrates the overall proposed MANGO framework.

Based on the effect of operating points on modal damping, MANGO can improve small signal stability through operating point adjustment. Simulation studies show that damping ratios can be controlled by operators through adjustment of grid operating parameters, such as generation redispatch, or load reduction as a last resort. At the same stress level (total system load), inter-area oscillation modes can be controlled by adjusting generation patterns to reduce flow on the interconnecting tie-line(s).

3.4.3.5.2 Management of Large Numbers of Intermittent Generation

Being able to monitor intermittent generation, such as renewable energy sources, in realtime is valuable to the management of the generation. EMS and DMS systems might not have sufficient monitoring capacity both in terms of monitoring points and signal sampling to include intermittent generation. Several dedicated WAMPAC systems can be deployed to perform such management functions. Under this scheme, WAMPAC can be used to manage the highly changing operating conditions, including intermittent power flows, autonomously. The same system can communicate with existing EMS or DMS system for receiving operator dispatch orders, if necessary.

3.4.3.5.3 Maintaining Power Quality

PMUs have been used in the past to monitor current and voltage harmonics. With WAMPAC, this information can be relayed and visualized in the network control room so that operators can resolve the power quality issue. WAMPAC time-synchronized data drive a host of monitoring applications that brings value to monitoring power quality in the control center and substations. Modern visualization tools such as those based on geographical information systems (GIS) can be made dynamic by layering the time-synchronized data of the power system captured by WAMPAC. Operators can then benefit from wide area visibility of power quality, allowing prompt preventive action to take place. For example, voltage sags or swells can be easily viewed in an animated fashion using WAMPAC visualization applications previously discussed.

3.4.3.5.4 Increasing Transmission Efficiency

WAMPAC systems can contribute to improving smart grid transmission efficiency in the following ways: (1) improved smart grid network management, (2) congestion management via stabilizing control, and (3) real-time optimization of grid operating parameters. Unlike traditional EMS/SCADA systems, WAMPAC systems can capture the accurate real-time state of the system. For example, up-to-date conductor temperatures calculated from phasor measurements can help determine additional power transfer capability of the transmission grid. Thermal margins in smart grid operations can be used by economic dispatch to ensure that most economical units are allocated, thereby minimizing total cost of power delivery. PMUs can also be used to enable stabilizing controls to mitigate potential destabilizing phenomena such as voltage collapse and angle instability. Traditionally, operators place flow limits on transmission lines to ensure that no destabilization takes place following a disturbance. It results in inefficient utilization of transmission assets. In most cases, stabilizing control will contribute to transmission efficiency by releasing extra transmission margins and/or allowing more economical generator dispatch. As discussed previously, deployment of PMUs enables an accurate estimation of the smart grid system model. This model can be used to calculate the most optimal set points for HVDC and FACTS devices. HVDC and FACTS devices can contribute to transmission efficiency if WAMPAC can modulate their operating set points according to the current smart grid conditions. For example, the objective can be expressed defined as optimal power flow with minimum transmission losses.

3.4.4 ROLE OF TRANSMISSION SYSTEMS IN SMART GRID

Renewable energy generation is a key topic of today's power systems, in all countries. Driven by the need to reduce CO_2 emissions to stop or at least reduce the global warming effect, new "CO_2-free" technologies are investigated to fulfill the energy requirements of the future. Based on the Kyoto protocol and its subsequent conferences, most countries have committed to specific CO_2 reduction and renewable energy targets within the next 10–20 years.

Large synchronous power grids, for example, in the Americas and in Europe, continue to develop in complexity and were not originally designed to serve the purpose they are expected to carry out nowadays, and this progression will continue into the future. Originally, the conventional power plants which are very easy to control were mostly built in the vicinity of cities and load centers and the grid around them was designed to provide the required capacity. The power demand was growing over the years and the ever-increasing amount of power capacity had to be brought from the adjacent grids over large distances. In addition to this, in the course of deregulation and privatization, a great number of power plants had to change their location; in the meantime plenty of volatile wind power has been installed in many countries, causing parts of the grid which may already be overloaded to become even more overloaded. For power grids, wind energy is the most difficult to process due to its inherent variability, whether it is located onshore or offshore. These fluctuations create great difficulties for the grids, for in this case not only the power flow, that is, the power supply, but also the voltage of the grids is affected. This results in fluctuations of both active and reactive power. This deteriorates voltage quality; the corresponding grid code can no longer be adhered to, and the adjacent loads as well as the grid itself are affected detrimentally. Moreover, in the event of grid faults, larger power outages referred to as "voltage collapse" can easily occur due to cascading tripping of wind or solar generators at low voltage levels. Due to this, in a large number of countries, the grid codes have been significantly tightened in order to fix the voltage within the exact, time-dependent ranges of tolerance and to protect the grid.

The security of power supply in terms of reliability and blackout prevention has the utmost priority when planning and extending power grids. The availability of electric power is the crucial prerequisite for the survivability of a modern society and power grids are virtually its lifelines. The aspect of sustainability is gradually gaining in importance in view of such challenges as the global climate protection and economical use of power resources are running short. It is, however, not a

Smart Grid Technologies

means to an end to do without electric power in order to reduce CO_2 emissions. A more appropriate way is to integrate renewable energy resources to a greater extent in the future (energy mix) and, in addition to this, to increase the efficiency of conventional power generation as well as power transmission and distribution without loss of system security. The future power grids will have to withstand increasingly more stresses caused by large-scale power trading and a growing share of fluctuating regenerative energy sources, such as wind and solar power. In order to keep generation, transmission, and consumption in balance, the grids must become more flexible, that is, they must be controlled in a better way. State-of-the-art power electronics with HVDC and FACTS technologies provide a wide range of applications with different solutions, which can be adapted to the respective grid in the best possible manner. DC current transmission constitutes the best solution when it comes to loss reduction for transmitting power over long distances. The HVDC technology also helps control the load flow in an optimal way. This is the reason why, along with system interconnections, the HVDC systems become part of synchronous grids increasingly more often—either in form of a B2B for load flow control and grid support or as a DC energy highway to relieve heavily loaded grids.

HVDC technology allows for grid access of generation facilities on the basis of availability-dependent regenerative energy sources, including large on- and offshore wind farms, and compared with conventional AC transmission, it suffers a significantly lower level of transmission losses on the way to the loads.

Based on these evaluations, Figure 3.89 shows the stepwise interconnection of a number of grids by using AC lines, DC B2B systems, DC long-distance transmissions, and FACTS for strengthening the AC lines. These integrated hybrid AC/DC systems provide significant advantages in terms of technology, economics, as well as system security. They reduce transmission costs and help bypass heavily loaded AC systems. With these DC and AC ultrahigh power transmission technologies, the "smart grid," consisting of a number of highly flexible "microgrids," will turn into a "super grid" with bulk power energy highways, fully suitable for a secure and sustainable access to huge renewable energy resources such as hydro, solar, and wind, as indicated in Figure 3.90. This approach is an important step in the direction of environmental sustainability of power supply: transmission technologies with HVDC and FACTS can effectively help reduce transmission losses and CO_2 emissions.

Despite a significant share of wind power, the stability of the grid has to be maintained that is, grid access solutions are needed, which provide both sustainability and security of electric power supply. This can be made possible by means of power electronics with dynamic fast control, which makes the grid more flexible and subsequently able to take in more regenerative and distributed energy sources. The solution of choice to tackle this complex task is FACTS and HVDC technology for they can be controlled on demand which takes a conventional grid to the "smart grid."

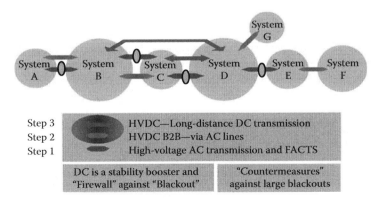

FIGURE 3.89 (See color insert.) Hybrid system interconnections—"Super Grid" with HVDC and FACTS. (© Copyright 2012 Siemens. All rights reserved.)

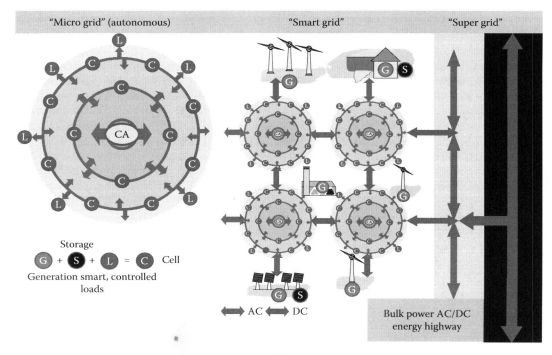

FIGURE 3.90 Prospects of smart transmission grid developments. (© Copyright 2012 Siemens. All rights reserved.)

3.5 DISTRIBUTION SYSTEMS

3.5.1 Distribution Management Systems

Stuart Borlase, Jiyuan Fan, and Tim Taylor

3.5.1.1 Distribution SCADA

Supervisory control and data acquisition (SCADA) systems are a relative mature technology for the management of distributed asset systems. While they have long been used for the management of generation and transmission systems, they are increasingly being employed for the monitoring and control of distribution systems. Technology advances that will aid in the deployment of SCADA technologies are still occurring, particularly in the communications area. These are described in other chapters in this book.

Figure 3.91 conceptually illustrates the major components of a SCADA system. The SCADA master hardware and software is typically located centrally at the control center. The control center consists of the SCADA application servers, the communications front end processors, a data historian, interfaces to other control systems, operator work stations, and other supporting components. The primary SCADA system is often redundant, with a local backup system and/or remote backup at another site. Other system environments are often installed by the utility for testing and quality assurance, development, and training. Various types of communications links to the remote terminal units (RTUs) are used. These communications links are now becoming more IP based using open protocols.

In the application of SCADA for distribution systems, the costs of the additional sensors, IEDs (intelligent electronic devices), RTUs, communications, and SCADA master station must be considered relative to the benefits that are realized. It is rarely economical to monitor and control an entire distribution system with SCADA points. Distribution organizations typically choose to

Smart Grid Technologies

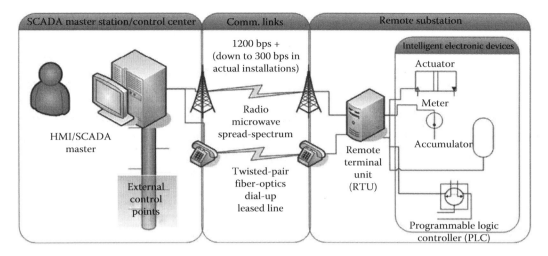

FIGURE 3.91 Major components of a SCADA system.

apply SCADA only to equipment that provide them with adequate return on investment in terms of improving reliability, volt/VAr control (VVC), situational awareness, remote control, or other business benefits. Monitoring and control of large distribution substations is usually always beneficial, but monitoring and controlling equipment further down the network on distribution feeders is not widespread, at least in the United States and other utilities with geographically large distribution systems. Figure 3.92 shows typical equipment types that can be part of a SCADA system applied on overhead distribution systems.

The most common equipment monitored and controlled in distribution SCADA include recloser controllers, switch controllers, voltage regulator controllers, and switched capacitor bank controllers.

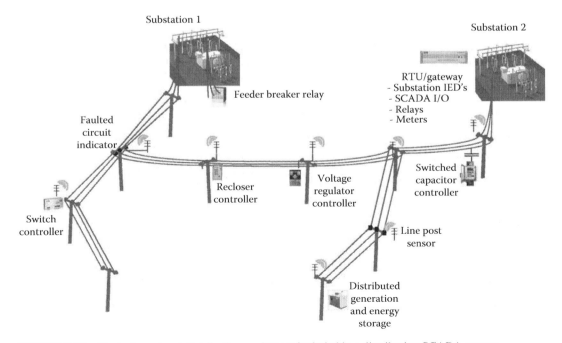

FIGURE 3.92 Typical overhead distribution equipment included in a distribution SCADA system.

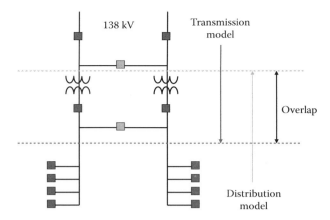

FIGURE 3.93 Possible overlap in separate transmission and distribution SCADA systems.

In many cases, IEDs and associated CTs and PTs are installed at these devices on the feeder, and adding the communications capability is only an incremental cost. The status and analog values monitored at these points provide operators with valuable visibility of the network operations further down the distribution system. In addition, if remote control is enabled for these devices, then reliability can be improved from the control center (through the recloser controllers and the switch controllers), and VVC can be improved (through the voltage regulator controllers and the switched capacitor bank controllers.).

In underground distribution systems, SCADA can be applied to equipment such as the network protectors in network transformer vaults, automatic throwover equipment, and the ring-main units that are used in many parts of the world for protection and switching. In these cases, the status, analog, and control points are similar to those for the overhead distribution system.

With the extension of SCADA to the distribution system, an important consideration is the best way to manage the SCADA within the distribution substation, both from a technology viewpoint and a business process perspective. If the transmission system SCADA and the distribution system SCADA are handled by the same utility operators, then it is greatly simplified. But in many organizations, distribution operations and transmission operations are separate. In such cases, coordination between the two organizations, for work flows such as switching, tagging, and control, must be established. Development, maintenance, and coordination of the two network models must also be addressed.

Figure 3.93 shows a typical distribution substation. An area of overlap exists between a newly defined distribution SCADA and an existing transmission SCADA/EMS. It shows the area of overlap between transmission and distribution, as well as the extent of their respective network models.

3.5.1.2 Trends in Distribution SCADA and Control

For master station developments, one of the key trends in the industry is the increase of bandwidth from the substation to the control center and also from the monitoring and control points on the distribution network to the control center. This increase in bandwidth enables the proliferation of thousands of low-cost sensors to be deployed on the network to increase the monitoring and measuring capability of SCADA, which will enable the applications at the master station to have a more complete view of the network and increase the accuracy of calculations and predictions—enabling more automated operations to take place.

The exponential rise in the number of real-time points means the old fixed capacities of SCADA systems have to be left behind and modern systems need to be able to scale up while maintaining and improving upon accepted performance standards. This is aided by the power of CPUs and the relatively cheap availability of large amounts of RAM, but must also be inherently supported by the design of the software systems processing the information. Additionally the more accurate modeling of the distribution network will enable optimization algorithms to run, reducing peak load and deferring investment in transmission and distribution assets. Many localized fault locators will be able to be deployed to accurately locate faults and enable restoration to occur quickly. Significant changes will also be seen in the area of database management and the reduction of configuration costs. IEC 61850 will greatly improve communications between devices. For the first time, vendors and utilities have agreed upon an international standard protocol. This will allow an unprecedented level of interoperability between devices of multiple vendors in a seamless fashion. The self-description feature of IEC 61850 will greatly reduce configurations costs. Also because of a single standard for all devices, training, engineering, and commissioning costs will be greatly reduced. IEC 61850 supports both client/server communications as well as peer-to-peer communications. The IEC process bus will allow for communication to the next generation of smart sensors. There is currently work under way to harmonize the EPRI CIM model and enterprise service bus IEC 61968 standards with the substation IEC 61850 protocol standards. Bringing these standards together will greatly reduce the costs of configuring and maintaining a master station through plug and play compatibility and database self-description. Advances in GUI/HMI interfaces will also be greatly improved. Moves toward browser-based displays will become more prevalent. New improvements that enhance the user experience will be developed especially in the area of safety.

Control systems already contain more and more intelligence and that trend will continue. Users are used to operating on an exception basis, for example, responding to a feeder lockout alarm only after local auto reclose schemes have completed. In the future there will be a lot more information available to the system, which in turn means that additional intelligence must be applied to that information in order to present the operator with the salient information rather than simply passing on more data. Taking the example of a fault on a distribution feeder further, an example would be that instead of presenting the user with a lockout alarm, accompanied by associated low volts, fault-passage indications, battery alarms, etc., leaving it up to the operator to drill down, diagnose, and work out a restoration strategy, the distribution control system will instead notify the operator that a fault has occurred and analysis and restoration is in progress in that area. The system will then analyze the scope of the fault using the information available, tracing the current network model; identifying current relevant safety documents, operational restrictions, and sensitive customers; and locating the fault using location data from the field. The system will automatically run load flow studies identifying current loading, available capacities, and possible weaknesses, using this information to develop a restoration strategy. The system will then attempt an isolation of the fault and maximum restoration of customers with safe load transfers, potentially involving multilevel feeder reconfiguration to prevent cascading overloads to adjacent circuits. Once the reconfiguration is complete, the system can alert the operator to the outcome and even automatically dispatch the appropriate crew to the identified faulted section.

Control systems will not only be able to present information to operators for their consideration but they will be able to advise the operator on how to best deal with a situation. Systems are able to propose and validate switching for planned and unplanned work. Faults can be automatically isolated and partially restored automatically as in the aforementioned example, or more likely in the short term, the system can carry out the analysis and present the proposed strategy to the users to carry out interactively. This mode of operation also supports the current reality that most distribution switching is manual; however, a control system can be set up to operate a "first pass restoration" via available SCADA devices to maximize customers on supply, then a second wave of manual switching coordinated by the operator. The system conditions can be monitored and checked against expected future loads and contingencies without the operator taking action.

3.5.1.3 Current Distribution Management Systems

Distribution management systems (DMSs) started with simple extensions of SCADA from the transmission system down to the distribution network. A large proportion of dispatch and system operations systems in service today rely on manual and paper-based systems with little real-time circuit and customer data. Operators have to contend with several systems and interfaces on the control desk ("chair rolls") based on multiple network model representations. The experience of operators is the key to safe system operation. With an increase in regulatory influence and smart grid focus on advanced technologies, there is a renewed interest in increasing investment in distribution networks to defer infrastructure build-out and reduce operating and maintenance costs through improving grid efficiency, network reliability, and asset management programs.

As distribution organizations have become more interested in increasing asset utilization and reducing operational costs, advanced DMS applications have been developed. These include load allocation and unbalanced load flow analysis; switch order creation, simulation, approval, and execution; overload reduction switching; and capacitor and voltage regulator control.

Two specific examples of advanced applications that reduce customer outage durations are the fault-location application and the restoration switching analysis (RSA) application.

Various DMS applications are commonly used today.

Fault detection, isolation, and service restoration (FDIR) is designed to improve the system reliability. FDIR detects a fault on a feeder section based on the remote measurements from the feeder terminal units (FTUs), quickly isolates the faulted feeder section, and then restores the service to the unfaulted feeder sections. It can reduce the service restoration time from several hours to a few minutes, considerably improving the distribution system reliability and service quality. The fault-location application estimates the location of an electrical fault on the system. This is different than identifying the protective device that operated, which typically is done based on the pattern of customer outage calls or through change in a SCADA status point. The location of the electrical fault is where the short-circuit fault occurred, whether it was a result of vegetation, wildlife, lightning, or something else. Finding the location of an electrical fault can be difficult for crews, particularly on long runs of conductor not segmented by protective devices. Fault location tends to be more difficult when troubleshooters or crews are hindered by rough terrain, heavy rain, snow, and darkness. The more time required to locate the fault, the more time customers are without power. DMS-based fault-location algorithms use the as-operated electric network model, including the circuit connectivity, location of open switches, and lengths and impedances of conductor segments, to estimate fault location. Fault current information such as magnitude, predicted type of fault, and faulted phases is obtained by the DMS from IEDs such as relays, recloser controls, or RTUs. After possible fault locations are calculated within the DMS application, they are geographically presented to the operator on the console's map display and in tabular displays. If a geographic information system (GIS) land base has been included, such as a street overlay, an operator can communicate to the troubleshooter the possible location including nearby streets or intersections. This information helps crews find faults more quickly. As business rules permit, upstream isolation switches can be operated and upstream customers can be reenergized more quickly, resulting in much lower interruption durations. The DMS fault-location application uses the electrical DMS model and fault current information from IEDs to improve outage management.

RSA is an advanced application that improves reliability performance indices. This application can improve the evaluation of all possible switching actions to isolate a permanent fault and restore customers as quickly as possible. The application recommends the suggested switching actions to the operator, who can select the best alternative based on criteria such as number of customers restored, number of critical customers restored, and required number of switching operations. Upon the occurrence of a permanent fault, the application evaluates all possible switching actions and executes an unbalanced load flow to determine overloaded lines and

Smart Grid Technologies

low-voltage violations if the switching actions were performed. The operator receives a summary of the analysis, including a list of recommended switching actions. Similar to the fault-location application, the functionality uses the DMS model of the system but improves outage management and reduces the customer average interruption duration index (CAIDI) and SAIDI. The RSA application is particularly valuable during heavy loading and when the number of potential switching actions is high. Depending on the option selected, the application can execute with the operator in the loop or in a closed-loop manner without operator intervention. In closed-loop operation, the RSA application transmits control messages to distribution devices using communications networks such as SCADA radio, paging, or potentially AMI infrastructure. Such an automated isolation and restoration process approaches what many call the "self-healing" characteristic of a smart grid.

Integrated volt/VAr control (IVVC) has three basic objectives: reducing feeder network losses by energizing or de-energizing the feeder capacitor banks, ensuring that an optimum voltage profile is maintained along the feeder during normal operating conditions, and reducing peak load through feeder voltage reduction by controlling the transformer tap positions in substations and voltage regulators on feeder sections. Advanced algorithms are employed to optimally coordinate the control of capacitor banks, voltage regulators, and transformer tap positions.

The topology processor (TP) is a background, off-line processor that accurately determines the distribution network topology and connectivity for display colorization and to provide accurate network data for other DMS applications. The TP may also provide intelligent alarm processing to suppress unnecessary alarms due to topology changes.

Distribution power flow (DPF) solves the three-phase unbalanced load flow for both meshed and radial operation scenarios of the distribution network. DPF is one of the core modules in a DMS and the results are used by many DMS applications, such as FDIR and IVVC, for analyses.

Load modeling/load estimation (LM/LE) is a very important base module in DMS. Dynamic LM/LE uses all the available information from the distribution network—including the user transformer capacities and customer monthly billings, if available, combined with the real-time measurements along the feeders to accurately estimate the distribution network loading, for both individual loads and aggregated bulk loads. The effectiveness of the entire DMS relies on the data accuracy provided by LM/LE. If the load models and the load values are not accurate enough, all the solution results from the DMS applications will be useless.

Optimal network reconfiguration (ONR) is a module that recommends switching operations to reconfigure the distribution network to minimize network energy losses, maintain optimum voltage profiles, and balance the loading conditions among the substation transformers, the distribution feeders, and the network phases. ONR can also be utilized to develop outage plans for maintenance or service expansion fieldwork.

Contingency analysis (CA) in the DMS is designed to analyze potential switching and fault scenarios that would adversely affect supply to customers or impact operational safety. With the CA results, proactive or remedial actions can be taken by changing the operating conditions or network configuration to guarantee minimal number of customer outages and maximum network reliability.

Switch order management (SOM) is a very important tool for system operators in real-time operation. Several of the DMS applications and the system operators will generate numerous switch plans that have to be well managed, verified, executed, or rejected. SOM provides advanced analysis and execution features to better manage all switch operations in the system.

Short-circuit analysis (SCA) is an off-line function to calculate the short-circuit current for hypothetical fault conditions in order to evaluate the possible impacts of a fault on the network. SCA then verifies the relay protection settings and operation and recommends more accurate relay settings or network configuration.

Relay protection coordination (RPC) manages and verifies the relay settings of the distribution feeders under different operating conditions and network reconfigurations.

Optimal capacitor placement/optimal voltage regulator placement (OCP/OVP) is an off-line function used to determine optimal locations for capacitor banks and voltage regulators in the distribution network for the most effective control of the feeder VArs and voltage profile.

Dispatcher training simulator (DTS) is employed to simulate the effects of normal and abnormal operating conditions and switching scenarios before they are applied to the real system. In distribution grid operation, DTS is a very important tool that can help the operators to evaluate the impacts of an operation plan in advance or simulate historical operation scenarios to obtain valuable training on the use of the DMS. DTS is also used to simulate conditions of system expansions.

3.5.1.4 Advanced Distribution Management Systems

Distributed energy resources (DER) on the distribution network will be from disparate sources and subject to great uncertainty. The electricity consumption of individual consumers is also of great uncertainty when they respond to the real-time pricing and rewarding policies of power utilities for economic benefits. The conventional methods of LM and LE in the traditional DMS are no longer effective, rendering other DMS applications ineffective or altogether useless. The impact of demand response management (DRM) and consumer behaviors may be modeled or predicted from the utility pricing rules and rewarding policies for specified time periods, which can be incorporated into the LM and LE algorithms; this requires a direct linkage between the DMS and the DRM applications. When the DRM application attempts to accomplish load relief in response to a request from the independent system operator (ISO), it will need to verify from the DMS that the DRM load relief will not result in any distribution network connectivity, operation, or protection violations. The high penetration of distributed generation will require the load flow algorithm to deal with multiple, incremental, and isolated supply sources with limited capacities, as well as a network topology that is no longer radial or is weakly meshed. In a faulted condition, the distributed generation will also contribute to the short-circuit currents, adding to the complexity of the SCA, RPC, and FDIR logic.

A number of smart grid advances in distribution management are expected, as shown in the Figure 3.94.

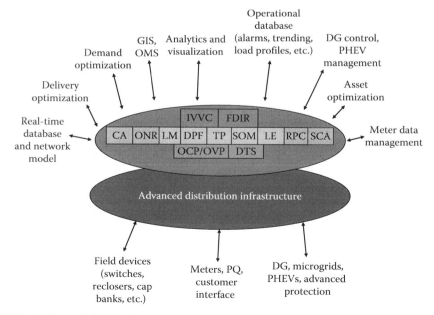

FIGURE 3.94 Advanced distribution management for the smart grid. (Fan, J. and Borlase, S., Advanced distribution management systems for smart grids, and *IEEE Power and Engineering.* © Copyright March/April 2009 IEEE.)

Monitoring, control, and data acquisition will extend further down the network to the distribution pole-top transformer and perhaps even to individual customers by means of an advanced metering infrastructure (AMI) and/or demand response and home energy management systems on the home area network (HAN). More granular field data will help increase operational efficiency and provide more data for other smart grid applications, such as outage management. Higher speed and increased bandwidth communications for data acquisition and control will be needed. Sharing communications networks with an AMI will help achieve system-wide coverage for monitoring and control down the distribution network and to individual consumers.

Integration, interfaces, standards, and open systems will become a necessity. Ideally, the DMS will support an architecture that allows advanced applications to be easily added and integrated with the system. Open standards databases and data exchange interfaces (such as CIM, SOAP, XML, SOA, and enterprise service buses) will allow flexibility in the implementation of the applications required by the utility, without forcing a monolithic distribution management solution. For example, the open architecture in the databases and the applications could allow incremental distribution management upgrades, starting with a database and monitoring and control application (SCADA), then later adding an IVVC application with minimal integration effort. As part of the overall smart grid technology solution or roadmap, the architecture could also allow interfacing with other enterprise applications (such as a GIS, an outage management system (OMS), or a meter data management system (MDMS) via a standard interface. Standardized web-based user interfaces will support multiplatform architectures and ease of reporting. Data exchange between the advanced DMS and other enterprise applications will increase operational benefits, such as MDM and outage management.

FDIR will require a higher level of optimization and will need to include optimization for closed-loop, parallel circuit, and radial configurations. Multilevel feeder reconfiguration, multiobjective restoration strategies, and forward-looking network loading validation will be additional features with FDIR.

IVVC will include operational and asset improvements—such as identifying failed capacitor banks and tracking capacitor bank, tap changer, and regulator operation to provide sufficient statistics for opportunities to optimize capacitor bank and regulator placement in the network. Regional IVVC objectives may include operational or cost-based optimization.

LM/LE will be significantly changed where customer consumption behaviors will no longer be predictable but more smartly managed individually and affected by distribution response management.

With a significant increase in real-time measurements available from more widespread installations of field IEDs on feeders and meter data from end users and AMI systems, distribution state estimation (DSE) will play an important role in monitoring the overall grid operation condition and situation awareness, as well as in supporting IVVC and other distribution optimization functions. More accurate estimation of distribution system voltages extending from the substation down the feeders to end user locations will allow IVVC to precisely control the voltage profiles along the feeder and at the end user to realize more economic benefits.

TP, DPF, ONR, CA, SCA, and RPC will be used on a more frequent basis. They will need to include single-phase and three-phase models and analysis, and they will have to be extended down the network to individual customers. Moreover, distribution optimization functions such as FDIR, IVVC, and ONR will be more effectively integrated for real-time and look-ahead operational support. Distribution optimization functions will also be coordinated with consumer demand management and DER optimization.

Distributed generation, microgrids, and customer generation (such as plug-in hybrid vehicles [PHEVs]) will add many challenges to the protection, operation, and maintenance of the distribution network. Small generation loads at the customer interface will complicate power flow analysis, CA, and emergency control of the network. Protection and control schemes will need to account for bidirectional power flow and multiple fault sources. Protection settings and fault restoration algorithms may need to be dynamically changed to accommodate changes in the network configuration and supply sources.

The development of new technologies and applications in distribution management can help drive optimization of the distribution grid and assets. The seamless integration of smart grid technologies is not the only challenge. Also challenging is the development and implementation of the features and applications required to support the operation of the grid under the new environment introduced by the use of clean energy and distributed generation as well as the smart consumption of electricity by end users. DMSs and distribution automation applications have to meet the new challenges, requiring advances in the architecture and functionality of distribution management, that is, an advanced DMS for the smart grid. Expect to see an evolution of traditional distribution management to include advanced applications to monitor, control, and optimize the network in the smart grid, that is, an advanced DMS for the smart grid.

Databases and data exchange will need to facilitate the integration of both geographical and network databases in an advanced DMS. The geographical and network models will need to provide single-phase and three-phase representations to support the advanced applications. Ideally, any changes to the geographical data (from network changes in the field) will automatically update the network models in the database and user interface diagrams. More work is required in the areas of distributed real-time databases, high-speed data exchange, and data security.

Dashboard metrics, reporting, and historical data will be essential tools for tracking performance of the distribution network and related smart grid initiatives. For example, advanced distribution management will need to measure and report the effectiveness of grid efficiency programs, such as VAr optimization, or the system average interruption duration index (SAIDI), the system average interruption frequency index (SAIFI), and other reliability indices related to delivery optimization smart grid technologies. Historical databases will also allow verification of the capability of the smart grid optimization and efficiency applications over time, and these databases will allow a more accurate estimation of the change in system conditions expected when the applications are called upon to operate. Alarm analysis, disturbance, event replay, and other power quality metrics will add tremendous value to the utility and improve relationships with customers. Load forecasting and load management data will also help with network planning and optimization of network operations.

Analytics and visualization will assimilate the tremendous increase in data from the field devices and integration with other applications, and they will necessitate advanced filtering and analysis tools. Visualization of the data provides a detailed but clear overview of the large amounts of data. Data filtering and visualization will help quickly analyze network conditions and improve the decision-making process. Visualization in an advanced DMS would help display accurate, near-real-time information on network performance at each geospatially referenced point on a regional or system-wide basis. For example, analytics and visualization could show voltage magnitudes by color contours on the grid, monitor and alarm deviations from nominal voltage levels, or show line loading through a contour display with colors corresponding to line loading relative to capacity. System operators and enterprise users will greatly benefit from analytic and visualization tools in day-to-day operations and planning.

Enterprise integration is an essential component of the smart grid architecture. To increase the value of an integrated smart grid solution, the advanced DMS will need to interface and share data with numerous other applications. For example, building on the benefits of an AMI with extensive communication coverage across the distribution system and obtaining operational data from the customer point of delivery (such as voltage, power factor, loss of supply) help to improve outage management and IVVC implementation.

Enhanced security will be required for field communications, application interfaces, and user access. The advanced DMS will need to include data security servers to ensure secure communications with field devices and secure data exchange with other applications. The use of IP-based communications protocols will allow utilities to take advantage of commercially available and open-standard solutions for securing network and interface communications.

3.5.2 Volt/VAr Control

Stuart Borlase, Jiyuan Fan, Xiaoming Feng, Carroll
Ivester, Bob McFetridge, and Tim Taylor

VVC relates to switching of distribution substation and feeder voltage regulation equipment and capacitor banks with two main objectives: reducing VAr flow on the distribution system and adjusting voltage at the customer delivery point within required limits. An effective VVC approach combines, coordinates, and optimizes the control of both VAr flow and customer voltage. Components of VVC are as follows:

VAr control, VAr compensation, power factor correction

Substation and distribution feeder capacitor banks are used to minimize VAr flow (improve power factor) on the distribution feeder during all load levels (peak and base). Reduction of VAr flow reduces distribution system losses, which reduces load on the substation and distribution feeders.

Conservation voltage reduction

CVR (*conservation voltage reduction*) is the control of substation transformer LTCs (load tap changers) and distribution feeder voltage regulators to reduce customer delivery voltage within specified and safe margins at the customer service point during peak periods of load, which may result in a reduction of customer load and, in turn, result in load reduction on the substation and distribution feeders. CVR may also be implemented during base loading periods. Voltage control is not only exercised for CVR but also to comply with normal operation and regulatory compliance.

Integrated Volt/VAr control

IVVC is the coordination of VAr flow and CVR to reduce distribution feeder losses and control the voltage profile on the feeder, which may reduce system losses and may improve service voltage to the customer. Other possible benefits may include reduction in capacitor bank inspections and capacitor bank troubleshooting.

Volt/VAr optimization

VVO (*volt/VAr optimization*) is the capability to optimize the objectives of VAr (loss) minimization and load reduction (with voltage constraints) using optimization algorithms and well-defined control objectives subject to various system constraints through centralized or decentralized decision makings.

The discussions that follow on voltage and VAr control reference the voltage levels and operation of the U.S. electrical system as an example.

3.5.2.1 Inefficiency of the Power Delivery System

Electric utilities have two concerns when it comes to transmitting electricity from the generator to the customer. First, it must get there safely and reliably. Second, the majority of what is generated must make it to the customer in order for the utility to be profitable—efficiency of power delivery to the customer. For a utility to maximize profits, it must minimize the amount of electric losses on the system during the transfer of electricity from the generation site to the customer. Electric losses are mostly a result of the heating effect (I^2R losses) of current passing through power delivery equipment. These are known as resistive losses. Other electric losses, known as reactive losses, are a result of losses in magnetic flux coupling in transformers and other inductive equipment, including transmission and distribution lines themselves. Power is transmitted over long distances at high voltages in order to reduce losses. Utilities have increased both transmission voltages and distribution voltages in order to reduce losses.

Traditionally, 115 and 230 kV were considered the primary transmission voltages, but lately 500 and 745 kV have become more dominant. The same is true at the distribution level, where 4 kV distribution lines are replaced with 12, 25, or even 34.5 kV lines. The purpose of the distribution system is to reduce the voltage to a lower level and deliver power at the required voltage level to the customer. In the United States, residential customers interface to the system at 120 or 240 V, not the 115–500 kV found on the transmission lines. It is cost prohibitive to have 115 kV–120 V transformers at the many customer service points. So, from the generator to the customer, the voltage is transformed multiple times and carried over hundreds of miles. Each transformation causes losses. Likewise, each mile the power flows causes additional losses in lines and cables. Since the distribution system operates at a lower voltage and a higher current than the transmission system, the distribution system will produce more losses per mile of equivalent impedance compared to the transmission system.

Another factor that can reduce electrical losses is a reduction in the distance from generation to end use. In the early days of electricity, power plants were built close to the customers in large cities. Since then, several trends have occurred that are placing the power plants further away from the customers. First, more customers are moving away from the city centers and living greater distances apart. Power plants are also moving further away. This is due to several factors. First, customers do not want power plants close to their homes. Next, with the advent of gas peaking generation plants, power plant sites were selected based on the availability of gas pipelines, water and transmission line capacity as the major factors, and not where the main load centers were located. This has resulted in an increase in the distance electricity is transmitted and distributed, which increases electrical losses on the system. In the future, with the advent of distributed generation, smaller plants that are closer to the end customer will help to reduce electrical losses.

Electrical losses occur at every level of the transmission and distribution system due to the electrical impedance (resistive and reactive) of the equipment: from the step-up transformers at the power plants and the transmission and distribution grid (lines and transformers) down to the customer end delivery points. VArs in the system are caused by current flowing through inductive equipment on the system, such as transformers and lines, and also by the type of load. VArs in the system increase the current flowing in the system, which results in an increase in energy delivery losses. To reduce the VArs, capacitance in the form of capacitor banks is added to the system.

3.5.2.2 Voltage Fluctuations on the Distribution System

Electric utilities are required to deliver voltage to the customer at a nominal voltage within a specified operating range. In the United States, for single-phase, three-wire service, the nominal service voltage is specified under ANSI (American National Standards Institute) standard C84.1 [2] as 120 Vac with an acceptable range (Range A) of ±5% of the nominal voltage. Therefore, in the United States, any voltage between 114 and 126 Vac is deemed to be an acceptable voltage or, as the standard states, a "favorable voltage." The voltage is allowed to enter an acceptable zone of 110–127 Vac for short durations. Acceptable delivery voltages for U.S. customers are listed in Table 3.8. Range A is considered the favorable zone where the occurrence of delivery voltages outside these limits should be infrequent. Range B is considered the tolerable range and includes voltages above and below Range A limits. Corrective actions must be undertaken for sustained voltages in Range B to meet Range A requirements.

If the distribution system voltage is too high, it can damage power delivery equipment, such as transformers, as well as consumer equipment, such as appliances and electronic equipment. High voltages can also reduce the life of lighting products. If the incoming voltage is too low, lighting will dim, motors will have less starting torque and can overheat, and some equipment, such as computers and TVs, will power down. As a general rule, lower voltages result in more damage to the load on a distribution system and higher voltages cause more permanent damage.

TABLE 3.8
Acceptable Delivery Voltages for U.S. Customers

Nominal Service Voltage	Range B Minimum	Range A Minimum	Range A Maximum	Range B Maximum
Percent of nominal	91.7%	95%	105%	105.8%
Single phase				
120/240, three wire	110/220	114/228	126/252	127/254
Three phase				
240/120, four wires	220/110	228/114	252/126	254/127
208Y/120, four wires	191/110	197/114	218/126	220/127
480Y/277, four wires	440/254	456/263	504/291	508/293
2.4–34.5 kV% of nominal	95%	97.5%	105%	105.8%

Source: Voltage ratings for electrical power systems and equipment, American National Standard ANSI C84.1-1989.

Power flows from generators through transmission and distribution lines and several transformers before it reaches the end customer. Transmission and distribution lines and transformers all have electrical impedance (resistive and reactive) and the current flowing through the impedance results in a voltage drop. Therefore, the main factors affecting the amount of voltage drop are the load (the amount of current), the types of load, and the distribution system impedance.

If the voltage at the customer closest to the distribution substation is below 126 and the voltage at the customer furthest from the substation is above 114, then there may be no need for any voltage corrective action. This is typically not the case. Without any means to compensate for the reduction and the continual change in distribution voltages, customers closest to the substation will experience highest voltage levels and customers furthest from the substation will experience lowest voltage levels. To regulate the voltage levels at the substation and along the distribution feeder and ensure the voltage levels are within limits at the customer, substation power transformers and distribution feeder voltage transformers (voltage regulators) are equipped with means to actively change the turns ratio (taps) of the transformer while energized. The tap changing equipment on power transformers are referred to as LTCs.

3.5.2.3 Effect of Voltage on Customer Load

There are two main types of load on the system: constant resistive load and constant power load. With constant resistive load, when the voltage is decreased, so does the current. This causes a reduction in electrical losses on the distribution system and a further reduction in voltage drop due to the losses. With constant power load, when voltage is decreased, current increases. Resistive load, mostly lighting and heating, as well as appliances powered via nonswitching power supplies, has been the predominant type of load on electric systems in the past. Constant power load, such as fluorescent lighting, appliances with switching power supplies, and heat pumps, is becoming more dominant. Motors are the worst type of load for voltage fluctuations because as the voltage decreases excessively, the current increases in a linear manner. If the voltage at the motor load is too low, the motor stalls and this causes an exponential increase in load current.

Studies have shown that a reduction in voltage typically results in a reduction in load when the voltage is first reduced. The level of load reduction achieved with voltage reduction is highly dependent on the type of load on the distribution feeder [3]. The effects of the voltage reduction diminish over time as certain loads require additional current at the lower voltage to complete the task.

For example, with hot water heaters, once the water temperature falls below the desired setting, the water heater turns on. At lower voltages, the water heater does not produce as much heat and therefore has to run longer to heat the water to the appropriate temperature.

To decrease the load, either during emergencies or when generation is not available or too expensive, utilities can only reduce the voltage to the point where the customer at the end of the feeder is still above 114 V. This means that customers closest to the substation are served power at a higher voltage. If this is the case, the full benefit of voltage reduction cannot be achieved. A constant level (or "flat") voltage profile from the substation down the distribution feeder to each customer would allow for maximum voltage reduction benefits. Large reactive loads and impedance in the distribution system result in an uneven voltage profile that sometimes cannot be entirely compensated by adjusting the substation transformer and distribution feeder voltage regulator taps. In this case, reducing the VArs flowing in the distribution system will help maintain a flatter voltage profile along the distribution feeder.

3.5.2.4 Drivers, Objectives, and Benefits of Voltage and VAr Control

Not all utilities have generation, transmission, and distribution systems as part of their operations, so the drivers, objectives, and benefits of VVC differ for utilities. Whether utilities have generation plants or purchase energy from other utilities or power pools, by reducing the losses in the system through VAr control, the utility can obtain more revenue for the same amount of electricity generated or purchased. However, the grid has a limited capacity. As load continues to increase, the capacity of the grid has to increase with it, meaning more power plants, more transmission lines, and more transmission and distribution substations. This is a high cost to the utility, both financially and politically. Customers consider electricity a commodity and are not always willing to pay for the costs that it takes to deliver it. Customers want to keep the cost down but are not always willing to allow the construction of the plants, lines, and substations in their neighborhoods. By reducing losses through VAr control and customer demand through voltage control, utilities can increase the capacity of the grid and avoid or defer system expansion projects. Time is also a factor that has to be reviewed. It typically takes upwards of a year to build a substation, several years to build transmission lines, and in the case of nuclear power plants, up to 10 years to bring a new power plant online. With the current increase in demand, it may not be possible to keep up with the customer demand, even if customers are willing to help with the costs. If the amount of power generated can be reduced by eliminating losses, less CO_2 will be generated from the power plants. Reducing the emissions from the plants can be a tremendous benefit for both the utilities and the public in general.

Another advantage of VVC has been driven by deregulation and trading of electricity on the open market. Utilities generate electricity in multiple ways: hydropower plants (power plants that use water to generate the electricity), fossil power plants (plants that use coal, natural gas, or oil to generate electricity), nuclear power plants, and the green power from the sun or wind. Each of these plants has different costs associated with producing the electricity, with hydro and nuclear the cheapest to produce, followed by the fossil fuel plants, and finally the green plants. Utilities maximize their profits by generating electricity from the most cost-effective plants, using the remaining plants for peak or emergency power only. Sometimes it is actually cheaper for the utility to reduce load rather than run an additional power plant for a few hours to support the temporary increase in load. Weekday load typically starts to increase in the morning when customers are getting ready for work and commercial and industrial businesses start for the day. Load then typically peaks during the day before reducing to the lowest level overnight as shown in Figure 3.95. The distribution feeder and substation load profile depends on the type of customer and load. Other factors can change the load pattern, such as weather. Residential loads typically peak during the hours of 2 and 6 pm in the summer in areas with high temperatures due to air-conditioning load. Irrespective of the load pattern, there will always be times when the load on the system is higher or lower than

Smart Grid Technologies

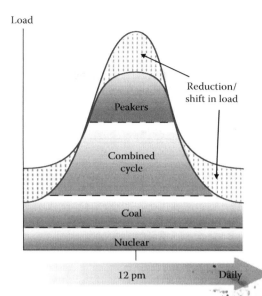

FIGURE 3.95 Effect of voltage control on customer load.

average. Ideally, a constant load on the system would help utilities operate the most cost-effective base generation, such as fossil or nuclear power plants. However, peak loading requires utilities to operate more expensive, fast-responding generation, such as gas turbines. Utilities are also required to have reserve generation capacity on immediate standby as contingency for any loss of generation on the system or any possible system reconfiguration due to equipment failures. As discussed earlier, lower voltages will, for at least short periods of time, reduce the power consumed by customer loads. The opposite is true for higher customer delivery voltages, causing a general increase in load. Customers on the same distribution feeder with exactly the same loads will have slight differences in their electricity bill, with the customers closer to the substation having a slightly higher bill due to the fact that their incoming service is at a higher voltage than those of customers located at the end of the line. Therefore, changing the voltage can change the loading and therefore the revenue. One would think that a utility would always want to run the voltage on the high side to increase revenue but this not always the case. There are several operating conditions that benefit the utility to run the system at lower voltages.

While VVC has been implemented over the many years to varying degrees, smart grid initiatives now bring renewed focus on the implementation considerations and measured benefits of voltage and VAr control to utilities. Current utility installations have shown that IVVC can reduce distribution feeder losses up to 25% and the peak load by up to 3% (see Table 3.9), depending on the load characteristics.

Another key benefit of VVC is that of security and reliability. As the growth of customer load outpaces the supply, the utility power delivery reserves are dwindling, making the system more susceptible to brownouts (suppressed voltage conditions) and blackouts (loss of power). VVC helps increase the available capacity of the power delivery system.

An emerging trend in generation is that of CO_2 emissions with carbon taxes and credits. If a utility can reduce the amount of generation required, especially with the coal and fossil plants, the utility can reduce the CO_2 levels, thus reducing the taxes and possibly actually building up credits.

While VVC brings significant benefits to the utility, the downside is that more equipment is required on the distribution system, such as voltage regulators and capacitor banks, as well as the means to monitor and control the devices. However, smarter monitoring and control can minimize

TABLE 3.9
Typical Load Reduction Due to CVR*

Percent Voltage Regulation	Load Reduction at Unity Power Factor	Load Reduction at 0.9 Power Factor
2	1.5	0.5
4	3.0	2.0

Source: M.S. Chen, R.R. Shoults and J. Fitzer, Effects of Reduced Voltage on the Operation and Efficiency of Electric Loads, EPRI, Arlington: University of Texas, EL-2036 Volumes 1 & 2, Research Project 1419-1, 1981.

* Empirically, load reduction can be in the range of 0.6%–0.9% for each percentage of voltage reduction on a distribution feeder. The trend with recent smart grid initiatives has included some form of field pilot or verification phase in order to verify the implementation requirements and costs of distribution automation applications before deciding on any system-wide deployment strategies. More importantly, the focus has been on measuring and verifying the expected benefits of the applications, such as from volt/VAr control.

the number of total operations of the equipment and therefore reduce maintenance costs. Smarter monitoring can also identify failed equipment such as fuses and pole-top transformers or capacitor banks to allow for quicker service restoration. Besides detecting outages faster, the smart grid will enable utilities to restore faster by remotely changing the configuration of the grid. This change in configuration can add complexity to the role of volt/VAr implementation which will be discussed in upcoming sections.

3.5.2.5 Volt/VAr Control Equipment inside the Substation

The majority of electricity delivery losses occur at distribution voltages. Distribution substations are typically fed from two or more transmission lines (e.g., 115 or 230 kV). There are typically one or more power transformers in the distribution substation that step the voltage down to between 4 and 25 kV. The low side of each transformer will connect to a bus that feeds multiple distribution feeders to distribute the power to the end customers. Some substation configurations allow the interconnection of the low-side busbars so that one transformer can feed multiple busses if another transformer fails or is removed from service for maintenance. The distribution power transformers and low-side busses in the substation are the primary point in the distribution system for voltage regulation.

3.5.2.5.1 Power Transformers

A power transformer has fixed winding ratio connections (taps) and, as discussed earlier, may also have variable taps to actively change the turns ratio of the transformer while energized (LTCs). The fixed taps allow for an adjustment of the voltage on the low side and there are typically settings for 0%, ±2.5%, and ±5%. This means when the rated incoming voltage is present, the secondary voltage can be at rated voltage (0% fixed tap) or 2.5% or 5% higher or lower than rated voltage. The fixed tap setting can only be changed when the transformer is out of service. If the transformer is equipped with an LTC, the low-side voltage can be varied with the transformer in service. The LTC typically allows for a 10% variation in voltage in either the raise or lower direction by having 16 taps, 8 taps that lower the low-side voltage and 8 taps that raise the low-side voltage. The LTC has very little effect on the high-side voltage. When in the neutral position, the LTC has no effect on the low-side voltage.

The LTC is a single-phase sense, three-phase operate device. It monitors one phase of voltage and current to make decisions but then acts on all three phases. For the LTC to operate correctly, all three phases of load need to be fairly well balanced. If one phase has significantly more load than another, the voltage drop will be greater. If it is not sensing that phase, the customers on that phase

FIGURE 3.96 Photograph of a distribution substation power transformer with an LTC.

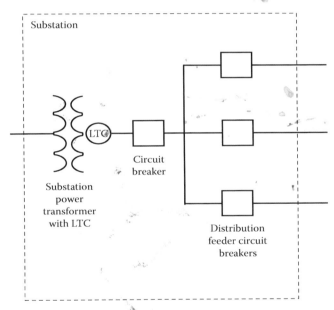

FIGURE 3.97 Single-line representation of a distribution substation power transformer with an LTC controlling voltage on three distribution feeders.

may have low voltage near the end of the line. If the phase with the highest current is the phase monitored, the customers on the other two phases may have high voltage near the source. Therefore, in order to have correct three-phase operation while only monitoring one phase, an assumption is made that all three phases have close to the same load. A photograph of a typical distribution substation power transformer with an LTC is shown in Figure 3.96 and a single-line representation of a distribution substation power transformer with an LTC controlling voltage on four distribution feeders is shown in Figure 3.97.

The first advantage of an LTC is that of cost. If the transformer is feeding more than two or three distribution feeders, it will be less expensive to use the LTC than to use single-phase regulators on each distribution feeder. Second is the cost of property. The LTC takes up the smallest amount of real estate. There are several disadvantages of LTCs. First, because it is a single-phase sense, all three phases have to be fairly well balanced. This is easy to do in urban areas where customers are condensed in small areas and distribution feeders are short, typically less than 4 miles. It is not so easy to do in rural areas with distribution feeders that are over 15 miles, most of which is single-phase runs. The next disadvantage is that a failure by either the LTC or the LTC controller can cause over- or undervoltages. Because the LTC is regulating the voltage to all the customers on all three phases of all feeders attached to the transformer, a failure affects many more customers. Another disadvantage is one of maintenance. To maintain the LTC mechanism, the entire transformer has to be taken out of service. This typically requires many hours for switching and limits the maintenance to off-peak load times such as evenings and weekends.

For smart grid applications, LTCs make life much more difficult. Because one device is affecting the voltage on many feeders, it is difficult to achieve a flat voltage profile. Each distribution feeder attached to the LTC will have a different length and, therefore, different impedance and also different loading conditions. Therefore, the voltage drops (and profile) along each feeder will be different, making it difficult for the LTC alone to attempt to correctly regulate all the feeders.

3.5.2.5.2 Substation Bus Regulation

Substation bus regulation attempts to regulate the voltage on the low-side bus of the distribution substation power transformer before the individual distribution feeders. There are two approaches to bus regulation, single-phase regulation and three-phase regulation. With three-phase regulation, the approach is similar to that of the LTC with one exception: the LTC is separate from the power transformer. The advantage of doing this is that maintenance of the three-phase regulator can be performed without taking the transformer out of service. The disadvantage of a separate bus regulation LTC over the transformer LTC is that the bus regulation LTC is more expensive and the installation footprint is much larger. This approach was popular many years ago, but few implement and manufacture three-phase regulators today.

The second approach to bus regulation is the use of three single-phase voltage regulators. One advantage of using three single-phase regulators is that with each phase being individually sensed and controlled, the loads do not have to be balanced for effective phase regulation. Another advantage is that a failure of the regulator or regulator controller now only affects the customers on that phase. The primary disadvantage to single-phase bus regulation is that of equipment size. The single-phase regulators cannot handle as much current as the three-phase bus regulators and therefore the substation transformer size and loading must be limited when using single-phase voltage regulators. From a cost standpoint, if the transformer is 20 MVA or smaller, single-phase bus regulation can be very economical. For this reason, single-phase bus regulation is very popular in rural substations, which tend to have fewer distribution feeders and less load as customers are more dispersed. A photograph of a set of three single-phase bus regulators is shown in Figure 3.98 and a single-line representation of a distribution substation with bus voltage regulation is shown in Figure 3.99.

Bus regulation, whether single phase or three phase, still poses many of the same limitations of the LTC, mainly one device, or set of devices, regulating the voltage of multiple distribution feeders.

3.5.2.5.3 Single-Phase Voltage Regulators

With single-phase regulation, each distribution feeder is regulated separately prior to leaving the substation. In this approach, the transformer and low-side bus are left unregulated. This approach requires the most space in the substation for installation, and is typically the most expensive, but offers the most flexibility and reliability. For example, if the substation transformer is attached to a bus feeding four distribution feeders, then 12 single-phase regulators and controls would be required.

The reliability of this voltage regulation approach is higher than the approaches discussed earlier since a failure of a single control or regulator only impacts the customers on that phase of that distribution feeder. The flexibility comes from the fact that each feeder is independently regulated. This allows for different distribution feeder lengths and loads to be accommodated independently.

FIGURE 3.98 Photograph of three, single-phase bus voltage regulators.

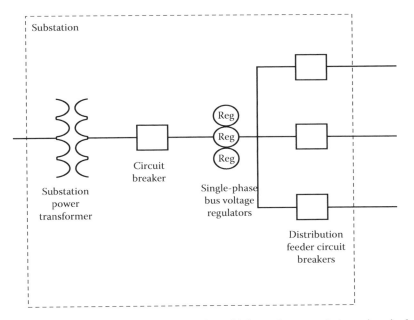

FIGURE 3.99 Single-line representation of a substation with bus voltage regulation using single-phase voltage regulators.

For this reason, many utilities are now designing new substations with single-phase voltage regulation even though it has the highest installed cost. A photograph of single-phase voltage regulators on distribution feeders in the substation is shown in Figure 3.100 and a single-line representation of a distribution substation with single-phase voltage regulators on each of four distribution feeders is shown in Figure 3.101.

FIGURE 3.100 Photograph of single-phase voltage regulators on distribution feeders in the substation.

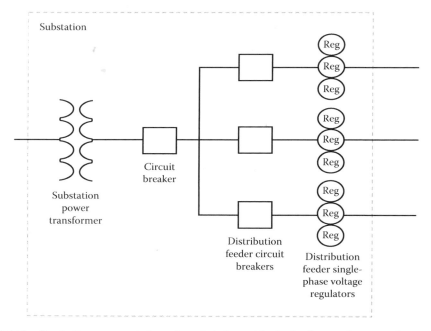

FIGURE 3.101 Single-line representation of a substation with single-phase voltage regulators on each of three distribution feeders in the substation.

3.5.2.5.4 Substation Capacitor Banks

Placing capacitor banks at the substation bus level is sometimes done in order to both regulate the bus voltage and supply VArs to the distribution system and load (VAr compensation). Use of capacitor banks at the substation level has disadvantages. First, the size is typically fairly large, so when the capacitor banks operate, they have a larger effect on the secondary voltage. This can cause LTCs and regulators to operate in response. It is not recommended to use substation capacitor banks with LTCs as it is typical to see the LTC having to operate 5–8 times per operation of the capacitor bank. As discussed earlier, it is difficult to perform maintenance on LTCs so it is not a good idea to increase the number of operations. The other negative effect to the use of substation capacitor

Smart Grid Technologies

FIGURE 3.102 Photograph of a substation capacitor bank. (© Copyright 2012 Siemens. All rights reserved.)

banks is that the use of the bank provides VArs at the substation and does not compensate the VArs flowing down throughout the distribution system. Therefore, station capacitor banks do not reduce losses in the distribution system. Capacitor banks for VAr compensation to reduce losses are more effective if distributed throughout the distribution system and applied close to inductive load.

While voltage regulators, whether LTCs, three phase or single phase, can adjust the secondary or load voltage dynamically, capacitor banks can also perform a similar function. When capacitor banks are added to a point in the power system, there is a resulting increase in system voltage at the capacitor bank. Therefore, capacitor banks provide both VAr and voltage support in the system. There are two primary differences between capacitor banks and regulators regarding voltage control. First, regulators are unidirectional in that they only affect the voltage on the load side of the regulator. Capacitors are bidirectional in that when they operate, the voltage on both sides of the capacitor is affected. Next, regulators can control the voltage in smaller increments, whereas capacitors banks do not have multiple steps of voltage control as a bank is either on, which will cause the voltage to increase, or off causing the voltage to decrease. The effect that the capacitor has on the secondary voltage is a combination of the rating of the bank (kVAr or MVAr) and the location of the bank in the distribution system. A photograph of a typical distribution substation capacitor bank is shown in Figure 3.102.

3.5.2.5.5 Summary

As can be seen, some form of voltage regulation is typically required inside the substation. Transformers with LTCs or bus regulators can be used in service areas with short distribution feeders and balanced load, but from a smart grid standpoint, ultimate flexibility is achieved with individual distribution feeder voltage regulation. Even with this equipment in place, it is typically not possible to adequately regulate the voltage along the length of an entire distribution feeder from within the substation. For this reason, devices on the distribution feeder can be added to provide additional voltage support as well as VAr compensation. Control of these down-line devices has to be coordinated with the substation devices in order to achieve the overall IVVC goals.

3.5.2.6 Volt/VAr Control Equipment on Distribution Feeders

There are two main components used to aid in the voltage regulation and VAr compensation outside the substation down the distribution feeder: the single-phase line regulator and the pole-top capacitor bank. Most utilities use pole-top capacitors to flatten the voltage profile along the distribution feeder and then regulators to adjust the voltage levels.

FIGURE 3.103 Photograph of three single-phase voltage regulators down-line on a distribution feeder.

3.5.2.6.1 Single-Phase Line Regulators

On feeders of considerable length or load, the regulating device at the substation may not be adequate due to the excessive amount of voltage drop along the entire feeder. If the voltage difference between the customer closest to the substation and the customer furthest away is more than 10–12 V, additional voltage correction will be required by placing single-phase line regulators somewhere between the substation and the end of the feeder ("down-line" of the substation). The single-phase voltage regulators used along the distribution feeder are similar to the single-phase voltage regulators used in the substation but are typically smaller in size and rating.

Substation regulation will regulate the voltage of the entire feeder, with primary voltage control of the section of the feeder up to the set of line regulators. The line regulators will control the voltage on the section of the feeder below the regulators. Coordination of the operation of multiple sets of regulators is required, usually the substation regulation acting first, then the line regulators. Coordination is achieved through time delay settings, with each set of regulators having a longer time delay than the set on the source side of it. A photograph of three single-phase voltage regulators down-line on a distribution feeder is shown in Figure 3.103 and a single-line representation of a distribution substation with single-phase voltage regulators down-line on three of the four distribution feeders is shown in Figure 3.104.

From a smart grid standpoint, additional real-time information from customer revenue meters on the feeder can better help in determining the placement of the line regulators and sizing that will be required to handle both normal operations and emergency operations.

3.5.2.6.2 Pole-Top Capacitor Banks

There are two types of pole-top capacitor banks: fixed and switched. Most utilities will use fixed capacitor banks to compensate for the minimum or average amount of VAr support required on the distribution system. VAr flow on the distribution system varies daily, and, therefore, fixed capacitor banks cannot effectively compensate VAr loads continuously over the load profile, and in some cases, fixed capacitor banks with the inappropriate rating and location on the distribution feeder can contribute to increased VAr flow on the distribution system. A photograph of a pole-top switched capacitor bank is shown in Figure 3.105.

Switched capacitor banks are similar to fixed banks except they have additional switches and controls, which allow the capacitor banks to be switched on or off either remotely via SCADA or an IVVC controller or locally via an automatic sensing control. Capacitor banks are similar to LTCs in that they use a single-phase sense with a three-phase operate. If multiple capacitor banks are deployed on the same distribution feeder, each bank is typically connected to sense-alternating

Smart Grid Technologies

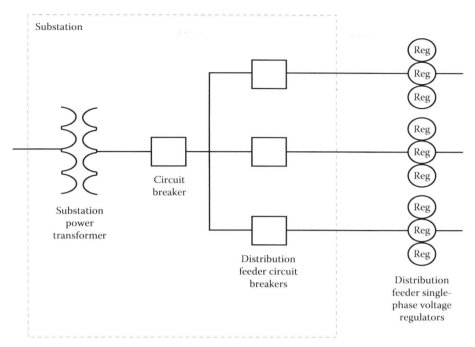

FIGURE 3.104 Single-line representation of a substation with single-phase voltage regulators down-line on three distribution feeders.

FIGURE 3.105 Photograph of a pole-top switched capacitor bank.

phases of the feeder. The sense on the given phase can be voltage, current, or both depending on the type of control selected. Coordination of capacitor bank control is done exactly the opposite of regulators. Coordinated control is required between the capacitor banks and the regulators in order to avoid conflict between the control schemes. Coordinated control between voltage regulation and capacitor bank switching is the premise of IVVC.

The main reason to use a voltage regulator in the distribution system is for voltage correction. There are two possible reasons for applying capacitor banks: power factor correction and voltage correction. Some utilities use capacitor banks for voltage correction. Switched capacitor banks are less expensive to purchase and install than single-phase line voltage regulators, and the long-term maintenance costs are also less. The theory is to apply many small capacitor banks on the distribution feeder so that the voltage can be regulated in smaller steps. The capacitor bank controls used for this type of system are typically voltage controls as it is easier to coordinate substation LTCs or feeder voltage regulators with switched capacitor banks if they are all using the same measured parameter. If capacitor banks are used mainly for VAr compensation, then coordination is not as critical between the capacitor banks and the LTCs and regulators. The capacitor bank controls will typically be based on power factor or VAr measurements and include voltage overrides in the control that require coordination with the LTCs and regulators.

3.5.2.7 Volt/VAr Control Implementation

3.5.2.7.1 Voltage Control

Regulator controllers are becoming much more advanced. Some controllers include functions such as voltage sag and swell detection, flicker detection, CBEMA (Computer and Business Equipment Manufacturers' Association) power quality violation detection, fault detection, and harmonic measurements. They also have predictive maintenance features such as motor current monitoring and incorrect tap position alarming. These data can help customer service engineers responding to customer complaints on incoming service. As regulator controllers become more intelligent, changes to the construction of the regulator will follow. Some controllers have the capabilities to detect faults on the feeder. This can be useful in providing exact fault locations to field personnel for quicker restoration of service. Currently, protection relays in the substation can estimate fault distances on distribution feeders, but accuracy of the estimation depends on the distribution feeder design and configuration since the feeder may contain numerous subcircuits (taps or laterals). Therefore, more fault information from equipment down the distribution feeder, such as voltage regulators, can provide more accurate operational data to the utility.

Smart grid can aid in the deployment of voltage reduction in many ways. First, typically it is difficult to reduce the voltage outside the substation because there is no communications to regulators on the distribution feeders. With the added communications networks being built for distribution automation and AMI, utilities can now extend communications to equipment and controllers down the distribution feeders. This additional monitoring and data collection of operations of the distribution systems can also help determine the optimal location of voltage regulation equipment on any given feeder and the optimal control of voltage to meet the objectives of combined voltage and VAr control.

3.5.2.7.2 VAr Control

There are many different types of capacitor bank controllers. In the earlier days, most controls operated on either time of day or temperature. Now, more advanced controls are being used that monitor voltage, current, power factor, VAr flow, or a combination of all. Twenty years ago, switched capacitor bank controller did not include remote control capabilities, but today the majority of controllers are communicating, either via one-way or two-way communications. The amount of intelligence in a capacitor bank controller depends on whether communications to the controller is provided. If communications is not provided, then all the intelligence must be in the controller, but if communications is included, the controller can have varying degrees of intelligence.

Time-based controls were popular 20 years ago because of the cost. A time-based control only required voltage to power the capacitor bank switch, so no additional sensing equipment was needed. The theory behind the time control was that for residential feeders, load would peak during certain times of the day, in the morning as people prepared for work and in the early evening as they came home and cooked dinner. The control could be programmed to have the capacitor closed during the peak hours and open during the remaining portion of the day. For commercial and industrial feeders, the load would be greatest when the businesses were open or plants were manufacturing and then would drop off when the business was closed. The control could even be programmed to take into account the weekends and holidays. Time-of-day controls have lost favor for several reasons.

First, load in general is not as predictable as in the past with the advent of 24-h businesses. Second, the time clocks in the capacitor bank controllers could deviate, causing the banks to operate at the incorrect time. The controllers lose power every time there is a feeder outage, and this power loss would cause the control to lose time. Therefore, batteries are used to keep the time when the control was without power. This causes a maintenance nightmare as personnel have to inspect and replace batteries. Other concerns such as daylight savings time can also create difficulties. So, while the time-based control is the least expensive to install, it can create large maintenance costs and may not always operate correctly.

Some capacitor bank controllers are based on temperature where the controllers monitor the ambient temperature and switch the bank on or off at predefined temperatures. The theory is that, particularly in warm climate on hot days when air conditioners are running, there is a need for the capacitor banks to compensate the considerable increase in VAr current from the air-conditioning load. The advantage of temperature controls is that they require no additional sensors and are thus very inexpensive, as with time-based controls. The temperature-based controllers, unlike the time controls, do not require a battery backup to keep the time and therefore require less maintenance. The problem with temperature-based capacitor bank controllers is that not all load predictably follows the temperature. For this reason, temperature controls are not used very often.

Voltage control for capacitor bank switching has gained popularity for several reasons. First, as in the time and temperature controls, voltage control requires no additional sensing equipment and is therefore relatively inexpensive. There are two basic types of voltage controls: absolute voltage control and delta voltage control. Both use voltage as the sense to decide operation, but they are done in different manners. Absolute voltage control is based on the actual measured voltage level of the distribution feeder, as in the control of voltage regulators. Delta voltage control is based on the change in voltage level measured on the distribution feeder. The theory of delta voltage control is that the impedance of the system is primarily inductive and therefore any voltage change is due to a current change and the majority of the current would be reactive.

More advanced capacitor bank controllers using current control, power factor control, and VAr control all require monitoring of the current flowing in the distribution feeder. Since these types of controllers require a current input, the location of the capacitor bank is limited. The capacitor bank must be placed on the main distribution feeder and not off any of the taps or laterals (subcircuits, usually single phase). This is because, while the voltage on the tapped sections of the line is comparable to the main distribution feeder, the current seen at any tap will only be the amount of current flowing through that tap, not the total current flowing through the main feeder. The VAr control is typically the most popular of the three, the obvious reason being that the capacitor bank should be switched on to provide reactive support when there is a lagging power factor.

While time- and temperature-based capacitor bank controls are not extensively used as much as the primary controlling function, many utilities may still employ an override feature using either time or temperature. A primary control function such as voltage or VAr flow is now more common but may include a temperature or time control override. For example, the capacitor bank may be switched on or off at different voltage levels, but if the temperature exceeded a certain level, the capacitor bank would be closed to provide more VAr compensation.

3.5.2.7.3 Volt/VAr Control Approaches

IVVC can be implemented in several ways. Each approach has its own strengths and weaknesses. Utilities need to analyze their existing infrastructure and the goals of VVC to determine the best approach to deploy VVC.

3.5.2.7.3.1 Local Intelligence Approach The first basic approach relies on the individual transformer LTC, voltage regulator, and capacitor bank controllers to control the voltage and VArs, where the transformer LTC and voltage regulator controllers are set to properly coordinate with the capacitor controllers as discussed earlier. With this approach, the transformer LTC, voltage regulator,

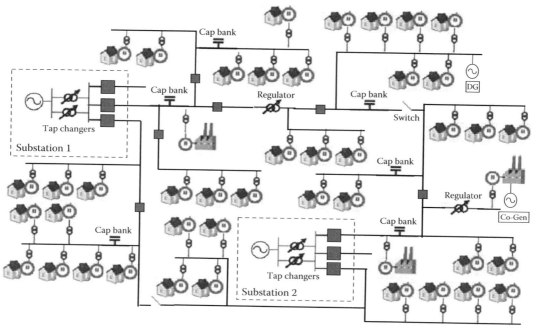

FIGURE 3.106 Local intelligence approach to VVC.

and capacitor bank controllers are typically not monitored and controlled remotely, although some utilities will still use communications to monitor overcurrent and to obtain metering quantities as a means to verify proper operation by the local controls. The capacitor bank controllers are run in the automatic mode and may use time, temperature, voltage, current, or VArs as the determining factor for operating the capacitor bank. Figure 3.106 shows the basic architecture of the local intelligence approach to VVC.

The capacitor banks are usually coordinated so that the one furthest from the substation closes first and opens last. This is implemented by varying the time delays in each control. The capacitor controls are also set to operate before the transformer LTCs and voltage regulators, again by the use of time delay settings. There are several types of capacitor controls that are used with this approach. The least expensive approach is to use a time, temperature, or voltage control as these require no additional inputs. The voltage is already required to operate the switch. Typically a delta voltage, or change in voltage, algorithm provides the best algorithm. The time-of-day and temperature controls are used with predictable loads, typically residential, and are usually only effective in certain climates and during certain times of the year. The voltage control is easy to use if the capacitor bank is being used more for voltage support than VAr support as it is easier to coordinate the control of transformer LTCs and single-phase voltage regulators with down-line capacitor banks if both are using voltage measurements to make decisions. A simple voltage control may not be appropriate for installations where the capacitor bank is used for power factor correction only.

While this approach to VVC is typically not supervised directly with communications, it can be supervised at the distribution feeder level with metering from the substation. Metering data, from protection relays, substation panel meters, or single-phase voltage regulators, can allow the VVC scheme to monitor the voltage, power factor, and VAr levels at the distribution feeder level to verify proper operation of the downstream devices.

3.5.2.7.3.2 Decentralized Approach This approach is similar to the previous approach, but with the addition of some level of integrated and coordinated control logic at the substation level

Smart Grid Technologies

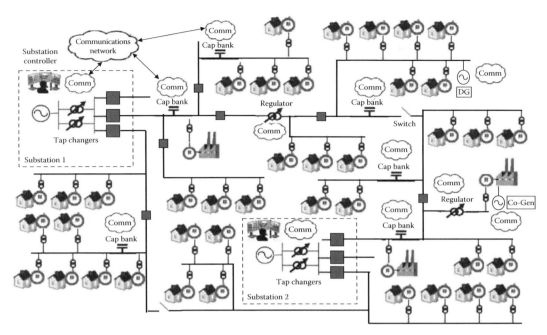

FIGURE 3.107 Decentralized approach to VVC.

remotely communicating with the transformer LTC, voltage regulator, and capacitor bank controllers. The volt/VAr controller in the substation provides "localized" monitoring and coordinated control of the volt/VAr scheme on a feeder-by-feeder basis. Communications with the field controllers provides the ability for the distribution operations control center to override the scheme remotely. Two-way communications with the field controllers can also help detect abnormal operating conditions and alarms, such as blown capacitor bank fuses, which eliminate the need for routine inspection trips to the field and may pay for the additional cost of adding the communications. Figure 3.107 shows the basic architecture of the decentralized approach to VVC.

3.5.2.7.3.3 Centralized Approach with No Local Intelligence The centralized approach is currently the most prevalent approach implemented. The centralized approach consists of central and master control logic (usually implemented on the distribution management control system [DMS]) communicating with the controllers in the field. With this approach, the transformer LTC, voltage regulator, and capacitor bank controllers do not require local intelligence for VVC actions. The capacitor bank controller tends to be an inexpensive switch with little, if any, decision-making capabilities. The capacitor bank controller has control outputs to turn the capacitor bank on or off and communications for remote control. The main cost of implementing this approach lies within the communications network and the centralized intelligence. The communications networks employed for this type of VVC approach have traditionally been one-way paging systems in order to reduce implementation costs. The centralized master sends out a command to a transformer LTC, voltage regulator, or capacitor bank controller and expects that the controller received the command and acted accordingly. More advanced control schemes also check metering quantities at the feeder level in the substation to determine whether the VAr load or voltage changed by the expected amount. If not, an alarm is generated to inspect the capacitor control, switch, capacitors, and fuses for a possible problem. Figure 3.108 shows the basic architecture of the centralized approach to VVC with no local intelligence.

The centralized VVC scheme is usually responsible for monitoring and coordinating control over a distribution service area with numerous substations. This allows the implementation of control schemes that have the objective of maximizing the benefits over a large service area, and not on an

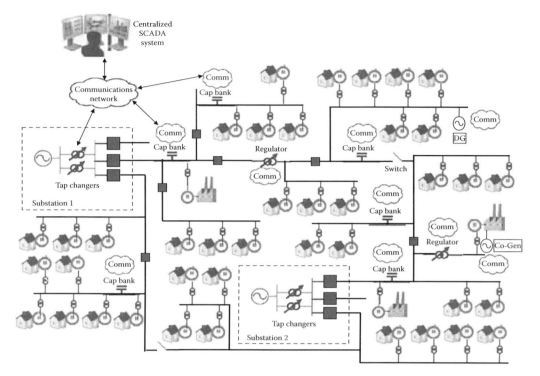

FIGURE 3.108 Centralized approach to VVC with no local intelligence.

individual distribution feeder or substation basis, where the best control decision for each feeder or substation may not maximize the benefits of VVC for the service area.

3.5.2.7.3.4 Centralized Approach with Local Intelligence This approach is similar to the previous approach, but the one-way communications with the transformer LTC, voltage regulator, and capacitor bank controllers is replaced with two-way communications, typically digital cellular or an unlicensed mesh-network radio system. By implementing two-way communications networks, the centralized approach receives additional information from the field controllers, such as confirmation of control operations and measurements of VArs and voltage on the distribution system at the field controller locations. This further enhances the capabilities of the centralized approach.

Also, with this approach, the transformer LTC, voltage regulator, and capacitor bank controllers have local intelligence for VVC actions in order to safeguard against loss of communications or issues with the central, master VVC logic. The field controllers are typically operating individually with their local control logic, but the central master monitors and controls setpoints in the field controllers to bias and enhance the overall operation of the VVC scheme. Therefore, if communications is lost or the central master fails, the field controllers will still be operating locally, although the control actions may not be optimal.

Any additional sources of field operations data will help enhance the VVC scheme, for example, some utilities are adding end-of-the-line voltage sensors to the feeders so that the centralized application can determine how low the voltage can be reduced without affecting customers at the end of the feeder. While capacitor bank controllers typically monitor only one voltage phase, these end-of-line monitors are typically monitoring all three phases. Some utilities are also deploying midfeeder monitors that also report back VArs on each phase. Some utilities are also adding communications to line regulators, reclosers, and sectionalizing switches further down the distribution network to provide additional system data.

The centralized control approach is further complicated by reconfigurations of distribution feeders due to switching or outages, which changes the load, VAr flow, and voltage along the distribution feeder and at the substation. This drives the need for the integration and coordination of control actions and objectives among distribution automation applications in a smarter grid.

From a broader view than just the control, the centralized VVC approach typically provides the best flexibility and operation of the distribution system but comes with additional cost. The cost can be offset if there is an existing communications infrastructure that can support communications to the field devices. The cost can also be offset if the communications system is shared with other applications, such as FDIR for communications to fault locators, down-line reclosers, and sectionalizing switches. As part of smart grid communication deployments, some utilities are also considering communications networks that can support distribution automation field communications as well as automatic meter reading.

3.5.2.7.3.5 Hybrid Approach with Local Intelligence The hybrid approach to VVC is a combination of the centralized and decentralized approaches and leverages the advantages of both approaches. The basic principal is to implement a hierarchical control scheme with intelligence in the local field controllers, in the substation, and in the central control master. One advantage of the hybrid approach over the centralized approach is in the communications network, where the hybrid approach can use different communications technologies without depending on a single, system-wide communications network for centralized VVC. Also, a failure of the communications network only impacts a limited area of the control system instead of the entire system. The hierarchical approach can also distribute the logic and processing power required for the VVC scheme, while maintaining the objectives of the control over a large service area. The hierarchy approach is designed to have the capability to default to the lower levels of control if there are any issues with communications or control at the higher levels in the hierarchy. The hybrid approach is cost effective where utilities have some form of substation controller that can be easily upgraded to include VVC in the substation. A major drawback of the hybrid approach is the additional hardware, software, programming, configuration, and coordination of the levels of control. Figure 3.109 shows the basic architecture of the hybrid approach to VVC.

3.5.2.8 Volt/VAr Optimization

3.5.2.8.1 Optimization versus Control

In simple terms, VVC is the capability to control the voltage levels and reactive power (VArs) at different points on the distribution grid by using a combination of substation transformer LTCs, feeder voltage regulators, and capacitor bank controllers. A distribution system is complex in that an operation on any single control device can result in considerable changes in multiple aspects of the system. For example, when a capacitor bank is energized, a certain amount of reactive power is injected into the system, which will affect the voltages, VAr flows, and power factors along the distribution feeder and at the distribution substation and, in turn, will affect the distribution system energy loss and load demand.

IVVC is the capability to coordinate the control actions of the substation transformer LTCs, voltage regulators, and the capacitor bank controllers such that the interaction of the control actions are integrated and optimally coordinated during the decision-making process. VVO optimizes the objectives of VAr (loss) minimization and load reduction (with voltage constraints) using optimization algorithms and well-defined control objectives subject to various system constraints through centralized or decentralized decision makings.

3.5.2.8.2 Selecting the Right Optimization Objective

The control objective of VVO should be of business and engineering significance. People sometimes minimize an indirect objective instead of the real objective by assuming that the two are equivalent.

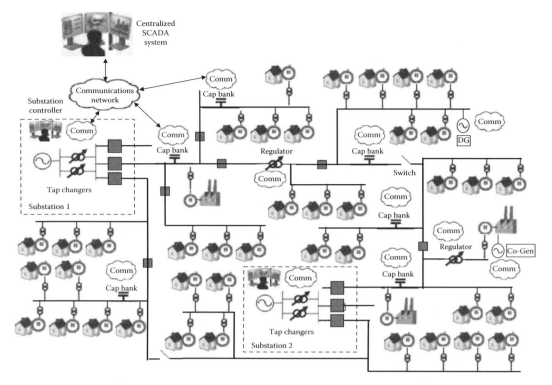

FIGURE 3.109 Hybrid approach to VVC with local intelligence.

Intuitively, such an assumption makes a lot of sense and appears to match the operational experiences. However, close examinations on the results often indicate otherwise.

For example, some utilities use capacitor switching to bring the distribution feeder power factor close to unity because a unity power factor can be a perfectly legitimate business objective, although the unstated business objective is actually to minimize feeder losses, while it is believed that power factor correction is the means to reduce feeder losses. If the real business objective is power factor correction and the measured distribution feeder power factor is lower, then the correct action is to switch in more capacitor banks no matter where they are located along the distribution feeder. However, if the real objective is to minimize losses, the correct action is never that simple and depends on the locations of the capacitor banks and the different VAr flows on the distribution feeder.

The whole concept of CVR is based on the premise that voltage reduction will result in energy or demand reduction, which implies that, in the aggregation, the loads on a feeder will respond to voltage reduction with demand reduction. Consider a simple example of two loads on the same distribution feeder. If the voltage is reduced at both loads and the first load decreases but the second load does not change, then the second load causes an increase in load current at lower voltage (constant power load). Therefore, while the objective was to reduce the voltage, some load on the feeder was reduced, but the losses on the feeder may not be minimized. A better understanding of the system load characteristics would help to optimize VVC. Therefore, in this hypothetical example, instead of indiscriminately reducing the voltage for the entire distribution feeder, a more effective approach may be to reduce voltage at the first load and increase the voltage at the second load to achieve maximum load and loss reduction. This example shows that minimization of losses does not always occur when the voltage profile along a distribution feeder is flattened. The difference between CVR and VVO is that VVO does not presume that the voltage reduction or increase is the correct solution. VVO determines the correct actions for different

Smart Grid Technologies

parts of the network based on load characteristics and the available measurements and controls. The objectives of VVO must be achieved while maintaining acceptable voltage profiles along the distribution feeder under dynamic operating conditions. Although the differences between ordinary VVC and optimal control are quite obvious, it is not unusual in common practice that they are either not well recognized or are ignored. Understanding the differences will enable the utilities to select the right technologies consistent with their business objectives and realize the maximum benefits.

3.5.2.8.3 Volt/VAr Optimization Approaches

As discussed earlier, there are many different approaches to the practical implementation of VVC, some requiring communications between field controllers and decentralized substation controllers or between the local controllers or the substation controllers and a central master, while some others do not. In VVO, communications between the local controllers and the central controller is a "must." It is not possible to achieve optimal control among a group of geographically dispersed controllers without communications.

As with VVC, there are several approaches to implementing VVO; however, VVO requires the integration and exchange of data with other utility applications, such as DMS, OMS, and GIS, and the levels and capabilities of optimization will evolve in the smart grid realm.

The model hierarchy of the substation-based decentralized versus centralized VVO approach is shown in Figure 3.110. As with IVVC, the decentralized approach is implemented at the substation level of the distribution system down the feeders associated with each substation, whereas the centralized approach optimizes voltage and VArs among substations, within operating regions and across the entire distribution system.

3.5.2.8.4 Decentralized VVO Approach

The major advantages of the decentralized VVO solution, as with the decentralized IVVC approach discussed earlier, are that it can be deployed incrementally, and one decentralized system's failure leads to very limited downtime for a small portion of the entire system, resulting in higher reliability for the

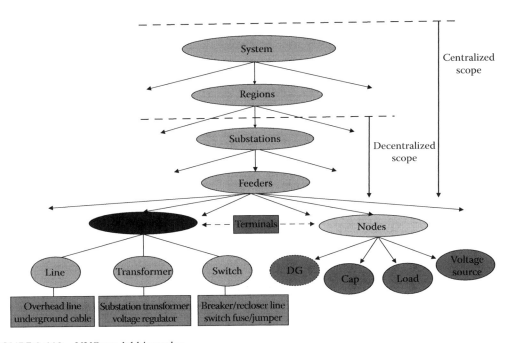

FIGURE 3.110 VVO model hierarchy.

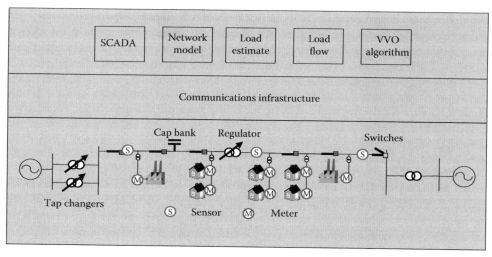

FIGURE 3.111 Centralized VVO approach.

overall system. This can make it easier for a utility to deploy VVO or distribution automation projects with smaller budgets and multiyear plans. However, as discussed earlier, the operation of breakers or switches in a distribution system will change the feeder configuration, which is a challenge in the decentralized control solution because it has only a partial model of the overall system. A certain level of coordination among the neighboring substations is required for proper operation of decentralized VVO.

3.5.2.8.5 Centralized VVO Approach

The centralized VVO approach uses a model of the entire distribution network, while the decentralized VVO approach uses a model of only a portion of the system. The following components should be included in the centralized VVO system (Figure 3.111):

- Distribution SCADA system for acquiring and processing real-time measurements from the field devices
- Short-term LE/forecast for look-ahead optimization
- Unbalanced three-phase load flow function for the distribution network operation validation and optimization
- Controllable devices (voltage regulators, capacitor banks, as well as other dispatchable energy resources)
- Sensors (voltage and current transducers)
- Substation controllers/RTUs
- Communications infrastructure
- VVO control algorithms in the control centers

The short-term load forecast and estimation, and the unbalanced three-phase load flow functions are used in the optimization to identify optimal switching plans and evaluate how the control strategy will perform with respect to the objective function and their impacts on other operation constraints, such as voltage limit and line rating violations. Based on the actual system operation conditions from the SCADA system, the optimal control strategy is updated on a continual basis.

3.5.2.8.6 Model-Based Approach

A recent approach to VVO utilizes a dynamic operating model of the distribution system in conjunction with mathematical optimization and power engineering calculations to optimize the volt/VAr performance of the distribution system within a given operating objective. The source of the dynamic operating model of the distribution system is typically the distribution connectivity model in a distribution organization's GIS. The model is adjusted in the control room of the utility on a near-real-time basis. This is done either through manual operator action on the model or through an interface to the SCADA system, which transmits changes in the status of system components. The model reflects changes in the status of distribution breakers, switches, reclosers, fuses, jumpers, and line cuts.

With model-based VVO, the "as-operated" state of the system, including near-real-time updates from SCADA and the OMS application, impacts the precise level of voltage control required for distribution companies to implement CVR without violating regulatory-prescribed voltage limits. For CVR, the optimum settings for each transformer LTC, voltage regulator, and capacitor bank depend on the following:

- The spatial distribution of load throughout the system
- The phases on which load is connected
- The connection type for three-phase loads (i.e., delta or wye connected), as well as transformer connections and parameters
- The voltage-dependent characteristic of loads (constant impedance, constant power, and the percentage mix of the two)
- Network topology and characteristics

The optimal settings of the controls in model-based VVO are determined by evaluating status and real-time data from equipment. The recent advances in technology required for distribution organizations to implement VVO include GIS-based network models, two-way communications to distribution substations and line equipment, and improvements in computing resources and architectures.

With model-based VVO, an objective function is specified, subject to a set of nonlinear equality and inequality constraints. The constraints consider the thousands of equations and state variables used for unbalanced load flow analysis.

One model-based VVO algorithm is summarized as follows:

"*Minimize* (real power demand and/or real power losses) *subject to* the following engineering constraints":

- Power flow equations (multiphase, multisource, unbalanced, meshed system)
- Voltage constraints (phase to neutral or phase to phase)
- Current constraints (cables, overhead lines, transformers, neutral, grounding resistance)
- Tap change constraints (operation ranges)
- Shunt capacitor change constraints (operation ranges)

Using the optimization control variables:

- Switchable shunts (ganged or unganged)
- Controllable taps of transformer/voltage regulators (ganged or unganged)

One significant difference between model-based VVO and other VVC methods is that model-based VVO can use power engineering calculations and analysis as its solution. These include load allocation and per-phase, unbalanced load flow analysis to compute the electrical state of the network. For many distribution systems, unbalanced load flow is required rather than balanced load flow, because of single-phase loads and laterals, unsymmetric component impedances, various unbalanced transformer models, and single-phase and unganged operation of some voltage regulators and switched capacitors. With this modeling and analysis capability, model-based VVO is also able to model the voltage dependency of customer loads in terms of the percentages of constant impedance and constant power components. This influences the calculation of capacitor switch states, as well as optimal LTC and voltage regulator settings.

The development of model-based VVO, accompanied by GIS modeling and two-way communications on the distribution system, provides distribution organizations with capabilities that were not previously available.

1. The true optimal state of voltage and VAr control equipment is calculated—nonoptimal rules of thumb are not used.
    a. Optimization algorithms, such as mixed integer, nonlinear programming, can be employed. Such algorithms insure that reduction in power, energy, and customer demand is maximized.
    b. In addition to execution based on predetermined intervals, model-based VVO can be configured to run based on events such as feeder reconfigurations and changes in feeder loading.
    c. A copy of the online model can be used to perform off-line studies, so operations planning can study different scenarios and configuration.
2. The algorithm uses the "as-operated" network model.
    a. The as-operated network model reflects the actual connectivity of feeders, loads, and voltage/VAr control devices. This is important due to the very dynamic nature of distribution system switching and outages that routinely occur. As SCADA and operators make changes to the as-operated network model, the volt/VAr network model in inherently kept up to date without the need for a separate update process.
    b. All distribution applications, including power flow, fault location, FDIR, switch orders, and VVO, can use the same model and network view.
3. For distribution organizations implementing multiple advanced distribution applications that use the same as-operated model, a common platform reduces computing hardware and ongoing maintenance costs.
    a. Model-based VVO leverages the investments made in SCADA, OMS, and other DMS applications, minimizing duplication in the costs in computing infrastructure and communications environment.
    b. A common distribution network model used for all applications results in no synchronization issues, either in real-time or in incremental updates from the GIS, between different models being maintained in different applications.
4. A three-phase unbalanced network model is utilized.
    a. A detailed, unbalanced three-phase system model is utilized to provide accuracy.
    b. Subtransmission and secondary voltage levels can be included in the analysis. VVO can be executed on one feeder, all feeders from a substation, several substations, or an entire system.
    c. In addition to radial systems, networked and looped systems can be analyzed.
    d. Voltage and thermal limitations on the network components are calculated.
    e. Voltage unbalance on the system is included, as are ganged and unganged controls.
5. Distribution system loads are modeled, since they impact the optimum settings.
    a. The voltage-dependent component of customer loads is modeled to represent load variation in real and reactive power as a function of voltage.
    b. Load characteristics such as location, size, and type (percentage mix of constant impedance and constant power) impact the optimal LTC and voltage regulator settings, particularly for CVR.
    c. Node voltages at load points, as well as all nodes throughout the circuits, are calculated and compared to operating limits before control actions are taken.
    d. Customer load profiles can be utilized in the load allocation and load flow applications. This enables a temporal representation of customer loads, which can also impact settings.

3.5.2.8.7 Volt/VAr Optimization and the Smart Grid

VVO will also enable distribution organizations to operate their systems as new complexities are being introduced. These complexities include increased renewable generation located at distribution voltage levels, increases in automated fault location and restoration switching schemes, increased system monitoring and asset management processes, and an electric vehicle charging infrastructure.

Smart grid initiatives are now providing the means to share data among enterprise applications. For example, voltage readings of customer revenue meters from AMI systems can be shared with a centralized VVC master in order to monitor lowest customer service point voltages and ensure that the voltage profile from the substation to the customer is as uniform as possible. All the components, except for the VVO control algorithms, are usually available in modern distribution system control centers. VVO control algorithms have been undergoing rapid advancement in recent years and are at the stage to move from R&D into feasible field deployments. With the introduction of DER and the deployment of consumer demand management, the integration of VVO with the control and optimization of these resources in the distribution system is becoming a new smart grid challenge to the industry in practice. The increasing penetration of renewable generation and energy storage in the coming years with smart grids will introduce great challenges and also opportunities to VVO. Nondispatchable distributed renewable energy resources, such as PV and wind, are intermittent and unpredictable in operation. They also increase the likelihood of overvoltage conditions in the distribution system. This means that the controllable voltage and VAr resources on the distribution system need to be controlled more frequently and more accurately in order to match the stochastic output profile of the renewable energy sources. Energy storage systems, such as battery storage, will affect the power flows on the distribution system. The variable power flow from energy storage will need to be taken into account for VVO, such that the VVO becomes volt/VAr and watt optimization (VVWO). Future regulatory and business models will profoundly affect the way DERs will be owned and operated. Considerable uncertainty and R&D, however, still remain in this area.

VVO will also provide the flexibility in handling network reconfigurations. For example, a permanent fault in a distribution feeder will result in switching operations and topology changes. The power flow directions will change along with the topology change. VVO can use the updated power flow analysis to determine the optimal VVC actions based on the new configuration of the distribution system. Therefore, centralized VVO in DMS can fully take advantage of the global distribution network and load models, system configuration, and the full complement of remote measurements from SCADA, as well as data from other applications, such as load forecasting/estimation, AMI, demand response, etc. Smart grid advances in VVO should drive toward the support of more advanced features, such as look-ahead optimization under dynamic operation conditions with planned and unplanned outages and maintenance schedules.

3.5.3 Fault Detection, Isolation, and Service Restoration

Witold P. Bik, Christopher McCarthy, and James Stoupis

Traditionally, electric utilities use the trouble call system to detect power outages. Initially, distribution system faults are interrupted and cleared by a fuse, recloser, pulsecloser, or relayed circuit breaker. Once the faults are isolated and customers experience power outages, they call the utility and report the power outage. The distribution system control center then dispatches a maintenance crew to the field. The crew first investigates the fault location and then implements the switching scheme(s) to conduct fault isolation and power restoration. This procedure for power restoration may take several hours to complete, depending on how quickly customers report the power outage and the maintenance crew can locate the fault point and conduct the power restoration [4]. Thus, one of the main drivers for smart grid is the possibility to enhance and optimize the reliability of the distribution system, which is being pushed strongly by utility regulatory bodies such as public utility commissions (PUCs).

With the recent push in smart grid, utilities have deployed more feeder switching devices (e.g., reclosers, pulseclosers, circuit breakers, switches) with IEDs for protection and control applications. The automated capabilities of IEDs, such as measurement, monitoring, control, and communications functions, make it practical to implement automated fault detection, isolation, and service restoration (FDIR). As a result, the power outage duration and the system reliability can be improved significantly. The IED data can be transmitted via communications between the IEDs themselves or back to a substation computer or a control center.

In addition to the FDIR application, other reliability-related issues arise in the normal day-to-day operations of a distribution utility, such as the failure of key distribution system assets, power quality issues caused by utility and customer equipment, and protection miscoordination. Equipment monitoring and diagnostics is a key technology that will be significant in smart grid due to its capability to prevent (and potentially to predict) the failure of assets vital to the operation of the distribution system, such as substation transformers and circuit breakers. Power electronics devices are gaining more attention due to their capability to reduce power quality issues. Adaptive protection schemes will also play a key role in modifying the substation and feeder IED protection and control settings in realtime for optimal device and system performance during faults.

3.5.3.1 Faults on Distribution Systems

The majority of faults that occur on distribution systems can be linked to a partial or complete failure of electrical insulation. The result is an increase in current, causing much stress on the overhead conductors or underground cables along the feeder. Of the faults that occur on medium-voltage overhead networks of utilities across the world, approximately 80% of the faults are transient (temporary), and 80% of the faults involve only one phase to ground [5].

Most distribution feeders today are radial in nature, meaning that power flows in a hierarchical fashion, from the distribution substations out to the loads. Most feeders have a three-phase main line, which forms the backbone of the power delivery system. It is typically an overhead line, which allows for clearing of temporary faults, as well as easier permanent fault location and repair. Single-phase and three-phase laterals, both overhead and underground, are fed from the main line and typically protected by fuses for fault isolation. Single-phase and three-phase sectionalizers and reclosers for overhead circuits, as well as underground fault interrupters, could also be used on the laterals where heavy loads are connected. A fault on the main line causes the substation circuit breaker to operate to isolate the fault. If automatic reclosers or sectionalizers are also used on the main line, and a fault occurs downstream of one of those devices, then the effect of the fault can be isolated to the downstream portion of the feeder only.

Due to the radial nature of most distribution systems today, no backup source is available, so customers are susceptible to a power outage even when the fault occurs several miles away. A basic grid is formed when the capability to transfer loads to adjacent circuits is added. A smart distribution grid emerges when the switching points between circuits, as well as several points along each circuit, have the intelligence to reconfigure the circuits automatically, either directly themselves or when receiving control commands from a substation computer or control center, when an outage occurs. More intelligent switching points yield more options to reroute power to serve the load, and communication between or to those points makes self-healing a practical reality.

Thus, in the future, it is envisioned that the distribution system will be more meshed than radial, especially when considering the connection of distributed generation and energy storage systems. This means that multiple sources will be connected to the same load. This reality makes fault management and maintaining reliability great technical challenges. Advanced protection and FDIR schemes, along with advanced sensing and high-speed communications, will be required to quickly isolate a distribution fault and restore unaffected customers.

3.5.3.2 Drivers, Objectives, and Benefits of FDIR

The major drivers for FDIR and other similar initiatives are improved reliability, enhanced system operation, and improved system efficiency. These factors all contribute to restoring unaffected customers faster after a disturbance, reducing the number of affected customers significantly, thus increasing customer satisfaction. The improved reliability is tied directly to utility reliability metrics. In most cases, utilities are under pressure from regulatory bodies, such as PUCs, to improve reliability, and the reliability metrics are used to determine distribution circuit performance.

The main reliability indices that are predominantly used throughout the world are the following IEEE Standard 1366 metrics [6]:

- SAIDI = Sum of all customer interruption durations/Total number of customers served
- SAIFI = Total number of customer interruptions/Total number of customers served
- CAIDI = SAIDI/SAIFI = Sum of all customer interruption durations/Total number of customer interruptions
- Momentary average interruption frequency index (MAIFI) = Total number of customer momentary interruptions/Total number of customers served

The majority of the world uses the IEEE reliability indices as is or calculated in a similar way but with a different name. Some other metrics used by utilities worldwide include average cost per outage, energy not supplied, customer minutes of interruption (CMI), and average interruption time. It should be noted that each utility may define their own metrics to assess reliability, that is, the IEEE indices are the closest thing to a standard on reliability but by no means the only metrics used.

The major benefit related to smart grid applications is the improvement in the reliability metrics, which ultimately results in an improvement in customer service. The need to meet goals related to the reliability metrics has motivated many utilities to install more automated switching devices out in the distribution system, which has reduced the duration of the outages due to the faster response time for the isolation of the fault and the restoration of the unaffected customers. With the installation of these devices as part of automated FDIR schemes, most utilities have established a target restoration time for fault disturbances, such as 1 or 5 min. In the future, the proliferation of more automated switching devices with communications and control capabilities will lead to even faster restoration times, leading to an even higher level of reliability. The automation of feeder switching devices also benefits other distribution automation applications. For example, the deployment of multifeeder reconfiguration or load balancing schemes enables a utility to transfer load from one feeder/substation transformer to another feeder/substation transformer, in the cases of transformer failure or peak loading conditions.

Other methods that can be used to improve reliability include more frequent tree trimming programs, the deployment of faulted circuit indicators, and the deployment of reclosers and sectionalizers instead of fuses, as well as fuse-saving schemes to only interrupt customers for permanent faults.[3] Also, circuit topology and load density of distribution feeders have a large effect on the frequency of faults and, thus, the duration of outages. Longer circuits typically lead to more interruptions. Shorter circuits, especially urban networks that form a meshed network, have been found to be more reliable. Also, utilities with higher load densities tend to have better SAIFI indices [6]. These factors are a key issue for future utility reliability, because as city areas and suburbs expand, the circuits will get longer and, thus, less reliable.

3.5.3.3 FDIR Equipment

3.5.3.3.1 Substation Circuit Breaker

A circuit breaker is a switching device typically located in the substation that can make, carry, and break currents under normal and short-circuit conditions [5] by opening and closing contacts.

FIGURE 3.112 Outdoor substation circuit breaker. (© Copyright 2012 ABB. All rights reserved.)

Most distribution feeders have a substation circuit breaker as the most upstream protective device, feeding the medium-voltage conductor wires that leave the substation. For FDIR schemes, each substation circuit breaker has a protection and control relay/IED that can communicate with the local substation automation devices, transmitting data and receiving control commands. Many circuit breakers today contain vacuum interrupters with magnetic actuators to operate a drive shaft, with encapsulated poles for protection from the weather and external elements. Figure 3.112 shows an example of a substation outdoor circuit breaker.

3.5.3.3.2 Manual Switch

A switch is a switching device that can make, carry, and break currents under normal conditions (not short-circuit conditions) by opening and closing contacts. Manual switches are typically located out on the distribution feeders, although some manual disconnect switches are deployed by utilities in distribution substations. Manual switches are typically not used in automated FDIR schemes, only in manual FDIR schemes when maintenance crews are dispatched to perform the switching. Most the switches today contain air blade-type contacts that open and close in such a way to give visual indication to the maintenance crews.

3.5.3.3.3 Remotely Operable Load-Break Switch

A remotely operable load-break switch is a switching device typically located outside the substation that can make, carry, and break currents under normal conditions (not short-circuit conditions) by opening and closing contacts. Remotely operable switches allow for the utility operations department to operate the switches via communications from the control room or the substation, typically through a SCADA or a substation computer interface. For FDIR schemes, each remotely operable

Smart Grid Technologies

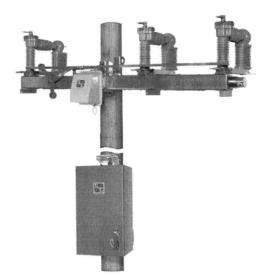

FIGURE 3.113 Overhead load-break switch. (© Copyright 2012 S&C Electric Company, Chicago, IL. All rights reserved.)

switch has a control IED that can communicate with the control room SCADA system in a control center-based FDIR scheme or a substation computer in a field-based FDIR scheme, transmitting data and receiving control commands. Most remotely operable load-break switches today contain either air blade-type contacts or vacuum interrupters with magnetic actuators as the operating mechanism. Figure 3.113 shows an example of an overhead load-break switch.

3.5.3.3.4 Automatic Sectionalizer

Automatic sectionalizers are essentially manual or remotely operable load-break switches located out on the feeder with added intelligence in their IEDs. The added intelligence allows for a local control decision to be made based on local voltage and current measurements. The sectionalizer IED counts the number of overcurrent events and/or voltage drops below a threshold when a fault on the connected feeder occurs. When the sectionalizer reaches its preconfigured count number, it opens during the dead time of an upstream circuit breaker in the substation or recloser outside the substation. Sectionalizers can be incorporated as part of FDIR schemes which would typically allow for local control decisions to be made for the fault isolation, and subsequent restoration decisions made by a master device or peer as part of a multipoint communication scheme. It should be noted that in some cases, single-phase automatic sectionalizers are used on laterals to isolate a single-phase fault that may occur, so that the rest of the distribution feeder can remain energized.

3.5.3.3.5 Automatic Recloser

The automatic recloser is similar to the automatic sectionalizer, except that it can make, carry, and break currents under normal and short-circuit conditions. Thus, instead of counting the number of overcurrent events occurring downstream on the feeder, the recloser will actually trip for those events. Like sectionalizers, automatic reclosers are also placed outside the substation out on the feeder. In some cases, single-phase automatic reclosers are used on laterals to isolate a single-phase fault that may occur, so that the rest of the distribution feeder can remain energized.

A recloser protection and control IED is typically coordinated with the upstream substation circuit breaker relay and downstream recloser IEDs and fuses, for coordination during fault disturbances. During the reclosing sequence, one or more fast trips are typically used in an attempt

FIGURE 3.114 Pole-top recloser. (© Copyright 2012 ABB. All rights reserved.)

to clear a temporary fault, followed by slower trips if the fault is permanent. Like sectionalizers, automatic reclosers can be incorporated as part of FDIR schemes which would typically allow for local control decisions to be made for the fault isolation, and subsequent restoration decisions made by a master device or peer as part of a multipoint communication scheme.

Many reclosers today contain vacuum interrupters with magnetic actuators to operate a drive shaft, with encapsulated poles for protection from the weather and external elements. Figure 3.114 shows an example of an outdoor overhead recloser.

3.5.3.3.6 Source Transfer Gear

Source transfer equipment typically consists of pad-mounted switchgear connected to multiple feeders and to the connected loads. Voltage sensors are used on each feeder to determine the presence of voltage on both the primary and secondary sources. If the voltage on the primary source feeder drops below a predetermined threshold and the voltage on the secondary source feeder remains above a predetermined threshold, then the three-phase switch connected to the primary source is first opened and subsequently the three-phase switch connected to the secondary source is closed. In an industrial park configuration, the source transfer switchgear is wired as part of a multiloop system, where one feeder acts as the primary source for a first set of loads and the secondary source for a second set of loads, and a second feeder is the primary source for the second set of loads and a secondary source for the first set of loads. If one feeder loses voltage upstream, then all the loads are switched to the healthy feeder. Figure 3.115 shows an example of a piece of source transfer switchgear.

3.5.3.3.7 Sensors

Sensor devices come in many forms and perform various different functions. Sensors typically used for the FDIR application are typically in the form of clamp-on fault current indicators (FCIs); however, data from other types of sensors (e.g., line post sensors, temperature sensors) could also

FIGURE 3.115 Source transfer switchgear. (© Copyright 2012 S&C Electric Company, Chicago, IL. All rights reserved.)

be applied. All FCIs measure current, and some also measure voltage, which can be useful for other distribution applications. Most FCIs have a visual indication that fault current has passed through the device or have a wireless signal for short-range communications, so that the utility maintenance crew can receive the data via drive-by. More recently, sensor companies are transmitting these data for long-range communications to a local data collector or even directly back to the substation or control center, for use in the FDIR application and outage management.

3.5.3.3.8 New Technology

3.5.3.3.8.1 Single-Phase Dropout Recloser This device is a single-phase cutout-mounted fault interrupter with a two-operation sequence—one timed overcurrent trip on a fast TCC curve and then one on a delayed TCC curve. Utilities will often implement a "fuse-blowing" scheme that coordinates the substation breaker with the lateral fuse so that the fuse will clear any downstream fault within its rating, not the breaker. Most feeder customers experience no power interruption, but the lateral customers get a prolonged outage—a bad result if the fault is temporary. A "fuse-saving" scheme has the first trip of the substation breaker intentionally set to operate faster than the fuse to clear a temporary fault downstream of the fuse. Often, the fault will be cleared during the open time interval, so that when the breaker closes back in, service is automatically restored and there is no prolonged customer outage. The second breaker trip is slower, so that if the fault is permanent, the lateral fuse will operate to clear the fault and isolate that section. A downside is that all feeder customers experience a momentary interruption for a lateral fault when the breaker trips before the fuse, regardless if the fault is temporary or permanent. Overall, this can have a negative impact on customer satisfaction since the vast majority of faults occur on taps off the main lines. A single-phase dropout recloser, used instead of a fuse, provides the best of both scenarios. The fast trip prevents a permanent outage for a temporary fault, and the substation breaker is spared from any trips even if the fault is permanent.

3.5.3.3.8.2 High-Performance Fault Testing After a conventional recloser or relayed circuit breaker opens to interrupt a fault, it typically recloses into the fault several times to determine if the fault is still present. Pulseclosing is a new technology for overhead distribution system protection that tests fault persistence without creating high-current surges that cause feeder stress. The pulsecloser device very rapidly closes and reopens its contacts at a precise point on the waveform to send a very short low-current pulse down the line then analyzes the pulse to determine the next course of action. If the pulse indicates a persistent fault, the pulsecloser will keep the contacts open, wait a user-configurable interval, and pulse again. This process can repeat several times until the pulsecloser determines that the line is no longer faulted. It then closes to restore service. However,

FIGURE 3.116 Pulsecloser. (© Copyright 2012 S&C Electric Company, Chicago, IL. All rights reserved.)

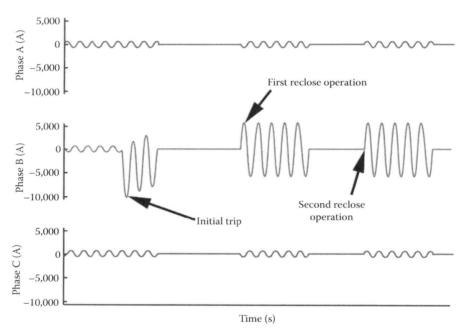

FIGURE 3.117 Conventional reclosing in response to a permanent fault. (© Copyright 2012 S&C Electric Company, Chicago, IL. All rights reserved.)

if the fault persists for the duration of the test sequence, the pulsecloser will lock out to isolate the faulted section (Figure 3.116).

Figure 3.117 shows a typical current waveform pattern that would result from a conventional recloser or relayed circuit breaker operating in response to a permanent single-phase-to-ground fault. The random point-on-wave closing often results in asymmetric fault current, significantly increasing peak energy into the fault. When the pulsecloser clears a fault, however, it tests for continued presence of the fault using pulseclosing technology, closing at a precise point on the voltage wave. Figure 3.118 shows how a pulsecloser would respond to the same permanent fault. Note that both the positive and negative polarities are tested to verify that high currents are due to faults, and not transformer inrush currents.

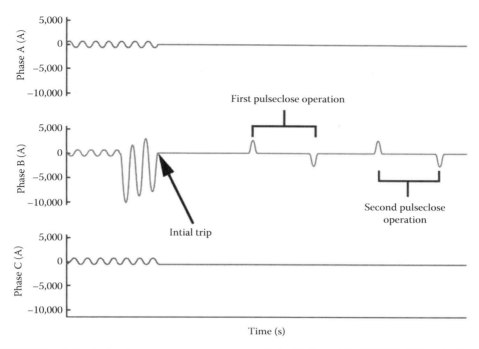

FIGURE 3.118 Pulseclosing in response to a permanent fault. (© Copyright 2012 S&C Electric Company, Chicago, IL. All rights reserved.)

3.5.3.4 FDIR Implementation

FDIR can be implemented in several ways. Each approach has its own strengths and weaknesses. The most appropriate approach will depend on a utility analyzing their existing infrastructure and the intended goals of reliability optimization, as well as the speed of response to the fault conditions required by the utility.

3.5.3.4.1 Field-Based FDIR Schemes

3.5.3.4.1.1 Substation Breaker with Fault-Detecting Switches Basic feeder protection is implemented with a substation breaker that coordinates with fuses on the feeder. When customers report an outage, a crew is dispatched and uses the customer outage reports to find the blown fuses. The fault is isolated when the crew opens a manual switch. Faults can be located more quickly if fault-passage detectors with a visible indicator have been installed on each phase at the switches. FDIR response times can be further improved when switches report fault current to SCADA and can be opened remotely by a SCADA command, then service can be restored to all customers on the substation side of the switch isolating the fault. The problem with using this approach is that during breaker operation, the whole feeder is subjected to momentary outages, and if the fault is permanent, the entire feeder will be locked out until the fault is located and manual switching performed. The restoration time in this case can take hours.

3.5.3.4.1.2 Substation Breaker with Midpoint Recloser About half of the customers on a feeder can be spared an outage when a recloser is installed at the midpoint of a radial feeder. Like the substation breaker, a recloser can interrupt fault current. The zone of protection is expanded because a midpoint recloser will sense current for a fault near the end of a feeder more accurately. If a fault occurs on the load side of the recloser, it is coordinated to open and isolate the fault before the substation breaker operates. So customers on the substation side of the recloser will not experience loss of power. The advantage of this scheme is that it can limit the extent of outage by effectively splitting the feeder into two sections. However, upstream customers are still subjected to voltage dips caused by reclosing, which is measured by the system average RMS variation frequency index (SARFI). In the case of permanent faults, SAIDI can be improved by 50%.

3.5.3.4.1.3 Substation Breaker with Automatic Sectionalizers Automatic sectionalizers sense the passage of fault current and then open on a predetermined loss-of-power count (substation recloser operations). This coordinates the operation of multiple sectionalizers. Because only a few reclosers can be coordinated in series, more automatic sectionalizers can be installed on the feeder, and more customers will avoid a permanent outage whenever a fault occurs near the end of a radial feeder. The advantage of this approach is that better segmentation can be achieved and the impact of the outage reduced within seconds. On the negative side, the entire feeder is subject to momentary outages during fault testing.

3.5.3.4.1.4 Fault Hunting Loop Schemes with Reclosers (No Communications) A loop scheme can automatically restore service with a normally open tie to a nearby feeder. Initial sectionalization occurs by the responses of various coordinated overcurrent protective devices to a feeder fault. Reconfiguration to restore power to the unfaulted feeder sections occurs by using a combination of timers and fault interruption, and no communication is required. The feeder is returned to normal configuration manually by opening the tie device and closing the midline devices. Figures 3.119 and 3.120 show the circuit topology for a three-recloser loop and a five-recloser loop, each with a normally open recloser at the tie point. Simple reliability calculations for loop systems assume a constant fault incidence rate in all feeder segments, equal segment lengths, even customer distribution, and a constant restoration time throughout the system. The benefit of a three-device loop system over two radial feeders is a 50% reduction in SAIFI and SAIDI compared to a breaker only. Expanding to a five-device loop improves the reliability indices even further.

Conventional loop schemes use loss-of-voltage timers to set the order of device operations. Closing into a fault is the only way to know if the fault is still present. The first reconfiguration

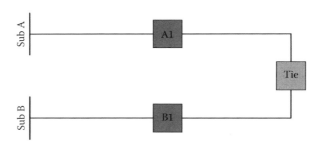

FIGURE 3.119 Three-recloser loop scheme.

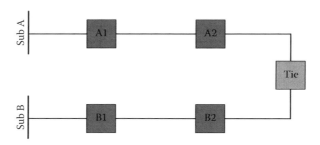

FIGURE 3.120 Five-recloser loop scheme.

Smart Grid Technologies

action to occur, after the fault has been interrupted and isolated by the breaker and/or the midline reclosers, is when the normally open tie recloser closes based on expiration of a timer that initiates upon loss of voltage on either side. A three-recloser loop system with equal segment lengths has a one-in-two chance that the tie recloser will sense loss of voltage due to a fault in the adjacent line section. When the tie recloser closes, fault current flows through the entire previously unfaulted feeder until the recloser times on its TCC curve and locks open. For a five-recloser loop system, the faulted section is found when the tie recloser closes into the fault or when the next midline recloser subsequently closes into the fault. A loop scheme application is commonly limited to the use of two specific sources, and each source must have capacity to supply the combined load of both feeders.

3.5.3.4.1.5 Loop Restoration with Pulseclosers A pulseclosing device used at the tie point, in an otherwise conventional recloser loop scheme, will use a pulseclose to test for faults before closing in. This avoids putting a fault on the otherwise unfaulted feeder. Pulseclosing benefits are compounded when multiple pulseclosing devices are used in series, since pulseclosing devices will properly sectionalize a system without using TCC coordination. A loop system with automatic noncommunicating restoration can be expanded to include any number of pulseclosing devices to provide desired segmentation and improve reliability for critical customers or a problem area. Furthermore, the entire restoration process can be completed without ever reintroducing the fault to either feeder.

3.5.3.4.1.6 Distributed Systems with Peer-to-Peer Communications The optimal self-healing system uses a mix of decentralized fast-acting local response with a centralized system for oversight. Local clusters of automated feeders function independently of the central control to isolate problem areas and minimize disruptions quickly. The feeders may be in a reconfigured state for several hours until the crews locate and repair the fault, so the distribution operators may want to shift load, switch capacitors banks, or modify voltage regulation to optimize efficiency. Peer-to-peer communications is used as the basis for such distributed restoration systems.

Distributed logic is also capable of handling multiple events, looking for alternate sources to restore unfaulted sections that are without service. This is especially useful during strong storms that sweep across a service territory and cause multiple outages. Reconfigurations can occur simultaneously at more than one location, and by accounting for real-time loading, it ensures that a circuit will not pick up more line segments than it can handle. It's a great advantage to have the distribution system automatically do the best restoration possible, quickly and efficiently, and report the *final* reconfigured state to the dispatchers. A smart restoration system will minimize the required excess source capacity, because it dynamically monitors load and can pick up more load with existing resources.

Figure 3.121 is an example of how a four-source, 12-switch deployment with distributed logic is divided into teams. All the switching points that bound a given line segment form a team. Switching points can be load-break switches, reclosers, pulseclosers, or breakers. The controls for each switching point communicate directly with all other controls in the team—which is why it is called peer-to-peer communications. Each team can share information with adjacent or remote teams, facilitating the deployment of large-area distributed logic functions such as protection and service restoration. Applications are scalable since additional teams of switching points can be added as necessary.

3.5.3.4.1.7 Substation Computer-Based Schemes Some utilities prefer to deploy substation computer-based FDIR schemes, especially when they have a mix of new and legacy control IEDs at the switching points outside the substation. In this case, it is simpler to add communications to the feeder switching devices and a substation computer with its own logic and to retrofit the older legacy control IEDs out on the feeder switching points. The substation computer typically has a simple connectivity model of the connected feeders with automated switching devices and receives

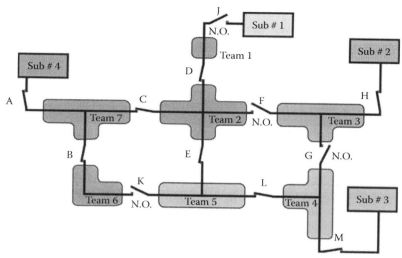

FIGURE 3.121 Example of distributed logic with teams of switches. (© Copyright 2012 S&C Electric Company, Chicago, IL. All rights reserved.)

data from the feeder and substation IEDs to make intelligent switching actions after a fault has been isolated upstream. After a recloser or the substation circuit breaker has operated to isolate the fault, the substation computer processes the IED data, determines the downstream isolation switching device that must be opened, and then determines the normally open switch that must be closed to restore power to unaffected customers.

The restoration algorithms of substation computers vary, but typically a capacity check is at least performed to ensure that alternate feeders can pick up the excess load. Some substation computers are also capable of supporting multisource multibackfeed restoration when more than two alternate sources, and thus at least two normally open switching devices, make up the distribution system. Figure 3.122 shows an example of this type of system. These devices also typically support restoration when multiple faults occur in the system, return-to-normal switching after the fault, and can be disabled by a single virtual "button" in the graphical user interface.

3.5.3.4.1.8 Source Transfer Applications Source transfer makes two different sources available to supply a specific customer load. If one source is lost, the customer is switched to the alternate source. When power is restored, the customer load is returned to its primary source. Sophisticated controls monitor source quality and availability and automatically switch to the best or only source.

3.5.3.4.1.9 New Technologies Large-Scale energy storage. High-power batteries, efficient inverters, and sophisticated switching make energy storage a practical new alternate source. There are a small but growing number of installations of 1–4 MW energy storage systems on utility systems using sodium-sulfur (NaS) batteries, among other technologies. The intelligence in the control system charges the batteries during off-peak times and then supplies energy during peak times. This creates several opportunities for economic justification, such as the ability to make full use of intermittent renewable sources regardless of the time of day or present loading, the ability to shave peak load, and the deferral of substation capacity upgrades. In another case, energy storage will result in a vast improvement in electric service reliability in the case of a town located a long distance from the substation and supplied power by only one feeder. The stored energy can power the town as an islanded network for several hours while the feeder is out of service.

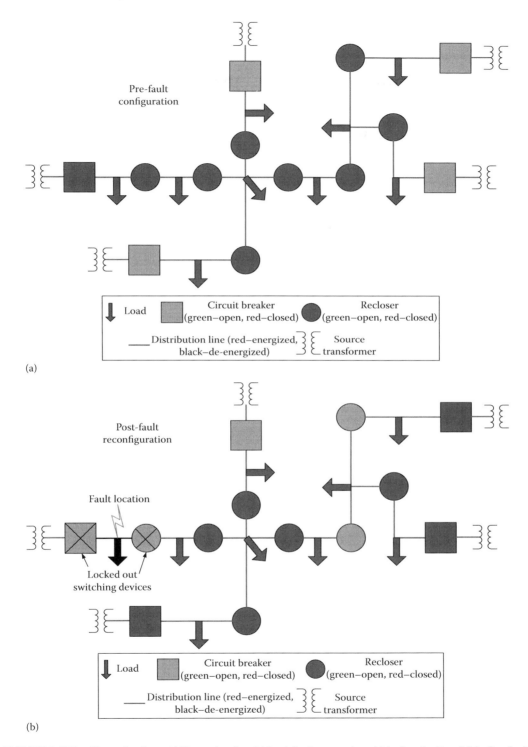

FIGURE 3.122 (**See color insert.**) Example of multi-backfeed restoration: (a) before fault and (b) after fault. (© Copyright 2012 ABB. All rights reserved.)

3.5.3.4.1.10 Closed-Loop Fault Clearing Systems A high-speed fault clearing system can use two parallel redundant circuits to protect an important customer area from outages. Customers can be supplied from either circuit, and directional overcurrent protection relays with fiber-optic communication are used to keep response time fast. Faults can be cleared in 3–6 cycles, with power immediately restored to unfaulted sections, and only faulted sections will have a power interruption. With a closed-loop circuit configuration, load can be supplied while a fault is being cleared and only a few customers will experience a voltage dip. These systems offer fault clearing with no outage when deployed on URD or circuits with no load connected between switches.

3.5.3.4.1.11 Deployment Considerations Because the field-based FDIR schemes vary significantly depending on what scheme is used, there are a wide range of deployment aspects that must be considered, including the use or nonuse of communications devices and fault-passage indicators and the deployment of automated reclosers and sectionalizers. These devices help to determine the fault location quicker, thus enhancing reliability. By automating the field switching devices with communications and intelligent control devices, the reliability indices will be reduced significantly, compared with the noncommunicating, nonautomated schemes previously described. Cost is the major issue with the deployment of these devices. Each utility typically sees a savings associated per customer with each minute of the outage reduction. Thus, there is a trade-off between the cost to deploy the automation devices and the observed savings. The utility will see a return on investment only once the savings exceeds the deployment costs. Hence, the utility must seriously analyze the number of automated devices deployed and would be wise to deploy these devices in stages in order to ascertain the reliability improvement and cost savings as the level of automation increases across their system.

In many states in the United States, the PUCs encourage utilities to report action plans for improving their worst-performing feeders. In this fashion, reliability is considered in the utility rate case decisions, providing a financial incentive for reliability improvement.

Some countries, particularly in Europe, have large monetary incentives in place that motivate utilities to improve service reliability. The incentives often take the form of both financial penalties to the utility for poor performance or capital reimbursement funds for improved results. The external influence of positive or negative cash flow based on service reliability provides a more quantitative environment for cost/benefit calculations.

The system topology must also be considered when determining the type of field-based FDIR scheme that is deployed. For basic radial systems, simpler schemes can be deployed, such as the midpoint recloser. For the more complex multibackfeed (meshed) systems, more automated devices are required, driving the deployment cost up, but the level of reliability improvement and cost savings will also increase due to the available alternate backfeed sources. For these systems, the more advanced distributed logic solutions based on peer-to-peer communication or substation computer-based schemes should be deployed.

3.5.3.4.2 Control Center-Based FDIR Schemes

3.5.3.4.2.1 Manual Switching Using SCADA/DMS and Remotely Controlled Switching Devices Historically, when customers lose power, they call a utility's automated answering system, which enters the outage data into the utility's OMS, which in many cases is part of the DMS. Outage data are then displayed on the operator interface (see Figure 3.123). As more phone calls are answered, the OMS tries to determine the cause of the outage, for example, if a switching device or fuse in the field operated to clear a fault or if a transformer or other component failed. The operator then uses the interface to coordinate isolation and restoration of the feeder by dispatching crews to conduct the switching operations.

If the utility has automated switching devices that directly or indirectly communicate with their SCADA, then they can remotely control the feeder switching devices for faster isolation and restoration.

Smart Grid Technologies

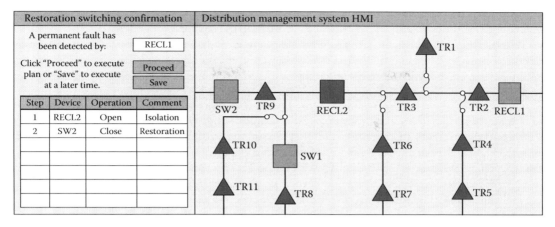

FIGURE 3.123 DMS-based restoration. (© Copyright 2012 ABB. All rights reserved.)

3.5.3.4.2.2 Automatic Switching Using SCADA/DMS and Remotely Controlled Switching Devices With an integrated and automated control center-based restoration scheme, the fault detection occurs in the field based on the IED-sensed network events, and the SCADA software is automatically informed, subsequently sending these data to the DMS. When the DMS receives this information, it will run a restoration switching analysis (RSA) with respect to the outage area and generate power restoration schemes or switching plans.

The RSA is based on the detailed network model and load flow analysis of the network model to ensure that the postrestoration network does not have current and voltage violations. In North America and other parts of the world, it is a requirement that the RSA supports unbalanced grounded systems, which are typical since the loads are split between the phases due to the way in which the loads are dispersed. In Europe and other parts of the world, a balanced load flow analysis is sufficient to support the three-phase ungrounded distribution systems in place. The RSA performs network topology analysis techniques that typically support both lightly loaded and heavily loaded network conditions. If the loading of the network is light, most likely a single-path restoration is sufficient. If the loading of the network is heavy, either a multipath restoration or multilayered feeder segment restoration has to be used. Figure 3.123 shows a control room operator interface with proposed switching actions.

Whether or not a switching plan is sent to the field devices for execution immediately after the RSA is executed is based on the operator's preferences. The application has three types of restoration control settings: (1) fully automated control mode, where the operator is not involved in the switching plan execution process, that is, the best switching plan is automatically selected; (2) semiautomated mode, where the best switching plan is automatically selected and the operator performs a one-click confirmation-based switching plan execution; and (3) semiautomated supervisory mode, where the operator selects the best switching plan based on information provided as a result of the RSA.

3.5.3.4.2.3 New Technologies The AMI and sensor technologies are having a significant impact on control center-based FDIR schemes. These technologies are being used to help enhance the outage analysis process of the DMS, including outage verification, faulted line segment location, and restoration verification. The AMI data are typically retrieved by the DMS via an MDMS, where the data are stored after retrieval from the field. By deploying communicating meters and sensors in the field, the utility operator can pinpoint the general line segment location where the fault occurred in the DMS graphical user interface. These real-time data allow the utility to dispatch field crews for repairs before customers even call the utility trouble call system to report an outage. Once the repair work is completed and power is restored to the affected customers, the meter and sensor data can be used to verify that the power has been restored to the customers.

Because the control center-based FDIR schemes require data via direct or indirect communication to field devices, other field data could be used in the outage/fault management process, as well as other everyday processes. For example, by obtaining the current fault magnitude from a substation circuit breaker IED after a fault has occurred, the DMS can run an algorithm to estimate potential fault locations. Coupling these results with the AMI data provides a very powerful new method to improve reliability and to significantly reduce customer outage minutes. In the future, it is envisioned that utilities may call affected customers immediately after a fault has occurred, indicating estimated time to restoration. Another example is the availability at the control center to meter-level power quality data, which can be analyzed to determine the effect of disturbances at the system and customer levels. The key message is that the more data the utility control center has, the more effectively it can run its operations, becoming more proactive rather than reactive.

3.5.3.4.2.4 Deployment Considerations When deploying control center-based FDIR schemes, the main issues that must be considered are the communications system, supervisory modes, and the integration of AMI and sensor data. For the communications part, some utilities prefer to deploy communications to the feeder devices directly from the control center SCADA system, bypassing substation computer and gateway devices. However, the advantage of getting access to data through these devices is the proximity of the substations to the feeder devices. In some cases, the utility may also prefer to bypass the SCADA system itself, sending field data directly to the DMS, due to the data bottlenecks that sometimes occur with SCADA and the access to nonoperational data that SCADA does not typically collect.

Another issue to be determined during deployment is the setting of the operation mode of the control center-based FDIR scheme. Most utilities prefer to employ a supervisory mode for these FDIR schemes, allowing the operator to have some level of control of the field switching. However, some utilities prefer the automatic mode to expedite the management of the fault, allowing restoration to occur much more quickly.

As discussed previously, there is great value in integrating AMI and sensor data into the FDIR schemes to improve outage/fault management. However, this is a large undertaking that will consume many utility (and vendor) resources to complete. Thus, it may be wise to first integrate these data in sections of the distribution system where reliability issues are the gravest.

3.5.3.5 Reliability Needs in a Smarter Grid

With the future smart grid, there are many challenges that must be addressed to reap the operational benefits. Deploying FDIR schemes alone does not ensure that reliability will be optimized. Coordinating FDIR with other control functions, such as VVC and optimization schemes, demand response programs, DER dispatch, and load balancing, will result in more effective distribution grid management. Coordination with volt/VAr schemes alone will help to increase efficiency and minimize losses after a fault has occurred on the distribution system. Load balancing is another future function that will allow for dynamic reallocation of load to adjacent feeders to ensure reliability during overload conditions.

DER devices and demand response programs will have a big effect on utility operations and devices, as well as the way in which consumers receive and use power. Due to the availability of alternate generation, DER devices will help to enhance reliability, especially when "basic FDIR" results in unaffected customers in specific areas of the grid losing power for a significant period of time after a fault. The proliferation of DER devices on the grid will require better protection algorithms and schemes, due to the two-way flow of power and the fact that they will contribute to faults. Demand response is another resource for grid management, where loads can be shaved during peak periods of operation to reduce overall system demand. By coordinating demand response with FDIR, restoration can be achieved over larger areas of the grid due to the lower demand, again enhancing reliability.

Reliability is becoming much more of an issue in the smart grid, due to technologies becoming available that can make the deployment of reliability functions such as FDIR and load balancing more realistic. Many PUCs are also pushing for higher reliability from the utilities over which they monitor. As the level of reliability increases in other industries, especially consumer industries (e.g., smart phones and laptops with "on demand" features), the expectations on reliability of the electricity supply from consumers will only increase as a result. These realities will make reliability optimization of the utility distribution grid a mandatory requirement in the future.

3.5.4 OUTAGE MANAGEMENT

Stuart Borlase, Steven Radice, and Tim Taylor

Modern computer-based OMS, utilizing connectivity models and graphical user interfaces, has been in operation for some time now. OMS typically includes functions such as trouble call handling, outage analysis and prediction, crew management, and reliability reporting.

Connectivity maps of the distribution system assist operators with outage management, including partial restorations and detection of nested outages. Outage management was originally based on receiving calls from customers and did not include a connectivity model of the system, including the connection points of all customers. Manual data recording and the use of paper maps were used to estimate the location of outages.

With the modern OMS, system connectivity information is typically stored in the GIS. Network data from GIS (and/or other data sources) are imported to the OMS database using a network data interface. This interface extracts data from GIS and performs a data model conversion based on business rules and data model mapping. The interface initially populates the database with all network data, including connectivity information, system components including protection and switching device types and locations, and distribution transformers. This is referred to as the "bulk network data load" or bulk load. The interface can also be periodically run to transfer the subset of data that has changed since the last update. This process is referred to as the "incremental network data update" or simply incremental update. A screen capture of an OMS system model in a large metro area is shown in Figure 3.124.

FIGURE 3.124 (**See color insert.**) OMS system connectivity model in a large metro area. (© Copyright 2012 ABB. All rights reserved.)

The data extracted from the GIS will capture necessary network data to support OMS and DMS operation. The required type data can be provided from other sources or entered manually. The following tasks are usually performed for the interface specification:

- Review and determine the data inputs/outputs and data flow, identify application modules, and identify all unique key attributes that will need to be maintained and used by the OMS and DMS applications.
- Review source data models in order to verify data requirements.
- Determine mapping of source data objects to OMS/DMS objects.
- Determine mapping of source data attributes to OMS/DMS attributes.
- Develop a data mapping spreadsheet including transformation rules as appropriate.

Another key to successful outage management predictions with OMS includes an accurate representation of customer connectivity on the system. When a customer calls in to report an outage, or an AMI meter sends an outage notification or restoration notification, the system has sufficient information to know where the customer is connected on the system. Customer connectivity is typically maintained in either the GIS or CIS. By evaluating report locations and the as-operated topology of the network, an OMS can identify probable outages that may constitute a single customer out, a single transformer out, a protective device operation, or a de-energized source.

Outage engine algorithms use the connectivity model, location of customer calls, and statistical parameters, such as the ratio of affected customers to total customers, the number of distribution transformers with calls, and the number of downstream protective devices with predicted outages, to determine probable outage location. These parameters can be combined in various ways to achieve optimum prediction accuracy for the network. System operators are then able to track outages using dynamic symbols on the geographic maps, such as the one in Figure 3.125, as well as in tabular displays.

In recent years, OMS has become more automated. Outage prediction—the process of analyzing outage events such as trouble calls, AMI outage notifications, and SCADA-reported status changes—has improved. Interfaces to interactive voice response systems (IVR) permit trouble call entry into an OMS without call-taker interaction and also permits the OMS to provide outage status information to customers and provide restoration verification callbacks to customers who request them.

OMS systems have also become more integrated with other operational systems such as GISs, customer information systems (CIS), mobile work force management (MWFM)/field force automation (FFA), SCADA, and AMI. Integration of OMS with these systems results in improved workflow efficiency and enhanced customer service.

Today's OMS is a mission-critical system. At some utilities, it can be utilized simultaneously by hundreds of users. It integrates information about customers, system status, and resources such as crews, providing a platform for operational decision support.

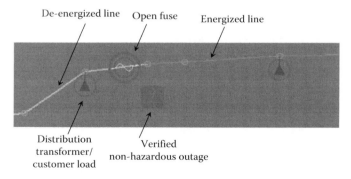

FIGURE 3.125 Example of outage representation using dynamic symbols in OMS.

Three ways in which outage management is changing in smart grid implementations are the following:

1. Integration of AMI data in OMSs
2. The use of advanced DMS applications for supporting outage management
3. The integration of SCADA with DMSs and OMSs

As distribution organizations have become more interested in increasing asset utilization and reducing operational costs, advanced DMS applications have been developed. These include load allocation and unbalanced load flow analysis; switch order creation, simulation, approval, and execution; overload reduction switching; and capacitor and voltage regulator control.

Two specific examples of advanced applications that reduce customer outage durations are the fault-location application and RSA, sometime called FDIR application.

The fault-location application estimates the location of an electrical fault on the system. This is different than estimating the protective device that actually opened, which typically is done based on the pattern of customer outage calls or through change in a SCADA status point. The location of the electrical fault is where the short-circuit fault occurred, whether it was a result of vegetation, wildlife, lightning, or other cause.

Finding the location of an electrical fault can be difficult for crews, particularly on long extents of conductor not segmented by protective devices. Fault location tends to be more difficult when troubleshooters or crews are hindered by rough terrain, heavy rain, snow, and darkness. The more time required to locate the fault, the more time customers are without power.

A DMS fault-location algorithm uses the as-operated electric network model, including the circuit connectivity, location of open switches, and lengths and impedances of conductor segments, to estimate fault location. Fault current information such as magnitude, predicted type of fault and faulted phases are obtained by the DMS from IEDs such as relays, recloser controls, or RTUs.

After possible fault locations are calculated within the DMS application, they are geographically presented to the operator on the console's map display and in tabular displays. If a GIS land base has been included, such as a street overlay, an operator can communicate to the troubleshooter the possible location including nearby streets or intersections. This information helps crews find faults more quickly. As business rules permit, upstream isolation switches can be operated and upstream customers can be reenergized more quickly, resulting in much lower interruption durations.

A second advanced application that improves reliability performance indices is RSA. This application can improve the evaluation of all possible switching actions to isolate a permanent fault and restore customers as quickly as possible.

Upon the occurrence of a permanent fault, the application evaluates all possible switching actions and executes an unbalanced load flow to determine overloaded lines and low-voltage violations if the switching actions were performed. The operator receives a summary of the analysis, including a list of recommended switching actions. Similar to the fault-location application, the functionality uses the DMS model of the system but improves outage management and reduces the CAIDI and SAIDI.

The RSA application is particularly valuable during heavy loading and when the number of potential switching actions is high. Depending on the option selected, the application can execute with the operator in the loop or in a closed-loop manner without operator intervention.

In closed-loop operation, the RSA application transmits control messages to distribution devices using communications networks such as SCADA radio, paging, or potentially AMI infrastructure. Such an automated isolation and restoration process approaches what many call the "self-healing" characteristic of a smart grid.

3.5.5 HIGH-EFFICIENCY DISTRIBUTION TRANSFORMERS

V.R. Ramanan

With the ever-growing global population and an ever-increasing global demand for energy consumption, sustaining our power-hungry world calls for energy-efficient products and reliable grids. The future smart grid needs to be not only "smart" but also highly efficient and environmentally sustainable.

Although there is a tendency to take them for granted, transformers are key components in the electrical power distribution grid, playing a significant role in the efficiency of that grid. In spite of the fact that they have high efficiencies, a large total loss of energy results due to the large number of distribution transformers. These losses are estimated to be approximately 2%–3% of the total electric energy and represent approximately 25 billion dollars annually for the United States.

A testament to the importance of energy saving programs and energy efficiency requirements is present in several global and local initiatives. Examples of mandates or standards requiring high efficiencies in distribution transformers are the following: the U.S. Department of Energy's mandated National Efficiency Standard, Australia's Hi efficiency 2010, India's 4 and 5 Star programs, China's SH15 standard, and the $A_k A_0$ standard in Europe.

Transformers in operation incur two types of losses: no-load loss, P_0, occurring in the transformer core which is always present and is constant during normal operation, and load loss, P_k, which occurs in the transformer electrical circuit, including windings and components, and is a function of loading conditions. Since most transformers are rated to handle peak loads which only happen at certain intervals during the day, distribution transformers can remain lightly loaded for significant portions of the day. So, specifying as low a no-load loss as possible reduces energy consumption and goes hand in hand with increased efficiency.

More end users now choose low-loss transformers based on criteria other than pure short-term profitability. The prevalent criterion is total ownership cost (TOC), which considers the future operating costs of a unit over its lifetime, brought back into present day cost and then added to its total purchase price. In calculating TOC, the losses are accounted by their financial impact, capitalized for an expected payback period for the transformer:

$$TOC = C_t + (A \times P_O) + (B \times P_k)$$

where
   C_t is the transformer purchase price
   A, the no-load loss factor, and
   B, the load loss factor, are the assessed financial value (e.g., USD/W) for no-load loss and load loss, respectively

The most optimal selection would be the design with the lowest TOC as calculated earlier. Simply put, the customer/user will obtain a practical balance between investment and reward, reflecting the continuous change in global and local business conditions at any time. TOC provides the true economics in evaluating a transformer purchase.

With low-loss, high-efficiency transformers, a higher material cost anticipates a higher first cost. However, this will be compensated by reduced running costs from lower losses. Beyond a certain time, the lower losses will give a net financial saving from reduced energy costs. If higher loss transformers are replaced with new low-loss transformers, this saving becomes even greater. Furthermore, lower losses result in cost avoidance derived from elimination or deferral of extra generation and transmission capacity additions.

The development of amorphous metal core distribution transformers (AMDT) is an important step in this direction. Amorphous metal (AM) enables a significant reduction in no-load losses

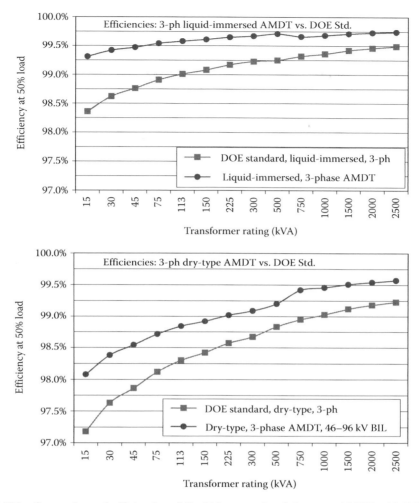

FIGURE 3.126 Comparison of efficiencies of liquid-immersed and dry-type AMDT with DOE National Efficiency Standards. (© Copyright 2012 ABB. All rights reserved.)

of transformers by up to 70%, as compared to conventional grain-oriented silicon steels (RGO). A quick back-of-the-envelope calculation highlights the energy savings potential from the deployment of AMDT. Assuming that about 1% of the installed U.S. generating capacity of 1.4 TW is lost in distribution transformer no-load losses, a 70% reduction of these losses from the use of AM cores suggests a potential annual energy saving of about 85 billion kWh.

Figure 3.126 compares the efficiencies of liquid-immersed and dry-type AMDT with the mandated minimum efficiency standards from the U.S. DOE across a wide range of transformer ratings. The improved energy efficiencies from AMDT are quite clear.

Figure 3.127 compares the TOC of liquid-immersed 1000 kVA transformers having RGO and AM cores, wherein the loss capitalization factors are as follows: A = 10 USD/W and B = 2 USD/W. The various components comprising the TOC are individually highlighted.

In summary, energy efficiency is the name of the game for electrical power distribution systems in the future. Reduction of losses due to transformers in the grid is an important first step, and developments are under way to specify and design lower loss, higher efficiency transformers. AM core transformers represent the ultralow loss, highest efficiency solutions.

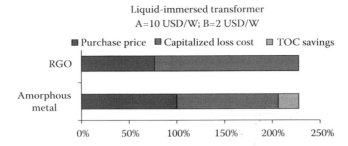

FIGURE 3.127 TOC comparison of RGO and AM transformers. (© Copyright 2012 ABB. All rights reserved.)

3.6 COMMUNICATIONS SYSTEMS

Harry Forbes, James P. Hanley, Régis Hourdouillie, Marco C. Janssen, Henry Jones, Art Maria, Mehrdad Mesbah, Rita Mix, Jean-Charles Tournier, Eric Woychik, and Alex Zheng

3.6.1 COMMUNICATIONS: A KEY ENABLER OF THE SMART GRID

Never before has the electric utility industry experienced a technology revolution as transformative as the smart grid. One of the major aspects of this transformation is the addition of an integrated and pervasive communications network that will touch every part of the grid from generation, transmission, distribution down to the consumer and will support the automated intelligent transactions that will make the new grid "smart." Many utilities will end up with communications network infrastructures that will rival the size and scale of the telecommunications companies.

Several major forces have influenced the development of wireless solutions resulting in changes to the interfaces among traditional electric industry domains. For the last 100 years, electric utilities have mostly operated in three separate domains of generation, transmission, and distribution. In the twentieth century and before, the separation of these domains enabled major development to occur. The result was to enable electricity to be generated, transmitted, and distributed to hundreds of millions of homes and businesses across the nation. In the last part of the twentieth century, technological and societal forces have changed the dynamics of these domains and how they interact. Two new domains have emerged that provide services and allow consumers to interact, not merely as customers, but also as suppliers of electricity and to directly participate in electricity supply and demand markets.

During the last 10 years, there has been an increasing regulatory and consumer interest in wireless solutions for electrical distribution and automated metering infrastructure in the United States. This movement gathered strength as a result of the northeast blackout of 2003 and, to a lesser extent, brownouts in 2000 in California. These events underscored the increasing vulnerabilities with grid reliability, which provided regulatory impetus for the passage of the 2005 Energy Policy Act that mandated states to study smart metering solutions. In 2006, North American Reliability Corporation (NERC) defined new rules to protect the bulk electric system, including components of an electronic security perimeter to protect the critical infrastructure of the national electric grid.

Since 2006, there have been several initiatives to define wireless communications architectures and their associated protective measures including funding for Advanced Metering Infrastructure (AMI) security initiatives by Congress in the American Recovery and Reinvestment Act (ARRA), which provided a portion of US$4.5 billion in funding for energy efficiency and reliability initiatives.

While AMI developments have been one of the main drivers of additional communications networks sought by electric utilities in smart grid initiatives, utilities are realizing the potential of an integrated system of technologies and communications solutions in a smart grid architecture—the merging of technology and communications in a smart grid.

3.6.2 Communications Requirements for the Smart Grid

3.6.2.1 AMI Communications

Technological advances in the areas of telecommunications network coverage, speed throughput, privacy, and security have enabled the implementation of more encompassing and capable AMI systems. AMI networks enable utilities to accomplish meter data collection, customer participation in demand response, and energy efficiency and support the evolution of tools and technology that will drive the smart grid future, including integration of electric vehicles and distributed generation.* Without the collection of AMI (interval) metering data, it is difficult to determine when customer consumption occurs in time, what customers do in response to grid management needs, and the value of customer response. Smart meters and related submeters that form the end points in the AMI architecture provide two critical roles. One is access to more granular interval usage data (e.g., last 15 min rather than last 30 days). Second, a durable communications link that is bidirectional (two-way) to deliver messages/instructions to the meter. While smart meters offer the potential of substantial benefits to electric utilities and consumers alike, the electric distribution company faces a number of possible deployment challenges. First is the need to establish and manage a communications network that is sufficiently flexible to reach most meters in the service area and is adaptable enough to change as customer and business needs change. Second, the deployment must be justified in terms of its cost and must provide for revenue recovery to satisfy both regulators and utility management. Third, customers must be educated about the benefits of smart meters and the related services that will be enabled through AMI. Fourth, the AMI architecture must embed systems and software that fully address cybersecurity and privacy needs.

Communications for smart grid AMI and demand management deployments should include

- An open-standard architecture to enable interoperability among systems, flexibility in communications choices, and future innovations from third-party technology providers
- Two-way communication to every meter to enable advanced control capabilities as well as remote device configuration and firmware updates
- Wireless in-home networking for demand response and load control devices, such as smart thermostats, smart appliances, in-home displays, and load controllers
- Advanced service switch for remotely connecting, disconnecting, and limiting service
- Positive outage notification and restoration verification
- Data management separate from network management, to enable customers to gain the benefits of more granular energy use data on timely basis

The purpose of an AMI communications system is to provide electric utilities with a communications network permitting connectivity between grid devices such as electric meters and a head-end system. AMI communications network options are numerous: they can be power line carrier (PLC), satellite, cellular (2G, 3G, or 4G), WiMAX, RF mesh, etc. PLC and cellular technologies (general packet radio service [GPRS]) have been traditionally used in Europe, whereas the United States has generally favored wireless technologies (cellular, RF mesh).

Several OFDM†-based PLC technologies are being standardized in Europe (PLC-G3, Prime). Wireless solutions are becoming more prevalent around the world. New technology standards are emerging, such as 802.15.4g (the PHY specification for the ISM [industrial, scientific, and medical] radio band for smart utility networks) and 802.15.4-2006, the frequency-hopping spread spectrum

---

* See, for example, Itron, Open way: Smart metering for the smart grid, 2011, https://www.itron.com/PublishedContent/OpenWay%20Overview.pdf
† Orthogonal frequency-division multiplexing (OFDM) is a method of encoding digital data on multiple carrier frequencies. OFDM has developed into a popular scheme for wideband digital communication, whether wireless or over wires, used in applications such as digital television and audio broadcasting, DSL broadband Internet access, wireless networks, and 4G mobile communications.

MAC. Furthermore, the AMI communications system may require access point, remote, and backhaul radios. The AMI communications system leverages the earlier technologies to provide an industry standard, reliable, scalable, and secure system for AMI applications.

The choice should be made based on edge device density, network performance, and AMI application requirements. Smart grid communications must support the use of multiple transport technologies seamlessly integrated to cost-effectively deliver the best combination of reliability, security, and functionality.

The scope of communications, beyond current grid interface and control, will expand dramatically with the smart grid. Grid intelligence is closely related to the extent to which the "decision platform" extends beyond current capabilities to more fully integrate grid components with customer premises equipment and response. This integration of communications and grid functionality enables operational visibility into the power delivery process, from the generation of electrical power down to the energy consumer. This visibility can only be achieved through platforms that support the exchange of data across the interfaces in the power system. This requires capabilities to reliably exchange large amounts of information in short periods of time. Reliable, fast, and secure communication is the cornerstone of any modern and smart grid power system.

The smart grid aims to assure secure, reliable, and optimal delivery of electrical power from generation to consumption point, with minimal impact on the environment through

- Better coordination of the energy supply and demand to reduce the overall need for generation, transmission, and distribution, while minimizing system losses
- Increased integration of dispersed renewable and distributed energy sources
- Enhanced security of energy delivery, optimal use of all network assets, and minimal environmental impact

3.6.2.2 Communications for Smart Grid Operations

Table 3.10 summarizes smart power network attributes and related communications requirements that will enable advanced operations in smart grid deployments. Critical to this is a viable platform for enhanced usage of information and communications technologies. Operational communications are already used extensively in specific parts of the power system (e.g., transmission) but are less developed in others (e.g., distribution and customer interface). Figure 3.128 presents some of the operational communications domains that can be used to further integrate the power delivery system.

3.6.2.2.1 Operational Applications

Protection relays require real-time transfer of electrical measurements, signals, and commands between substations to ensure the protection of the power system and its assets. This is a critical operational communication need in the substation and constitutes a basic building block for the network's "self-healing" capability. Stringent data transmission time requirements dictate the use of telecommunications grade communications devices and circuits. This requires that the utility either build out a dedicated communications infrastructure or procure advanced communications services. Dedicated communications is typically required between substations for exchanging data between protection relays.*

Automation in the "smart" substation at the feeder, bus bar, or substation levels exchanges information through different levels of Ethernet LANs.† Although automation at this level is generally

---

* In most cases, this can be accomplished through transparent dedicated circuits over SONET/SDH (multiplexing transport protocols), dedicated fiber, or HV PLCs. Ethernet transport is employed in some cases at distribution voltage levels.
† These are defined by IEC 61850. Serial (RS232 & RS485) communications links are still widely used in the substation and will be part of the legacy interface requirements in smart substations.

TABLE 3.10
Communications Requirements to Support Smart Grid Operations

Power Delivery Attributes	Operational Requirement	Smart Application	Communications Services
Self healing	Prompt reaction of the power network to changes through a coordinated automation system and "network-aware" protection schemes for rapid detection of faults and power restoration	Protection relay, networked automation, wide area protection	Low-latency, time-predictable communications channels, time-controlled Ethernet wide area network (WAN)
Enhanced visibility and grid control	Enhance visibility of the power flow and the network state across the interconnected, multiactor, competitive market to achieve an increased level of security of supply	EMS/DMS/SCADA, wide area monitoring systems	Resilient IP connections for SCADA and WAMS. Secure communication for inter-control center communications
Enhanced control of the power flow	Enable decoupling of networks and control of the power flow by utilizing power electronic devices in the network (HVDC, FACTS)	Wide area control	Time-controlled Ethernet local area network (LAN) within the perimeter of the plant
Empower consumers	Incorporate consumer equipment and behavior into the design and operation of the grid. Demand response and peak shaving through information exchange with the energy consumer (and potentially the consumer's electrical appliances)	Smart metering and AMI	Two-way "real-time" wireless or wireline communication from the service provider to the individual consumers
Resist and survive physical and cyber attacks	Mitigate cybersecurity issues in the power delivery information system covering the control center, substation, and the communications network in-between. Remotely monitor unmanned installations and assure physical integrity of power utility critical sites	Centralized cybersecurity monitoring facilities. Video surveillance and access control	Secure and redundant communications for remote security barriers and intrusion detection systems. Communication to video and security access systems
Maintain service and recover from natural disasters	Enable the operational system to resist disasters and be prepared to reestablish operations when they do happen	Backup control centers. Fast deployment, mobile control platform	Automatic failover communications to back up facilities, sites, and staff dispersed across the network
Accommodate clean power	Secure integration of the dispersed power generation, mainly large wind farms, but also "energy-producing consumers" (solar, wind, etc.)	Remote generator unit monitoring, supervision, and connect/disconnect	Reliable two-way communications with dispersed generators to automate switching, routing, and storage of energy from alternative sources
Optimize asset usage and life-cycle management	Use power system assets at full efficiency, to their real end-of-life, without disruption of service due to failure or environmental risks, through remote monitoring and right-on-time corrective action (no unnecessary preventive replacement)	Asset condition monitoring (circuit breakers, transformers, etc.)	Secure IP communication from the monitoring device associated with the power asset to the monitoring application platforms

Source: © Copyright 2012 Alstom Grid. All rights reserved.

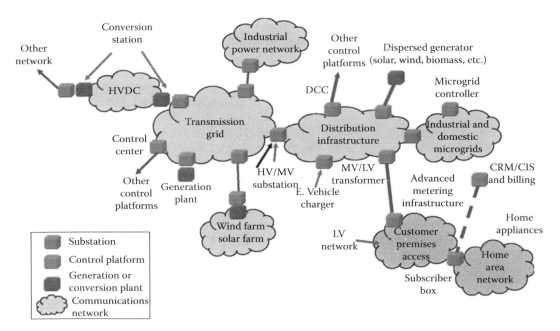

FIGURE 3.128 (See color insert.) Operational communications domains in the electric utility. (© Copyright 2012 Alstom Grid. All rights reserved.)

constrained to the substation perimeter, external communications beyond protection relays is also necessary.* Real-time data exchanges beyond the substation are required for incorporation of feeder automation and interfaces to distributed energy resources.†

Energy management and control center communications include a number of different applications for the "enhanced visibility and grid control" component of smart grids:

- *SCADA* communications are evolving from serial RS232 links into network-centric IP communications protocols. For example, the IEC 60870-5-104 protocol allows the RTU to connect directly through an Ethernet LAN port at 10 or 100 Mbps, although the bandwidth allocated to each RTU remains often around 10 kbps. This is rendered possible in many cases through SONET/SDH and/or gigabit Ethernet optical networks that carry IP traffic.
- *Inter-control center* communications to provide data exchange over WANs between utility control centers, utilities, power pools, regional control centers, and nonutility generators are supported through an IP network using the inter-control center protocol (ICCP or IEC 60870-6/TASE-2 [telecontrol application service element]). The required bandwidth ranges from less than two up to tens of Mbps. Cybersecurity is in general the major issue in implementing these interconnections.

Wide area monitoring, protection, and control (WAMPAC) systems enable accurate visibility of transmission system power flow across multiple interconnected power networks. They constitute GPS-synchronized measurements of the power system bus voltages, line currents, etc., using synchrophasors or phasor measurement units (PMUs) collecting time-tagged measurements every 5–20 ms as defined by the IEEE C37.118 standard. The required bandwidth is in the range of

---

* These are mainly non-real-time TCP/IP exchanges such as supervision and system monitoring, configuration downloading, and parameter setting and confirmation.
† Ethernet VLANs with priority assignment and adequate restoration mechanism are appropriate in this case.

10–100 kbps per PMU device and few hundred kbps for PDC (phasor data collector) communications.* The communications requirements depend upon the application:

- Display systems (voltage, phase, power swing, line loading, etc.) and monitoring applications (frequency and voltage stability, power oscillation, line temperature) can tolerate a relatively large time latency (tens of seconds)
- Applications related to online analysis of network stability have a time constraint similar to SCADA (i.e., few seconds)
- Wide area protection systems and emergency situation control require much faster responses (10–100 ms)

Energy metering for settlement and reconciliation at the HV substation delivery point is an important component of the new deregulated power system. This is another potential IP-based communication of the HV substation with a bandwidth that depends upon the frequency of data capture and transfer.

Condition monitoring and asset management of primary components of the substation (circuit breaker, power transformer, etc.) generate condition monitoring data collected for maintenance, loading and stress analysis, and life-cycle management. All substation intelligent electronic device (IED) as well as telecommunications devices also need to support remote management. An asset monitoring network can be implemented across the communications infrastructure using web services with servers residing in the substation or at some other location. Monitoring in the substation should also include environment monitoring to protect substation assets and premises (e.g., temperature monitoring, fire detection). While this type of data is not as time critical as, for example, SCADA data, it potentially includes a large amount of data from equipment all over the T&D grid and over various communications networks.

Security, surveillance, and safety communications is driven by national authorities with increasing focus on the mitigation of security risks on "critical infrastructures" including those related to electric utilities. Authorities are setting specific security standards that concern not only the security of information and communications but also the physical security of electrical installations and process sites across the power delivery system. The deployment of these applications require broadband IP connectivity as enormous quantities of information need to be transported in realtime from a large number of dispersed sites to centralized security monitoring stations.

Video monitoring of unmanned substations and night surveillance of premises can be performed via widely used intelligent video over IP surveillance cameras, advanced video analytics, and automatic alarm triggering based on motion detection.

Substation nonoperational data collection includes event recorder data and analog waveforms captured continuously and uploaded for postincident analysis. The volume of data and its non-realtime nature often dissociates it from SCADA. Communications requirements related to these applications are covered through IEC 61850 and constitute part of the TCP/IP network load.

Mobile workforce communications need to incorporate new field work practices using multimedia-centric command control and dispatch communications solutions, replacing or complementing the traditional switched telephone network and/or wireless voice facilities. In-house and contractor maintenance staff require remote access to online maintenance manuals, maintenance applications, substation drawings and plans, accurate maps, pictures, and timely communication of work orders to carry out their tasks. Broadband IP network access with reliable wireless connectivity is required in order to meet an acceptable transactional time performance considering the relatively large volume of data to be handled (file transfer, multimedia group communication, instant messaging, video streaming, etc.).

---

* Dedicated circuits using SONET/SDH (multiplexing transport protocols) or time-controlled Ethernet may be used for these applications.

3.6.2.2.2 *Operational Communication Constraints*

In order to provision communications services for operational applications, whether through procurement of telecommunications services or through implementing a dedicated network infrastructure, a number of issues must be addressed. Apart from the most basic need, which is the coverage of the operational zone, these issues imply requirements to be predictable, robust, error-proof, and future-proof.

Many critical power process-related applications require predictable behavior in the related communications service. Predictability in this sense can be defined as follows:

- Deterministic information routing—This means that both in normal time and in presence of anomalies and failures, one can precisely determine the path taken by the communication. Fixed or constrained routing limits the operation of network resilience mechanisms into a predefined scheme in which every state taken by the network is previously analyzed. Deterministic routing is not a natural instinct of the network designer who is tempted to employ every resilience capability of the employed technology. However, it constitutes one of the bases for fault tolerant design and for a predictable time behavior.
- Predictable time behavior—This is the capability to determine the time latency of the communications link for an application. This attribute is essential for applications such as protection relaying and WAMPAC and needs, as a prerequisite, deterministic information routing.
- Predictable network transit time, requiring dedicated network resources including backhaul network and back-end connection resources, to prevent competition for resources with other network traffic.
- Predictable time behavior must also take into account the time required to restore service in the event of a network anomaly.
- Predictable time behavior is assumed to harness "measuring tools" to monitor the "time latency" for every critical service.
- Fault tolerance is the capability of continued service in the event of a communications network fault, achieved through the predictable behavior of the system, for example, normal and back-up services without use of common resources, equipment, link, power supply, fiber cable, etc.

Robustness is a system capability to resist to the severe environment in which it must operate. Concerning the operational communications services, different aspects must be considered:

- Reliable and stable hardware and software—The duplication of critical modules and subsystems and, in certain cases, of the whole equipment or platform increases the availability of the system. Availability is a statistical parameter that must be estimated across the whole chain and must be coordinated among the different constituents of the system. It complements but cannot replace fault tolerance which is a deterministic concept.
- Power autonomy—An essential attribute of operational communications is their continuity in the event of AC power supply interruption for a specified duration ranging from few hours to few days depending on operational constraints. This is often a major drawback for using public communications facilities that generally lack sufficient autonomy. Adequately dimensioned DC batteries and backup generators allow utility telecommunications infrastructure to remain operational for restoring the power system.
- Mastering cybersecurity risks—The robustness of the communications service depends also upon the degree of invulnerability of the network infrastructure to security risks in particular for increasing numbers of IP-based smart grid applications. Proper isolation of different services (through VLANs or VPNs) and a security policy covering not

only the control center but the whole distributed intelligence system incorporating the communications infrastructure and monitoring of cyber access are some of the aspects of security risk mitigation.

3.6.2.3 Home Area Network

Home area networks (HANs) provide the means for electric utilities to communicate to individual consumer load devices (mostly residential consumers) in support of demand management applications. The HAN is the local communications network in the house to communicate among the various demand management devices, such as in-home displays, home energy management devices, programmable communicating thermostats, and smart appliances, and is the interface to communicate with the electric utility. The electric meter is considered the obvious choice for the communications interface to the residential consumer; however, other means of communications to the customer premise include cable TV, phone lines, and commercial wireless networks. Within the customer premise, several communications options are available to integrate demand management devices—wired and wireless. From the utility side, there is momentum to use PLC and wireless solutions, for instance, HomePlug GreenPhy (wired) modulating data over the home electrical wiring and ZigBee (wireless), a constrained mesh network. HomePlug GreenPhy is intended for low-cost energy management and home automation and allows for less expensive hardware modulation by sacrificing bandwidth but still allowing for coexistence with the high-speed version called HomePlug AV. ZigBee leverages IEEE 802.15.4 technology and a network, transport, and application layer as well as security that is currently tied to ZigBee called SEP 1.x (smart energy profile 1.x.).

The consumer side is tending to lean toward home automation through the use of ZigBee home automation profile (which unfortunately does not coexist with SEP 1.x) and low-power Wi-Fi. Although home automation provides some rudimentary energy management, it is not nearly as complete as SEP clusters.

The HomePlug consortium and others have worked with the ZigBee Alliance to create a link layer agnostic version of SEP (version 2.0) that has separated the SEP v1.x layers out and leverages IPv6 and TCP/UDP for the networking and transport, off-the-shelf certificate technologies for security, and HTTP for services. In the case of ZigBee and applicable constrained networks, there is also a requirement for 6LoWPAN (IETF RFC 6282), which performs header compression of the IPv6 network layer and the UDP/TCP transport layer, and a requirement for CoAP (constrained application protocol—draft-ietf-core-coap), which performs compression of HTTP server and client headers. In addition to being link-layer agnostic, going to an Internet-based network layer and off-the-shelf certificate management allows for SEP 2.0 devices and the next specification for the home automation profile to coexist at the link and network layers.

Other technologies used on the consumer side that will give opportunity for the HAN-based energy management include low-power Wi-Fi and Bluetooth low energy, which can each coexist at the link layer of their associated full-power implementations.

Implementing demand management systems requires utilities to take into account flexibility and scalability of networks. Utilities can take advantage of multiple network protocols, topologies, and potentially carriers to implement DR systems. In addition, utilities can also potentially implement their own proprietary networks in geographic localities where coverage is poor or when it otherwise makes sense in order to implement and support demand management systems.

But energy efficiency opportunities are not limited to simply the domain of infrastructure owned and controlled by the electric utility. Home energy management is an area that is just beginning to come into focus, but some forms will involve the networking of appliances possessing "intelligence" with applications that monitor a broad variety of devices consuming energy within the home and optimize consumption based on consumer preferences and knowledge of the marginal cost of the energy being consumed. This capability will require new access to data from the electric supplier (marginal price, current consumption, load curtailment signals) as well interactive capabilities and action notices to the home owner (e.g., to permit override of planned/automated actions). More informed and efficient consumption decisions on the part of the consumer will require gathering and storing information that has the potential for misuse. Thus the participant in this sector, the electricity, energy management application, and communications service providers will be faced with new demands to assure the privacy of the data.

For the average ratepayer and electric consumer, wireless fourth generation networks will enable consumers to access smart grid network applications across utilities from a great number of mobile platforms, homes, and other networks without degradation of service and fast access to Internet-based and web-enabled services. It also means that smart grid applications will be extended to mobile devices enabling consumers to have availability to a wide range of applications and controls, for example, controlling electric consumption, monitoring appliance usage, and interfacing with utility's back-office applications (such as billing) from anywhere in the world.

3.6.3 Wireless Network Solutions for Smart Grid

3.6.3.1 Cellular

The adoption of cellular wireless networks as the communications choice for electric utilities started more than 20 years ago. As cellular technology became more readily available, utilities began the migration of certain applications to wireless networks run by the telecommunications companies, but the adoption path has been slow at times as the cautious, risk-averse world of electric utilities has lagged the rapid growth and constant innovation in the mobile wireless telecommunications industry.

Cellular networks are familiar to a majority of the world's population today. The ubiquitous personal mobile phones and the cellular towers dotting the landscape have become so commonplace that most people take their existence for granted. However, these networks are complex systems that have been refined over decades of development and use. A cellular network is a radio network made up of small low-powered transceivers (also known as "cell phones," "cellular radios," or "mobile phones"), a network of powerful fixed, geographically distributed transceivers (also known as "base stations," "cell sites," or "cell towers"), and an infrastructure to tie the base stations together and to the public telephone system. Though a full description of cellular technology is beyond the scope of this book, the details necessary for an assessment of their role in the development of the smart grid are included here. Figure 3.129 provides a representative depiction of a basic cellular network.

The first use of cellular technology in the electric utility industry was the adoption of AMPS (advanced mobile phone system) starting in the late 1980s for automated meter reading applications, particularly for commercial meters or for very hard-to-read residential meters. AMPS (considered to be a first-generation or "1G" cellular technology) was an improvement in many ways over the previous methods of meter reading for the commercial and hard-to-read environments in that it did not require a field visit, nor did it require that a POTS ("plain old telephone service") line be run to the meter and maintained by the customer. Though there were growing pains with the low-bandwidth (9–24 kbps) wireless technology, utilities began to recognize its benefits. When products based on paging networks became available as competitors to the AMPS networks in the late 1990s, pricing pressures and improved electronics led to further adoption of wireless technologies across the utility industry.

However, a challenge to broader commercial wireless adoption for years to come would then follow—the decommissioning of the AMPS networks by their owners that started in 2002 and was complete by 2008. These actions, which were due to a combination of federal pressure, wireless carrier economics, and the inefficient use of valuable spectrum by the AMPS technology, led many utilities to question the wisdom of relying on systems and networks not only beyond their control, but under the control of a commercial entity with a much broader set of business objectives than just

Smart Grid Technologies

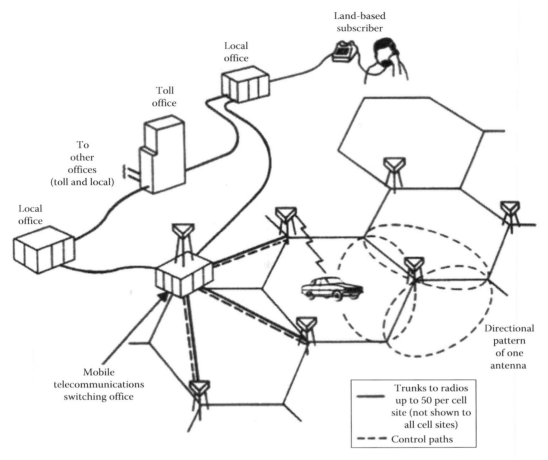

FIGURE 3.129 Basic cellular network.

keeping utility communications networks intact. This concern, set off by the experience of "losing" AMPS, would linger for years.

3.6.3.1.1 2G Networks

The second-generation ("2G") wireless systems that began to be deployed in the 1990s have slowly repaired the relationship between utilities and commercial cellular network providers. The 2G network rollouts coincided with the massive adoption of cellular technologies worldwide, making the networks more ubiquitous, reliable, and cost effective. The cellular world split into two main camps at this point: those carriers who chose to utilize the global system for mobile communications (GSM) system and its data protocols of GPRS and enhanced data rates for GSM evolution (EDGE) and those who chose the CDMA2000 family and its one times radio transmission technology (1×RTT) data protocol. The North American carriers that chose GSM included Cingular/AT&T Wireless, T-Mobile, and Rogers Wireless, while Verizon, Sprint, and Bell Canada chose CDMA2000.

The 2G systems could provide much greater bandwidth and better pricing than 1G, and advanced electronics made it simpler to create full meter reading systems that could use the new 2G networks. As a result, acceptance for its use in grid applications grew worldwide. This speed was doubled in 2000 with the introduction of commercial GPRS networks in the United States that provided speeds of about 28 kbps—a major leap forward—but only comparable to dial-up speeds experienced by users connected via wireline networks. The introduction of EDGE and similar technologies boosted speeds from about 28 kbps to about 150 kbps of expected average user throughput, which enabled another class of applications to enter the smart grid arena. While 28 kbps was sufficient for transmitting point-to-point metering data containing a few hundred bytes of data, the introduction of EDGE allowed file transfers and support of applications requiring much greater bandwidth. This was a fivefold increase in throughput speeds, and as commercial carriers invested heavily in coverage, reliability of the network significantly increased. The largest AMI deployments through 2009 were in Europe, where hundreds of thousands of meters with GPRS radios inside were installed, though pilots for deployments on a similar scale are underway in North America as well. EDGE allowed an entire new class of utility applications to be implemented. For example, utilities were able to implement wireless wide area network (WWAN) routers to provide backup and continuity of service to traditional wireline circuits. They were also able to extend networks with reasonable throughput to geographic areas where wireline DSL connectivity was not available—for example, remote generation, distribution facilities, and service centers in remote locations.

Perhaps most importantly, a number of other industries began adopting 2G wireless systems for their remote communications needs, creating applications in security, oil and gas drilling, fleet management, point-of-sale terminals, and vehicle tracking. The large numbers of applications and devices have attracted greater attention by the carriers, and they have subsequently created business units and pricing plans to further grow this high-margin and low-turnover component of their business. The carriers have recognized that these devices do not require significant spectrum to support and have consequently begun to provide long-term service-level agreements (SLAs) to large data-only customers who are concerned about the longevity of the 2G networks. This has further increased the installed base of 2G devices in the field, and thus 2G deployments are expected to continue to grow due to this "snowball effect." The attractive economics, profitable spectrum use, and business agreements have combined to make an "AMPS-like" decommissioning of the 2G systems unlikely without significant economic fallout for the carriers.

3.6.3.1.2   3G Networks

In the mid-2000s, the first of the third-generation or "3G" networks were deployed. The GSM-based carriers such as AT&T and T-Mobile deployed the GSM variant of 3G (known as high-speed packet access or HSPA) while Sprint and Verizon deployed the CDMA variant, evolution-data optimized (EV-DO). These networks provided substantially greater bandwidth to the mobile device—approximately 10 Mbps (peak speed). The uptake of 3G technology by consumers was greater than anticipated, in part due to breakthrough devices like Apple's iPhone, and, as a result, the commercial carriers have invested billions of dollars in the infrastructure necessary to support the greater bandwidth requirements, with billions more yet to come. The increase of average throughput speed from 150 kbps to 1.5 Mbps was significant. Most smart grid applications (such as AMI) today place a premium on low cost and ubiquitous coverage rather than greater bandwidth, which has led to further adoption of the 2G technologies that have been deployed already. The introduction of 3G allowed AMI solutions to transmit large volumes of data using data collector units connected to the utility meter data management systems and back-office servers using 3G cards.

In addition to the boost in wireless throughput speeds, 3G technologies provided significant enhanced security capabilities, which enabled utilities to address privacy and security concerns (for instance, GPRS and HSPA allow the implementation of authentication mechanisms). In addition, 3G technologies introduced a newer, stronger, and more robust algorithm named Kasumi that addresses the vulnerabilities associated with earlier generation cryptographic algorithms. These capabilities have and will continue to affect the type of smart grid network elements and the network architecture of future systems.

3.6.3.1.3   4G Networks

The changes in network speeds and capabilities are just beginning—all major carriers have decided to converge on the long-term evolution ("LTE") set of protocols as their future 4G infrastructure.

This has the potential to further accelerate technological innovation in the cellular field, as it will be the first time since the 1980s that a single technology platform will be shared among the vast majority of cellular users worldwide. These networks are scheduled to be deployed beginning in 2011, which will enable utilities to have peak speeds up to 100 Mbps and average user throughput speeds of about 10 Mbps—faster than existing commercial WANs and enabling wireless utility data transmission capabilities that approximate the speed of DSL wireline networks currently in existence. 4G networks use packet switching technologies. One of the key features of 4G in contrast to 3G is that it is expected to be all IP-based.

To date, little has been said about what comes *after* the 4G technologies such as LTE and WiMAX. However, based on the history of cellular development thus far, the degree to which the desire for bandwidth is insatiable, and the amount of money worldwide that flows through the industry to seek a competitive advantage, something will likely come along for deployment in the 2017–2020 timeframe. Based on the trends that have been set by the industry so far, the bandwidth would likely be on the order of optical fiber today (1 Gbps or more), with increased reliability and reduced cost per bit. The impact of this level of capability on the smart grid of 2020 and beyond is hard to predict, but it is likely that these trends will make cellular technology more attractive to utility customers as the broadening gap with utility-owned wireless networks becomes clear.

3.6.3.1.4 Strengths and Weaknesses of Cellular Communications
Cellular currently plays some part of the communications infrastructure for a large number of utilities, typically for their commercial and industrial customers or for hard-to-read residential meters. A number of vendors have focused on this part of the overall smart meter market, for example, SmartSynch, Metrum, Trilliant, Elster, and Comverge. With the rapidly dropping prices for data and electronic components, the residential market as a whole has now become addressable by cellular technology. For example, the largest single deployment of cellular-based smart grid technology to date was the first major AMI deployment in the Province of Ontario by Hydro One, who selected SmartSynch to provide and install 20,000 residential meters using 2G technology (GPRS) in 2006. As for 4G deployments to date, Grid Net has been the market leader in promoting the use of WiMAX for the smart grid and has an early market win in Australia. Gridnet is now also supporting LTE [1]. Ausgrid in Australia have announced intentions to migrate their network from Wimax to LTE [2].

The widespread global use of cellular networks—with over five billion mobile phone connections worldwide in 2010 [3]—certainly implies that there might be broad benefits to using the networks that could also be applicable to smart grid use. Perhaps the most significant benefit is the enormous ecosystem that supports the global cellular industry and those five billion customers. Utilities who utilize cellular networks are leveraging billions of dollars per year in technology innovation, infrastructure deployment, and engineering education. The intense competitive environment in the telecommunications industry—not only for the carriers acting at the retail level but also for infrastructure providers, software companies, and consulting firms—leads to continuous cost savings and service improvements at every link of the communications chain. The U.S. carriers alone invest billions of dollars each year just in infrastructure, which is subsequently invested by their vendors to improve performance and so on. The result, as the world has already seen in the last few years, is an accelerating pace of innovation and progress. With the number of cellular users increasing daily, the cumulative benefit to utilities will be trillions of dollars of investment in their communications choice, regardless of whether the utility itself takes any proactive steps to invest in that technology or not. No other utility communications option can match even a significant fraction of the investment being made in the cellular industry.

Another positive attribute of cellular networks for the smart grid is that they are already deployed throughout a very large part of the developed world. This gives utilities the ability to choose where they want to deploy the actual grid-specific components of the smart grid such as sensors and meters without having to deploy a communications network first. Even when factoring in the extremely remote areas of a utility's service territory, using inexpensive cellular for those areas where coverage exists leaves the utility with more funds to target the remote areas where cellular networks have not been deployed.

Alternatively, cellular systems are a much simpler and cheaper solution for provisioning in developing regions or countries, as the infrastructure cost associated with green-fielding a wired POTS system is intractably more costly than a wireless cellular system.

Finally, a great strength of the choice of cellular for utilities is the widespread base of expertise that exists to build, operate, troubleshoot, and optimize these networks, developed by virtue of the numerous networks that have been deployed for many years. Most university electrical engineering departments now include an option in their design programs to learn about cellular networks, and, perhaps more importantly, virtually all students are intimately familiar with the capabilities and challenges of daily cellular use.

Cellular is not without its weaknesses as it relates to the smart grid—if not for these factors, cellular would certainly see a higher penetration rate than it does today. The technical issue cited most often by utilities as a weakness of cellular is the restriction on any network's coverage footprint. "I couldn't get coverage there" is a lament that has a first-order impact on the ability to use cellular at a given location. In the past, utilities have not been able to significantly influence placement of cellular towers due to business case issues. Commercial carriers have improved the situation in certain circumstances by using higher-powered cellular radios, purchasing and provisioning sub-GHz spectrum (which has better penetration in buildings containing cement and rebar), providing external antennas with long runs, or installing repeaters and micro/femto-cells.

Yet, utility organizations are made up of individuals who likely use a cellular phone regularly, either for work or personal use. They use them at all times of the day in a variety of conditions, and consequently these individual experiences mold opinions about the apparent reliability of cellular communications. "Dropped calls" and unclear connections are familiar experiences for most cellular phone users, and they assume that these issues will occur for cellular-based smart grid communications as well. In reality, any wireless network will have some low-fidelity and lost connections, and for low-latency high-quality communications like voice calls, these issues will be highlighted. In the case of smart grid applications, the data networks—which perform differently in many material ways—are used exclusively. However, given that most power-system cellular communications equipment will be directly connected to power and not need to rely on maximizing handset battery, equipment vendors can use cellular radios with higher transmit power and more sensitive antenna to achieve data communications links in areas that would be considered "dead zones" for cellular voice communications. In addition, cellular systems have a number of algorithms in place at the tower and the device to enhance the reliability and security of data connections. Nonetheless, any concerns regarding the perceptions of reliability and daily performance must be addressed directly by smart grid equipment vendors who utilize these networks.

Another significant hurdle for cellular-based smart grid business cases is the relatively higher bill of materials cost of the radios used to communicate on the networks. The cellular protocols and infrastructure were designed to accommodate secure and reliable communications even when the handset is moving at high speeds, with the need to switch towers frequently, in environments with high amounts of electromagnetic noise. The result is a very capable but complex radio that inevitably costs more than a radio without the same security, noise, bandwidth, and protocol requirements.

While the aforementioned positives and negatives are fundamental traits of commercial cellular networks, some factors regarding their use within the smart grid are changing rapidly and thus are the subject of debate. One of the most contentious is the utilities' desire to have some sense of "control" of the networks to minimize operational risk, which manifests itself in two important ways—the level of counterparty risk (i.e., the risk that a wireless carrier would default on its responsibility to provide a reliable network to the utility) assumed by the utility and the risk that technology evolution would drive wireless carriers to make the utility communications equipment obsolete. These are legitimate and serious concerns.

Sophisticated utilities consider the counterparty risk within the context of the overall risk environment, including the counterparty risk for the alternatives. The risk of having a nationwide wireless carrier default on its responsibility, given the size of those companies and very large number of other parties with whom they have counterparty arrangements (including a large consumer population and government agencies), and combined with the level of internal risk to the utility if the carrier defaults (the utility only needs to change out the end devices to a competitor network) are small relative to the actual counterparty risk of having a utility-only communications provider default, as these companies are typically smaller, less diversified, and with very few counterparty arrangements, and their defaults put at risk the entire utility network infrastructure (not just end devices) without future support. Some utilities, such as Ausgrid in Australia, have decided to install, own, and operate their own cellular communications networks.

With regard to obsolescence, utilities expect equipment deployments that do not need to be revisited for many years, if not decades, and expect a very high level of consistency of performance throughout those years, while carriers must balance the desires of consumers, nationwide spectrum holdings, and network operations cost. On one side of the debate, cellular would not seem to be a good candidate: carriers are subject to economic and market considerations that have led to the decommissioning of first-generation networks and the deterioration of the paging networks, and the wireless industry as a whole moves so quickly that some utilities believe they are better off not participating directly. On the other hand, when looking at decisions made today from the vantage point of a few years in the future, utility-specific technologies will almost certainly find that they cannot keep up with the rest of the world, creating frustrated utility customers as well as regulatory commissions and utilities who have the albatross of an inadequate communications network on their balance sheets. The reality for utilities in the twenty-first century is that technical obsolescence is a fact of life, and the wisest choice is to minimize their exposure to the ongoing technology upgrade cycle by focusing on only their equipment and not the network infrastructure as well. A solid middle ground perhaps exists in which carriers address the concerns of utilities through SLAs, long-term commitments to utilize the networks, and products that embrace the utilities' and the carriers' concerns.

Other arguments revolve around how regulatory commissions allow cost recovery of capital equipment deployed. So far, regulatory commissions have not widely allowed utilities to recover operating expenses associated the operation of the network and the subsequent meter reading expenses. This creates a large operational expense (OpEx) that can be difficult for some utilities to accept based on the regulatory incentives. The alternative for utilities is to create and build their own networks and capitalize the cost (CapEx) of these proprietary networks, increasing the rate base and passing the cost to consumers and ratepayers. When considering the overall cost of these deployments, industrial and consumer groups are pressuring regulatory bodies to allow cost recovery of expenses associated with commercial networks. Utilities will have to find an optimum financial equilibrium point between using commercial networks and deploying their own proprietary networks based on these factors. Utilities have a choice: in 5–10 years, will they have to address the costs of upgrading the devices on their network to satisfy smart grid needs, or will they have the costs of the devices *and* the network itself that they are responsible for upgrading—and in which direction will their regulating commissions be amenable to cost recovery?

Lastly, the impact of natural disasters on both the electric and telecommunications networks is an important area for constructive discussions. Electricity generation, transmission, and distribution systems in the developed world are a modern marvel, and its consumers have grown accustomed to the availability of ample electricity whenever it is needed. As a result, modern societies have become almost completely dependent on electricity for normal life, and this dependency is never more apparent than after a natural disaster. Energy—specifically electricity and fuel—has joined food, water, and shelter as necessities of life. Utilities are therefore inclined to take whatever steps they can to ensure that they can restore electricity as soon as possible after a widespread outage, and the "smart" elements of the grid are no exception. Some utilities believe that it is thus necessary to build and operate their own communications networks, since they will then have control over which aspects of the network are brought online first to assist in restoring electricity overall. Wireless carriers argue, however, that the emphasis on building a proprietary network for this reason arose during a different time, when only a small part of the population was interested in wireless networks. In addition, since the introduction of 2G networks, commercial carriers have configured the EDGE, HSPA, and soon-to-be-introduced LTE networks to allow data throughput even in cases where voice channels are saturated. This voice/data resource segmentation ensures utilities that their data will still get through the commercial networks even during periods of high voice utilization or limited network infrastructure availability. The long-term path forward probably involves a much closer relationship between the electric utilities, communications providers, and emergency response personnel so that the highly connected world is restored in the best interests of the population as a whole. This is especially true for municipalities that may push for a single telecommunications infrastructure for providing electricity as well as water, gas, electricity, video, etc.

The worldwide impact of cellular technologies has been significant since it was first introduced over 20 years ago, and its effect on the adoption of the smart grid will surely be significant as well. The nature of that effect, the timing of the adoption, and the evolution of the relationship between commercial wireless carriers and electric utilities will all be interesting to watch.

3.6.3.1.5 Role of Cellular Communications in the Smart Grid

Twenty-first century smart grid solutions should be founded on architectural principles and not on specific technologies because of the accelerated pace of technological change. In general, three architectural principles should be followed when considering cellular communications for smart grid deployments: (1) rapid expandability, (2) integrated stratums, and (3) transparent commonality. These three architectural principles allow utilities to implement infrastructure elements that can be upgraded in a modular manner which enables a longer life for smart grid elements consistent with regulatory commissions' desires to reduce cost for ratepayers and systems subscribers.

Rapid expandability: Electric utilities should implement network elements and smart grid components that are based on technologies that can be expanded rapidly. The concept of expandability allows utilities to respond to changes in regulatory climate and ratepayers' and consumers' demand and provides utilities with the ability to respond to technological network changes in a cost-effective manner. While the full realization of the smart grid vision may take many years to be fulfilled, utilities should implement communications network capabilities in a rapidly expandable manner in order to support various forms of smart grid deployments—such as AMI—in the next 3–5 years. In addition, utilities must also implement networks that can respond to technological changes and rapid evolution in the world's wireless ecosystem. An example of this rapid evolution in the GSM architecture during the last 10 years is the introduction of WWAN technologies every 18 months beginning with GPRS in 2001, EDGE in 2003, UMTS in 2005, HSPDA in 2007, HSUPA in 2008, and HSPA 7.2 in 2009, and now LTE. The evolution of this ecosystem has enabled utilities to support legacy network implementations while at the same time introduce newer and faster network capabilities as needed. For example, utilities are using the faster 3G network capabilities to support data collectors that manage large number of meters in mesh networks. Solutions implemented should be rapidly expandable across smart grid components and systems and should also have the ability to transcend individual data technologies.

Integrated stratums: At the same time, smart grid communications systems should be built upon *an integrated stratum approach* where one layer can be upgraded or changed without disturbing other layers of the smart grid model. For example, specific wireless modems supporting smart

Smart Grid Technologies 281

meters should be able to be upgraded to LTE without affecting the AMI head-end system or meter data management system application layer.

Transparent commonality: While rapid expandability and integrated stratums are important principles, wireless solutions also depend on their ability to integrate across multiple platforms and wireless systems. For example, an electric utility may have more than one wireless commercial carrier and more than one type of WWAN solution that supports sending data to an integrated meter data management system. These systems must be capable of interfacing with a variety of application solutions and carriers to effectively collect data and interface to other back-office core utility systems. Therefore, transparent commonality should be a key component of any communications solution implemented in smart grid systems. Systems that are integrated rely on transparent and common application interfaces to communicate across stratums. The term transparent commonality is used to denote the ability to implement architecturally similar systems in order to reduce the initial cost of capital and ongoing operational expenses. Transparently common systems allow utilities to implement a supportable architecture over their service territory minimizing change. In other words, utilities should minimize the types of networks and technologies deployed across their smart grid systems and ensure that networks are implemented in response to consumer and business requirements. For example, a utility may choose to implement one type of AMI architecture to support urban and high-density consumers while another AMI architecture to support consumers in rural areas. By creating these categories, utilities can deploy common and transparent systems across their service territory while optimizing the cost of implementation. Thus, *transparent commonality* is an important component of the smart grid that reduces overall costs associated with smart grid architectures and allows components to communicate among each other in a transparent manner. Commonality requires smart grid elements to use published and open interfaces and that these interfaces only provide exposure to a limited set of system capabilities allowed for security purposes. Thus, applications do not need to know the internal structures of other network elements and application systems, but they do need to know how to interface in a transparent manner. The principle of transparent commonality reduces the level of complexity associated with smart grid components which if left unmanaged can become unreasonable.

3.6.3.2 RF Mesh

Radio-frequency (RF) mesh technologies compose the communications backbone of numerous existing AMI deployments today. RF technologies are simple, cheap, and widespread. RF relies on RF wavelengths to communicate between devices and back to an access point that then is connected to the utility via a backhaul network. RF technologies can generally provide bandwidth of approximately 100–200 kbps, which is usually deemed sufficient for typical smart grid field applications, but not for backhaul networks that are expected to aggregate data. Examples of major RF mesh communications providers include Silver Spring Networks, Trilliant, Itron, Landis & Gyr, and Elster.

A key feature of RF mesh technologies is the ability to form a "peering" network. In this configuration, each device is capable of communicating with nearby peers and then sending information via those peers to an access point that has a direct communications path to the utility. A simple way to think of this is that every mesh node acts as a router—the advantage of this method is that not every device has to have a direct communications path all the way back to the utility, they only have to have a communications path to a peer. This saves on costs and power in that the communications chips in each device can be considerably less sophisticated and use less power in their signals. This peer-to-peer type network can also repair itself if a connection to a given peer is interrupted, as long as there are other peers in communicating range (usually line of sight). Thus, coverage is easy to roll out in a phased manner. In an RF mesh configuration, an access point may cover several blocks in a neighborhood, as opposed to a single-cell tower that must cover several miles. Thus, while more access points are required, the power and size of each access point is smaller than of a cell tower.

Another advantage of RF mesh networks is that they provide a greater degree of network control for utilities. Utilities can closely specify the operating characteristics, extent, and cost of the network. Utilities are less concerned about the impact of emerging technologies and the eventual phase-out of legacy technologies if their network is under their own control.

The disadvantage of RF mesh technologies is that they are typically based on proprietary technologies with single-vendor sources. This is a risk for utilities looking for long-term sustainability and vendor support where the vendors supplying the technology may go out of business. In addition, some RF mesh solutions are based on the unlicensed ISM frequency that is susceptible to interference. Licensed spectrum is narrowband and is vendor specific via arrangements with spectrum owners or municipalities. This limits the applications and device ecosystem compared to cellular networks.

3.6.4 Communication Standards and Protocols

The key word in smart grid development is information. In order to operate a smart grid, accurate information about the power system, its current status, its trends, its historic information, and its applications, is necessary. The access to information is granted by communication, and this is why communication standards and protocols play a key role in smart grid developments around the world. Through standardized communications interfaces that describe the functionality, the information is not only accessible, but it is also interoperable allowing a cost-effective implementation of the required functionality across domains.

Communication is the core of smart grid applications. It enables the exchange of information between the different elements of the grid, such as feeder equipment, substation equipment, and network control centers, by providing a common set of rules for data representation and data transmission. In a utility environment, the exchange of information handled via communications protocols that, depending on the application, have to satisfy different constraints. For example, protection applications have more stringent requirements on real-time and reliable information delivery compared to monitoring applications. Similarly, the cybersecurity requirements may vary. For example, the cybersecurity requirement can be of higher importance for applications interacting with customer meters than for monitoring applications local to a substation.

The discussion that follows is an overview of the major communications protocols considered for smart grid applications. The goal is not to give an exhaustive list of communications protocols but rather to identify the main areas of application.

From a smart grid point of view, communication can be classified into two main categories:

- Communications systems specific to smart grid applications such as IEC 61850 [3–6] and IEC 61968-9 [7] and communications protocols such as DNP3 [8], IEC 60870-5 [9], IEEE C37.118 [10], ANSI C12.19 [11], ANSI C12.18 [12], ANSI C12.21 [13], ANSI C12.22 [14]
- Auxiliary protocols playing a major role in smart grids but not limited to this application domain, such as IEC 62439 (HSR/PRP)* [15], IEEE 1588 [16], NTP [17], and the widely used Ethernet [18], IP [19], and TCP [20]/UDP [21]

IEC 61850, and more precisely the mapping to the communications protocols as defined in IEC 61850-8-1 and IEC 61850-9-2, was originally designed for use within substations at transmission and distribution levels. However, through the latest extensions, IEC 61850 now provides a communications solution for substations, between substations, to control centers, for hydro-electric power plants, for distributed energy resources and wind farms, and it is expected that more domains will be added to the standard in the near future.

---

* High-availability seamless redundancy (HSR) and parallel redundancy protocol (PRP) are the latest additions to the IEC 62439 standard for high-availability industrial Ethernet networks. Designed for mission-critical and time-sensitive applications such as those found in protection and control applications, HSR and PRP provide guaranteed behavior under failure conditions and increased network reliability.

DNP3 and IEC 60870-5 are protocols that are mainly used at the transmission and distribution levels to exchange data between substation equipment and the network control center.

IEC 61968-9 specifies the interaction with customer meters and therefore targets low-voltage applications. Similarly, ANSI C12.19, ANSI C12.18, ANSI C12.21, and ANSI C12.22 define a set of standards for the data exchange between a customer meter and a meter reader over multiple media such as an optical port, a modem, or a network.

HSR and PRP are two reliable communications protocols for industrial automation suited for protection and control applications, while IEEE 1588 and NTP can handle the time synchronization constraints required by the devices on the network. IEEE 802.3 (Ethernet), IETF RFC 791 (IPv4), and RFC 761 (TCP) and RFC 768 (UDP) are the basis of most of the smart grid communications systems and protocols. IPv4 born out of DARPA (Defense Advanced Research Projects Agency) in the 1970s, which is used to this day as the backbone for the modern Internet has been so successful, it has started to outgrow itself—the network address blocks have been completely depleted to the point that IANA (Internet Assigned Number Authority) and RIRs (Regional Internet Registries) since April of 2011 have been in a constant churn of reclaiming, redistributing, and reallocating network space. This is not so much an issue for the utility sector, but is for the Internet, and was seen as a problem on the horizon for the rest of the Internet as early as the 1990s. A great amount of research leads to a solution that not only expanded the address space but also solved fundamental problems associated with IPv4. Born out of that research was RFC 2460 (IPv6)—a network protocol that would solve issues of address space, network configuration, network discovery, neighbor discovery, routing redundancy, mobile routing, and network security. At the transport layer, TCP and UDP have been used for reliable streaming and datagram support for applications, an alternative which may be more suited for the utility sector is the use of RFC 4960 (SCTP [Stream Control Transmission Protocol]) which allows for reliable transport of packets out of order. It should be noted that the Internet Protocol Suite (IPv4/v6 & UDP/TCP) combines layers that are defined in the ISO OSI 7 layer model (Figure 3.130) into four layers—link, Internet, transport, and application. For a better understanding of Ethernet, IPv4/v6, and TCP/UDP, readers should refer to the abundant literature for Internet protocols such as [22] or [23].

3.6.4.1 IEC 61850

The standard IEC 61850 "communication networks and systems for power utility automation" defines a communications system for interoperability between equipment from different manufacturers. The standard introduces several features that impact the design of systems, such as the use of communications services for the exchange of time-critical information between IEDs, such as protection relays. With that, the hard wiring of signals between the relays used to implement the protection schemes can effectively be eliminated. Other impacts of IEC 61850 are related to hardware and software design of equipment, the design, commissioning and testing of a system, and the required training for engineers.

The standard IEC 61850 Edition 1 was published by IEC between 2003 and 2005. Since then, new editions of the standard are under development, and the scope of the standard has been extended extensively to include many new domains as indicated earlier. The purpose of IEC 61850 is to provide all the necessary specifications required to achieve interoperability between the equipment of an integrated system. To achieve that, the standard defines communications services based on TCP/IP and Ethernet, and object models describing data visible to the other equipment. It further defines a language to exchange engineering information between tools. More information on IEC 61850 can be found in the Smart Substation section of this book.

3.6.4.2 DNP3 and IEC 60870-5

DNP3 and IEC60870-5 were developed in the 1990s and are two commonly used protocols for the communication between field devices, residing either inside or outside a substation, and the network

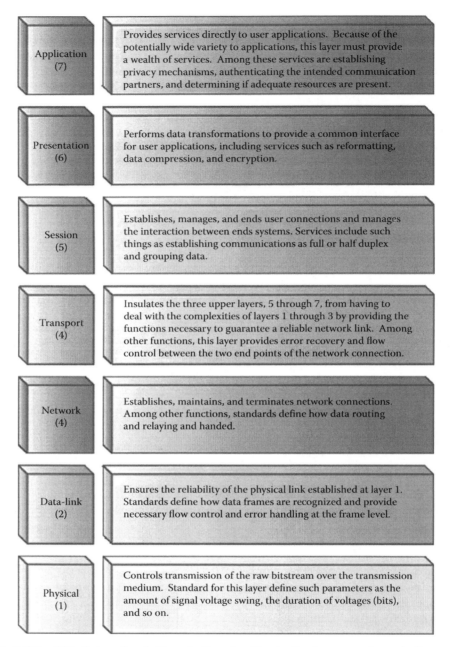

FIGURE 3.130 ISO (International Organization for Standardization; www.iso.org) Open Systems Interconnect (OSI) seven-layer communications model (http://www.novell.com/info/primer/prim05.html).

control center. They are also sometimes referred as "SCADA protocols" as they are intended to standardize the communication of SCADA systems. Their main applications are monitoring and controlling field equipment. Both protocols use a simplified version of the open system interconnection (OSI) seven-layer reference model [24]. It is referred to as the enhanced performance architecture (EPA) and only includes the physical, link, and application layers. However, a fourth layer is sometimes considered for DNP and targets the transport layer (DNP transport). While DNP3 and IEC 60870-5 were primarily intended to be based on RS-232 or RS-485, they evolved to now use

Ethernet and TCP/IP. In many respects, DNP3 and IEC 60870-5 are quite similar; however, they have significant technical differences and are not compatible. Moreover, DNP3 is mainly used in North and South America, while IEC 60870-5 is predominant in Europe and the Middle East.

3.6.4.3 IEEE C37.118

The IEEE C37.118 standard defines the exchange of synchronized phasor measurements used in power system applications. It was first published in 1995 and was revised in 2006. A synchronized phasor measurement, or synchrophasor, is produced by a PMU and represents the magnitude and phase angle of a waveform. PMUs distributed across the electric grid produce synchronized measurements, that is, measurements taken at the same time. As of today, IEEE C37.118 is primarily used for WAMPAC applications.

The IEEE C37.118 standard defines the communication rules for a single PMU, or a PMU data aggregator, called PDC (Phasor Data Concentrator). Due to the distributed nature of the measurement points across the network, IEEE C37.118 has to be implemented on top of a routable protocol such as TCP/IP or UDP/IP. To ensure the success of applications based on IEEE C37.118, two main challenges have to be addressed: (1) synchronization of the measurement points or PMUs and (2) transmission delays. The synchronization aspect is currently handled through the integration of a GPS receiver directly into the PMU. In the future, IEEE 1588 should be able to provide the required synchronization needs through the communications network and therefore remove the need for a GPS receiver in PMU devices. The transmission delay challenge depends highly on the network communication topology, that is, number of switches, length and type of links, etc. However, a common practice is to sacrifice the reliability characteristic of TCP by implementing IEEE C37.118 over UDP.

3.6.4.4 IEC 61968-9 and MultiSpeak

The IEC 61968-9 standard, issued in 2009, defines the interfaces for reading, monitoring, and controlling meters installed at a customer site. As such, IEC 61968-9 is not a full communications protocol since only the seventh layer of the OSI model is covered. The standard defines a set of XML schemas for the different operations applicable to meters. As an example, such operations can be *read load curve*, *read or write contract parameters*, *read technical maintenance data*, *read meter values*, etc. Even though IEC 61968-9 defines only the XML schemas, it implicitly imposes some constraints on the lower-level protocols. The usage of XML implies that most messages will be truncated to be transmitted over the physical media. Therefore a protocol with fragmentation capabilities, such as IP, is required. Moreover, due to the implicit fragmentation, each fragment must be sent reliably; that is, a fragment cannot be lost otherwise the message is unreadable, and therefore a protocol such as TCP rather than UDP is preferred.

Similarly to IEC 61968-9, MultiSpeak is a specification defining messages to interact with electrical meters. However, there are some important differences between the two standards. First, MultiSpeak is focused to meet the needs of electric cooperatives in the North American market, while IEC 61968-9 is focused toward all utilities in the international marketplace. Second, IEC 61968-9 is transport independent while MultiSpeak requires SOAP messages using HTTP and TCP/IP socket connections. Third, message headers can be easily mapped between MultiSpeak and IEC 61968-9, but the mapping of message content between the two is more complex. MultiSpeak has a longer history than IEC 61968-9 and had its third and latest version in 2008. It is worth noting that an ongoing harmonization effort was started in 2008 and will eventually lead to a complete mapping of MultiSpeak messages to IEC 61968-9.

3.6.4.5 ANSI C12.19, ANSI C12.18, ANSI C12.21, and ANSI C12.22

ANSI C12.19 initiated in the 1990s by the American National Standards Institute is the main standard used in North America for data exchange between gas, water, and electricity meters and utilities. It provides a data model of the meter through the specification of a set of common data structures referred as "tables" to read, write, and configure a metering device. However, ANSI C12.19 only provides a model of the meters and is not as such a complete communications protocol. ANSI C12.18, ANSI C12.21, and ANSI C12.22 define the underlying protocols used by ANSI C12.19 to transport the data over various communication media.

The ANSI C12.18 standard is written specifically for meter communications via an ANSI Type 2 Optical Port. It is a complete point-to-point communications protocol covering the seven layers of the OSI model. Additionally, it details the criteria required for communications between an ANSI C12.18 device and an ANSI C12.18 client via an optical port. The ANSI C12.18 client may be a handheld reader, a portable computer, a master station system, or another electronic communications device. It is mostly used for manual meter reading using the infrared optical port currently in use by most North American meters.

Similarly, the ANSI C12.21 standard details the criteria required for communications between a C12.21 device and a C12.21 client via a modem connected to the switched telephone network. It is also a point-to-point communications protocol, but compared to ANSI C12.18, it allows the remote reading of a meter. It also includes authentication.

ANSI C12.22 is the designation of the latest standard being developed to allow the transport of ANSI C12.19 table data over networked connections. C12.22 is intended for use over already existing communications networks just as C12.21 is intended for use with already existing modems. Examples of such communications networks covered by C12.22 include TCP/IP over Ethernet, SMS over GSM, or UDP/IP over PPP over serial communications links. ANSI C12.22 is suited for automated reading of meter devices.

3.6.4.6 High-Reliability Protocols

IEC 62439-3, published in 2003, standardizes several protocols for industrial communication with a strong focus on reliability aspects. From a smart grid point of view, two protocols are particularly of interest: PRP (parallel redundant protocol) and HSR (high-availability seamless ring). Compared to other protocols, PRP and HSR provide an instantaneous recovery time in case of a link failure, which is a crucial feature for real-time applications, for example, differential protection application based on IEC 61850-9-2. Moreover, PRP and HSR are primarily intended for LANs and impact only the layer 2 of the OSI model, which make them good candidates for substation automation applications. Other protocols require at least several milliseconds to recompute a new route after the occurrence of a link or switch failure.

The principles of PRP and HSR are simple and can be summarized in three points: (a) each device is redundantly connected to the network through two independent network interface controllers (NIC) and two independent links, (b) the messages issued by the sender are duplicated over the two connections and sent simultaneously, and (c) the receiver transmits the first received message to the application (e.g., a protection function or a TCP/IP stack) and discards the duplicated message. From an application point of view, PRP and HSR are transparent and therefore do not require any modification. Moreover, failure of a link between the sender and the receiver does not introduce any delay since the messages are duplicated and transmitted simultaneously. The use of PRP versus HSR depends on the network topology: PRP is applicable for a point-to-point topology, while HSR is only applicable for a ring topology. PRP can be implemented entirely in software (at the driver level) and only requires an additional NIC on the device, while HSR requires the HSR switch functionality implemented by each device participating in the ring.

3.6.4.7 Time Synchronization Protocols

Time synchronization over communications networks is mainly achieved through NTP/SNTP (Network Time Protocol, Simple Network Time Protocol) or IEEE 1588 also called PTP (Precision Time Protocol). While NTP was defined back in 1985, IEEE 1588 is more recent and was first published in 2002 and revised in 2008. Besides the technical differences of the two protocols, their main differentiator is the accuracy they can provide: SNTP can provide an accuracy of tens of milliseconds across a WAN, while PTP can provide submicrosecond accuracy on an LAN. SNTP is mainly intended to run over a WAN but can also be run on an LAN in which case the accuracy can improve to a couple of hundreds of microseconds under ideal conditions. Conversely, PTP is restricted to an LAN and requires some specific hardware to achieve high accuracy. From a smart grid point of view, SNTP is mainly used for control and monitoring applications, while PTP is mostly used for protection applications.

SNTP and PTP are based on a similar mechanism involving the exchange of messages between a reference time source and a device. The purpose of the message exchange is to transmit the value of the reference clock and then to evaluate the transmission delay. While SNTP assumes a symmetric delay between the reference time source and the device, which is never valid in a WAN because of the switched nature of the network and the unpredictable delays introduced by switches and routers, PTP precisely evaluates the transmission delay by requesting the switches to report the residence time, that is, the delay a message is held by the switch. Therefore, for high accuracy, PTP requires some specific features implemented by the switches to support the residence time calculation.

3.6.5 Communications Challenges in the Smart Grid

3.6.5.1 Harnessing Technology Complexity

Modern operational applications in the smart grid environment and the corresponding communication access systems are propagating network intelligence to hundreds of substations spread across the grid. IP routers and Ethernet switches, VPN coding devices and firewalls, web servers, service multiplexers, and communication gateways require a great amount of parameter setting, which is generally expert oriented.

Furthermore, unlike serial communications links that did not operate with incorrect configurations, the present IP networks and Ethernet LANs generally "find a way to deliver information" even with incorrect parameters. Latent "setting errors" in the substation communications can cripple the communications network's performance, availability, capacity, and security. Communication and network devices installed in the substation environment must have "substation user-oriented" interfaces converting substation parameters into telecommunications network technical parameters to allow error-free configurations and operation by staff with limited communications network expertise.

3.6.5.2 Legacy Integration, Migration, and Technology Life Cycle

Telecommunications is a fast-moving technology driven by an enormous mainstream market and competition. Power system technology, on the other hand, evolves orders of magnitude slower and is deployed to fully replace an older technology over many years. As an example, Ethernet and IP networking that are being gradually introduced into the substation communications environment have been mature technologies for a long time in the mainstream commercial market, while serial RS232 communications links, which are still currently used in many power systems, have virtually completely disappeared in the telecommunications world. The increasing introduction of electronic intelligence into the smart grid will therefore produce two extremely unequal life cycles inside the same system: circuit breakers and power transformers do not have the same "technology life cycle" as their associated protection, monitoring, control, and communications devices. It should be assumed that in future, not only the power network but often every single substation may incorporate different generations of information and communications technology installed at different times. "Legacy integration" becomes an important part of any power system communication plan or project.

Operating with some older generation components in the system is not a temporary transitional state but the permanent mode of operation of the power system communications network: by the time that the older generation equipment is dismantled, the "once new generation" equipment itself has become obsolete and "legacy." The master plan for the operational communications network must include a preestablished migration strategy that stipulates not only how a new technology can be introduced into the network but ideally also how it can be removed from the network in a smooth manner without jeopardizing the whole power system. Excessive functional integration may present an attractive cost advantage at the time of deployment but may also be a major concern when one part of the integrated system needs to be replaced.

In general, the communications system refurbishment can be partial and performed by layer or service according to requirements (e.g., upgrading or replacing the transport core but not the substation multiplexing). The communications network architecture must be layered in order to allow such layered refurbishment and replacement of one technology without causing major network disturbances and service disruption. Similarly, all sites of the power network are not constructed, equipped, or refurbished at the same time and through the same project. This results in a multivendor and multirelease environment inside the same functional layer of the network. The power system communications network is therefore implicitly multivendor, multirelease, and multitechnology but still should operate as a single network.

3.6.5.3 Communications Service Planning and Evolution Trends

Most of smart grid operational applications that constitute the basis for the "power network of the future" already use to some extent communications solutions in today's existing and field-proven telecommunication industry. Therefore, future prospects in terms of power system telecoms are more about estimating power system application requirements than about predicting telecommunications technology evolutions.

When an operational telecommunication network is being planned for deployment or rehabilitation to enable smart grid applications, the following points require particular attention:

1. Ethernet ubiquity and SONET/SDH bandwidth allocation
   Ethernet is the dominant access interface for almost all smart grid operational applications, the standard local network technology and the optimal transport technology in the operational environment of the electric utility providing low connection cost, bandwidth flexibility, and a wide variety of topologies and transmission media (copper pair, fiber, wireless, etc.) [25]. Converters and coordination between many types of communications interfaces are gradually disappearing. However, legacy interfacing shall remain a major issue still for a long time. Terminal servers and interface conversion remain the solution to many legacy issues and allow the encapsulation of many non-Ethernet services in order to benefit from Ethernet flexibility and wire-saving properties. Moreover, Ethernet transport being the underlying network for many time-sensitive operational applications, it is essential to allocate reserved bandwidth resources as well as the flexibility of virtual separation (VLAN). Ethernet over SONET/SDH is a particularly efficient manner of implementing time-controlled Ethernet connections. SONET/SDH over optical fiber is used to implement multiple independent Ethernet transport connections with individually allocated bandwidths together with some small capacity dedicated to multiplexed circuits for protection relay communications and legacy applications.
2. Multiple secure IP networks
   The great majority of smart grid operational applications rely on the capability to connect network sites to control, monitor, or support platforms through an IP network. Many of these applications need segregated bandwidth to assure predictable network behavior,

guaranteed time performance, or security. Although applications may be grouped according to their requirements into a number of networks, a large multiservice IP network with VPN separation and IP-based Quality of Service (QoS)* control may not fulfill the operational requirements. VLANs over separate SONET/SDH bandwidth or separate wavelengths are currently employed to implement separate IP networks and can be scaled up through a technology such as VLAN trunking prioritization with 802.1p, MPLS (Multiprotocol Label Switching),[†] or DSCP (Differentiated Services Code Point) when the numbers get too large. These distinct IP networks can, for example, be allocated to EMS/SCADA, to asset monitoring, to site surveillance and facility management, and to support voice and data services (refer to Figure 3.131).

3. Service separation through wavelength multiplexing
   Another telecommunications technology which is increasingly used in utilities for separating multiple networks is wavelength-division multiplexing (WDM). WDM is becoming a secure and affordable way for separating traffic between the following:

   - Operational and corporate networks over the same fiber
   - SONET/SDH multiplexed network and MPLS/gigabit Ethernet networks
   - Protection relay communications and other communications

4. Use of wireless technologies and procured telecommunications services
   Distribution utilities often have a much larger number of sites to cover and very small traffic requirements associated with monitoring and control of each of these sites. Moreover,

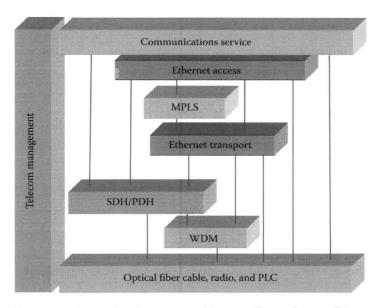

FIGURE 3.131 Communications network overlay architecture (Power System Telecommunications—A new landscape, M. Mesbah, Areva T&D Application Note, January 2009). (© Copyright 2012 Alstom Grid. All rights reserved.)

---

* Quality of Service (QoS) is the ability to provide different priority to different applications, users, or data flows or to guarantee a certain level of performance to a data flow.
† MPLS is a mechanism in high-performance telecommunications networks that directs data from one network node to the next based on short path labels rather than long network addresses, avoiding complex lookups in a routing table. MPLS can encapsulate packets of various network protocols.

distribution relays do not always intercommunicate, and when they do, they have less severe time constraints. Distribution companies also have larger mobile workforce on the operational side due to little or no staffed installations.

These challenges often favor different wireless technologies for communications across distribution systems: private mobile radio (PMR) operated by the utility or wireless services procured from a service provider (GPRS, VSAT, WiMAX, etc.). When the use of a telecommunications service provider is considered, it is important to assess the capability of the operator to continue service provision at times of major disaster, for example, in the case of a prolonged power outage.

3.6.5.4 Cybersecurity for Wireless Networks

In 2008, the Federal Energy Regulatory Commission (FERC) approved eight new critical infrastructure protection (CIP) reliability standards designed to protect the nation's bulk power system against potential disruptions from cybersecurity breaches. These standards were developed by the NERC and provide a cybersecurity framework for the identification and protection of critical cyber assets. The eight cybersecurity standards address the following areas: critical cyber asset identification, security management controls, personnel and training, electronic security perimeters, physical security of critical cyber assets, systems security management, incident reporting and response planning, and recovery plans for critical cyber assets.

A key concept associated with these NERC requirements is the establishment of an electronic security perimeter to protect smart grid network elements. These perimeters allow utilities to define transport network paths for data delivery.

One way of implementing these electronic security perimeters in WWANs is through defined standards for security in mobility networks using defined access point names (APN). These standards enable utilities to transport data from the AMI to core information technology (IT) infrastructure using authorized and encrypted capabilities. The implementation of APNs in 3G networks provides linkage from the wireless network to the utilities' core IT infrastructure using either frame relay circuits or MPLS connectivity and provides multiple levels of security, access controls, and encryption that many electric, natural gas, and water utilities find beneficial. For example, all data traffic from mobile devices using the radio access network is encrypted and subsequently tunneled from the core network serving nodes to core network gateways that provide connectivity to utility enterprise systems. Custom APNs segment traffic in a layer 2 VLAN at the gateway layers in the commercial carrier core network. The traffic then enters an MPLS virtual routing facility (VRF) using connectivity routers to maintain traffic separation to the customer's enterprise system.

Utilities should be aggressive in identifying and correcting vulnerabilities and exposures associated with smart grid network elements. In the context of smart grid deployments, security must be (1) encompassing, (2) circulative, and (3) aggressive.

Encompassing security: In order to address security requirements of smart grid components, utilities must consider exposures and vulnerabilities within network domains and supporting smart grid infrastructures that include intranets (premise domains), Internet connectivity, fixed communications links, and wireless connectivity. An encompassing approach to identifying these cross-domain vulnerabilities is required because an exposure in any of these domains can (if not properly isolated) lead to exploitation of smart grid network elements residing in other domains.

Circulative security: Because of the potential large number of network elements and components associated with the smart grid, utilities must deploy automated and centralized security event management and control systems that should be centralized and have a broad encompassing situational awareness. This implies that utilities must implement integrated security management and incident reaction controls. These security controls would provide wide spread alarms that flow to a central location where critical incident and crisis action teams can react and take proper action. Without this

Smart Grid Technologies

type of circulative situational awareness and centralized control, vulnerabilities can be exploited, and security events associated with smart grid network elements can occur without the appropriate response. Security of smart grid network elements must also be pronominally current and up to date meaning that operating systems, applications, browsers, network interfaces, access control lists, and other smart grid network elements must be constantly updated and protected accordingly. Thus, keeping network elements protected cannot be a one-time or periodic event for electric utilities but rather a continuous process where smart grid network elements are updated and protected proactively before events occur.

Aggressive security: Utilities must also be very aggressive in the care and resources dedicated for the protection of smart grid network elements. Electric utilities must take the initiative in complying and exceeding security standards by implementing knowledge-based and artificial intelligence systems that are able to detect network security events before they actually occur. By the time a smart grid security event occurs, the damage is already done, and it is too late. After the incident, utilities must deal with a consequence management effort as opposed to a security event. Therefore, it makes good policy and business sense for electric utilities to be proactive about security and invest in aggressive measures that will see trouble before it occurs.

3.6.5.4.1 Functional Domains

In order to plan and deploy wireless smart grid systems that are secure, it is helpful for utilities to have a functional view of the network and logically segment the smart grid WWAN elements into five domains. These domains are consistent with the 3GPP [26] standard body view of cellular networks and include the following: (1) smart grid mobility devices, (2) smart grid airlink interfaces, (3) carrier core networks, (4) connectivity to utilities' core systems, and (5) utilities' enterprise domains.

The first domain of the smart grid network view includes wireless end-point devices or mobile stations (such as electric vehicles) that provide services to smart grid consumers. In 3GPP terminology, mobile devices are termed user equipment or UEs. These mobile devices have a subscriber identity module or SIM that serves as the identity mechanism of the device. The SIM contains pre-programmed identification information such as an International Mobile Station Identifier or IMSI, the SIM key, authentication algorithms, and home short message service (SMS) numbers. The SIM can also store a list of subscriber names, numbers, and received short messages.

Devices that encompass this domain are the only components in the entire smart grid security architecture that users can directly tamper with because they reside in their homes and/or they have control over them. If a device is not properly secured, users can access protected data, perform unsafe software downloads, disable local store encryption, turn off local authentication, remove or disable virus protection, and potentially retrieve sensitive user names and passwords. For these reasons, policies and cybersecurity measures to protect, monitor, enforce policy, and control smart grid devices by electric utilities are critical. In this domain, it is worth mentioning the trend toward LTE and M2M (machine-to-machine deployment). In particular, the ongoing work on M2M communication identity module (MCIM) implementing a "software SIM."

The second domain of this functional view of smart grid networks is the airlink interface, which is composed of base transmitting stations in WWANs. In the GSM/HSPA environment, these base stations are called Node Bs (radios) and are attached to radio network controllers (RNCs). The combination of Node Bs and RNCs is referred by 3GPP as the universal terrestrial radio access network or UTRAN. The UTRAN works in close coordination with the serving support node and mobile switching centers described earlier to provide the basic functions of smart grid data connectivity for consumers and ratepayers. The UTRAN maintains and manages the air interface protocols for smart grid devices attached to wireless networks as well as specific call-sustaining procedures (power control and handover management). The UTRAN interfaces with core serving nodes in order to authenticate users and provide airlink encryption.

The wireless industry term "airlink" refers to the radio transmission of voice or data from the wireless device to the network base station, and from there to the other network segments for authentication and transport. The airlink segment is a separate functional domain and is not usually included in the carrier segment because, when the user is roaming, the airlink segment is not under the carrier's control. For example, mobile wireless devices with a SIM that belongs to one wireless carrier but roams to another geographic area that is serviced by another carrier. These carriers may have roaming agreements, but security for the UTRAN belongs to each respective carrier.

Securing the air interface is a critical responsibility either for the utility operating their own WAN or for the commercial network carrier working together with its roaming partners. While utilities today operate at national levels, it should be noted that over 90% of the world commercial wireless carriers comply with 3GPP air interface security standards.

An important security mechanism that protects smart grid data transmissions is encryption which can occur at the air interface layer and at the application layer. In the air interface layer, smart grid data are encrypted between the base stations or Node Bs and the mobile smart grid wireless end-point device.

At a high level in 3G networks, following authentication and key agreement, the network and smart grid wireless end-point device calculates a one-time (once per session) 128 bit encryption key by applying a key-generating algorithm known as Kasumi. Once the encryption key is derived, communication between the wireless end-point device and the 3G network is encrypted using the Kasumi algorithm. It is stronger than earlier 2G GSM proprietary algorithms. Authentication of wirelessly enabled smart grid devices in 3G networks is a two-step process in which devices such as electric meters first authenticate *the network* and then the commercial carrier serving the meter authenticates *the user* to the network.

The third domain of this functional view is the core mobility network or carrier segment. Once a smart grid device has authenticated to the WWAN, a data session or a packet-data-protocol (PDP) session is established in the Gateway Support Node or GGSN. The GGSN serves as the gateway between commercial wireless carriers and the utilities' core enterprise network routable network. All packets between the air interface UTRAN and core network interface exit through gateways called GGSNs. The GGSN provides IP services to the utility enterprise information system network via a number of connectivity options including Internet connections, secure network-to-network VPN connections, dedicated frame relay circuits, or MPLS cloud connectivity. Commercial carriers supporting smart grid networks are responsible for securing customers' confidential data as they move through or are stored in the carrier's network. This includes logged and archived data and all customer personal data such as billing information.

The connectivity segment of this model includes network elements and network circuits that link the commercial carrier core network and the utility enterprise network which resides outside of the carrier's control. As mentioned earlier, common connectivity segments include the Internet, either with or without VPN, frame relay circuits, and links through MPLS clouds.

The enterprise segment includes the utilities' back-office and core IT systems located inside the enterprise. Utilities are responsible for perimeter defense and other security systems within this segment.

3.6.5.4.2 Application Domains

While breaking down smart grid wireless communications into five functional domains is helpful in order to understand a functional view of the network, a different type of analysis is required in order to understand the network application architecture of smart grid network elements. In order to facilitate this analysis, an eight-layer architectural model was originally created by Todd Allen and refined for the utility industry by Art Maria of AT&T. This architectural model which was originally developed for AT&T's Wireless Reference Architecture has been adapted for the utility industry. The Allen/Maria model addresses the architectural elements that exist between the smart grid users and devices on one side of the communications link and the utility's enterprise applications the user wishes to access on the other side.

The first layer of the Allen/Maria model is the *user layer*. Because wireless end-point users are the weakest link of this model, utilities must implement policies that enforce security of wireless end-point devices. Without the enforcement of these policies, security becomes merely guidelines and not rules. In most cases, wireless telemetry end-point devices such as smart meters should never allow end users to access these devices. In other cases where devices such as smart meters provide gateway access into the home, end-user policies must be carefully crafted and implemented in order to enforce strict security standards.

The second layer of this architectural model is the *device layer*. If strong security controls are not implemented, devices such as electric meters, thermostats, and other demand response units represent the greatest security risk in a wireless smart grid application. If a device is not properly secured, it can provide access to data stored on the device or to data in the utility's enterprise systems that the device is connected to. Therefore, utilities must evaluate what type of security software must reside in the wireless end-point device. Depending on the end-point device capability, additional layers of user authentication, encryption, and device management must be implemented. These security controls are in addition to those provided by wireless carriers.

Taking human factors into account is an important part of the successful deployment and adoption of a wireless security architecture. Security policies can either drive or impede adoption, depending on the circumstance. The more intrusive a security policy is on users, the more they will attempt to circumvent it, which makes policy enforcement even more important.

The third layer of the architectural model is the *network layer*. Authentication and encryption elements of this layer enable utilities to establish electronic security perimeters.

The fourth layer of the architectural model is the *presentation and interface service* layer, which provides management of smart grid network elements. For example, wireless meters are connected through the wireless network to one or multiple gateways. These gateways are considered head-end systems (HESs) and manage individual devices. Head-end system interface with meter data management systems (MDMSs), which can manage multiple head-end systems and provide overall system management, control, and data collection of the AMI for the utility. HES and MDMSs are a significant portion of the presentation and interface service layers in the wireless smart grid architecture. Head-end systems and meter data management systems must provide additional layers of authentication, application controls, and encryption to reduce the risk of security exposure.

The fifth layer of the architectural model is the *business service layer* that provides management control services for smart grid network elements connected to the utility. Elements of this layer most likely reside in network control centers and provide logical interfaces between the fourth layer HES and MDMS and the utility's back-office support systems.

The final three layers of the architectural system can be tightly coupled and form the infrastructure of the utility's back-office and SCADA control systems. They include the application service layer, the data service layer, and the data source layer.

The application service layer includes systems such as the AMI servers and other applications that interface with meter data management systems. This layer also includes application service hosts such as those supporting outage management systems and customer billing. Application servers can be web-based servers, which use standard web service interfaces, or they can also rely on middleware servers. They communicate with the data service layer via applications programming interfaces (API) such as net, HTTP, MV90, SOAP, and ODBC. The data service layer provides a repository of data where customer information is stored.

As the electric utility addresses the pressures of creating and maintaining a secure environment, it will need to transition from a relatively closed system to one where millions of end points have the potential to send information to or require information from the electricity provider. This heightens the need for assuring security and privacy. The risks come from a variety of sources and motivations, and the result is disrupted business processes and higher costs. The defense is not to remain frozen in the twentieth century; rather the solution is to apply best practices of privacy and cybersecurity.

3.6.5.5 Management and Organization Challenges

Providing communications services to the whole spectrum of new smart grid operational applications in the power utility represents a change of scale in terms of management and organization. The requirements are indeed very different depending on the mode of service provisioning.

In a procured service mode, this represents a much larger scope of contract and therefore new grounds for negotiation with the provider, but also the opportunity to redefine SLAs* regarding the availability and continuity of communications services. It may also require new ways to measure the quality of the delivered service and the assurance that the contracted SLA is met.

In a utility-operated dedicated telecommunications network environment, a significant increase in the number of communications services requires the reorganization of the telecommunications delivery structures. If previously, service management was nothing more than a few phone calls between the telecommunications O&M team, SCADA supervisor, and substation staff, a sharp increase in the number of concerned parties may imply a fundamentally different "service user/service provider" management model in which the tasks of service management need to be explicit and formal.

The first step toward this change of scale is the formal definition of a two-level architecture separating core communications services from different application networks using core communications resources. The management of the core network infrastructure then becomes the responsibility of the "core service provider" with SLA obligations toward each power system application network. The core service provider notifies service users of the availability and performance of the communications services through "service dashboards" constituting the basis for service "situation awareness."

3.6.6 Communications in the Smart Grid: An Integrated Roadmap

In the myriad of changes that face the domain of communications for smart grid, it is difficult to know where to start. Utilities have to manage short-term requirements (deployment of AMI enforced by national regulations, adapt to the increasing number of small renewable power generators) and long-term requirements (anticipate future communications needs, provide security all along the value chain). Utilities are often enticed to make heavy investment in AMI without a clear idea of the long-term benefits. They may be pushed along the tracks of mainstream telecommunications technologies without always understanding what is at stake. In order to build a reliable, resilient, and secure communications infrastructure, utilities need to embrace a holistic approach starting by the definition of their smart grid roadmap.

Communication changes impact all stages of the power delivery chain, and different solutions are already deployed, exist, or are being developed to fulfill new requirements. Communications requirements differ from utility to utility depending on a great number of factors, including

- Density of communication end points—The geographical area to cover and the distances between the sites greatly differ depending upon the segment of the power delivery system (generation, transmission, distribution), the geographical spread and scale of the country (e.g., European versus American scale), and population distribution (e.g., urban, semi-urban, rural).
- Topology of the power network—Where some distribution networks may be extensive and serve a large number of customers, such as a densely populated urban area, a two-tier

---

* An SLA (Service Level Agreement) is a part of a service contract where the level of service is formally defined. In terms of communications, this may include minimum data rates, availability of the service, etc.

communications system with LV PLCs to a concentrator at an MV/LV transformer and a backhaul system from that point to the central platform may be appropriate communications solution. On the other hand, for a distribution network out in a rural serving only a few customers over a large geographical area, a one-tier communications architecture using wireless solutions may be more attractive.

- The condition of the power network assets—Depending on investments made over the years, the need for condition-based asset monitoring may be more crucial in certain networks than in others, creating the conditions for a strong requirement on communications to the sites where these assets are located.
- The amount of distributed generation currently in the power network and planned in predictable future—In many European countries, the current legislation has greatly encouraged the deployment of small renewable generators (solar panels, individual wind generators, small biomass plants). Bidirectional communication to these sites is becoming essential not only for metering purposes but also for the safe and secure operation of the power system.
- Population and industry growth and resulting power network extension plans—Residential and industrial complexes built from the ground up in many parts of the world are today based upon the principles of energy autonomy (e.g., microgrids) with centralized intelligent energy exchange with the outside world and therefore heavily rely upon communications.
- Regulatory context and legislation—Enhanced human safety, critical infrastructure site security and surveillance, liability on nondelivery of power, disaster readiness with auditable recovery schemes and tools, as well as environmental hazard monitoring are some of the hot legislation topics in many countries which necessitate increased "smartness" of the network and therefore more extensive communications.

For numerous utilities, smart grid is currently narrowed down to smart metering as the local regulators are often pushing to deploy AMI. However, AMI is only one out of several utility communications domains in which intelligence is to be shared. Table 3.11 summarizes major smart grid applications and corresponding data communications requirements and potential communications technologies.

Wireless solutions are becoming more and more widespread. One of the major advantages of wireless communications is that, in many cases, solutions can be deployed more easily and at a lower cost than wired solutions. For instance, it would be very costly to install an optical fiber to each existing secondary substation of a distribution grid. From a technical point of view, new technologies, such as LTE, also allow communications systems to provide higher-bandwidth and lower-latency requirements that were not possible before.

The communications network in smart grid must enable a set of functionalities across the power system to facilitate interaction across the grid and with customers [27]. First, it must make use of advanced sensors that are integrated with real-time communications to enable modeling and simulation computations. These functions must be provided in visual forms to enable system operations and administration. Second, the smart grid communications system must reinforce the transmission and distribution systems in ways that enhance data transfer, control center operation, and protection schemes. Third, communications functionality must facilitate the relief of congestion on the grid, from generators to customer premises, to enable increased power flow, enhanced voltage support, and greater reliability. Connectivity to customers to enable value-added services is also expected, as the consumer is the final user of the smart grid. This must take the concept of wholesale market settlement—attribution and accounting of power transactions—down to the retail level. Bridging wholesale and retail transactions to ensure dollar settlements is critical for the smart grid.

TABLE 3.11
Communications Requirements for Smart Grid Applications

Smart Grid Application	High-Level Communications Requirements	Candidate Communications Technologies
Home area network	Connect meters and home appliances for demand response and energy efficiency applications	Broadband PLC, Wi-Fi, WSN mesh, WPAN ZigBee
AMI (last mile)	Connect customer meter to a concentrator forwarding data to a customer relationship management platform and distribution management system (outage management)	NPLC, GPRS (low requirements) broadband PLC, fiber to the home (FTTH) for backhaul
Customer premises access (not via meter)	Home gateway or energy box for energy services (connect to server)	Public Internet
Distribution network automation	Connect MV switches and control platforms (SCADA)	Wireless, satellite, fiber, license-free spread spectrum radio
Microgrid management	New specific architecture for local microgrid (LV or MV) management and DR/DG local optimization	Wireless, PLC
Distribution network asset monitoring	IP connectivity at MV/MV and MV/LV devices (substation)	Broadband PLC, wireless
Dispersed generation (solar, wind, biomass, etc.)	Treated as a SCADA RTU or via meter if residential	Broadband PLC, wireless, satellite, license-free spread spectrum radio
Substation automation	Substation digitalization	Real-time process bus, IEC 61850
Large renewable plants (off-shore wind farms)	Voice, video, control, monitoring, SCADA	Fiber, microwave, UHF, etc.
Transmission asset monitoring	IP and web service in the HV substation	Robust Ethernet and IP router with security architecture
Transmission network wide area monitoring	IP connectivity from substation to PDC	Robust Ethernet and IP router, specific communications architecture
Transmission network automation (W. A. Protection and Control)	Time-predictable wide area Ethernet	Ethernet over SONET/SDH and time control
HVDC communications	Long-distance, low-capacity communications	Long-range optical links

Source: © Copyright 2012 Alstom Grid. All rights reserved.
PLC, power line carrier; WSN, wireless sensor network; WPAN, wireless personal area network (IEEE 802.15); NPLC, narrowband power line carrier.

3.7 MONITORING AND DIAGNOSTICS

Mike Ennis and Mirrasoul J. Mousavi

Monitoring and diagnostics in smart grids require three fundamental elements in the broadest sense: data, intelligent algorithms, and communications. Data are provided by sensors and sensor systems including intelligent electronic devices (IEDs) and switch controllers. Intelligence is provided by digital processors, which are instructed to perform certain operations on sensor data based on specific algorithms. Communications are required to deliver the derived monitoring and diagnostics intelligence to the right person/device, in the right format, and at the right time. These three elements are also the building blocks of current control and monitoring systems, but, in the era of smart grids, a dramatic boost is needed in functionality, performance, and coverage across the power delivery chain down to the last mile including the end customers. More importantly,

Smart Grid Technologies

expect to see an infusion of intelligence into every system and device, coupled with an integral means for communications, ranging from a local HMI to a broadband IP or fiber optic network.

In recent years, considerable progress has been made in measurement and instrumentation due largely to the progress in integrated circuit technology, the availability of low-cost analog and digital components, and efficient microprocessors [1]. Consequently, the performance, efficiency, and cost of sensors and sensor systems have seen much improvement. The emergence of local and international standards coupled with advancements in communications technology paves the road for more advancements (e.g., wireless sensor networks) and application areas such as monitoring and diagnostics for smart grids.

Smart sensors are composed of many processing components integrated with the sensor on the same chip. These sensors have intelligence of some form and provide value-added functions beyond passing raw signals, leveraging communications technology for telemetry and remote operation/reporting. Increasingly, local devices will have to report information rather than data, since otherwise data bottlenecks will ensue, compromising the ability of transmission and distribution grids to deliver value to their operators. Automated, reliable, online, and off-line analysis systems are needed in conjunction with sensors/sensor systems supporting smart grid monitoring and diagnostics applications.

3.7.1 Architectures

Smart sensor technologies enable condition monitoring and diagnosis of key substation and line equipment including transformers, cables, breakers, relays, capacitors, switches, and bushings. These sensors use digital data for monitoring and diagnostic purposes that, depending upon the type of the asset and monitoring requirements, may include conventional and nonconventional voltage and current measurements and temperature readings. A fault passage indicator on a distribution line is an example of a smart sensor that senses the overcurrent condition and communicates the passage of the fault current to a local or remote human or machine operator (see Figure 3.132).

These sensors, empowered by a central processing unit, offer functionalities beyond conventional sensors through fusion of embedded intelligence to process raw data into actionable information that can trigger corrective or predictive actions. It is this combination of sensors, intelligence,

FIGURE 3.132 Example sensor system with embedded intelligence and communications. (© Copyright 2012 GridSense, Inc. All rights reserved.)

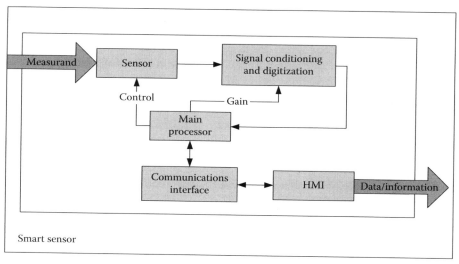

FIGURE 3.133 Block diagram of a smart sensor.

and the communication of information, rather than mere data, which earns them the description "smart." They may perform a number of functions based on the level of sophistication [1]. These functions, depicted in Figure 3.133, may include

1. Basic sensing of a physical measure
2. Digitization and storage
3. Raw data processing and analysis by the central processing unit
4. Local and remote communications
5. Local and remote HMI

These sensors may be stand-alone devices or integrated into multifunctional IEDs. To deliver the best value, these sensor systems may be deployed in three tiers depending upon the available architecture and application requirements.

3.7.1.1 Tier 1: Local Level

All smart sensor functions including sensing and analysis are local to the asset they are monitoring. The sensor is a stand-alone device with embedded intelligence for local data processing and local/remote communications (Figure 3.134). A visual fault indicator on a terminal pole is an example of a sensor product in this tier. A transformer monitor used solo inside a substation is another example.

Information and data from these sensors may be loosely or tightly integrated into feeder or substation automation systems. When fully integrated, these sensors make up an integral part of the automation solution. By far, this is the most common architecture for smart sensors, and fault indicators are the most common sensor in use today. The future trend is a higher level of integration of these sensors with operations and automation systems.

3.7.1.2 Tier 2: Station/Feeder Level

Monitoring and diagnostics at this level involve smart sensors that are in fact distributed systems with remote access to sensor measurements outside the substation environment. In these systems, sensor functions are distributed among system components that may physically be located apart. A common architecture involves sensing and measurements that are polled into a computing environment (e.g., station computer) for analysis and interpretation as shown in Figure 3.135.

Smart Grid Technologies

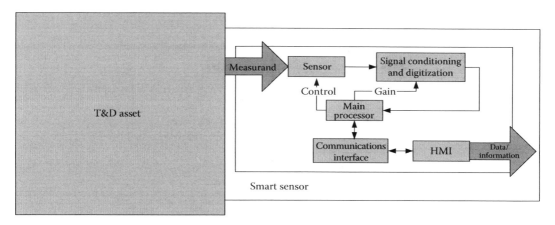

FIGURE 3.134 Tier 1 monitoring and diagnostics.

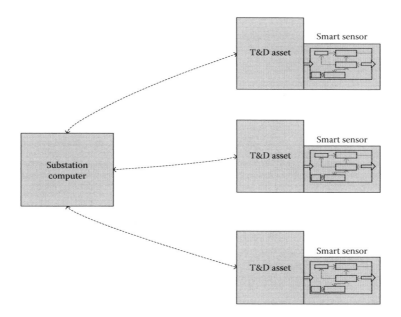

FIGURE 3.135 Tier 2 monitoring and diagnostics—hierarchical topology.

Since smart grids will contain both hierarchical and distributed sensors, the topology of Figure 3.136 is also likely. The substation computer in Figure 3.136 might take on either a supervisory or gateway role, but it is equally likely to envisage that such a mesh topology might relate to third-party equipment, operated by, for example, energy service provider. In this type of mesh topology, communications occur between peers based on system needs rather than polling cycles from a master controller.

Feeder monitoring through peer-to-peer communications or via a substation computer is an example of this tier solution. Often these functionalities are integrated into protection and control IEDs and systems forming multifunctional devices and systems.

3.7.1.3 Tier 3: Centralized Control Room Level

System-wide monitoring and diagnostics applications require architecture at the control room level where information and/or data from field sensors are pooled into a central repository to support

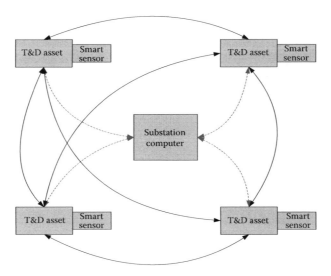

FIGURE 3.136 Tier 2 monitoring and diagnostics—meshed topology.

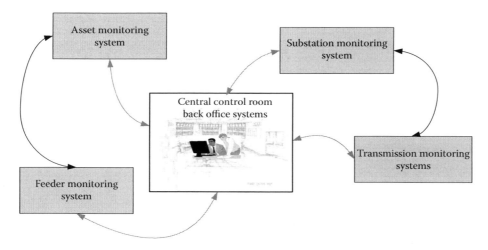

FIGURE 3.137 Tier 3 monitoring and diagnostics. (© Copyright 2012 ABB. All rights reserved.)

real-time and back-office applications as presented illustratively in Figure 3.137. Integrated substation monitoring application with enterprise network security systems, including both cyber and physical security, is based on such architectures.

For enterprise T&D asset management applications, data repositories are needed for data update, retrieval, and reporting via a stand-alone application with server inside client firewall with interfaces to external test data systems (i.e., DOBLE, POWERDB, Hydran), mobile computers, and handhelds.

This architecture empowers operations and maintenance departments with decision support based on real-time and historical data. Such decision support systems provide asset condition diagnostics function by utilizing pattern recognition and intelligent algorithms, enabling asset data management system to perform reliability-based as well as predictive maintenance. Such architecture should be designed for seamless integration into other enterprise systems such as work management system, inventory system, financial system, and regulatory compliance.

Smart Grid Technologies

3.7.2 Wireless Sensor Networks

The majority of the T&D assets are located inside transmission or distribution substations, but other system components are outside or extended from the substations. Regardless of the location, the sensor or sensor system installed on these assets has to have communication ability to send and receive signals or enable remote diagnostics.

The communications and interfacing technology deployed depend upon the application and requirements of the specific sensor or sensor system. Most systems are equipped with LCD screens, RS232, RS485, and Ethernet interfaces [1]; wireless communications are also developed and utilized in some applications with continued research and development to improve performance. Bundles of lead wires and fiber optic cables are common in most hard-wired sensors, but the trend is changing. Significant installation cost, long-term maintenance cost, and limited number of deployed sensors are impediments to the widespread use of wired sensors, but wireless sensor networks are now eliminating these constraints and offering attractive sensor solutions.

A wireless sensor network is typically composed of a number of sensors that are linked to each other through a base station or gateway or through peer-to-peer connections forming a star or mesh network. The data are collected at each sensor node, possibly preprocessed, and forwarded to the base station directly or through other nodes in the network. The collected data are then communicated to the system via the gateway connection. Recent advancements in wireless sensor networks offer a single sensor package integrating sensors, radio communications, and digital electronics into an integrated circuit. This compact design results in substantial cost reduction and enables low-cost sensors to communicate with each other using low-power wireless data routing protocols [2].

The radio link in a wireless network is the largest power-consuming component that can be characterized in terms of the operating frequency, modulation scheme, and hardware interface to the system. There are many low-power proprietary radio chips on the market, but the use of a standard-based radio interface enables interoperability among networks from different vendors.

The existing radio standards include IEEE 802.11x (LAN), IEEE 802.15.1 and 2 (Bluetooth), IEEE 802.15.4 (ZigBee), and IEEE 1451. Public carrier telecom networks are also now beginning to open up and become viable for middle-mile communications.

For short-range wireless sensing applications, IEEE 802.15.4 has a number of features which can be used as a benchmark for other wireless solutions. The IEEE 802.15.4 standard specifies multiple data rates of 20, 40, and 250 kbps for transmission frequency bands of 868 MHz, 902 MHz, and 2.4 GHz, respectively. The 2.4 GHz band being essentially license-free worldwide is the most appealing band. By accommodating higher data rates, it reduces the transmission time and consequently lowers the power consumption level of the radio. This provides for a long term and potentially maintenance-free network for monitoring applications in many areas including smart grids.

A number of companies are working together to develop reliable, cost-effective, low-power, and wirelessly networked products. The ZigBee Alliance, for example, promotes the use of wireless networks for home/building monitoring and control applications using an open global standard (IEEE 802.15.4). As smart grid initiatives are rolled out, wireless sensors networks will be an integral and vital part of many application areas related to grid monitoring and diagnostics including the consumer space. In this rapidly evolving area, however, new solutions have to be always borne in mind; DASH7 (ISO 18000-7), Wibree (Bluetooth low power), and UWB PHY applications of ZigBee have the potential to open up new niches within low-power, short-range wireless transmission.

3.7.3 Diagnostics

Transmission and distribution asset monitoring and diagnostics applications extensively utilize sensors and sensor systems for various functionalities ranging from basic alarming to online and nondestructive condition assessment. Transformers, load tap changers, regulators, circuit breakers, reclosers, HV and MV vacuum/SF6 switchgear, underground cables, overhead lines, switched capacitors, reactors, surge arresters, insulators, shunt devices, batteries, battery chargers, and power electronics interfaces are the major power system assets that may be equipped with some kind of sensor or sensor systems for continuous monitoring, diagnostics, and real-time asset management. The total cost of the power equipment and its failure risk usually determine the need, complexity, and features of the installed monitoring systems. Some components of the power system may not have a monitoring system installed, but every piece of equipment participating in smart grid communications should be equipped with sensors for measuring, monitoring, and/or control applications.

Figure 3.138 shows some of the most common sensors and application areas. The sensors and sensor systems supporting monitoring and diagnostics applications range from conventional CTs and VTs to state-of-the-art optical and acoustic sensors. These sensors are used to measure and sense physical attributes such as electric current and voltage, temperature, gas-in-oil, moisture-in-oil, acoustic wave, vibration, pressure, weather parameters, UHF and RF waveforms, water, thermal

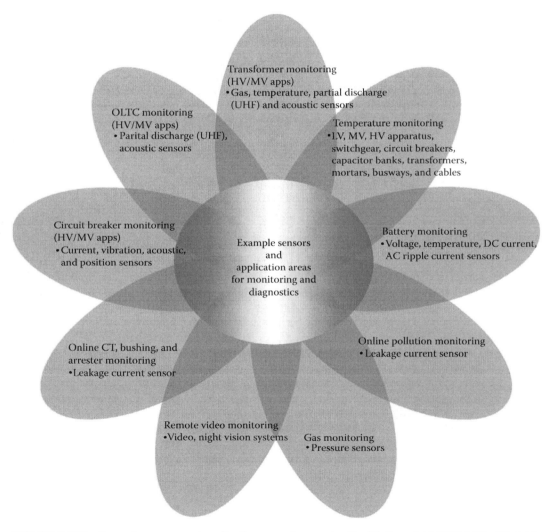

FIGURE 3.138 Example sensors and application areas for smart grids. (© Copyright 2012 ABB. All rights reserved.)

Smart Grid Technologies

profile, motion, proximity, x-rays, displacement, and erosion. Constant improvements in performance and cost are expected to continue and accelerate full-scale deployment of various sensors to enable smart grids.

These sensors support the following applications in particular:

1. Cable diagnostics and prognostics
2. Water penetration monitoring of high-voltage cables
3. PD detection, localization, and monitoring of power transformers and cables
4. Power and distribution transformer monitoring
5. Transmission grid monitoring
    a. Real-time monitoring of conductor temperature
    b. Asset health monitoring
    c. Voltage instability monitoring of transmission corridor

In addition to monitoring and diagnostics areas, the data collected from these sensors may be utilized to support other system functions such as VAr management, design improvements, real-time control applications, dynamic loading of transformers, triggering advanced diagnostics, and off-line applications.

In the era of smart grids, ubiquitous sensors and measurement points will enhance situational awareness and monitoring of system components down to the last mile. This will in turn mean more data and increased processing needs. The utility environment of today is overflowed by the amount of data already collected by existing systems. Thus the addition of new data points will simply exacerbate the situation, unless the data-to-information conversion process is considered in each step of the way giving rise to more automation and emergence of proactive system health management and auto-notification systems [3].

The ability to proactively address T&D system problems and respond as quickly as possible to outages and asset failures, along with movement toward predictive maintenance, will be a significant contributor to fulfill the promise of smart grids. Typically, maintenance schedules for the assets are set on a preprogrammed basis without specific intelligence about the asset condition and health. In a self-healing smart grid, with automated analysis of sensor data and predictive maintenance technologies, the Operations and Maintenance departments will have the ability to respond quicker to outages, send the right restoration/repair crew, assess the risks, and proactively address system problems. The planning department will in turn have access to better information for upgrades and long-range reliability enhancement projects.

The intelligent machine algorithms used for monitoring and diagnostics in smart grids may reside at different levels in the supervision and control hierarchy. Protection and control IEDs can host such algorithms to detect anomalies in the power system behavior and identify an abnormal situation (such as emerging faults). They can take appropriate action (such as tripping appropriate circuit breakers) based on this local analysis or just forward the fault/event information to the higher entity in the hierarchy, which can be a substation computer or a full-fledged health management system (see Figure 3.139). This essentially minimizes the amount of data that should be retrieved by the substation or control center computers for analysis and decision making giving rise to reduced communications bandwidth required.

The substation computer may host the intelligent algorithms; this can provide system monitoring capability at Tier II, which is unavailable with conventional sensor systems without remote communications. Along the same lines, the intelligence can be hosted at the control center level. Each deployment option comes with its own benefits and limitations, which need to be evaluated carefully to achieve optimal net benefit from the chosen architecture. These solutions can be tailored to the application based on the complexity, scalability, cost, communications options, and customer preference requirements.

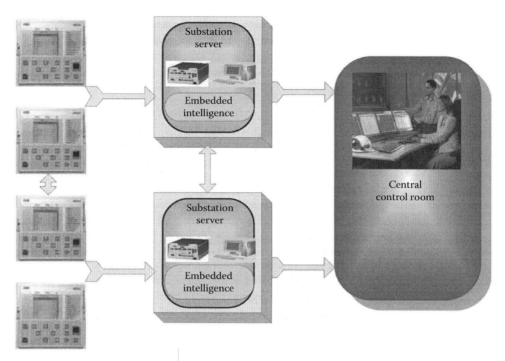

FIGURE 3.139 Monitoring and diagnostics data flow [3]. (© Copyright 2012 ABB. All rights reserved.)

3.7.4 Future Trends

Many attributes of smart grids for monitoring and diagnostics require sensors and sensor systems that are reliable, scalable, and integrated into automation systems. Although there has been significant progress in the recent decades on sensors and sensor systems in general, there is room for continued improvement for smart grid applications. These improvement areas include cost per node reduction, managing power requirements, expanding communications capabilities, reducing footprint, retrofitability, ease of installation/configuration/calibration, accuracy, scalability, reliability, interoperability, and finally security aspects. Future trends will involve efforts to continue to make sensor systems cost-effective, accurate, scalable, fault tolerant, interoperable, secure, self-powered, remotely available, and maintenance free, all in an integral part of utility automation infrastructure.

3.8 GEOSPATIAL TECHNOLOGIES

Stephen Byrum and Paul Wilson

3.8.1 Technology Roadmap

The business of an electric utility is inherently spatial in nature. Managing power flows over a large geographic area requires detailed information about the vast network of wires and equipment that composes the grid—and much of that information is spatial.

Since the genesis of widespread electricity distribution in the 1880s, there has always been a need for geospatial information to help manage the grid. The electrical grid is inherently spatial, rooted in the geography of the service territory served by the utility. It is a complex network of wires, supported by devices that control the flow of electrons through those wires. To build and manage that network, the utility has to know the location of all those components and how they are connected. Managing this locational and topological data, and providing users with methods to view and use it, requires technology that is designed to handle large amounts of geographic data.

Smart Grid Technologies

A great deal of utility work has a high level of "where" content, reflecting the spatial nature of the grid. For any operations function, much of the day-to-day work requires access to location-based facilities data. Where are my facilities? Where are my customers? Where is the device that controls this circuit?

For field crews—the "tech in the truck" that makes up a large part of the utility workforce—there are additional spatial questions at the heart of their daily work. Where am I? Where do I need to be for my next assigned job? Where is the switch that controls this line?

3.8.1.1 Age of Paper

For almost a century, the mechanism for storing all of this spatial data was the paper map (or, for permanent records, a more durable equivalent such as vellum). The "data" were created and maintained by manual drafting. It was distributed by making copies of map books for each person (or field crew) needing the data.

Edison's first distribution network, the Pearl Street project, covered a very small geographic area—several blocks of lower Manhattan. Even with a territory that almost disappears within the territory of a modern utility, however, a map was needed to show the spatial extent of the network (Figure 3.140).

As the size of utilities' service territories grew, the scope of the mapping effort grew as well. Recording changes became more of a problem as data volume rose dramatically. Each utility

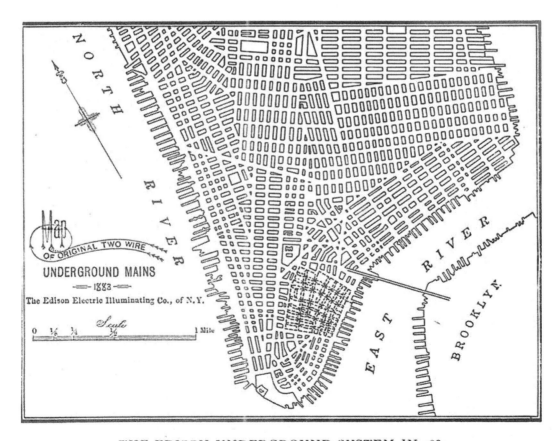

FIGURE 3.140 Pearl Street project in Manhattan. (Courtesy of Consolidated Edison, New York.)

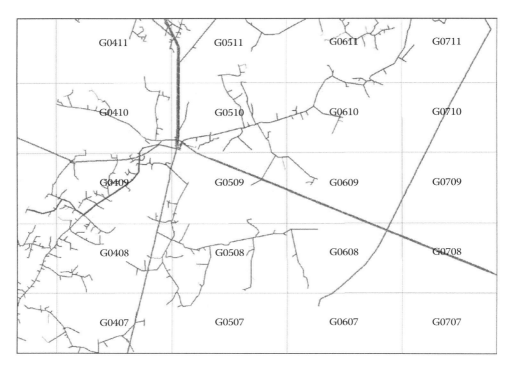

FIGURE 3.141 Typical utility map grid. (© Copyright 2012 General Electric. All rights reserved.)

developed a system for organizing and cataloging maps. The service area was typically divided into a map grid—a series of tiles, where each tile corresponded to a defined geographic area and was represented by a map sheet (Figure 3.141). This mapping structure often made its way into the field, as numbers based on the map grid were stamped onto poles and other equipment. As paper maps became more congested, the grids had to be split and redrawn into fourths and sixteenths to provide a workable resolution, adding to the cost and effort of maintaining, publishing, and distributing map books.

3.8.1.2 Emergence of Digital Maps

Utilities, like most other businesses, first used computers for back-office functions such as payroll, billing, and accounting. Starting in the late 1960s, some utilities (notably Public Service of Colorado) started to experiment with harnessing this computing power for representing maps.

The mainframes were, by today's standards, very limited and primitive tools. Even with the limitations, however, it quickly became apparent that the growth in computing technology offered utilities a new way to handle spatial data. As we moved through the 1970s, there was clearly a new age of geospatial technology in utilities: the sheer volume of map data had overrun the ability of paper-based methods to keep up with changes, and digital tools were increasingly seen as a potential answer.

Over the next two decades, almost every large utility invested heavily in computer infrastructure. Massive data conversion projects were necessary to turn paper maps, often decades old and with questionable cartographic accuracy, into usable data. This required tying points and lines on the old maps to a common coordinate system (latitude/longitude, state plane, or UTM*).

---

* UTM coordinate system, based on the Universal Transverse Mercator map projection, is a planar locational reference system that provides positional descriptions accurate to 1 m in 2500 across the entire earth's surface except the poles.

Although at this stage the storage of spatial data started to move from physical to digital, the communication of spatial knowledge still relied on paper. After all, the mainframe was not readily accessible to the people involved in managing the grid. Interactive, on-screen graphic capability was fairly primitive and very costly. The emphasis, then, was still on producing paper maps. The data used to produce the maps may have been stored digitally, and the maps might have been generated by a digital plotter, but the end result was still a paper map. In most cases, the goal was to reproduce—in a more efficient way—what had been used for decades. The paper map products from a digitally stored map had a more appealing look and feel with the capability of more detail and consistency, but the information that the map contained still had to be communicated through paper to humans with no interface or tools to make use of all the captured map data. Map content and symbology mirrored the standards in use at each utility but also carried with them the accuracy issues of the original paper maps.

3.8.1.3 From Maps to Geospatial Information Systems

The next stage in the evolution of geospatial technology shifted the emphasis from maps to applications. The graphic representation of facilities in two- (2-D) or three-dimensional (3-D) space—the map—was still important, but the data behind the map began to be used in different and more powerful ways.

Early systems were used to store geographic data and communicate it through maps. The next big step was to treat data about the grid as not just a map but as a collection of objects that have location, attributes, and topology. Adding attributes makes it possible to retrieve related information (what are the voltage ratings on that transformer?) and to search (where is pole B45806?). Common database functions allow for complex queries across data sets (where do we have 500 kV oil-filled pad transformers installed within 1000 yards of the Chesapeake Bay management area?). Establishing topology (the relationship of features to each other) enables network connectivity, supporting models of current flow. This added a whole new dimension to the data available from the mapping database, a powerful resource for utility network planning and operations.

This change marked the transition from automated mapping to true geospatial information systems (GIS). Together, these characteristics support analysis of asset attributes and much more. CAD (computer-aided design) systems also played an important role during this period, adding intelligence to the process of designing new facilities on the grid.

Even though these systems enabled profound changes in the way that map data were stored and managed, the direct effect on utility operations back then was minimal. Access to GIS tools and applications was limited to professionals with extensive training in the technology. Frontline users (the crews in their trucks) were, for the most part, still using paper maps. Even though the maps represented digital data and were generated by plotters, the users were still constrained by the limitations of having to use the paper format of the geospatial information.

3.8.1.4 Across the Enterprise

A fourth stage of the geospatial grid started to emerge in the 1990s. Desktop computers proliferated, network infrastructure grew, and (in the late 1990s) mobile computers rugged enough to survive field conditions were deployed. Paper was replaced by computer applications that could search for objects, display attributes, and even trace through the network to identify trouble spots. A rich set of applications made GIS capabilities accessible to planning and operations managers and also field crews.

Geospatial technologies have evolved to become a true enterprise system, extending from meager map digitization, to meaningful GIS systems, to a valuable data resource that crosses many enterprise applications. GIS for utilities has become a business-critical technology, supporting operations as the "system of truth" for the grid. Interoperability has allowed it to become the integration point for other utility enterprise data—asset databases, sensor and monitoring equipment, customer information systems, work management, compliance records, as well as third-party and public map sources. It is now common for a utility to visualize load information, assets, protection schemes, workforce locations, and public/commercial maps and photography all at the same time and through the same interface.

Interoperability has magnified the need for accurate data in all systems. Reliability is tied to properly functioning applications that are dependent on accurate and up-to-date data. Smart devices that report load information can dictate a demand response application, but if there are inaccuracies in the asset management system and geospatial representation, the application will be ineffective.

In many ways, this drive to automation in the utility industry has mirrored technology trends in other sectors, where the platform for computing has moved steadily closer to the user's place of work. It parallels ubiquitous mobile platforms and social networking, which have brought computing power to the hands of almost everyone. (And, in developing countries, it is outstripping conventional desktop computing as the dominant platform.)

Delivering geospatial tools to the field is essential to the operations of a utility company because much of the work has to be done outside the office. The assets and the customers are all outside in the field, spread across the service territory. Consequently, much of the utility workforce is also outside the office. The field personnel jobs are inherently mobile, moving around the grid "in the geography" to job locations that change rapidly (Figure 3.142).

This spread of geospatial technologies to the field is worth emphasizing because of its profound impact on how utilities do their work. While in the early digital age, a utility might map the orders and track the location of field workers, the push to provide this capability to the field has been driven by the demand for enterprise information by the field worker. The field technician is usually the front line of work with the grid and is increasingly a frequent point of contact with the customer. There are several forces driving this spread of technology to the field workforce:

- Fewer people, more work
- Growing complexity of work
- Increased safety and security standards
- The increased cost of outages
- Higher expectations for customer service
- The expectation of technology by the younger workers accustomed to social networking

The aging utility workforce has a major impact here. Most utilities are faced with the prospects of replacing a key cadre of workers that represents much of the organizational knowledge. This group, in effect, carries the system maps in their heads. As this segment of the workforce nears retirement age, it will be essential to support less experienced workers with strong geospatial tools.

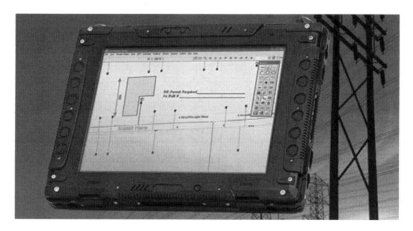

FIGURE 3.142 Extending maps to the field. (© Copyright 2012 General Electric. All rights reserved.)

Smart Grid Technologies

Mobile applications often show a short return on investment. By taking technology to the work site, these systems can close the loop and digitize work processes from beginning to end. This eliminates many sources of errors and speeds up processes that were once paper-bound. For safety and efficiency, much of the supervisory team is also in the field close to the work being performed. Often the supervisor is the most qualified to make an assessment and network decision, but now the data streaming from the system is required to make those decisions. Extending the data set from GIS and related enterprise applications to the field improves field work efficiency and safety and provides a synergistic return that is often overlooked and hard to measure by traditional standards.

Over the last decade, we have seen a major transformation in mobile computing technology. The rapid development of consumer technology has helped drive acceptance of smartphones and tablets into commercial markets. Ubiquitous personal and business improvement applications are now used in almost every company. Because of the field-centric nature of much utility work, mobile systems play a large role in operations. Field applications for a utility are job-critical and time-critical. A breaker or regulator that is bypassed for maintenance must be accurately identified and modeled in the GIS for the systems that rely on them to function properly. The most reliable current source for this information is the worker performing the action. Therefore, field applications have to work wherever and whenever they are needed.

3.8.1.5 Developing World

The technology evolution described earlier has been fairly consistent in North America, Europe, Australia, and many parts of Asia. In the developing world, technology for managing the grid has taken a different shape.

Part of this difference is due to the grids being different. In some emerging economies, large power grids have not been as common as in the developed world. Therefore, with developing countries, there is an opportunity to take advantage of modern tools for the smarter grid in the expansion phase rather than having to deal with issues of retrofitting the older grid. The technology aspect is different too. By starting later on the GIS curve, some parts of the world are avoiding the sometimes uneven evolution of hardware and software systems over the last four decades.

In much of the developing world, where large landline communications infrastructures are lacking, mobile phones are rapidly becoming the tool of choice for both businesses and consumers. A newer technology has replaced the need to build out an older (and more expensive) infrastructure. Similarly, the predominant computing platform is not the desktop but mobile devices. Tablet computers and smartphones provide utility employees with enterprise-wide strategies from the onset, rather than having to later add mobile applications to office-bound systems.

A significant benefit of the late implementation of geospatial technologies is skipping much of the data conversion process. Rather than dealing with the painful and expensive projects to convert paper maps of old facilities into digital form and to correct the cartographic error of paper products that were generated over time, a utility that is now expanding into new areas can capture designs and as-built drawings electronically as part of the construction process.

Capturing this data electronically not only allows for more accurate asset and grid inventory but also provides a means for spatially accurate records and correct connectivity. These data can be captured at the time of installation to improve accuracy and provide a shorter database posting cycle. This allows the GIS to be as accurate as the actual facilities it represents as quickly as possible.

3.8.2 Changing Grid

Throughout the first four stages of geospatial applications, the technology has changed dramatically, but the electrical grid has remained largely the same. (It is often said that Thomas Edison, looking at today's grid a century after his Pearl Street project, would easily recognize what he saw: a one-way, fairly static network where a flow of electrons was created at a small number of power generation plants and distributed to customers.) The electrons, for all practical purposes, flow one way. There is little system-wide information flow. SCADA systems are sometimes used to monitor overall flows through the system backbone, but this capability rarely reaches all the way to the customer. Each customer has a meter—a device that measures the usage at the customer point so that billing can take place. Almost all of these characteristics change with the smart grid. The old grid, with its static, one-way flow, becomes a much more complex and dynamic system.

Much of this added complexity has a geographic dimension. To begin with, take generation: the old paradigm of a few power plants, all controlled by the utility, gives way to a system that may have numerous power sources. Wind farms and solar installations are often privately owned, so the utility has a challenge in adding them to the network data model. And since they are subject to weather factors that neither the owner nor the utility can control, managing system flows become far more complex. Information about the "whereness" of weather, which varies over space and time, can help balance the complexity of the generation mix in the utility system.

It is a similar story on the customer side. Most utilities have not included details about customer locations in their spatial data. The GIS data model often extended only to a distribution transformer, sometimes with links to data about the customers fed from that transformer. Does the smarter grid, with smart meters and perhaps smart appliances, require that the GISs capture location beyond the transformer?

The increased penetration of electric vehicles (EVs) will add yet another dimension. Although charging points are static, the vehicles themselves move around and might connect to the grid at different locations.

There is also an impact of the utility's crews. These frontline employees, who have to build the system and resolve any operational problems, are faced with a more complicated job, needing far more data and new tools to analyze this data.

Clearly the changing grid will increase the demand for more geospatial data and the need to integrate the geospatial data across numerous business and operational applications in the utility enterprise.

3.8.3 Geospatial Smart Grid

Now we are on the edge of a fifth "age" of the geospatial grid. This time, the changes are driven not by gains in geospatial technology but by the transformation of the grid itself: the emergence of the smart grid.

How does geospatial technology contribute to planning, building, and operating the smart grid? In this section, we will examine the importance of these tools, reviewing a number of applications in the utility sector. One key in planning business-critical applications is to ensure a consistent base of geospatial data. The GIS is typically seen as the platform for managing these data—the "system of truth," which is synchronized with local data requirements for other enterprise systems. Close attention to interoperability is required. The stringent requirements of a smarter grid with constantly updated data may challenge the traditional abilities of GIS to continually exchange data with other applications.

3.8.3.1 Core Spatial Functionality

It used to be easy to equate geospatial applications with geographic information systems. After all, GIS was the tool of choice (and often the only tool) for any functions that required spatial data. That has changed dramatically; today, applications in virtually every part of the utility automation sector manage and display map-based data.

Here, the focus is on software applications. Although the division is somewhat arbitrary, the applications can be divided into categories that reflect how they are used.

The first group of functions includes those that are central to spatial data handling. They are traditionally the core components of the GIS.

3.8.3.1.1 Managing Spatial Data: The System of Truth

Geospatial tools, at a basic level, provide a common source of information for operating the grid. Since the operating system of a utility is spread out over a large geography, the data necessary to run it are spatial in nature, and managing these data spatially is critical to the business. This has been a major driver in the adoption of GIS within the utility sector. GIS today is often viewed as the "system of truth," the single trusted source for any data that are spatial. For many years, this viewpoint was hard to challenge. Almost any application of geospatial data was handled by the GIS software and managed by the utility's GIS group. It was clearly the single source of data because the data were not used anywhere else.

That has changed, however. As more operation functions have been automated, spatial data have made its way into other applications. OMS/DMS applications rely heavily on a spatial view of the grid. Who would have thought in the early days of a GIS "map" that we would see it presenting and interacting with SCADA? Work management systems now include map-based views of how work and crews are distributed over the service territory. Even planning and marketing groups in the utility employ geospatial data, using maps of current infrastructure in conjunction with demographic and land use data. It is due to this data codependence that the savvy spatial data manager recognizes the importance of data accuracy and strives for perfect data.

These applications must therefore use the same source of spatial data. They are, after all, covering the same geographic area. And they should reflect the same "reality" of the physical grid. For many years, all spatial data—anything with coordinates attached—were strictly the province of the GIS. It was argued that "spatial is special" or that the unique nature of geospatial data meant that only dedicated GIS systems were capable of handling and displaying these data. Now, however, many of these other systems have evolved to include spatial tools. So where are the geospatial data? Which data sets reside in which system?

Advances in hardware and software technologies make these questions more difficult. Abundant storage means that keeping multiple copies of spatial data (perhaps in slightly different forms) is not cost prohibitive. The concept of cloud storage even eliminates the "where is my data" question—although it does raise other questions, like how to maintain security for critical infrastructure data. On the software side, database software commonly used by other applications may now include tools to manage this type of data (e.g., Oracle Spatial).

This spread of spatial functionality into other systems clearly has great advantages. It does raise a data management dilemma, however. If every application that uses geospatial data stores a copy of that data, how do we synchronize the systems to ensure that they are all operating on the same "truth"? A common approach is to keep the base data in the GIS and then feed to other systems as needed. As we will see in a later section, the changing requirements of the smart grid may make that method more difficult.

3.8.3.1.2 Geovisualization

Maps are used for a reason—they are the best means of communicating certain types of information. For the electric grid, this means a spatial view of the relationships between network, customers, and field crew locations.

We can refer to this process as geovisualization. It is a way of communicating spatial information in ways that support human decision making. If done well, presenting a clear view of operating data supports situational awareness and improves decision making.

One of the beauties of spatial technology is that the same data can be used in so many different ways. It can, in effect, produce a near endless series of maps. At one scale, the data produce a wall map, an overview of a large area. The same data can also generate a series of larger-scale maps (or even schematics) with details for smaller areas. By managing scale, geospatial technology can produce the map that is most appropriate for the job at hand.

Today's computing technology can offer ways of visualization that go far beyond the static, 2-D paper map. GIS tools can quickly render views based on an almost endless combination of geographic and thematic filters.

FIGURE 3.143 LiDAR point cloud. (Courtesy of LiDAR Services International, Calgary, Alberta, Canada.)

The emergence of 3-D viewing also adds exciting possibilities. Most existing facility data, since they were created by converting paper maps, are 2-D. This has limited the use of 3-D viewing tools. New data collection methods such as LiDAR create point clouds that can be processed to build 3-D facility databases (Figure 3.143).

3.8.3.1.3 Queries and Reporting

Early automated mapping systems utilized special file structures to handle XY coordinates and attributes. In the late 1970s, the emergence of general-purpose relational databases offered a new storage paradigm. Soon most GIS vendors offered databases as a way to manage the increasing volume of noncoordinate data. Over time, the ability to manage and manipulate geospatial data has become widespread in commercial databases (e.g., Oracle Spatial).

With this underlying database structure, it is, of course, very straightforward to perform queries and generate reports. The added geospatial element enables spatial filters that add to the power of data retrieval:

- A pure data query—"list all of my transformers"—has usually too much information.
- Adding a filter by a landbase polygon—"list all of my transformers is this district"—is more useful but may still be too much information for most tasks.
- Adding a filter by proximity from a linear landbase feature—"list all of my transformers within 1000 ft of this road"—starts to focus an important subset of the data.
- The real power of spatial data may come from a filter based on connectivity—"list all of my transformers between these two points on this circuit."
- More complex queries—"list the customers served by the transformers selected earlier"—can be designed to pinpoint data that are key for a certain task.

This query/report capability, combined with geovisualization, is often used to extend spatial data to settings where computers may not be appropriate. For example, work packets for a vegetation management crew can combine lists of tree trimming work with maps that illustrate the work in a geographic context. This capability may also be useful as a way to provide data outside the company, such as contractor crews that may be validated for access to live data.

3.8.3.2 Planning and Designing the Grid

GIS and CAD systems have a long history of supporting the plan/design/build processes in electric utilities. A number of commercial systems are available for these tasks. While the spatial aspect of

Smart Grid Technologies

laying out facilities is native to these applications, the details of structural and electrical analysis and even work management have to be considered for a design tool to be effective.

3.8.3.2.1 System Planning

Prior to detailed engineering design, utilities often have to perform long-range planning for service territory expansion or system improvements. This may involve projections of population growth used to predict future system needs or an analysis of environmental factors for a construction project.

Defining a transmission corridor is a classic example of this type of project. The need for a new line may be established based on current demand and on projections of future demand. Once it is determined that a new line is needed to connect a generation source with an area of demand, there may be an array of corridor choices that must be analyzed. This analysis will include factors such as terrain, environmental impact, current land use patterns, and cost. The selection process includes a bewildering mix of political and public interest actors—another case where geovisualization tools can have a major impact by communicating the spatial context.

While much of the data used in this process are spatial, they likely do not reside in the utility GIS. Land use and population data may come from local governments, while terrain, weather and wildlife patterns, and other geospatial technical data sets may be available from federal agencies. It is almost inevitable that multiple data sources, with data in multiple formats, will be needed. The tools used for utility system planning must be capable of handling this combination of data sources.

3.8.3.2.2 Grid Design

Detailed design of the electrical network is a category at the heart of the geospatial smart grid. These analysis and optimization systems, often including DLTs (design layout tools), are critical components for designing robust networks (Figure 3.144). When used well, they can also achieve significant cost savings.

Applications in this category have to handle the entire spectrum of the utility's facilities:

- Both transmission and distribution
- Overhead as well as underground

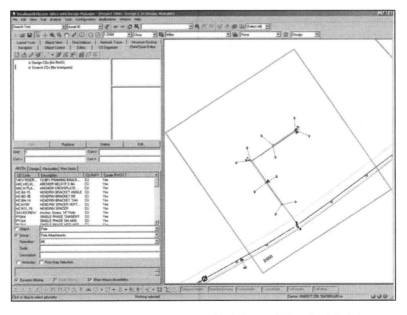

FIGURE 3.144 Grid design application. (© Copyright 2012 General Electric. All rights reserved.)

- Linear UG facilities (ducts, trenches, conduits)
- UG structure nodes (manholes, handholes, vaults, pads)
- Substation internals

Fundamental capabilities of these applications include tracing by phase and circuit, schematic layout creation, and the ability to handle multiple levels of detail (e.g., showing a switch as a single element and the related internal view).

A large component of the design system capability is optimizing conductor and transformer sizing. Flexibility is important. Results can be based on customer class data or spot load models. Tools have to consider load growth and check for allowable transformer overloading settings and potential voltage drop and flicker problems. Conductor sizing relies on both power factor and quality, considering both overloading and underloading.

Along with design of the core electrical network, these applications also may have tools for corridor management (right-of-way, vegetation, dam inundation), joint-use pole management, and streetlight layout.

3.8.3.2.3 Communications Network Design

One of the primary changes with smart grid is the addition of a communications network to the electric grid: a truly smart grid is as much about information flows as electron flows. This requires tools that enable efficient design of the communications network.

Communications design tools have to support an integrated view of the entire network. This includes both inside and outside plants, and both physical and logical networks (Figure 3.145).

The communications physical network model includes all of the ducts, cables (both underground and overhead), and support structures (street cabinets, manholes, splice closures, rack-mounted equipment) that compose the system. One of the challenges is the need to manage both extensive geographic areas and the details of buildings (including floor plans, rack locations, down to the communications port).

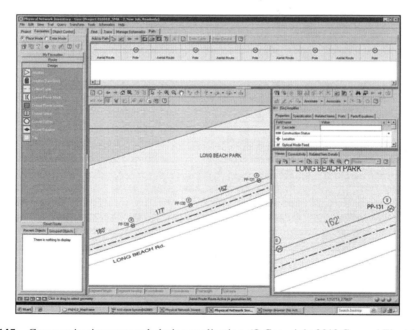

FIGURE 3.145 Communications network design application. (© Copyright 2012 General Electric. All rights reserved.)

Smart Grid Technologies

Communications design applications also have to manage the logical network (active network elements, customer circuits, and bearer circuits).

Utilities have been utilizing grid design software for many years. What is different now is the need to manage the rollout of large communications networks. And, clearly, the key is integration of the engineering design of both electric and communications networks so that together they help manage a smarter grid.

3.8.3.3 Operating and Maintaining the Grid

Once the smart gird is designed and built, the emphasis changes to operating and maintaining it. Geospatial technologies are a core component of operational processes and applications.

3.8.3.3.1 Network Analysis

As the complexity of the grid increases, power system analysis tools will play an even larger role. These tools are used to manage circuit configuration, direction of flow, voltage, and phasing.

A major part of this functionality is transformer load management. By aggregating data from summer and winter peak loads for each customer (gathered from CIS billing data) and adding information about the performance specifications of individual transformers, these analysis applications can identify overloaded or underloaded transformers (Figure 3.146).

3.8.3.3.2 Outage Restoration

The previous section described analysis tools that are used in the day-to-day operations of an electric utility. Another set of tools comes into play when things go wrong; the lights are out, impatient customers are waiting for answers, and the utility is facing significant monetary losses (both in lost revenue and penalties).

Outage management and work management systems are described in more detail elsewhere in this book. Here, we will just mention the key role of geospatial data in several parts of the outage process.

Much of the restoration process is driven by the utility's field crews. Here, it is vital to have a coordinated view of repairs between the OMS and work management systems. Communications

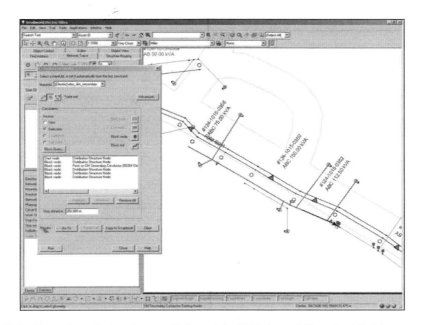

FIGURE 3.146 Network analysis software. (© Copyright 2012 General Electric. All rights reserved.)

between dispatchers and the field are very location-based; getting the right skills and equipment to the right place requires a detailed view of the network. As repairs are made, it is also important to record changes to the facility base and communicate those changes back to the GIS.

Responding to storms and other outages is a perfect example of the need for high performance in geospatial systems. It is truly a "high-stress GIS" situation. There are huge financial stakes in restoring critical infrastructure more quickly. Public perception, fueled by high-profile outages in the last decade, plays an increasingly important role. The GIS cannot be a roadblock, so it is crucial that it can manage large volumes of rapidly changing data.

3.8.3.4 Mobile Geospatial Technologies

As noted earlier, one of the more recent steps in the evolution of utility geospatial technology is the ability to move map and facility data out of the back office and to the field worker. This trend has been enabled by major advances in mobile computing and related technologies such as GPS and wireless communications.

All field applications, of course, have to link closely to back-office systems. As noted in a later section, managing the data flows between office and field is a difficult but necessary element of geospatial design.

3.8.3.4.1 Map Viewing

A fundamental part of field capability revolves around viewing geospatial data. It is giving the field user answers to many of the "where" questions described earlier.

Early field automation systems were aimed at replacing paper. There are both productivity and cost advantages in eliminating the paper map books that utilities had relied on for decades. It is not hard to beat the functionality of paper. Instead of dealing with fixed scales, users can easily zoom in and out, getting the level of detail they need for the task at hand. Symbology can change with scale so it is more easily read. And by grouping data into different layers, each with a display range tied to zoom levels, it is possible to reduce clutter and improve visibility.

There may even be useful view modes that take advantage of attributes linked to geometry, such as the ability to render circuits by color rather than a default mode of showing conductor color and line thickness by voltage or phase (Figure 3.147).

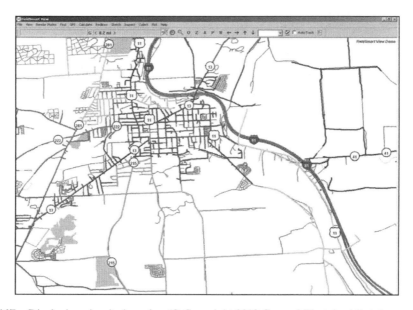

FIGURE 3.147 Displaying circuits by color. (© Copyright 2012 General Electric. All rights reserved.)

Smart Grid Technologies

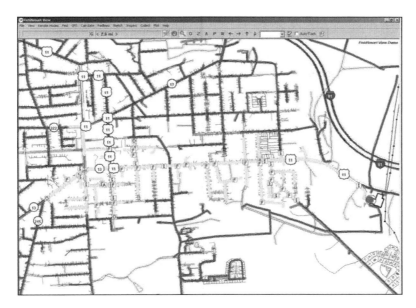

FIGURE 3.148 Circuit trace. (© Copyright 2012 General Electric. All rights reserved.)

One of the primary advantages of a digital map viewer is the ability to search. Finding a specific facility on a paper map can be very time consuming, even if some assets, such as line poles, used a numbering system linked to a grid based on map sheets. Searching on objects other than the grid facilities is important too. Landbase features (streets, intersections, points of interest) are useful in helping a crew navigate to work assignments—especially if they are in storm recovery mode and working in an unfamiliar area.

Searches can extend to external databases that can be linked to location. Customer data are a good example. Even if the customer (meter) coordinates are not part of the GIS data model, most utilities do link customer records to a transformer. This lets viewers zoom to a location that is in close proximity to the meter and to even show lists of other customers served by the same transformer.

Most mobile viewers today include some analysis functions like circuit tracing (Figure 3.148), which takes advantage of the connectivity data present in the GIS. Tracing an electric circuit is a huge productivity gain in the field when field troubleshooting. The ability to designate an underground route through a building complex or urban setting with the connection to protective devices is impossible on paper. Similarly, the field crew cannot gain a clear picture of affected areas from an out-of-service protective device using a paper map.

3.8.3.4.2 Workforce Management

It is easy to forget the people element of smart grid. After all, much of what we hear is about the totally automated, self-healing nature of the future electric network. It is described as almost an "untouched by human hands" system. Although a great goal, we know that this will not always be the case. Smart meters will not be very smart when they are lying in the rubble of a house destroyed by a tornado, and the intelligent network will fail if vital components are damaged by falling trees. The bottom line is that people will always be a key part of running a utility.

Even though workforce management software has traditionally been a separate category, managed by a different group in the utility and provided by a different set of vendors, it is included here because of the strong geospatial link. The essence of these systems is to get the right people to the right place at the right time, so the space/time aspects of geospatial technologies are an essential component.

Work management systems come into play in almost all aspects of utility field work. These systems may, for example, schedule distribution designer field visits with customers and then manage the resulting construction work. After the system is built out, work management systems play a vital role in managing both the daily service work of the utility and the stressful periods of outages. Workforce systems may also play a role in special projects such as AMI deployment.

These applications typically have both back-office and field components. The back-office system manages the overall field workforce, tracking crews and equipment. As work is needed, the system creates service requests. It then assigns the task to a specific crew based on a complex mix of factors including crew/vehicle capability, current locations, and expected task completion times. This drives the scheduling and dispatch of a given crew to each work task.

As the work is completed, the system tracks the progress, looking at current status and estimated time for completion. It may also manage parts inventory based on the materials used in each job.

The field component of workforce management takes a different perspective. Instead of managing multiple crews, the focus is on the assigned work of a single crew. Communicating with the back-office system is important to update assignments and job status. Although routing from one job to another is often handled in the office system, the ability to update routes in the field is useful since traffic conditions may affect the original path.

For many years, commercial work management systems tended to focus on a specific type of work. They were designed to handle either short-cycle (service) crews or long-cycle (construction) work. Today, as the utility workforce has evolved, much of that distinction has disappeared, and these systems use a "work is work" philosophy and take a unified approach to the entire mobile workforce (both in-house and contractor crews).

A relatively new capability in workforce applications is predictive analytics, forecasting how the utility's future workloads are most likely to be distributed over time and over the geography of its service area. This functionality utilizes historic trends and projected needs to help balance future demand (the amount and distribution of required work) and supply (crews, vehicles, materials).

3.8.3.4.3 Inspections

All utilities are required to periodically inspect facilities. Some inspections are self-imposed, and others are mandated by regulators. These inspections may focus on a specific facility type such as transformers, conditions that affect facilities like vegetation growth, or they may look at all facilities in a given area (e.g., a circuit or a substation).

Special purpose categories include pole audits (looking at either the utility-owned poles themselves or updating foreign attachments), vegetation surveys, or storm damage assessment work (Figure 3.149).

The back-office part of an inspection system schedules the work and manages historical data for the relevant facilities. The field application provides form display, validations, and editing capabilities, along with markup or sketching functions and attachments such as photographs. An additional advantage of digital inspections is the incorporation of GPS. GPS can be used with inspections to improve the "where" of mapped facilities as well as validate that the inspector was actually at the correct site when performing the inspection.

3.8.3.4.4 Routing and Navigation

As consumer navigation systems have proliferated, it is increasingly common to see navigation capabilities as part of a field automation suite (Figure 3.150). This can be an important capability even if a back-office system generates a preferred route as part of a work order since traffic or other real-world conditions might require changes in the original route.

The underlying technology is familiar: after the user defines a destination, the system uses GPS to determine current location, calculates point-to-point routing over an intelligent street network, and highlights the route with turn-by-turn directions. A utility setting adds some special requirements. For example, some bridges and underpasses might have restrictions that would prevent certain utility trucks from using them; the generated route has to take these restrictions into account. Similarly,

Smart Grid Technologies 319

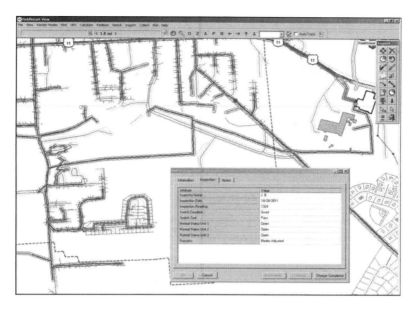

FIGURE 3.149 Field inspection application. (© Copyright 2012 General Electric. All rights reserved.)

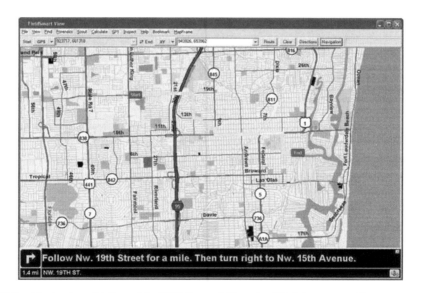

FIGURE 3.150 Navigation. (© Copyright 2012 General Electric. All rights reserved.)

some states constrain the use of driving directions on computer screens in certain vehicles, so the routing app has to communicate with verbal driving directions.

3.8.3.4.5 Data Collection and Update

Mobile applications that enable field data collection take advantage of the fact that utility field crews are in a good position to update the geospatial database. They are, after all, knowledgeable about the facilities and are often placed in close proximity to the objects in the field.

These capabilities can be used throughout the facility lifecycle. In some cases, they focus on collecting data that are not in the GIS such as dangerous trees or dig-in damage. They are also used as part of the construction process, capturing the differences between as-designed and as-built facilities. One of the more frequent uses of this capability lies in the ad hoc data updates that arise from a field crew seeing a discrepancy between what they see on the screen and what they observe in the real world.

There are several flavors of mobile data collection tools. Some are simple drawing tools, letting someone in the field draw on top of the map and submit the sketch to a mapping group for interpretation. Other redlining tools provide the ability to add notes and more complex drawing capabilities (text, symbols, and annotation) over the existing map. The most complex applications link to compatible unit databases and include data validation tools to help ensure that data collected in the field is usable in related systems.

Even if mobile applications support the field update of facility data, the back-office components must be able to capitalize by providing a rapid and secure process for inserting changes into the corporate GIS. The slow update process has been a source of frustration for many utilities. There are, of course, valid reasons for insuring the validity of data before the "system of truth" is changed. In many cases, the legacy of paper mapping systems remains a roadblock. Given today's technology and the example of "crowdsourcing" tools (see next section), there is little excuse for an update cycle that is measured in weeks or months.

3.8.3.5 Engaging the Consumer

The applications discussed earlier are all aimed at the employees and contractors of the utility—the people who build and maintain the grid. What about the consumer, the end user of smart grid?

In most cases, there is no legitimate need for the consumer to access the detailed facility data in utility geospatial systems. Even if there is interest, security is a real concern. There are cases, however, where the customer would find it useful to have a spatial view of the grid. In a major outage, for example, many utilities post a web page showing the extent of current outages. These data should, of course, reflect the more detailed view of current status that the utility is using internally.

The emergence of EVs may yield other examples. The driver of an EV, dealing with range limitations, has a vital need for updated locations of charging stations and perhaps even a count of the available outlets. If these data are present in the utility GIS, it should be in sync with what the consumer is seeing. These data need to be made available to the driver in the vehicle through an onboard system.

Other consumer-facing applications will undoubtedly emerge as we move into the smart grid era. Geospatial data will often provide a framework for these applications, serving as a common view of network assets and status.

3.8.4 Smart Grid Impact on Geospatial Technology

In the previous section, we looked at how geospatial technology will help support the growth and management of smart grid. There is another interesting angle to that relationship: How will the emergence of the smart grid impact geospatial?

Today, GIS has been implemented in most utilities and is considered a successful example of technology implementation. Many in the industry, however, believe that the new grid will challenge current systems. After all, most current technology was designed to support the business processes that were rooted in the old grid—processes that, in some cases, may date to a century ago.

How does the shift to a smarter grid impact the geospatial systems now being used by utilities? Perhaps the most obvious change is that everything scales up—there are more data, tied together in more complex ways, and the need for speed and accuracy is dramatically increased. These factors will challenge almost every aspect of geospatial system design, forcing major changes in how spatial data are managed and distributed. The following sections describe potential problem areas and suggest how geospatial system design may evolve to handle these issues.

3.8.4.1 Coping with Scale

One of the clear differences with the geospatial smart grid is the change in scale—everything gets bigger. Even now, a spatial database for a large utility that contains detailed landbase and complex facility data for its entire service territory commonly exceeds 100 GB. That does not count related data from even larger systems (customer information, asset tracking).

Many observers estimate that there will be at least a thousand times as much data with smart grid deployment. Not all of that data, of course, will be managed by the geospatial systems, but we can anticipate a significant rise in volume.

For a large utility, the data volume is driven by the need for a lot of detail. This volume is multiplied by the large area that has to be covered, resulting in tremendous amounts of data. The GIS data models typically cover both the transmission system (power plant to substation) and the distribution network (substation to customer). In most cases, the model (or at least the populated database) does not extend to the actual customer premise (the meter) but ends at a transformer that may handle dozens of customers. Adding intelligence at the customer site will require handling data about the meter and the characteristics of the customer. (Are there solar panels on the roof? Is there an EV?) Not only are these data tied to a premise, but much of them are associated to the consumer (e.g., details on EVs and smart appliances). The historical aspect of this consumer data, as well as the need to keep it updated as the consumer moves or replaces items, all add to the data volume.

The smart grid will require more objects and more attributes for those objects that are in the facility "layer" of the utility GIS. It may also demand new sources of data (e.g., weather) and new analysis tools to dissect complex relationships among objects.

The key question here is whether current GIS architectures are optimal for these larger volumes. Data structures that have worked reasonably well in the current environment may not meet performance expectations as data needs scale rapidly.

3.8.4.2 Moving to Realtime

Even though a typical utility GIS now has thousands of changes every day, the system can be viewed as relatively static. New or changed facilities are reported through a review process and then validated by GIS staff before being added to the database.

GIS is often seen as a spatial data warehouse, a "system of truth" that is used as a trusted reference for the grid. Currently, the practice is to do periodic extracts for applications like outage management or mobile data; GIS is seen as a data source that feeds other more time-critical applications. These operational systems are important for maintaining the grid and responding to power outages, and there are critical safety considerations when crews are doing work on the lines.

When the lights are out after a storm, data can start to change very quickly. For example, as the system is reconfigured to reroute power and resolve outages, the status of switches may change. The load on devices and conductors fluctuates, changing their working capacities. Some of the information about the grid is less likely to be valid if there is damage. Repairing the system takes precedence over recording the details of how those repairs were done. And public demand to get the lights back on adds to the time pressure; making this a true "high-stress GIS" situation.

The current slow change GIS cycle may not be adequate for a grid that is highly dynamic. In today's grid, even when object attributes and electrical flows change, location is almost always static. With EVs, that is no longer true; an EV is an active part of the network that can change location during the day. With solar generation rates changing as clouds pass across the sky and wind turbine farms coming on- and off-line based on wind speed, energy storage that absorbs system capacity can also move throughout the day and night. Managing this data will require the ability to handle moving objects in the spatial database.

Unless the GIS can support these near-real-time demands, it will not be able to retain its position as the single reference for the grid.

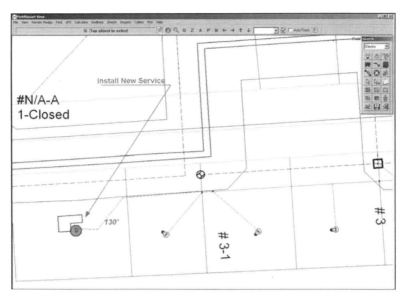

FIGURE 3.151 Geocollaboration. (© Copyright 2012 General Electric. All rights reserved.)

3.8.4.3 Supporting Distributed Users

One of the paradoxes of the smart grid is that even though the system is touted as being self-healing, its deployment puts even more burden on the people who maintain it. We not only need more information but need it faster; we need to get it to the people who can act on the information. In a utility, that means the field crews—often spread out over a very large area and often (in situations like storm recovery) without consistent communications capability.

Data communications is a crucial design parameter for mobile applications. Utility crews have to be able to work anywhere within the service area, which means that there are almost always limited coverage areas for any large utility. Systems also have to allow for "no comm" situations, where storm damage to the grid may be accompanied by communications outages (or, at best, limited bandwidth). The bottom line is that field applications must be designed to support base functionality without any wireless connectivity. Critical applications have to work whenever and wherever they are needed.

There are many cases in grid maintenance where multiple crews, along with supervisors in an office, are working together to handle a situation like a large outage. These users would benefit from a higher degree of interactivity with the back-office GIS (data input or sketches) to communicate changes; this can be seen as a need for geocollaboration (Figure 3.151) on a large scale.

3.8.4.4 Usability

The changes noted earlier—more data, changing faster, with distributed users—will lead to another challenge: how do users interact with the smart grid geospatial system? In an era with a more complex grid, user categories may be less distinct. Instead of a separate group of GIS professionals who maintain and control the system, we may find more of an operational bias—electrical engineers viewing the system as a platform for applications.

An additional degree of difficulty is present in field settings. It is a challenging work environment; the display screen is typically smaller than is common in the office, and viewing conditions are rarely ideal. In events like storms, there is a great deal of pressure on the user (and consequently the system) to work quickly.

Smart Grid Technologies

A design goal is to hide the complexity of GIS and CAD systems. The details of the application user interface should disappear. Ideally, users view the field application as a tool: something that helps them do their work and has an easily understood function.

There are several key questions that must be addressed in usability. How can we help users pick out key information in a system with more potential for clutter? Is it possible to define key data based on context (location, time of day, current user activity)? What design strategies are needed to support high performance?

3.8.4.5 Visualization

One of the design elements that relates to usability is visualization. One key advantage of a geospatial system is its ability to render maps in many different ways depending on the audience and the intent. A GIS screen can act, if needed, like a wall map, giving operations staff in the control room a quick overview of system conditions, such as power flows and bus voltage levels by color combinations over a large service territory. It can, alternatively, display a detailed schematic of a transformer vault to help a crew safely make repairs.

It is important to remember that, in field settings, viewing maps is made more difficult by system constraints (typically a smaller screen size) and environmental conditions (glare from sunlight).

The increasing complexity of the grid will demand new forms of visualization. We may need to extend the GIS toolkit to take advantage of advances in other fields of computer graphics like 3-D entertainment systems.

One of the key aspects of visualization is helping the user quickly focus on what is important. How can we help users pick out critical information in a system that has more data and consequently more potential for clutter? Symbology design and color will play important roles in this area. Another key to managing data display is the use of multiple layers. Given the vast amount of data that relate to the grid (especially if other networks, such as gas, water, or communications, are also present), the user can be overwhelmed by visual clutter. We can have the "fog of data" like the confusion of the "fog of war." By selecting groups of data elements that are job-specific and giving the user easy ways to select what they want to see based on the task they are performing, we can minimize information overload.

This problem can be mitigated with intelligent filtering of the data:

- Functional layers (e.g., landbase vs. facilities)
- Setting visibility (zoom) levels for each object type
- Symbology changes based on display scale
- Highlighting key objects

These techniques can be used together to create thematic views, where the goal is to present only that data related to the user's current task, using visualization techniques to highlight critical objects. For example, a field user may need an overview of an electrical circuit, where the conductor would be rendered with a thicker line and devices like switches would be represented with larger symbols.

Other aspects of data presentation drive from the environmental factors noted earlier. For example, using a light color line style to denote a high-voltage line may work fine in a controlled office environment but is likely to cause problems on field devices.

3.8.4.6 Standards

Given the vision of the smart grid as a vast interconnected network, having standards for all of the components is essential. Interoperability of software components moves from a goal to a requirement.

Several organizations, notably NIST (National Institute of Standards and Technology) in the United States, have led the way with grid standards. Although there has been a considerable amount of work with GIS standards by OGC (the Open Geospatial Consortium) and others, the perception is that geospatial technology is less advanced in this area than some of the "engineering" disciplines. There does seem to be momentum around ideas like Common Information Model (CIM), and discussion of GIS/grid standards appears to be growing. OGC has been involved in Smart Grid Standards Roadmap Workshops organized by NIST and the Electric Power Research Institute (EPRI).

As we have seen in earlier sections, the need for spatial data is not confined to GIS. Many applications are driven by spatial data. It is vital, then, that these systems use common structures for managing and visualizing geospatial data.

3.8.4.7 Data Quality

What are the key characteristics of usable data for the geospatial grid?

- It must be complete, covering all the relevant data types over the right geographic area. For example, the extents of the data must include all relevant parts of the electric grid that could be included in a network trace.
- Objects must be positioned accurately. This is often a challenge for utility data derived from old paper maps, which typically were not surveyed and were mapped in the pre-GPS era. Positional accuracy problems become more obvious when facility data from paper maps are overlaid with other data sources that were derived from imagery, GPS, or other more accurate methods.
- Objects have to be classified accurately and have to include attributes that support the utility's business processes (e.g., size of transformers, phases of electrical lines).
- The data must be reasonably current. The update cycle varies according to use. In some cases (e.g., switching status of operating devices on the grid), outdated or inaccurate data pose safety hazards.
- For data types that would be part of a network trace, topological relationships must be accurately captured. Connectivity may be explicit (i.e., driven by a table) or geometric (based on drawn location and proximity).

Information that passes these tests can form the basis of useful smart grid applications. Smart grid adds an element of uncertainty to current GIS practice. Where does the data model stop? Current distribution network traces, for example, cover the range from substation to transformer. As intelligence is added at the edge of the grid, other objects may come into play. Distributed generation sites and microgrids add complexity to power flows. On the customer side, do intelligent appliances and EVs need to be included?

In general, we can assume that there will be higher requirements for data quality to operate the smart grid than is true with the current grid. Control is more data-driven and that requires complete and accurate data.

An ESRI survey of electric utilities published in August 2010 found that most companies were not ready for the increased demands of smart grid data. Only 15% of respondents reported high confidence in their GIS data accuracy. A common issue highlighted in the ESRI report is the long lead time to update the GIS with data from the field. Lag times of several months are not uncommon. Part of the problem is that update processes in many utilities are still based on the age of paper maps, where a specific group in the utility has the sole responsibility for final changes to the database. This slow process creates problems today and could be a major issue in the faster changing world of smart grid. As suggested in a later section, concepts like VGI (volunteered geographic information [VGI]) may offer a solution to this slow update cycle.

3.8.4.8 More Open: Sensors and Other Data Sources

IEDs (intelligent electronic devices), RFID (radio-frequency identification) tags, and other types of sensors are being used more often in utilities. As the grid becomes smarter, more of its components will be digitally accessible and identifiable. The grid will be, in some ways, a perfect example of the "Internet of things."

Although most of these devices are deployed to monitor and communicate specific measurements, their location—in terms of XY coordinates and relative to the topology of the grid—is an important element. In addition to these inputs that are controlled by the utility, there will be a need to integrate external data sources. Current and forecast weather data, for example, will be an important tool when trying to predict power flows from geographically dispersed solar and wind sites.

Research is needed to determine whether available GIS architectures can handle the flow of data from sensors and integrate with other data sources, such as weather feeds, into grid management applications.

3.8.4.9 More Closed: Security

Utility GIS has always been a very closed system. Part of this, of course, is based on security concerns, given the need to keep the grid (clearly an example of critical infrastructure) safe from malicious actions. This issue is an even more visible concern with smart grid since automating control of power flows means that physical security is not sufficient.

Another reason for the closed nature of utility systems has been a concern for data integrity. In most utilities today, GIS management has restricted change access to the database to trained personnel in the belief that it is necessary to maintain accuracy. A field user reports a change (which could result from an actual construction change or an observation that the existing map is incorrect) through a structured (and often time-consuming) process. A trained GIS professional validates the update and then makes the actual database change. This multistage process is frustrating to operations staff, since field employees are usually in a better position to compare the map with the real-world features that they are seeing.

How can these security concerns be addressed? There is considerable activity around security standards, made more challenging by the rapid pace of development in hardware and software technology. Because the stakes are so high, we can assume that security concerns will be a major filter on adoption of new approaches (such as cloud computing and crowdsourcing, covered in a later section).

NERC (North American Electric Reliability Corporation) in the United States has implemented a CIP (Critical Infrastructure Protection) program to establish a set of standards for all facets of utility security, including cyber assets. The IEEE is also very active in defining standards for grid data and applications.

3.8.4.10 More Closed: Privacy

The previous section looked at security from the perspective of the utility. It also works in the other direction: concerns about individual consumer privacy.

Given recent high-profile cases of hacking and identity theft, it is no surprise that many people have issues with any technology that collects and manages data about individuals. Smart grid, with its emphasis on data collection from tracking consumer usage habits to smart meters and sensors, fosters a growing concern about how to protect privacy rights.

The geospatial nature of much of the data adds another complication to the issue. Even if great care is taken to protect the most obvious aspects of individual identity (name, address, phone numbers), location information can be used as a link to find these data.

Legal challenges around the use of map-based tools (e.g., Google Street View) have resulted in the term "geolocation privacy." Safeguarding individual data becomes more difficult as data sources proliferate. Any utility applications that utilize geospatial data have to take these privacy concerns into account.

3.8.5 FUTURE DIRECTIONS

Geospatial technology—even when the technology was ink on paper—has always been an essential part of the electric grid. Tools have improved dramatically over the last decade, and GIS technology has become a business-critical force in helping utilities cope with a changing business environment. Utility geospatial tools will continue to evolve. They have moved from mapping to designing to managing in a relatively short period of time. As the GIS has grown in capability, we have seen the use of spatial data and spatial analysis tools in other applications. And now, as we enter a new era, we can see more clearly that the smart grid is about data—and much of that data are spatial. The need for these geospatial tools will continue to grow. At the same time, we see the need for new and better tools. As the electric network undergoes major changes, it seems clear that the technology being used today will not be adequate to support the increased size and complexity of a smarter, more connected grid. As the grid becomes smarter, systems that help manage the grid will have to be smarter too. We can look forward to the continuing evolution of geospatial technology to meet these needs.

The previous sections, while describing the key role of geospatial technology in running the grid, have also pointed out several gaps—areas where there are needs for future development to support a smarter grid. Here, we will look at several areas of development that may prove useful.

As the application of geospatial tools to electrical networks becomes more widespread, we are seeing an interesting convergence of technologies. There are four major threads:

- Vendor-based GIS systems that have been the core of the growing geospatial industry for four decades
- Ancillary technologies, such as GPS receivers and navigation systems, which focus on the value of location
- Consumer-oriented mapping tools such as Google Maps, offering toolsets for displaying data on map backgrounds
- Open-source tools and data (OSM, Ushahidi, the Map Kibera project), initially used for data collection but now extending into visualization and analysis

3.8.5.1 Architecture

Looking at the geospatial grid automation picture as a whole, what is likely to be the predominant future architecture? Will this thing called a GIS continue to be the central repository for spatial data, feeding other applications as needed? Or will GIS disappear as other applications add the ability to store and manipulate spatial data?

It is likely that we will see something in between. With the rise of ubiquitous spatial capabilities, it seems clear that the preeminent role of corporate GIS will decline. Some planning functions, like grid design, are likely to remain within the domain of GIS systems and GIS vendors. Other operational tools, with the need for more real-time capability, will store and manipulate spatial data internally. In effect, the GIS will be embedded in multiple places.

The key is data integrity. It is vital that there is a single accurate and consistent view of the grid across all applications, and GIS still seems the most appropriate place to manage that "truth" for the utility.

3.8.5.2 Cloud

Cloud computing is one of the hottest areas of technology today. The appeal, in many ways, is obvious: forget the details of managing hardware, storage, and software and view applications as a service.

Does the cloud have a role in utility geospatial? (There is some irony in the fact that the cloud computing concept is often defined as a utility and is described by comparing it to the electrical grid.)

Security concerns are often cited as a barrier. At this point, it is hard to envision a scenario where utilities turn over the management of critical infrastructure and operations data to a third-party

Smart Grid Technologies

storage provider. Some utilities are further constrained by regulations that govern the physical location of data storage. (Private clouds, utilizing the same concepts but maintaining physical custody of vital data, are more likely.) The tools to support the design and creation of the GIS data, however, may reside on a public cloud to reduce utility costs and allow the company to better leverage shared and contractor labor.

As we have seen with the grid, however, data needs extend far beyond the facility objects that represent the utility's assets. Other geographic data from other sources are needed for planning and design. Renewable energy is subject to weather variations, so real-time weather data are needed for grid operations. These external spatial datasets are good candidates for the cloud approach.

3.8.5.3 Place for Neo-Geo

In the last few years, we have seen rapid progress in the "neo-geo" arena—companies and individuals using consumer tools such as Google maps (or even open-source spatial toolkits) to solve significant real-world problems.

A great example is crowdsourcing or VGI. Fueled by grassroots efforts to help disaster recovery efforts in Haiti and elsewhere, VGI has proven to be a valuable tool. By supporting a larger group of users, it has been far more agile and more responsive than many of the traditional GIS efforts.

In some ways, communications and connectedness are the essence of a smarter grid. Can neo-geo play a part? There are security concerns and other barriers, but it does appear that these tools can help apply "people" intelligence to some utility processes, like collecting damage information after a storm.

3.9 ASSET MANAGEMENT

Catherine Dalton, Soorya Kuloor, Tim Taylor, and Steve Turner

Utility executives have turned to asset management as an organizational model that creates operational and financial success by reducing the dependence on capital spending. Federal and state regulatory agencies and utilities themselves have raised the expectations for power system reliability. Asset management is a common approach or tool that ties these objectives together into an actionable methodology. An effective asset management program can help utilities to maximize the rate of return per O&M and capital dollar spent, evolve to a competitive culture, invest in training, base decisions on sound business principles, and learn the impact of expenditures on quality of service.

Asset management is managed maintenance of generation, transmission, and distribution assets by means of acquiring data from these assets to execute actionable intelligence on their behalf.

One of the main drivers for smart grid is the possibility to optimize the reliability of the distribution system, which is being pushed strongly by electric utilities as well as their regulatory bodies. Therefore, asset management is no longer business as usual. Rather, new developments in ways to approach asset management are driving smart grid activities. At the same time, smart grid activities are driving new developments in approaches to asset management.

With the recent push in smart grid, utilities have deployed an increased number of intelligent electronic devices (IEDs) for protection, monitoring, control, and metering applications. The functionality built within these IEDs allows for very robust asset management tools to be implemented on the electrical distribution system that enable real-time asset management or managed maintenance of existing and new electrical equipment. The addition of intelligence optimizes the delivery of electricity by allowing utilities to operate electrical systems at maximum capacity at all times.

Asset management for electric utilities can be considered the process of

- Optimizing system performance, profitability, and business growth
- Balancing stakeholder interests
- Positioning for long-term viability

- Scheduling replacement of capital assets based on specific criteria
- Scheduling the addition or replacement of fixed assets required to maintain current or anticipated levels of service
- Enabling reliability-centered maintenance for substations
- Inspection and performance monitoring of assets
- Tracking of asset data
- Prioritizing of investment decisions

Successful asset management programs require a balanced perspective that takes into account not only reliability but also safety, financial, and regulatory perspectives. Managing and balancing these critical perspectives is the key to future success in today's environment. The electric utility and the consumer will both benefit from using asset management philosophies. Benefits to successful asset management include gaining ability to manage infrastructure and understand necessary resource requirements, enhancing customer satisfaction, reducing costs to improve return to investors, and improving reliability data and reporting to regulators.

3.9.1 Drivers

3.9.1.1 Safety

The safety of the public and the electric utility employee are nonnegotiable. There is zero tolerance for human mistakes made with respect to daily operations and maintenance of the electric utility system. Safety is among the highest priorities of electric utilities. Given that fact, electric utilities proactively invest in training that includes safety, equipment, installation, operations, commissioning, testing, settings, design, and many other areas of expertise. Utility managers encourage teamwork. Central organizations develop plans with their field operations groups. These plans encourage involvement at all levels of the organization. No one should be afraid to speak up if he or she feels there may be room for improvement in safety processes. Utility managers also empower field operations employees and make them accountable. Field personnel should be rewarded and encouraged for a job well done. They have clear accountability in an end-to-end process. Management should clearly delineate who is responsible for what process and/or procedure and train personnel accordingly. Management should be held accountable as well as field personnel when it comes to safety. Assets not only include power system equipment but also the employees themselves, and safety is a critical component in managing and safeguarding a utility's assets.

3.9.1.2 Reliability

Electric utilities' knowledge of the regulatory environment may be limited. The utility may be located in numerous jurisdictions with different regulatory requirements. The utility has to expend time and effort to become familiar with regulatory requirements in each of those jurisdictions. Mandated expenditures and penalties from regulatory agencies may force utilities to prescribe to methods about which they need to become more adequately educated, such as new IT systems and processes they may need to implement. Prescribed standards and continuing demands for improvements in reliability, or at least no degradation of reliability, place continuing pressure on electric utilities. The increased scrutiny on reporting methods and consistency require utilities to dedicate resources that create additional overhead costs. Loss of credibility in the eyes of the regulatory agencies creates even more pressure for the utilities to implement improvements in their electrical systems and supporting systems. Utilities must balance capital and O&M investments with reliability requirements and pursue rate cases to ensure adequate equitable arrangements in cost recovery.

There are some actions that utilities can take to help balance the reliability perspective. They can establish local reliability "owners" who understand the goals of reliability, participate in identifying and suggesting reliability improvements, and coordinate day-to-day reliability activities of a district or region. They can support IT systems that will enable tracking of assets and their associated parameters. Moreover, they can learn state mandates regarding reliability, safety, etc. They can maintain service based on performance information such as customer demand, revenue, and cost by segment and operating history. And finally, they can balance analyses with experience from field operations.

3.9.1.3 Financial

Electric utilities face several operating and maintenance challenges. There may be underutilized assets within the utilities' service area. "Smart" devices will assist utilities in identifying geographical areas or individual pieces of equipment for immediate action and ensure equipment is being utilized to its maximum capabilities. Utilities can assess the variable profitability of service territories, since an electric utility's profitability depends on allowed rate of return by a state regulatory agency. Acceptable rates of return vary from state to state. Utilities can also provide for investment in new customer or system improvement without adequate incremental cash flow. If adjustments are not permitted in an electric utility's rate base, the utility may not be able to afford to invest in system improvements. Utilities struggle with low allowed returns on investment (ROI). Depending on the state and its regulatory environment, some states allow higher ROI than others. Utilities want to maximize ROI in the states in which they are allowed to do so.

Asset management tools can also assist in work prioritization. With "smart" devices, utility employees will have up-to-date information reporting or predicting probable equipment failures and possible consequences. This information will allow utilities to prioritize their workload. Similarly, smart tools will allow utilities to accurately forecast operational and maintenance budgets, capital forecasts, and required resources. In addition, it is very challenging to predict future revenue with so many unknowns. Will costs be recovered in a rate base? Will load growth trends continue in certain areas? Asset management tools can assist in more accurate revenue forecasts. Most importantly, utilities will be able to foresee the consequences of lack of maintenance and make informed business decisions and develop effective contingency plans that are actionable because they will be based on real information and not only on assumptions.

3.9.1.4 Regulatory

There are numerous regulatory challenges faced by electric utilities, mainly with respect to electrical power distribution system reliability. There exists a complexity of multiple regulatory bodies, with different measuring and reporting criteria for electrical distribution reliability indices, such as SAIFI, SAIDI, CAIDI, MAIFI, and others. Even the same reporting criteria may vary in its formula from state to state. Also, obtaining accurate, complete, and timely information for reporting purposes is a challenge. Furthermore, managing distributed generation and renewable energy sources, momentary interruptions, and other operational challenges requires complex technical and regulatory insight. Some actions that utilities can take to help implement a more effective asset management program include being proactive with regulators in order to help regulators understand the challenges of regulation in a smarter grid environment.

3.9.2 Optimizing Asset Utilization

Electric utilities can optimize assets by means of measuring efficient operation and improved performance. These measurements are made possible with the use of smart grid technologies. There are numerous ways that electric utilities perform optimal asset utilization from the power plants all the way to the electrical consumers' homes. They include the following:

1. Electric utilities can operate the assets they already own for as long as possible. They can retire inefficient equipment and maximize energy throughput of existing assets. Electric utilities can spend capital funds to build new revenue-producing assets and invest maintenance funds to support the legacy system. Some older equipment may not accommodate

new technologies that can enable predictive maintenance. These assets will still need to be maintained, but manually. A communications infrastructure may make financial sense to implement for asset management considering the benefits of the increased amount of asset monitoring data available. Electric utilities should understand that the age of revenue-producing assets should be irrelevant, but the condition of the assets is key. They should understand how often equipment fails in service and understand how often a power plant or power delivery system is out of service due to unplanned events.

2. Utilities should have a solid standards strategy for work practices, design, construction, materials, and equipment specifications.
3. Utilities should have an effective supply chain that includes consolidated supplier arrangements.
4. Utilities should perform long-range system planning. System planners need to know the options available to optimize asset loading. Asset data are a knowledge base for system planners and engineers. These data can show different ways to optimize loads that would allow more efficient load management. Examples of these data include an energy supply plan, an infrastructure risk analysis (provides information on over- or underutilized assets and information for maintenance and replacement planning), and a delivery expansion plan (provides information for targeted economic development).
5. A human asset plan should be in place for workforce optimization. This plan would contain a workload assessment that would ensure that employees are fully utilized and maintain the proper work/life balance, while maintaining safety. This plan would also contain a skill gap assessment that would allow for augmenting system training with targeted training in order to address specific gaps in skill sets. Moreover, a training and development plan would be part of the human asset plan. It would provide appropriate tools and equipment to employees, as well as position employees to challenges, technologies, and common issues they may face. It would also identify how long after training it takes an employee to reach proficient productivity levels.
6. Maintenance procedure performance should be evaluated. Repeatable and measurable processes should be put into place that minimize and manage maintenance procedures. Tools, such as performance metrics, balanced scorecards, work activity analysis, process improvement tracking, and cost/performance models and correlations, can be used to further enhance the management of assets. Performance metrics can be tied to processes. Balanced scorecard data can be tracked, trended, and measured among various criteria such as financial, operational, safety, and employee metrics. Work activity analysis can reduce exposure to potential accidents by minimizing manual maintenance. Less manual maintenance could mean less exposure to potentially risky situations. Improvement in these processes can result by tracking specific data associated with specific processes and analyzing this data. Furthermore, by developing cost models and process performance models, correlations can be made that may have not been considered in the past due to the availability of more data from "smart" devices.
7. Utilities can set customer-based targets to measure customer satisfaction. Some targets can include measurements around transaction surveys, customer advocacy, perception of reliability, voice of the customer, and customer phone calls. Utilities can use transactional surveys, which periodically survey their customers to identify areas of improvement from the customer perspective. Utilities can act as customer advocates. For example, if a customer calls into a call center with an issue, utility call center employees should work as customer advocates and promote customer needs in the utility. Moreover, utilities can track the perception of reliability. They can ask customers about their power interruptions. For example, how often does a customer believe he or she is suffering inadequate electric service? They can listen and respond to the voice of the customer. The customer should be periodically queried via surveys or focus groups so that utility management can realistically assess customer needs and requirements. Utilities can improve customer satisfaction

through common approaches and processes and empower personnel with knowledge to work more effectively and confidently with a customer. Most importantly, utilities can quickly provide accurate and courteous customer service.

3.9.3 Asset Management Implementation

There are five steps for implementing an efficient asset management program.

The first step is to deploy focused and effective predictive maintenance. Identifying and prioritizing defective conditions that might lead to interruptions is the key to lower cost reliability. Intervening in the deterioration process with low-cost maintenance avoids the cost of replacement and continually reduces future dependence on construction spending. Preventative maintenance is the key to lower cost and is the foundation to an effective asset management culture.

The second step is to provide lower-cost options to the cost of equipment replacement. Providing the organization with alternative choices that maintain or improve reliability, but at a lower cost than replacing the asset, requires innovation, creativity, and organizational discipline. Some ways to provide lower-cost options include the following: learning the impact of type of expenditures on quality of service and developing plans to optimize; setting targets for per unit cost of work; tracking overhead costs and setting targets for reduction; allocating capital for replacements and maintenance based on territory, asset types, and regulatory environment. For example, a utility should look at its expenditures versus its improvement in reliability indices (such as SAIFI, CAIDI), for a particular state, or in other words, the cost/reliability relationship.

The third step is to assess the condition of assets using data from smart devices. Managers can be empowered with asset condition data as a way to create accountability for making better decisions and using lower-cost options. Without detailed asset condition information, the organizational reflex is to build something new or do nothing. In order to assess the condition of assets, the asset manager needs to (1) identify asset age, condition/health, and time to failure prediction; (2) monitor performance based on key measures that include safety, reliability, customer satisfaction, and ROI; (3) develop a maintenance strategy using consistent and cost-based maintenance practices; and (4) plan any required additions, plan to address underutilized assets, and plan for asset replacements and life extensions.

The fourth step is to implement a balanced scorecard evaluation. A balanced scorecard can be used to track the performance improvements based on the asset management approach. Some possible measures that can be used in each area include the following:

1. Financial performance: Performance measures can include added shareholder value and O&M per line mile.
2. Customer service: Performance measures can include percent of customers very satisfied, number of public utility commission (PUC) justified complaints, number of PUC violations, and CAIDI.
3. Operational performance: Performance measures can include SAIFI, percent of jobs completed within plus or minus percent of estimate, percent of meters read, and percent of calls answered within a certain time period.
4. Employee performance: Performance measures can include motor vehicle incident rate, average employee satisfaction survey, lost time incident rate, and unscheduled hours off per employee.

Finally, the fifth step is to have management commitment and controls in place. Changing a culture that evolved from decades of managing in a cost plus environment is a daunting challenge. Therefore, leadership support for key initiatives is imperative for successful implementation. A clear vision, continual communication, and actions that support the plan are all requirements for success. Controls should be in place to restrict the old activities that add little value to the organization.

Low-cost, efficient, and measurable reliability improvements are needed. Success takes ownership, passion, and commitment from senior management, and it requires accountability from everyone. Success takes leadership, and success takes time.

3.9.4 WHERE SMART GRID MEETS BUSINESS: THE ELECTRIC UTILITY PERSPECTIVE ON ASSET MANAGEMENT

Utilization of assets can be maximized by efficient operation of existing assets. Such an operation involves the following steps:

1. Identification of stressed assets
2. Improving the operation of these stressed assets by relieving the stress on these assets
3. Working around aging assets

In order for performance to be improved, an electric utility must ask itself numerous questions related to its operations. These questions should be asked in terms of capacity planning, investment planning, maintenance and replacement planning, design, regulatory strategy, customer service processes, and work and resource management.

3.9.4.1 Asset Condition Monitoring

The direct way to monitor asset condition is with the use of sensors to measure power flow and other conditions (such as oil temperature) of the asset. Such sensors may be monitored in real-time through a SCADA system or off-line by periodically collecting the measurements accumulated by the sensors. The information from these sensors is then analyzed to analyze asset utilization and stress.

3.9.4.1.1 Asset Loading

Asset loading is used to measure the stress on any given asset, such as a transformer or an underground cable. The percentage loading is expressed as

$$\text{Percentage Loading} = \frac{\text{MVA or Amps Flow}}{\text{MVA or Amps Rating}} \times 100$$

Assets are stressed when their percentage loading is high, which causes degradation in asset life expectancy when the assets are loaded above 100% (i.e., above their rated value). Identifying system loading and preventing overloading contribute significantly toward increasing the life of assets.

3.9.4.1.2 Improving Load Factors

Typically the total system load peaks in the evening (Figure 3.152). Transmission and distribution (T&D) systems are designed to handle this peak load demand. Effective management of this peak load has significant impact on the utilization of system assets. The system load factor defines the ratio between the average system load and the peak load. Load factor is defined as

$$L_f = \frac{L_{avg}}{L_{peak}}$$

where the average load L_{avg} is expressed as

$$L_{avg} = \frac{\text{Total kWh}}{\text{Number of hours}}$$

Smart Grid Technologies

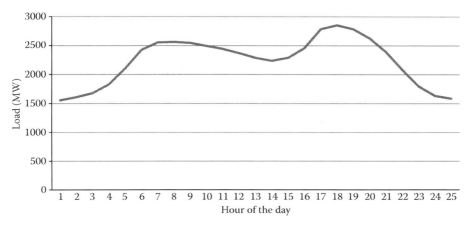

FIGURE 3.152 Typical daily electric load curve of an electric utility.

Load factors can be calculated for the entire system, or for individual assets, such as transformers. Load factors measure the average utilization of any asset or a group of assets. Assets that have low load factors are considered underutilized. Assets that have high load factor are better utilized assets.

Reducing peak loads on assets improves their load factor and hence their utilization. For example, consider a scenario where the peak load on an asset is reduced by 10% while keeping the total kWh loading the same:

$$L'_f = \frac{L_{avg}}{L'_{peak}} = \frac{L_{avg}}{0.9 L_{peak}} = 1.1111 L_f$$

As shown in the earlier equation, this results in load factor improvement of over 11%.

Present utility planning and operational practices are based mainly on peak load. Heavily loaded assets need to be replaced or upgraded. This upgrade or replacement cost can be avoided if the loading on the asset is reduced. By using smart grid technologies and the information obtained through systems such as AMI and advanced distribution automation, a much better understanding of loading and loading durations can be obtained. By reducing system and feeder peak load and flattening load profile, the load factor of the existing system assets can be significantly improved. Thus the system can be used to supply more demand.

Several approaches can be used to reduce asset loading:

1. Load shifting: Shifting load from higher loaded parts of the system to lower loaded parts of this system, for example, on distribution systems by switching operations to change the configuration of feeders. This can be performed on a seasonal, weekly, or daily basis depending on loading patterns.
2. Phase balancing: Balancing the loading on phases, more typically of distribution feeders. This has dual benefits:
   a. Provides better balancing of the phases, which can reduce I^2R losses on a feeder and address potential peak demand issues.
   b. Moves load from loaded phases to less loaded phases, hence better asset utilization.
   c. Dynamic rating: Maximum loading limits of power equipment such as transmission lines and transformers vary based on how long the loading levels are sustained. Short-term overloading of these devices is acceptable provided that these load levels do not occur frequently and enough time is provided for the power equipment to cool down. Similarly,

external temperature and weather conditions have an effect on how much an asset can be loaded. Dynamic rating of devices takes these into consideration during system operation.
  d. Demand management: Demand management can be used to effectively manage system peak loads and improve asset utilization. Smart technologies, such as the use of distributed energy sources, battery storage, or load management, can help manage peak demand loading at specific points of the power delivery system.

3.9.4.1.3 Lowering System Losses

Electrical losses in line sections and transformers occur due to the current flowing through them. For any given electrical element that is carrying an AC current, the active power electrical loss is expressed as

$$P_{loss} = I^2 R$$

where
I is the current flowing through the element in Amperes
R is the resistance of the element in Ohms
P_{loss} is expressed in Watts

As seen from the equation, the loss varies as the square of the current flowing through the element. Therefore, as current increases, the amount of losses also increases. The electrical loss is wasted as heat in the element. When electrical assets are overloaded, their losses significantly increase, thus causing heat-related damage. Managing the flow of current through an asset improves the health of the asset. Reducing losses also reduces waste and has direct financial benefits.

Reducing system losses can have significant revenue savings for a utility. For example, assume a utility with a peak load of 2500 MW, a load factor of 52%, and an average generation cost of 6 c/kWh. Assume an average loss for distribution systems around 5%. If this loss is reduced by around 10%–4.5% of the total system load, the annual saving for the utility is calculated as shown:

$$\text{Annual cost of power} = 2,500 \times 1,000 \times 52\% \times 8,760 \times 6/100 = \$683,280,000$$

$$\text{Annual cost of losses} = 683,280,000 \times 5\% = \$34,164,000$$

Savings due to reducing losses by a factor of 10% = $3,416,400

Reducing system losses has direct impact on asset loading. Reduced losses reduce the power flow on the assets. In some cases, the reduction of losses may avoid or delay the need for reconductoring lines and feeders to increase the available power delivery capacity.

3.9.4.1.4 Reducing Outage Frequency and Duration

Improving outages has the following direct benefits to asset management:

- Improved system reliability (SAIDI, SAIFI, CAIDI, and other reliability indices)
- Better utilization of assets
- More loads served and hence improved revenue

Typical approaches used for improving outages include the following:

- Fault detection, isolation, and service restoration (FDIR): Installing automated switches and control devices in the system to automatically restore as much of the system as quickly as possible. (Covered in more detail in this book.)
- Vegetation management: Managing vegetation around overhead lines will minimize the frequency of related faults.

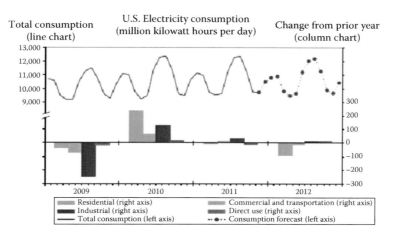

FIGURE 2.7 U.S. outlook for electricity consumption in 2011–2012. (From U.S. Energy Information Administration (EIA), Short-term Energy Outlook, November 2011.)

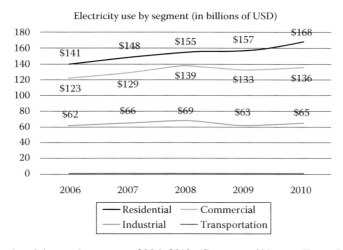

FIGURE 2.8 U.S. electricity use by segment 2006–2010. (Courtesy of Newton-Evans Research—based on DOE EIA source information.)

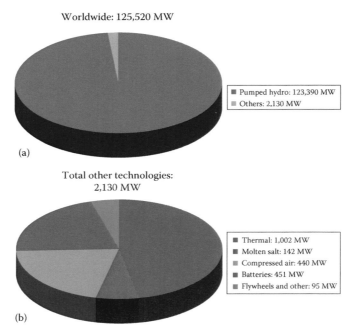

FIGURE 3.7 Installed grid-connected energy storage technologies worldwide as of April of 2010. (From Current Energy Storage Project Examples, California Energy Storage Alliance (CESA), http://www.storagealliance.org/presentations/CESA_OIR_Storage_Project_Examples.pdf)

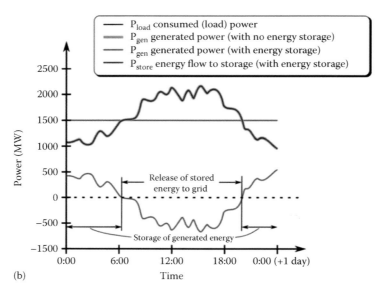

FIGURE 3.9 Conceptual description of grid energy storage. (b) Energy storage and release cycles. (From Wikipedia, Grid Energy Storage, http://en.wikipedia.org/wiki/Grid_energy_storage)

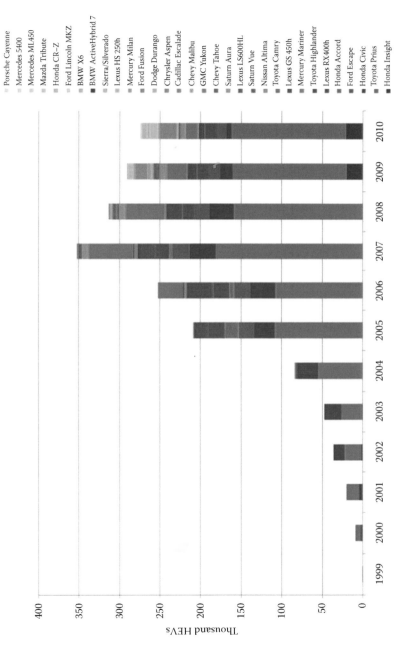

FIGURE 3.12 U.S. HEV sales from 1999 to 2010. (From U.S. DOE Alternative Fuel Vehicles (AFVs) and Hybrid Electric Vehicles (HEVs) http://www.afdc.energy.gov/afdc/data/vehicles.html#afv_hev)

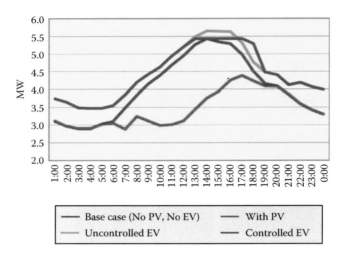

FIGURE 3.21 Example of feeder load under PV and EV penetration scenarios. (From Agüero, J.R., *IEEE Power Energy Mag.*, September/October 2011, 82–93.)

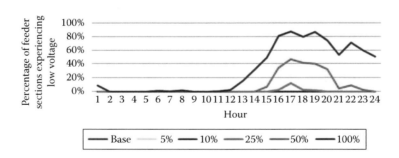

FIGURE 3.22 Example of percentage of feeder sections experiencing low voltage for various PEV penetration levels. (From Agüero, J.R. and Dow, L., Impact studies of electric vehicles, Quanta Technology, Raleigh, NC, 2011.)

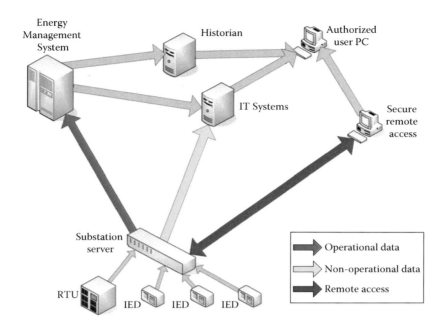

FIGURE 3.27 Substation data flow.

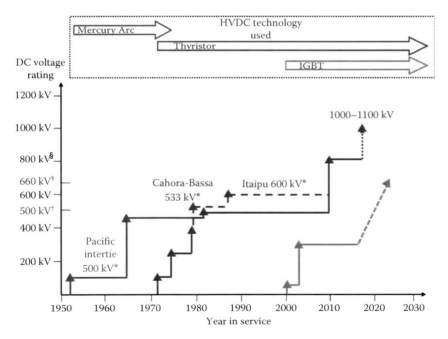

FIGURE 3.66 Evolution of HVDC voltage rating and technology. *Multiple bridges per pole; †500 kV becomes *de facto* standard for single 12-pulse bridge per pole; ‡660 kV used as "standard" in China; §800 kV used as "standard" in China and India. (© Copyright 2012 Siemens. All rights reserved.)

FIGURE 3.81 Swissgrid web page showing current European PDCs links (January 2012). (© Copyright 2012 Swissgrid AG. All rights reserved.)

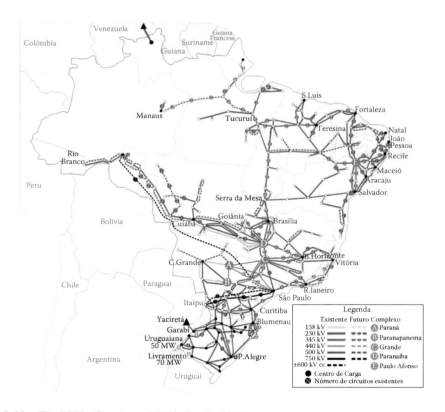

FIGURE 3.82 The BIPS. (Courtesy of ONS, Brazil, 2010.)

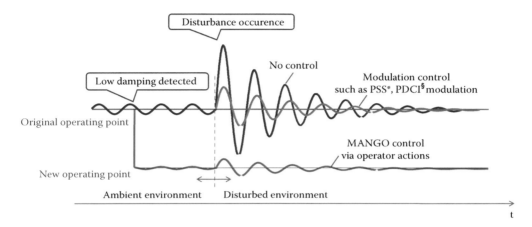

FIGURE 3.87 MANGO versus modulation stability control. (* Power system stabilization; § Pacific DC Intertie damping).

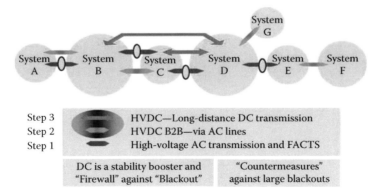

FIGURE 3.89 Hybrid system interconnections—"Super Grid" with HVDC and FACTS. (© Copyright 2012 Siemens Energy, Inc. All rights reserved.)

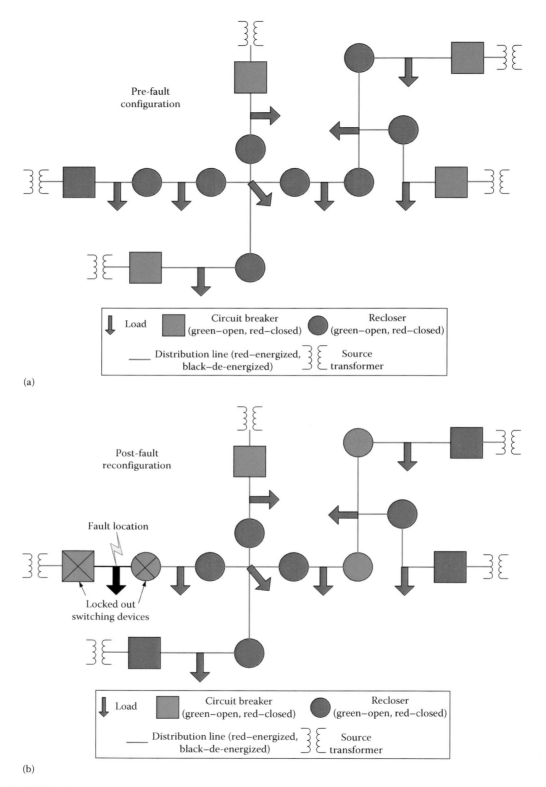

FIGURE 3.122 Example of multi-backfeed restoration: (a) before fault and (b) after fault. (© Copyright 2012 ABB. All rights reserved.)

FIGURE 3.124 OMS system connectivity model in a large metro area. (© Copyright 2012 ABB. All rights reserved.)

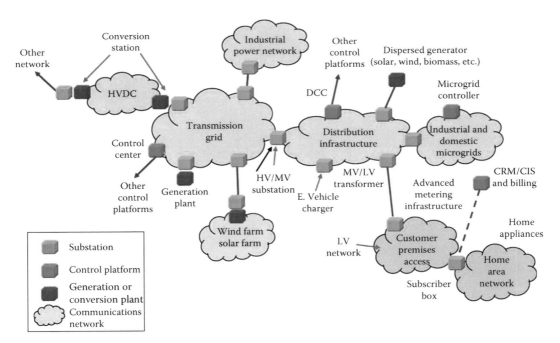

FIGURE 3.128 Operational communications domains in the electric utility. (© Copyright 2012 Alstom Grid. All rights reserved.)

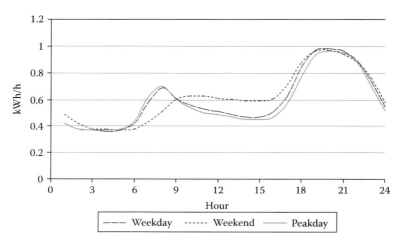

FIGURE 3.168 Example diurnal load shape. (© Copyright 2012 Pacific Northwest National Laboratory. All rights reserved.)

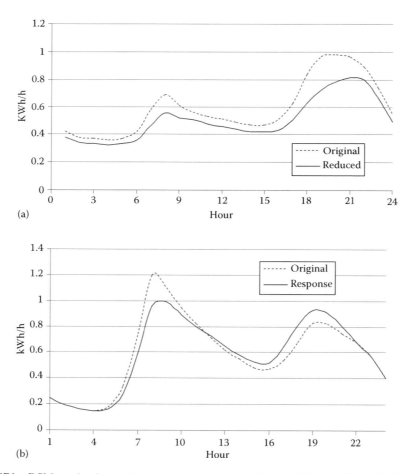

FIGURE 3.171 DSM mechanisms: (a) energy conservation and (b) peak load shifting. (© Copyright 2012 Pacific Northwest National Laboratory. All rights reserved.)

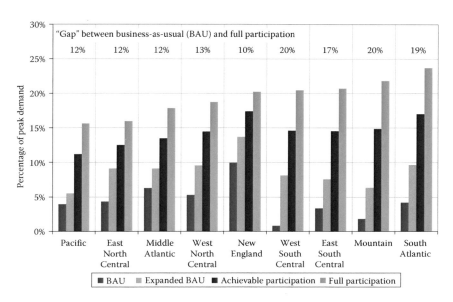

FIGURE 3.173 Gap in demand management participation in the United States. (From National action plan on demand response, Federal Energy Regulatory Commission, June 17, 2010, p. B-1, http://www.ferc.gov/legal/staff-reports/06-17-10-demand-response.pdf)

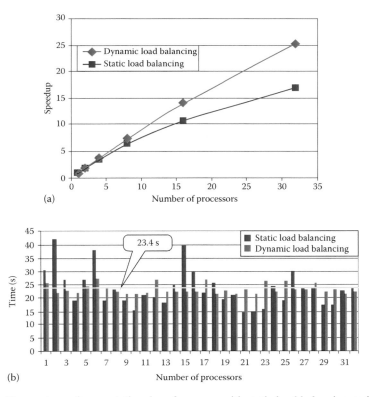

FIGURE 3.198 Comparison of computational performance with static load balancing and single-counter-based dynamic load balancing: (a) speedup performance comparison and (b) computational evenness comparison. (From Huang, Z. et al., Massive contingency analysis with high performance computing, *Proceedings of PES-GM2009—The IEEE Power and Energy Society General Meeting 2009*, Calgary, Canada, July 26–30, 2009.)

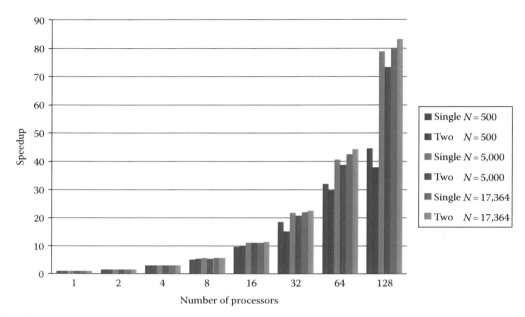

FIGURE 3.199 Comparison of computational performance with dynamic processor load balancing using single counter and two counters in a slow computer networking environment. (From Chen, Y. et al., D., Performance evaluation of counter-based dynamic load balancing schemes for massive contingency analysis with different computing environments, *Proceedings of the IEEE Power and Energy Society General Meeting 2010*, Minneapolis, MN, July 25–29, 2010.)

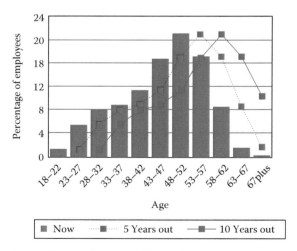

FIGURE 4.4 Electric and gas utility employee age distribution.

Smart Grid Technologies

3.9.4.1.5 Improved Investment Planning

Asset investments are required for increasing capacity and for maintenance and replacement programs. Project expenditures could include increased tree-trimming budget, increased pole inspection and maintenance, lightning protection upgrades, infrared inspection programs, animal remediation programs for substations, tap fusing programs, etc.

Utilities should determine per unit capacity costs and volumes for

- The cost per installed MW or capital dollars to serve a MW of electricity
- The cost per delivered MW or O&M dollars spent to serve a MW of electricity
- The cost per MW transformed or O&M cost of transformers per MW of electricity

Utilities need to take into account the existing infrastructure before developing optimal investment strategies for smart grid implementations. The investment analysis is challenging because of the large number of technologies, operational practices, and enterprise application software within the electric utility, where some large investments typically require more than a decade to fully implement. Smart grid investment planning and analysis should include a short-term as well as long-term, comprehensive, enterprise-wide, detailed cost/benefit analysis. Over time, utilities must make good use of their resources and determine relationships or parameters that have the greatest impact on investment returns.

3.9.4.2 More Effective Management of the Workforce

The organization must be in alignment in terms of its vision and values. People need to understand how they contribute to the success of the business and be rewarded for their efforts. Communication by top leadership is essential. People, integrated systems, and best in class processes make up the foundation of a solid asset management program, but people make the difference.

With an effective and efficient asset management program in place based on smart grid innovations and technologies, grid maintenance activities in the field can be achieved with much greater ease and accuracy. An efficient work schedule can be determined for field services, design, engineering, and other important groups within the electric utility. Also, the work schedule will be more accurate and allow timely completion of work since it will be based on real-time information on the processes. In addition, it will be easier to maintain organizational alignment and identify groups accountable for organizational processes or functions. Integrated asset management programs will help institute required performance incentives and prepare employees for organizational changes.

3.9.4.2.1 Mobile Workforce Management

Geospatial Information Management (GIS) is a long-accepted technology that has proven its value repeatedly to thousands of utilities worldwide. Likewise, mobile workforce management (MWFM) (also known as field force automation [FFA]) has enjoyed similar success and utilization over many years. The former originates essentially in an office environment, with paper maps then produced and used by mobile crews and technicians. The latter occurs primarily in the field, used also by mobile crews and technicians but also by dispatchers and managers. Over the last 5 years in particular, these two technologies have been converging in ways that are powerfully useful in contributing to both improved workforce productivity and more effective asset management. To better appreciate this convergence and its likely direction and development over the next 5 years, it would be helpful to first review how MWFM has developed to where it is today.

Utilities conduct business with a unique set of operating characteristics that have evolved over decades, beginning with paper-based operations and evolving into complex IT-driven business processes. Aiding in this transformation is the arrival of improved workforce management (or MWFM) technologies and systems that have helped utilities more efficiently keep the lights on, the water running, and the gas flowing.

MWFM represents the evolution of work and workforce management in a utility environment and the associated technologies and strategies for managing the three main divisions of work (i.e., customer and meter services, inspection and maintenance, and construction). As a result, current business trends and economic and environmental drivers are pushing utilities to more creatively manage all phases of work from an enterprise perspective in order to achieve the company's ultimate objectives.

There are four overarching divisions of work that any utility must manage:

1. Customer and meter services (outage)—This division of work includes everyday field work typically completed within a shorter duration (hours, days); it is often considered "unplanned" and "undesigned" work that the utility must manage. Examples include responses to customer inquiries, such as a new service hookup, gas or water leak, etc.
2. Outages—This division of work includes responses to outage conditions, such as a downed power line or broken main.
3. Inspection and maintenance—This division of work includes fieldwork around managing and maintaining assets, such as regular maintenance inspection of distribution assets/equipment and dispatching work crews to repair or replace key distribution assets.
4. Construction—This division of field work includes complex work that must be planned, designed, scheduled, and executed over longer periods of time (i.e., days, weeks, months). Long-cycle work is often resource-intensive, involving multiple stakeholders and the management of compatible units. Examples include construction of new infrastructure, including mains, points, and spans, and replacement of aging infrastructure and corresponding assets.

MWFM from its origin has focused on work type #1 mentioned earlier, customer and meter services, including outages. As such and given the available technologies of the day, there was limited, if any, GIS support available for this type of work.

The challenges facing utilities today can, in part, be addressed by holistically managing the three divisions of work described earlier. While aging workforce and aging infrastructure issues are accepted as major challenges by utility management, the utilization of MWFM technology to help efficiently address these business process changes can be a distinct advantage.

When addressing the dual threat of aging assets and infrastructure in the field, utilities must leverage integrated asset and workforce management technologies to capture and transmit data across the enterprise. The advantage of all field users living in a single system stretches across the three divisions of work. For example, when a field technician repairs a downed power line, he or she enters data via a mobile device, which is shared with customer service, dispatch, scheduling, and other departments. When it comes time to inspect and/or maintain that same power line, the field crew can access the prior repair information on-site. In turn, this data can be leveraged when planning and preparing for larger-scope work (long cycle). It is easy to see how access to this information, especially in geospatial form, can ultimately be a cost-saving tool when it comes to deploying both human capital and physical resources.

A recent analysis of the workforce management industry indicated that, given various market drivers in play, mobilizing the asset-oriented/long-cycle work remains an untapped opportunity for utilities. Most utilities do not have a viable option across short- and long-cycle work and often manage mobility with multiple MWFM systems versus an enterprise application. Also, an aging workforce continues to create challenges that highlight the importance of enterprise-wide technologies to support business process execution. In addition, aging infrastructure issues are driving companies to focus resources on asset-intensive work execution. These combined forces set the stage for companies to transition their focus toward improving field productivity through MWFM.

A clear strategy for achieving both business effectiveness and peak operational performance is the deployment of an enterprise-wide MWFM system. Most solutions in existence today provide ample functionality to manage singular parts of the three divisions of work; however, the majority of technologies handle only specific pieces, such as customer/meter service and inspection and maintenance, or are unable to handle construction work in conjunction with inspection and maintenance. Companies need a MWFM solution that can support utilities in managing all three divisions of work, but this solution to be truly effective must be geospatially enabled across the board.

Significant functionality is required to handle all three key divisions. Many utilities have carved off only part of the MWFM solution. For instance, some utilities only use mobile devices for deployment of customer work, or they schedule and dispatch inspection and maintenance work. With an enterprise MWFM solution with full GIS capability, a utility can operate a whole enterprise application, where mobility, scheduling, dispatch, and data transparency are carried throughout every layer of work.

A key opportunity for organizations will be the enterprise MWFM functionality that can link the customer, meter, outage, inspection and maintenance, and construction work with GIS support in the field to decrease overall IT spend and increase companywide productivity.

Utilities are continually looking to improve service, react effectively to emergencies, and identify opportunities in the field to improve service and operations. By definition, work consists of the day-to-day operations that define a utility. In the future, the work schedule will be intrinsic to the holistic operational plan. Responses to customer service inquiries, such as a new service hookup or a gas or water leak, can be combined and balanced with inspection and maintenance work.

For example, responses to customer care can be combined with managing other important assets in the field that are in close proximity to the customer care request. A mobile system with GIS can provide visibility into the work and work type, along with the crew and contractor information. This increased visibility will result in better customer service, workforce efficiency, and, ultimately, cost savings.

As much of the utility infrastructure is aging in parallel with a maturing and retiring workforce, utilities are entering a proactive phase of managing both assets and people with improved technology. For years, most work has been managed by silos of IT systems supporting only short-cycle (customer and meter services) work. With modern MWFM systems, utilities can now more directly accommodate, plan for, and have visibility into the status of inspection and maintenance duties.

The next focus of development will be the successful management of construction in addition to improving customer, inspection, and maintenance works. Installing the next mile, managing new neighborhood designs and implementation and maintaining this infrastructure will require enhanced MWFM capabilities with even stronger GIS components. MWFM combined with full-featured GIS capabilities is going to be a major field force transformative process that utilities and communications companies will face in the twenty-first century.

It meets the challenges of increased customer service with lower costs that is and will be a hallmark of field operations, and at the same time provides significant configurable flexibility at the point of service provision in the field. Through a combination of software solution implementation and professional services on the organizational and procedural aspects of field operations, this MWFM/GIS solution can produce measurable and repeatable benefits to virtually all utility companies who deploy large and active field service organizations.

3.9.4.2.2 Dual Threat: Aging Infrastructure and Aging Workforce Call for Integrated Asset and Workforce Management

Few people in the utility T&D business need convincing that the above and below ground asset infrastructure, be it electric, gas, or water, is a critical component that is today showing unmistakable signs of age. And most utility managers understand that the field workforce tending to that infrastructure—and the customers attached to it—is also aging and retiring at an increasing rate. These issues of aging infrastructure and aging workforce are often examined independently in articles, papers, and presentations to industry forums. Although richly worthy of attention in their own right, it is the interaction of the two, the dual threat of asset infrastructure aging simultaneously with the workforce maintaining it, that should be a particular concern of utility management, regulators, employees, and customers.

When a utility field technician or crew is dispatched to complete a series of inspections and maintenance orders at, for example, a substation, one can assume that they are experienced and fully trained to do so. With the median age of utility industry employees in the United States currently at 49, this would certainly be a valid assumption. Indeed, for the utility industry, there is a large group of well-experienced, well-trained current employees in the 45–54-year-old range. In executing the various procedures and tests involved in inspections and maintenance work, these employees are drawing directly on the considerable expertise and familiarity with the assets they have developed over some 25 years of work in the field.

Although inspections and maintenance procedures are well documented and the subject of frequent refresher training for crews, the personal knowledge component is just as important. Knowing the peculiarities of a given asset type, even down to the model number, is a valuable additional aid to properly conduct an inspection and, if necessary, make repairs. As these lead technicians and crew chiefs age, however, an increasing number are taking advantage of utility retirement packages to depart on or, in some cases, before their scheduled retirement dates. When such individuals leave, their individualized knowledge goes with them. And with strict cost controls in place at most utilities, hiring replacements is a lengthy and demanding process. Even once hired, the time needed to achieve the same level of personal knowledge is measured in many years, not months. With the increasing use of contractors (also exposed to the aging workforce factor), utilities might not even be in a position to hire new employees who will eventually be developing the in-depth skills needed for field work.

So when a relatively new lead technician or crew goes out next year to do that same inspection series at a substation, the personal expertise brought out to the job with them may be significantly lower than in prior years. No doubt they will exert their best efforts, will follow documented procedures, and will be subject to on-site personal review by a field supervisor. But that innate ability to sense what the status really is for a given piece of the asset infrastructure will be lessened.

At the same time as this aging workforce factor is coming into play, those same assets and infrastructure being maintained are also aging, thus requiring more and lengthier inspections and maintenance procedures and, in some cases, replacement. Newspaper headlines and TV news stories the last 5 years have all too often featured a failure, sometimes spectacular, of a given piece of electric, gas, or water infrastructure that at the very least led to service interruptions or, more dramatically, produced significant damage and human injury as a result of its failure. Much of that infrastructure is composed of operating assets that utilities of all types must continually inspect, maintain, and replace to ensure reliable performance. And while a good portion of the nation's utility asset infrastructure is in satisfactory condition, an increasing percentage of it is nearing (or exceeding) its planned operating life and therefore requiring field work to maintain or replace it. In such work, utility employee personal knowledge of individual assets as well as the overall T&D network is critical.

This work is typically generated in some form of utility asset management or work management system. Such a system consists of technology, certainly, as well as business processes, procedures, and even accounting practices, all of which come into play in operating a satisfactory asset management program. For many years, such systems were "siloized" by the type of asset being installed or maintained; thus, all meters would be managed by one system (often a spreadsheet), all transformers and related substation equipment managed in another, and so on. In the last 5 years, however, there has been an increasing trend toward maintaining the history of all assets in a single enterprise-wide asset management system of record.

Ironically, this same silo approach also has applied to the utility field workforce, with meter and customer service personnel managed by one workforce management system, distribution personnel in another, and outage technicians in still a third. Like asset management, however, the trend in recent years has been toward a common workforce management system covering the entire enterprise field force. With both of these consolidation trends, therefore, the ability has arisen to take a more holistic and integrated approach to asset management and workforce management, both as separate disciplines and now also as a useful synergistic point of operational excellence. Nowhere is this outcome more helpful than in addressing the problems posed by an asset infrastructure aging as fast as the workforce that maintains it.

How would such synergy appear in a normal field environment? Consider the example from earlier of a technician or crew going out to a substation for a series of inspection and maintenance tasks or orders. The very composition of the crew is a good starting point, in that the enterprise workforce management system would have facilitated the creation of a crew with the most appropriate skills to the orders at hand and structured the assignment and scheduling of the work to achieve maximum efficiency.

The orders to be accomplished would be generated by the enterprise asset management system, using a variety of triggers (time and condition based in addition to regulatory requirements) to determine the specific asset items to be inspected and maintained. These orders would be passed electronically to a mobile data terminal (MDT) carried by the technician or crew. On its screen would be details of the orders and assets, giving the utility employees a clear view of what needed to be done, but not how. That question would in the past have been left to the combination of training and personal knowledge that exists today in most utilities. But with the aging workforce and asset infrastructure issue, the "how" may not be as clear today and in the future as a utility would prefer.

This illustrates the new abilities and benefits of an integrated approach to asset management and workforce management. In this scenario, besides passing down the order details, the upstream system would also send to the MDT a hyperlink or a file attachment for viewing by the technician or crew. The hyperlink would be a clickable reference to the specific detailed inspection and maintenance procedure for that asset, which would be resident on the device hard drive. If wireless communications was available, one would see an internal website where details on the procedure were presented. Besides text information, the hyperlink can also provide graphical guidance, with detailed schematics of a given asset model, for example. Using the MDT's zoom in capability, the technician can see the diagram in great detail as needed. The technician thus has at his or her fingertips specific and always up-to-date guidance on how to achieve the work order objective in detail.

Whereas a hyperlink reference would be largely generic to a group of similar assets, another possibility that is even more specific is the attachment capability. In this case, detailed information relevant to that particular asset and order, such as a digital photograph taken yesterday by the previous crew working on it, would be made available on the MDT. Any changes made that day can be captured via another digital photograph and sent back up as a file attachment as part of the order completion process. The difference between hyperlinks and attachments in this example, therefore, is largely one of specificity to the asset and order in question.

Another possibility would relate to a maintenance procedure (such as replacement of an asset grouping) that a crew might need to undertake following an inspection. Using software on the MDT and the wireless capability, the crew would be able to interface to other third-party applications that are resident on the device itself or accessible via the web. For example, compatible units (CU) are a major component of new construction for the maintenance replacement of assets. With this third-party reference tool, the crew can access a library of available compatible units for the order in question and determine what materials and labor should ideally be expended in the procedure. Whereas in the past, this information would be on paper and well known to experienced utility or contractor employees, with this capability, they can easily and quickly reference compatible unit definitions that are always up to date and complete.

Through all of these means, useful information is made accessible at the place and time where it is most needed to properly perform the work, certainly a tool and capability previously impossible. This tool, as a result, greatly helps augment the personal familiarity of the utility workers on the job with the asset and work to be done. The foregoing are two examples of where a properly thought out and integrated solution architecture approach can greatly assist a utility in its daily work as well as meeting its long-range objectives of reliability. The overall capability, achieved only through the proper integration and coordination of enterprise asset management and enterprise workforce management, thus can play an instrumental role in helping utilities deal with the dual threats of aging infrastructure and aging workforce.

3.9.4.2.3 Forecasting

Forecasting typically utilizes historical data and other inputs (such as known blocks of work) in a three-layer neural network demand model to calculate workload demand up to a year in advance. Input data can then be manipulated through a series of "what-ifs" in an iterative process that eventually matches resource supply with work order demand. Output from this system can then be fed into subsequent scheduling processes.

Forecasting allows a planner to, among other things,

- Deliver highly accurate workload/workforce forecasts
- Forecast over variable days, weeks, months up to a full year ahead
- Operate with a user-friendly, intuitive interface
- Automatically import needed historical data to produce forecasts
- Input special data reflecting known workload programs
- Recognize complex patterns and relationships in input data

3.9.4.2.4 Scheduling

Using scheduling, a utility can do the job right with the right resources the first time. Appointment scheduling is automated and immediate, using predefined scheduling parameters and rules. Call center representatives can initiate service requests and schedule service appointments, assuring the right technician at the right place and time. The MWFM scheduling algorithms enable both the initial optimization and reoptimization of assignments and schedules to ensure the choice of a field resource to do a given job is and remains the most efficient possible.

Scheduling allows a utility to, among other things,

- Define and adjust technician/crew availability
- Negotiate and reserve customer appointments
- Automatically assign work orders to technicians
- Reoptimize the schedule when the operational environment changes
- Sequence and route work assignments based on the street network
- Make manual adjustments to the schedule when conditions change

3.9.4.2.5 Dispatch

Dispatch offers personnel a high-level and/or detailed look at the entire field service program in realtime, in both Gantt chart (timeline) views and tabular windows. Dispatchers are given the flexibility to make modifications to assignments for technicians or crews, with the ability to adjust when there are conflicts or other conditions requiring their attention. With dispatch, the system performs most of the mechanical dispatch functions needed during a shift, leaving dispatch personnel time to focus on those exceptions requiring personal intervention.

Dispatch allows a utility dispatcher to, among other things,

- Monitor technician and order progress and status
- Be alerted to conditions that require a dispatcher's attention and action
- Monitor the position of the fleet using dispatch maps and GPS
- Enter status and complete orders on behalf of techs who do not have operational devices
- Monitor the location of field resources compared to open orders
- View the GIS representation of assets involved with open orders

3.9.4.2.6 Mobile

Mobile transforms the way all field workers process orders. Instead of receiving work orders on paper or over a voice radio, technicians obtain and manage all of their work orders wirelessly on a mobile device.

Mobile provides field workers with the tools, information, and wireless capabilities to receive work orders, complete the work, and send status updates and order completion information back to the enterprise—wirelessly and in realtime. With integrated GPS-based mapping and street-level routing, mobile ensures optimal routes for traveling between assignments to minimize distance and time.

Mobile allows a utility field worker to, among other things,

- Receive and work orders in the field both in and out of wireless
- Provide real-time status updates to dispatch and the back-office
- Request supplementary information from the back-office
- Review and work orders on the mobile map, along with optimal driving directions
- Notify dispatch of potentially dangerous situations
- Complete, edit, and approve timesheets

3.9.4.2.7 Reporting

Another big step toward optimizing field service operations is possible with a performance reporting system within MWFM that provides critical visibility and insight into operational status, both at a tactical real-time level and at a strategic trending level. Easy web access to multidimensional analysis and detailed performance reports allows a utility to uncover the real drivers of workforce performance and make adjustments to maximize efficiency.

Reporting typically comprises a knowledge warehouse, operations dashboard, and decision support capability. These integrated components provide a comprehensive, consistent view of operational data, addressing daily field tactics and providing in-depth views of past performance to assist with strategic analysis, planning, and forecasting. Input from reporting can be fed back into the initial step in the work cycle of forecasting, thus creating a self-referential loop of optimized field operations.

Reporting allows a utility to, among other things,

- Obtain long-range trending views based on weeks and months of data
- Obtain short-term, tactical views of current operations in the field
- Develop its own reporting structures based on the historical database
- Distribute reports automatically to target audiences
- Access all reporting through an easy-to-use web interface

3.9.4.2.8 End-to-End and Top-to-Bottom Benefits

Utility service organizations today operate in an ever-changing environment of rising costs, globalization, and mergers and acquisitions. Utilities at the heart of the T&D operation continue to strive to meet customer expectations and comply with increasingly complex industry regulations and infrastructure requirements. To many, these "new field" requirements and technologies are starting to cover all work types, all workers, and support each functional step in the overall field operation process. As a result of managing many of these new field requirements, T&D organizations are on the heels of realizing some amazing new benefits (Figure 3.153).

To gain these new benefits, an MWFM solution now must fully and truly support "end-to-end" processes and "top-to-bottom" functionality. If it does not, then the new gains outlined in the following will become much more of a challenge and delay the required return on investment.

End-to-end means that, in doing field work orders, there is a natural progression of necessary functional steps. These begin with the process of forecasting both the workload and work resources needed over a projected period of time. Work orders must then be scheduled and assigned to specific

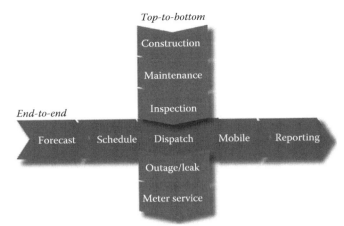

FIGURE 3.153 End-to-end and top-to-bottom benefits of MWFM. (© Copyright 2012 ABB. All rights reserved.)

resources; those orders should then be dispatched. Technicians and crews must then receive that work as they operate in a mobile state in the field. Finally, the results of such activity immediately and over time may be reported and analyzed.

As such, end-to-end and top-to-bottom capabilities available in Service Suite extend the range of the enterprise workforce management solution to include forecasting, scheduling, dispatch, mobile, and reporting. Such capabilities result in superior functional processes that support all construction, maintenance, inspection, outage/leak, and meter services work order types. In short, virtually everything a T&D utility's workforce does and requires in the field is supported by such an approach to MWFM.

In the early days of utility workforce management, the primary order types supported were related to customer and/or meter services. Doing these high-volume, short-duration orders through MWFM soon proved to have powerful paybacks in terms of customer service and field workforce productivity. But this affected only part of the utility field workforce; there were still a substantial number of field technicians and crews doing more complex work in the field that were using paper orders and voice radio only.

The solution now supports—from top to bottom—all of the work order types conducted in the field by a T&D utility. Construction, inspections, maintenance, outages/leaks, as well as customer and meter services work orders are all now equally supported. To provide this scope, the MWFM industry solution has been greatly expanded to cover the complexities and often lengthy job duration of asset infrastructure work orders conducted in the field.

Easier access to and improved accuracy of data

- Automate field collection and validate work results
- Deliver critical enterprise and customer and asset information to the field
- Provide real-time feedback on work progress
- Validate data in the field to ensure correct information in back-end systems
- Increase data accuracy with user-friendly graphical interface, context-sensitive drop-down lists, and auto-fill fields

See more and control more in the field

- Measure and analyze operational performance
- Access an enterprise-wide view of workforce and its workload
- Access web-based, real-time tactical reporting tools

Smart Grid Technologies

- Continuously optimize scheduling as daily changes occur
- Allow the user to configure, deploy, and modify the system

Consolidate work and save time

- Eliminate manual work order sorting, bagging, and entry of completion information
- Consolidate geographically dispersed dispatch centers
- Reduce number of field technicians required
- Scale the system in response to changes in the organization
- Be flexible and adaptable through open, distributed architecture that supports an enterprise workforce management solution

Increase customer satisfaction

- Meet more customer appointment commitments
- Ensure workers have the right equipment, skills, and parts to finish the job in one visit
- Enable service agents to confidently offer narrower service appointment windows
- Provide real-time service order feedback to customers
- Meet regulatory requirements for customer appointments

Decrease costs

- Reduce travel time and distance driven with geospatially optimized routing
- Reduce employee overtime
- Decrease operating expenses and overhead
- Optimize use of contractors
- Increase the number of technicians supported by each dispatcher

Increase efficiency

- Eliminate voice and paper-based processes
- Increase volume of work orders and completion rates
- Improve service response times
- Improve monitoring and management of field resources
- Significantly reduce time needed for generating, reviewing, and changing schedules through optimized scheduling

3.9.4.3 Examples of Utility Asset Management Applications

3.9.4.3.1 Transformer Monitoring Systems

Transformer monitoring systems use direct instrumentation of critical transformers in the system to accumulate measurements on power flows and transformer oil condition such as temperature, moisture level, and combustible dissolved gas levels. These measurements are then fed into analytical and statistical tools to determine transformer condition and alert utility personnel about transformer health.

3.9.4.3.2 Transformer Load Management Systems

Transformer load management (TLM) systems are used to estimate the loading and peak loads on distribution transformers. This is done by combining maximum consumption information for all customers connected to each distribution transformer. Monthly customer kWh consumption is generally obtained from their meter reads. If the end customer meters provide interval meter reads, significantly more information is calculated in the TLM system including peak loading and load factors of the transformers.

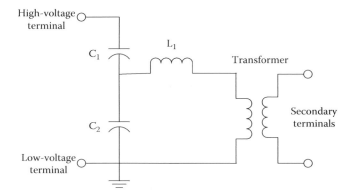

FIGURE 3.154 Capacitor coupled voltage transformer (CCVT) components.

3.9.4.3.3 Circuit Breaker Monitoring

Circuit breaker monitoring helps with the scheduling of circuit breaker maintenance by monitoring the number of breaker open and close operations and magnitude of current interrupted. The function calculates an estimate of wear on the breaker contacts by measuring the duration and magnitude of current during the interruption period. The individual operations are accumulated and then compared to a user selectable threshold value. An alarm is sent when the threshold is exceeded.

3.9.4.3.4 Pending Voltage Transformer Failure

The capacitor coupled voltage transformer (CCVT) consists of three parts (Figure 3.154):

- Two capacitors across which the transmission line voltage is divided
- An inductive element to tune the device to the line frequency
- A transformer to isolate and further step down the voltage for input to the protection relay

CCVTs typically have a life expectancy of 20–25 years. The capacitors are usually stable until close to the end of life but can then suffer a catastrophic failure.

Three separate CCVTs are applied for a three-phase bus or line voltage measurements. A failing CCVT can be detected early by monitoring the zero-sequence voltage ($V_A + V_B + V_C$) with a sufficient time delay to avoid alarming on system transients such as switching operations.

3.9.4.3.5 Through-Fault Monitoring

Through fault is an ideal means of performing asset management for distribution transformers. In the southeastern United States, lightning storms are common during the hot summer months. A transformer can be subjected to multiple through-fault current surges during a storm if lightning strikes the distribution system fed by the transformer. Fault current flows through the transformer fed by the grid to the fault on the feeder. Through faults are high current events that can cause both mechanical and thermal stress to the winding insulation. It is possible to perform just in time maintenance by monitoring the number and magnitude of through faults a distribution transformer experiences over time. A transformer protective relay can count the number of through-fault events and send an alarm when the number of operations exceeds a selectable limit. An electric utility may want to send out a crew to examine the transformer after a certain number of through faults have occurred.

3.9.4.3.6 Fault Location

When an electrical disturbance such as a lightning strike occurs on an overhead high-voltage transmission line, high magnitude current flows through the line conductor and associated power equipment to the point of the disturbance. The high current magnitude can quickly damage the line conductor and power equipment, such as a transformer bank.

Protective relays can detect the presence of a disturbance on overhead transmission lines and send commands to open the circuit breakers at each end before any damage occurs. Accurate fault location helps utility personnel expedite service restoration, thereby reducing outage time. If the distance to fault is known, the utility can quickly dispatch line crews for any necessary repair and save a lot of time and expense that would otherwise be lost in having to patrol the overhead line for possible damage.

3.9.4.3.7 Just-in-Time Maintenance

Maintenance of physical assets has always been a challenge in the electric utility industry due to the harsh operating environment and the need for safety. The best maintenance practice makes significant contributions to an electric utility company's bottom line, which, in turn, justifies maintenance expenditures in terms of return on investment.

Electric utilities in the past practiced preventive maintenance, which is often referred to as periodic maintenance. Preventative maintenance is maintenance work performed at regularly scheduled intervals (i.e., time based). Maintenance departments may perform work that is not required and replace parts that are still in working condition. The majority of equipment failures are not related to the number of hours they were operated. Therefore, periodic maintenance is not always cost effective.

Some electric utilities only perform reactive maintenance, which is simply to only fix or replace existing equipment when it fails. This practice leads to both unplanned downtime and production delays, which, in turn, result in lost revenue. Repairing equipment only after it breaks down usually also leads to secondary damage, which increases maintenance costs.

JIT, or just-in-time, maintenance is condition based. The electric utility monitors the health of the grid assets and sends out crews to perform maintenance when an alarm is raised. Maintenance decisions are made on the analysis of the monitored data. Ideally, the condition of an asset is continuously monitored by online sensors and some form of data analysis.

As part of the smart grid programs, electric utility companies are making initial investments in sensors and monitoring equipment to regularly monitor critical equipment and predict problems before they occur, preventing downtime and preserving resources. JIT maintenance places the emphasis on both corrective and preventive approaches. Condition-based monitoring ensures that the appropriate maintenance work is performed at the right time by identifying the root cause(s) of the problems. This proactive managed maintenance approach, a form of asset management, improves profitability.

3.9.5 Where Smart Grid Meets Consumer Reality: The Consumer Perspective on Asset Management

Commercial, industrial, and residential consumers are finding ways to be engaged in energy use reduction. Consumers have explored demand-side smart grid technologies that can help optimize the use of their assets while helping reduce energy bills with minimal impact on their way of doing business or life style at home. Consumers feel the need to become more engaged and educated in terms of energy and usage based on global warming issues, political issues, and economic incentives, and more public awareness of smart grid initiatives in their local regions.

3.9.5.1 On-Site Generation

Whether it is for financial or environmental reasons or both, a growing number of electric consumers are now considering generating their own renewable energy. With generous state and federal incentives, there has never been a better time to consider solar and other alternative energy options. Some electric utilities are committed to increasing the amount of renewable energy generation on their systems to reduce dependency on fossil fuels and foreign oil and address climate change, so they are eager to help consumers with alternative energy resources.

Net metering: If an electricity consumer generates power with wind, hydro, solar, fuel-cell technology, biogas, or biomass, electric utilities' net metering services will allow the consumer to offset the cost of the electricity used from the electric utility with the energy the consumer generates. The consumer will get credit in his or her electric utility account for any power produced and not consumed. Net metering rate schedules are typically provided by electric utilities that offer such services.

Solar power initiatives: Electric utilities are working to supply the growing needs of our nation's energy future with clean renewable energy. They have assisted hundreds of self-generating customers who supply the grid with their excess power through net metering service, which allows consumers to receive payment or credit for energy supplied back into the system. Electric utilities have also partnered with public agencies and private businesses on many innovative solar initiatives, such as solar highways, solar rooftop installations using thin film solar panels, etc. For example, the electricity generated by a solar highway is operating highway lights and reduces the state's reliance on fossil fuels. The solar panels produce electricity during the day, supplying power onto the grid, and the electric utility returns an equivalent amount of power at night to light the interchange, through a net metering arrangement with the state. Utilities are helping to grow solar energy in a variety of ways, including recruiting solar manufacturers to their states and creating partnerships to implement solar electric installations.

3.9.5.2 Managing Energy Demand and Consumption

It is virtually impossible to store large amounts of electricity. It must be produced exactly when people need it. Peak periods of energy usage correspond to consumer living and work routines and weather influences. Some utilities experience peak demand periods during summer months with building cooling load, some utilities during winter months with building heating loads. To account for the peak loading hours, which may only occur for less than 5% or 10% of the total generating hours in a year, utilities must build enough power plants to meet the peak demand. Most of the new power plants built today burn fossil fuels, which contribute to air pollution. By reducing peak-hour energy use, the consumer can help delay the need for new power plants. If the consumers can shift energy use away from on-peak and mid-peak periods, then it reduces the stress on the utility generating and power system assets. Utilities have implemented, or are considering implementing, time of use or real-time pricing in order to reflect the true real-time cost of supplying electricity to consumers. This will help drive the behavior of utilities to consumer less electricity during peak times. This consumer-driven approach to asset management will allow consumers to have insight and control over their home energy usage and understand their costs. Consumers will be able to gather information, manage usage, and have access to real-time knowledge that empowers them to make smarter energy choices. The consumer will be able to know the best times to perform high energy consumption tasks. Perhaps over time, consumers will be informed with recommendations by smart devices to better manage their energy usage based on trending data. Residential electric loads that use the most electricity in a typical month are electric water heaters and heating and cooling systems (depending on season). Costs to operate these appliances can account for as much as 50% of a consumer's average energy use. Installing programmable thermostats on heating and cooling systems and a water heater timer on water heaters can help control when energy is used and therefore reduce the cost of running these appliances. For larger industrial and commercial consumers, improved asset management practices may include better use of cooling and heating equipment, such as precooling buildings to avoid peak heating or cooling loads. Some industrial and commercial consumers have backup generators that can be used to generate electricity back on the system during peak hours (distributed generation)—these are not renewable energy sources but typically diesel- or gas-powered generators that can be better utilized by the industrial and commercial consumers, as well as help utilities manage peak demand on the system and improve usage of T&D assets from the utility perspective.

Smart Grid Technologies

FIGURE 3.155 Integrated asset management process. (© Copyright 2012 ABB. All rights reserved.)

3.9.6 CENTRALIZED, DATA-DRIVEN ASSET MANAGEMENT

Much of the asset condition data at most electric service providers exist in silos. For example, the oil lab group has the DGA (dissolved gas analysis) and general oil data, the electrical testing group has the power factor and other electrical testing data, the substation group has all substation inspection data, the infrared group has the infrared inspection results, and so on. Given this situation, there is great difficulty in gaining the best insight into the condition of the equipment upon which these measurements have been made. An approach to T&D asset management that emphasizes an efficient, data-driven business process is shown in Figure 3.155. The process is divided into four continuously integrated stages: data collection, integration and analysis, decision making, and work execution. These are described in the following.

3.9.6.1 Data Collection

A necessary starting point for an effective asset management program is to have a description of the installed equipment base in a centralized, accessible format. Having such a database makes it easier to plan for spare parts and know where potential problem equipment might be. For example, if a particular make and model of transformer bushing is creating problems, the centralized database can be used to determine where all those transformers are located. It is also important to have inspection and maintenance data, including inspection results, testing results, operator logs, and maintenance scheduling, all in the same database. As mentioned earlier, in many organizations, if these data exist at all, they are scattered in many different locations and formats. Good decisions require good inspection and maintenance records. For example, in the bushing problem described earlier, the database can be used to determine when power factor testing was last performed or thermographic images were last taken on the suspect bushings, and what those results were. This assists in making the proper decisions about the bushings.

Most electric service providers now have an installed base of real-time systems such as online monitors, SCADA, EMS/DMS, and outage management systems (OMS). These systems provide valuable information on the data-to-day operation of equipment in the field. Key parameters from these systems—such as peak loading, number of switch operations, number of tap-changing operations, etc.—can be integrated into the asset management database to further enhance decision

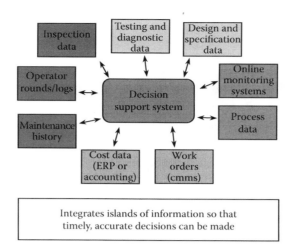

FIGURE 3.156 Central data repository for assets. (© Copyright 2012 ABB. All rights reserved.)

making. An integrated asset decision-support system, which leverages data from multiple sources, is shown in Figure 3.156.

3.9.6.2 Integration and Analysis

The right data and information, in the right format at the correct location, permits improved decision making. Equipment condition assessment involves interpretation of inspection/test results, analysis of operating data, trending of monitored points, cross-correlation of data, and the use of experience, guidelines, and rules-of-thumb to judge the condition of a piece of equipment. The data must be integrated for maximum effectiveness.

Reporting sounds simple but is extremely important. An asset manager must be able to produce reports that give details about particular assets, or summaries of all assets, with respect to condition, major problems or trends on the system (such as escalating failure modes), effectiveness of testing and diagnostic procedures, cost benefits of various activities on the system. Reporting also summarizes the data and analyses in preparation for the next phase, decision making.

3.9.6.3 Decision Making

The decision-making process takes available data and information from the first two steps; evaluates various alternatives and scenarios; and makes the best decision, considering all stakeholders. System planning considers load growth, the capabilities of the existing system, and system performance targets such as reliability to develop a capital investment plan. Risk analysis is an inherent part of this—at the various possible levels of capital expenditures, what are the risks of not being able to meet system performance targets.

Maintenance and inspection policies, as well as O&M budgets, have a large impact on system performance. In recent years, electric service providers have evaluated how condition-based maintenance (CBM) and reliability-centered maintenance (RCM) can be applied for more effective maintenance processes. CBM and RCM processes can only be applied with the proper data and information, as furnished by the first two steps of the cycle. Ultimately, asset managers would like the ability to optimize CapEx and OpEx simultaneously. That is, they would like to get the maximum system performance out of the available dollars. Decision making also involves work order generation (what condition status warrants the issuance of work orders), determination of spare equipment/parts requirements, and development of contingency plans for equipment with deteriorating condition.

Smart Grid Technologies

3.9.6.4 Work Execution

Work execution simply means doing the work. This involves the performance of maintenance, new construction, and new customer connections; retrofit projects; equipment replacements; etc.

Elements of this work, or even all of this work, can be outsourced to companies, which have advantages due to technologies, process efficiencies, economies of scale, or broad customer bases. The last step is monitoring, which involves continuously monitoring the performance of the system and its assets, as well as the costs of achieving the performance. It is really part of data collection. Without monitoring, there are no grounds from which to make continuous improvements.

An ideal asset management system integrates inspection information from many different sources, including testing results, predictive maintenance tests (oil analysis, infrared thermography, sonic/ultrasonic and vibration analysis, etc.), operator logs, on-line system monitoring data, real-time data (SCADA), maintenance histories, and failure data into a single database. This gives the asset manager the ability to use all available asset information in analyzing the condition of assets on the system.

Email- and page-alarming features within some of these applications can be setup to notify the responsible persons when alarms exist or condition status change. This feature is designed to "push" the existence of potential problems to the responsible people instead of them having to continually review volumes of data to uncover problems. This is called threshold alarming: high, low, in range, out of range. Various engines can be designed to perform specific high-level manipulations of the data. For example, a trend engine can look at long-term changes in parameters, with the ability to compare similar pieces of equipment or cross-trend with other data points. Other data management capabilities can help an asset manager determine the effectiveness of each testing technology and maintenance activity. Such systems provide information needed to implement a condition-/reliability-based maintenance strategy or reduce the frequency of time-based maintenance schedules.

3.9.7 GEOSPATIAL INTEGRATION FOR ASSET MANAGEMENT

Particularly for organizations with geographically dispersed assets, such as those in electric utilities, an operational view of assets in a spatial context supports informed decision making for operations and maintenance. Some enterprise asset management (EAM) systems have the capability to display work and asset information on maps through standards-based integration with leading geospatial information systems from vendors like ESRI, GE, and Intergraph. Out-of-the-box integration enables users to display, query, and update the location of assets from a map view. Users can make spatial inferences from the relationship between assets—poles and distribution transformers poles, for example—that have not been possible with traditional, tabular views.

Mobile users can leverage map-based data as well to locate and view assets on their devices. The data can invoke location-based services on the mobile device for enforcing job plans, safety compliance and locations, and time-based data collection.

To make a spatial view possible, the EAM can expose asset data as first-class geospatial features using standards-based, GIS vendor independent formats—making the EAM system a virtual GIS data store. Thus there is no need to copy EAM data into a GIS tool, eliminating concerns about version control or analyzing what data to copy and when to copy it.

With an open and vendor-neutral approach to geospatial integration, GIS users can view the latest asset and work order data, since the EAM data being accessed through the GIS system reside in the EAM. Likewise, users can view and search asset locations in map views in the EAM environment as part of a unified interface, as shown in Figure 3.157. For example, the geographic locations of asset management entities such as work orders can be created, updated, or deleted from within the map view.

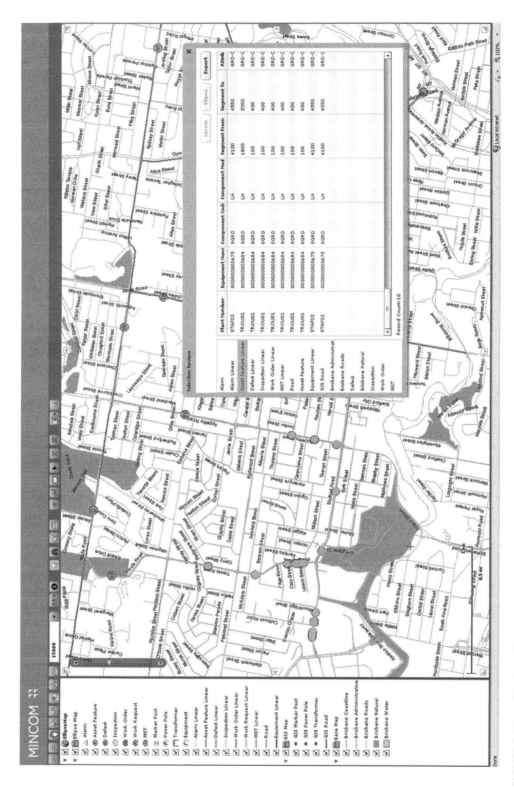

FIGURE 3.157 GIS view of asset management. (© Copyright 2012 ABB. All rights reserved.)

Smart Grid Technologies 351

3.9.8 Advanced Asset Management for the Smart Grid

The proliferation of disparate IT systems throughout T&D organizations has created a need for business intelligence solutions in asset management. With business intelligence solutions, data are extracted from the multiple systems, summarized, and aggregated. Users are provided with multiple tools to convert the data into actionable information, including key performance indicators (KPIs) and graphs, dashboards, ad hoc querying and reporting. An example of an architecture for a business intelligence solution for a T&D organization is shown in Figure 3.158.

The digital dashboards with their associated KPIs, "drill downs" and "drill through," analytics, and reports represent the presentation layer of information. The ETL (extract, transform, and load) engine rapidly extracts the data from customers' disparate systems, cleanses "dirty" data, and then loads the data into the BIM (business intelligence module). The integrated development environment and administration environment allow administrators to easily create and manage KPIs, queries, and dashboards specific to their unique requirements and provide proper levels of security within each business area.

The business intelligence solution can be used for applications such as tracking asset reliability performance. By integrating points within an organization's operations, work management, SCADA, and financial systems, an asset reliability performance module can provide the asset manager with detailed information regarding asset performance by identifying assets that are performing below (or above) expectations. The organization is able to determine which metrics meet their asset performance criteria. Relevant metrics are matched with an organization's targets and goals to automatically provide visual identification of at-risk, poor performing, and failing assets. A dashboard for tracking distribution asset performance is shown in Figure 3.159.

Another example of how business intelligence solutions can assist with asset management is an asset health calculation. An Asset Health Index can provide a centrally located repository that displays aggregated health information. The health information can be extracted from various sources and used to analyze asset condition and performance information to identify population condition, risks, and impacts of the assets' condition. The Asset Health Index can support the ability to modify algorithms and formulas that are employed to calculate an asset's overall health score. Asset managers can be supplied with various dashboards to develop plans for replacement of degrading assets before failure. These dashboards can also enable an organization to optimize the activities to sustain the life expectancy of the particular asset (Figure 3.160).

The electric utility industry, and indeed the entire commercial and consumer customer base, still has much to learn about what the smart grid world may look like 5, 10, or 20 years from now. As always, much of what is now assumed will change, technology will evolve, and people will adapt. But even at this early stage, it is clear that this new smart grid will not be able to meet or exceed expectations for it without the involvement of key supporting systems existing today such as asset management, network management (NM), and workforce management. These systems together can go a long way toward gathering the right and necessary smart grid data and making it available and actionable to obtain the benefits everyone is demanding.

Asset management should be able to

1. Predict potential failure of a smart asset based on measurements received from SCADA or other operations systems; from an asset management program standpoint, this enables the utility to obtain the last gasp of work from assets, affording JIT retirement and replacement
2. Optimize maintenance timing and processes, with received distribution data used to determine maintenance priorities; additionally, asset management would create an actual work order to be executed by MWFM
3. Support near real-time asset decision making, with predictive asset-modeling capabilities built on real-time data

352   Smart Grids: Infrastructure, Technology, and Solutions

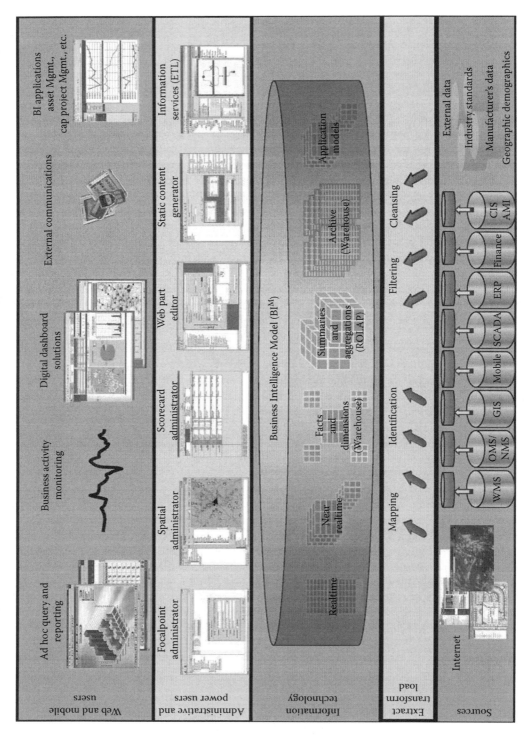

FIGURE 3.158 Utility business intelligence architecture. (© Copyright 2012 ABB. All rights reserved.)

Smart Grid Technologies

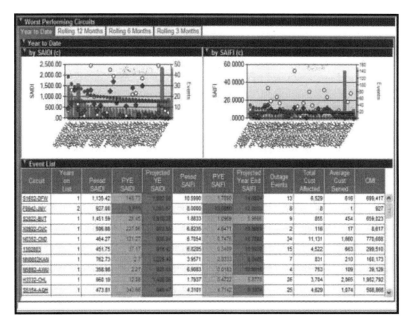

FIGURE 3.159 Dashboard for tracking distribution asset performance. (© Copyright 2012 ABB. All rights reserved.)

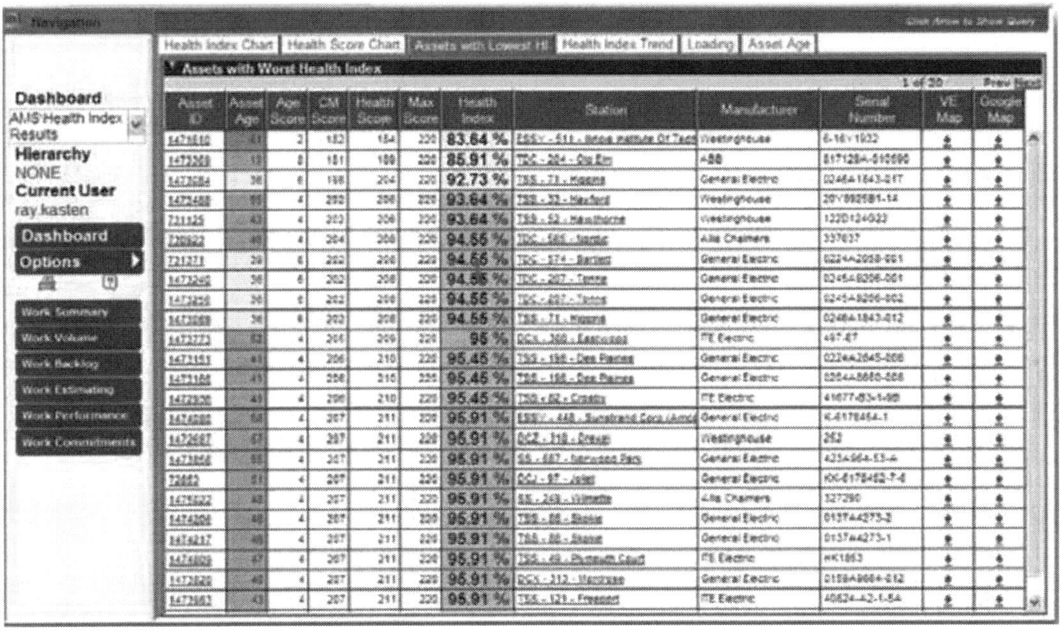

FIGURE 3.160 Dashboard for tracking asset health. (© Copyright 2012 ABB. All rights reserved.)

4. Enable tagging and tracking of all smart assets involved in the smart grid, using asset management to help manage the rollout of these assets as well as integrating with other asset reporting mechanisms to gain better predictive and analysis capabilities on smart asset performance
5. Enable smart assets to provide greater reliability and performance, with fewer and less lengthy outages, shorter restoration times, and lower overall costs; customers and regulators will expect nothing less
6. Provide more detailed and precise asset location data, from purchase through retirement, including accurate latitude/longitude GPS positioning

Workforce management (MWFM) should be able to

1. Increase productivity in the field by eliminating or greatly reducing the need for crews to search for an outage cause location
2. Enable proactive maintenance in conjunction with asset management, where crews can perform work on other assets (smart or not) while at a location
3. Perform reliable field root cause analysis, where network issues can be pinpointed and diagnosed, with results sent back to NM
4. Provide smarter customer-related services, including more precise appointments, self-service web portals, and immediate back-office connection to and registration of new network assets
5. Conduct more accurate and analytical crew performance measurement
6. Better utilize a greatly increased volume of data in the field to perform work and reduce restoration time

NM should be able to

1. Pinpoint and diagnose smart grid issues and automatically direct corrective action to be taken in the control center or field through asset management or MWFM, or both
2. Determine the state of the network in realtime
3. Portray in realtime the asset connection status and topology for use by downstream systems and personnel
4. Conduct failure notification for appropriate assets to downstream systems and personnel
5. Provide remote access monitoring and reporting of smart assets

Now more than ever, consumers and regulators insist on responsive and high-quality service from their utility service provider. To succeed, T&D utilities need to optimize their field service delivery processes and improve the reliability of critical assets, while making efficient use of field workers and equipment.

Most MWFM systems in place today automate at least part of the field service workflow—from scheduling and optimization of order assignment to dissemination of work to the field, mobile order processing, and performance measurement. In the face of constantly changing investments and formidable field operations challenges, MWFM today does provide a platform for managing work across much of the enterprise—regardless of the enterprise application from which the work is generated—and enables seamless, real-time data and communications flow between certain enterprise systems and certain mobile workers. However, are the systems that are in place today going to help build the workforce operation of tomorrow?

Clearly there remain many obstacles to distributing work across organizational or geographic boundaries. As a result, companies continue to look to further increase operational efficiencies and heighten productivity, while optimizing performance, improving internal controls, and reducing costs.

Smart Grid Technologies

T&D outfits will continue to look toward their MWFM systems as the primary platform for addressing new obstacles and gaining new benefits. Challenges include

- Improving end customer service and responsiveness
- Driving out operational and supply chain costs
- Meeting tougher regulatory compliance requirements
- Delivering system reliability in light of an aging field workforce and an aging asset infrastructure
- Supporting a progressive field asset management program
- Unifying fragmented technologies and lowering IT support costs
- Developing and enforcing utility-wide process standardization
- Increasing field force productivity
- Reducing the carbon footprint of field operations

Asset management is a balancing act among the electric utility, its customers, its shareholders, its employees, and its regulators. First and foremost, the electric utility must deliver value and performance to the electric utility customer. It must also create meaningful metrics, provide service options, and move to a customer-driven organization. The utility must achieve a reasonable rate of return for shareholder and maximize share value. It must manage assets prudently, optimize costs, and increase asset utilization. The utility must do more with less and develop skills, identify expectations and accountability, tie performance to compensation, and create competitive culture. Regulators also need to change their thinking around regulation and smart grid drivers and technologies, such as decoupling rates from return on investment to support energy usage reduction through loss reduction and demand response programs, representing the customer, and setting standards for electric utility performance.

Within the electric utility's organization, asset managers and smart grid committees drive asset excellence, strategic asset planning, holistic consideration of spending, and performance accountability. Utilities should focus on performance of all assets (financial, physical, human, etc.), economic life cycle, manage risk, and leverage information technology for accurate and timely information. They should balance financial, technical, and sociopolitical components of asset investments. Utilities should use a balanced scorecard to evaluate asset management programs to optimize their value and identify areas in which assets (including human assets) are under or over performing based on smart grid technology enablers.

In summary, what can a successful asset management approach do for an electric utility, as well as commercial, industrial, and residential electricity consumers?

1. Utilities can save money while improving reliability.
2. Utilities can avoid mandates and penalties.
3. Utilities can enhance all systems, especially electrical, while reducing expenditures.
4. Utilities can retain and enhance the value of their infrastructure for the future.
5. End use consumers can consider solar and other alternative energy options.
6. Utility net metering will allow the consumer to offset the cost of the electricity with the energy the consumer generates.
7. Consumers can monitor energy demand and consumption and save money by responding to time-of-use pricing and shifting load to off-peak hours.

In the United States, reliability problems have appeared due to underinvestment in the electrical infrastructure concurrent with growing demands. In Europe, there is more concern with the integration of renewable energy resources, which places different demands on the electrical system. The challenge has been to connect these renewable sources to T&D networks that are not designed for the additional power flow, which is also a concern in the United States as more renewables are added to the system. In China, the grid is developing quickly in order to meet the growing electricity demand. In Japan, there is less legacy infrastructure and fewer players in the market. Although smart grid means different things to different people, the objectives are the same, that is, to optimize the use of assets and ensure reliable delivery of electricity to the consumer. Asset management is an important tool or enabler that is driving smart grid enhancements, while at the same time, smart grid initiatives are also driving asset management capabilities and functionality within intelligent products.

The International Electrotechnical Commission (IEC) is coordinating smart grid efforts around the world. The IEC is developing many international standards that help operate the electric grid. IEC standards allow smart grids to cross national borders, ensure safe and efficient interoperability, and allow electrical grids to integrate renewable energy sources. Just as they are doing in North America, smart grid initiatives are strengthening electrical distribution system reliability around the world. From an end use consumer standpoint, smart grid initiatives are saving consumers' electricity and money. Consumers are taking an active role in controlling their energy consumption worldwide through many smart grid initiatives.

3.10 SMART METERS AND ADVANCED METERING INFRASTRUCTURE

Stuart Borlase, Mary Carpine-Bell, James P. Hanley, Chris King, Eric Woychik, and Alex Zheng

AMI stands for advanced metering infrastructure (AMI), a term coined in 2002 in a California PUC (CPUC) proceeding; today, AMI is synonymous with a smart meter system. Many industry experts consider smart meters as the foundation for smart grid, because smart meters measure energy, demand, and power quality at the endpoints of the grid. As the name implies, AMI is a technology that involves not only a meter but also an infrastructure of communications, software applications, and interfaces related to exchange of data between the electric utility, the meter, the consumer, and authorized third parties in some cases.

AMI communications systems vary depending on the location, geography, the utility company preference, and the technology choices available. The various forms of communications are discussed in an earlier section of this book. This technology of remotely and automatically collecting diagnostic, consumption, and current status from the meter and then transporting that information to a central database for billing, analyzing, and operations is changing not only the way utilities do business but the way that consumers behave. It is a change far greater than any seen in the utility industry since its inception. Metering technology has remained largely unchanged until the introduction of advanced metering during the past decade.

Smart meters are defined by their functionality. A typical set of smart meter functions is as follows:

- Two-way communications between the utility and the meter
- Recording of usage intervals of 15 or 60 min
- Sending of data to the utility at least daily
- Internal switch to disconnect or disconnect power
- HAN interface
- Recording of power quality information such as voltage and outages
- Functionality to ensure reliable and secure data communications

3.10.1 Evolution of the Electric Meter

Use of commercial electric meters spread in the 1880s in order to properly bill customers for the cost of energy. The electricity meter is a device that measures the amount of electric energy consumed

Smart Grid Technologies

FIGURE 3.161 Typical electromechanical electric meter. (© Copyright 2012 GE Energy. All rights reserved.)

by a residence, business, or an electrically powered device. Electricity meters are typically calibrated in billing units, the most common one being the kilowatt hour (kWh), which is the amount of energy consumed by a 1 kW load in 1 h. Billing of customers relies on periodic readings of the electric meters to determine energy used during that period or cycle. Most residential meters are of the single-phase type, whereas most commercial and industrial customers require three-phase metering. The electric meters measure the units billed according to the type of customer: in most cases, residential customers are charged a flat energy (kWh) rate, and commercial and industrial customers are charged an energy rate as well as a maximum demand rate (kW).

Electric meters fall into two basic categories: electromechanical and electronic. The most common type of electric meter currently in use today is the electromechanical induction watt-hour meter. An example of an electromechanical meter is shown in Figure 3.161.

The electromechanical induction meter operates by counting the revolutions of an aluminum disk, which is made to rotate at a speed proportional to the power through the meter to the load. The number of revolutions is thus proportional to the energy usage. In addition to showing continuous power use via the rotating disk, these meters also show cumulative energy usage via the dials at the top of the meter. The cumulative energy usage is the total energy used since the meter was energized, which utilities use to determine how much energy has been used by the customer. In order for a utility to read these electromechanical meters, they must send someone to physically go to each meter and write down the readings on the meter. The meters are read on a regular basis (monthly, quarterly, or annually depending on usage and population density)—the difference between the two readings is how much energy (kWh) was used by the customer between the readings. The readings are then entered manually into some type of customer database for billing purposes.

There are several drawbacks to the electromechanical meter:

Inefficient: Sending utility employees to walk house to house to manually read each meter and record each customer electricity usage is labor intensive and can be expensive depending on the location of the utility and the country.

Inaccurate: While the meters themselves are prone to inaccuracies, the manual meter reading and manual data entry into the billing system is also a source of error. Any errors required a significant amount of time to resolve, usually involving a separate visit to the customer to read the meter again. This also led to the development of a software application called VEE (validation, estimation, and editing). These applications take into account historical patterns of use and validate readings to ensure that they were in the expected range and edit or correct readings accordingly.

Tamper prone: Tampering with electromechanical meters to reduce the amount of energy registered by the meters can be one of the major sources of energy losses in a utility T&D system. Electromechanical meters are relatively easy to bypass or manipulate to alter energy usage readings.

No remote monitoring or control functionality: As part of the process of dealing with customers moving residences (typically more frequently for apartments or high-turnover housing, such as college campuses) and people not paying their electricity bills, utilities have to send someone out to the customer to turn the power off and back on for new customers. Also, if a customer reported loss of power, the utility would need to send someone to the customer to verify if the power supply on the utility side of the meter had failed or if it was an issue on the customer side of the meter (sometimes referred to as "OK on arrival," when loss of power is not the fault of the utility).

No consumer visibility of energy usage: Unless consumers physically read their own meters on a regular basis and more frequently than the billing cycle, they have no visibility of their energy usage until they receive the bill. Without detailed or real-time information on energy usage and time-based energy pricing, consumers have much less ability to reduce or time-shift energy usage and save money on their bills.

As the needs of electric utilities progressed and technology advanced, the next generation of meters were developed using microprocessors, which included communications interfaces and data storage capability. These electronic meters brought the benefits of digital technology with no moving parts, more accuracy, much more functionality, and also software programmable features. Smart meters are electronic meters that also have a communications interface for communicating with the utility. Most smart meters also contain an interface for communicating with the premise and a service disconnect switch. Today there are a multitude of electronic meters available and each with the capability to support numerous types of communications systems. An example of a smart meter is shown in Figure 3.162.

Electromechanical meters are still the most predominant type of electric meter currently in use around the world. The electromechanical meters are lower in cost than smart meters and represent a significant investment in terms of replacing them with smart meters. Some electromechanical meters have been is service for numerous decades. Electromechanical meters have a long service life. Utilities need to have a strong business case to replace electromechanical meters with smart meters.

The advent of electronic metering gave way to a more economical manufacturing process over electromechanical meters, mostly due to the electronic content, but also since one electronic meter could handle multiple meter configurations through firmware or software changes compared to the different types of electromechanical meters required for different applications.

FIGURE 3.162   Example of a digital smart meter. (© Copyright 2012 Trilliant. All rights reserved.)

Smart Grid Technologies

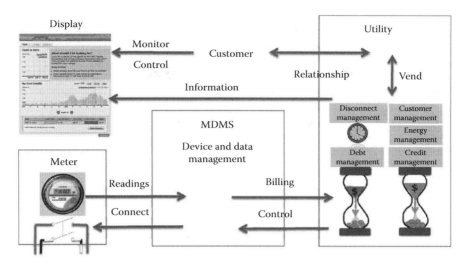

FIGURE 3.163 Data flows in the "central wallet" prepayment approach. (© Copyright 2012 eMeter, a Siemens business. All rights reserved.)

As the smart meter communications capabilities evolved to support two-way communications, so did the functionality of the meters. Some of these functions include the capability of remotely connecting or disconnecting power to the customer at the meter, notifying the utility when the meter loses power (outage notification), power quality monitoring, remote diagnostics capabilities, communicating time of use rates (pricing signals) to customers, and supporting various communications technologies—home area network (HAN)—interfaces that allow the meters to communicate not only to the utility but also to the customer. Smart meters are deployed not only in electric utilities but also in gas and water utilities.

The standard business model of electricity retailing involves the electricity company billing the customer for the amount of energy used in the previous month or quarter. Some utilities offer prepayment metering as an alternative to requiring deposits from customers with poor credit or customers who wish to budget their electricity spending more closely; in a few countries—notably the United Kingdom—customers who have not paid their bills are required to have prepayment metering [1]. This service requires the customer to make advance payment (prepayment) before electricity can be used. If the available credit is exhausted, then the supply of electricity is cut off at the prepayment meter. In the United Kingdom, mechanical prepayment meters used to be common in rented accommodations. The disadvantages of these included the need for regular visits to remove cash and risk of theft of the cash in the meter. Modern solid-state electricity meters, in conjunction with smart cards, have removed these disadvantages and such meters are commonly used for customers considered to be a credit risk. In the United Kingdom, one type of prepayment system is the PayPoint network, where rechargeable tokens can be loaded with the amount of money that the customer has available. In South Africa, Sudan, and Northern Ireland, prepaid meters are recharged by entering a unique, encoded 20 digit number using a keypad. This makes the tokens, essentially a slip of paper, very low in cost to produce. Smart meters enable the provision of prepayment service without any tokens at all, such as via a "central wallet" approach as shown in Figure 3.163. This approach further reduces the cost of prepayment metering [1].

3.10.2 Evolution of Meter Reading

With microprocessor-based electronic meters and the capability to communicate with the meters, utilities first started deploying systems that could remotely read meters, known as AMR (automated meter reading). The first AMR schemes allowed the meter reader to use a handheld computer to access data from the meter via a magnetically coupled or infrared communications interface. This was done primarily to capture energy usage interval data because such data could not be read on a simple display. The meter readings in the handheld computer were then later downloaded to a data collection or billing system when the meter reader returned to the utility office. This still required the meter reader to walk to each meter. AMR progressed to the use of a short-range radio frequency (RF) system to communicate with the meters. This allowed the meter reader to either "walk by" or "drive by" the meters without the need to enter the customers' premises. This approach allowed utilities to save on the cost of meter reading by increasing the speed of meter reading and avoiding issues with "hard to read" meters, either difficult for the meter reader to access the meter or the meter reader was unable to enter the customer premise due to locked gates, vicious dogs, etc. Despite the improved accuracy from the perspective of the meter reader and the savings in labor, there was still the cost of using the vehicle for "drive-by" meter reading, plus AMR offered no additional functionality; it simply replaced human meter readers.

AMR evolved to include one-way fixed communications networks from the meter to a central data processing system without the need to "walk by" or "drive by" the meters. This provided utilities with significant operational savings in reading meters and integrating the meter readings directly into their billing system. It eliminated the vehicles and enabled the reading of meters every day. It also added alerts, such as theft alerts and outage alarms. Later, two-way communications with the meters allowed the utilities to send messages and controls to the meters, such as to remotely connect or disconnect service to a customer. The advanced capabilities of the meters and the communications system, and the integration of the meter data into the utility billing and customer information systems (CISs) became known as the AMI.

A typical AMI system is shown in Figure 3.164 and is composed of an AMI/communications network headend; an integration platform; a meter data management system (MDMS); the AMI communications network utilizing various protocols, including mesh, cellular, WiMAX, PLC, and WiFi/Ethernet; and the AMI endpoints including AMI meters. The communications link to the customer is via a "home area network"—or "HAN"—an interface that is usually provided as part of the smart meter and which also supports commercial customer applications. The meter HAN interface

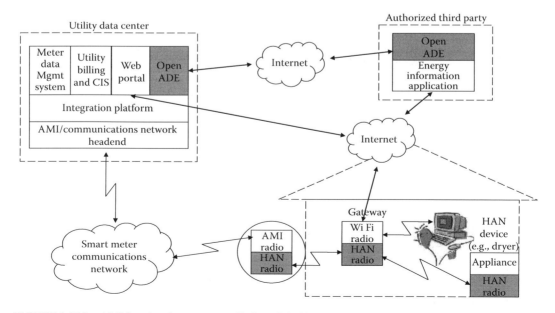

FIGURE 3.164 AMI functional components. (© Copyright 2012 eMeter, a Siemens business. All rights reserved.)

Smart Grid Technologies

TABLE 3.12
Advanced Metering Penetration in the United States (end of 2011)

State	Advanced Metering Penetration
California	10,017,553
Texas	4,934,632
Georgia	2,096,453
Florida	1,638,500
Pennsylvania	1,403,000
Arizona	1,215,000
Alabama	1,205,066
Oregon	832,643
Nevada	750,000
Idaho	492,000

is used to communicate with consumer thermostats, appliances, and other devices for consumer demand management, as well as support communications to EV charging stations.

The network headend monitors, controls, and manages the communications protocols and exchange of data on the communications network, including the provisioning of the meters and HAN devices on the communications network. Provisioning is the process used to identify and validate devices on the AMI communications network. The headend also enables the utility's operational employees to remotely monitor and maintain the health of the AMI network.

The MDMS remotely collects metering data from the meters in the AMI network and performs long-term data storage and management for the vast quantities of meter data. The MDMS also integrates the headend with the utility's CIS and other enterprise-relevant applications. The headend feeds the received endpoint meter data, via an integration platform, to the MDMS system, which will typically perform VEE on the data before making it available for billing and analysis. The MDMS also provisions the smart meters on the data network, that is, end to end from the meter to the CIS, and activates the meters for billing, once reliable and accurate data from the meter have been verified. The CIS captures a broad set of customer, location, service, asset, and financial information within a single application. The AMI system maintains a cross-reference of endpoint identification with the CIS, as does the MDMS, with the MDMS synchronizing the three applications. Apart from billing, the MDMS is also the data source—the system of record for metering data—for other utility enterprise applications, such as consumer online web content, demand management systems and outage management systems.

As of late 2011, advanced metering represents 18% of total installed meters in the United States, with existing utility plans and commitments raising that total to 40% by 2019 [2]. However, the penetration varies significantly by state. Table 3.12 lists the top 10 states by number of installed meters as of the end of 2011 [3]:

3.10.3 AMI Drivers and Benefits

Smart meters and AMI are considered to be one of the early drivers of smart grid and are commonly associated with the term "smart grid"—people typically thought of AMI as the first step toward smart grid. U.S. smart meter and AMI initiatives are a result of state initiatives (e.g., California and Texas), utility proactive steps (e.g., Oregon, Idaho, Florida, Georgia, and Alabama), and Smart Grid Investment Grants in the 2009 American Reinvestment and Recovery Act (e.g., District of Columbia and Nevada).

A major benefit of AMI is reduction in costs related to manual meter reading and the manual connection and disconnection of customers. Other benefits include the following:

- Valuable outage detection data that can be integrated with outage management systems.
- Tamper and theft detection and notification help in utility revenue recovery.
- More accurate meter reads and improved billing, and a reduction in customer complaints.
- A key facilitator for demand management programs. Consumers can receive price signals and more frequent and timely energy usage information so that they are aware of how much their energy use is costing and the value of conservation or efficiency investments.
- More customer information for the utility: The meter data that continuously stream in to the utility provides valuable information that can be used to make real-time operational decisions, as well as long-term decisions about planning and maintenance.
- The AMI communications link can also be used for other services to the customer, such as Internet access. Such access, along with real-time access via in-home displays that connect to the smart meter HAN interface, has been shown to reduce total consumption by an average of 8.7% [4].
- Some utilities are also trying to leverage the AMI communications system for other utility applications, such as communications for distribution automation.

Some potential, but so far unproven, disadvantages of AMI include the following:

- Loss of privacy since more detailed energy usage information is available from each customer meter
- Greater potential for monitoring or interception of customer data by unauthorized third parties
- Increased security risks from network or remote access

There has been much discussion recently regarding the advantages and disadvantages of smart meters (see examples of smart meter case studies at the end of this section). The business case for smart meters and AMI will continually evolve and be under intense scrutiny for years to come.

An interesting use case for real-time AMI is that of the requirements for the United Kingdom in 2014, where a newly formulated data communications company (DCC) for all utilities will be formed to collect 15-min interval data for all residential and small business customers. This is similar to centralized meter data systems in Ontario (Canada) and Texas, as shown in Figure 3.165.

3.10.4 AMI Protocols, Standards, and Initiatives

3.10.4.1 ANSI C.12.18 and C.12.19

Given the number of meter manufacturers and the utilities' desire not to be "locked in" or single sourced to a specific meter manufacturer, a standardized configuration/interface protocol and description standard was developed for meters. ANSI C.12.18 and C.12.19 were formulated to standardize the configuration, data gathering, and link layer communication to the meter. ANSI C12.19 is the American National Standard for Utility Industry End Device Data Tables and was established in 1997. ANSI, NEMA, and IEEE have adopted this standard for electronic metering products. This standard defines a set of flexible data structures for use in metering products. In particular, this standard defines a table structure for utility application data to be passed between an electricity meter and a typical handheld device carried by a meter reader or a meter communication module, which is part of an automatic meter reading system. The standard organizes metering device data and operating criteria into standard tables, which contain data common across all manufacturers, and manufacturer tables, which contain manufacturer-specific data.

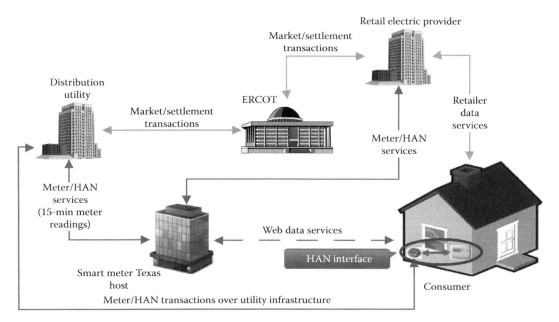

FIGURE 3.165 Smart Meter Texas portal. (© Copyright 2012 eMeter, a Siemens business. All rights reserved.)

3.10.4.2 IEC 61968-9 Common Information Model

The IEC 61968 series have been devised to facilitate interapplication integration for various distributed software applications supporting the management of utility electrical distribution networks. IEC 61968 supports the integration of disparate applications (legacy and new). IEC 61968 is intended to support applications that are event driven. It supports loosely coupled applications that have different run-time environments with heterogeneity in languages, operating systems, protocols, and management tools. Specifically, Part 9 of the IEC 61968 standard specifies the information content of a set of message types that can be used to support many of the business functions related to meter reading and control.

3.10.4.3 IEC 62056 DLMS-COSEM Standard

IEC 62056 (DLMS/COSEM)—Device Language Message Specification (DLMS)/Companion Specification for Energy Metering (COSEM)—is a suite of standards that specifies object models to represent the functionality of a meter, an identification system for all metering data, a messaging method to communicate with the model, and a transport method to carry information between metering equipment and data collection system. It is a standard that is growing in adoption in the IEC markets.

3.10.4.4 North American Electric Reliability Corporation: Critical Infrastructure Protection Security Requirements

The North American Electric Reliability Corporation: Critical Infrastructure Protection is the standards used to secure bulk electric systems. A section of the Reliability Standards for the Bulk Electric Systems of North America describes security requirements for critical cyber assets. The bulk electric system standards provide network security administration and support best practice industry processes.

3.10.4.5 National Institute of Standards and Technology

The National Institute of Standards and Technology (NIST) is developing key technical standards for an intelligent power distribution grid. ANSI C12.19 and IEC 61968 are part of the initial Smart Grid Interoperability Standards Framework released to date [5].

3.10.4.6 Smart Energy Profile

Zigbee Alliance's Smart Energy Profile (SEP) Specification was included in the NIST Framework and Roadmap for Smart Grid Interoperability Standards. SEP is intended as a choice for implementing AMI and smart metering on HANs and was in use in Texas, Nevada, and Oklahoma—and in pilots in California—as of the end of 2011.

3.10.4.7 Common Information Model

The Common Information Model (CIM) is defined and published by Distributed Management Task Force (DMTF). The CIM includes the CIM Infrastructure Specification and CIM Schema. The schema provides the actual model descriptions, while the specification defines the details for integration with other management models. CIM is also included in NIST's Smart Grid Framework and Roadmap.

3.10.4.8 802.16e

The 802.16e-2005 standard is known as Mobile Broadband Wireless Access System. 802.16e was an amendment to 802.16.2004 standard. 802.16e addressed mobility, better support for quality of service and of scalable OFDMA (orthogonal frequency-division multiple access). 802.16e-2005 is sometimes called "Mobile WiMAX."

3.10.5 AMI SECURITY

AMI security must take in consideration the costs of mitigating control techniques compared to the likelihood and risk scenarios of occurrence. While no security process, control, or solution can provide a 100% guarantee, utilities must identify and focus on the highest threats compared with the highest risks and impact occurrences. From this, methods to best identify security anomalies quickly, react accordingly, and segment where feasible are all controls needed to lessen the impact, outages, and risks.

3.10.5.1 Strategy

As outlined in the U.S. Department of Homeland Security, National Infrastructure Protection Plan, 2009, the Components of Cybersecurity Strategy focus on the following:

1. Prevention: Actions taken and measures put in place for the continual assessment and readiness of necessary actions to reduce the risk of threats and vulnerabilities, to intervene and stop an occurrence, or to mitigate effects.
2. Detection: Approaches to identify anomalous behaviors and discover intrusions and to detect malicious code and other activities or events that can disrupt electric power grid operations, as well as techniques for digital evidence gathering.
3. Response: Activities that address the short-term, direct effects of an incident, including immediate actions to save lives, protect property, and meet basic human needs. Response also includes the execution of emergency operations plans and incident mitigation activities designed to limit the loss of life, personal injury, property damage, and other unfavorable outcomes.
4. Recovery: Development, coordination, and execution of service- and site-restoration plans for affected facilities and services and reconstitution of smart grid operations and services through individual, private-sector, nongovernmental, and public-sector actions.

3.10.5.2 AMI Security Requirements

The following specific topics are considered key components that are needed to address the areas of compliance as outlined earlier. The smart grid security requirements have been developed to exceed industry security standards and recommendations identified by government organizations such as DHS, DoE, NIST, and NERC, as well as industry groups such as NEMA and OpenSG (AMI-SEC). The following smart grid security requirements address the following cybersecurity-related domains:

- *Time synchronization*: Time synchronization is a critical function to nearly every component of the smart grid system. Accurate time synchronization across multiple systems is required to ensure proper record keeping for both control system events and security events. Further, accurate time synchronization is essential to the proper implementation of certain cryptographic protocols and authentication schemes.
- *Logging and auditing*: Logging and auditing are common system functions that are required by many components of the smart grid system. There are three possibilities for a shared logging system. The first option is a centralized logging system, where all log records are transmitted to a central system on an immediate or "as soon as possible" basis. The second option is a decentralized logging system, where each system maintains logs separately, but the smart grid system provides an interface for querying remote system logs from a centralized location. The third option is a hybrid system, which allows each system to maintain its own logs of record, but those logs are transmitted on some periodic interval to a centralized logging service. The most appropriate option is ultimately based on the goals of the system.
- *User management*: User management is the process of managing users of the smart grid systems.
- *Role management*: Role management is the process of managing roles (or groups) that control access to various smart grid system functions. It includes but is not limited to creation, modification, and deletion of roles, assigning users to roles.
- *Resource management*: Resource management is the process of managing various resources such as HTML or other pages (JAP, ASP, etc.) or resources such as EJB, etc. and which roles can access those resources.
- *Authentication*: Authentication is the process of proving that the requestor of an action is valid. A requestor may be a person but also may be another system or service. This is typically accomplished through such mechanisms as passwords, token authentication, or certificates.
- *Authorization*: Authorization is the process of proving that the validated requestor of an action is authorized to perform that action. A requestor may be a person but may also be another system or service. This is typically accomplished through such mechanisms as role-based access control or directory group membership and can include transaction level security.
- *Password management*: User's password to the smart grid systems should be managed so that various complexity requirements can be met without having to redeploy any component. Password management process will provide the functionality to configure various complexity requirements for any or all users and will provide functionality to reset or change passwords.
- *System account management*: System accounts that are used within the smart grid systems such as username and password to connect to database, certificate key store password, etc., should be configurable both at deployment and at any time during the lifetime of the smart grid system.
- *System and information integrity*: Smart grid systems exchange data within and outside of system boundary and may contain business critical or confidential messages. It is highly important to protect the integrity of the data during transit.
- *System and information confidentiality*: Encryption technology is used to secure data while in motion or while at rest. Data in motion are considered particularly vulnerable, especially when passing from trusted systems to less trusted systems. All system components that will use encryption to secure information in transit will adhere to the following system requirements. Encryption of data while at rest (in storage) is not addressed in these requirements.
- *Cryptographic key management*: Cryptographic keys are used widely in smart grid systems, either symmetric encryption keys or public/private keys of the certificates.

The cryptographic keys typically have an expiry date and sometimes mandated to be changed either at regular intervals or when there is any breach.
- *Firmware management*: Numerous smart grid components will contain firmware as an integral part of the component. How this firmware is managed represents a significant portion of the security related to the lifecycle of hardware systems.
- *Interoperability standards*: Security architecture of the smart grid systems embrace the various interoperability standards set by organizations such as W3C, Oasis, IEEE, IETF, etc. By adopting standards wherever possible, it differentiates our products from competitors and provides value for the customer as they can leverage either built in security or can integrate with their existing security systems. The requirements in the following will describe various standards that smart grid systems shall support wherever applicable.

3.10.5.3 AMI Security Threats

AMI systems are implemented over a wide variety of infrastructures. Designs include both wired and wireless communications as well as a mix of public and private networks. The applications, which run through these infrastructures, are capable of implementing risk-appropriate security strategies in order to mitigate the impact of a range of threats. Security for information exchange between applications is supported by both the endpoints that either receive or provide the information, as well as any middleware. This is essential because the AMI as a whole may be exposed to several types of threats:

- Compromise of control (i.e., system intrusion)
- Misuse of identity or authority to gain inappropriate depth of access
- Exposure of confidential or sensitive information
- Denial of service or access
- Breach of system, import of errors (integrity)
- Unauthorized use (authorization)
- Unidentified use/misuse (authentication)
- Manipulation and destruction of records (auditability/proof)
- Delayed/misdirected/lost messages (reliability)
- Loss due to system (loss/damage)

Because of the number of products and technologies used in AMI deployments, typically layered security architectures are designed and implemented. Such architectures allow for a blending of different cost-effective technologies with suitable risk mitigation techniques, including using compensating controls in the case that some system components are not inherently secured.

3.10.5.4 Applying Security Specification to AMI

There are three predominant integration environments to consider for security purpose:

1. Utility mission critical operational systems and data access/exchange
2. Utility internal enterprise front and back office applications and data/exchange
3. Utility external application and data access/exchange

Security measures are implemented in AMI implementations to support the specific needs of these environments.

3.10.6 AMI Needs in the Smart Grid

Smart meters and AMI aggregate data collected from the meters and provide the means to communicate data to the customer. Within this context there are many requirements that must be supported by the smart meter and AMI as smart grid advances beyond the pilot phase.

3.10.6.1 Meter Data Reads

The systems must maintain a default schedule for uploading scheduled meter reads and also respond to on-demand reads for all authorized systems within a defined timeframe.

3.10.6.2 Internal Device Management

Each endpoint device within AMI systems has the ability to perform internal self-monitoring and trigger alarms and/or AMI notifications based on configurable thresholds. The devices are also required to maintain an onboard history of log information that can be delivered to all authorized systems on demand.

3.10.6.3 Remote Configuration

Each endpoint device must support configuration for internal settings and processes. Most AMI meters receive and process meter configuration files over the AMI network that provide configuration for all tables.

3.10.6.4 Firmware Upgrades

Each endpoint device will need to support the ability to handle remote firmware upgrades delivered from all authorized systems. The upgrade process should minimally impact the production operation of the AMI system.

3.10.6.5 Time Synchronization

Each network device is required to maintain time synchronization with the NMS that utilizes an NTP server for consistent time.

3.10.6.6 Local Connectivity

The AMI meter must allow connectivity from all authorized devices, which include support for local handheld devices that are used by utility personnel in the event of a network disturbance.

3.10.6.7 Testing and Diagnostics

Each endpoint device must perform self-diagnostic processing to identify both actual and potential issues and report these to the NMS in either realtime or on demand depending on the criticality of the event.

3.10.6.8 Other Functions

The AMI system also provides transport for several utility back office functions and event-driven processes that include the following:

- Service/load control
- Outage detection
- Service switch
- Service limiting
- Tamper detection

3.10.6.9 Supporting the Customer Interface

AMI will need to support demand response functionality by responding to NMS-delivered events and enabling delivery to HAN devices within the customer premise. These will include the scheduling, cancelation, and rescheduling of demand management events. AMI must also be able to support electric vehicle billing. Take, for example, roving PHEVs on the utility's

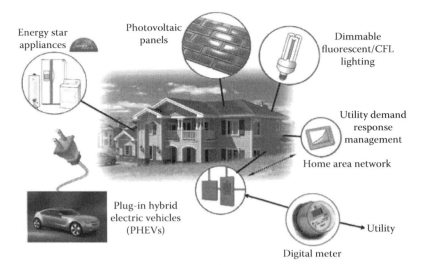

FIGURE 3.166 Customer interface challenge. (From Fan, J., Borlase, S., Advanced distribution management systems for smart grids, *IEEE Power Engineering*, March/April 2009. Copyright 2009 IEEE.)

(or another utility's) distribution network. Point-of-use metering and energy charge or credit must be managed and tracked on the distribution network. This is a challenge in terms of not only the additional load or potential supply (and related protection and control issues) but also the tracking and accounting of energy use or supply at various points on the distribution network or on a neighboring utility's distribution network. This is a huge challenge for utilities and is leading to a significant change in data management and accounting away from the once-a-month meter reading and billing of customers. The customer interface challenge is illustrated in Figure 3.166.

3.10.6.10 Integration with Utility Enterprise Applications

AMI must be able to integrate seamlessly with other smart grid enterprise applications in order to realize the full benefit of exchanging data between the AMI, OMS, and DMS in addition to meter-related applications of billing and the CIS. This is achieved by the integration platform shown in Figure 3.167.

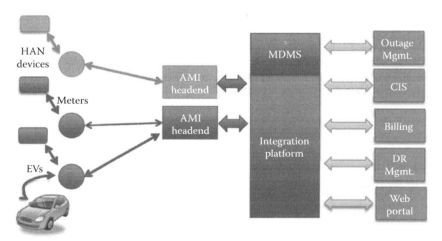

FIGURE 3.167 AMI integration with enterprise applications. (© Copyright 2012 eMeter, a Siemens business. All rights reserved.)

ADDRESSING PUBLIC OPPOSITION TO SMART METERS IN CALIFORNIA

California's utilities, which have recently ramped up efforts with regard to AMI, have nonetheless had their share of hurdles to overcome. As of late 2011, the three major investor-owned utilities—Pacific Gas & Electric (PG&E), San Diego Gas & Electric (SDG&E), and Southern California Edison (SCE)—had completed nearly 90% of their planned combined deployment of approximately 10 million electric smart meters.

A small but very vocal group of customers has opposed the deployments (New York Times: "New Electricity Meters Stir Fears", 30 January, 2011). Some, who already had the smart meters, claimed the new meters were inaccurately measuring their power use, resulting in abnormally high bills. The CPUC officially received 1100 such complaints about higher rate. After following up with these concerns, however, the CPUC found that none of the complaints were due to faulty meters and that all of them were in fact due to misunderstandings about the rate structure or an unexpected spike in electricity usage.

As part of its investigation, the CPUC ordered a third-party audit of the advanced meter deployment to determine if there were billing problems (San Francisco Chronicle, "Customers say new PG&E meters not always smart," October 18, 2009). In this case, the issue was not technical, but in fact a public relations problem. Since then, the utilities have taken a greater eye to managing public relations, investing millions more in public education and outreach programs. The CPUC has also allowed customers to opt out of smart meters, provided the customers pay the utility's added costs associated with maintaining manual meter reading. Fewer than 1 in 1000 customers have chosen to opt out.

CASE STUDY: ENEL SEEING REAL AMI BENEFITS IN ITALY

ENEL, the largest electricity distributor in Italy, has deployed advanced meters to almost all of its 30 million customers, at least 24 million of which are being read remotely. As the first utility in the world to embark on such an aggressive, large-scale deployment of advanced meters, ENEL has served as a positive example for other utilities around the world. The installation came at a cost of about €2 billion, equal to approximately €70 per customer. Benefits ENEL realized through its AMI deployment include better revenue collection, lower operating expenses, increased operational efficiency, and improved service to customers. The operating savings to ENEL have been estimated at €500 million per year by the utility, suggesting a payback period of 4–5 years [5]. The primary drivers of these savings are reductions in field labor.

3.11 CONSUMER DEMAND MANAGEMENT

Stuart Borlase, David P. Chassin, Gale Horst, Salman Mohagheghi,
Eric Woychik, and Alex Zheng

The management of energy demand at the consumer has been the focus of research and debate for several decades. Consumer demand management takes many forms driven by utility incentives for consumers to reduce overall energy usage or to manage energy usage patterns. Widely used terms for consumer demand management include energy efficiency, energy conservation, demand-side management (DSM), demand response (DR), and DR management (DRM).

A major challenge for a utility regarding supply and demand of electricity is that the load on the system is not constant and the utility must try to efficiently dispatch generation to meet the load on the system at various times during the day. Utilities typically have various generation sources, such as coal-fired plants, gas turbines, hydroelectric power, or power purchased on the open market, that cause the utility to incur different costs. In reality, the cost to supply electricity on the grid is related to the change in load over time and the changes that occur in electrical supply. Changes in supply include outages of generation and transmission and changes in supply from energy sources such as wind and solar photovoltaic (PV). During peak energy demand or major changes in supply, higher cost generation sources are used such as gas turbines, which result in a higher cost to supply required energy grid needs. During the lowest demand periods, lower cost generation such as nuclear, hydro, and coal-fired power plants are the primary sources of electricity. A traditional way to bill customers for electricity is to charge the average price to supply electricity throughout the year and to measure energy consumption in terms of non-time-differentiated energy (kWh) use over a period or weeks of months.

It may seem counterintuitive for a utility to implement measures that reduce consumer consumption since it reduces the utility revenue. However, in some cases, the reduction in energy consumption during peak load periods can provide significant operating and financial benefits to a utility. This may translate to avoiding grid congestion, or the ability to supply electricity to customers during contingencies, such as when generation is unreliable. This shift of peak electricity may result in significant reductions in capital investment and operating costs, while maintaining grid reliability. Many retail electricity suppliers currently implement specific usage rates or charges for industrial and commercial customers that include some type of maximum demand component and corresponding fee or penalty for consumption beyond agreed-upon levels.

3.11.1 Demand Management Mechanisms

Demand management is generally based upon actions on the consumer side of the meter that reduce consumer load, invoke energy efficiency, DR, distributed generation (DG), or energy storage. Use of the full set of demand management options is also called integrated DSM (IDSM) [1]. To distinguish, energy efficiency measures alone generally focus on reducing total energy consumption of consumer loads, such as lighting, space conditioning, appliances (e.g., refrigerators, washing machines, dishwashers), variable speed motors, etc. Instead, IDSM is an approach to offer customers a full suite of demand-side opportunities.

One part of this, DR, typically refers to reducing consumer electricity usage at specific times during the day, week, or season. Historically, DR has been used to reduce electric usage during peak load times. Reduction in the peak load typically results in deferral of the energy used to an off-peak period. Alternatively, peak load reductions can effectively reduce the load on the system altogether. In recent times, DR has taken a number of forms such as voluntary curtailable load, behavioral-based response, price-based response, or event-based response.* An early form of DR is direct load control where utilities could remotely control consumer loads such as water heaters and air conditioners to turn them off during peak load.

The idea that the demand side of the grid can be managed and demand can respond to information and to signals from the utility is not new. Still, in most markets, demand remains unresponsive to price signals from suppliers or the suppliers may not yet provide demand signals. Both direct load control and price response are increasingly being considered or piloted in various locations. To illustrate, consider the impact of higher gas prices on the amount of air travel or the type of car you consider purchasing. If prices go up a few cents per gallon, you might not care, but when gas prices double, or triple, or even go up by an order of magnitude, your behavior is more likely to change. Presumably most consumers will shift their behavior or respond in some way as prices change.

---

* On the one hand, voluntary customer curtailment can reduce loads but is usually paid only an energy price ($/kWh) for such reductions. This behavior is neither certain nor predictable. On the other hand, sophisticated electronic controls enable rapid dispatchable load reduction at a specific location. These certain, predictable actions to lower loads may be in response to network operating events or changes in generation pricing.

Smart Grid Technologies

As an alternative approach, customers may be offered incentive payments in lieu of high prices as a method to encourage demand reductions.

If load reduction measures can be encouraged through customer incentives, information, pricing, and technology, the costs to provide the capacity needed for reliable electric service can be significantly reduced. In these ways, a portion of the burden of reliability and the risks with power outages can be shifted to the consumer who will either directly or indirectly realize the benefits. As consumers more cost-effectively manage their consumption through load response, overall system costs can be reduced. For example, with customers enrolled in peak load response in ISO-NE, the system reserve margin may be lowered 10% or more [2]. Correspondingly, this is expected to reduce the probability that peak load would exceed power availability by between 10% and 50%. This reduction in peak capacity bodes to reduce ISO-NE supply side costs by as much as 8.5%. Thus, with an incremental amount of demand management, significant electricity cost reductions are possible.

DG installed at the consumer site is another effective means to reduce load demand on the grid seen by the utility. DG generally includes any generation at the customer premises, such as solar PV, wind turbines, fuel cells, combined heat-and-power, and diesel generation or microturbines. Some of the DG is considered to be *must-take* as it is simply put on the grid whenever electricity from these sources is produced. In other cases, it has become dispatchable* by either the utility or the ISO who can determine when consumer sources are connected to the grid. When DG provides power to the grid, it may be *net-back-metered* (credited at retail rates), paid or credited at wholesale prices, or valued at some other contractual or tariff regulated rate. If the DG is dispatchable by the utility or the ISO, consumers may be paid for both availability of the distributed energy resources (DERs) as well as additional incentives when the resources are actually called upon. Larger-scale nondispatchable DG such as wind power or PV electricity are generally paid an energy value but are not considered to provide capacity benefits or avoided transmission and distribution (T&D) benefits as they are variable in nature and thus are uncertain resources. This is changing, however, as some renewable resource deployments (e.g., wind and PV) are integrated with battery storage to create a reliable source of energy.

Energy storage devices at a consumer site can take power from the grid (to be charged) and provide power back to the grid (discharge) at critical times. Accordingly, storage can be dispatched to gain market benefits and take advantage of low-cost power (for charging) at times when grid reliability and costs are less consequential. The ability of storage to perform market arbitrage, as with DR, depends on the speed of the response and the availability of the storage when grid needs and market prices are greatest. Storage may be controlled through voluntary (manual) behavioral response or automatically with use of event- or price-based triggers. This suggests that for storage to be of greater value, advances in control and grid interface technology are a high priority. Energy storage can serve a number of purposes to meet peak loads, variability of renewable generations, or other grid dispatchability needs. Increased use of storage is expected to provide capacity availability, energy, voltage support, and frequency regulation. It is considered a flexible resource with significant market opportunities. Batteries can be utilized in various formats. Larger batteries are being piloted in substations and to support large PV or wind installations. Smaller batteries are being installed as community energy storage (CES) and technology is being developed that can utilize energy backfeed from vehicle batteries.

3.11.2 Consumer Load Patterns and Behavior

At the same time, every day as people get up and go to work and as industry and commerce begins, electric power systems ramp up in order to meet demand. The electric grid sees predictable changes

---

* Dispatchable resources are those available to be turned on and synchronized to grid frequency or can be turned up or down to vary generation capacity, in response to grid operator instructions. Nondispatchable resources cannot respond to grid operator instructions, so are considered *must-take* or may be *baseload* resources.

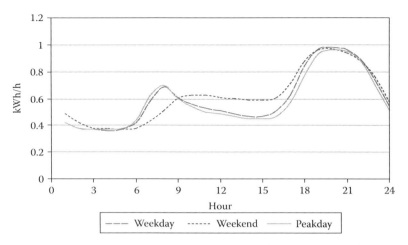

FIGURE 3.168 (**See color insert**) Example diurnal load shape. (© Copyright 2012 Pacific Northwest National Laboratory. All rights reserved.)

in load over the course of each day depending on the type of load and various other factors, such as the day of the week, temperature, etc. This results in a series of daily, weekly, seasonal, and annual cyclical changes in load and load peaks. In each case, the utility must provide sufficient generation and transmission capacity to ensure that demand is fully met in all circumstances.

During the hottest hour of the hottest day in the summer, some utilities experience an enormous demand for electricity, caused by air conditioning units, that the utility must meet to avoid a local brownout (low-voltage condition) or a system blackout (complete loss of voltage). In order to meet these few hours of "peak demand," the utility must build or have access to enough generation and transmission capacity to meet the system peak. However, much of that capacity remains unused during the remainder of the year. In fact, for much of the year, electricity demand does not approach the level of the annual peak.

Load shapes describe the changes in load on a daily or seasonal timescale. Most load shapes are typically represented as an average hourly energy use, for example, kWh/h (Figure 3.168). However, seasonal load shapes are sometimes represented as peak values, for example, MW, even though sometimes that peak value is obtained from the maximum of an average diurnal load shape, for example, MWh/h. In either case, the load is effectively a power value and not an energy value.

There are some important characteristics of load shapes that must always be considered when they are used. First, the difference between summer and winter lighting load shapes is greater the further the load is from the tropics. This means that any load control system that is affected by diurnal phenomena, such as temperature or insolation, is going to vary more seasonally the further the location is from the equator. Second, for the same outdoor temperature, air-conditioning loads are typically higher in humid climates than dry climates. This means that air-conditioning control strategies may tend to yield greater benefits in humid climates than they do in drier climates. Third, higher-income regions typically have higher loads per capita than less affluent areas. Fourth, commercial loads may be less dependent on climate and weather than are residential loads. Commercial buildings' cooling systems are more dominated by internal heat gains from lights, computers, and people, and they have less exterior surface area per square foot of floor than do residential buildings. Fifth, industrial loads are also sensitive to economic conditions. When the economy slows, the first things to slow are typically the factories. Sixth, agricultural loads are highly seasonal and sensitive to weather. Water pumping and refrigeration are driven by the growing season in any given region. Seventh, load shape data can change significantly over time because of evolving energy efficiency standards and consumer purchasing habits. Much of the load shape data from the 1980s are still

Smart Grid Technologies

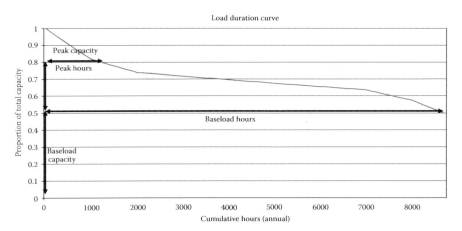

FIGURE 3.169 Typical load duration curve. (© Copyright 2012 Alex Zheng. All rights reserved.)

being used because of a lack of newer better data. But the penetration of consumer electronics into the residential market has changed substantially since then, even though the efficiency of appliances has improved significantly at the same time. Eighth, the composition of the load has changed over the years, particularly due to an increase in consumer electronics or inductive motors. And finally, because lighting efficiency programs have been successful in recent three or four decades and new loads have emerged, lighting load has become a smaller fraction of the total load.

One way to visualize utility peak loads is to take all the hours in a year and the corresponding maximum load in each hour and then rank them by the load demanded. Graphing this from the hour of the highest load to the hour of lowest load produces a "load duration curve," which is shown in Figure 3.169.

The figure immediately makes evident that the slope of the line does not remain constant. Rather, the slope is steeper at the left end and at the right end of the graph and less steep in the middle. This signifies rapid change in power demand as you consider the top 1000 load hours and lowest 1000 load hours of the year. Figure 3.169 labels a few notable features of the load duration curve. In addition to the peak shown in this curve, the response of consumer loads can also add value when utilized to compensate for variations in output from renewable energy sources. The most common use of a load duration curve is for planning studies, when planners estimate the number of hours per year for which a system resource must be allocated. The load duration curve can also be used to estimate the maximum amount of load that that should be curtailed during a certain period.

While the load shape describes the amount of load that is present at any given time of day and day of year, that description is *not*, by itself, sufficient for the study of loads in the context of the smart grid. The load ramp time, duty cycles, and periods of those cycles are also very significant when load control strategies seek to modify them. It is possible to measure directly both the duty cycle and the period of loads using end-use metering technology. However, end-use metering is quite expensive, and the data collected about any given device are often the superposition of the device's natural behavior and other driving functions such as consumer behavior. So it is often very difficult to clearly identify the fundamental properties of each load. Furthermore, for any time-domain models where the load aggregate is a consideration, not only the duty cycles or on-probabilities of devices must be considered, but also their periods and state phases, as shown in Figure 3.170.

In such cases, the state of a single device is the position of the device in relation to one complete cycle. Hence, it may be more complicated to aggregate devices with differing usage patterns and functional power cycles. This is important in demand management where utilities need to take into account the diversity of consumer load patterns when modeling loads.

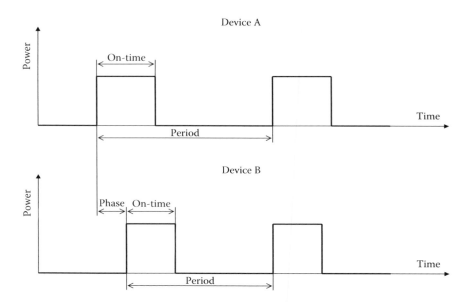

FIGURE 3.170 Load aggregation. (© Copyright 2012 Pacific Northwest National Laboratory. All rights reserved.)

3.11.3 Conserved versus Deferred Energy

Load managed by DR can be separated into two parts—conserved energy and delayed energy. In some instances, the load that DR turns off at a designated time is not "recovered" or deferred entirely for use at a later time. This permanent reduction in energy use due to short-term reduction in demand is referred to as conserved energy, because the net impact is equivalent to having never used that energy in the first place. However, not all of the consumer load reduction is completely conserved. For example, although loads from lighting may not need to be made up for at a later time, shutting off a hot water heater or air conditioner for a short amount of time may result in additional loading at a later time. This "bounce-back" effect can sometimes result in secondary peaks later in the day, when the entire curtailed load comes back online. This energy consumption is known as deferred energy, or the "snap-back" effect, because it still occurs but at a later time. Consumer load reduction needs to be managed carefully to ensure that it does not create artificial peaks that are costly to manage as a result of deferred demand for energy returning later in the day. Utilities have a variety of methods for managing deferred energy, but most methods are essentially different ways of staggering the return of consumer consumption to full load.

From the perspective of load behavior, demand management has two mechanisms to offer utilities and customers (Figure 3.171). The first is energy efficiency or energy conservation. These strategies reduce the total electric energy consumed by a customer. Typically these include (a) reducing total runtime, for example, by lowering a thermostat; (b) reducing load during operation, for example, by retrofitting higher efficiency equipment; and (c) substitution of fuel sources, for example, replacing central fossil-fueled electricity with distributed renewable sources. The primary benefit of energy conservation is to allow the utility to avoid the cost of acquiring new sources of energy to meet growing demand. Although a utility's primary incentive is to increase revenue from energy sales, a significant fraction of a utility's costs include the acquisition and financing of new energy sources. New energy resources can often be so expensive that the effect on rates is too great for consumers to bear. For example, if a utility forecasts a 50% growth in demand over the next 10 years that would result in a 25% rate increase, but it can implement an energy conservation program that reduces

Smart Grid Technologies

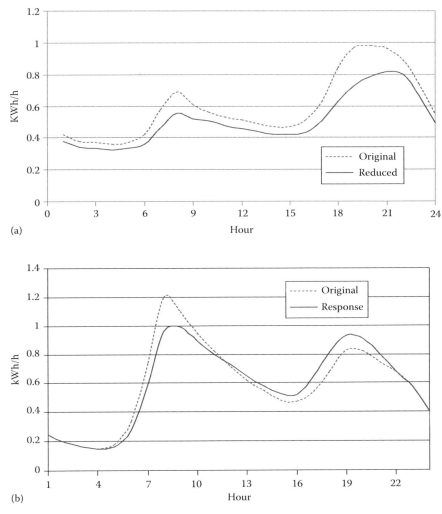

FIGURE 3.171 (See color insert.) DSM mechanisms: (a) energy conservation and (b) peak load shifting. (© Copyright 2012 Pacific Northwest National Laboratory. All rights reserved.)

that growth to less than 10% over the same period without a rate increase, the obvious choice is the conservation program.

The second is peak load shifting. These strategies reduce the peak load on the system by shifting coincident demand to non-coincident times, for example, by using energy storage. The primary benefit of peak load shifting programs is that they allow the utility to avoid the need to build new system capacity that does not come with a corresponding increase in energy sales revenue. Adding capacity is typically a very capital-intensive proposition for a utility, so any program that can move load off peak without reducing revenues from energy sales is attractive.

As a general rule, strategies that reduce energy consumption are supported by existing utility DSM programs. These programs are not generally considered smart grid programs in today's sense of the word because (a) they are already widespread and (b) they do not require information technology (IT) to realize the majority of their benefits. In contrast, peak load shifting programs require accurate and timely information to operate effectively, particularly if incentive signals, such as prices, are to capture all possible opportunities and reconcile any contradictory signals that may persist. One exception is conservation voltage reduction (CVR). CVR can be used to maintain voltage levels between allowed maximum and minimum ranges, through use of smart grid technology to monitor, manage, and reduce voltage, producing an overall conservation effect.

The focus of most advanced load modeling research is on those load behaviors that are affected by or can directly participate in smart grid technologies. Hence load models primarily address load shifting behavior and other behaviors that respond to relevant signals from utilities or from the bulk system.

3.11.4 Supply Side of the Equation

Three basic types of power generation are used meet the varying characteristics of electric loads: baseload units, intermediate units, and peak load units. These three categories of power generation largely reflect trade-offs between capital cost and operating cost, where each type of generator serves a different role, and in combination, they lower total costs and meet reliability needs.

Baseload units: Refer to Figure 3.169. The area labeled baseload capacity represents generation that essentially operates continuously at the same level. Baseload generation typically includes nuclear, coal, or very efficient gas-fired plants. The generation from these plants is used during all hours of the year. Note that in some regions, hydro may also be used to supply base load when it is in ample supply.

Intermediate or responsive and load following units: These generation plants are generally more responsive to load changes than baseload units, are intended to ramp (up and down) to supply varying load. Many units in this group may be considered load following units. Moving up the load duration curve from baseload, these units help the grid supply electricity load demanded by consumers on the timescale of minutes or hours, and some of these plants can respond to load changes in seconds. Responsive and load-follow generation capacity is often provided by combined-cycle natural gas power plants and hydro generation. Intermediate generation units are not always needed on the grid, such as at night, and when they are, it is not always at full capacity.

Utilities or system operators pay a premium for intermediate and responsive load following power over baseload units because such plants can be quickly dispatched as needed at a relatively low variable cost. As they must be flexible enough to be ramped up and down as a regular part of operations, the operating efficiency of these plants is usually less than baseload generation which translates to somewhat higher variable costs.* These price premiums help compensate plant owners for the reduced production that occurs, to make up for the fact that they do not generate power for as many hours as baseload plants. In exchange for this higher cost, intermediate plant owners make assurances that they will be available to provide power when needed.

Peaking units: Peaking units help utilities serve loads during the hours of highest demand. The steep slope of the left end of the load duration curve in Figure 3.169 indicates that for a few hours in the year demand for power is much greater. In this example, approximately 20% of the total system generation capacity must be available to serve less than 10% of the hours in the year. As with responsive and load following units, peaking units operate even fewer hours per year, and the owners of these plants must receive premium payments in order to stay profitable.

The pattern of generation use and the types of generation available have major cost and environmental consequences. Less efficient units tend to emit greater pollution and are more expensive to operate, but they are responsive when needed. Responsive, load following, and peaking units many times sit idle until the few hours per day or per year they are needed. While utilities and grid operators would undoubtedly prefer not to have to pay for these expensive units, they represent the only

---

* This also reflects the distinction between energy payments and capacity payments. Energy payments are for electricity produced ($/MWh) in an hour. Capacity payments are for the availability to respond as needed to major changes in load or generation on the grid and provide a specific level of output ($/MW). Baseload units are considered to provide energy supply. Responsive and load-following units provide energy and capacity and thus in some markets get paid for both.

supply-side tool to meet large spikes in demand to respond to contingencies that occur with other plants and the T&D system.

Because utilities and system operators have minimal influence and control on the load or demand side of the grid, they are required to build-out redundant supply-side capacity in order to ensure a high level of reliability. With significant peak demand by consumers, utilities must procure generation portfolios that are largely overbuilt. As a result, most power systems have low-capacity factors. To illustrate, in 2006 in the ISO-NE region of the United States, 25% of the entire generation fleet ran 2.92% of the time or less, and 15% of the entire generation fleet ran 0.90% of the time or less. These calculations exclude the idle generation that was kept available to meet the required 10% planning reserve margin [3]. Put another way, 20% of ISO-NE's total generation capacity was used to deliver only 0.34% of the annual energy use, and 30% of its total generation capacity was used to deliver only 1.63% of annual energy use [4]. In most other parts of the United States, generation asset utilization is similarly low, and, on average, the entire U.S. generation fleet operates at a 50% capacity factor.

These illustrations show that building plants to meet the highest peak loads is quite cost-inefficient, and much of the power plant fleet is unused over many hours of the year. In other words, the system is significantly overbuilt in order to compensate for varying loads. Currently these units must still be paid for by the utility or power market and ultimately by consumers. The electric power industry has followed this paradigm for many years. However, smart grid technology is now able to provide efficient alternatives to supply-side only solutions.

3.11.5 Consumer Side of the Equation

Utilities and power providers have for the most part found that management of consumer loads provides the lower financial returns. As a result, consumer loads today are relatively unresponsive to changes in system conditions. To utilities and other grid operators, consumer loads must be met regardless of how much power is demanded. Utilities and grid operators have assumed complete responsibility for meeting consumer demand, regardless of the pattern of the demand or how much it costs to provide the electricity.

There are significant pilots and research examples that indicate that consumers may be willing to change their demands in response to incentives, information, and prices. Recent studies have shown that something as simple as an in-home display that shows peak hours or peak prices can help shift consumer demand. Recent advanced technology allows appliances to automatically change power use based on the grid needs or in response to different electricity prices in ways that minimize the impact to consumers. These advances suggest that the consumer mass market can change their demand, which opens up a whole new industry for DSM technologies and services. This has resulted in the development of a cohesive set of product technologies, programs, standards, and consumer devices for consumer demand management. Measurement and validation of demand management participation at the consumer is required in addition to a means of financial settlement, both of which can be enabled by *smart meters*. Communication is also a key component over both the utility service area as well as within a customer premise. AMI infrastructures can provide communications to the consumer for demand management, although other communications technology options are also being explored and piloted.

Assume that consumer loads can be divided into two categories, so-called critical or nondiscretionary loads that cannot be disconnected from the grid at any time and noncritical or discretionary loads that are not significantly impacted if disconnected from the grid. Critical loads might include hospitals, critical telecommunication infrastructure, security systems, and emergency response sites. Noncritical loads might include residential customer washers and dryers, hot water heaters, decorative lights, and some part of air conditioning load. Discretionary loads could include dimming certain lights, reduced heating/cooling needs, and altering less critical certain business processes.

Beyond the distinction between critical and noncritical loads, it is useful to think of loads across a spectrum of values based on the customer's willingness to alter usage patterns. Consider a potential scenario where each load is prioritized compared to other loads, and the importance of each load is reflected in the price the consumer will pay to retain the service provided by that end-use device. The load (importance) ranking for every load in every house, neighborhood, or city could be put in line to receive power based on how much value it provides, rather than treating all loads as having equal importance. If each load can be controlled on or off based on the current price of electricity and the value placed on the service by the consumer, one could envision a prioritization of each load in terms of a specific electricity price level. A similar approach is to couple specific consumer electricity uses with specific incentive levels that the customer will accept. Both market pricing and customer incentives aim to provide a value proposition for consumer responses to specific loads, which reflect the customer's value to sustain or curtail each load.

3.11.6 Utility–Customer Interaction

There are a number of ways that customers can use demand management. Generally, consumers need to know what loads they want to reduce and utilities needs to provide customer participation options that may include advanced metering and market pricing data. With a smart meter in place and advanced electricity pricing that is communicated to consumers, automated DR can be employed to directly trigger load reductions, also called *auto*-DR, when specific price levels are exceeded. A major aim of auto-DR is to enable DR through a preprogrammed response, for example, to reduce appliance loads at specific price levels, so that consumers can directly participate with minimal effort and gain the benefits of DR. This back and forth between market prices, customer incentive, the meter, and consumer loads will enable a more complete electricity market, particularly as loads can increasingly respond to prices as much the supply-side responds to price. This fully participatory DR market will provide the needed compliment to the supply-side of the electricity market, resulting in greater market efficiency.* In contrast, today's largely supply-only market leaves electricity prices unresponsive to loads, and consumer response is not a factor. Not only will a more dynamic consumer load and supply-side market be more efficient and increase reliability (reduce blackouts and brownouts), it will reduce market price volatility and the potential for market power manipulation.

Figure 3.172 illustrates a set of steps and information flows between the utility or market and the consumer, focusing on the use of price signals. The electricity market takes energy information and provides price signals, which are sent to consumers by advanced smart meters or other communications mechanisms such as cable TV, phone line, wireless networks, etc. Consumers can

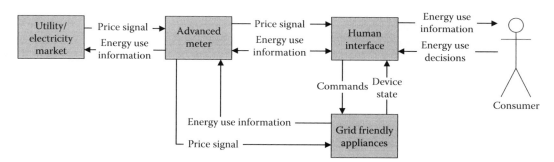

FIGURE 3.172 Example of information flow between utilities and consumers in DR. (© Copyright 2012 Alex Zheng. All rights reserved.)

---

* This is known as a dual, supply-demand market equilibrium, as compared to the predominantly supply-side only electricity market equilibrium. See Ref [5].

respond extemporaneously or through automated technology to direct incentives and market prices. Smart consumer devices, such as lighting and appliances that can respond to price signals to turn off, reduce, or defer load, will provide the basis for consumer DR enablement and help respond to market prices and customer information. One example could utilize home area network (HAN) technology and open standards. The smart meter, in addition to its digital time-based metrology, can provide advanced information flow to the utility, the market, and to customers. This can be especially useful for the electricity market and the grid operator to verify the availability of DR as it prepares to respond to system contingencies and to reduce loads in response to prices.

3.11.7 Value of Demand Management

Utilities benefit in many ways from demand management programs: (1) avoiding dispatch of expensive peak generation units, (2) deferring long-term capital investments in excess generation and T&D capacity, (3) reducing carbon footprint by using more efficient units, (4) increasing system reliability, and (5) savings from lower energy use during high-cost times, (6) additional tools to support the growth of renewable energy (e.g., wind and solar) for which the power output varies.

The value of demand management is usually compared to opportunities to defer or fully avoid supply-side alternatives that are constructed. The resources and related costs deferred or avoided include electricity generation (power plants), transmission lines, distribution, customer costs, and environmental pollutants, including SO_X, NO_X, and CO_2. Where competitive markets exist for deferred or avoided resources, market prices can be used to value demand management.

Demand management provides grid support functions such as contingency response, reserves, and frequency control. Demand resources can be called upon to respond to disturbances on the system to prevent and mitigate outages. In addition, by providing reserves and frequency regulation, demand resources can enhance the stability of system operations without the need for additional generation capacity. In California, for example, if only 20% of the state's retail demand in 1999 was subject to time-based pricing, and with only a moderate amount of price responsiveness, the state's electricity costs would have been reduced by 4% or $220 million. The following year, in 2000, electricity prices were more than four times as high, and the same amount of DR would have saved California electricity consumers about $2.5 billion—or 12% of the statewide power bill. (any reference for these numbers??) These, and other estimates of benefit potential, were presented to the U.S. Congress in a senate-requested report by the General Accounting Office in 2004. The PJM Interconnection estimated that during the heat wave of August 2006, DR reduced real-time prices by more than $300 per megawatt-hour during the highest usage hours, estimated to be equivalent to more than $650 million in payments for energy. Many utilities such as PG&E and Southern California Edison have been able to use demand management programs to help justify recovery on extensive AMI rollouts.

Demand management can also make important contributions to addressing climate change and other environmental issues. One way that it does this is by enhancing and reinforcing customer energy efficiency, the accepted cornerstone of emission reduction policies. With DR technologies, customers will receive information on their electricity usage that they have never had before and receive it in a timely manner such that it acts as feedback to reinforce their energy management efforts. Furthermore, they will have price and rate options that will stimulate them to be more efficient energy consumers. DR technologies will be the answer to the question "how can you manage what you cannot measure?" A report in 2007 from the Brattle Group [6] has shown that even where customers are not on time-differentiated rates, they may reduce their electricity usage by 11% just as a result of being more informed and understanding better how and when they are using electricity. The report suggested that if DR were implemented nationwide in the United States using only existing, cost-effective technologies, that peak load could be reduced by 11.5% (assuming nationwide consumer acceptance of such a program). The study concludes that a more conservative nationwide DR program would result in a peak load reduction of 5%, which would correspond to nationwide savings of $3 billion each year, or $35 billion over the next two decades. This figure does not include other benefits such as lower wholesale electricity prices, improved reliability, or enhanced customer service.

SIDE BAR: BENEFITS REALIZED: DEMAND RESPONSE SAVES THE TEXAS GRID FROM BLACKOUT

On February 26, 2008, the Texas grid suffered a significant increase in demand (4.4 GW) due to colder than expected weather coupled with a decline in wind power (1.4 GW) and an underdelivery of power from other power sources. This sudden strain caused a drop in system frequency, which triggered emergency grid procedures into action. Because of the system-wide lack of generation capacity, the Electric Reliability Council of Texas (ERCOT) turned to their demand response (DR) program, also known as Load acting as a Resource (LaaRs), to help bring demand back in line with supply. These loads consisted of large industrial and commercial users who signed up in advance to curtail their electricity use for payment during grid emergencies. The cost of dispatching these resources is significantly lower than dispatching peaking gas turbines, whose costs can be as much as an order of magnitude higher. This program enabled an estimated 1.1 GW of DR resources within a 10 min period, helping to stave off a blackout. Most of these loads were restored after an hour and a half [7,8].

The economic benefits of demand management have historically been based on grid capacity needs and demand management operational capabilities. In many places, demand management has been used only during system emergencies when generation capacity was scarce. It is increasingly accepted that demand management can reduce the need to purchase high-cost, capital-intensive infrastructure (e.g., generation and transmission capacity) that is used to preserve reliability, and also reduce uncertainties in loads and system conditions. This contrasts with earlier versions of demand management programs that had limited availability and uncertain response times when called. Still, these earlier demand management programs did offer significant operational certainty to ensure that specified load reductions occur.

In the last three decades, demand management has been primarily viewed solely to bolster grid reliability during emergencies. More recently, demand management is viewed as a flexible resource to respond to a full set of market needs, to mitigate price and congestion needs, and to respond to a series of needs for specific reliability and energy-based services. The Federal Energy Regulatory Commission (FERC) in the United States has provided rules to enable demand management to be treated comparably with supply-side resources, which means that demand management can be used, and compensated, in the same specific ways as supply-side resources. Demand management, in response to price or reliability needs, is no longer just for emergency peak load management. Demand management can now be used for the full set of market opportunities, on the one hand to respond to variations in renewable energy supplies and on the other to reduce fuel costs in power plants.

The long-standing goal of many in the demand management industry has been to reduce the peak loads and to increase loads during minimum load times, in order to increase power plant fleet utilization, that is, increase the fleet *capacity factor*. With greater use of demand management, daily load curves would have lower peaks resulting in lower average electricity costs. With a flatter load profile, grid operators and utilities can use the more efficient plants more hours per year.

Demand management can provide major wholesale benefits and is increasingly used to derive benefits that are monetized in organized electricity markets. When electricity market generation is scarce, load reduction from demand management is valuable. Many demand management resources can participate directly in organized markets, though energy efficiency is largely the exception.

The basic competitive wholesale market services that demand management, DR, and energy storage may participate in are as follows:

1. Resource adequacy (planning reserve), which can be defined on a locational (subregional) basis (e.g., 15% of total planned load)
2. Operating reserves, including spinning and non-spinning generation reserves that must be available online within 10 min (e.g., 7.25% of current hourly load)
3. Frequency control or automatic generation control to ensure that regional frequency (on a subsecond basis) is maintained
4. Emergency capacity, which may include capacity market requirements (e.g., in PJM and ISONE)
5. Energy, on a zonal, nodal, *instructed*, or distribution circuit basis, including providing supplemental energy needed to "back-fill" operating reserve requirements
6. Congestion management for locational "out-of-merit" or "out-of-sequence" conditions
7. Energy price mitigation, particularly on a locational basis, to reduce energy prices such as when scarcity conditions exist

Organized competitive markets provide most of these services in separate markets for day-ahead, hour-ahead, and "real-time" trading and scheduling. Power generation plants and demand management services that comparably satisfy necessary conditions can operate in many different markets on a given day. A major increase in the need for ramping capacity for grid balancing is in response to the large amount of variable renewable resources on the grid, particularly wind generation and PV generation. Demand management can be also be used for load management at specific locations on the distribution system. The greatest value from demand management is when it can serve multiple purposes—such as for the high-voltage grid, customer needs, and the distribution grid. The flexibility of demand management to serve these purposes depends on the hours of availability and the trigger(s) used to activate and thus harness the demand management.

The potential contribution that demand management can make to renewable energy development should be noted. In the case of wind energy, a particular geographic wind resource may not be available during peak demand periods. By matching that wind resource with DR during the period that wind is nonavailable, the wind resource may become more viable. In order to support high levels of intermittent renewables on the grid, demand management enables load to follow generation, instead of the traditional model of generation following load. The result is a greater opportunity to replace less environmentally friendly resources with a combination of wind (or solar) and DR.

The calculation of benefits for demand management usually depends on (1) whether it can be dispatched (in contrast to voluntary response), (2) the certainty (predictability) of availability, (3) the response times when it is called, and (4) the ability to verify its availability and its dispatchability when called upon.*

The installation of interval metering has been one of the main enablers of demand management and has brought about a greater certainty of demand management availability and performance. Many states in the United States have begun to address demand management cost-effectiveness.† California has decided that dispatchable demand management qualifies as resource adequacy, which allows it to be more valuable.‡ This also means that dispatchable demand management can

---

* The recent DR cost-effectiveness protocol highlights the following operational factors for DR: availability, notification time, trigger, distribution, and energy price. These factors do not directly reflect the requirements to qualify for specific CAISO markets or to provide distribution load management, which seem essential for cost-effectiveness.
† For example, see CPUC ALJ Hecht's August 27, 2010, ruling in Rulemaking 07-01-041 (DR OIR) to provide guidance on the scope and contents of the utilities' DR applications. This ruling emphasizes a set of related topics: use of price-responsive DR, resource adequacy (planning reserve margin) requirements, integration with the wholesale market, integrated demand-side management, load impact estimates, and cost-effectiveness metrics.
‡ Resource adequacy has also been defined as long-term planning reserves, which are needed when other plants and transmission lines do not operate, most typically because of "forced outage."

qualify to provide grid ancillary services called operating (spinning and non-spinning) reserves and emergency capacity.* Fast responding demand management may also qualify to provide *instructed energy*,[†] which is usually paid for at the highest energy market prices.

> **CASE STUDY: PNNL OLYMPIC PENINSULA PROJECT SAVES UTILITY AND CONSUMERS MONEY**
>
> In 2004, Pacific Northwest National Lab (PNNL), in partnership with the Bonneville Power Administration, started the Olympic Peninsula GridWise demonstration project, which equipped over 100 households with advanced meters, as well as thermostats, water heaters, and dryers that could respond to communications signals from the meter. The software used in the pilot program enabled homeowners to customize devices in terms of choice of the desired level of comfort or economy, to automatically optimize the level of power use based on dynamic electricity prices that changed every 5 min. This DR demonstration project yielded average electricity bill savings of 10% for participants [9].
>
> The project also provided benefits to the utility by reducing transmission congestion during peak hours and the need to build additional transmission. This pilot project showed that the Bonneville Power Administration could successfully defer additional transmission investment for at least 3 years. Called "GridWise," this demonstrated that intelligence-enabled appliances can reliably and economically be used to alter load profiles, to reduce the need for peaking plants and additional capacity expansion.

3.11.8 Demand Management Enablers in the Smart Grid

A smart grid uses digital technology to improve the reliability, security, and efficiency of the electricity system. The vast number of stakeholders and their various perspectives have brought about an ongoing debate regarding the definition of a smart grid, a debate that addresses the special emphasis each participant brings to the anticipated transformation of the utility system.

A significant number of technical areas are included in any reasonable electricity system taxonomy that covers the areas of concern to smart grid technology advocates. Progress made in moving toward a smart grid is often described in terms that focus on the interfaces between elements of the system and the systemic issues that reach beyond any single area. The areas of the electricity system that cover the scope of a smart grid include the following:

Area, regional and national coordination regimes: A series of interrelated hierarchical coordination functions exists for the economic and reliable operation of the electricity system. These include balancing areas (BAs), independent system operators (ISOs), regional transmission operators (RTOs), electricity market operations, and government emergency-operation centers. Smart grid operations in this area include collecting measurements from across the system to determine system state and health and coordinating actions to enhance economic efficiency, reliability, environmental compliance, and response to disturbances.

DER technology: This is arguably the largest "new frontier" for smart grid advancements; this area includes the integration of DG, energy storage, as well as demand-side resources for participation in electricity system operation. Smart appliances and electric vehicle chargers are becoming important components of this area, as are renewable generation components such as

---

* Operating reserves are considered to be short-term reserves to be used within 10 min, typically when generation or transmission outages occur. Operating reserves come in two forms, spinning or "hot" reserves and nonspinning or "cold" reserves.
† Instructed energy is provided by the CAISO's electronic dispatch, which requires the generator to be available and to respond, and either to rapidly increase or decrease generation output as needed.

Smart Grid Technologies

those derived from solar and local wind sources. Aggregation mechanisms of DERs are also under this rubric.

Delivery T&D infrastructure: This is the delivery part of the electricity system. Smart grid items at the transmission level include substation automation, dynamic limits, relay coordination, along with the associated sensing, communication, and coordinated action. Distribution-level items include distribution automation such as feeder-load balancing, capacitor switching, and restoration, as well as advanced metering such as meter reading, remote-service enabling and disabling, and demand-response gateways.

Information networks and finance: IT and pervasive communications are cornerstones of a smart grid. The capabilities and performance requirements for information vary widely depending on the application area. Nonetheless many common IT issues fall under this rubric. Examples include interoperability, the ease of integration of automation components, as well as cybersecurity concerns. IT-related standards, methodologies, and tools also fall into this area. In addition, the economic and investment environment for procuring smart-grid-related technology is an important part of the discussion concerning implementation progress.

Figuring prominently in any discussion of the smart grid is the role of the consumer. Smart grid enables informed participation by customers, making it an integral part of the electric power system. With bidirectional flows of energy and coordination through communication mechanisms, a smart grid should help balance supply and demand and enhance reliability by changing how customers use and purchase electricity. These changes are expected to be the result of smarter consumer choices and shifting patterns of behavior and consumption. Enabling such choices requires new technologies, new information regarding electricity use, and new pricing and incentive programs.

Having smarter consumers allows a smart grid to add consumer demand as another manageable resource, together with power generation, grid capacity, and energy storage. From the standpoint of the consumer, system management in a smart grid environment involves making economic choices based on the variable cost of electricity, the ability to shift load, and the ability to store or sell energy. From the standpoint of a smart grid operator, system management in a smart grid environment involves sending the price signals necessary to stimulate the right load shift or utilization of energy storage at the right time.

Consumers who are presented with a variety of options for purchasing power and consuming energy are given the ability to do at least two things. First, respond to price signals and other economic incentives to make better informed decisions regarding when to purchase electricity, when to generate energy using DG, and whether to store and reuse it later with distributed storage. Second, consumers need to make informed investment decisions regarding more efficient and smarter appliances, equipment, and control systems.

System engineers must be able to understand and incorporate models of the devices consumers use and the patterns of their use. The models must include all the salient features of the devices that support the smart grid, so planners and engineers can quantify the financial benefits and the operational impact of the smart grid on the overall electric system.

3.11.8.1 Beyond Peak Shifting

The benefits of demand management go beyond mere peak shifting and many times include lower overall levels of consumption—a conservation effect—as well as very large price reduction effects. It is difficult to convey the economic impact of these benefits as they are complex to estimate and are specific to each electricity control area. FERC in the United States reports over 50,000 MW in peak load reduction potential from its 2010 survey [10], which is broken down as follows:

- Wholesale commercial/industrial DR potential increased from 12,656 MW in 2008 to 22,884 MW in 2010.
- Utility commercial/industrial DR potential increased by 23% from 2008 to 2010.
- Wholesale and commercial/industrial segments are over 80% of the DR potential.

- Residential DR potential is estimated to be over 7000 MW.
- Four DR programs (emergency response, interruptible load, direct load control, and load as capacity resource) account for 79% of total U.S. peak load reduction potential.

This FERC report also summarizes the market barriers to greater use of DR that regulatory reform may remove or significantly mitigate [10]:

- Disconnect between wholesale and retail markets—pricing is not consistent.
- Measurement and verification challenges with establishment and use of base line levels of DR.
- Lack of real-time information sharing—retail entities do not share wholesale or distribution system-level information with customers.
- Ineffective DR program design—retail DR programs do not reflect wholesale market realities and benefits.
- Disagreements on cost-effectiveness analysis—the benefits attributable to DR.

The FERC's estimate of national DR potential broken out by region is shown in Figure 3.173. The gap between business as usual (BAU), achievable participation in DR, and full participation in DR is estimated. These differences reflect varying degrees to which barriers to DR adoption reside.

While demand management at the industrial level has been in place in different forms for decades, at the commercial and residential levels, it is relatively new. This is because industrial loads tend to be larger and more concentrated, making them ideal candidates for curtailment during peak hours. Utilities can easily cut large loads by making only a few calls. But curtailing such high-value loads

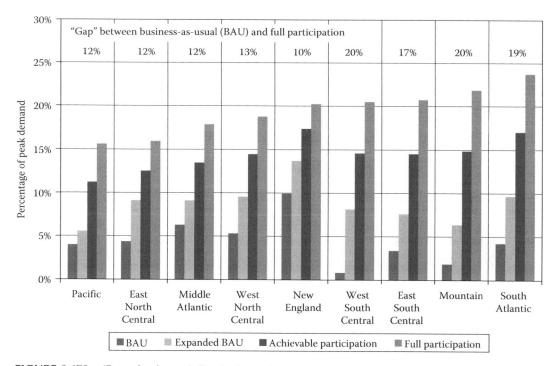

FIGURE 3.173 **(See color insert.)** Gap in demand management participation in the United States. (From National action plan on demand response, Federal Energy Regulatory Commission, June 17, 2010, p. B-1, http://www.ferc.gov/legal/staff-reports/06-17-10-demand-response.pdf.)

Smart Grid Technologies

also risks greater economic harm. Demand management would allow low-value, noncritical commercial and residential loads to be turned off with less economic impact.

There are several evolving trends in the industry that hold promise for wider and more active participation from commercial and residential consumers:

- *Improved human interfaces*: Among the many in-home displays recently released or in development, an example is Intel's Intelligent Home Energy Management Proof of Concept. This features a vibrant OLED screen and integrates traditional thermostat features with energy cost management (through connection with ZigBee compatible advanced meters), home security system monitoring, tasks reminders, and media functions such as video memos [11].
- *Lower-cost sensors and communications devices*: Lower-cost sensors and communications devices will make it cost effective and faster to communicate between the utility and the consumer through a variety of channels.
- *Better energy management software*: Google's PowerMeter and Microsoft's Hohm energy management software may serve as predecessors leading to new tools for customers to manage their energy use in an online, highly visual environment.
- *Greater public awareness*: Increased public education and concern about environmental and energy issues, especially in the past decade, have driven greater involvement in personal energy use management and demand for information and control over emission sources for both commercial and residential applications.

3.11.8.2 Utility Demand Response Management
3.11.8.2.1 Load Modeling and Forecasting

For most of the twentieth century, loads were quite simple to understand and represent, even as they grew increasingly unpredictable over the years. The philosophy of power systems engineering is that purpose of the bulk power system is only to satisfy the load and not to question why it is present or attempt to manage it directly. Consequently load was treated largely as a boundary condition. In fact, very early electric system planners generally anticipated only three types of loads: constant impedance loads from devices such as incandescent light bulbs, inductive loads from motors, and system losses from cables and transformers. The times and quantities that were present were quite easy to anticipate, and the system could easily meet the need to balance supply and demand by having generators follow the load using voltage and frequency feedback signals.

As a result, with few exceptions, power engineers tended to downplay the significance of load behavior. Load was treated largely as a static boundary condition, even when numerical simulations made it practical to do otherwise. It made perfect sense to keep load a constant parameter in the context of steady state flow solutions or subminute dynamic simulations because (a) that is how the system was actually operated, and (b) whatever uncertainty was present was both small enough and random enough to be readily addressed by good dispatch practices and the existing feedback controls. At every time scale in between, load behavior simply was not complex enough to warrant much consideration beyond basic forecasting needs.

The first hint of the potential significance and complexity of load behavior came with the cold load pickup problem. This problem arises after prolonged power outages where all thermostatic loads have settled well outside their normal control hysteresis bands and load patterns have lost diversity. When the power is turned back on, all these devices turn on simultaneously, causing a surge in demand that can far exceed the demand prior to the power outage and even exceed the maximum capacity of the system (which is why it can take several days to restore full service to all customers after a major system outage). This phenomenon is also observed in load control rebounds, which are similar in nature although not generally as severe.

But, it was not until the advent of smart grid technology that power engineers came to seriously consider the potential role that loads could play in meeting system needs. It was realized rather quickly that loads exhibit behaviors that are not simply detrimental to the system, such as cold-load pickup or load control rebound. The same phenomena that give rise to adverse behaviors might also be used productively to support strategies such as bulk system underfrequency load shedding at the end-use level (instead of at the neighborhood level), distribution system undervoltage support, or support of intermittent renewable generation, such as wind and solar.

Economies of scale, regulatory barriers, customer expectations, and a strong preference for centralized command and control in vertically integrated utilities have made it far easier to govern a few large generators than many small loads. So for more than one hundred years, the system was controlled exclusively from the supply side. Understanding load behavior was unnecessary, and with remarkably few exceptions,* it remained largely an academic question.

The introduction of smart grid technology, the growth of intermittent renewable generation resources, and the advent of intelligent load controls have converged to make loads potentially an equal partner in the electric system's physical and economic operation. As a result, load modeling has become an increasingly important consideration in the design and operation of smart grid technologies and in the debate about how to implement renewable resources en masse.

To understand how loads respond to changes in circumstances, utilities must begin by understanding how loads behave in general. Load behavior analysis and modeling requires subdividing loads into the main classes that influence the kinds of behavior observed. Primary among these taxonomies is the economic nature of the load, viz., residential, commercial, industrial, and agricultural. For smart grid purposes, residential and commercial loads are the most challenging and interesting. In contrast, industrial loads are not the focus because they have individual characteristics that can be difficult to model, and agricultural loads because they tend to be fairly simple in comparison.

In addition, at least from the perspective of electric system modeling, it is vital to identify the degree of electrification of the load itself. Electrification is typically characterized by end use, that is, the type of device that meets an electric customer's needs. Not all devices use electricity to satisfy the demand for goods and services, and the fraction of those that do varies according to factors such geography, demographics, regulatory policy, and long-term expectations for energy prices for the fuels, if any, needed by the various prime movers, for example, steam, water, wind, sun.

Finally, intermittent availability of lower-cost fuels, particularly those that are relatively uncorrelated with electricity prices, can make multifuel systems economically advantageous. From the standpoint of load behavior, these can either be implemented as direct-delivery systems, such as solar water heaters, or be mediated by electricity delivery systems, such as rooftop PVs. The availability and behavior of these systems can influence load behavior as well.

Smart and efficient execution of demand management commands rely strongly on smart measurement and analysis of the consumer and market data. Different portions of end-user consumption level are qualified for various demand management programs, such as interruptible load, direct load control, or demand dispatch. Metering of individual consumer load consumption versus an aggregate measurement of the consumer total load provides detailed knowledge on the consumer habits and consumption patterns and enables a more appliance-oriented DR approach with a higher chance for success. This would require smart and efficient analysis of the raw data in order to extract meaningful information from it for DR purposes. Various estimating and forecasting techniques can be used to develop a reasonably accurate model for consumer demand and provide a forecast for its demand during the future time intervals. The sensitivity of the consumer to the electricity prices

---

* One notable exception was the advent of DSM programs focused on energy conservation. These programs address the problem of load growth through conservation measures such as efficient appliance retrofits and consumer education. To this day, DSM programs remain effective at controlling the net rate of load growth.

can be incorporated into the model to reflect the response of the consumer to demand management events. More complicated econometric models can be developed to account for the qualitative data, such as the personal habits of the individual consumer toward DR events. These models can determine a probability value by which an individual consumer may comply with a demand management event issued by the utility, and if proved reliable, they become part of statistical reliability. This information proves to be very useful in validating the demand management event and determining whether extra measures are necessary. Matching this information with electricity market prices, from the full set of separate market elements, is an additional challenge.

Traditionally, DR programs have been offered to the customers as a set of fixed options with preset terms and conditions. The customer would then pick the program that fits its needs the best. There are several attributes that are directly associated with each program such as the maximum duration of the demand management event, the maximum number of times a demand management event may be issued in a year/month, and the maximum number of consecutive days a demand management event may be issued. Other attributes are more related to the consumer, for instance, the minimum advance notice for the event. All these attributes could make a difference in the comfort level of the customer participating in the corresponding programs and therefore impact the acceptance and success of the program.

With proper modeling of the customer load patterns and habits, it is possible to customize the programs to tailor these attributes to fit customer needs and habits. This will create a wider and more flexible set of program terms and incentives for selection by the customer. More choices can lead to higher customer program participation. With additional choices, a consumer may view the process of selecting the suitable programs confusing, making the selection difficult or even risky. This issue could be buffered via a mechanism that utilizes consumption patterns and habits of individual consumers to propose an optimal program selection to the consumers to create mutual benefits for both parties.

3.11.8.2.2 Load Responses

Load response can be separated into two distinct behaviors, one which is a one-time irreversible behavior, for example, a person turns off the lights when leaving home, and one which is reversed later, for example, a person defers doing a load of laundry until the next day. Some load responses have both. For example, lowering a thermostat by 2°F for 8h every day will reduce the heating energy consumption, but the strategy will also introduce a recovery period during which some of the savings received are lost when the thermostat is raised back up. A comparison of the reversibility of different residential load responses is shown in Table 3.13.

Irreversible load responses reduce overall energy consumption by the amount of load that responds. Irreversible load response is simply a change in the load shape, which results in a net reduction of both energy and maximum power, as illustrated in Figure 3.171a. Most irreversible load response requires a one-time investment, and the benefit is typically enduring. However, some load responses such as consumer awareness programs may appear to be irreversible over the short term, but in fact decay in the long term.

TABLE 3.13
Residential Load Response Reversibility

Irreversible Responses	Reversible Responses
Lighting controls	Heating and cooling thermostat schedules
Cooking/heating fuel switch	Washer and dryer deferral
Heating and cooling thermostat setback	Refrigeration and freezer defrost deferral
Energy efficiency retrofits	

Reversible load response is a change in the load shape that results in a change in maximum power but no net change in energy consumption, as shown in Figure 3.171b. This behavior is typical for thermostatic loads, such as heating and cooling systems. These responses are often called load shifting or deferral. Preheating and precooling are also reversible load responses, but with the opposite sign (i.e., load increase precedes load decrease).

Reductions in consumer real power consumption are typically associated with energy efficiency programs; that is, reducing real power consumption reduces energy consumption for nonthermostatic loads such as motors and lights. In some cases though, reduction of real power can also reduce peak load, which, from a smart grid perspective, may result in deferred capacity expansion. One caveat is important for thermostatic loads: reduction in real power typically results in increased duty cycle or run time and does not result in reduced energy consumption. This can affect the saturation load (the load at which diversity disappears) and contribute to increased adverse load behavior associated with loss of load diversity, such the onset of load rebound and price instability.

3.11.8.2.3 Price Signals

Consumer DR can be controlled using price signals. Different rates at different times, implemented effectively, can drive desired end-user consumption behavior. However, unlike the availability cost of other services such as airfare or hotel rooms, currently the typical electric price to the residential customer is "one size fits all." Such a price offers no reduction for conservation and no premium for consumption during peak periods. While utilities across the country have used these types of pricing mechanisms to differing extents for some time now, AMI technology paves the way for much greater adoption of dynamic rates and pricing signals for all customer classes—"smart meters" enable "smart rates." As with any control system, utilities can employ either an open-loop strategy or a closed-loop strategy. An open-loop price-based control strategy generally relies on time-varying prices with the expectation that higher prices lead to lower loads. Several types of open-loop pricing strategies are commonly found and are distinguished by the rates or tariffs they use. Each rate is designed to elicit a different response from customers.

Time of use (TOU): TOU rates have existed in some countries for many decades because they are simple to implement and meter. The peak-time schedule is typically determined seasonally, and it sometimes has a "shoulder" rate that is an intermediate rate between the off-peak and on-peak rates.

Critical-peak pricing (CPP): CPP is similar to TOU pricing, but instead of having a daily schedule, the CPP is declared only on the few days the utility expects peak conditions to prevail. For this reason, the price on critical peak is usually very much higher than the standard rate. It is not unusual to find the CPP rate is more than 10 times the standard rate.

Peak-time rebate (PTR): PTRs are similar to CPP rates, but instead of charging customers more, the rebate works by refunding customers who reduce load on peak. Unfortunately, it is often difficult to determine the savings precisely for any given customer, and solving this problem can lead to complexity in the program implementation.

Dynamic pricing (DP) or real-time pricing (RTP): DP or RTP works by sending customer prices that reflect to some extent the variations in prices seen at the wholesale level. Unfortunately, because the fluctuations in wholesale prices can be unpredictable and vary as frequently as every 5 min, RTP can be difficult for customers to respond to without special hardware. RTP is a closed-loop price-based DR control strategy. The implementation of RTP can be difficult to understand, but its flexibility and scalability are very important attributes that have led to growing interest in its use. RTP systems may require DR equipment to be installed in the customers' homes. The fact that prices change in periods as short as 5 min means that customer cannot be expected to respond all the time. Some may contend that customers will not accept price change more frequently than hourly. For this reason, RTP systems include devices that can respond to prices and interact automatically on behalf of the customer. One very important caveat for RTP in particular is that a customer's subscription must be voluntary. Some customers may have load shapes that are

Smart Grid Technologies

particularly ill-suited to RTP because the unresponsive part of their peak load is highly coincident with peak price. In contrast, other customer may have highly response loads on peak and be able to provide a load of flexible DR to the utility. At the other end of the scale, some customers may not have enough demand on peak for the cost of the RTP system to be justified, and utilities must retain the ability to exclude certain customer from using RTP when they would essentially be free-riders or not viable as a DR resource.

Regardless of the specific pricing mechanism proposed by a utility, there are two fundamental implications of these changes:

- More active involvement on the utility's part in helping customers understand what they can do to reduce their usage
- More active involvement on the customers' part in what they use and when they use it

Smart grid is more than simply new technology. It will have a significant impact on a utility's processes. Perhaps more importantly, it is also about the new information produced and made available by these technologies and the new customer–utility relationship that necessarily emerges as a result of these technologies. A critical element of this new relationship is "decoupled rates." Decoupled rates break the linkage between what a utility charges for power delivery and how much energy the end user actually consumes. While the costs of generating power are clearly a function of usage, the cost a utility incurs to provide a power delivery system has little to do with how much an individual customer actually uses. Public, regulatory, and local government interest in renewable energy sources and DSM programs has never been higher. However, until new pricing mechanisms such as "decoupling" become more common, utilities will continue to have a financial disincentive to encourage their customers to use less of their product. Greater alignment between the end user and the utility will result in greater reductions in energy consumption (and emissions).

3.11.8.2.4 Consumer Behavior

Different rates elicit different behaviors from consumers. Consumer behavior affects many aspects of utility's operations, sometime in very complex ways. Specifically, utilities are often concerned with one or more business performance metrics, such as the rate of return on capital investments, exposure to short-term wholesale price fluctuations, minimizing operating costs, controlling net revenue, or maximizing earnings. Consequently, utilities are challenged to not only design the different rates that elicit needed behaviors from customers, but they must also determine what fraction of their customers would ideally have to be on each rate in order to meet any of these business performance metrics.

One approach to this challenge lies with a method developed as part of the capital asset pricing model used in modern portfolio theory. The concept of an efficient frontier (dotted line) is illustrated in Figure 3.174, where the expectation of an outcome for a mixture of two rates is plotted against the uncertainty of that outcome [12].

The concept of the efficient frontier applies in this situation because a utility would try not to choose a mixture of rates that does not lie on the dotted line shown in Figure 3.174. Suppose a utility proposed to place all its customers on the RTP rate and none of the TOU rate. The expected earnings would be low, but the uncertainty would also be quite low. However, for the same uncertainty, the utility could realize significantly higher expected earnings by choosing a more balanced mixture of customers on each rate. Thus, for any outcome the utility wishes to maximize, only the mixtures of rates that lie on the top of the curve would be efficient, and all other mixtures would be suboptimal. Similarly, for any outcome the utility wishes to minimize, only mixtures that lie on the bottom of the curve would be efficient, and all other mixtures would be suboptimal. Typically utilities have more than two rates that are being mixed, so the mixing regions between each of the rates may overlap as all combinations of mixtures are examined. However, the frontiers remain either the upper or lower boundaries of these regions.

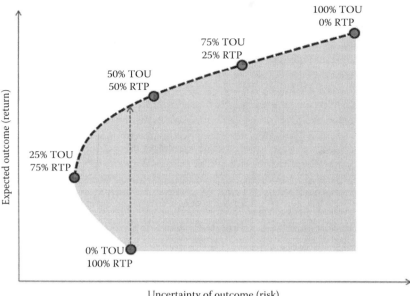

FIGURE 3.174 Portfolio theory applied to utility rate design. (© Copyright 2012 Pacific Northwest National Laboratory. All rights reserved.)

In practice, utilities must collect data on consumer behavior in response to the rates. This data can be used to establish both outcomes for each business performance metric under each rate as well as the uncertainty of those outcomes in the face of uncertainty about costs and operating conditions. Therefore utilities may wish to continuously update the analysis at least annually, perhaps more frequently, to determine the objective rate mixtures that the utility's DR programs seek to achieve. The rate mixture objectives evolve over time in response to changing demographic conditions, the seasons, and perhaps even wholesale market conditions. The portfolio analysis method can be thought of as a long-term closed-loop control process that the utility uses to continuously optimize the performance of its DR systems.

A demand management event issued by the utility is successful if it can attract sufficient participation by end-use consumers resulting in the amount of demand reduction desired by the utility. Clearly this creates a dynamic environment where the consumers and the utility interact through a system of incentives and agreements in order to achieve the target demand reduction. Large numbers of consumers, with their stochastic nature of energy consumption patterns, make it difficult to model the problem in a deterministic way. The behavior of consumers is affected by various market-driven factors such as energy prices as well as personal habits and consumption patterns that vary from one individual to another.

When a store runs a sale advertisement, they learn over time their expected response rate from direct mail or other media. Customers do not tell them, "If you put this item on sale, I will come into your store and make a purchase." However, the response rate from the sale ad is learned over time and can become very predictable. This ability to learn the response reliability over time based on data management experience is referred to as statistical reliability. Certain customer responses to smart grid stimuli will fall under this category of statistical reliability. Over time, the grid and utility operations will learn that when a specific signal is sent out, the response will be a predictable amount based on their historical learning. Statistical reliability is already applied in grid operations, utilizing inputs such as experience with the load curves, weather predictions, and utility experience to predict peak days and peak energy consumption. Learning to apply this concept to predict the response to demand management will save a considerable amount of cost and will enable inclusion of additional consumers and devices. This way, responses that cannot be foretold via electronic means or where the customer or device manufacturer does not support via communicating device technology can be included under this category of DR.

Throughout the years, most demand management applications have adopted solutions based on control and response models of aggregating individual consumers into groups of end users connected to a specific substation, feeder, or service transformer. This way, the volatility of the individual consumer behavior is reduced, and the problem can move further toward a probabilistic problem where the uncertainties can be treated as random variables. However, DR necessitates the introduction of a higher level of granularity where the stochastic models of individual customers are accounted for. On one hand, these models should consider the behavioral and financial aspects of individual customers; on the other hand, they must incorporate the impact of the terms and conditions of the demand management programs into the decision making by the consumers. Therefore, grid entities prefer to have an accurate indication of the current availability of the dynamically changing components of the demand management environment.

Identifying the behavioral patterns of consumers when it comes to electricity consumption is essential for ensuring sufficient participation following issuance of a demand management event. On one hand, the event must be issued at times and locations when there is enough electricity consumption available; on the other hand, it must not contradict with individual consumer lifestyles. In other words, there is a clear tradeoff between achieving demand reduction and consumer inconvenience. From the utility perspective, an accurate model/prediction of consumer behavior and consumption patterns is critical to a successful demand management program. From the consumer perspective, there must be the ability to opt out of any program or specific instance that conflicts with a consumer lifestyle or specific schedule. Consumer device manufacturers, concerned about the satisfaction of their customers, require the opt-out feature before allowing their products to participate in demand management automation. This will tend to induce an element of variability into consumer demand management programs. In the general sense, the demand pattern of consumers can be analyzed and estimated from two aspects:

Consumption habits: This is related in part to the individual habits for using various household electric appliances, for instance, washer/dryer, dishwasher, etc. The number of times a week/day that each appliance is used and the duration of each usage are likely to reflect a specific pattern for each consumer. This portion of demand reduction is what qualifies for demand dispatch, where the utility tries to shift the consumption from peak-load to off-peak hours. This can be done either manually by proposing timeframes for usage of the various appliances or automatically by remote activation/deactivation of appliances according to utility needs and customer acceptance parameters. A currently less common source of demand reduction, which is likely to grow, is associated with charging electric vehicles. This is perhaps one example where most consumers are very flexible as to what time of the day the charging phase should take place (as long as it is done automatically). Another major portion of demand reduction is related to the temperature settings of air conditioner units in winter and summer seasons, where consumers show different levels of sensitivity to heat/cold and show various degrees of flexibility to deviations from their habitual comfort zone as a result of demand management.

Elasticity to electricity prices: A rather crucial assumption behind DR assumes that consumers are willing to temporarily forgo their convenience to avoid higher electricity prices. While this is perhaps true for a sizeable portion of the consumers, the degree to which they are willing to give up their comfort level varies from one individual to another and may be impacted by the financial incentives offered. The consumer sensitivity to the electricity prices is utilized by the utility by introducing real-time pricing tailored toward peak-hour needs.

Incentives offered to consumers in order to encourage them to participate in a demand management event play an important role in its success. For demand dispatch applications at the residential level, to shift certain loads from peak to off-peak hours, the value of the incentives is not extremely critical, especially if the shifting is done automatically and may not even be detected by the consumer. Examples of this are electric vehicle charging and water heating. However, for other residential applications, for instance, air conditioning usage, where the comfort level of the consumers is affected most, the role of the incentive payments is higher. For commercial buildings, air conditioning usage can be shifted—rather than curtailed—by preheating/precooling the building during off-peak hours, for example, early in the morning before arrival of the occupants, and turning off the air conditioning unit during peak-load hours. Such practice becomes more difficult in the residential market where consumer lifestyles and schedules are more dynamic in nature. The financial incentives a utility is able to offer have limitations dictated by the financial calculations by the utility relative to their cost structure and level of vertical integration.

3.11.8.2.5 Demand Response Management Systems

Key to the implementation of demand management programs is an integrated set of applications that enable both the utility and consumers to understand and have control over energy consumption. At the utility level, economic DR programs are implemented that offer economic incentives to customers to contain and/or shift their energy demands to better match system-level demand with system supply resources. At the consumer level, home energy managers (HEM) provide real-time energy consumption information and optimize operation of appliances and energy management devices on the HAN. The primary objective is not only to provide consumer applications and HEM devices that empower and allow better management of energy consumption without adversely compromising lifestyle, but to also allow utilities to maximize rate structures, reduce reserves, and allow greater economic flexibility in the near and long term. Demand management may also be used to mitigate unexpected supply disruptions or overload conditions. During such conditions, customers may be asked to curtail consumption on short notices or utilities may use direct load control to shed consumer loads.

Demand management should not be limited to solutions that are able to suppress or shift demand and alleviate peak load. The vision should be to provide a solution that goes beyond peak shifting and provides flexibility and increased efficiency in managing overall demand. The solution should match generation resources with demand from electricity consumers in an efficient, predictable manner and incorporate the control, integration, and optimization of renewable and other DERs as the adoption of these technologies increases.

The demand response management system (DRMS) is the utility application that manages DR capabilities from the utility down to the consumer. The DRMS also interfaces and operates with other utility operational and information systems, such as EMS, DMS, OMS, customer information systems (CISs), and billing. To be an end-to-end solution, a DRMS needs to have this level of functionality at a minimum in order to provide real value to a utility.

First generation DRMSs typically focus on functionalities needed to support peak shifting by obtaining load information from the meter and statistically estimating load availability based on customer enrollment and consumer historical data compared to the load forecast from EMS. Once a DR event is selected, the DRMS sends a basic signal to preestablished groups of customers based on the estimated amount of MWs required. In-home enabling technologies, such as smart thermostats, in-home displays (IHDs), HEMs, and smart appliances, receive the signals and perform the load management activities based on the consumer's preferences. Two-way communications allow the utility to measure the effect and verify consumer participation in demand management.

As load management technologies advance and the DRMS is integrated with more utility enterprise applications, the DRMS enables the use of DR to support emergency response and virtual generation capability in the EMS. Through aggregation, it estimates demand in near realtime and is able to dynamically select customer groups based on electrical nodes and resource availability. The DRMS also incorporates and accounts for distributed resources such as energy storage, wind, solar, and PHEVs.

Smart Grid Technologies

The operations component of the DRMS contains all the critical applications for a utility to manage and maximize resources for DR events. These applications include response estimation, dispatch, aggregation/disaggregation, measurement and verification, and reports and analysis.

3.11.8.2.5.1 Consumer Response Estimation Many existing direct load control programs do not enable utilities to accurately estimate how much load reduction they will obtain when an event is initiated. As a result, a common strategy is to send a signal to a larger subset of the population in order to ensure the necessary reduction is met. The impact of the event is evident at the system level; however, there is no direct feedback from particular premises and very little learning on the impact from one event to the other, making the planning and execution of demand management very inefficient. The lack of feedback also makes estimation of the potential rebound effect at the conclusion of the demand management event more difficult, making the grid vulnerable to a subsequent rebound peak or operational instability. The response estimator function in the DRMS determines the amount of MW and MWh available for DR over a time frame of interest, including the estimated rebound effect. In addition, this estimation can be tied to existing load forecasting tools since there is direct correlation between the two. The response estimator evaluates the likely response from participating homes, as well as their associated probabilities of participation.

3.11.8.2.5.2 Demand Dispatch With few exceptions, utilities today rely on manual processes, spreadsheets, and independent software applications to decide if, when, and how much demand resource is needed to support forecasted demand requirements. The demand dispatch application in the DRMS is a decision support tool that provides utilities with recommendations as to when to initiate a DR event and how many customers to include in the event. The demand dispatch tool determines the optimal schedule and resource mix, taking into account generation costs and the impacts of the rebound effect, when providing recommendations for how much DR to request for each given time period. The demand dispatch toll should take into account optimal dispatch of demand across multiple customer types and pricing programs.

3.11.8.2.5.3 Aggregation/Disaggregation Aggregation is a necessary component of the response estimator application. The aggregation function determines the total DR available based on customer participation and availability. The aggregation function collects up-to-date metering data from each of the applicable premises to enable as accurate an assessment as possible of the current load state and potential for DR. The disaggregation function identifies the participating customers for each pricing event.

3.11.8.2.5.4 Measurement and Verification As utilities initiate demand management events, there is little feedback to measure the extent to which an event is successful. Customers may have participated in their demand management program, or they may not have participated due to an endless range of possibilities. This function in the DRMS calculates baseline customer load profiles according to contractual terms and verifies reductions/changes in load from their profile for billing purposes. This information can be tied into a utilities' CIS to facilitate accurate billing and rewarding for participation in demand programs. The application also validates the probabilities of participation, expected load change, and anticipated rebound effect as estimated by the response estimator application.

3.11.8.3 Empowering Consumers

3.11.8.3.1 Delivery of Real-Time Information
In order to look at an electric load as a resource, there must be an architecture enabling management of the load. As the wholesale price of electricity fluctuates, there is a desire to be able to reflect this fluctuation in the retail electric rates. Most people have become accustomed to watching the cost of fuel for their car. Their buying decision may be accelerated or delayed in accordance with the price.

In a similar way, the price of electricity can impact consumer purchases of the electricity product. In the past, there was no mechanism to inform the consumer of the current price of the electricity product. But when the price of these products starts to change rapidly, the price became a key piece of information the consumer needs to know prior to the purchase, or in the case of electricity, consumption of the product.

With the trend toward a variety of time-based pricing rates, electric customers need access to the price information they have not had to deal with before. Compared with auto fuel, the consumption of electric power has additional layers of complexity. One could easily determine the miles-per-gallon (MPG) in a vehicle. But in my home, it is similar to having multiple vehicles with a very wide range of MPG ratings. Furthermore these "vehicles" may operate concurrently or in any combination. Some of them operate without our knowledge and without a reasonable method to control their utilization. For example, consumers do not know exactly when their refrigerator will operate. Other than unplugging it completely, there is little control over its operation. To motivate a change in electricity use, customers will need more information prior to purchase.

Unlike other commodities purchased as consumers, the electric information has several caveats to address. Getting the information to the customer via an adequate mechanism may also have a dependency on how often they need to have the information updated. Can the price change every month, week, day, and hour or even in a shorter block of time? How much advance notice of the price is needed? As noted in the MPG discussion, price is not enough information for a consumer to manage consumption in an environment where usage is not known. The customer needs to know the quantity they expect to consume.

This leads to the two core requirements for smart display of information: price and consumption. Consumers can also benefit from additional information such as when the price will change again and whether it will be more or less expensive in the future. Knowing how much electricity will need to be purchased must also be known to make good financial decisions.

It is also necessary to communicate information customers have not previously had to understand. A kilowatt-hour is not a term in the average daily vocabulary nor is its meaning. Some manipulation of the data is required before it is presented to the customer. Several types of display devices have been designed to do this. The first devices for demand management applications were mostly independent display devices (IHDs) that, using their own sensors or meter access methods, could be located or mounted according to consumer preferences. The IHDs generally displayed the key pieces of information to answer the questions, how much has my electricity cost me this month (or during some selectable period of time) and at what rate am I purchasing electricity now? An IHD may indicate the current electricity price, the rate at which it is being used, and the cost per hour. For example, the display may indicate that at the current rate of energy consumption, the cost is $0.37 per hour. The display could show additional computations for the consumer. These could include the projected cost of the current month at the current rate of energy consumption, a comparison with last month, a graph of usage by day or month, or any of a number of other potential calculations depending on the amount of energy usage history the device is able to store.

In the 2009–2010 timeframe, NIST (National Institute for Standards and Technology) in the United States was approaching a task handed to them via legislation. Recognizing the need to make this type of information available, a priority action plan was initiated to help create a standard method to communicate electric consumption and usage information in a standard format. With the introduction of these standards, a variety of methods of reporting this information to the consumer was enabled, and the open market could look at the best way to relay the information to the consumer. The standards also led to the introduction of the "Green Button," which utilizes this usage* information to display energy consumption to consumers in new creative ways via a variety of personal devices that include dedicated devices in the home, the Internet, and personal smart phones.

---

* Based on UCAIug OpenADE and NAESB PAP10 standards ratified in October 2011.

Smart Grid Technologies

The smart meter is one device that can be enabled with communications technology to provide consumption information to the consumer. The newer communicating electric meters, often referred to as the "smart meter," are designed to calculate consumption at programmed intervals. To provide this information to the customer, several methods are available. One is to design communication electronics in the meter that will transmit this information into the home/premise. Another method is where the utility uses the Internet to forward real-time data back to the home/premise.

An advantage of routing the real-time consumption through the Internet is that it would enable a third-party firm to contractually agree with the utility and consumer to have access to the data. This third party could provide the service of displaying the data in a very advanced graphical format that is accessible via a number of devices including the computer, Internet, PDA, phone, TV, text message, and any other available means. These third-party service providers could also provide technology to assist the consumer in managing the energy inside the premise. A disadvantage is the dependency on other nonutility- and nonconsumer-owned systems and devices that may exist in the communication and control path. There may also be more concerns with data security and privacy when the data pass through more systems.

One advantage of having a smart meter capable of transmitting the data directly to devices inside a premise is that the route is more direct. The information may be available sooner and more reliably due to fewer points in the pathway. Privacy concerns are easier to manage since the data do not pass through third-party systems.

3.11.8.3.2 Smart Loads and Appliances

Past approaches to controlling large residential loads have included ways to limit electric use in water heating, pool pumps, and air conditioners. The basic approach is to control a switch to turn on or off the load remotely. For certain loads, such as the water heater or pool pump, this can be done typically without consumer objection or knowledge of when activation has occurred. The cost of adding this type of switch required on-site installation at a total cost nearing the cost of the device being controlled. Manufacturers are starting to include this switching ability in core product lines that makes the addition of this type of control possible at a small fraction of the cost of previous methods. These advances will likely pave the way to a simple consumer installable add on that is also utility trackable and verifiable. For control of air conditioning, several approaches have been tested. These have included control of the compressor itself in some pilots. Another approach is to wire the control between the thermostat and the AC unit to effectively mimic the thermostat control without having to enter the premise for installation. Other more sophisticated approaches involve smart thermostats able to receive demand management messages that provide both control and the interface to the consumer. The thermostat messages from the smart grid could include messages used for other methods of impacting consumption.

In addition to the energy display mechanisms, the same data can be utilized anywhere the capability exists to receive the information and relay it to a customer. As other in-premise devices advance, they continue to have better hardware to communicate with the consumer, and additional places become available for the display of energy information. One distinct advantage of this display advancement is that the consumer could use the display to decide when to operate an appliance, such as an oven, dishwasher, or washing machine. This provides an opportunity to impact the use at the decision-making time for these process-oriented devices (e.g., cooking and cleaning) that interact directly with the consumer. In addition to optionally displaying energy information, a device can respond by changing or limiting energy consumption in an automated manner with full knowledge of the best way to limit, delay, or optimize performance over a specified period of time. In considering this capability, information can be transmitted into a home or premise for the purpose of impacting energy consumption in parallel with an informational display and ability to manage consumer preferences.

Devices that can receive communicated energy information and respond by altering energy consumption are often referred to as "smart devices" or "smart appliances." For example, a drying appliance could reduce the amount of heat applied and lengthen the drying cycle. A product utilizing refrigeration may have a variable-speed component able to scale back the use of electricity in an acceptable way over a temporary period of time without actually turning off the device. The microprocessor controlling the device, based on detailed internal knowledge, can determine what it can do and for how long while maintaining safety and success of the process being controlled.

By automating the process of demand reduction/curtailment, smart appliances can help smooth execution of DR events with minimum effort from the consumer. The key to designing smart appliances is to reduce the amount of consumer interaction needed in decision making by putting energy management and interface logic into the device controls. The appliance can perform necessary actions to both meet the utility needs as well as accommodating the customer preferences. These preset rules and conditions, updated by the consumer according to individual needs and preferences, remove the burden of decision making from consumers. A demand management event issued by the utility is followed to the variable extent that matches requirements and options set forth by the consumer. The actions taken afterward, turning off a device, reducing the load, shifting the load to a different time, or ignoring the request, can then be implemented automatically. The smart appliances already have an interface with the consumer and are well qualified to present the energy configuration options to the consumer for their selection. For simple devices such as water heaters, thermostats, or even remote switches, the device may be turned on/off entirely. In more complicated designs, these can be intelligent controls or responses to intelligent controllers that react in accordance with the price of energy, or demand management events issued and also considering the preferences of the local consumer.

3.11.8.3.3 *Consumer Energy Management*

As smarter load controls evolve with communications technology, it is possible to integrate and manage the control of loads to more effectively respond to DR signals. The smart loads exchange data over a local communications network (wired or wireless) in the home or consumer premise, commonly known as an HAN.

Such devices that integrate and manage the control of consumer loads are commonly referred to as HEMs. An HEM can determine the operating status of all loads and optimize the control and scheduling of loads based on consumer preferences along with demand and price signals from the utility or grid. More advanced HEMs will include the capability to manage consumer renewable generation and even electric vehicles to provide estimation and historical data to help consumers make more informed decisions about managing their energy usage.

3.11.8.3.4 *Customer Education and Participation*

Successful smart grid implementation requires educating consumers on the benefits of the technologies and enlightening them on the easiest ways to enjoy the benefits without having to change their lifestyle, thereby ensuring consumers are voluntarily engaging in the programs offered by their respective utility. For a demand management program to be effective, the consumer benefits must be clearly understood and sought after by the consumer. The value of the smart grid investment increases significantly as consumer participation increases, and in the long run, increased participation could drive down the cost of electricity for everybody.

It will be important for customers to understand how the cost of the DRM program will be recovered, especially if it is tied to a smart meter deployment, and ensure that customers do not associate implementation of smart meters and smart grid with increased personal energy costs. Additionally, through effective education, consumers will "opt-in" to utility programs and continue to be engaged about how much they are saving—both themselves and the environment.

Customer education is a key to the success of DR. Without proper information, consumers might consider DR as an action that leads to inconvenience and a disruption of their lifestyle. Customers also need help to determine their most effective course of action in impacting their energy consumption and cost [13]. The incentive payments for subscribing to DR—specifically for the residential customers—might not be high enough to provide financial justification by itself. Clearly, DR can lead to beneficial short-term impacts on the electricity market that increase as the number of customers participating in the demand management program increases. However, more efficient results could be achieved by focusing on the benefits to individual customers. These include

- Individual financial savings: In addition to receiving incentive payments and discounted rates, a customer participating in a demand management event, for example, by shifting the noncritical portion of demand from peak-load hours to off-peak hours, could also benefit from savings in monthly electricity bills.
- Second to personal cost savings, consumers have a growing concern about the environment [13]. Consumer engagement may be increased by making sure they understand the environmental benefits of their proposed response to grid conditions.
- Avoiding uncontrolled loss of service: By participating in a demand management event, for instance, through direct load control program for air conditioning units, a customer can help the utility achieve a controlled load reduction where power will be restored after the preset duration of the event is passed. Lack of sufficient participation in the long run could lead to weakening of the distribution network during peak-load hours, which in turn could lead to an uncontrolled loss of supply.

3.12 CONVERGENCE OF TECHNOLOGIES AND ENTERPRISE LEVEL INTEGRATION

Stuart Borlase, John Chowdhury, Greg Robinson, and Tim Taylor

Smart grid is a system of systems. In order to realize the full potential of the benefits of smart grid, a new set of "smart" applications and significant enhanced or new utility business processes will be required. Much of this converges into the operations domain of smart grid. These new applications and business processes will require a holistic solution to achieve seamless and secure interoperability of new and existing information sources. Today, many of the utility operations centers (transmission and distribution [T&D] control rooms) have a wide variety of software systems that are silos—islands of information. Compounding this problem, many of the software systems in the operations domain at utilities are often from different vendors, each with separate, independent data models, user interfaces, and incompatible and nonstandard security measures.

It is imperative that these "smart" applications leverage (not replace) critical operations information and services provided by existing legacy systems and be highly flexible and scalable to a new generation of information sources while preserving the system quality attributes (performance, scalability, availability, security, etc.) that is mandatory to management and control of our most critical national infrastructure—the electric power grid. A systems approach to smart grid is required to ensure interoperability and provide consistent set of measures to guarantee against cyber-related vulnerabilities and potential coordinated targets of attack on the nerve center to T&D operations.

Some major challenges emerge in this context. The first set of challenges refers to the adoption of these new technologies, which are, for the most part, untested and unproven in the field. The second set of challenges refers to the lack of industry standards, protocols, and operational guidelines that should be required in an industry that has very high requirements for reliability and control. The final set of challenges refers to the existing gap between technical operations and business decision making within the utility's organization. This gap reflects on the business process, information technology (IT) infrastructure and integration, workflow consistency, and timely interaction.

While standard services are imperative for vendors to agree on the semantics behind interoperability of respective systems, the implementation and underlying security of these services in an operation-centric environment require an appropriate software infrastructure. Many in the industry believe that an enterprise service bus (ESB) or operational service bus, as sometimes referred to in this industry, and the use of a set of standards-based and secure domain services are a key element in the integration architecture.

3.12.1 Synergies of Integrated Applications

The smart grid initiative uses technology building blocks to drive toward a more integrated and long-term infrastructure that is intended to realize incremental benefits in operational efficiency and data integration while leveraging open standards. Utilities should consider the synergies of smart grid technology deployment and sharing of infrastructure costs when planning smart grid programs. A well thought out smart grid roadmap leveraging several technologies is also a good strategy for any regulatory approval.

The smart grid will help move the thinking away from tactical implementations of technologies to address specific issues or gain isolated benefits to longer-term integrated and interconnected strategic solutions. The fully interconnected, integrated nature of a smart grid provides additional benefits that cannot be realized from isolated, tactical solutions.

Consider Figure 3.175 that highlights the potential of smart grid functions to deliver value to utility business processes.

In addition to synergistic benefits, cost savings are also realized among smart grid technology deployments. For example, an AMI communications network with two-way, high bandwidth and open connectivity can also support communications to distribution automation devices in the field and provides opportunities for enhanced customer service solutions (such as Internet access, through a home area network [HAN]) and therefore a more attractive return on investment.

There are three key areas in which business systems need to be integrated:

- IT/back office to IT/back office
- Operational/realtime to operational/realtime
- IT/back office to operational/realtime

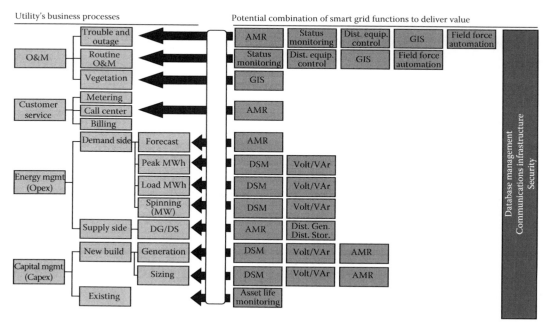

FIGURE 3.175 Synergies of smart grid functions delivering value to utility business processes. (© Copyright 2012 GE Energy. All rights reserved.)

Smart Grid Technologies

The integration of IT systems and operations technology (OT) systems is vital to a successful implementation of new technologies under the smart grid umbrella. IT systems are mostly software applications for commercial decision making, planning, business processes management, and resource allocation. IT applications can include, among others,

- *Enterprise resource planning—ERP*. For managing financial and human resources, materials, and assets
- *Enterprise asset management—EAM*. For supply chain, inventory management, work and asset management
- *Customer information systems—CIS*. For managing customer data, metering data, settlements, and invoicing
- *Energy portfolio management—EPM*. For energy planning, portfolio optimization, scheduling, energy trading and risk management, market analysis, retail management, price and load forecasting, ISO bidding, settlements, and postanalysis
- *Geographic information systems—GIS*. For mapping and geographic information

OTs are software applications that provide operational control of assets in the electric network in realtime (or near realtime). OT applications should be integrated in order to present a complete and current view of the network to the utility business and can include, among others,

- *Supervisory control and data acquisition—SCADA*. For real-time data acquisition
- *Distribution management systems—DMS*. For managing and control of distribution networks. Includes advanced applications such as fault detection, isolation, and service restoration (FDIR), volt/VAr optimization, state estimation, outage management systems (OMS), etc.
- *Energy management systems—EMS*. For managing and control of generation and transmission systems
- *Outage management systems—OMS*. For identifying distribution system outages and helping with service restoration activities
- *Mobile workforce management—MWFM/field force automation—FFA*. For managing mobile field crews, mapping, work scheduling, and optimization
- *Advanced metering infrastructure—AMI*. For gathering and managing metering data (interval and noninterval). Includes remote reading and possibly remote control
- *Demand response management—DRMS*. For managing demand response programs and Virtual Power Plants (VPP)

The following list provides examples of operational/real-time system integration:

- OMS–SCADA/EMS: Real-time SCADA data providing accurate outage information in realtime.
- OMS–DMS: Real-time DMS switching programs and network applications cross-referenced with outages based on customer information.
- OMS–MWFM: Outage analysis and management integrated with mobile solutions for automatic/intelligent dispatch, crew management, and outage restoration. Avoids the control room bottleneck when faced with extreme outage conditions as manual intervention is reduced significantly while maintaining the integrity of the network.
- OMS–AMI: Real-time meter data indicating customer outages. Pinpoints scope of outage and can be used to detect nested faults through ping functions.
- DMS–EMS: Accurate real-time data shared between T&D networks allowing dual control, sharing of real-time data where applicable, and coordinated switching and tagging between T&D networks.

- DMS–SCADA: Real-time distribution SCADA data allowing utilities to automatically restore the distribution network, perform network analyses such as load flow, FDIR, and VVO, as well as standard monitoring and control features using an integrated network model.
- DMS–MWFM: Mobile solutions enabling field crews to reflect field switching directly in the network model in realtime as performed in the field.
- DMS–AMI: Metering data provides accurate load profile model for use by power analysis and demand management applications. Integration is an enabler for partial/selective load shedding functionality.
- EMS–SCADA: Real-time transmission SCADA data allowing utilities to use advanced EMS applications to monitor and control the transmission network.

The aforementioned integration results in improved customer experience and productivity levels of call center, dispatch, network operators, and field crews.

The following list provides examples of IT/back office to operational/real-time system integration and the resultant benefits to the utility business:

- EAM–EMS/DMS/OMS: EAM data in EMS/DMS and OMS applications provide more and more data for intelligent network operation, including ratings, limits, internal diagrams, and other asset data. Real-time systems can also provide EAM systems with operational usage data in order to improve maintenance regimes and asset performance statistics.
- GIS–DMS/OMS: GIS network model data are increasingly being considered the sole source of network data for the enterprise. Having DMS/OMS models integrated with the GIS allows the real-time network to be constructed from a single common data source.
- SCADA–EAM: SCADA can provide EAM systems with operational history on and assist utilities with condition-based monitoring and outage prevention.
- WFM–EAM–ERP: Through tracking of assets utilized during a project via mobile solutions, project accounting can be more accurately recorded and forecast.

To be as effective as possible, an integrated solution relies on the following data to be current and accurate:

- SCADA I/O
- Customer fault information
- Network topology
- Resource positioning and workload
- Asset data
- Power flow calculations

3.12.2 Examples of Converging Technologies

3.12.2.1 Integrating Distribution Operation Applications

Integration of distribution operations systems will be critical to the success of the smart grid. The isolated operational systems of the 1980s and 1990s have become increasingly integrated, through various interfaces such as file transfers, application programming interfaces (APIs), middleware, and most recently, web services. As the smart grid progresses, and interoperability standards advance, the sophistication of operational systems will continue to increase.

As electric utilities look to the future with the intense pressure to improve reliability, operational efficiencies, and customer satisfaction, utilities will require advancements in DMS and operational management systems to meet the growing demand. Evolving business and regulatory challenges

Smart Grid Technologies

have resulted in utility demands to use DMS and OMS tools seamlessly to manage OMS processes such as unplanned outages and crews, while also managing complex and heavily loaded distribution networks with advanced distribution applications.

Utilities are experiencing an increase in automation and the amount of data collection points being applied to customer premises and utility networks. The ability of network operators to proactively manage large and complex networks requires new advanced tools that can turn network data into useful information in order for safe, timely, and effective operational decisions can be made. Together, DMS and OMS will provide a comprehensive and integrated approach to managing networks and outages during routine and high-volume periods.

Figure 3.176 shows an integrated distribution operations environment using a single network model and integrated operator functionality. An architecture in which OMS functionality and DMS network applications use a single, common distribution network model has the advantage of only requiring one network model that needs to be maintained. Such an architecture is also modular, in order to effectively meet the needs of different distribution organizations that are in different phases or have different strategies of smart grid implementation. This modular distribution operations platform permits organizations to incrementally add particular modules, such as OMS capability and advanced DMS network applications, as their business needs change.

The components of the integrated distribution operation environment include the following.

3.12.2.1.1 SCADA

Distribution SCADA infrastructure is shown at the bottom of Figure 3.176. SCADA collects analog and status data from RTUs and IEDs and provides functionality such as control, alarms, events, and tagging. The OMS and advanced DMS network applications can utilize SCADA

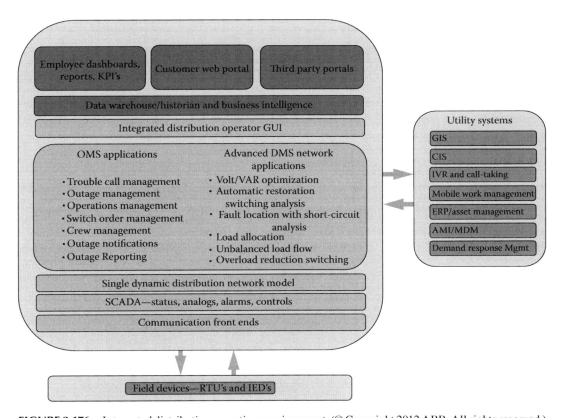

FIGURE 3.176 Integrated distribution operations environment. (© Copyright 2012 ABB. All rights reserved.)

from a single vendor in an integrated platform, which facilitates the integration for a distribution organization, or they can be integrated with a third-party SCADA through ICCP or other types of interfaces.

3.12.2.1.2 Single Dynamic Distribution Network Model

As shown in the center of Figure 3.176, this architecture utilizes a common distribution network model for DMS applications and OMS applications. It greatly simplifies system maintenance, as it eliminates the need to build, maintain, and synchronize multiple data models. An additional benefit is coordination of planned outages and unplanned outages due to temporary lines, line cuts, manually dressed device operations, and subtransmission operations. The single dynamic distribution model also means that the advanced DMS network applications always utilize the as-operated state of the distribution network, with the present connectivity and state of switching devices, capacitor banks, and customer loads.

3.12.2.1.3 Advanced DMS Network Applications and OMS Functionality

This architecture includes integrated OMS functions and advanced DMS network applications. The OMS functionality includes the identification and resolution of outages throughout the distribution network, the management of crews, and the tracking of corrective work, including temporary cuts and line jumpers. Advanced DMS applications that have been developed include load allocation and unbalanced load flow analysis; switch order creation, simulation, approval, and execution; overload reduction switching; and capacitor and voltage regulator control. Other applications such as fault location and restoration switching analysis permit operators to reduce customer interruption durations during outage management. Applications such as load allocation, unbalanced load flow, and line unloading permit improved asset utilization and operation closer to equipment thermal limits.

3.12.2.1.4 Integrated Operator Graphical User Interface

The integrated architecture provides a consistent user interface across the operational functions. This can include distribution SCADA, DMS applications, OMS, and even transmission SCADA if required. Operator workstations consist of tabular displays, in combination with geographic and schematic displays, that provide fast response during storms. The result is improved operator effectiveness and flexibility, as well as reduced maintenance and training costs.

3.12.2.1.5 Packaged Business Intelligence for Distribution Organizations

Packaged business intelligence solutions are now available to distribution organizations. This enhances an organization's reporting, situational awareness, and business intelligence needs, and allows individuals across the distribution organization to understand what is happening through the use of standard KPIs, dashboards, and reports that come out-of-the-box. With minimal training, custom pages can also be designed. This provides operations, management, and others across the organization an improved picture of situational awareness. It also provides customer service representative information that they require to be very responsive to customers inquiring about service issues.

3.12.2.1.6 Integration of SCADA, DMS, and OMS

Integration of DMS/OMS with SCADA is an increasing trend. While the inclusion of SCADA breaker-open operations in OMSs has been used for outage detection for years, recent business challenges have driven a more comprehensive integration between the systems. Available functionality now includes the transfer of status/analog points from SCADA to the DMS/OMS, the sending of supervisory control and manual override commands from the DMS/OMS to the SCADA, an integrated user interface running on the same operator console, and integrated single sign-on for users.

Smart Grid Technologies

Benefits of integrating SCADA with DMS/OMS include the following:

- Improved operations by close integration of DMS applications with distribution SCADA
- Increased operator efficiency with one system, eliminating the need for multiple systems with potentially different data
- Integrated security analysis for substation and circuit operations to check for tags in one area affecting operations in the other
- Streamlined login and authority management within one system
- One network model for OMS and DMS analysis
- Consolidated system support for DMS/OMS and distribution SCADA

3.12.2.1.7 Integration of OMS and Mobile Workforce Management

Integration between OMS and MWFM is often done to improve work flows in the outage management process. Functionality that is typical includes the following:

- Transmittal of outage assignments directly to the mobile data terminal (MDT) from the OMS
- Receiving of crew assignment status updates from the MDT (en route, arrived)
- Updating of assignments automatically to MDT as the OMS outage engine repredicts outages
- Verification and completion of outages in the OMS from the MDT
- Display of crew login status in the OMS, as entered from MDT

There are other advantages of closer integration between the control center and field operations. Switch order management and execution can be more efficient. Resources can be dispatched and managed more efficiently. Figure 3.177 illustrates benefits from improved integration between control center and field operations functions.

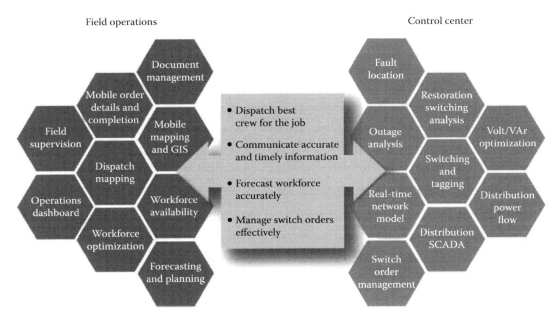

FIGURE 3.177 Integrated network management and field operations. (© Copyright 2012 ABB. All rights reserved.)

3.12.2.2 Integration of AMI into the Distribution Operations Environment

AMI is a key component of many distribution organizations' smart grid plans. Business cases are being built on the premise that AMI systems with the right functionality can improve system operations. Interfaces between AMI/meter data management (MDM) and the OMS already have been developed. Work continues to enhance the functionality, but already there are several ways AMI data can improve the outage management process (Figure 3.178).

First, if the AMI meters and communications are so equipped, the OMS can receive a last-gasp or outage notification message from the meter when it loses voltage (i.e., a customer outage event has occurred.). That way the OMS is notified of any customer outages, even if the customer does not report it. Receiving outage notification messages is in addition to phone calls from customers reporting outages. These messages are particularly useful when no one is at a property where an outage occurred or when people there are asleep. The outage notification message can provide the granularity of identifying individual customer outages at the meter and help with outage management algorithms to more accurately predict the source of the outage and therefore reduce customer interruption times and result in a more efficient dispatch of repair crews.

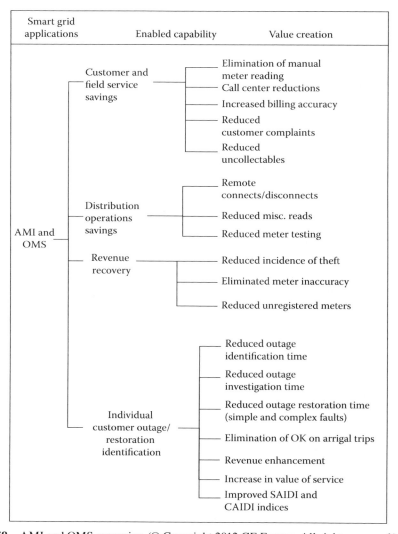

FIGURE 3.178 AMI and OMS synergies. (© Copyright 2012 GE Energy. All rights reserved.)

Second, with the proper interface between the OMS and AMI system and the right communications infrastructure and meter, a message can be sent from the OMS to query if a meter is in service. This is sometimes referred to as "pinging the meter." The meter can be pinged directly, assuming the AMI communications permit it, or the MDM can be pinged to determine the status of a meter. The meter can be pinged either by a customer service representative or an operator. The value in pinging the meter is that many customer outage reports are results of problems on customer sides of meters. Utilities commonly report that 50%–67% of single-customer-call outages are results of problems on the customer side of meters (OKA—OK on arrival) and not the responsibility of distribution organizations. By remotely interrogating the meter, the utility can inform customers if any loss of service issues requires further investigation by the customer on the customer side of the meter. If personnel can ping a meter to determine it has voltage despite a customer's reports of no power, responding troubleshooters and crews could save time and vehicle miles. Utilities can also take advantage of other measurements available in the smart meter, such as power quality (e.g., sag and swell) data, and identify and respond to any service issues well before the customer calls to complain. An additional value in meter query is in the ability to potentially perform outage scoping or define the outage area by pinging select meters. This can lead to a faster definition of the outage area.

A third area in which interfaces between OMSs and AMI systems can provide value is through restoration notifications. They provide confirmation to distribution operators that customers have been restored downstream of a particular protective device. Restoration notification can be done through a restoration notification message transmitted from the restored meter to the OMS or through pinging of meters that presumably have been restored. The value of restoration notifications is that when all customers have not been restored because of a nested outage within the larger outage area, field personnel can be notified of additional problems before they leave the area. This reduces the need to redispatch crews during multiple outages and reduces crew costs and travel time. The smart meter can therefore further reduce the duration of extended outages as part of the improvement in customer service.

Improved ways of using AMI infrastructure for improved outage notification still are being explored. In addition, the use of other AMI data in DMS applications, such as interval demand data and voltage violations, is being investigated. Methods are being developed in which actual customer load profile information, obtained from AMI data, is being used in load flow programs, instead of typical customer-class load profiles that are developed for typical day types. This means that more accurate loading information can be applied to each distribution transformer before load flow calculations on the system are performed. The results are improved understanding of the actual power flows and voltages on a distribution systems.

Voltage violation alarms that are generated in the AMI system can also be sent to the DMS system so that system operators are notified of system voltage problems. Volt/VAr control can then be adjusted to compensate for the voltage problems, either by the operator or directly by the automated volt/VAr control application. This will be especially important for utilities who aim to reduce customer demand by operating their systems in the lower levels of the permissible voltage bandwidth, a concept commonly known as conservation voltage reduction.

3.12.2.3 Multiple Smart Grid Functions Benefit Outage Management

In recent years, improved tools for FFA, including the use of mobile dispatch and GIS, have improved the productivity of field crews and reduced outage durations. The implementation of distribution feeder FDIR can also reduce customer outage duration on unfaulted feeder sections. Other technologies, such as asset monitoring and diagnostics and vegetation management, will also provide incremental benefits to outage management (Figure 3.179).

Geographic information system (GIS)—Typically a GIS serves as the distribution organization's master repository for distribution asset data and system connectivity. The distribution operations environment utilizes the asset data and connectivity information, which is periodically updated in the

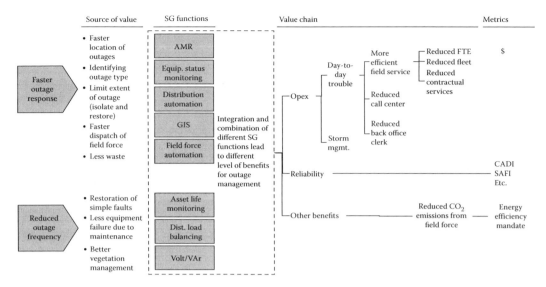

FIGURE 3.179 Multiple smart grid technology benefits outage management. (© Copyright 2012 GE Energy. All rights reserved.)

GIS. Data can be extracted from the GIS via XML files, comma delimited text files, or shared database tables. A bulk load process is typically used when a distribution organization first implements a distribution operations environment with an OMS and/or DMS. An incremental update process is frequently used afterward to transfer changes from the GIS to the distribution operations environment.

Customer information system (CIS)—Customer data, customer-to-transformer ties, connects/disconnects, and trouble calls can be received by the OMS from the CIS. Typically customer data are updated either on a nightly basis or on a near real-time basis. Trouble calls can be fed into the OMS, and feedback to the caller can be provided in realtime.

Interactive voice recognition (IVR)—The distribution operations environment can be integrated with both incoming and outgoing IVR systems for trouble call entry, call back, and planned outage notifications.

Mobile workforce management (MWFM)/field force automation (FFA)—DMS/OMS are frequently configured to create orders in an MWFM system for routing to mobile data terminals. The status from the trucks can be displayed (en route, arrived, etc.), as well as outage completion status and estimated restoration time.

Global positioning system (GPS)—Outage management systems can support receipt of periodic vehicle location coordinates, from GPS or AVL (automatic vehicle location systems). The coordinates can be used to display the vehicle position on a geographic world map.

ERP/asset management/work management—In many cases, an interface is implemented between the DMS/OMS system and the asset management system. This can be done for switch orders, or to create work orders and work requests in work management systems for follow-up work.

3.12.2.4 Integrating Workforce, Asset, and Network Management Systems

There are certain aspects of a working smart grid that are often not at the forefront in design and deployment activities. One of these overlooked aspects relates to the underlying systems infrastructure that will be needed to keep the utility network running effectively post-implementation. Specifically, there are operating assets (indeed, a whole new class of such assets), along with the data produced from those devices, that will need to be monitored, managed, reported, repaired, and, at times, replaced. All of this work will need to be done, and at the same time the existing grid infrastructure requires and receives ongoing maintenance, repair, and operation. Utilities can approach these systems challenges, utilizing existing technology solutions firmly linked in to the intelligent distribution network. Asset management, network management, and MWFM systems all can, and should, play a big role in a utility's planning, implementation, and operation of the smart grid. Smart grids will require significant supporting operational solutions to achieve their full promise.

There will be many parts in a fully evolved and operating smart grid. Spanning the energy flow from generation and transmission to distribution, these include smart meters, smart substations, distributed generation, renewable energy, and many other related parts. These components include functions needed for smart grid operation such as real-time simulation and contingency analysis, distributed generation and alternate energy sources, self-healing wide area protection and islanding, demand response and dynamic pricing, and energy market participation, among others. The design objectives and means of a traditional grid asset and network management approach, supported by an MWFM framework, must address a wide spectrum of network management, asset management, and workforce management requirements. Taking an asset-centric view (as opposed to customer-centric, which is equally important), utilities must be able to develop assets, operate assets, and maintain assets efficiently and effectively. Each of these has very different challenges, as seen in Figure 3.180.

Developing assets is typically a complex, multitask undertaking involving carefully planned work, extended periods of time, multiple crew types and skills, and, potentially, contractors. Operating assets, however, is typically a high-volume, short-duration activity, with mostly unplanned orders for individual field workers. Maintaining assets is an amalgam of the two, with work that is planned and scheduled involving individual technicians or crews working varying durations. Supporting smart grids involves these very same breakdowns of work, yet the tools, scheduling, skills, and durations can significantly vary from what many utilities are accustomed to. What is particularly new is the amount and types of data these new assets produce, and the time immediacy they will require.

"System of record" is a term often used in the utility industry. EAM, supported by a GIS, is the system of record for the asset, SCADA/DMS/OMS is the system of record for actions taken on the

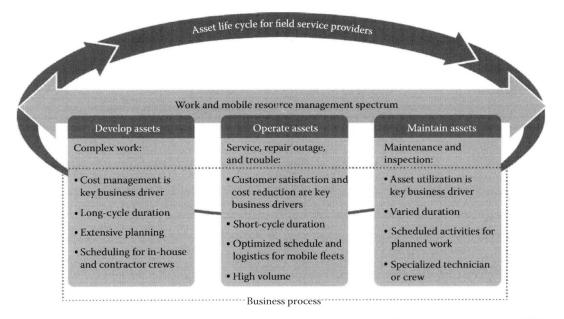

FIGURE 3.180 Challenges of developing, operating, and maintaining utility assets. (© Copyright 2012 ABB. All rights reserved.)

grid, and MWFM is the system of record for the work performed and workforce that did it. Any of these systems alone is a powerful tool, but properly integrated, these systems enable a new level of operations and asset management efficiency.

There is great value to be gained in addressing the challenges of building, operating, and maintaining a smart grid. Better managing expensive smart grid assets can lead to extended life of that equipment, and thus a utility's distribution asset acquisition demands can be lowered by 10% or more, and on-hand inventory levels reduced 10%–20%. With smart asset diagnostics and repair capabilities combined with more efficient crew routing and scheduling, regulatory fines can be reduced and workforce productivity improved as much as 20%.

The concept of integrating the SCADA/DMS/OMS, Enterprise Asset Management (EAM) system, and the Enterprise Workforce Management (EWFM) system is shown in Figure 3.181. Work to be done can originate in either the EAM or SCADA/DMS/OMS systems and immediately (without intervention) flow to the EWFM system as field actionable. The results can flow back to the originating system and to the other host system as well. After it acts, data are returned to the upstream systems now positioned as objects describing that grid asset.

Shifting that focus from the assets to the customers, it is also clear that the demands of a smart grid customer will differ in some ways from a traditional customer. With intelligent devices in or near their homes and businesses, these customers will expect quicker response and restoration times. Furthermore, they will need better and more detailed explanations of what is occurring in an outage, for example, and what the utility is doing about it. They will also expect tighter appointment windows and more explanatory billing. In short, the smart grid customer will want more information and greater control. To achieve these demands, the utility will have to provide greater integrated workflow communication between the control center and the field operations workforce.

The control center today already has plenty to do. Responsible for switch order management, fault analysis, real-time network modeling, and many other activities, the center is at the heart of efficient grid operation. For its part, field operations are equally busy as it fulfills its duties for outage responses, customer and meter service work, and grid infrastructure inspections and

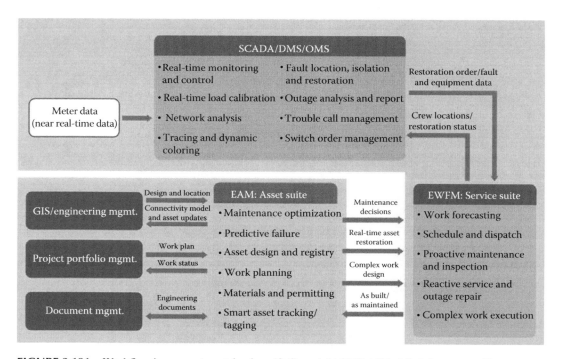

FIGURE 3.181 Workflow in an asset-centric view. (© Copyright 2012 ABB. All rights reserved.)

Smart Grid Technologies

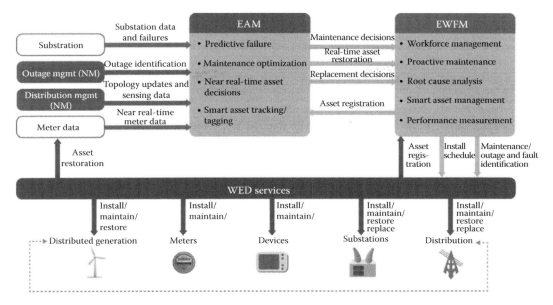

FIGURE 3.182 Workflow in a customer-centric view. (© Copyright 2012 ABB. All rights reserved.)

maintenance work. All of these and many other duties combine to execute control center directions in the field and on the ground or in the air. Figure 3.182 shows the workflow components that are vital in a smart grid world where these two groups must closely work together, emphasizing four areas in particular:

1. Dispatching the best crew (skills, location, equipment, and materials) for the job
2. Communicating accurate and timely information about the order, especially data arising from smart assets that field techs must know
3. More accurately forecasting the expected workload by geography and skills, to ensure field resources are available where and when they are needed
4. Managing the process of planning, executing, and documenting switching orders in outage management

3.12.3 Enterprise Integration

"Integration" is defined as the act or approach of making two of more independently designed things (e.g., systems, databases, or processes) work together to achieve a common business goal. There are many styles of integration. The types of "things" in question determine the category or style of integration that applies. Some example integration includes the following:

- People: To function at an optimum level, human-to-human and human-to-process interaction requires integration throughout the various organizations and not limited to the end users. Business partners, customer, and employees are all important resources to the value chain provided within the smart grid. Think of renewable integration for a utility where utilities are engaging third-party providers, customer through net metering, and utility-owned services.
- Process: Recurring elements (security, service level, monitoring, etc.) can be shared across applications to provide horizontal services to decouple these reusable application components. The use of service-oriented architecture (SOA) or ESB to implement these processes will facilitate more rapid changes in these processes.

- Application: Utilities have invested enormous resources and capital into custom-designed and off-the-shelf applications. The application integration goal is to leverage, rather than replace, these assets by providing ways of connecting, routing, and transforming the data that are stored or shared among them.
- Systems: Systems manage, process, and deliver data to the people and applications in the utility environment. Smart grid operating environment will require the system to be transparent to the elements that interact with it.
- Data: Data are the primary business element of a system. The data are the source of the information and can more easily be shared through the adoption of the standards specification.

Today's typical utility IT systems were installed over several years, and they were expensive to deploy and are costly to maintain. They are often too complicated and often difficult to use. These systems were built within the enterprise business operations silos, and they were built for the specific purpose. These systems largely ignored the informational needs of the surrounding functional areas. They are also not easy to integrate causing inefficient process within the utilities.

Within the last several years, as utilities looking to implement a new and improved business processes, they expect new systems to provide a revolutionary application platform that includes open source, standard-based solutions, XML, and SOA. With smart grid, the new systems are ideally expected to be easy to use, highly adaptable, and responsive to the often changing needs. They are expected to have significantly lower total cost of ownership than the legacy systems. These systems are built to integrate with existing or future applications. As utilities are looking into these new systems, focus should be in configuration rather than coding of the new systems.

3.12.4 Data Integration versus Application Integration

Within the utility enterprises, the majority of integration activity can be classified as either data or application integration, but data and application integration overlap. Bringing together data from two independently designed databases to create third database is an example of data integration. Creating a composite system by leveraging the functionality of two independently designed applications is an example of application integration. A common requirement is to ensure that two or more applications contain consistent data such that users interacting with these applications will see a consistent state of critical business entities (such as customers or products). While the end result of this activity is that the applications involved share the same perspective (i.e., application integration), achieving this state involves synchronizing the data between them (data integration). According to Gartner, enterprise integration is "Making two or more independently designed things (applications, systems or DBs) work together to achieve a business goal." An example of such integration can be seen in Figure 3.183.

3.12.5 Enterprise Service Bus

The enterprise information bus (EIB) or ESB provides an infrastructure view, in terms of the middleware layer that supports and manages the run-time business and infrastructure services, and facilitates their creation from existing assets. The EIB or ESB contains the enterprise integration components. Typical enterprise integration component includes

1. *Message transport*: Moves data from the application to broker and vice versa
2. *Transformation engine*: Translates messages from one format to another
3. *Process management*: Provides ability to apply business logic to events
4. *Routing*: Route requested data to the application or events
5. *Application adapter*: Forms the bridge between integration layer and application layer

These components are typically managed by a concept called enterprise SOA.

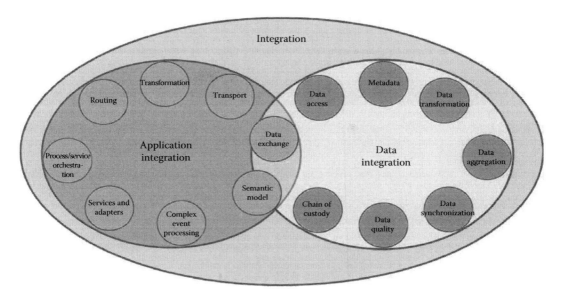

FIGURE 3.183 Conceptual integration of data and applications.

3.12.6 Service-Oriented Architecture

Enterprise SOA is a business-driven software architecture that allows a utility to share and reuse services within the company and with strategic partners. With enterprise SOA, business functionality is represented through reusable enterprise services. Complex processes can be broken down into smaller process steps, which are represented in software as reusable enterprise services. Though SOA, utility IT organization can support business change with increased adaptability, flexibility, and cost efficiency. With increasing smart grid adaptation, a utility can compose new business processes for business innovation. Enterprise services—including those from strategic business partners—can be created, combined, and rearranged to compose new business processes to support innovative business strategies. A utility also can quickly modify business processes to achieve operational excellence. When change is required, utility IT staff only needs to modify the enterprise service for a particular process step, so the IT organization can respond more quickly and cost effectively to changing business requirements.

SOA can assist utilities to automate processes to empower information workers. An organization can utilize enterprise services to automate many process steps that currently require human intervention. With automated process information, workers spend less time on routine tasks, freeing them to contribute to the strategic goals of your business.

3.12.7 Enterprise Information Management

Enterprise Information Integration (EII) and Enterprise Application Integration (EAI) are not new concepts [1]. They are processes and tools applied in other industries to integrate and share data among multiple sources and multiple applications. The challenge with the utility industry and other real-time mission-critical systems is the integration of data that are frequently updated and required within short time periods (seconds) with data that are not required to be updated as frequently (days).

The main show-stopper for large-scale integration is that data reside in thousands of incompatible formats and cannot be systematically managed, integrated, or cleansed. Integration infrastructures suffer from data errors and ambiguities that arise with different interpretations about the intended meaning of information exchanged between applications, such as (a) confounding conflicts, where information appears to have the same meaning but does not; (b) scaling conflicts, where different reference systems measure the same value; and (c) naming conflicts, where naming schemes differ significantly.

With the smart grid, the types and also the volumes of data are growing dramatically. Traditional approaches, already strained with existing systems, will simply not support information exchange requirements of utilities implementing smart grid systems. Enterprise Information Management (EIM) is of strategic importance because it establishes data and information as business enablers rather than inhibitors. EIM does this by resolving information to a single version of the truth, thereby reducing the risk of misinformation and increasing the efficiency of system and human interactions.

Many tools and industry standards exist to facilitate process integration and business intelligence. An EIM-based methodology embraces these tools and standards as a way to resolve the semantic differences that make data difficult to exchange, analyze, and understand. In an IT environment where utilities have dozens of representations of the same information, resolving these differences by creating common meaning across the enterprise is of utmost importance. This methodology manages business semantics as the foundation for supporting common understanding, transparent flow, and usage of information across an enterprise. It decouples applications so that data can be used throughout the enterprise as a true information asset. This not only insulates applications from changes to data sources and application interfaces, but it also enables utilities to dramatically reduce redundant data sources, migrate systems much more easily, and free utilities from vendor lock-in.

3.12.7.1 Establishing an EIM Framework

The challenges with integrating systems are many and begin with the way systems are procured. When a project procures applications, vendors are driven by the procurement process to meet user requirements at lowest cost. Each of the procured systems has its own unique mixture of platform technologies, databases, communications systems, data formats, and application program interfaces. While utilities prefer products that support industry standard interfaces, another high priority is for product vendors to supply application interfaces that remain relatively stable across product releases. In that fashion, once an application is interfaced to the utility's enterprise integration infrastructure, incorporation of future product upgrades will be easier, charges for custom interface development will be decreased, and the risk of errors will be reduced during installation and maintenance of each product release. Success will depend on how well EIM is defined before the procurement process begins, and then used as proposals are evaluated/selected and then ultimately through implementation.

A well-designed EIM strategy requires business units and IT to look at enterprise data and information as assets in an effort to understand the nature of the information and how it is used and controlled. This effort includes addressing key issues around data definition, quality, integrity, security, compliance, access and generation, management, integration, and governance. These issues are interrelated and systemic in nature and require business units and IT to work together to understand and catalog information and data to build a robust model that can be leveraged across the enterprise. This is an iterative process that requires a holistic and evolutional EIM strategy and framework to ensure a consistent and effective approach.

EIM frameworks and strategies provide a clear roadmap for utilities to establish the necessary governance and technology solutions. EIM is not only complementary to SOA but also required for the business to drive and enable the convergence of OT and IT, which are key parts for the ultimate realization of a smart grid. An example of such a framework can be seen in Figure 3.184.

At the core of an EIM methodology is the development and use of an enterprise semantic model (ESM), which serves as the logical representation of the information assets an enterprise uses to manage and facilitate business processes. When governance, for example, is established by a utility in the absence of an overarching semantic life-cycle design, governance will provide little return on investment because the opportunities for effective reuse of project artifacts will be greatly diminished.

Smart Grid Technologies

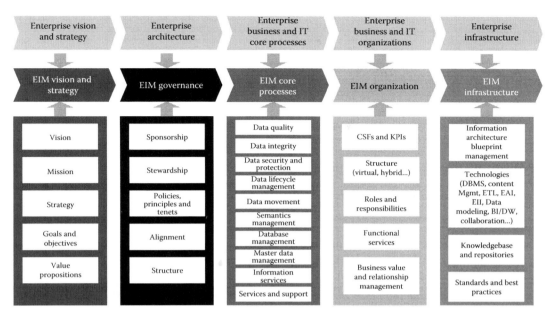

FIGURE 3.184 EIM framework. (© Copyright 2012 Xtensible Solutions. All rights reserved.)

An effective EIM methodology allows utilities to embrace industry standards such as IEC common information model (CIM) and, more importantly, enables utilities to create their own information model by organizing metadata and models from their existing applications. This is a critical component of an enterprise strategy for creating reusable data services that would otherwise not be achieved with SOA investments [2].

Basic ESM goals include the following:

- *Enterprise information driven*: The ESM utilizes internal models, metadata, and terminology already in use in the enterprise. Existing models and common vernacular, whether or not they are documented, are the most important sources for ESM development.
- *Enterprise owned*: The enterprise owns the ESM, including its terminology, semantics, and implementation. The enterprise has the final word on if and when externally controlled semantics are introduced.
- *Stable*: The ESM must be stable, keeping established semantics clear and unambiguous.
- *Nonstatic*: ESMs are nonstatic in nature and must allow semantics to evolve toward greater clarity as existing business information evolves or new information is introduced.
- *Openly accessible*: The ESM must provide open access to business-critical information about semantics, data restrictions, entity refinement, and constraints targeting specific business contexts.
- *Semantic traceability*: Semantic traceability and lineage are important to enterprises that require or desire traceability and correlation across internal information or to non-ESM semantics.
- *Industry standards aware*: By providing mechanisms to systematically take advantage of applicable industry standard models, data types, and code lists as input, a robust ESM incorporates standard and broadly adopted semantics.
- *Multiple standards capable*: An ESM must be capable of incorporating multiple external reference models, even allowing for referencing more than one standard from a single business entity.

- *Concise enterprise semantics*: By benefiting from both internal sources and available industry standards, an ESM provides concise enterprise semantics appropriate for business information across the enterprise.
- *Business context capable*: The ESM must support data exchange and information sharing within a particular business context.
- *Leverage available methodologies*: When appropriate, any existing modeling methodologies should be used in order to avoid reinvention or introduction of proprietary concepts. Proprietary methods limit choices of service providers, which indirectly will drive up maintenance and enhancement costs.

3.12.7.2 Role of an Enterprise Semantic Model

A utility needs to resolve semantics across information sources scattered around the enterprise to support consistent system development, integration, and analysis. As depicted in Figure 3.185, a common approach to resolving enterprise semantics is to map information sources to each other. Key challenges with this approach include

- Difficulty arriving at common agreement of semantics across all uses
- Varying formats and change rates of mapping sources (i.e., inconsistencies due to revisions, upgrades, and replacements)

As a means for resolving these issues, industry standard information models are often employed (refer to Figure 3.186). However, at least initially, this actually adds standard terminology to enterprise semantics. Therefore, key challenges with this approach include

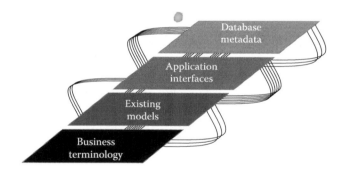

FIGURE 3.185 Typical sources of enterprise information. (© Copyright 2012 Xtensible Solutions. All rights reserved.)

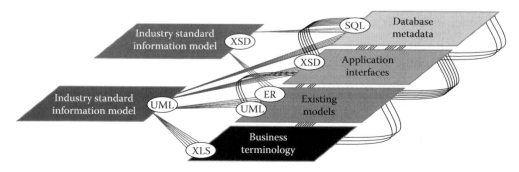

FIGURE 3.186 Introducing industry standard information models. (© Copyright 2012 Xtensible Solutions. All rights reserved.)

Smart Grid Technologies

- Additional semantic mapping to develop and maintain
- Complexity of understanding and using multiple standards
- Differences in format of mapping sources
- Possible internal model vulnerability to external model changes

Utilities are compelled to create more flexible businesses and are weary of plans being hindered by their complex and brittle IT systems. These utilities are now planning their investments in a way that will draw more knowledge and value from the data collected by their smart grid devices, especially the AMI. Intelligently using the data wherever it is needed throughout the enterprise requires an architecture and strategy to deal with how utilities collect, use, and act upon data and information. This data goal leads the utilities to an integration strategy that is based not only on standards such as the CIM but also on EIM concepts to give the standards a proper context. Gartner states,

> Enterprise Information Management (EIM) is an organizational commitment to structure, secure and improve the accuracy and integrity of information assets, to solve semantic inconsistencies across all boundaries, and support the technical, operational and business objectives within the organization's enterprise architecture strategy.

The ESM provides the means to leverage commonly defined business semantics. See Figure 3.187. The ESM serves as the logical model on which all semantically aware design artifacts are then based, such as those for integration services, data warehouses, portals, databases, business process automation, business process activity monitoring, and reporting.

In reference to Figure 3.188, the first step is to identify existing enterprise concepts through analysis and collaboration. Semantics already in use in the enterprise, but not necessarily documented or formalized, are agreed upon. As with all steps in this process, the information analyzed should only cover areas of interest and go as deep as initially required. This step may also be "seeded" with terminology from a reference model, that is, CIM. The output of this step can be as informal as a glossary or as formal as a thesaurus. The content represents terminology agreed upon by business stakeholders that is implementation agnostic, meaning that the terminology can be represented in any modeling tool or development environment.

Once a set of enterprise terms and basic definitions is identified and chosen, it needs to be formalized to whatever degree is determined appropriate for the project. A project team must iterate

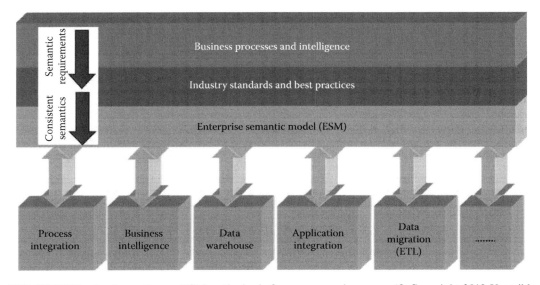

FIGURE 3.187 Implementing an ESM as the basis for a common language. (© Copyright 2012 Xtensible Solutions. All rights reserved.)

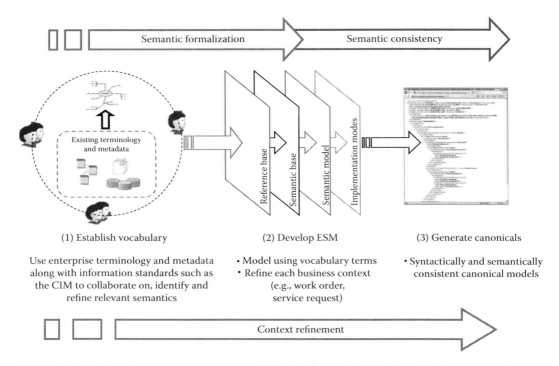

FIGURE 3.188 Developing and applying an ESM. (© Copyright 2012 Xtensible Solutions. All rights reserved.)

the enterprise terminology to diminish ambiguity. It is only after ambiguity is eliminated from enterprise terminology that effective use of external sources such as standard reference models can be accomplished. In this formalization process, synonyms between enterprise terminology and reference models are identified, and preferred terms are established.

The formalized semantics are then implemented in a modeling environment suitable for transforming the formal semantics into a form for implementation by projects consuming or referencing the ESM. Based on the requirements of typical utilities' enterprise landscape, the default modeling environment is UML. However, UML is not required. The appropriate level of formalization for a given project is modeled semantic concepts without excessive rule sets or context specific entities. (Context specific business entities will be defined as part of actual implementation projects.)

Implementation by projects means making the UML model available to technology teams working in the enterprise. If the requirements of a particular project can be addressed with UML, then deployment may simply mean modeling the project-required business entities and implementation models in order to generate deployment artifacts. Other projects may have different deployment requirements.

3.12.7.3 ESM Architecture

A key goal of EIM is to enable the information to be used in a semantically consistent way across disparate technologies and business functions. This is depicted in Figure 3.189. Whether starting at the top with business process models or at the bottom with interfaces to systems, the goal is to represent each concept precisely in the ESM. Concepts may then be utilized properly and consistently wherever they show up, such as in business process automation, business activity monitoring, business-to-business integration, business intelligence, decision support, and inter-application integration. To avoid carrying today's baggage into tomorrow's systems, transformation is used where necessary to interface ESM-based canonical models to legacy and other nonconforming systems.

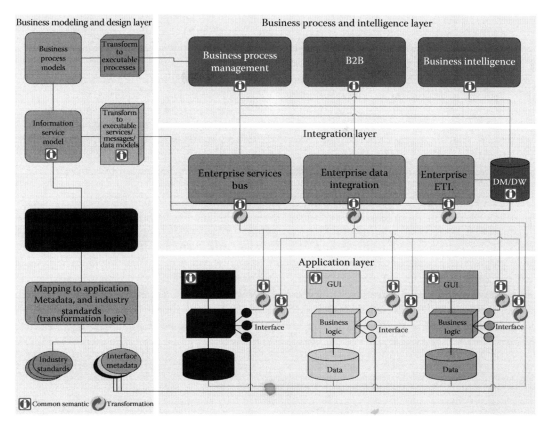

FIGURE 3.189 Architectural view of how an ESM binds everything together. (© Copyright 2012 Xtensible Solutions. All rights reserved.)

Figure 3.189 depicts four major components of how to introduce consistent semantics into the enterprise architecture at both design and run times:

1. Business modeling and design layer: Typically, business process management and design are done on a project-by-project basis, governed, if available, by a corporate IT life-cycle process. What is often missing is how to introduce and manage consistent business semantics at design time. The business modeling and design layer shows that business process models will drive information service models, which are supported by an ESM. The information service models are collections of the services, operations, and messages utilized for information exchange. The ESM is developed through a combination of industry standards, internal application metadata, and business terms and definitions and is defined using UML constructs. This model is transformed into WSDL and XSD definitions for transaction message exchange or DDL for database design and data integration. The output of the process and information service models will drive the run-time environments in the three layers on the right.
2. Application layer: With the increasing amount of commercial off-the-shelf (COTS) applications being implemented at utilities, the ability to dictate how application internal data and information are modeled and represented is eliminated. Utilities can enforce consistent semantics on applications within an enterprise that need to exchange information and provide services outside of the application boundaries. On top of that, applications today are capable of being configured with fields that represent how a utility wants to see their data, thus enforcing consistent semantics at the GUI and reporting levels.

3. *Integration layer*: In today's enterprise, several integration technologies coexist. For example, the ESB for process and services integration and EDI/ETL for data integration often coexist in an enterprise. The key to introducing consistent semantics is to have an ESM to drive both the design of integration services (typically in WSDL/XSD format) and the design of the data services (ETL transformations) and database models (DDL). This ensures that what is exposed to the enterprise is a consistent representation of the data and information.
4. *Business process and intelligence layer*: At the business process level, there are needs for orchestrating multiple applications to accomplish process automation or process management. There is also the need to exchange data with applications or users outside of the enterprise (B2B), as well as to present business data in a way that business intelligence can be mined. All these speak to the necessity of a consistent representation of business meaning (semantics) [2].

3.12.7.4 ESM Information Sources

An effective EIM methodology allows utilities to embrace industry standards as well as enabling them to create their own information model by organizing metadata and models from their existing applications. This is a critical component of an enterprise strategy for creating reusable data services that would otherwise not be achieved with SOA investments. While many industry models may be helpful, at the core of the ESM for most utilities will be IEC's CIM, which was designed for the purpose of integrating disparate utility applications (IEC 61968 and IEC 61970 series of standards). However, ERP and supply chain are largely outside of the scope of the CIM, so additional industry or proprietary models will be employed for these aspects. For communicating with intelligent electronic devices, IEC 61850 contains a rich model that can be incorporated into the ESM.

As the CIM is a large information model and requires that a systematic and model-driven methodology be followed to achieve desired benefits, there are common misconceptions and concerns about using the CIM, some of which are summarized as follows:

1. *The CIM is too large.* For the common system language, the CIM can be thought of as the unabridged dictionary. It is important to note that projects only use the portion of the dictionary relevant to their implementation. But as the dictionary is much richer, there will be consistency and congruity for other areas that the implementation must interface with.
2. *The CIM inhibits innovation.* Because people do not have to waste time reinventing things that have been well vetted in the community, they can leverage the existing dictionary while focusing more energy on their innovative concept. Not only is this more efficient for the innovator, but it is also much more efficient for the people the innovators want to share his ideas with. The community is already educated on how to use the well-vetted language.
3. *The CIM is too slow.* This is like saying the English language is slow; it is based on the speakers command of the language and the choice of media used. When a person communicates with someone, they must (a) articulate the information for the receiver to comprehend it and (b), as a person, may then provide this information through the U.S. Postal Service, through e-mail, through phone calls, etc.; the system may provide this information over many types of middleware (messaging, file transfers, database, etc.).
4. *The CIM is too abstract.* This quality enables the CIM to continue to be relevant and valid even as technology continuously changes. The ability to properly convey unambiguous information primarily boils down to one's skills in applying the common system language.
5. *The CIM is not a best practice data model.* For any individual purpose, one can always invent a model that is superior to any other existing model. The CIM has not been developed for only one functional area but rather by a wide range of domain experts for integrating disparate applications. So for interapplication integration purposes, a superior model does not exist and would be difficult to achieve.
6. *The CIM is too hard to implement.* Specialized models are often biased for a particular implementation and used with a specific implementation technology. If the implementation

Smart Grid Technologies

never had to interface with other systems, this would be easier. As the CIM is an information model that is technology neutral, using the CIM does require following a process that restricts the general information model for particular contexts and then generates the appropriate design artifacts. The CIM community has been doing this for some time, and many tools are available for automating the process [3].

As evidenced by the aforementioned concerns, it is time consuming and resource intensive for project teams to fully understand all of the necessary details to correctly and optimally apply relevant industry standards. So leverage is needed as a means of lowering both costs and risks and is often found through their vendors and in industry user groups such as those of the Utility Communications Architecture International Users Group (UCAIug).

3.12.7.5 Developing and Implementing an EIM Master Plan

Figure 3.190 highlights a 10-step approach with key activities, sequences, and focuses, leading toward an end goal of a shared organizational understanding and commitment of value as well as investment needed for EIM.

While the approach is strategic, the execution of this approach needs to be tactical and flexible to support the short- and long-term needs of the utility. Therefore, it is suggested that steps 1 and 2 be conducted first, then findings evaluated, and then followed by necessary adjustments to the remaining steps [4].

3.12.7.6 EIM Benefits

EIM enables raw data to be turned into information, intelligence, knowledge, and wisdom. As information systems are becoming critical to the success of business, information management must be dealt holistically. In summary, EIM

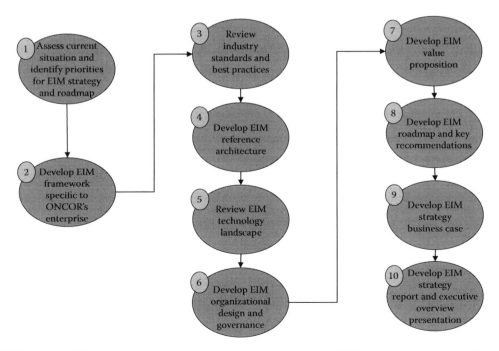

FIGURE 3.190 EIM strategy and roadmap approach. (© Copyright 2012 Xtensible Solutions. All rights reserved.)

- Enables utilities to take ownership, responsibility, and accountability for the improvement of data quality and information accuracy and consistency
- Enables utilities to establish single version of truth for data over time
- Improves utility process and operational efficiency and effectiveness
- Provides a strategy and technique to mitigate the risks as well as maximize the value of implementing commercial packaged applications
- Reduces the number and effort of integration over time
- Enables the control of unnecessary data duplication and proliferation
- Enables a more flexible and scalable process integration
- Improves the data quality, integrity, consistency, availability, and accessibility over time
- Maximizes the return on investment of SOA technologies
- Establishes a critical component of the enterprise architecture
- Provides guidance and services and enables consistent implementation of SOA and information management across major programs

3.12.7.7 Integrating OT and IT Systems

One of the key issues faced in data integration projects is locating and understanding the data to be integrated. Often, one would find that the data needed for a particular integration application are not even captured in any source in the enterprise. In other cases, significant effort is needed in order to understand the semantic relationships between sources and convey those to the system. Tools addressing these issues are in their infancy. They require both a framework for storing the metadata across an enterprise and tools that make it easy to bridge the semantic heterogeneity between sources and maintain it over time.

There is a current trend toward the integration of operations applications and the integration of "enterprise" or "back-office" applications, but they remain as separate integration solutions as shown in Figure 3.191.

Historically, IT and OT reside in different parts of the organization. The operations side of the utility is responsible for execution, monitoring, and control of the electric system, making sure the network is operating within the allowed ranges of reliability, quality, and cost set by the regulations and parameters of the corresponding agencies (i.e., in the United States, NERC, Public Utility Commission, FERC, etc.).

Utility operations groups have control over the assets and infrastructure that is part of the electric network: power generation units, transmission systems, substations, distribution networks, feeders, meters, etc. This control and monitoring are executed via control and protection devices such as relays, circuit breakers, switches, voltage regulators, capacitor controls, and feeder protection. Due to the nature and properties of the electric power systems, speed and precision are fundamental elements within the operations groups (electrons travel very fast and require a balanced equation between load, losses, and generation) in order to keep the system "live," basically complying with Ohm and Kirchhoff's laws, among many others. The decisions that the operations group make are aimed at (in priority order)

1. Protecting the network—Prevent a failure that can damage or destroy expensive equipment and infrastructure
2. Keeping the "lights on"—Prevent outages and blackouts by ensuring that electric demand is met
3. Reducing cost of operation—Ensure that demand is met in the most economical way

These goals are achieved through the use of effective OT.

The business side of the house is responsible for decision making, energy planning, operations planning, resource and asset allocation, and support of any activities required to facilitate the tasks of the operations group such as trading, fuel nomination, field crew dispatch, customer service, etc.

Smart Grid Technologies

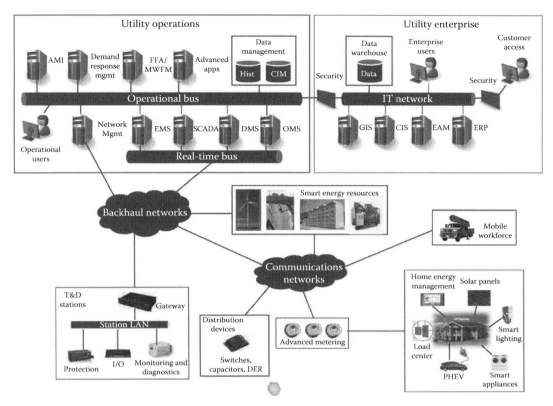

FIGURE 3.191 Dichotomy of utility operations and IT systems. (© Copyright 2012 GE Energy. All rights reserved.)

Decision making at the enterprise level usually involves (directly or indirectly) multiple departments within the utility; for example, sending a field crew to repair a transformer will involve field operations (personnel and vehicles), finance, human resources, inventory/warehouse, and customer service (if the maintenance task will affect end customers). The involvement of multiple departments on any given task or decision-making process demands a tight integration of systems and applications at the enterprise level that must be provided by a solid and consistent IT infrastructure.

IT plays a major role in the success of effective decision making at the utility. Data and application integration, business intelligence, hardware capabilities to run complex algorithms and display mapping features, workflow coordination, and reporting are some of the elements that IT facilitates to the business groups for efficient operation.

The following paragraphs describe the key business drivers, benefits and obstacles of an integrated approach to utility operational and IT business systems (Figure 3.192).

3.12.7.7.1 Benefits

Reduced operating expenses: Integration of operational and business systems allow utilities to benefit from reduced operating expenses through optimal process design, proactive maintenance programs avoiding emergency asset costs, improved technology investment programs, reduced theft, and significant gains in employee productivity.

Reduced working capital expenses: Integration of systems can assist utilities to lower working capital through improved inventory management targeting end-of-life components, slowed peak demand growth, support for distributed generation and improved performance of existing assets.

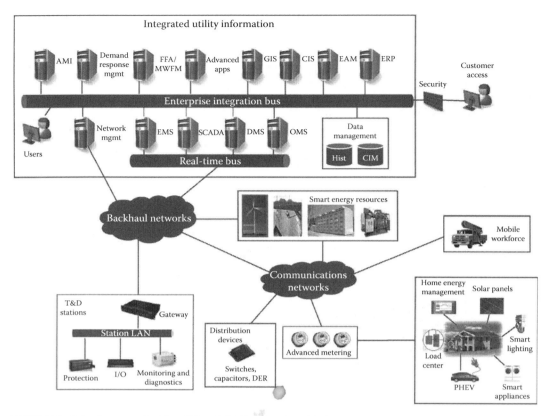

FIGURE 3.192 Integrated utility OT and IT. (© Copyright 2012 GE Energy. All rights reserved.)

Increased reliability and operating revenue: Through integration of systems, utilities are positioning to increase operating revenue through better customer experience, improved forecasting on asset utilization, increased capacity, and by minimizing unplanned network downtime/blackouts.

Integrated vision fosters teamwork: Integrated solutions allow management, operational, planning, and backroom operations staff to truly operate as a team, with each having access to the same pool of information. This provides visibility of the goals of the entire business and how well they are performing at any given time.

Promotes best practice and high productivity: Enterprise data are available to the key business users with minimum human intervention. The business is provided with a complete and current view of the network and company assets with easy access to all functions and data. This enables high-quality decision making and high productivity.

Consistency: With integration between systems, users are presented with a consistent three-dimensional view of the current network conditions allowing informed real-time decision making alongside strategic planning and network optimization. A common data model serving an integrated application suite provides improved data maintenance and integrity of data. This has direct productivity benefits for the business support system staff.

Customer expectations: Customers demand a reliable and consistent high-quality service. An integrated solution supports and facilitates these expectations, which can be maintained even during severe storm events. Customers affected by supply failures are less upset when

Smart Grid Technologies

provided with substantive information relating to the fault—especially its likely duration. Integrated systems can help ensure that accurate information flows quickly between modules.

Quality of service: Integration of tools such as volt/VAr control can assist utilities with improving the quality of supply delivered to customers as well as improving network performance and reduced losses.

Regulatory performance: A fully integrated system is key to monitoring exposure to guaranteed standard payments by monitoring the progress of all network incidents and their potential impact on guaranteed standard payments to customers. Key to this is systems providing an accurate and auditable means of measurement, which can only be achieved through provision of data from a range of systems. In addition, it is becoming increasingly important for utilities to track performance against both a specific property and individual. Systems need to be able to collect and report against such data.

Changes to business structure: Systems should be designed to manage change in business structure and accommodate mergers, acquisitions, organic growth, and separations. By implementing flexible, productized, and open systems, utilities can position themselves for the organizational turmoil of the utilities marketplace. For example, by implementing a DMS that can accommodate numerous legacy SCADA protocols, a newly combined business can continue to operate legacy SCADA systems without any impact on the control/operational user interface.

Media attention: With systems becoming strained to breaking point, high-profile failures are drawing increasing media and hence public attention that can impact not only utilities but also the GDP of large industrialized nations. Through better decision support tools making use of increasingly voluminous and better quality data, utilities are better placed to avoid unwanted media attention through controlled management of both planned and unplanned outages.

3.12.7.7.2 Obstacles

Complexity of installation: Integrating the entire suite of business systems is a complex problem that requires open yet secure applications that are productized yet flexible to global customer requirements. The foundation of an intelligent grid must be an open-systems-based comprehensive reference architecture that can integrate intelligent equipment and data communications networks into an industry-wide distributed computing system.

Adoption of standards: Although there has been a recent move toward standardization of technologies, protocols, and architectures, adoption of such standards has been variable. Even where standards may exist, interpretation can still differ. For this reason, it makes sense to purchase a solution from a single vendor, because this ensures a consistent interpretation of standards across the product suite.

Performance and availability: The performance and availability requirements of real-time control systems have historically differed from back office applications. With integration between the back-office applications and real-time control systems, consideration should be given to availability of data from back office applications. In the longer term with advances in hardware performance and availability, such issues are likely to become less important.

Business process: Business process has often been at odds with technology, and where technology is an enabler, business process can often lag behind. A common example is the single network model concept that requires the master data repository to be maintained in time to ensure that the real-time network model always accurately reflects the actual construction in the field.

Volume and rate of data acquisition: With dramatic changes to the communications networks forecast and vastly increased volumes of data becoming available, the challenge will be to optimize usage of the data in assisting the system to make rapid and intelligent decisions.

Change management: Possibly the most important yet least addressed issue in integrating systems is managing organizational change. With safety at the core of every control room decision, utilities have historically been conservative in their approach to adopting new technologies and processes. The concept of change management describes a structured approach to transitions in individuals, teams, and organizations that moves the target from a current state to a desired state. Stated simply, change management is a process for managing the people side of change.

With the advent of smart grid, two major issues arise:

1. The need to integrate new types of assets/agents to the electric network and make them "operational ready," taking into consideration all the complexities of operating interconnected electric systems. These assets can be electric vehicles, demand response programs, HAN, distributed generation (including solar panels), and large-scale renewable generation (particularly volatile generation such as wind).
2. The need to manage very large quantities of "new" data in near realtime that will be available to the operations and business groups within the utility. Data will come from new devices and sensors spread throughout the T&D networks, metering devices (AMI), recharging stations (for electric vehicles), and HAN, among others.

Both the business and operations groups will face significant challenges in terms of infrastructure, communications, business processes, and coordination when trying to deal with the two issues outlined earlier. Now, for the first time, there is a real need to integrate real-time operations with business decision-making processes and applications. An effective integration of these two groups not only will solve the problems discussed earlier but, more importantly, will transform the utility industry like never before.

The smart grid concept promises to increase operational efficiency, reduce costs, and be more environmentally friendly by enabling new assets/agents to be part of the electric network. Real integration of IT and OT not only helps fulfill that promise but enhances the opportunities to add more value and effectiveness to the energy value chain.

The challenges of having consistent data flowing among systems are many and begin with the way systems are procured. Few companies define an effective enterprise-wide architectural framework before plunging into product selections. Ironically, the least of the costs is usually the actual purchase price of the product. A poorly designed or ad hoc architecture is a hidden liability that substantially increases overall costs with each application system that is procured. When a project procures applications systems, each vendor is driven by the procurement process to meet user requirements at lowest cost. Each of the procured systems will have its own unique mixture of platform technologies, databases, communications systems, data formats, and application interfaces. As a result, the same information is stored and maintained on many separate systems throughout the utility's enterprise. This results in integration anarchy, which is a chaos of duplicated logic, duplicated data, duplicated effort, newly acquired integration difficulties, lack of ability to easily create new application functionality from services, and lack of ability to support business processes with applications. This integration anarchy will therefore result in higher costs and an inflexible system of systems. True integration is achieved by systematically managing information as an enterprise asset.

The smart grid has brought about a whole new level of lateral thinking in terms of the huge amounts of data available data and the need to share the data among enterprise applications. It is clearly apparent that synergies in sharing and exchanging operating data can significantly enhance the integration of applications and drive toward a common platform of technologies. To date, sharing of data among applications has been mostly implemented as "point-to-point" solutions; software applications are still independent systems, and complexity of integration increases drastically as more systems are integrated. As smart grid moves forward, there will be a new generation of highly integrated applications supported by enterprise platforms and integration frameworks. There will always be some need to integrate new systems with legacy systems using some form of standardized integration adapters and services in order to reduce integration efforts.

The lines between what are seen as traditional operational or real-time applications, such as SCADA, DMS, OMS, etc., will start to blur as the applications merge into a single, highly integrated enterprise solution. Similarly, nonoperational and what are currently considered "back-office" systems will be able to easily integrate with utility operational systems so that data can be freely exchanged across all utility operational and business applications. Much synergy in the sharing of data and an increase in the functionality and performance of applications can be achieved with the use of common standard data models and an integration framework. This enterprise level of integration will have a tremendous impact in the way a utility is operated and managed in the future—this is considered the ultimate nirvana of a smart grid realization.

3.13 HIGH-PERFORMANCE COMPUTING FOR ADVANCED SMART GRID APPLICATIONS

Yousu Chen and Zhenyu (Henry) Huang

The power grid has been evolving over the last 120 years, but it is likely to see more changes over the next decade than it has seen over the past century. In particular, the widespread deployment of renewable generation, smart load controls, energy storage, and plug-in hybrid vehicles will require fundamental changes in the operational concepts and principal components of the grid. Encouraged by aggressive public policy goals, such as the U.S. State of California's push to generate 33% of its energy from renewable sources by 2020, this evolution will continue at an accelerated speed. In the next 10–15 years, it is estimated that more than 15% of electricity will be supplied by intermittent renewable sources, and more than 15% of loads will actively respond to grid situations and incentive signals. In addition, distributed generation will supply power directly to the distribution grid, and electric vehicles (EVs) will be able to serve as both mobile and roving electricity consumers and electricity suppliers. The traditional one-way power flow, from generation to transmission to distribution and to consumer, will be fundamentally changed to a two-way power flow. This will result in stochastic operating behaviors and dynamics the grid has never seen nor been designed for. To plan and operate such a grid with sufficient reliability and efficiency is a challenge.

As a smart grid evolves and the number of smart sensors and meters on the grid increase by orders of magnitude, the information infrastructure will also need to drastically change to support the exchange of enormous amounts of data. With the significant increase in the number of data sources and the amount of data, smart grid applications will need the capability to collect, assimilate, analyze, and process the data. In particular, operation applications of the smart grid will need to process data at rates that satisfy the real-time requirements of functions that monitor, control, and protect the grid. Not only will smart grid sensors and meters generate an unprecedented volume of data (estimated at a million times more data), they will have distinct attributes that differentiate them from traditional power grid measurement devices. For example, sensors that use high sampling rates (30 samples per second or higher) provide an opportunity to more accurately monitor transmission grid dynamics. Other types of sensors will become more prevalent in monitoring the distribution system and the end users, providing transparency into the demand side of the grid. In the traditional scenario of one-way power flows, only limited digital information is required to operate and manage the traditional grid. In contrast, the future grid will have two-way information flow to monitor and control the two-way power flows. This requires not only fundamentally changing the way information is generated, transferred, and managed, but also (and more importantly) how the information will be utilized. The challenge is how to take advantage of this information revolution in terms of capturing the large amounts of data and processing it. Without addressing this challenge, the grand aspirations of smart grid evolution will remain unfulfilled.

New techniques and computational capabilities are required to meet the demands for higher reliability and better asset utilization, including advanced algorithms and computing hardware for large-scale modeling, simulation, and analysis. High-performance computing (HPC) is considered one of the fundamental technologies in meeting the computational challenges in smart grid planning and operation. HPC involves the application of advanced algorithms, parallel programming, and computational hardware to drastically improve the capability of handling data analysis, modeling, and computation complexity in software applications.

In addition to being driven toward HPC from the complexity due to internal smart grid development, power grid applications are also driven toward HPC by external forces in computer development. Computer processor hardware has been significantly improved over the last decade from about 300 MHz in 1997 to about 4 GHz today. However, the single processor speed (i.e., clock frequency) is reaching a plateau and no longer follows Moore's law [1], due to thermal limitations with the current CMOS processor technologies. To further increase the processor speed, computer vendors are offering multicore processors while the performance of each core remains relatively flat. With this trend, essentially all computers are going to be parallel computers ranging from a few cores in desktops to hundreds of thousands of cores in a large-scale high-end computer. Power grid software tools, designed traditionally as sequential codes for single-central processing unit (CPU) personal computers, are essentially running on only one core in the multicore computers. The easy gains of the past in which sequential applications simply got faster due to increased performance and clock frequencies on newer processors are long gone. Any performance gains for applications must be realized through the use of parallelism across multicore processors. It is imperative to redevelop power grid software tools with explicit parallelization for such a parallel computing platform. Only such parallelized software tools can take advantage of the multicore parallel computers.

3.13.1 Computational Challenges in a Smart Grid

Power grid planning and operation rely heavily on modeling and simulation. Due to the large size and high complexity of a power grid, experimentation with the system in realtime is very limited, and for most cases, infeasible. Simulation becomes almost the sole means to understand how a large-scale power grid would behave and how a control method or operation procedure would affect such a large system. Simulation requires accurate representative models, real-time data for situational awareness, and computational algorithms and hardware for timely execution. State estimation is a traditional example of such a simulation function that combines the use of system models and measured data to provide real-time knowledge of a power grid using computational methods. State estimation typically receives telemetered data from the supervisory control and data acquisition (SCADA) system every few seconds and extrapolates a full set of grid conditions based on the grid's current configuration and a theoretical power flow model. State estimation provides the current power grid status and drives other key functions such as contingency analysis, optimal power flow (OPF), economic dispatch, and automatic generation control (AGC), as shown in Figure 3.193 [2].

Mathematically, the functions shown in Figure 3.193 are built on complex algorithms and network theories in combination with optimization techniques (Figure 3.194) [3]. Given the sheer size of some power grids, all these mathematical problems require significant time to solve. The computational efficiency in grid operations is low, which can lead to the inability of grid operations to

Smart Grid Technologies

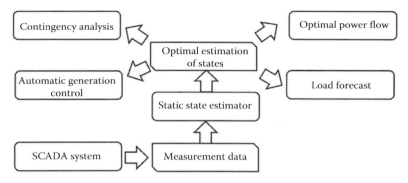

FIGURE 3.193 Functional structure of real-time power grid operations. (From Chen, Y. et al., An advanced framework for improving situational awareness in electric power grid operation, *Proceedings of the 18th World Congress of the International Federation of Automatic Control (IFAC)*, Milano, Italy, August 28–September 2, 2011.)

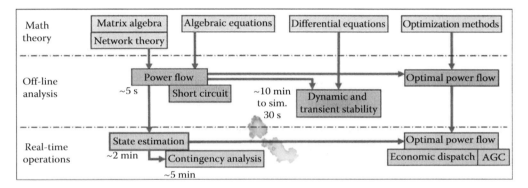

FIGURE 3.194 Grid computational paradigm. (From Huang, Z. and Nieplocha, J., Transforming power grid operations via high-performance computing, *Proceedings of PES-GM2008—The IEEE Power and Energy Society General Meeting 2008*, Pittsburgh, PA, July 20–24, 2008.)

respond quickly to adverse situations and eventual instability and voltage collapse of the grid within a matter of seconds [4].

Smart grid advanced applications will create an enormous computational challenge. On one hand, the information revolution offers opportunities for full transparency with tremendously larger real-time data sets made available through the information system. The challenge is that a computational platform needs to be in place to handle such data. On the other hand, the grid evolution introduces new dynamic and stochastic behaviors into the power grid. Simulating such a smart grid will require enhanced modeling techniques and computational tools to solve the models. There are several computational challenges that result from the complexity of data, modeling, and computation.

3.13.1.1 Data Complexity

The primary source of real-time operational data for today's grid is through the SCADA system. Typically the scan rate is once every few seconds. The future power grid is expected to contain millions of smart sensors and meters with various data transfer speeds ranging from once every few seconds to many samples per second. These new sensors and meters will result in significantly larger amounts of data compared to the traditional power grid data infrastructure. One significant category of sensors is the phasor measurement unit (PMU). PMUs generate high-speed time-synchronized measurements at a typical data rate of 30 samples per second. Currently there are approximately

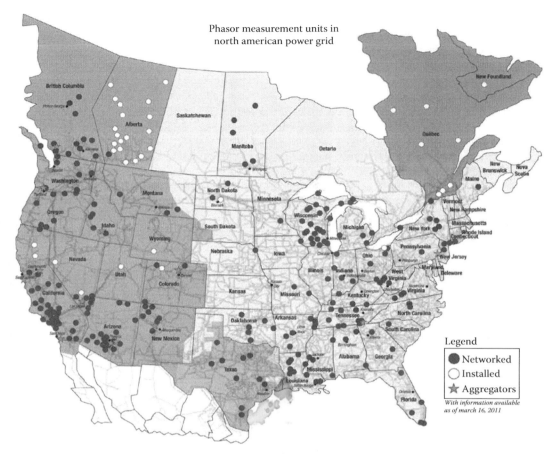

FIGURE 3.195 Installed PMUs in North American Power Systems. (From North American Synchrophasor Initiative (NASPI), Current PMU locations—Map of networked PMUs, https://www.naspi.org/site/Module/Resource/Resource.aspx)

250–300 PMUs installed in North America (Figure 3.195). Major phasor efforts are coordinated by the North American SynchroPhasor Initiative (NASPI) [5] supported by the U.S. Department of Energy (DOE) and the North American Electric Reliability Corporation (NERC). For the deployment of a full-scale phasor system in the future North American power grid, the phasor data volume is estimated to be at the order of terabytes per day. Consider the following:

$$N\_PMUs \times N\_Phasors \times N\_Bytes \times N\_Samples \times 24\text{ h} \times 60\text{ min} \times 60\text{ s} = 50{,}000 \times 8$$

$$\times 4 \times 30 \times 24 \times 60 \times 60 = 4.15 \times 10^{12} \text{ bytes/day} = 4.15 \text{ terabytes/day}$$

where
  N_PMUs is the total number of PMUs in the system, estimated at 50,000—the number of transmission-level buses in North American power grids
  N_Phasors is the number of phasors each PMU will generate (assumed to be 8)
  N_Bytes is the total number of bytes each phasor sample has (assumed to be 4)
  N_Samples is the data speed, assumed to be 30 samples per second

Smart Grid Technologies

TABLE 3.14
Estimated Amount of Smart Meter Data for 1 Year

Number of homes	100	1k+	10k+	100k+	500k+	1 Million
Compressed data size	2.5 GB	38.5 GB	366.3 GB	2.9 TB	13.6 TB	27.3 TB

GB = gigabyte, 10^9 bytes; TB = terabyte, 10^{12} bytes.

If considering other data such as timestamps and higher data rates, the data volume would be several times larger.

Another major category of data is the smart meter data. Smart meter data are available through technologies such as automatic meter reading (AMR) and automatic metering infrastructure (AMI). Beyond energy measurements for billing, smart meters increase visibility of the distribution network for planning and operation and help improve demand management. There are already millions of smart meters installed in the U.S. power grid. The American Recovery and Reinvestment Act of 2009 fostered significant investment in the United States through various Smart Grid Investment Grants and Smart Grid Demonstration Projects. Deployment of smart meters results in large volumes of data (Table 3.14) [7]. There are over 140 million households in the United States with electricity. If they all had smart meters, the resulting data volume could be at peta-scales over a time period of 1 year even at today's relatively low data acquisition speeds (e.g., one sample every minute) for smart meter data.

The challenges posed by the increase in data are twofold: high data volume and high data speed. Power grid planners and operators need to explore options for rapid retrieval and analysis of large volumes of data. Power grid operators rely on the data to make real-time decisions; hence the data need to be transferred and managed in very short time intervals over large geographical areas, and the analysis functions need to have high computational speeds to keep up with the large volume and the high data rates. Large-scale problems combined with requirements for real-time data collection, categorization, and processing pose a unique challenge for data-intensive computing.

3.13.1.2 Modeling Complexity

Modeling of grids today is typically performed on a first-principle-based* approach to describe the grid connectivity and associated parameters at the bulk transmission level (69 kV and above). The model size is typically on the order of 10^4 components for an interconnection-scale grid. The assumption when modeling only the bulk transmission system is that loads residing in the low-voltage-level grid (distribution system) are passive devices and the majority of their behaviors are predictable; the distribution system is at most approximated using simple models for special studies that involve identification of voltage stability behaviors.

This assumption is no longer true with the penetration of smart loads and distributed generation (especially small wind turbines and rooftop photovoltaic panels). The grid load is actively participating in and responding to grid dynamics and/or incentive signals, and the electricity flows bidirectionally between the transmission grid and the distribution grid, instead of one way as in the traditional grid paradigm. The need for modeling of some of the lower voltage levels in the grid becomes apparent. However, inclusion of modeling of the distribution system would increase the order of the model size significantly. Figure 3.196 shows the complexity of the power grid and the number of devices at different levels.

Given the sheer number of devices to be modeled and the complexity of each of them, it is apparent that the traditional first-principle-based modeling approach will no longer suffice. For example, it is not feasible to model air conditioning units based on the electrothermal conversion

---

* First-principle-based modeling relies on creating a block diagram model that implements known differential algebraic equations governing dynamics of the system.

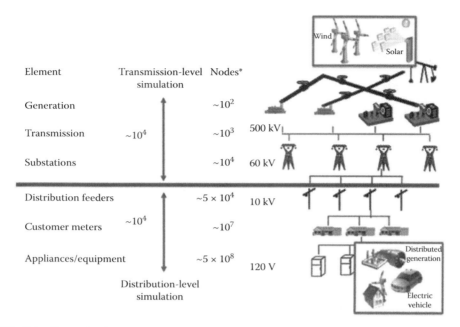

FIGURE 3.196 Example of power grid modeling complexity (*Western USA). (Courtesy of Pacific Northwest National Laboratory, Richland, WA.)

equations in more than 200 million homes and businesses in the United States. A fundamentally new modeling approach must be developed to achieve higher modeling resolution while retaining the feasibility of solving the models. There is a need to explore statistical approaches and behavior modeling to characterize the stochastic nature of smart grid devices and an increased level of human intervention. The model will be aggregated at a level that is feasible for mathematical and numerical solutions. Even with aggregation, the number of components in the model will inevitably increase, probably from 10^4 to 10^5. For example, the composite load model recently developed by the Modeling and Validation Work Group of the United States. Western Electricity Coordinating Council (WECC) extends the traditional one-component load model to a 10-component load model that includes a transformer, a feeder line, four motors of different types, an electronics load, a ZIP (constant impedance, current, and power) load, and two capacitors. Considering dynamic models of both traditional centralized power plants and distributed energy resources, the model size would increase by another order of magnitude to 10^6.

3.13.1.3 Computational Complexity

Computational complexity used in this context describes the order of computational operations required to solve a specific modeling equation. "Big-O" notation is used as the measure of computational complexity that reflects how much computational capacity is required to solve a specific problem. Note that it does not consider computer memory requirements.

Some key equations used in power grid modeling and analysis are given as follows, along with the computational complexity of the standard algorithms used to solve them. The equations are for power flow solution and contingency analysis, state estimation, dynamic simulation and dynamic contingency analysis, small signal stability, and dynamic state estimation. In the computational complexity expressions, the symbols are as follows:

N is the number of variables
M is the number of measurements

Smart Grid Technologies

m is the number of outer-loop steps
a is the degree of sparsity* in the underlying matrix

3.13.1.3.1 Power Flow Solution and Contingency Analysis

$$\begin{cases} P_i = |V_i| \sum_{j=1}^{N} |V_j|(G_{ij}\cos(\theta_j - \theta_i) - B_{ij}\sin(\theta_j - \theta_i)) \\ Q_i = -|V_i| \sum_{j=1}^{N} |V_j|(G_{ij}\sin(\theta_j - \theta_i) + B_{ij}\cos(\theta_j - \theta_i)) \end{cases}$$

$$O[m(aN^2 + N^2 \log N)] \tag{3.1}$$

where
  V_i and θ_i are the bus voltage and phase angle at bus i
  P_i and Q_i are real and reactive injections at bus i
  G_{ij} and B_{ij} are conductance and susceptance of line ij

3.13.1.3.2 State Estimation

$$z = h(x) + e$$

$$O[m(aN^2M + N^2 \log N)] \tag{3.2}$$

where
  z is the measurement vector
  x is the state vector
  h is the function that relates the measurement and the state
  e is the residual

3.13.1.3.3 Dynamic Simulation and Dynamic Contingency Analysis

$$\frac{dx}{dt} = f(x,y), \quad 0 = g(x,y)$$

$$O[N + m(aN^2 + N^2 \log N)] \tag{3.3}$$

where
  x is the state vector
  y is the algebraic vector
  t is time
  f and g are differential and algebraic equations, respectively

---

* In the field of numerical analysis, a *sparse matrix* is a matrix populated primarily with zeros.

3.13.1.3.4 Small Signal Stability (Calculation of Only k Eigenvalues)

$$\frac{d\Delta x}{dt} = A\Delta x + B\Delta y$$

$$O(aN^2 + kN^2) \tag{3.4}$$

3.13.1.3.5 Dynamic State Estimation

$$\frac{dx}{dt} = f(x,y), \quad 0 = g(x,y), \quad z = h(x,y) + e$$

$$O(N^2M + M^2N + M^2 \log M) \tag{3.5}$$

The aforementioned data rates, system size, algorithmic complexity, and required time to solution determine the computing power required. In particular, the types of real-time simulations and analyses discussed earlier drive the need for HPC power. Table 3.15 estimates the computing power required to solve the key equations to meet expected operational requirements with various model sizes and expected times to solution.

With the complexity of data, modeling, and computation, it is clear that advances in smart grid applications will present significant challenges in planning, analyzing, and operating the grid. The analysis tools employed in today's power grid planning and operation using serial computing* significantly limit the understanding of power grid behaviors and responsiveness to emergencies, such as the 2003 U.S.–Canada Blackout [3]. The information revolution provides an opportunity to enhance grid planning and operation functions and to enable new ones. But challenges exist in terms of large volumes of high-speed data, unprecedented modeling granularity, and large-scale computation. These challenges call for the exploration of HPC technologies to transform today's grid functions using improved computational efficiency and functionality.

TABLE 3.15
Computing Requirements for Modeling the Power Grid

Model size	10^4 (major transmission elements)	10^5 (+ major renewable and major loads)	10^6 (+ renewable, loads, distributed generation)
Time to solution	2–4 min	2–4 s	10 ms–1 s
State estimation	100 MFLOPS	10 GFLOPS	10 ExaFLOPS (dynamic)
Contingency analysis	100 MFLOPS	1 TFLOPS	100 TFLOPS
Dynamic simulation	1 MFLOPS (10× slower than realtime)	100 GFLOPS (10× faster than realtime)	10 TFLOPS (10× faster than realtime)
Small signal stability	10 GFLOPS	10 TFLOPS	1 ExaFLOPS

FLOPS = floating-point operations per second; MFLOPS = MegaFLOPS = 10^6 FLOPS; GFLOPS = GigaFLOPS = 10^9 FLOPS; TFLOPS = TeraFLOPS = 10^{12} FLOPS; PFLOPS = PetaFLOPS = 10^{15} FLOPS; ExaFLOPS = 10^{18} FLOPS.

---

* Serial computing uses only one CPU (processor) to perform all the simulation computation tasks sequentially.

3.13.2 Existing Functions Improved by HPC

State estimation and contingency analysis are two major functions employed in today's power grid planning and operation (Figure 3.194). As described earlier, state estimation applies a steady-state power grid model to a set of real-time SCADA measurements to estimate the states of the power grid, that is, bus voltages and phase angles. Based on the states, operation decisions would be made. The estimated states are used by contingency analysis for operational "what-if" studies to determine preventive measures for supporting grid stability. Contingency analysis is also used for longer-term planning studies. For large-dimensional equations and a large number of contingency cases, both state estimation and contingency analysis take minutes to solve. During grid emergency conditions, this low computational efficiency is not adequate for managing today's power grids. For future smart grids, such a low computational efficiency will further limit the capability of stable grid operation. Significant new grid dynamics will result from intermittency of renewable generation and randomness of load consumption, as well as new characteristics from large-scale storage operations and market trading. This requires more responsive grid planning and operation methods. HPC holds the promise to improve the computational efficiency and thus enhance the responsiveness of power grid functions. Two examples of applying HPC specifically to state estimation and contingency analysis are described as follows.

3.13.2.1 Parallelized State Estimation

The weighted-least-squares (WLS) algorithm, which involves solving a large and sparse system of linear equations at every iteration, is the most widely used solution method in state estimation. In each cycle of WLS state estimation, a large and sparse system of linear equations is being solved. The iterative conjugate gradient (CG) algorithm [8] is a candidate for solving state estimation because it exhibits a high degree of coarse-grained parallelism* and is often used on parallel computers for applications that require sparse linear solvers. The CG method was originally developed for solving symmetric positive definite systems of linear equations. Although they apply iterative solution methods, CG-type methods theoretically yield exact solutions within a finite number of steps. Discounting round-off errors, the CG method would guarantee convergence to the exact solution in at most N iterations, where N is the dimension of the linear system. With preconditioning, the system of equations would lead to solution convergence in far fewer than N iterations. Evaluation of the CG algorithm by modeling the 14,000-bus U.S. WECC system achieved a 5 s solution time for the full state estimation problem, a speed comparable with SCADA cycles [9–12]. The solution time can be further reduced if more parallel processors are used. Such scalability would be very important in future smart grids. Model sizes for a smart grid may be significantly larger due to the need to model the grid with more granularity in order to capture additional system dynamics and stochasticity.

3.13.2.2 Parallel Contingency Analysis

Contingency analysis assesses the ability of the power grid to sustain various combinations or scenarios of component failures based on state estimates. The outputs of contingency analysis, together with other energy management system (EMS) functions, provide the basis for preventive and corrective operation actions. Contingency analysis is also extensively used in power market operation for feasibility testing of market solutions.

Due to the high level of computations involved, contingency analysis is currently limited to mostly select "N-1" scenarios. Power grid operators manage the system in a way that ensures that *any single credible contingency (i.e., N-1) will not propagate into a cascading blackout*, which approximately summarizes the "N-1" contingency standard established by NERC [13].

---

* An application exhibits fine-grained parallelism if its subtasks must communicate many times per second; it exhibits coarse-grained parallelism if they do not communicate many times per second.

Although it has been a common industry practice, analysis based on limited "N-1" cases will not be adequate to assess the vulnerability of future power grids due to the high penetration of smart grid technologies. Intermittent renewable energy introduces to the power grid more-frequent, higher-amplitude energy imbalances; smart loads and plug-in hybrid vehicles cause new dynamics and random behaviors; new market development puts the grid close to its capacity and requires more frequent contingency analysis. NERC moves to mandate contingency analysis from "N-1" to "N-x" in its grid operation standards [15]. All this calls for a massive number of contingency cases to be analyzed. This would result in a huge number of contingency cases. As an example, the WECC system has about 20,000 elements. Full "N-1" contingency analysis constitutes 20,000 cases, "N-2" is roughly 10^8 cases, and the number increases exponentially with "N-x." It is obvious that the computational workload is beyond what a single personal computer can accomplish within a reasonable time frame. Parallel computers or multicore computers emerging in the HPC industry hold the promise of accelerating power grid contingency analysis. Unlike state estimation, contingency cases are relatively independent of one another, so contingency analysis is inherently a parallelizable process. The performance of parallel contingency analysis relies heavily on computational load balancing.

A straightforward load balancing of parallel contingency analysis is to preallocate an equal number of cases to each processor, that is, static load balancing. The master processor only needs to allocate the cases once at the beginning. Due to different execution times for different cases, each power flow run may require a different number of iterations and thus take more or less time to finish. The extreme case would be nonconverged cases which iterate until the maximum number of iterations is reached. The variations in execution time result in unevenness, and the overall computational effort is determined by the longest execution time of any of the individual processors. Computational power is not fully utilized as many processors are idle while waiting for the last one to finish. A better load balancing scheme is to allocate tasks to processors based on the availability of a processor, that is, dynamic load balancing. In other words, the contingency cases are *dynamically* allocated to the individual processors on demand so that the cases are more evenly distributed in terms of execution time by significantly reducing processor idle time. One implementation of the scheme is based on a shared variable (task counter) updated by an atomic fetch-and-add operation [14]. The master processor does not distribute all the cases at the beginning. Instead, it maintains a single task counter. Whenever a processor finishes its assigned case, it requests another task from the master processor, and the task counter is updated by one, as shown in Figure 3.197a. In contrast with the evenly distributed number of cases on each processor with the static scheme, the number of cases on each processor with the dynamic scheme may not be equal, but the computation time on each processor is optimally equalized.

This dynamic load balancing scheme can be further improved by extending it to multiple counters to avoid congestion for the master processor. Figure 3.197b shows a two-counter example. An equal number of cases is preallocated to the two-counter groups. Each group has its own counter. Within each group, the dynamic load balancing scheme is applied based on the availability of processors. When the preallocated tasks are finished in one group, the counter in this group *dynamically* "steals" tasks from the other group to continue the computation until all tasks are done. By implementing the multicounter dynamic load balancing scheme, counter congestion can be reduced, and further speedup is achieved. Such a multicounter scheme is particularly well suited for a computing environment with a large number of processors connected by relatively slow communications links [15].

Using a test system that has a size and structure equivalent to the U.S. Western Interconnection, the aforementioned load balancing schemes were tested and compared to demonstrate the performance of parallel computing. The test system has 2,748 generators, 14,000 buses, and 17,346 lines. Figure 3.199 presents the performance results with single-counter-based dynamic load balancing, in comparison with static load balancing. For this test, 512 "N-1" contingency cases, representing a typical scenario of today's practice, were selected. Obviously, much better linear scalability is

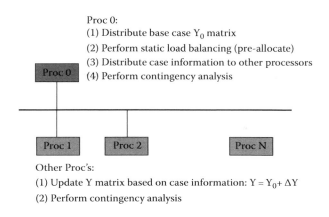

FIGURE 3.197 Framework of parallel contingency analysis: (a) single-counter dynamic load balancing and (b) multicounter dynamic load balancing with task stealing. (From Chen, Y. et al, Performance evaluation of counter-based dynamic load balancing schemes for massive contingency analysis with different computing environments, *Proceedings of the IEEE Power and Energy Society General Meeting 2010*, Minneapolis, MN, July 25–29, 2010.)

achieved with dynamic load balancing as shown in Figure 3.198a. Figure 3.198b further compares the processor execution times for the case with 32 processors. With dynamic load balancing, the execution time for all the processors is within a small variation of the average 23.4 s, while static load balancing has variations as large as 20 s or 86%. The dynamic load balancing scheme successfully improves evenness.

The single-counter dynamic load balancing scheme was further tested with massive numbers of contingency cases. Three large sets of contingency cases were computed using 512 processors: (1) 20,094 full "N-1" cases, which consist of 2,748 generator outage cases and 17,346 line outage cases; (2) 150,000 "N-2" cases, which randomly chose 50,000 cases from each of three combinations: double-generator outages, double-line outages, and generator-line outages; (3) 300,000 "N-2" cases, including 100,000 cases from each of the three combinations mentioned earlier. The results are summarized in Table 3.16, showing excellent speedup performance is achieved: about 500 times with the "N-2" scenarios and slightly less with the "N-1" scenario.

To demonstrate the performance of multicounter load balancing, a set of processors connected through slow-speed networking links equivalent to Ethernet were used to compute large sets of contingency cases. When the communication speed is relatively low, counter congestion is more likely to happen. As shown in Figure 3.199, when the number of contingency cases is large ($N = 17,346$), the two-counter scheme can improve the overall performance under the slow-speed networking environment. Very importantly, this is true when the number of processors is relatively large and the number of contingency cases is large enough. Such an Ethernet environment is of significant importance because it is more affordable, accessible, and available for many power companies.

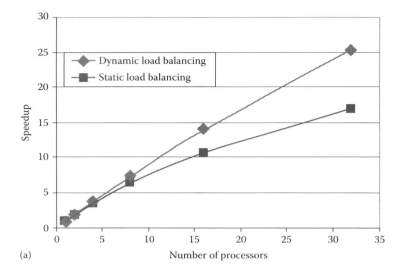

(a)

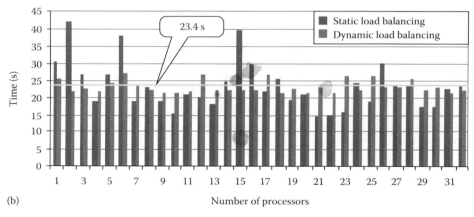

(b)

FIGURE 3.198 (See color insert.) Comparison of computational performance with static load balancing and single-counter-based dynamic load balancing: (a) speedup performance comparison and (b) computational evenness comparison. (From Huang, Z. et al., Massive contingency analysis with high performance computing, *Proceedings of PES-GM2009—The IEEE Power and Energy Society General Meeting 2009*, Calgary, Canada, July 26–30, 2009.)

TABLE 3.16
Summary Results of Massive Contingency Analysis

512 Processors Used	Total Time with Parallel Computing (s)	Total Time with Serial Computing (s)	Speedup
20,094 N-1 cases	31.0	14,322.2	462
150,000 N-2 cases	187.5	94,312.5	503
300,000 N-2 cases	447.9	227,085.8	507

Smart Grid Technologies

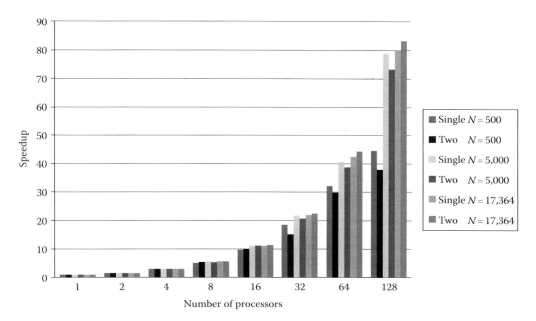

FIGURE 3.199 (See color insert.) Comparison of computational performance with dynamic processor load balancing using single counter and two counters in a slow computer networking environment. (From Chen, Y. et al., D., Performance evaluation of counter-based dynamic load balancing schemes for massive contingency analysis with different computing environments, *Proceedings of the IEEE Power and Energy Society General Meeting 2010*, Minneapolis, MN, July 25–29, 2010.)

The contingency analysis framework and dynamic processor load balancing schemes described earlier are generally applicable to more complex analyses. In the smart grid environment, renewable energy and other smart solutions will require faster and more extensive contingency analysis of larger systems. In addition, greater dependence on data and data networks requires contingency analysis to consider failures beyond the power grid. It needs to consider failures in the data network or manipulations of the data. Such data-induced contingencies will certainly increase the complexity and size of the analysis. HPC-aided contingency analysis is therefore essential.

3.13.3 New Functions Enabled by HPC

Over the years of developing computational methods for power grid analysis, certain assumptions and compromises were introduced to make sure the methods are practical within the constraints of limited computational resources. Typical in power grid planning and operation are modeling assumptions, quasi-steady-state assumptions, and compromises of efficiency for reliability.

As mentioned earlier, a modeling assumption is made in today's transmission models that loads are predictable with simple equations. This assumption reduces the size of the model and thus makes the computation feasible with standard single-processor computers. Another modeling assumption is the reduction of a substation model to a single-bus model, which results in the difference between an operational model (i.e., circuit-breaker-oriented model) and a planning model (i.e., bus-oriented model). This modeling assumption in turn results in a significant divergence of methods and tools between planning and operation. This assumption does significantly simplify the computation for planning studies; however, it is limited in that many scenarios cannot be easily studied in planning, such as breaker failure contingencies. A more profound impact is that planning and operation are typically in separate organizational silos, and the difficulties in communication between these two important functions cause significant grid reliability issues when emergencies arise.

A quasi-steady-state assumption is applied to operation studies. Today's operation is based primarily on a model that largely ignores dynamics in the power grid; electromechanical interaction of generators and dynamic characteristics of loads and control devices are not included in operational models. This assumption reduces the computation by several orders of magnitude and enables the operation studies on standard computers to be feasible within the required operational intervals. The problem with this assumption is that many studies cannot be performed in the operational environment. Development of the smart grid makes the grid much less quasi-steady-state, as compared with the power grid in the past. Furthermore, this assumption widens the divergence between planning and operation, as mentioned earlier.

One example of compromising grid efficiency for reliability is today's path rating studies. To support grid reliability, path rating studies are performed to determine transfer capabilities of transmission paths and apply a margin so the actual power flow will not exceed the predetermined limits. Path rating studies are typically performed in an off-line mode months ahead of time using the most conservative scenarios. This compromise makes the studies feasible with limited computational resources. The consequence is that the predetermined limits are conservative by definition; transmission capacity is underutilized, which results in reduced grid efficiency. This consequence will be more significant with smart grid development. Due to intermittency and stochasticity of new generation and loads, the most conservative case will be so conservative that it would render the path limits unreasonably low. On the other hand, with the increased complexity in the smart grid, there may be situations where the predetermined limits do not maintain adequate reliability in the grid.

These assumptions and compromises were reasonable and historically have contributed significantly to the development of power grid computational methods and tools. However, they no longer make sense in the much-evolved power grid. New energy resources and new load consumption devices make the power grid more dynamic and stochastic, which requires more granular analysis and more timely response.

With much-improved computational resources, there are no reasons to continue applying these assumptions in power grid planning and operation. Models for planning and for operation can be merged into one: operation can include dynamic models as in planning, and planning studies can be done in a more real-time manner to improve grid efficiency while still maintaining reliability. HPC will enable new functions that are much needed for smart grid development. Two examples of HPC-enabled new functions are presented in the following: dynamic state estimation and real-time path rating.

3.13.3.1 Dynamic State Estimation

Dynamic state estimation introduces dynamic models for real-time power grid operation. Traditional *static* state estimation only estimates static states, that is, bus voltages and phase angles. Power grids are inherently dynamic systems. A full dynamic grid view needs to estimate dynamic states, for example, generator speeds and rotor angles. This is possible with high-speed phasor measurements in the power grid [16]. Dynamic state estimation would then enable real-time dynamic simulation and dynamic contingency analysis to foresee the grid status with more transparency for better reliability and asset utilization. This is especially important given the increasing penetration of variable renewable energy sources and smart loads into the grid. However, bringing dynamic information into real-time grid operations is extremely challenging due to its computational demand. The dynamic model of a power grid consists of a set of differential and algebraic equations; solving such a set of equations in the time domain using numerical integration techniques is far more time consuming than solving static power flow equations.

One approach for dynamic state estimation is to formulate the problem as a Kalman filter process. Both extended Kalman filter (EKF) [18] and ensemble Kalman filter (EnKF) [17] techniques have been applied to dynamic state estimation. EKF and EnKF are capable of dealing with nonlinear equations [18] used in power system state estimation. Given a power system modeled as a set of

Smart Grid Technologies

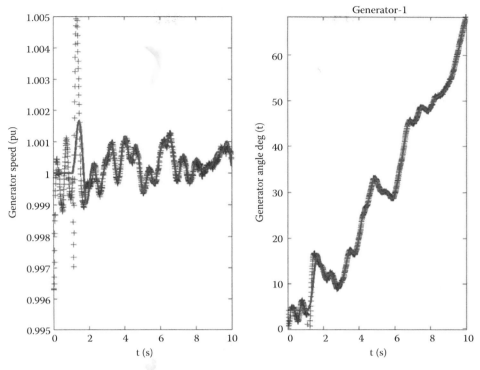

FIGURE 3.200 State tracking of generator 1 in the 16-generator system. The solid line is the true value from simulation. The scattered crosses are the estimated states from EnKF. (From Li, Y. et al., Application of ensemble Kalman Filter in power system state tracking and sensitivity, *Proceedings of the 2012 IEEE Power and Energy Society Transmission and Distribution Conference and Exposition*, Orlando, FL, May 7–10, 2012.)

difference equations, the dynamic state estimation is a two-step prediction/correction process [20]. The prediction step is a time update using the difference equation, which predicts the state variables of the next step based on the values of the previous step. The correction step compares the predicted values against actual measurements and uses the error to correct state variables.

Tests with a 16-generator system demonstrated good tracking performance of an EnKF-based dynamic state estimation. The test system consisted of 16 generators, 86 transmission lines, and 68 buses [19]. A 0.05 s disturbance of a three-phase-to-ground fault at a bus was simulated at 1.1 s with 3% noise added to all simulated voltages. The test results are shown in Figure 3.200. Initial errors in the state variables affect only the state variable tracking for the initial time period. After a short time of period, for example, about 1 s, there is no significant discrepancy for the state tracking. The disturbance does cause some deviation in the tracking. The Kalman filter can correct itself with continuous data and track the states again. Overall performance indicates that once the EnKF is locked into tracking, the performance is consistently good for the continuous tracking.

Achieving tracking accuracy is just the first step for the dynamic state estimation function. Performing the function fast enough for real-time purposes is another important aspect. The Kalman filter has high computation demands, especially for large algorithm solution systems such as the power grid. The EnKF computational complexity in "Big-O" notation is $O(M^2N)$. For large power grids, the computation is prohibitive for achieving real-time performance without HPC. For example, a 1081-bus system, representing regional power systems such as the U.S. California grid, would require 1.57×10^{11} FLOPS to complete one tracking step within 0.03 s (the phasor measurement cycle), and a 16,072-bus system equivalent to the WECC power system would require

5.60×10^{14} FLOPS. These FLOPS numbers translate to HPC computers with thousands to hundreds of thousands of processor cores. Implementing and parallelizing codes on such HPC computers is a fundamental challenge. An initial attempt (Figure 3.201) has shown a promising path forward to achieve the required computational performance. Figure 3.201a is the time used to compute one step of the EnKF dynamic state estimation. The parallel codes are based on the global array programming model [20,21]. Several computers are used to perform the test. Although the best time at 128 cores is still about 1000 times longer than 0.03 s, the execution time decreases consistently as more processor cores are applied. This is confirmed by the speedup curve in Figure 3.201b. The speedup curve also reveals potential saturation issues with parallelization. To prevent the overhead associated with parallelization that diminishes the benefits of using more cores, further improvements with the parallelization approaches are probably needed. With such improvements and larger-scale HPC, dynamic state estimation is very attainable, especially for regional power systems.

3.13.3.2 Real-Time Path Rating

Transmission grid path flow is limited by the lowest of three criteria: the thermal limit, the voltage stability limit, and the transient stability limit for all critical contingency scenarios. Figure 3.202 shows the U.S. California-Oregon Intertie (COI) ratings as an example [22]. The contingency rating is determined by stability studies using the worst-case scenario. The difference between the thermal rating and the contingency rating is significant, and it presents an opportunity for improvement if the analysis can be done in realtime with realistic power flow scenarios from online EMS snapshots. According to Bonneville Power Administration studies in the United States, a 1000 MW rating increase would generate $15 million revenue even if only 25% of the increased margin can be used for 25% of a year.

One fundamental challenge that limits conduction of path rating studies in realtime is the solution time that is needed to perform transient stability simulation. With today's commercial software packages, it takes hours or days to perform transient simulation for one path rating. The objective of path rating is to search for the maximum power transfer capability that will not violate stability criteria. One typical scheme is the binary search or its variations. As illustrated in Figure 3.203, at each loading level (1, 2, 3, 4,...), a set of contingency scenarios are simulated to test for stability limits. This is a serial process because the direction of the search depends on the results of the previous loading level.

The computational process is shown conceptually in Figure 3.204. A number of dynamic simulation cases need to be performed at each loading level. Since serial computing uses only one CPU to perform all the simulation sequentially, the total solution time increases when the number of contingencies or the number of search iterations increases. It usually takes a very long time. This serial computing is typical in current practice and renders the path rating studies to be off-line. Distributed computing has been applied to speed up the computation of multiple contingency cases by distributing the cases to multiple computers, but the solution time for each case remains the same, as shown in Figure 3.204b. The total solution time no longer depends on the number of contingencies, assuming enough processors are available. But due to the time it takes to solve each dynamic simulation case, the total solution time is still in the range of hours or more. It improves the performance, but it does not address the issue of computation time with the sequential search process. The key to further reducing the solution time is to parallelize the computation of individual simulation cases. Figure 3.204c shows the approach with parallelized computing. With enough processors and a scalable implementation of parallelized dynamic simulation, the total solution time can be kept short enough to enable real-time path rating analysis.

A version of parallel dynamic simulation was implemented on a 128-core HPC computer using OpenMP application program interface (API). This 128-core computer belongs to the shared-memory HPC family, which has a large block of memory accessible from multiple processing cores. Several test systems of sizes ranging from small to medium were used to test the parallel dynamic simulation implementation. Figure 3.205 shows results of speeding up the transient simulation in a research setting. The overhead required to coordinate the parallel computation among

Smart Grid Technologies

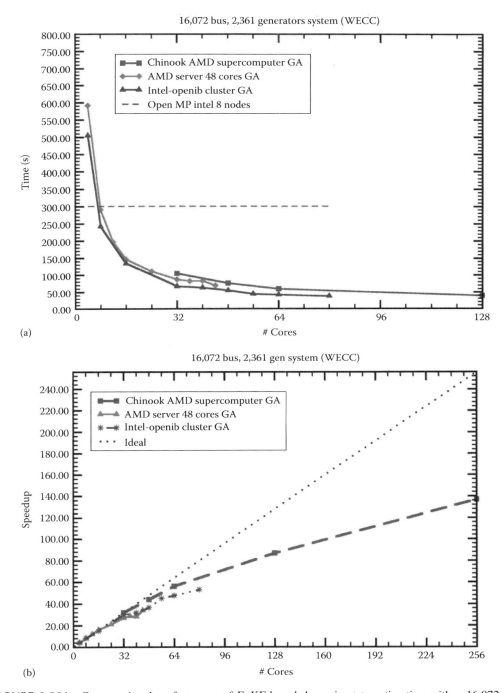

FIGURE 3.201 Computational performance of EnKF-based dynamic state estimation with a 16,072-bus system: (a) execution time and (b) speedup. (Courtesy of Pacific Northwest National Laboratory Richland, WA.)

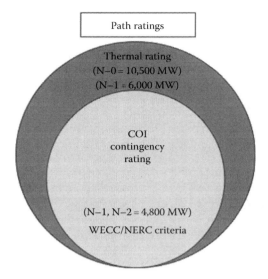

FIGURE 3.202 COI path rating—an example of underutilized assets. (From Western Interconnection 2006 Congestion Assessment Study, Prepared by the Western Congestion Analysis Task Force, May 08, 2006.)

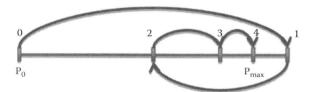

FIGURE 3.203 Binary search for maximum power transfer capability. (Courtesy of Pacific Northwest National Laboratory Richland, WA.)

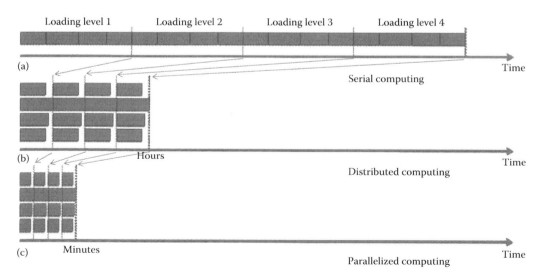

FIGURE 3.204 Conceptual computational processes of dynamic simulation for path rating with three different approaches (a) serial computing; (b) distributed computing; and (c) parallelized computing. Each block represents a dynamic simulation case for a contingency scenario. (Courtesy of Pacific Northwest National Laboratory Richland, WA.)

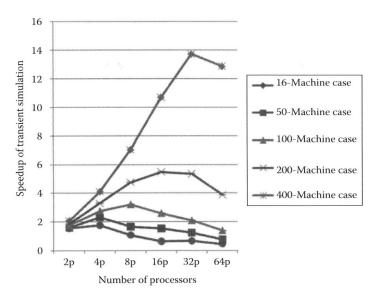

FIGURE 3.205 Speedup of transient simulation with an HPC computer. (Courtesy of Pacific Northwest National Laboratory Richland, WA.)

processors quickly dominates the simulation. Thus the speedup curve saturates at a relatively small number of processors. However, as the system size increases, better speedup performance is observed and saturation occurs at a larger number of processors. For example, the 400-machine test system achieved 14 times speedup with 32 processors. Speedup performance is expected to be even better with larger system sizes (e.g., the WECC system has 3000 generators). Figure 3.205 shows that HPC can run transient simulation much faster and enable real-time path rating for better grid asset utilization.

3.13.4 HPC in the Smart Grid

Driven by the deployment of smart grid technologies such as smart loads, renewable energy sources, and plug-in hybrid vehicles, the complexity of the power grid is increasing significantly. The complexity lies in three major aspects: data, modeling, and computation. The future smart grid will have significantly larger sets of data available from smart meters and sensors and require more granular models to describe the grid behaviors with new dynamics and stochasticity. Large amounts of data and complex grid models drive up the scale of computation in power grid functions.

Smart grid applications need to explore HPC technologies for two main reasons. First, the computational requirements cannot be satisfied by serial computing; the slow responses to adverse grid conditions lead to inability to prevent failures and grid instability. Second, power grid computation needs to adapt to the trend in the computer industry, which is undergoing a significant change from the traditional single-processor environment to an HPC era with multiprocessor computing platforms.

HPC has been shown to improve today's grid operation functions such as state estimation and contingency analysis. Study results show that these functions can be solved much faster when adapted to appropriate HPC computing platforms. Solution times can be reduced from minutes to seconds—comparable to SCADA measurement cycles. HPC also enables the development of new smart grid applications by integrating dynamic analysis into real-time grid operations and by moving off-line analyses to online applications. Dynamic state estimation and real-time path rating are two examples of new HPC-enabled applications. With HPC improving the computational performance of smart grid modeling and simulation, more studies and faster analyses will be possible, and that will generate many larger amounts of data in a shorter timeframe. To convert such big data into actionable information, especially for real-time grid operation, computer-driven visualization would be essential and would be another application that is enabled by and can take advantage of HPC technologies. Only when the information is used for decision making in controlling, operating, or managing smart grid devices, would HPC really exert its value in smart grid development.

HPC in smart grid is in its infancy. Significant further research and development are needed to adapt power grid computations to HPC platforms and eventually bring the HPC technology into power grid planning and operation.

3.14 CYBERSECURITY

Matt Thomson

Cybersecurity is a term that relates to technologies, processes, and measures taken to protect data, communications networks, information technologies, and computing systems against unauthorized access or attack. One of the most problematic elements of cybersecurity is the quickly and constantly evolving nature of security risks. The traditional approach has been to focus most resources on the most crucial system components and protect against the biggest known threats, which necessitated leaving some less important system components undefended and some less dangerous risks not protected against. Such an approach is insufficient in the current environment. To deal with the current environment, advisory organizations are promoting a more proactive and adaptive approach. The National Institute of Standards and Technology (NIST) in the United States, for example, recently issued updated guidelines in its risk assessment framework that recommended a shift toward continuous monitoring and real-time assessments [1].

The security of the smart grid is a heavily debated topic related to its growth. Everyone agrees that the smart grid should have a robust security model; but the challenge is how to address the changing requirements of security related to the smart grid and how to best apply the numerous alternatives that exist when securing a complex system such as smart grid.

3.14.1 Defining Security

Traditional security has relied on the three common pillars of confidentiality, integrity, and availability. The more modern Parkerian hexad relies on three more pillars: possession, authenticity, and usability. Within these six pillars can be found all of the concerns that exist with smart grid systems.

3.14.1.1 Confidentiality

Within confidentiality, there are the traditional concerns related to the transmission and storage of data related to smart grid operations. This kind of operational data is often deemed confidential, in the sense that if this data were known, it would have the potential to have a negative consequence on the security of the overall operation of the system. Alternately, this operational data may also be deemed confidential for competitive reasons. If the data were known to a competitor or other value-chain participant, it may allow that entity to extract an unfair advantage in a specific supply chain or within the overall market.

Additional new concerns related to confidentiality for smart grid are related to the privacy of consumer data that arise from smart grid technologies such as advanced metering and demand response programs. There are certainly consumer, as well as regulatory, expectations that the data regarding the use of electricity within a private residence remain confidential. Smart grid technologies such as advanced metering will enable greater data accuracy, and this enhanced accuracy will create new data that when combined with data mining capabilities has the potential to create significant privacy concerns.

The entry points that introduce the confidentiality risk are any location where data are stored (at rest) and any mechanism by which data are transmitted (in motion). For data at rest, whether on a smart grid device or within an operational data center, the data have the potential to be read, copied, and distributed to persons other than the intended recipients of the data. For data in motion, whether on a private, utility-owned network, on a service provider shared private network, or on a public network such as the Internet, the data have the potential to be intercepted, just like "tapping" a phone call, and then recorded, copied, and distributed.

The answers to solving the confidentiality problem are rooted in the security functions of encryption and access control. By providing the appropriate level of data encryption, data can be protected from observation by anyone who is not an intended recipient. Access controls are the technical measures that are put into place to protect information from those with access to the system from obtaining data that are not relevant to their job function, thereby further protecting the data from accidental or intentional disclosure. The functions of encryption and access control will be discussed in more detail in a later section.

3.14.1.2 Integrity

The integrity of the data that are fed into upstream systems is critical. If the data are able to be manipulated (intentional) or corrupted (unintentional) and upstream systems are fed inaccurate operational data, this could lead to the ability to negatively impact operations. In extreme circumstances, this could lead to the compromise of critical systems or lead to grid instability.

The entry points that introduce the integrity risk are any points where data are handed off from one system to another. The security of the exact mechanism of the handoff is important, but of more importance is how the system that is receiving the data ensures the validity of the message. If the data are able to be manipulated while it is in motion between two systems, the potential exists that the receiving system will take action based on this manipulated data. If the data are corrupted while it is in motion between two systems, the potential exists for an undesired response by the receiving system.

In either of these cases, the overall complex interdependency of the smart grid system could quickly escalate a message error as the data propagate throughout related systems. This makes ensuring the integrity of the data of critical importance to ensure stable operations. It could be argued that ensuring message integrity is the most critical operational security function that the smart grid must support. Most other operational security functions are done primarily either to ensure message integrity or depend on assured message integrity to achieve their goals.

The answers to solving the integrity problems are rooted in the security functions of auditing, authorization, nonrepudiation, and message-signing. These functions will be discussed in more detail in a later section.

3.14.1.3 Availability

Smart grid systems will be required to have similar reliability to the overall electric grid. These availability requirements may exceed the capability of some of the technology being used to create solutions. Reliability and availability are often used interchangeably but, in fact, mean two very different things. Reliability can be thought of as the dependability of the equipment. How often does it fail? How resilient is it? Availability, on the other hand, is the measure of the system being available to do the work it was designed to do, at the time it is needed. A portion of the system may be up, running, and processing, commands 100% of the time, and therefore very reliable, but if the performance is not adequate to the needs of the system, and critical command windows are missed or delayed, then the system cannot be said to be available.

The points that create availability risk are almost too numerous to list. Any system, any network, every communications-handling device, every message-handling process, and every service that is invoked, at every layer of the application and its underlying operating system when handling the delivery of a message or a command from one end of the system to the other is an availability risk. In the world of systems engineering, there is a popular saying, commonly attributed to Gordon Bell, a world-renowned computer engineer and researcher: "The most reliable components are the ones you leave out." In no system is this more true than in the highly complex system of systems that is the smart grid. Smart grid systems must be designed with full understanding of the overall system availability requirements.

Solving the availability risk is almost as complicated as identifying the components in the system that have the potential to impact availability. Most solutions rely on redundancy techniques, such as load balancing, clustering, or fail-over technologies. The cost and complexity of redundancy solutions grow as a factor of the number of failure points that are identified in the system. The more failure points there are in the system, the lower you may find the overall system availability to be. To adequately address this, one must perform careful analysis of the component-level MTTR and MTTF to ensure that the system itself is capable of meeting the target availability objectives.

3.14.1.4 Control

The ability to control information that needs to be secured is essential to ensuring its long-term usability and integrity. This has implications to systems which rely on control for various legal, regulatory, and business purposes. If information that is used for financial calculations, such as meter data, experiences a loss of control, through compromise of the systems that provide or transmit the data, then the provenance of that information can no longer be assured, and the reliability of the information decreases. At that point, a risk analysis must be completed to ascertain whether or not the information can still be acted upon or trusted for use in other systems.

3.14.1.5 Authenticity

The term provenance is often used to describe certainty of origin. In this sense, we also use the term authenticity. The process of verifying authenticity is similar to the process used for verifying integrity, ensuring that the source of the data, and the data itself, is authentic.

3.14.1.6 Usability

This aspect of the hexad is concerned with ensuring the data are usable. An encrypted communication stream stored in an encrypted form is very secure but can make it very difficult for the data to be useful. Usability is the factor that ultimately provides business value and therefore must be preserved as one of the highest priorities.

All of these confidentiality, integrity, availability, control, authenticity, and usability are predicated on there being a reason to secure the system at all. What is the risk to the overall smart grid system if the system itself, or any one of its components, is compromised? This risk, and the potential cost of this risk, drive the economics of security design decisions.

How risk acceptance is determined is a business decision, not a technical decision. A business choosing to employ a particular risk analysis methodology will have its own key input variables that drive the decisions, which are usually tied to a financial risk that the business assumes based on their own competitive analysis of the market or regulatory conditions under which they operate.

There are reasons other than financial ones that may drive risk acceptance. For example, a business using a strictly financial-based risk analysis model may be willing to accept the financial implications of their risk acceptance. However, industry regulations or law may dictate that a business may not accept that risk, or the industry governing organization may have the authority to assign penalties that drive behavior that may not otherwise be financially justified. These factors may drive security behavior even in the absence of a strong economic incentive to add additional security capabilities.

It is important to recognize that no security decision should be made without considering risk. Where are the risks in the system? How much will a failure of the system cost? How much will a

Smart Grid Technologies

failure of the security of a system cost? Not all risks are created equal. A business must understand the risk and must be able to make informed and intelligent decisions about how to address it.

3.14.2 Communications Model

The most important principle in overall smart grid design, as well as in the security of the smart grid, is related to open standards. This is especially important in security systems, because proprietary or incompatible security implementations will degrade the overall system security profile as all equipment is reduced to the lowest common denominator. To quote NIST directly, an open standard is

> "...developed and maintained through a collaborative, consensus-driven process that is open to participation by all relevant and materially affected parties and not dominated or under the control of a single organization or group of organizations, and readily and reasonably available to all for Smart Grid applications."

Therefore, the first architectural principle for security is that all security solutions for smart grid systems must be open. An architectural view for security for the smart grid is a challenging proposition. Just as there are widely varying ideas of what smart grid is, there are widely varying opinions on the nature of the architectures used to secure it. A popular mechanism for discussing architectures is the OSI* (Open Systems Interconnection) seven-layer communications model (see Figure 3.206). This model is challenging in the context of smart grid, due to the many overlapping components where a clear delineation between the layers is not always possible. However, it is useful to consider this model when discussing some of the various security standards that will apply to smart grid architectures. Additionally, some of the most important protocols, such as IPsec and SSL/TLS, operate at more than one layer.

Within the physical layer exist the base communications protocols necessary for communications. At this layer are found IEEE 802.11 (WiFi), IEEE 802.15 (ZigBee), and IEEE 802.16 (WiMAX), all

OSI Model			
	Data Unit	Layer	Function
Host layers	Data	7. Application	Network process to application
		6. Presentation	Data representation, encryption and decryption, convert machine-dependent data to machine independent data
		5. Session	Interhost communication, managing sessions between applications
Media layers	Packet/datagram	3. Network	Path determination and logical addressing
	Frame	2. Data link	Physical addressing
	Bit	1. Physical	Media, signal, and binary transmission

FIGURE 3.206 The ISO (International Organization for Standardization; www.iso.org) OSI seven-layer communications model (http://www.novell.com/info/primer/prim05.html).

---

* The OSI model is a product of the OSI (www.standards.iso.org) effort at the International Organization for Standardization (www.iso.org). It is a prescription of characterizing and standardizing the functions of a communications system in terms of abstraction layers. Similar communications functions are grouped into logical layers. An instance of a layer provides services to its upper layer instances while receiving services from the layer below.

of which appear regularly in various smart grid architectures. Protective measures at this layer are largely based on physical protection, but protocol deficiencies do occur.

The data link layer contains some of the most well-established protocols, such as Ethernet. The security architectures at this layer involve VLAN configurations, switch port protective measures, and network segmentation using firewalls.

The Internet protocol (IP) is clearly the dominant standard at the network layer. At this layer, measures designed to prevent the manipulation of data paths, or routes, and prevent the alteration of data packets. It is clear that IP will have a significant role in the development of smart grid despite the risk potential.

The transport layer contains IP's partners TCP and UDP. Certain functions of the SSL/TLS protocols are present at this layer as well. Mechanisms that control session establishment, such as firewall packet inspection, are employed at this layer.

The next three layers, session, presentation, and application, are nearly impossible to separate cleanly in the context of a large-scale system such as the smart grid. These layers also contain SSL/TLS functions, as well as the more well-understood protocols such as HTTP and NTP. Security at these layers requires the most wide-ranging security considerations.

3.14.3 Security Functions

So far we have identified a number of potential threats and identified examples of security functions that are important to complex interdependent systems like the smart grid. What we will now examine is one potential security architecture, encompassing all of the security functions that have been identified so far and endeavor to create an architecture that addresses the security vulnerabilities that have been discussed.

3.14.3.1 Layered Security Model

The concept of a layered security model (Figure 3.207), such as the protection ring model, is not a new concept. Operating systems and microprocessors have been using this architecture for decades to control access to privileged resources. It is not difficult to imagine this same concept applied to smart grid systems.

The ring model describes the mechanism by which communications between layers of a system is secured. In general, an outer layer ring is not permitted unrestricted access to resources on an inner layer ring. In addition, requests for resources should not be able to "skip" rings. A ring 3 resource should not be able to access a ring 1 resource directly. Access to an inner ring is done only through predetermined interfaces, and these interfaces would be designed to restrict the access sufficiently to ensure the security of the overall system.

In software systems, the ring security model is typically managed by the operating system, with the center, or ring 0, typically being the kernel of the operating system.

A well-designed smart grid system should also offer similar protections. Smart grid systems and resources at the edges of the system, such as home area network (HAN) devices or smart meters, should not have the ability to directly access the operations layer (ring 0) of the smart grid system. The data coming from these systems should pass through interfaces that ensure the integrity of the data before it is passed on to each lower layer.

The ring model is realized in smart grid systems through the use of network segmentation, system separation, and interface control. The exact mechanism deployed depends on the performance needs of the layer and the actual processing capability available at each layer. Regardless of the specific mechanism used to separate the layers, there are several key architectural principles that should be considered when passing data between layers.

First and most importantly, a failure at one layer should not result in a failure in a lower layer nor of any related system in the same layer. In the AMI example noted earlier, a failure of the meter at layer 4 should in no way impact any lower layer in the system. By impact, it is meant

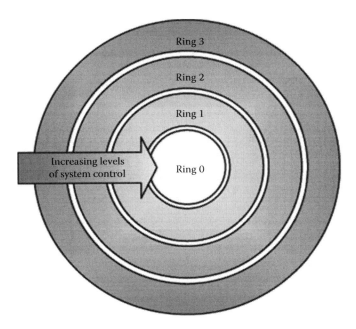

FIGURE 3.207 Layered security model.

that neither the confidentiality, integrity, nor availability of any lower layer should be affected the failure of the higher layer. It is worthwhile to consider that the converse cannot be true; it will very often be the case that a failure of a lower-level system will impact the availability of a system that resides at a higher layer. This is simply the nature of a hierarchical distributed system like the smart grid.

There are two key concepts in this architecture. The first is ensuring that the various levels of smart grid communications are sufficiently separated from one another. The second is ensuring that a failure at any one layer can be contained to that layer and not allowed to propagate to other layers.

Before beginning to discuss potential security architectures for smart grid systems, it will be useful to have a basic understanding of the different security functions that will be discussed. These functions will be the high-level security components that every system must address in one form or another. This section will not address all possible security functions, only those that will be used in later discussions.

3.14.3.2 Authentication

Authentication is the process of verifying the identity of the person or service requesting access to another resource and answers the question "who are you?" A robust authentication mechanism may require a number of subcomponents, each of which will serve a particular function in system authentication. We are very used to thinking of authentication in terms of user names and passwords, and this is correct. However, authentication is also between systems, or between applications, or between hardware components. Each of these may have a different way of achieving authentication, but it is a necessary process to ensure security.

3.14.3.2.1 Identity Database

An identity database is required for authentication. It can take many forms, such as a commercially available user directory, or database table, or other mechanisms. These data sources contain the information necessary to determine who, or what, is attempting to access a system. The identity database is likely to contain user names and passwords, but it will also hold information about devices and certificates. These certificates can be used in place of user names and passwords to facilitate machine-to-machine communication. Certificates will be discussed in more detail in a later section.

3.14.3.2.2 Identity Management System

An identity management system is designed to be a central repository of all resources that require authentication services. It might be compared to an asset management database, except that instead of physical equipment, the assets are identities. The identity management system tracks resources such as user accounts, systems, and permissions and is capable of tracking these resources throughout their life within the business environment.

For example, when personnel are first hired, they are entered into the identity management system. When their first e-mail account is created, it is also entered into the same system. When they are granted remote access, or access to any other specific systems, these requests are logged and possibly even processed by the identity management system. The same process is true for machine access, allowing machines and applications to be tracked as well as people. This creates a central place for access to resources to be tracked over the life cycle of personnel or systems within the organization. When an organization has invested in an identity management system, they will find that their ability to complete meaningful audits is greatly enhanced by the existence of this centralized repository.

3.14.3.3 Authorization

Authorization is the process of checking what the authenticated person or service is authorized to do within the context of the system, and answers the question "what are you allowed to do?" There are many approaches to authorization.

Simple systems often have two authorization modes: read-write and read-only. More complex systems might be capable of more specific authorization rules that control access on a very specific attribute basis, such as the ability to update one field, but not another, only at certain times of day, and only when logged in from certain machines. This is often referred to as granularity. There are benefits to a very granular authorization system, but it does come with the added complexity of managing the many potential roles and responsibilities across all the users of the system.

3.14.3.4 Auditing

The heart of any risk analysis and security program is its auditing program. Without periodic audits to validate the efficacy of the security controls that are in place, there can be no assurance that the system is secure. A good auditing program will be executed on a periodic basis and will test the controls that the business has determined to be essential to secure operation of the system. In some cases, regulations, such as Sarbanes-Oxley, or organizations such as NERC may dictate the frequency of the audits and the data that must be audited.

3.14.3.4.1 Logging

The generation and management of log files is not only essential for grid operations, it may even be required by regulation. In general, log files must record every access to a critical system, as well as record every change made to the system.

It is quite easy to imagine smart grid systems generating significant amounts of data in log files, and this represents yet another challenge with a system of the scale and complexity of the smart grid. When these log files are distributed across hundreds of different systems, there must exist a mechanism for centralizing the log file when needed. This can represent a challenge from a communications perspective, a data storage perspective, and an analysis perspective. Companies will need to have a process in place for log aggregation and analysis to ensure that audits can be done successfully, and so that analysis can be done after significant events.

3.14.3.4.2 Time Synchronization

As a minor subfunction to logging, but as a larger function to security as a whole, time synchronization is an essential function. Not only is time synchronization critical for event analysis, it plays an important role in other security functions, such as message validation and session management. In a unified smart grid system, it is important that all devices share an authoritative time source.

Time synchronization is especially important to other advanced smart grid devices, such as synchrophasors, which require a submicrosecond, or less than 1×10^{-6} s, level of precision in order for them to fulfill their function. While this is not strictly a security aspect of time synchronization, it serves as a representative example of why time synchronization is essential to smart grid systems.

3.14.3.5 Key Management

Key management is the process that governs how keys are issues to users, applications, and devices. These keys are used to establish identity and to ensure message integrity when sending commands between systems. When there are hundreds of devices, key management is a relatively simple task, but when there are hundreds of thousands or even millions of devices, key management becomes a significant challenge. Since smart grid end-point devices will require keys of some type, whether through X.509 certificates or other form of encryption key, these keys must be manageable. Every key in the system, whether application, device, or person, must be able to be updated or revoked on demand.

Unfortunately, a full discussion of public key infrastructure (PKI) systems and related infrastructure is beyond the scope of this chapter. The topic of key management is relevant to smart grid systems because it is almost certain that end-point devices will have keys that need a centralized management function.

3.14.3.6 Message Integrity

3.14.3.6.1 Signing

When a message is sent from one system to another, there is first an authentication process, to prove an identity, followed by an authorization process, to check what that identity is allowed to do. Once those two things have happened, messages may be exchanged between the two systems. To prevent communication tampering, another layer is recommended, and that is a process of message signing. There are two main reasons for signing a message. The first is to ensure that the message contents have not been modified while being sent between systems. The second reason is to verify the identity of the sender independently of the authentication process.

3.14.3.6.2 Nonrepudiation

When a message requires an acknowledgment, nonrepudiation is the term that is used to mean that the validity of the message has met all of the criteria necessary to be certain that the sender of a message received, processed, and acknowledged a message. In addition to guaranteeing that the message and its response have fulfilled all of the integrity checks, nonrepudiation also means that the message has been logged and that the log file is marked with the times of the transaction. This level of integrity checking ensures that the system can be extremely confident in the validity of the message and its ability to be audited.

3.14.3.6.3 Encryption

Whole books have been written on this topic, and it would be impossible to cover all of the technical intricacies of this incredibly complex topic here. Instead, we will summarize the purpose of encryption, which is to ensure that a message is not able to be read by any person or system for which the message was not intended. This can be accomplished by countless different encryption and hashing algorithms, all of which rely on preserving the integrity of the key that was used to encode the message. If the encryption key is compromised, then so is the message.

3.14.3.7 Network Integrity

There are a number of ways to ensure network integrity, with different devices meeting different needs. An appropriate device is chosen by understanding what the risk profile of the network. Every network has different requirements, and no one configuration is suitable for all networks.

3.14.3.7.1 Firewalls

A firewall is used to restrict traffic on a network to known set of rules. For example, it may limit traffic to a specific communications channel between one device and a known set of devices, blocking all other traffic. Firewalls can also log and record all traffic for audit purposes and can generate alarms as needed to alert an operator of violations of the firewall policies.

3.14.3.7.2 Intrusion Detection and Prevention

Intrusion detection and prevention devices monitor network traffic looking for specific data patterns that match signatures of known attacks. Certain viruses and certain forms of network attacks have predictable network traffic, and these devices can detect the signs of these attacks. Once one of these devices has detected an attack signature, it can take active steps to block the attack from reaching its intended target and generate an alarm for an operator.

3.14.3.8 System Integrity

3.14.3.8.1 Malware Protection

This refers to the general class of protection more commonly referred to as antivirus protection. Every system should maintain some form of malware protection strategy. For systems that are not capable of running traditional software, an alternate strategy, such as continuous auditing and file signature checking, should be employed. A robust malware protection strategy will not only block malicious files from being stored on the system but will also have the capability to alert an operator when malware is detected.

3.14.3.8.2 Configuration Management

The specifics of a how a system is configured can greatly impact its security. Configuration management is necessary to ensure that the system does not vary from a known baseline and that variations from the baseline are noted immediately.

3.14.3.8.3 Testing and Validation

The testing and validation, or quality assurance process, is essential to system integrity. Before any changes are made, and before new systems are deployed, these changes must be evaluated for impact to the total system. The changes need to be tested for security and to ensure that they do not impact the overall reliability of the system.

3.14.4 SECURITY THREATS

The choice and deployment of security functions are determined by the nature of the system and the risk presented by that system. In a system as large and complex as the smart grid, each of these will be necessary for some portions of the system. It is impossible to identify all of the possible security threats to smart grid systems. All we can hope to do is to use a considered risk management approach when identifying the weaknesses in the overall system design. In this section, we identify some of the most common security threats and discuss potential ways to control those threats. In a

Smart Grid Technologies

later section, there are several additional resources identified which go into much greater detail on the hundreds, if not thousands, of potential threats that exist in a system as potentially complex as the smart grid.

3.14.4.1 People

For all the attention paid to the technical parts of security, the human element remains one of the weakest links in any security strategy. In the course of normal operations, there must be a human who has the authority to perform an action, and that action may have the potential to have significant impact on the overall operations of the system. Even when the best possible controls are in place, there is likely to be at least one single, trusted, inside person who can be either the greatest threat to the system or the last, best line of defense for the system. In smart grid systems, these people could be control room operators, plant floor operators, field service personnel, or the information technology staff that keep the computer systems running.

3.14.4.1.1 Intentional Impact

The first and most obvious security concern when considering the human aspect is intentional abuse of the system. A single disgruntled employee with sufficient privileges in the system can cause havoc. This is why the concept of separation of duties and role-based access control is so important. No one person should have the ability to impact all aspects of the system. Each person who has access to the system should maintain unique access permissions and records of all actions that could impact the system should be kept.

For example, one common function of traditional distribution automation systems, and both current and future smart grid systems, is the ability to reconfigure distribution lines. A knowledgeable insider, with the desire to cause intentional harm to the system, could create a configuration that may be harmful to equipment or to grid stability. In an alternate scenario, a field engineer with access to substation automation equipment could upload malicious code on to a substation device or modify the firmware of a substation device.

Smart grid systems should be designed so that functions can be separated from each other, either logically separated, physically separated, or both. Within software systems, this can be accomplished with a robust access control mechanism. Within a network or other directly connected systems, this can be accomplished with the appropriate physical separation of networks. Most importantly, functions should be divided between personnel, so that no one person has full control over the entire system.

3.14.4.1.2 Accidental Impact

The second concern when considering the human aspect is that of accidental or unintentional damage to the system. These risks should be mitigated by sufficient separation of systems and by a mechanism for checks and balances between the various components of the system. Any form of human error might cause accidental impact. The goal is to ensure that the damage caused by human error is minimized to the greatest extent possible.

Consider one of the frequently cited sources of concern for smart grid advanced metering systems, the demand response function. This is the function that allows a load, such as an air conditioner, to be controlled with the intent of optimizing power consumption in a particular region. All of the security threats that will be discussed in this section will apply to this function. When examining the accidental impact of personnel, imagine the simple scenario where a human operator is empowered to issue a demand response command such as "turn off air conditioning for 30 min." The operator opens the application required for the procedure and accidentally enters 300 min instead of 30 min and submits the command. In a well-designed system, there would be a system of checks and balances to ensure that the data that were entered are within the expected normal range for requests of that type. Perhaps 300 min is an allowed value, but is it a typical value? In the cases where the entered value falls outside the normal range, there should be additional checks in place to prevent accidental impact. This is only one example; there are many more.

3.14.4.1.3 Prevention
The protection of the security threats related to people is as complicated as people are. The NERC CIP-004 Personnel and Training reliability standard offers excellent guidance in this regard. Personnel should be vetted through a process that meets the organization's risk avoidance policies and should be trained sufficiently in the appropriate procedures for their job function and also in general security awareness.

3.14.4.2 Process

Security threats that arise from process can be likened to those that arise from a lack of long-term situational awareness. Process is most closely related to security when it comes to activities such as requesting user accounts and requesting access to systems. An out-of-control process around either of these leads to a lack of knowledge about what users can access a system and what permissions are granted to them while they are using the system.

The next and perhaps the most important way in which process, or a lack thereof, can represent a security threat to a system such as smart grid is in auditing. Auditing is the foundation of long-term situational awareness. A comprehensive auditing program that reviews the risk profile of the system, combined with reviews of access records, accounts, and actions taken within in the system, can provide confidence that the system is operating as expected and is maintaining the appropriate level of security. There are challenges associated with auditing for the smart grid, which will be discussed in a later section.

3.14.4.3 Technology

After the less tangible threats associated with people and process, we come to a discussion of the threats associated with technology. These threats are the ones that you will find most often used as examples, simply because they are so numerous. As noted in an earlier section, as increasing amounts of technology are used in an already complex system, the number of failure modes increases with every new technology.

3.14.4.3.1 Hardware
The threats that may be considered uniquely within the hardware domain are those related to environmental operating conditions and physical tampering. Environmental threats can be man-made but are most often associated with the natural operating environment for the device, such as outdoors for a household meter, or inside a substation.

While environmental threats to hardware are generally easy to design for, since the operating conditions can generally be predicted, the threat of hardware tampering is much harder to protect against. Even so, this is the threat must be most carefully in a large-scale deployment of any piece of technology. The piece of smart grid equipment that is most likely to be susceptible to threats that are discovered through hardware tampering and reverse engineering is smart meters. The magnitude of their deployment, often scaling into the millions of devices, provides ample opportunity for interested attackers or simply curious tinkers to disassemble these devices and to try and figure out how they work. In the process of discovering how these devices work, flaws in their design may be uncovered. Whether related to their communications, their firmware, or their processors, there are many opportunities for a mistake.

There have been numerous examples in other industries ranging from mobile communications, satellite television, and in-home entertainment devices of hardware weaknesses being exploited to circumvent protections that may have been in place on the device. Smart grid devices that reach the same level of near ubiquitous deployment should expect to receive similar levels of scrutiny by both curious and nefarious persons.

3.14.4.3.2 Software

If there is something that can go wrong with the smart grid, it is more likely to be a failure of software than failure of hardware. Software attacks, especially against web applications, are perhaps the easiest to execute. It is for this reason that developers of software for the smart grid need to understand not only the industry in which their software will be used but also the unique challenges of maintaining security in that environment. In fact, there are so many potential threats that can arise from software, it is challenging to identify the threats that are most critical to the smart grid.

3.14.4.3.3 Input Validation

When any software component accepts data for processing, there is often an assumption that the data that are being sent are the data that it expects to receive. This assumption leads to many failure modes with names like "cross-site scripting" and "SQL injection," but ultimately they all reduce to a larger issue referred to as "input validation."

Input validation is simply the process that every software component should go through to verify that the data that are about to be processed meet its expectations in all regards. Are the data of the proper format and of the proper length? Do the data contain any extra data which I am not equipped to process? Is it numeric if I am expecting numeric, and text if I am expecting text? Does it contain any characters which have special meaning to the application, to the operating system, or to the server software on which the application is running? These kinds of questions should be asked any time data are accepted for processing. Unfortunately, these kinds of checks are usually not done or are insufficient for the task. It is especially true in batch data loads, or system-to-system data exchanges, that proper input validation is not performed.

Input validation should be done by the application receiving and processing the data. There are other mechanisms available that can protect against some attacks that use insufficient input validation as their means of attack, such as appliances commonly called "XML firewalls" or "application firewalls" that can monitor the traffic between two systems, identify data payloads that are indicative of this kind of attack, as well as other kinds of attacks, and block the traffic if desired.

3.14.4.3.4 Session Management

Session management refers to the processes and technologies required to maintain an active session between two communicating applications. Issues with session management can allow an attacker to usurp a session and begin to act as the previously authenticated end point. The danger here is of course that the attacker will send malicious commands to the other system, which believes that it is communicating with a previously trusted device.

Preventing session management attacks requires careful planning as to how communicating devices will establish a session, how long the session will remain valid, how session keys and session tokens are established, and how different applications share authentication information with each other. There are a number of technologies that can assist in this space. In the web service space, technologies such as SAML can provide managed session tokens to ensure session management objectives are met.

3.14.4.3.5 Operational Maintenance

Do not forget one of the most important significant threat sources for software: unpatched systems. Whether it is operating system software, application software, or equipment firmware, all of them represent a software threat if they are not maintained according to vendor recommendations. Mitigating this threat requires a process for identifying relevant software updates, performing a risk analysis, and then a change control process for rolling out these updates across the various components of the system.

What makes the process of operational maintenance so challenging for smart grid systems in particular is the complexity and high level of interdependency between systems. As the number of software components increases, the number of potential permutations for test case validation grows exponentially. It can be very difficult to test the downstream impact of a software update when there are several layers that may be impacted. Traditionally, the solution has been to update only when absolutely required and to leave a running system alone as much as possible. While this may be a viable solution in a tightly controlled and segmented environment, this will present issues in the future. As the evolution of the smart grid continues, dependency on external data sources, external systems, and external software applications will increase. This increased dependency on external systems may force internal systems to update more frequently than previously desired in order to maintain compatibility. It is often the case that software updates are recommended reasons of reliability or security. Since these are critical to overall grid operations, the operational maintenance process for software is an essential aspect of reducing the overall threat to a smart grid system.

3.14.4.3.6 Additional Information

The Common Weakness Enumeration dictionary can provide numerous examples of common issues with software systems. Links to the CWETM, and other resources, are provided in a later section. There are so many potential software vulnerabilities that are unique to any given operating environment that the risk analysis process should attempt to identify the critical few that have the potential for greatest impact in the system.

3.14.4.3.7 Communications

Within communications, there are a variety of well-known threats that need to be evaluated for their relevance to any particular smart grid solution. In this section, we will highlight some of the most critical, but as with any complex system implementation, there are many, many others. The resource guide at the end of the chapter can provide additional guidance. This is not intended to be an exhaustive list, only representative of the types of vulnerabilities with the largest risk potential.

3.14.4.3.8 Denial of Service

The easiest way to attack a communications channel is to block the channel entirely. A communications channel can be blocked in a number of ways, depending on the specific characteristics of the channel. If we consider a typical wired communications channel, the simplest and most effective attack is to just cut the wire. Depending on the design of the total system and on the functions of systems at either end of the broken wire, the effects can be minor or serious. Most systems designed for reliability include some form of communications redundancy to protect against both the intentional and unintentional failure of the wire.

What if the mechanism is not a physical severing of the wire, but instead the wire is flooded with so much data that legitimate messages cannot effectively use the channel to communicate? Picture a stadium full of people, all shouting commands with a megaphone to a player on the field. There is so much noise, and so much interference, that the poor player on the field is unable to hear anyone. This type of attack can have an impact not only on the communications channel but also on the devices that are the target of the attacking traffic, as it can cause those devices to slow down while they attempt to process all of the illegitimate data. These attacks apply equally to both wired and wireless communications channels.

Denial of service (DoS) attacks can be protected against in a number of ways. The most important mechanism is protection of the communications channel itself. Whenever possible, physical access to the channel should be secured. When that is not possible, the network communications equipment should take note of unauthorized devices being attached to the wire and take action to block the device or notify appropriate personnel. Network equipment, such as routers or firewalls, in the communications channel can also be used to detect and block attempts at DoS attacks.

When the channel is wireless, communications should be protected using encryption to ensure that unauthorized devices are not able to communicate on the channel. There is a challenge in securing the frequencies used from radio jamming. The only defense is to have a mechanism for changing frequencies, as with IEEE 802.11 devices. However, if the attacker jams all available channels, the only recourse is to locate and remove the source of the radio interference. When considering the use of wireless technology, it is important to consider the potential for radio interference as a mechanism for blocking communications.

3.14.4.3.9 Message Interception

Message interception is similar to listening in on a conversation. An attacker who is able to intercept a message may be able to discern information about the nature of the conversation. To exploit this vulnerability, an attacker positions himself on the wire or in the radio communications path for wireless communications, between two devices, and records the data being sent between them. If the data are not properly secured, through encryption or other means, he is able to see the data, much like recording a telephone conversation. After listening to the conversation for a while, the attacker may be able to "hear" enough words to learn the contents of the message. With this information, the attacker may be able to reverse-engineer the data stream and collect the data.

In a smart grid system, this data could be anything that relates to grid operations, from the consumer's HAN all the way through. This vulnerability may lead to the exposure of information that may provide additional information that will enable additional attacks on other parts of the system or allow for message tampering.

When considering message interception, it is important to consider the nature of the data that is being transmitted. If the data contain personally identifiable information about a consumer, then special care should be taken to ensure that appropriate safeguards are in place to adequately protect the privacy of an individual.

3.14.4.3.10 Message Tampering

Message tampering might be compared to using a voice changer on a telephone. If it were a good enough voice changer, you might fool the person on the other end of the phone into believing that you were someone else, and they may place a higher value on the message since it came from someone they trusted. To exploit a vulnerability related to message tampering, an attacker would modify a message while it was in transit between two systems. Similar to message interception from a previous section, the attacker would position himself on the wire between two communicating devices. Once there, the attacker has a number of options on how to tamper with the messages between those devices.

The easiest and most direct way to tamper with a message is to corrupt the message. The attacker may stop the message in its entirety, by blocking the communications between the devices, a condition referred to as a DoS attack. Alternately, the attacker may intentionally corrupt the message by deleting or inserting fragments of data. Since the message either never arrives or arrives in a format that is no longer able to be read by the receiving device, the message has effectively been corrupted. The effect of this attack can range from a minor impact, if the corrupted message is a notification requiring no action or a message that can be redelivered later, all the way to critical impact if the message contained an alert which required immediate, time-sensitive action.

Another form of message tampering is one that is referred to as a replay attack. As with previous attacks, the attacker is positioned in the communications path between two devices and records a conversation between two communicating devices. At some point in the future, the attacker replays the previous conversation over the communications channel with the intent of causing one or both of the communicating systems to act as if it is hearing a brand new conversation. As with other attacks, the severity depends on the nature of the conversation. In a worst-case scenario, the attacker may have recorded a conversation that requests that one of the communicating devices shuts down, resets, or changes operating state. If the appropriate protections are not in place, this could result in an undesired state change which could negatively impact the reliability of the system.

3.14.5 Cybersecurity in the Smart Grid

The topic of threats to the smart grid has been the subject of much research and documentation. There is a significant body of work that should be leveraged when doing additional research on the topic.

The NIST smart grid interoperability team, the SG-IP, through the Cybersecurity Working Group, has produced the document referenced as "NISTIR 7628, Smart Grid Cyber Security Strategy and Requirements." This document represents hundreds of hours of work by cybersecurity experts from all areas of both of the private sector and the energy industry.

For advanced metering in particular, there are two references that are of excellent value in helping to assess the devices security capability. The first is the document title "Wireless Procurement Language in Support of Advanced Metering Infrastructure Security," published by the Idaho National Laboratory National SCADA Test Bed Program, which provides excellent recommendations for utilities looking to choose an advanced metering solution. The second is the AMI-SEC document from UCAIug that covers these ideas and many more.

The NIST Framework and Roadmap for Smart Grid Interoperability Standards defines seven domains: markets, operations, service provider, bulk generation, transmission, distribution, and customer (Figure 3.208).

Each of these areas is described elsewhere. In the model diagram, note the solid lines, which indicate secure communications interfaces between each of the domains. There are several key security services which must exist in order for these secure communications interfaces to exist.

3.14.5.1 Authentication and Authorization Services

Authentication services will provide the resources necessary to validate the identity of any user or other system requiring access to an application, service, or system. A centralized authentication service will contain identity data gathered from multiple sources, including human resource databases, external user databases, and other sources of data related to identifying users, software systems, and equipment.

Authorization services will merge the data contained in a centralized authentication source with data from authorization sources. Authorization sources can be external application databases

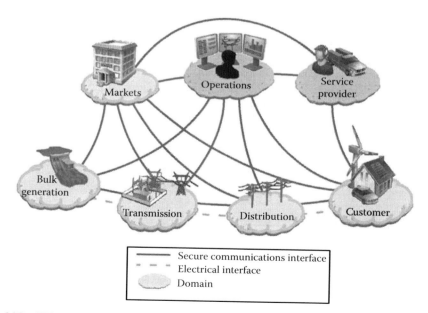

FIGURE 3.208 NIST framework for smart grid interoperability standards. (© Copyright NIST.)

Smart Grid Technologies 459

or other directory sources. Authorization may be maintained by direct definition, roles, areas of responsibility, groups, or by other mechanisms that meet the needs of the business.

Combined authentication and authorization services are often combined into an identity management system. With an identity management system at the core of a security architecture, it can act as the hub for all security-related requests, providing significant additional benefits for ensuring a robust security architecture.

3.14.5.2 Certificate Services

One of the significant challenges that faces a highly complex, highly distributed system with thousands of end points, like the smart grid, is how trust and identity are established between the components of the system. A detailed discussion of how certificates work is beyond the scope of this chapter, but in order to understand the requirements of the service, we will briefly discuss their purpose and use.

Certificates can be thought of as identification tokens that can be assigned to a user, a piece of hardware, or an application. When any of these want to trade data with another, they can use certificate services to validate each other's token, safely grant access, and to exchange encrypted information. How certificates are created, distributed, and verified is a detailed topic, so for the sake of this discussion, simply note the fact that certificates can be used as a means of verifying identity between two systems and encrypting communications between those systems.

Certificate services act in concert with an identity management system. The identity management system database or directory would typically store the certificates, which can then be retrieved as needed by the certificate services to validate the authenticity of a certificate that is presented by another system.

Certificates are not the only mechanism for achieving positive identification and encryption, but certificates are a very robust mechanism. There are deployment considerations when using certificates, such as the management of millions of unique certificates, the keys associated with maintaining them, and the processes for issuing and expiring certificates. These are all well understood, but thought must be given as to how to handle each of these deployment requirements.

3.14.5.3 Network Security Services

The separation of systems using networking technologies is an extremely strong tool in securing the network-dependent smart grid. The services available at the network layer are among the most well-proven security technologies in use today. These services include networking equipment such as routers and firewalls. The logical network separation of one set of systems from another can prevent a significant number of the vulnerability concerns. This logical separation can be with virtual LAN (VLAN) technologies or a physical break in the wire between critical systems and networks. In addition to VLAN technologies, smart grid solutions will make heavy use of protected network segments, commonly referred to as DMZ segments. These segments are used as a buffer zone between a secure network and a less secure network.

When these networking technologies are combined into a well-managed system, they provide an extremely strong defense against network-borne attacks. In addition, network-based security services can be tied to the other security services, such as authentication, authorization, and certificate services. When all of these services are tied together, the beginning of a true end-to-end solution for smart grid security is now in sight.

Like all other aspects of the smart grid, the development of improved security will be an evolutionary process. The technologies that are needed already largely exist; however, the challenge ahead lies in improving the standards and developing an interoperability profile for the smart grid industry. Without the creation of a security interoperability standard, smart grid solutions from many sources will be incompatible with each other, potentially requiring lengthy and costly system integration projects that will drive the cost upward and drive down the economic justifications for the investment in smart grid technologies.

As of the summer of 2011, the SGIP (Smart Grid Interoperability Panel) has identified five families of standards "necessary to insure smart grid functionality and interoperability." These five families are as follows:

- IEC 60870-6 series: Telecontrol protocols compatible with ISO standards and ITU-T recommendations
- IEC 61850: Communication networks and systems for power utility automation
- IEC 61968: Application integration at electric utilities-system interfaces for distribution management
- IEC 61970: Energy management system application program interface
- IEC 62351: Power system management and associated information exchange—data and communications security

Of these five, IEC 62351 is tied most closely to the needs of security for the smart grid but even more is needed to ensure true end-to-end security that can accommodate the full range of functions spanning from transmission operations to consumer home energy management functions. The work is ongoing and very promising.

Beyond the issues of security standards and interoperability are the next generation challenges facing security technology. Equipment used in the energy industry is typically long-lived, but the communications and security standards that will be used to implement the smart grid do not have a similarly long lifespan. Wireless technologies in particular are relatively short-lived compared to the expected life of utility-grade equipment, and security technologies and vulnerabilities change rapidly as researchers discover new techniques for penetrating computer systems. Each of these can be mitigated by a modular hardware approach that allows the wireless technology to be updated, and careful selection of other hardware technology, such as firmware, might allow new security weaknesses to be addressed through firmware upgrades. Any such upgrade would be constrained by the processing capability of the device and as such will limit processing to current state without a physical hardware change.

The future of smart grid all over the world will also be a matter of national security. A complex and interdependent system will, by its very nature, have a much larger risk profile than a small, self-contained system would have. The increasing reliance on automation, communications, and software technologies to drive efficiency and reliability will require that individual nations develop their own regulations on securing their electric grid from attack. This trend toward national security mandates runs the risk of fragmenting the security standards and interoperability process for smart grids across the globe. It is important that designers, implementers, and operators of the smart grid recognize that while many of the threats that have been discussed here are most often discussed in the context of traditional information technology solutions, they have just as much relevance to the complex technological solutions of the smart grid. We would be wise to leverage the years of security evolution that have already taken place in securing other complex information technology systems when looking for best practices for the smart grid.

GLOSSARY

CIP	Critical infrastructure protection
DoE	Department of Energy
DoS	Denial-of-service
FERC	Federal Energy Regulatory Commission
HAN	Home area network
IEC	International Electrotechnical Commission
MTTF	Mean time to failure
MTTR	Mean time to repair
NERC	North American Electric Reliability Corporation

NIST   National Institute of Standards and Technology
PKI   Public key infrastructure
SAML   Security Assertion Markup Language
SCADA   Supervisory control and data acquisition
SGIP   Smart Grid Interoperability Panel

3.15 SMART GRID STANDARDIZATION WORK

Erich Gunther

A fundamental factor in addressing the smart grid standards and technology problem is to find a way to decompose the power system and its underlying standards and technologies into coherent parts that allow targeted research to be applied, while exposing the barriers that have precluded solution development.

Four fundamental questions related to standards and technology for building a smart grid include the following:

- What new and emerging technologies are on the horizon that impact the smart grid of the future?
- How to avoid incompatible systems being fielded that result in costly replacements ahead of projections (i.e., stranded assets)?
- How to help foster open access, competition, and commercial growth of new and exciting technologies that offer energy consumers new ways to meet their energy needs while at the same time saving them money?
- Where government can help and where government should stay out (i.e., "what" vs. "how")?

3.15.1 Introduction to Standards and Technology

Standards development can be thought of as a continuum from single-company, proprietary specification or interface to a multiparty agreement, then a set of technology or market requirements, a national standard, and finally an international standard (see Figure 3.209).

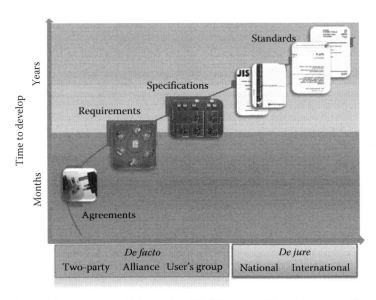

FIGURE 3.209 Standards continuum. (© Copyright 2012 Enernex. All rights reserved.)

Agreements, requirements, and specifications are defined by single or multiple persons, alliances, or organizations. Often proprietary in nature, they are quickly developed to design and measure elements, products, and systems, usually over a period of months.

A single-organization example drawn from the computing industry is the Microsoft set of application programming interfaces or APIs. These publicly available interfaces are used by developers in combination with the Microsoft Windows SDK (software development kit) to create applications intending to work under the Windows framework. Microsoft has mechanisms to incorporate feedback from developers and customers for their software, APIs and SDK, and also hosts conferences to increase awareness and information exchange among the entire ecosystem. Finally, there is a certification program (the "designed for" sticker/logo) to inform customers of the relative capabilities of hardware and software conforming to this set of practices.

Alliances are a group of entities and individuals that recognize the value of a particular technology and form a formal "interest group" to promote, for example, the codification of design and marketing of that technology. Multiparty in nature, the difference between an alliance and a formal standards group lies within both the rules and the work products. Since any number and balance of interested parties can form an alliance, the rules under which they operate vary widely. The ZigBee Alliance, for example, has a 15-member board (entitled "promoters") made up uniquely of technology vendors with other membership categories known as "participants" and "adopters." The work products of this alliance are known as "profiles" or agreed-upon specifications.

Since an alliance is not required to have a balanced membership, or in some cases to follow certain antitrust regulations, the work products must be submitted to a standards development organization (SDO) in order to become true *de jure* standards. Often, the alliance may actually promote specifications as "standards," before they are officially codified in a manner similar to a formal SDO. Also, one of the primary goals of most alliance efforts is demonstrable product interoperability within the framework of a defined certification program. Extension of the interoperability and certification efforts may even include joint work with other alliances to meet market needs. As an example, in August of 2008, the ZigBee and HomePlug alliances announced a joint effort on harmonization.

The following are well-known alliances related to the utility industry in the home area network (HAN) market space:

- HomePlug Powerline Alliance (www.homeplug.org)
- Wi-Fi Alliance (www.wi-fi-org/)
- Z-Wave Alliance (www.z-wavealliance.org)
- ZigBee Alliance (www.zigbee.org)
- U-SNAP Alliance (www.usnap.org/)

Another type of organization relevant to this discussion is a "user group." A differentiator between user groups, alliances, and formal standardization organizations is that user group rules often permit more free discussion between those actually using standards and specifications than those of the developing organizations. An example of the relationship is illustrated by the International Electrotechnical Commission (IEC) 61850 standards developing committee (Technical Committee 57, Working Group 10) and the UCA International Users Group (UCAIug) IEC 61850 committee. The IEC technical committee is made up of national experts, nominated and accepted by the IEC. Each committee follows a prescriptive process for producing IEC standards, in this case the IEC 61850 suite of standards. Part 10 of that suite addresses interoperability tests. The UCAIug IEC 61850 committee is composed of expert users that meet on a semiannual basis to discuss how the compliance of products to IEC 61850 is demonstrated. This is accomplished by that committee validating that the standard tests are applied in a consistent, transparent, and fair manner, and thereby conforming products meet the goals of the standard.

Also within the UCAIug are task forces such as OpenHAN (HAN) and AMI-SEC (Advanced Metering Infrastructure Security), which have published system requirement specifications. These "user group" specifications allow the customer and vendor communities to communicate with each other through a common document and to understand the complexities of what is needed to meet business objectives. These *de facto** specifications and requirements are also written in a manner that facilitates work by formal standards groups to develop *de jure* standards. Figure 3.209 depicts the "standards continuum" from the more proprietary to the more open with a relative development timeline.

SDOs operate under similar rules worldwide. In general terms, the members of the committees doing the actual development work are limited by antitrust rules or laws from engaging in anticompetitive behavior such as market division, pricing discussions, and the like. Also, intellectual property is treated as a potential source for standards language and requires disclosure by the holder [1,2]. A clear distinction from an alliance or user group is that strict control is maintained of the candidate voter pool for balloting to ensure a measure of fairness and balance. As an example, the American National Standards Institute (ANSI) has three categories, producer, user, and general interest, and for balloting purposes, no single category can exceed 40% of eligible voters.

Formal standards (and many specifications) may actually begin as *de facto* "standards," that is, enough commonality among enough producers to call the product/approach/protocol "standard." Beyond this, SDOs actually author *de jure* standards—those that are codified in a manner similar to laws. Given the careful attention to balloting balance, open rules, and open participation, standards may be adopted in place of laws in certain jurisdictions. To move beyond standards that are regional in scope, there are copublication pacts in place between the SDOs such as IEC and IEEE, ANSI and IEEE, and others. This is the first step toward true harmonization, whereby a standard is in place for a broader market.

The following are examples of relevant SDOs for the utility industry in North America:

- ANSI—American National Standards Institute (www.ansi.org)
- DIN—Deutsches Institut für Normung, German Standards Institute (www.din.de)
- IEC—International Electrotechnical Commission (www.iec.ch)
- IEEE—Institute of Electrical and Electronics Engineers (www.ieee.org)
- IETF—Internet Engineering Task Force (www.ietf.org)
- ISO—International Organization for Standardization (www.iso.org)
- ITU—International Telecommunication Union (www.itu.int)

Table 3.17 summarizes the standards continuum with respect to the elements described earlier.

3.15.2 Standards Development Organizations

There are a number of bodies established to organize and promote the development of technology standards, from formal SDOs to less formal alliances and user groups.

SDOs operate under similar rules worldwide. In general terms, members of standards committees who perform the development work are restricted by antitrust rules or laws from engaging in anticompetitive behavior, such as market division, pricing discussions, and the like. Intellectual property is treated as a potential source for standards—language—and requires disclosure by the holder. For balloting, candidate voters are carefully balanced with respect to the interests they represent in order to provide a measure of fairness and balance. Standards usually begin as de facto standards (i.e., sufficient commonality exits among a representative number of

---

* Journalist's Guide to the Federal Courts, Something that exists in fact but not as a matter of law.

TABLE 3.17
Standards Continuum Summary

	Level	Defined by	Recognized	Example	Time Frame
De Facto	Proprietary	One vendor or user	Market dominance	File formats, API	Months
		Two vendors or users	Market acceptance		
	Consortia	Group of vendors and/	Members of the	ASHRAE, DNP,	Months/years
	Industry	or users representing	alliance, consortium	EIA, IETF, ZigBee	
	Alliance	an industry or			
		market segment			
De Jure	National	National standards	Within one country or	ABNT, ANSI, CEN,	Years
		body	group of countries	CSA, DIN, JSA	
	International	International standards	Worldwide	IEC, IEEE, ISO, ITU	
		body			

Source: © Copyright 2012 Enernex. All rights reserved.

producers to call the product/approach/protocol—standard). Beyond this, SDOs compose de jure standards (i.e., those that are codified in a manner similar to laws). Given the careful attention to balloting balance, open rules, and open participation, standards may be adopted in place of laws in certain jurisdictions.

The National Institute of Standards and Technology (NIST) in the United States started a three-phase smart grid interoperability plan to expedite development of key standards that incorporate and align the results of efforts by industry players and research institutes until now. In particular, Electric Power Research Institute (EPRI) has participated in the NIST's plan, especially for consensus building among industry participants, government agencies, and research institutes. At the end of 2009, NIST had completed a testing and certification plan and issued the initial set of priorities, standards, and action plans. The Federal Energy Regulatory Commission (FERC) is in the process of institutionalizing the proposed standards.

In September 2009, NIST released a report expanding their initial 16 preferred standards to a list of 77 standards, which are all available for review in the full report [3]. However, there were about 70 other broad sectors of the smart grid where NIST has yet to come up with specific recommended standards. Of those, NIST has focused on 14 priority areas where key regulators—namely, the FERC—have said they need them sooner rather than later. Of those 14 "action plans," only one—a standard for upgrading existing smart meters—has been completed, for which responsibility is with the National Electrical Manufacturers Association (NEMA). In short, while there are steady moves toward establishing standards, much remains to be done.

3.15.2.1 Selected SDO Groups Focused on Smart Grid Standards

3.15.2.1.1 IECEE CB

The CB (Certification Body) Scheme of the IECEE (IEC System for Conformity Testing and Certification of Electrotechnical Equipment and Components) is a strong example of international cooperation; it encourages efficient international trade by slowly abolishing technical barriers and identifying national differences. The IECEE CB has obligated participating countries to conform to IEC standards for various electronic components, equipments, and products while providing evidence (CB Test Certificates) that applicants have successfully passed the requirements of IEC standards. If national standards do not adhere entirely to IEC standards, variances are permitted after formally declaring and detailing these to the IECEE.

3.15.2.1.2 IEC SG3

The IEC formed the Standard Member Board as a strategic group around smart grids to establish the interoperability of smart grid devices and systems. It has developed a framework for standardization that will help many countries take the first step toward addressing challenges in achieving energy efficiency. In addition, the IEC opened a global smart grid portal to provide a comprehensive catalog of standards for smart grid projects.

3.15.3 ALLIANCES

Distinct from SDOs are alliances, entities, and individuals that recognize the value of a particular technology and form an interest group to promote, for example, the codification of the design and marketing of that technology. The difference between an alliance and a standards group lies with both the rules and the work products. Because any number of interested parties can form an alliance, the rules under which they operate vary widely. One example is the ZigBee Alliance, which has 15 members (or promoters) made up of technology vendors and two other classes of membership, participant and adopter. The work products of the ZigBee Alliance are known as profiles or agreed-upon specifications. Since an alliance is not required to have a balanced membership and to follow certain antitrust regulations, the work products must be submitted to an SDO to become true *de jure* standards.

Each alliance combines different participants across the industry to meet their stated goals. Often, the alliance promotes specifications as standards, before they are officially codified in a formal manner. In August of 2008, the ZigBee and HomePlug alliances announced a joint effort to provide some harmonization to their efforts.

One of the primary goals of most alliance efforts is product interoperability, followed by some form of certification program to demonstrate that capability. In electronic and power technology—unlike physical technology—interoperability is at best an aspiration of the community that developed the standard. This highlights the need for a dedicated users' community to identify interoperability challenges and requirements, write tests to validate products, and certify results.

3.15.3.1 Example of an Alliance on Smart Grid Standards

3.15.3.1.1 GridWise Architecture Group

Another group laying the foundation for true interoperability is the GridWise Architecture Council (GWAC). In a partnership with NIST, the GWAC sponsors the Grid-Interop conference, which has the goals of achieving system-to-system interoperability, business process interoperation, preparing for a sustainable electricity system, developing policies for integrated smart energy and a holistic view of generation to consumption.

3.15.4 USER GROUPS

In addition to SDOs and alliances, a third important entity is a user group. User groups often permit more free discussion between those using the standards and specifications. An example of the relationship is shown by the IEC 61850 standards developing committee (Technical Committee 57, Working Group 10) and the UCAIug IEC 61850 committee. The IEC technical committee is made up of national experts, nominated and accepted by the IEC. Each committee follows a prescriptive process for producing IEC standards, in this case the IEC 61850 suite of standards. Part 10 of that suite is interoperability tests. The UCAIug IEC 61850 committee is composed of experts that meet on a semiannual basis to discuss how the compliance of products to IEC 61850 is demonstrated. The committee validates that the standard tests are applied in a consistent, transparent, and fair manner,

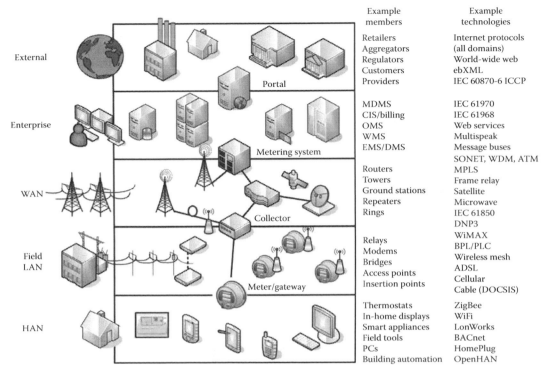

FIGURE 3.210 Smart grid domains of interoperation. (© Copyright 2012 Enernex. All rights reserved.)

and thereby conforms products meet the goals of the standard. Also within the UCAIug are task forces such as OpenHAN (HAN) and AMI-SEC, which have published system requirements specifications. These specifications allow the consumer and vendor communities to communicate with each other through a common document and to understand the complexities of what is needed to meet business objectives. The specifications are also written in a manner that facilitates work by standards groups to develop *de jure* standards that lead to products that meet the desired requirements.

3.15.5 Smart Grid Standards Assessment

Although the vast majority of engineers working on the smart grid understand the importance of standards in the deployment of the smart grid, a big concern is how long it takes to introduce standards and have them adopted by national or international standards bodies. Standards usually have a cycle of adoption from 2 to 4 years. The industry cannot develop smart grid standards using current methods and hope to achieve the goals of the stimulus plan. What then are the approaches to standards that the industry should take? First, luckily, there are already many standards already adopted by international standards bodies, which are fully applicable to the SG of today and tomorrow.

Figure 3.210 provides a high-level overview of the smart grid landscape. Ideally, each of the domains, members, and technologies would interact in a manner that empowered all actors (utilities, customers, etc.) to participate in meeting any of the business, technology, and societal goals. From the HAN, where consumers would be able to purchase and install monitors and controls, through the utility network formed of communications and power system infrastructure to the enterprise applications, data would be transformed into useful information wherever and whenever necessary.

Smart Grid Technologies

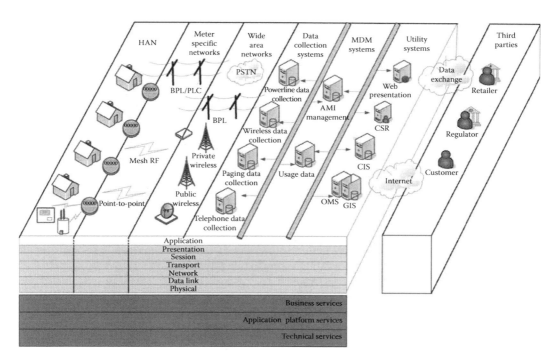

FIGURE 3.211 Interoperability hurdles using the AMI enterprise. (Adapted from G. Gray, AMI Enterprise, A Framework For Standard Interface Development. Available at http://osgug.ucaiug.org/sgsystems/OpenAMIEnt/Shared%20Documents/) (© Copyright 2012 Enernex. All rights reserved.)

In reality, each of the horizontal domains represents a zone of interoperation where there may be competing and complementary standards and technology, often focused on a specific technology or partner technologies and developed without consideration for the requirements of any other domains.

Figure 3.211 is a schematic representation of the hurdles to achieving interoperability (seamless information exchange) between devices and systems using the AMI enterprise as an example. One can easily trace from the HAN, through a local area network (LAN) and wide area network (WAN), and into the utility enterprise. Orthogonal to this view, there are seven layers displaying the open system interconnection, or OSI, layered communications model [4]: application, presentation, session, transport, network, data link, and physical. To be properly integrated, any two devices/systems must agree on standards and protocols for each of the seven layers.

Layer 1, the physical (PHY) layer, identifies the media (hardware) by which bits are exchanged. Layer 2, the data link, media access control (or MAC) layer, is where the exchange of frames (multiple bits) between hardware and software is performed. Layer 3, the network layer, is where paths are determined and logical addressing is applied for packets (multiple frames). Layer 4, the transport layer, is where end-to-end connections are formed and segments (multiple packets) are exchanged. Layer 5, the session layer, deals with interhost communication. Layer 6, the presentation layer, deals with data representation and encryption. Finally, layer 7, the application layer, is where data are used and information created and where users connect interface devices.

Communications protocols (the name given to a specific instantiation of the seven-layer OSI stack) may not explicitly define all seven layers of the OSI stack. These "short" stacks may only explicitly reference three or four layers; however, the functionality of all seven OSI layers is typically bundled somewhere in the protocol even though some layers are only implicitly defined. The OSI model is generally accepted as the most detailed practical representation of communications functionality.

3.15.6 SMART GRID GAP IDENTIFICATION AND DECOMPOSITION

Figure 3.212 shows the utility landscape decomposed into discrete domains, not intending to be all-inclusive. Those domains include the enterprise (integration), control center (LAN), WAN, substation (area network), field area network (FAN), residential customer (HAN), and distributed energy resources (DERs). Most of the existing utility business process systems have been designed, tendered, installed, and operated for a single application, in what are known colloquially as "silos." The most common application that may cross operational and communications domains is a supervisory control and data acquisition (SCADA) system, which is used by a control center to operate elements in the substation domain across a WAN. Despite this internal isolation, a significant amount of utility research and development effort has been devoted to advanced applications, usually with the presumption of some sort of high-speed communications network being available. Examples of these advanced applications are a broadly deployed power quality monitoring and control system or a phasor measurement monitoring system. The former often requires feature-rich monitoring devices (large quantities of data, to the oscillographic level) and extremely high-speed communications to leverage the system under operational conditions (subsecond to minutes). The latter may represent fewer data elements but still necessitates high-speed communications with low total elapsed time (latency) for its most useful applications of system protections (a few seconds) to system state calculation (a few minutes).

As much as the utility business is still composed of internal silos of operation, the product domain exhibits the same characteristics. It is rare for companies to have a broad portfolio of solutions for more than two of the domains. Even if one abstracts the problem further, one finds the standards

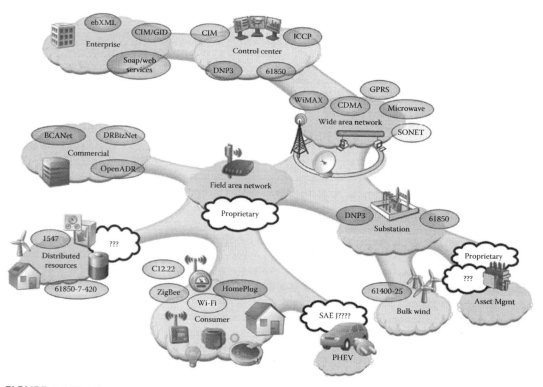

FIGURE 3.212 Communications domain decomposition and standards. (© Copyright 2012 Enernex. All rights reserved.)

and user communities also typically focus their efforts on one domain, one integrated solution, or one particular product or suite of products.

With the plethora of standards available in the utility space in each "network" domain as well as through the OSI layers, it is often beneficial to juxtapose existing standards against actual requirements, resulting in a "gap" analysis. A gap analysis is facilitated by preparing annotated requirements against applicable technologies. An approach that derives requirements from use cases* is one of the best methods to develop lists of functional requirements.

3.15.6.1 Generation

For generation power equipment, this is a well-defined space, with mature standards and regulations. However, technology in the generation area will always evolve, and new standards will always be developed.

3.15.6.2 Transmission

For transmission power equipment, this is also a well-defined space, with mature standards and regulations. For communications between utility operations applications and transmission equipment, the same is generally true. Technology in the transmission area will always evolve, and new standards will always be developed. Representative standards, specifications, and technologies are categorized by application domain in Table 3.18.

TABLE 3.18
Transmission Standards and Technology

Domain	Standard/Specification/Technology
Control centers	• IEC 61970 CIM
	• IEC 60870-6 Intercontrol Center Communications Protocol (ICCP)
	• National Rural Electric Cooperative Association (NRECA) MultiSpeak
Substations	• IEEE C37.1 SCADA and Automation Systems
	• IEEE C37.2 Device Function Numbers
	• IEC 61850 Protocols, Configuration, Information Models
	• IEEE 1646 Communications Performance
	• Distributed Network Protocol (DNP3)
	• Modbus
	• IEEE C37.111-1999 COMTRADE
	• IEEE 1159.3 PQDIF
Outside the substation	• IEEE C37.118 Phasor Measurement
	• IEC 61850-90 (in development)
	• IEEE 1588 Precision Time Protocol
	• Network Time Protocol
Security	• IEEE 1686 IED Security
	• IEC 62351 Utility Communications Security
	• NERC Critical Infrastructure Protection (CIP) Standards
Hardening/codes	• IEEE 1613 Substation Hardening for Gateways
	• IEC 61000-4 Electromagnetic Compatibility
	• IEC 60870-2 Telecontrol Operating Conditions
	• IEC 61850-3 General Requirements

Source: © Copyright 2012 Enernex. All rights reserved.

---

* A good discussion of use cases can be found in Wikipedia: http://en.wikipedia.org/wiki/Use_case

3.15.6.3 Distribution

For typical central generation supply and typical substation designs, this is a well-defined space, with mature standards and regulations. For legacy utility communications, the same is true. However, this is one of the areas where much of the innovation is expected to occur. Moving away from the central generation supply model, it is the distribution portion of the grid to which distributed resources (generation *and* storage) will be connected. Two standards bodies, IEEE SCC21 (standards coordinating committee) and IEC TC8 (technical committee), have groups focused on the technical standard aspects of attaching generation and storage to distribution systems on both the primary (higher voltage, up to 34.5 kV) side and secondary (lower voltage, down to 240 V) side. Examples of these standards are IEEE 1547, the IEC 61400 series for wind turbines, IEC 60364-7-712 for a building solar power supply, and the IEC 62257 series for small renewable energy and hybrid systems for rural installations. For treatment of applications of those technologies, much work has been performed defining information exchange standards with the goals of permitting model exchanges, load flow calculation exchanges, and driving toward operation and control data exchanges. Collectively, this work is known as the CIM standards. On the standard side, the IEC has several suites of publications in the 61970, 61968, and 61850 series. The first is known as the generic interface definition (GID) and the CIM; the second contains the CIM for business-to-business exchanges of information, while the third is for substation equipment monitoring, operation, and control. Other examples of protocols in this domain include the DNP3 and Modbus for what is known as distribution automation products, and ANSI C12.19 (table-based data model) and ANSI C12.22 (networked communications) for electricity metering products. A series of standards produced by the IEC in the 62056 series provides a competing set of metering protocols.

3.15.6.4 AMI Communications Technologies

Vendors are currently building "last mile" AMI communications solutions around five technologies: wireless star, wireless mesh, power line carrier (PLC), broadband over power line (BPL), and fiber optics. Wireless star technologies are available in both licensed (200, 900 MHz) and unlicensed spectra (900 MHz, 2.4 GHz). Advantages of licensed technology include greater allowable transmission power (2 W vs. 1 W) and blocking of interference sources. The principal disadvantage is the need to obtain a jurisdiction-by-jurisdiction license to operate. The desired frequency may also have been already allocated. Advantages of unlicensed technology are elimination of licensing requirements due to the use of the "free" spectra and more choices of a set of frequencies to use within the spectral bands. These two aspects often offset the potential interference and lower allowable transmission power. For wired technologies, the principal hurdle is propagation of the signal across power system equipment such as transformers. Transformers act as natural filters to the radio-frequency signal. Another difficulty is maximizing the bidirectional communications rate. For BPL technologies, the communications rate is solved by choice of the frequency band; however, power line communications equipment often interferes with other wireless communications technologies (amateur radio). None of the aforementioned limitations apply to fiber optic technologies. However, it is often difficult to cost justify "fiber to the home" for a single purpose use (such as advanced metering). Smaller utilities such as municipalities have sometimes successfully invested in this medium as they may then be able to offer cable television, phone service, and Internet service first (depending on competitive legal issues) with enough bandwidth still remaining for their utility operations.

The major disadvantage of wired technologies is that they are often incompatible with water and gas meters due to their use of the electrical distribution wires as the transmission media. A wireless technology is needed to reach any device not receiving electric service. Sometimes utilities will adopt a hybrid approach, utilizing different technologies to accomplish heterogeneous functional objectives.

3.15.6.5 Consumer

For the consumer, equipment standards are generally driven by product safety codes and regulations, especially for electricity-consuming products. Underwriters Laboratories (UL) and the Canadian Standards Association (CSA) are the main bodies that develop safety standards, with the National Fire Protection Agency (NFPA) responsible for the National Electric Code (NFPA 70), a version of which is used to judge installations. Another safety-oriented standard is IEEE C2, known as the National Electric Safety Code or NESC. For communications signals and interference, the Federal Communications Commission (FCC) has jurisdiction over products according to Code of Federal Regulations (CFR) Title 47 Part 15. Most consumer devices are tested and certified to accept any incoming interfering signals and continue operations and do not generate any interfering signals in a certain frequency band. Communications standards and specifications have historically failed to gain widespread application (and product adoption) due to the variety of products in this market space at all levels. Industry-facing efforts include the specifications developed by the ZigBee and HomePlug alliances, standards such as BACnet (building automation and control network communications protocol), LONWorks, and X-10. In certain cases, an alliance will coalesce around a standard to deal with the certification and marketing of products conforming to the standard. An example of this is the Wi-Fi (short for wireless fidelity) Alliance, formed to address the market needs of products conforming to the IEEE 802.11 series of standards.

3.15.6.6 Enterprise Integration

Enterprise integration identifies the connection of disparate applications needed to drive the utility business needs. This typically includes applications with "system" in their name such as outage management systems (OMS), graphical information systems (GIS), distribution management systems (DMS), energy management systems (EMS), customer information systems (CIS), meter data management systems (MDMS), or even an enterprise resource planning (ERP) system. Common practice is for each of these systems to be supplied by a different vendor, leading to difficulties in managing the data needed to run the utility business. The industry is moving toward common information model (CIM) development* and away from proprietary integration development.

3.15.6.7 NERC CIP Standards

The NERC CIP standards received the force of law in 2008 when the FERC approved their use. These standards dictate measures utilities must take in identifying and protecting critical cyber assets. However, jurisdictional issues complicate the matter for distribution-oriented smart grid technologies such as AMI. The primary point of contention over whether or not the CIP standards should apply to AMI has to do with one of the criteria explicitly delineated as requiring consideration for designation as a critical asset. CIP 002 states that utilities shall give consideration to "systems and facilities critical to automatic load shedding under a common control system capable of shedding 300 MW or more." If an AMI deployment includes integrated disconnects in the meters, this threshold is easily surpassed by distribution networks even in the 100,000 home range. For larger deployments with significant demand response program enrollment, it is possible this threshold may even be surpassed without disconnects in the equation. The take-away point here is that significant smart grid technology deployments such as distribution automation or advanced metering have the ability to dramatically affect load if compromised and used improperly or for malicious intent. Further, the automation capabilities of these systems may provide the ability to drop load suddenly or rapidly enough to impact overall system stability, including causing problems for transmission and generation, and possibly even causing a cascading blackout. This is a case of technological development having simply outpaced the relevant regulatory structures. By law, NERC has jurisdiction over generation and transmission as they frequently involve interstate commerce. On

---

* Example, IEC 61970 and IEC 61868 suites of standards.

the other hand, most distribution technologies including AMI do not involve interstate commerce and therefore fall under the purview of individual state utility commissions. As of this writing, the NERC CIP standards do not apply to AMI or any other smart grid technology deployed in the distribution domain.

3.15.7 Beyond Standardization

Standards attempt to meet the goal of creating a basic understanding of how to use a technology in a common manner. Unless interoperability tests or guidelines exist for a standard, at best a technology would be in compliance with the standard. In electronic and power technology, unlike physical technology (sizes), interoperability is at best an aspiration of the community that developed the standard. This highlights the need for a dedicated user's community tasked to identify interoperability challenges (requirements), write tests to validate products, and certify those results.

The relationship between the principal participants of successful standardization efforts is depicted in Figure 3.213. Utilities and vendors provide input into user groups and SDOs in the form of lessons learned, technical innovations, and application notes, and through direct participation. User groups, with early access to SDO drafts, can review and debate errors and propose other changes in a manner that provides a quicker consensus result. This results in the utilities and vendors more quickly receiving developed standards that address oft-desired implementation guidelines and test procedures, providing a higher degree of technology confidence and less "buyer's remorse."

A group that can be identified as laying the foundation for true interoperability is the GWAC. In a partnership with NIST, the GWAC sponsors the Grid-Interop [5] conference, which has the goals of achieving system-to-system interoperability, business process interoperation, preparing for a sustainable electricity system, developing policies for integrated smart energy and a holistic view of generation to consumption.

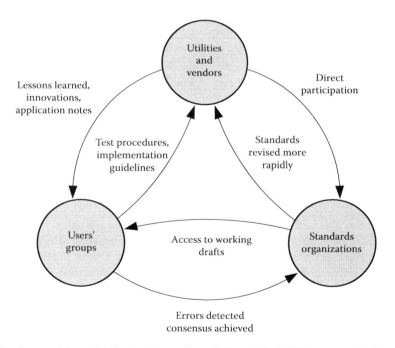

FIGURE 3.213 Successful standardization interaction. (© Copyright 2012 Enernex. All rights reserved.)

Another method to move beyond standards and technology driven by a single interest group is to use rules and regulations from governments and governmental agencies. The best of these are targets or guidelines with appropriate incentives and penalties without too much prescription. An example from another industry is the corporate average fuel economy (CAFE) standard for automobile manufacturers codified in the U.S. Code [6]. This is a target for the entire production, without specifying which particular models. The market of vehicles then signals to the manufacturers which vehicle and what volumes are needed, allowing them to develop technologies (engines, combustion techniques, etc.) to meet the government standard measurement.

The term "standard" is truly only applicable in certain situations. It is advocated here to reserve the use of "standard" for de jure standards, especially when employed without the "de jure" modifier. There may appear to be little harm in referring to de facto "standards" simply as "standards," but this actually dilutes and confuses the definition in the manner that the term "engineer" is often misapplied to functions requiring no engineering education or certification. It is recommended to employ the applicable term of "specification," "requirements," and "requirement specification" instead of "standard."

Finally, to move beyond standards that are regional in scope, there are pacts in place to allow copublication of standards between the IEC and IEEE, ANSI and IEEE, and others. This is the first step toward true harmonization, whereby a standard is in place for a broader market.

3.15.8 Key Issues

The electrical utility industry must overcome the following challenges to make the smart grid a reality.

3.15.8.1 Deployment of Technologies Still Under Development

Although most of the technologies necessary to build the smart grid already exist, most of them have not been mapped into the electric power domain.* Products for cost-effectively applying some of these technologies in the power system have only become available in the past few years. Due to the nascent state of product availability and pressures to deploy smart grid technologies, utilities right now often need to work in partnership with vendors to define requirements, provide design feedback, and evaluate prototypes. After downsizing and deregulation, many utilities do not have the research and development resources available to make this happen.

The pressures of rapidly deploying smart grid technologies and being first-to-market also generate particular concerns for security—both in the short and long term. Knowledge of how to develop secure embedded technology is a scarce and expensive commodity, and neither of these attributes meshes well with a high-pressure, get-it-done-now approach. As a result, many technologies are being deployed with numerous flaws that stem from a hurried and immature process.

Building the necessary controls for strong security into the production process takes time for the vendor and pressure from utility customers. The AMI-SEC Task Force has only recently finalized the first version of the AMI System Security Requirements document. Even once vendors start building product to the AMI-SEC guidance, it will take several months for this product to make it to the field and likely several more for the industry to work out any issues with interpretation of the guidance. In the meantime, vendors are still making products and putting them into the field, setting up a potential need for a significant replacement effort at some point in the future to remove points of vulnerability.

---

* A good example of domains, technologies, and power systems can be found in the NIST document entitled "Report to NIST on the Smart Grid Interoperability Standards Roadmap" found at http://www.nist.gov/smartgrid/InterimSmartGridRoadmapNISTRestructure.pdf

3.15.8.2 Lack of Market Power For Smaller Utilities

Deploying advanced technology is easier for bigger utilities for two reasons: first, they simply have more internal resources to apply to the project, and second, they must deploy to a larger number of sites and therefore can offer bigger incentives to vendors to implement the features they need. Smaller utilities do not have economies of scale, cannot offer large incentives, and therefore must often take off-the-shelf technology. This may mean their smart grid projects are "not as smart," or must be deferred because they are not yet cost-effective.

3.15.8.3 Interoperability Weak Spots

Several of the key smart grid communications standards, notably ANSI C12, IEC 61850, and IEC 61968/61970, follow a similar pattern:

- Committees have developed them over a long time, perhaps a decade or more, and the standards therefore represent heroic efforts on the part of multiple vendors to compromise. The fact that some of them exist at all is remarkable.
  - Nevertheless, the process in which they were developed means the standards contain options for most of the possible ways that vendors have implemented these utility applications over the years.
  - The standards therefore contain many implementation choices with few mandatory items, and implementations are difficult for utilities to specify without significant internal expertise.
  - Utilities use these standards in areas that have traditionally been dominated by single-vendor implementations, and for economic reasons, *unfortunately continue to be so dominated* despite the use of the standards. Therefore, significant multivendor interoperability testing in real-world situations is slow to arrive and sometimes painful when it does.
  - In some cases, such as ANSI C12, no organization exists even to provide certification testing. Although it is less effective than true interoperability testing, certification would at least represent a major step toward interoperability.
  - Devices implementing the standard typically can establish basic communications and exchange simple information very easily. However, when trying to deploy more advanced functions, utilities discover that vendors follow differences in philosophy that cause them to not work well together. The GridWise Interoperability Framework [7] would identify these philosophical differences as a lack of interoperability at the level of semantic understanding, business context, or business procedures.

    The traditional solution for these problems is to let more time pass and let the standard mature. Implementers discover the weak spots in the standard and utility users eventually begin to demand more mandatory items. The industry eventually develops guidelines for implementation that restrict the number of ways a vendor can implement the standard to a minimum set. IEC 61850, for instance, is about to release a second edition closing many of the "holes" in the first specification, 3 years after the first edition.

    However, the regulatory and economic realities of smart grid deployment mean that some utilities do not have the time to wait for the standards to mature. It may be necessary for groups of utilities to step in and impose guidelines for interoperability. Nevertheless, to do so, assume that someone knows what the best guidelines should be!

3.15.8.4 Enterprise Application Integration

This area represents a particular interoperability weak spot because the current state of the art is an extremely manual, labor-intensive process that is very dependent on the utility's existing information infrastructure and the utility's business practices.

Smart Grid Technologies 475

The two major standard players in this area, the IEC 61968/61970 CIM standards and MultiSpeak, approach these problems from different directions.

The CIM standards do not attempt to provide plug-and-play interoperability but instead define a "tool kit" that can be used to develop a set of essentially new protocols. An analogy for this process is that CIM defines a common set of words, but utilities must create their own rules for creating sentences from these words. One utility's implementation cannot talk to that of another utility if they have not worked together from the start of the project to define the same rules.

MultiSpeak, originally developed for and by smaller coop utilities through NRECA, takes the opposite approach. It strictly defines each application interface in a very clear manner through the command and payload semantics without a need to rely upon the possible variations of information infrastructure (e.g., "middleware") at the utility. This approach facilitates interoperability between vendors. Feature sets are scalable to better enable appropriate integration, and there are clear rules for extending the protocol.

The solution to the integration headaches lies somewhere between these two technologies, and both will likely migrate in that direction. In fact, there is a memorandum of understanding between the developers in both groups to find a manner to harmonize the two standards. The problem again is time.

3.15.8.5 Lack of Standard Distribution LANs, Especially Wireless Mesh and BPL

The distribution automation domain, also known as "access" or "last mile to the home," is dominated by proprietary solutions to a classic engineering problem: deploying millions of endpoints requires an extremely low-cost solution, but the area to be covered is huge with great variations in terrain that must be overcome. Furthermore, new distribution applications such as autorestoration and remote downloading of firmware are demanding lower latency and higher bandwidth.

Dozens of vendors have applied themselves to this problem space, producing a variety of solutions, each hoping to be the "better mousetrap" technology that takes the industry by storm. Unfortunately, none has (yet) emerged as a clear leader. As long as this is the situation, there is little incentive for vendors to work together to develop a common standard.

Standards do exist in this domain, and many of them are promising. ADSL and television cable are already deployed to a large number of premises, and a great variety of cellular phone technologies are available. The new wireless technology WiMAX is an open standard and seems particularly promising. However, utilities remain skeptical about these technologies. This reluctance to deploy standards in the last mile occurs for a variety of reasons: they may not provide sufficient reliability, sufficient bandwidth, or sufficient coverage. The most important barrier, however, is the cost of deploying thousands or millions of endpoints. These endpoints can be extraordinarily problematic as well, as utilities sometimes find themselves in the situation that the last 2% of the customer base represents 20% (or more) of the cost to deploy working technology under the obligation-to-serve paradigm. Utilities may deploy these open standards to the neighborhood or collector level, but the last mile typically remains proprietary.

In addition to creating interoperability issues, proprietary protocols also are a security concern. In the security realm, technology is never known to be unbreakable. The best one can hope for is that the security holds up long enough to support a useful life for the technology. Use of proprietary protocols often fosters a false sense of security when the developers of the security scheme are the only ones who have had significant exposure and interaction with the protocol. This is also called "security by obscurity" and is widely recognized as one of the weakest mechanisms for protecting valuable assets. The harsh realities of humanity mean that everyone is prone to mistakes and the limits of their own perspective. The process is better off leveraging the benefit of scrutiny and analysis by numerous industry experts early and often in the development lifecycle. This kind of exposure only takes place in open standards–based work.

The typical communications solution to a variety of proprietary physical layers is to use a common network layer or application layer to "bridge across" between technologies as illustrated in Figure 3.214. The clear leader in this area is the Internet Protocol suite, which has been doing just

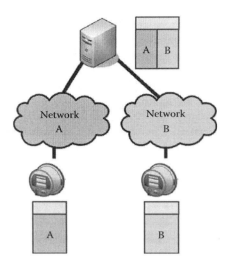

FIGURE 3.214 Using a common upper layer to bridge proprietary networks. (© Copyright 2012 Enernex. All rights reserved.)

this type of bridging for several decades now. However, most of the existing proprietary last-mile technologies were designed for applications with low messaging requirements and have traded off bandwidth for low cost. They are not capable of dealing with either the extra bandwidth required by an IP-based solution or with the underlying philosophy of the Internet technologies that permits any device to acquire bandwidth as needed at any time.

With this in mind, some vendors are offering field LANs with an open-standard application layer, namely, ANSI C12.22 (including ANSI C12.19 data models). In theory, this would permit a metering system using ANSI C12.22 to communicate over IP or variety of field LANs with any device that also supports ANSI C12.22. Unfortunately, in practice, no such multivendor metering system exists. The ANSI C12.22 standard permits so much flexibility in implementation that even metering systems that support ANSI C12.22 typically only support meters from the same vendor. Furthermore, using a metering standard, ANSI C12.22, as the only means of interoperability means that there are few nonmeter devices that can use these networks.

So the solution needed in the last-mile domain is more bandwidth at lower cost. This will permit the deployment of standard IP-based protocols to the customer premise, and as an added benefit, to the pole-top. There are some signs that a few vendors will soon be able to meet these requirements.

3.15.8.6 Too Many HAN Standards

In attempting to deploy smart grid applications such as AMI and DR to the home, utilities have necessarily entered the volatile home and building automation markets. These industries have implemented a huge variety of networking technologies, open and proprietary, wired and wireless. The leaders in this area include HomePlug 6LoWPAN and ZigBee, plus semiproprietary solutions like LonTalk, Insteon, and Z-Wave; more traditional (but costly and power-hungry) open standards like Ethernet and Wi-Fi; and popular legacy protocols such as X10. Such diversity and lack of interoperability have presented difficulties to the vendors in these markets previously, but the problem has been exacerbated by the deployment of AMI and other smart grid applications. Utilities would prefer to implement a single networking technology across every premise in their service area, a possibility that was previously very unlikely in this market.

There are a few promising efforts in this area. First among these is the OpenHAN working group, which has defined a set of common requirements for HANs intended to be used by utilities. The next step in the OpenHAN process will be a CIM that could be carried over a variety of networking technologies. Second was the swift creation of the "Smart Energy" profile, an application layer object model dedicated to electric utility functions, by the ZigBee Alliance. Another promising sign is the agreement between the ZigBee Alliance and HomePlug PowerLine Alliance to develop a common application layer across their respective wireless and broadband-over-power line technologies. All these efforts are being coordinated though the UtilityAMI organization.

However, any standards effort takes time. Utilities that must deploy customer-oriented smart grid applications right now are being forced to either commit to a particular technology, provide services only to customers who already have Internet access, or defer applications that require the use of an HAN until some future release. In the latter case, they may be faced with higher upgrade costs later.

3.15.8.7 Gateway Definition Between Utility and Premise

Although the EPRI and others have been studying the idea of a "consumer portal" for several years now, the technological shape of the gateway between the utility and each customer has never been well-defined. For instance, consider a system using ANSI C12 over the WAN and distribution LAN and the ZigBee Smart Energy profile on customer premises. There is no specification that clearly defines how the functions of these two technologies should be mapped to one another. For instance, which object or message on the HAN implements a demand response event expressed in ANSI C12?

3.15.8.8 Common Information Model (CIM)

The lack of a gateway mapping definition is a special case of the lack of a CIM that reaches across all smart grid domains. For instance, a CIM message to generate an energy price event, transmitted in the enterprise domain, must be translated to an equivalent ANSI C12 message to travel across the WAN and field LAN and then translated again to a corresponding ZigBee SE message before it reaches a thermostat in a consumer's home. Such translation is necessary because there is typically no common network layer used across all domains, and generally not enough bandwidth to carry an enterprise message verbatim down to the thermostat even if there was. There is even less agreement on how such a message might be translated and sent to a pole-top distribution automation device using, for example, DNP3.

At the moment, each of these translation steps is completely ad hoc and vendor-specific because there is no agreement on a common object model that works in all domains. The CIM, as a data model specific to the enterprise domain, is a good starting point, and various parts of UtilityAMI are beginning the first steps of harmonizing it with other technologies, but there is much work to be done.

3.15.8.9 Legacy Transmission and Distribution Automation

Utilities that attempt to deploy smart grid applications universally across their service area must first deal with the automation they have already deployed. Most utilities have several "islands of automation" in place, developed on a project-by-project basis over the years. Automation projects have tended to be "spotty" and incomplete due to a lack of a business case, especially in the distribution environment. Now that the business environment for widespread automation is improved, system engineers must find ways to incorporate these legacy systems into the new smart grid.

An important factor is that many of the technologies used in these legacy systems are becoming obsolete and are no longer supported. The "technology time warp" in the power industry is such that many technologies considered "advanced" by utilities are already considered to be aging and on the way out in general computing environments. Examples of such technologies are SONET, Frame Relay, 10 Mbit Ethernet, trunked radio, and even leased telephone lines. Many older technologies, such as Bell 202 modems, are now essentially *only* found in utility automation. Smart grid deployments must find a way to either integrate or replace these systems.

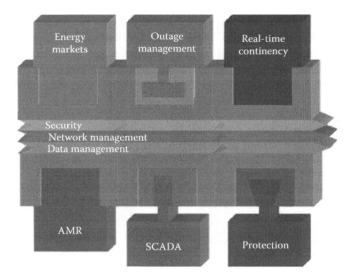

FIGURE 3.215 Holistic application integration. (© Copyright 2012 Enernex. All rights reserved.)

3.15.8.10 Poor Business Cases in Isolation, High Initial Investment When Integrated

In the metering and distribution automation environments, basic smart grid functions tend to have poor business cases by themselves. Examples of such functions are simple automatic meter reading or feeder fault location and autorestoration. While nobody can deny their intrinsic value, these functions typically do not provide enough return on investment to justify building the communications networks needed to deploy them over a wide area.

Many early adopters of smart grid philosophy have discovered, however, that when several smart grid applications are deployed together, the integrated business case becomes viable. For instance, when basic meter reading is combined with meter-aided outage management, theft detection, and prepayment and/or real-time pricing, it is easier to justify an AMI deployment. Similarly, when fault location is combined with phasor measurement and real-time state estimation, the business case for transmission or distribution automation becomes clearer.

Nevertheless, these more advanced features and holistic integration of security, network management, and data management (see Figure 3.215) require a larger up-front investment. The higher initial investment of integrating advanced smart grid applications may present a challenge to many utilities.

3.15.8.11 Merging Organizations

One of the biggest challenges facing utilities wishing to implement smart grid automation is that these applications force previously isolated organizations to communicate with each other and perhaps merge. Some examples include the following:

- *Merging SCADA and protection*: For several years now, it has become apparent that substation automation using intelligent devices and modern LAN technology requires the integration of substation functions. Protection devices must be capable of control and monitoring, and SCADA devices must take some part in protection. Similarly, the parent organizations of these devices must learn to communicate with each other.
- *Merging information technology (IT) and operations*: As utilities deploy enterprise bus technologies, the traditional separation between corporate IT and utility operations organizations must narrow, especially to address security issues.

- *Merging metering and distribution automation*: These two systems previously had nothing to do with each other although they shared a common geographic area of responsibility. Soon they will likely make use of a common, ubiquitous distribution communications network. Similarly, their respective organizations must now integrate operating procedures to realize some of the advantages of an integrated smart grid, such as advanced outage management.
- *Merging power and industry*: As more customer-centric applications like real-time pricing, distributed generation, and microgrids are deployed, utilities must take more of an interest in the industrial automation world. What was previously a one-way relationship must become a partnership as customers become active contributors to the operation of the power system.

3.15.8.12 Applying Holistic Security

Even if they applied to distribution-oriented smart grid technologies, the NERC CIP standards clearly specify that utilities must integrate information system security into all aspects of their automation systems: not just devices, computer systems, and technologies but also policies, procedures, and training. This is sound practice, and utilities would be well advised to follow this guidance even if the force of law does not apply. In order to be effective, utilities must apply these measures consistently and in an integrated fashion across their entire organization.

The smart grid vision of information flowing automatically throughout the utility provides great opportunities for efficiency, reliability, and cost-effectiveness. However, it also provides many more opportunities for attackers. The challenge for most utilities will be to build their smart grid systems in an evolutionary fashion, integrating security, network management, and data management into each smart grid application as it is deployed.

3.15.9 BEST PRACTICES

Within the scope of an AMI deployment, most utilities have rediscovered the value of a system engineering approach. Bringing together personnel from various groups impacted by the different aspects of metering (meters, communications, IT, system planning, system operation, customer interface, maintenance, reliability, and business operations, to name a few) to discuss on an abstract level, the use of such a system and capturing the responses via a "use case" is the first step. Included in the system engineering approach is the customer via a variety of mechanisms through direct workshop participation, consumer advocacy groups and focus groups, to name a few. From these use cases, it is possible to eliminate spurious actors and requirements (both functional and nonfunctional) through the documentation step. Functional requirements and nonfunctional requirements are then enumerated. Further discussion with utility personnel allows them to target high-impact (low-hanging) returns, making for a solid business case. Applying a ranking system to the requirements provides the utility personnel a tool with which to evaluate the relative maturity of any technology with respect to the cross-paired requirements and business goals. The vendors are then given a clear signal against which to develop (or modify) candidate elements and systems to meet the utility needs.

While a laborious process, both sides (utility and vendor) derive considerable benefit. The utility has several points along the process against which to test both their system and goals and candidate technology. The vendors have a relatively static target that facilitates development investment, and this static target also creates "vendor tension"—that is, competition—which the utility can leverage in either a performance or financial manner (or sometimes both).

The "smart grid" deployment will be a bit less concrete for many utilities than an AMI deployment. However, the same process will facilitate success in both grid operation and financial goals.

3.15.10 Legislation and Regulations

One area that is difficult to navigate is legislation and regulations. The utility space is rife with conflicting constraints, including obligation to serve, monopoly franchising, right-of-way obtention and defense, primary fuel source considerations, environmental, legal, rate of return, and many others. An example of this is provided by the attempt to mandate programmable, controllable thermostats (PCTs) in all new construction in California. These PCTs would enable the customer to determine a program that met their comfort and financial (and in some cases, environmental) goals, while providing utilities a touch point to limit peak demand (among other capabilities). Clearly nondiscriminatory in application (all new houses were required to have one), however, the perceived discrimination in operation (temperature and wind speed variance across a service territory might lead to a certain population "suffering" while another did not) provided a means to eliminate this law and regulation. One of the root causes of such vehement opposition is the flat consumption rate/tariff that is in place for the majority of electricity consumers. The fear of changing to another pricing model adds the potential for every single consumer to increase their cost of consumption. This is particularly unpalatable if one perceives others to be benefitting at one's expense, no matter what the "greater good" might dictate.

Clearly there was no lack of vision in this example. Rather, the legislation was defeated by good counter-marketing press. As demand response, smart grids, and even AMI are difficult concepts for the general public to grasp, fear of change drives efforts to understand, accept, and agree. A suggestion for strategy in the legislative and regulatory environment would be one of continual engagement with the general public at every stage. Communities of experts, such as those found collaborating in organizations such as GWAC and regulators in those such as the Mid-Atlantic Demand Response Initiative (MADRI), provide models for public engagement and leveraging the expertise in the industry.

As mentioned previously, cybersecurity also poses a particular challenge for legislation and regulation. While a uniform cybersecurity vision and strategy across the system may make sense from a technical standpoint, there is a hurdle in the way that our government is fundamentally structured. States still hold the trump card when it comes to legislation, and from a regulatory standpoint, the U.S. states operate more like 50 separate markets—each with their own rules and procedures. As long as state regulatory commissions determine how rates are figured, this issue is not likely to go away.

3.15.11 Advancing Smart Grid Standards

The following key items can help craft the smart grid standards landscape:

Standard field LANs: Especially where "Internet" technology is deployed, standard field LANs leverage existing network management tools and applications, allowing for reduced learning curves and minimizing special training. This would also allow utilities to create "vendor tension" to drive innovation.

Modems for field LANs: Whether wired or wireless, some locations will require a connection not limited by the existing communications architecture (e.g., wireless mesh, wireless star, Ethernet) and application(s) and dedicated for the "smart grid" purpose; this is also a reasonable backup to a primary communications technology.

More field bandwidth: Single-purpose networks (e.g., AMI, SCADA, phasor measurement unit monitoring) often are constrained where even a second application is precluded. It is rare that the realm of applications is known when hardware is deployed, and this modest investment would allow innovation to flourish.

CIM design framework: The IEC 61968 or CIM standards are a loosely defined semantic model for application-to-application integration. To date, only power system model exchanges have been tested for compatibility and interoperability. One of the challenges with this suite of standards is the choice of semantics is made by the user, within the limits of the standard. As such, only the fact that the messages conform to the semantic model can be verified. If different applications have employed different, yet compliant, choices, they may still be incompatible. The industry needs to be encouraged to tighten the allowable choices and offer an implementation guide. This would allow users to better develop interoperable applications.

CIM application security: The CIM standards are useless without applications and an integration layer across which information is exchanged. As such, there are no native security elements present. Instead, the native security of the applications and integration layer are used. Despite this, the standard's developers need to be encouraged to develop guidelines for verifiable CIM application security.

CIM/61850 harmonization: A gap analysis was performed, and set of recommendations for harmonization between the CIM and IEC 61850 has been developed. The user group or other certified entity should be encouraged to develop the recommendations presented in the latest investigation [8].

IEC 61850 outside substation: This protocol standard suite offers a compliance test and a user group committed to mediating and validating compliance and interoperability for substation communications equipment. The IEC working group has begun to extend the standard to communications outside of the substation (e.g., hydroelectric plants in 61850-7-410 and DER devices in 61850-7-420) and to work on substation-to-substation communications.

ANSI C12 guidelines: The ANSI metering protocol standards suffer from a lack of compliance tests as well as any means to demonstrate interoperable products, and this simply must change. Utilities should demand that the vendors in this market develop and demonstrate compliant and interoperable products.

Support of UtiliSec: The UtiliSec Working Group of the Open Smart Grid Subcommittee of the UCAIug is charged with developing security guidelines, recommendations, and best practices for the smart grid. The AMI-SEC Task Force has produced an AMI Risk Assessment and System Security Requirements document, with a component catalog being populated and the AMI-SEC Implementation Guide in final revisions. Ultimately these documents will provide additional assurance not previously available within the utility industry. While not directly measurable, the working group provides a focal point for industry discussions on security as it relates to the smart grid where the interchange of information and lessons learned between utilities on security-related issues is vital to the overall growth of the community.

Asset management: Already a key business element for many utilities, deployment of advanced communications infrastructure and smart grid devices will cause this problem to outgrow existing tools and management systems. However, that same infrastructure and smart devices will facilitate new systems in their efforts to perform true cradle-to-grave asset management. Whether certifying as-delivered functionality, monitoring the KPIs of the equipment during its operational life, or tracking maintenance and end-of-life information, this functionality is expected to greatly assist utility managers with minimizing operational and maintenance costs over the lifetime of utility assets.

DERs: Existing distribution systems are designed with a source-to-sink power flow in mind. Protection equipment and schemes, voltage support equipment and schemes, and other technical aspects assume a one-way power flow. DERs, which include generation *and* storage, require rethinking and redesign of both power equipment and analytical software applications as they contain energy supply or what some would consider "reverse" power flow. Standards are in the works for modeling, operation, and control of these devices, as well as protocols for defining them, but much of this may be revised once there is widespread deployment and integration of DER devices.

Plug-in hybrid electric vehicles (PHEVs): Another area of concern are PHEVs. This is one technology that completely shatters the central generation to end-consumer power system model that is in operation today. Most aspects of these load and storage devices (in technical terms) are yet defined. Policies and regulations are also not quite complete, and certainly business models for energy accounting are sorely lacking. There are several cross-industry groups assessing these gaps and proposing solutions. Utilities are attempting to prepare by obtaining advanced metering devices which are able to perform bidirectional energy accounting, by developing strategies around HANs and information models, and investigating financial accounting strategies, but much of this work is incomplete.

Three overarching strategies will prove useful when addressing the standards and technology. The first is a commitment to drive stakeholders from use of open standards to demonstrations of interoperability. Stagnation at the standards developer level can ruin a good standard (see IEEE 802.11.n and P1901). If compliance tests are not defined and moderated, the standards themselves are both too strict and too interpretable for each implementer. If interoperability tests are not defined and moderated, it is nearly impossible to provide any assurance of field compatibility.

A holistic approach (technical, communications, environment, regulation, security) is difficult but feasible. Clearly defined requirements in all of the areas allow system designers and integrators to develop, test, and deploy solutions that have a chance for success.

Finally, simple single-entity rate of return is often a poor measure of project value. As an example, AMI projects are built to leverage benefits beyond meter control and data collection. It may be the case that a utility can jump from all human-read to fully automatic reading, piggy-back on a significant existing communications infrastructure, or take advantage of favorable population density or terrain to minimize equipment investment. Looking outside of core metering aspects leads a utility to enhance outage management programs and perhaps improve on customer relations, to name a few areas. In the same manner, smart grid projects should look to provide benefits beyond core system operations improvements.

The systems engineering approach assists a utility with finding multiple benefit streams for a project deployment and helps to eliminate the siloed or "my project" syndromes that plague many large projects in many industries. An excellent domain-specific system engineering reference is the IEC Publicly Available Specification (PAS) 62559, which includes the IntelliGrid methodology for project definition decomposed into a five-phase project approach. The PAS also includes a use case development guide with three different (transmission synchrophasors, distribution automation, consumer) domain examples. Complementary to this open standard, third-party metrics such as the GridWise Architecture Interoperability Checklist and the Smart Grid Scorecard [9] provide an independent means for project teams to self-evaluate systems and products, leading to more transparent benefit derivation.

For large systems such as AMI and "smart grid" deployment, it is not longer the case that a "boutique" approach is viable: purchasing the latest and "greatest" components and stitching them together to construct a system. The process is reasonably clear for determining and acting upon requirements:

- Legislation and regulations (and markets) will determine targets for utilities.
- Utilities will use specifications and standards (and proprietary information) to procure and deploy equipment from vendors, within the legislative and regulatory constraints.
- Vendors will use specifications and standards (and utility information) to develop equipment.

None of this approach is unknown; however, the final steps are not always followed:

- Appropriate parties should construct compliance tests (e.g., vendors and third parties).
- Appropriate parties should construct acceptance tests (e.g., utilities and vendors).
- Appropriate parties should construct interoperability tests (e.g., vendors, third parties, and utilities).
- Appropriate parties should construct security tests (e.g., independent third parties and utilities).

Standards groups are ill-equipped to partake in these efforts as often the scope of the group or its ground rules preclude the open communication necessary to properly expose issues with technology as it is measured against the standard. However, the vendors participating in those groups are, with proper moderation and oversight, well-equipped to perform exactly these tasks. The other

Smart Grid Technologies

parties need to resist the "single vendor, end-to-end is better" which, while true initially, leads (and has lead) to incompatible products, poorly defined systems, and widespread buyer's remorse in the utility industry.

Research should be conducted to accelerate the development of standards that fill existing gaps including those for security, smart grid network and device management, information privacy management, and FAN interoperability. Appropriate private and public sector funding and organizations are needed to address these needs. Security, network, and device management and field interoperability touch on both technology and policy in that order. It is not enough to develop secure, manageable, and interoperable technology if policies, rules, and regulations inhibit its use in the power system.

Information privacy management also touches both policy and technology, though likely policy first. Best practices on developing policies are most likely not centered on the "terms of use" model from websites: long, legalese-filled missives that often inhibit understanding by the casual user. For example, this is an area where some discomfort may be found in the answer where the end consumer is the owner of the data collected and permits the utility, and only that utility, to use that data to provide the service. Permission is not granted to release the information to any entity, regardless of situation. This is not the current model nor is this a proposed solution, rather an example of an issue regarding information privacy management. This particular subject area requires legal research as well to examine what liability limits each party owns.

There are many mature standards and best practices already available that can be readily deployed to facilitate smart grid deployment. The main problem with adoption seems to be a lack of awareness of those standards and best practices by those involved in designing smart grid systems at a high level and regulatory guidelines for applying them.

Key recommendations include the following:

- Regulations should be developed that encourage utilities and product vendors to support standards-based technologies over proprietary solutions.
- Regulations should avoid mandating specific standards or technologies where possible in favor of specifying desired outcomes and important characteristics of the standards to be employed ("what" vs. "how").
- Research should be conducted to accelerate the development of standards that fill existing gaps including those for security, smart grid network and device management, information privacy management, and FAN interoperability.

GLOSSARY

6LoWPAN	IPv6 over Low-power wireless personal area networks
ADSL	Asymmetric digital subscriber line
AMI	Advanced Metering Infrastructure
AMI-SEC	AMI Security
ANSI	American National Standards Institute
ATM	Asynchronous Transfer Mode
BACnet	Building automation and control networks data communications protocol
BAN	Building Area Network—synonym for HAN
BPL	Broadband over power line
CAFE	Corporate Average Fuel Economy
CDMA	Code division multiple access
CDPD	Cellular Digital Packet Data
CFR	Code of Federal Regulations
CIM	Common Information Model

CIP	Critical infrastructure protection
CIS	Customer Information System
COMTRADE	Common format for transient data exchange
CORBA	Common Object Request Broker Architecture
COSEM	Companion Specification for Energy Metering
CSA	Canadian Standards Association
CSR	Customer service representative
DER	Distributed energy resource
DIN	Deutsches Institute für Normung
DLMS	Device Language Message Specification
DMS	Distribution Management System
DNP	Distributed Network Protocol
DOCSIS	Data Over Cable Service Interface Specification
DOE	Department of Energy
DRBizNet	Demand Response Business Network
ebXML	Electronic Business using eXtensible Markup Language
EMS	Energy management system
EPRI	Electric Power Research Institute
ERCOT	Electric Reliability Council of Texas
ERP	Enterprise resource planning
FAN	Field Area Network
FCC	Federal Communications Commission
FIPS	Federal Information Processing Standard
FRCC	Florida Reliability Coordinating Council
GID	Generic Interface Definition
GIS	Graphical Information System
GPRS	General packet radio service
GWAC	GridWise Architecture Council
HAN	Home area network
HTML	Hypertext markup language
HTTP	Hypertext Transfer Protocol
HTTPS	Hypertext Transfer Protocol over Secure Socket Layer
ICCP	Intercontrol Center Communications Protocol
IEC	International Electrotechnical Commission
IEEE	Institute of Electrical and Electronics Engineers
IPSec	Internet Protocol Security
IPv6	Internet Protocol version 6
ISO	International Organization for Standardization
IT	Information technology
ITU	International Telecommunication Union
LAN	Local area network
LEED	Leadership in Energy and Environmental Design
MAC	Media access control
MADRI	Mid-Atlantic demand response initiative
MAS	Multiple Address Systems
MDMS	Meter data management system
MPLS	Multiprotocol Label Switching
MRO	Midwest Reliability Organization
NEC	National Electric Code
NERC	North American Electric Reliability Corporation

NESC	National Electric Safety Code
NFPA	National Fire Protection Agency
NIST	National Institute of Standards and Technology
NPCC	Northeast Power Coordinating Council
NRECA	National Rural Electric Cooperative Association
OLE	Object Linking and Embedding
OMS	Outage management system
OPC	OLE for Process Control
OSGi	Open Services Gateway initiative
OSI	Open Systems Interconnection
PAN	Premise area network—synonym for HAN
PAS	Publicly Available Specification
PC	Personal computer
PCT	Programmable, controllable thermostat
PHEV	Plug-in hybrid electric vehicle
PKI	Public-key infrastructure
PLC	Power line carrier
PMU	Phasor measurement unit
PQDIF	Power Quality Data Interchange Format
PSTN	Public switched telephone network
PTP	Precision Time Protocol
RFC	Reliability First Corporation
RFID	Radio-frequency identification
RMON	Remote Network Monitoring
SAE	Society of Automotive Engineers
SAN	Substation area network
SCADA	Supervisory control and data acquisition
SDH	Synchronous digital hierarchy
SDO	Standards Development Organization
SERC	SERC Reliability Corporation
SNMP	Simple Network Management Protocol
SOAP	Simple Object Access Protocol
SONET	Synchronous Optical Networking
SPP	Southwest Power Pool
SQL	Structured Query Language
TLS	Transport Layer Security
TRE	Texas Regional Entity
UCAIug	UCA International Users Group
UL	Underwriters Laboratories
USB	Universal Serial Bus
USC	U.S. Code
VLAN	Virtual local area network
VPN	Virtual private network
WAN	Wide area network
WDM	Wavelength division multiplexing
WECC	Western Electricity Coordinating Council
Wi-Fi	Wireless fidelity
WiMAX	Worldwide Interoperability for Microwave Access
WMS	Work Management System
WPA2	Wi-Fi Protected Access 2

Annex: Technology Enumeration

A nonexhaustive list of technologies that straddle the identified decomposition is enumerated as follows:

Standard	Enterprise	LAN	WAN	SAN	FAN	HAN
Internet Protocol version 4 (IPv4)	Y	Y	Y	Y	Y	Y
Internet Protocol version 6 (IPv6)	Y	Y	Y	Y	Y	Y
Multiprotocol Label Switching (MPLS)	Y	Y	Y	Y	Y	Y
X.509 Public-Key Infrastructure (PKI)	Y	Y	Y	Y	Y	Y
Federal Information Processing Standard (FIPS) encryption	Y	Y	Y	Y	Y	Y
Federal Information Processing Standard (FIPS) authentication	Y	Y	Y	Y	Y	Y
Internet Protocol Security (IPSec)	Y	Y	Y	Y	Y	Y
Transport Layer Security (TLS)	Y	Y	Y	Y	Y	Y
Common Management Information Protocol	Y	Y	Y	Y	Y	Y
OSI (Open Systems Interconnection) network management	Y	Y	Y	Y	Y	Y
IEC 62351 Security	Y	Y	Y	Y	Y	Y
Simple Network Management Protocol (SNMP)	Y	Y	Y	Y	Y	Y
Remote Network Monitoring (RMON)	Y	Y	Y	Y	Y	Y
IEEE 1588 Precision Time Protocol (PTP)	Y	Y	Y	Y	Y	Y
IEC 61334-4-41 DLMS (Device Language Message Specification)			Y		Y	Y
IEC 62056 COSEM (Companion Specification for Energy Metering)			Y		Y	Y
ANSI C12.18 Optical Port and Protocol Specification for Electric Metering						Y
ANSI C12.19 Utility Industry Data Tables			Y		Y	Y
ANSI C12.21 Telephone			Y		Y	
ANSI C12.22 Networking			Y		Y	Y
ANSI C12.23 Testing			Y		Y	Y
Building automation and control networks (BACNet)						Y
HomePlug						Y
IEEE 802.15.4 with ZigBee						Y
IEEE 802.11 b/g "Wi-Fi"						Y
IEEE 802.15.1 "Bluetooth"						Y
Radio-frequency identification (RFID)						Y
IEEE 802.11i Wi-Fi Protected Access 2 (WPA2)						Y
IEEE 802.3 Ethernet						Y
LonWorks						Y
X10						Y
6LowPAN (IPv6 over low-power wireless personal area networks)						Y
Z-Wave						Y
Insteon						Y
WirelessHART (Highway Addressable Remote Transducer Protocol)						Y
Open Services Gateway initiative (OSGi)						Y
IEEE 802.1Q Virtual LANs (VLANs)						Y
Fieldbus			Y	Y	Y	
Profibus			Y	Y	Y	
IEEE 1390 Telephone Meter Reading			Y	Y	Y	
Cellular Digital Packet Data (CDPD)			Y	Y	Y	
IEEE 802.16 WiMAX			Y	Y	Y	

Standard	Enterprise	LAN	WAN	SAN	FAN	HAN
Multiple Address Systems (MAS)/Trunked Radio			Y	Y	Y	
IEC 60870-5-101/104 Telecontrol			Y	Y	Y	
Modbus			Y	Y	Y	
DNP3			Y	Y	Y	
IEC 61850 Substations			Y	Y	Y	
2G Wireless (1×RTT, GPRS)			Y		Y	
HomePlug Access BPL (Broadband over Power Line)			Y		Y	
X-Series Networking	Y		Y	Y		
Frame Relay	Y		Y	Y		
Synchronous optical networking (SONET)	Y		Y	Y		
Synchronous digital hierarchy (SDH)	Y		Y	Y		
Asynchronous transfer mode (ATM)	Y		Y	Y		
Wavelength division multiplexing (WDM)	Y		Y	Y		
Virtual private networks (VPNs)	Y		Y	Y		
IEC 60870-6 Intercontrol Center	Y					
IEC 61970 Common Info Model	Y		Y	Y		
IEC 61968 Distribution Interfaces	Y		Y	Y		
OpenGIS (Open Geographic Information Systems)	Y					
MultiSpeak	Y		Y	Y		
HTTP/HTML	Y					
Common Object Request Broker Architecture (CORBA)	Y					
Web services	Y					
Structure Query Language (SQL)	Y					
OPC (Object Linking and Embedding for Process Control)	Y					
Web services security	Y					
HTTPS (Hypertext Transfer Protocol over Secure Socket Layer)	Y					
IEC 62325 Energy Markets						
ebXML (Electronic Business using eXtensible Markup Language)						
Point-to-point microwave	Y		Y			
Licensed point-to-multipoint radio			Y	Y	Y	Y
Unlicensed point-to-multipoint radio			Y	Y	Y	Y
Licensed mesh radio network			Y	Y	Y	Y

REFERENCES

3.1 Technology Drivers

1. Fan, J. and S. Borlase, Advanced distribution management systems for smart grids, *IEEE Power and Engineering*, March/April 2009.
2. United States Department of Energy, Smart grid, http://www.oe.energy.gov/smartgrid.htm

3.2 Smart Energy Resources

1. Enslin, J. Grid impacts and solutions of renewables at high penetration levels, www.quanta-technology.com, June 2009.
2. Enslin, J. Dynamic reactive power and energy storage for integrating intermittent renewable energy, Invited Panel Session, Paper PESGM2010-000912, *IEEE PES General Meeting*, Minneapolis, MN, July 25–29, 2010.
3. Current Energy Storage Project Examples, California Energy Storage Alliance (CESA), http://www.storagealliance.org/presentations/CESA_OIR_Storage_Project_Examples.pdf

4. Electricity Storage Association (ESA), Utility Support, http://www.electricitystorage.org/technology/technology_applications/utility_support/
5. Wikipedia, Grid Energy Storage, http://en.wikipedia.org/wiki/Grid_energy_storage
6. Agüero, J. R. Steady state impacts and benefits of solar photovoltaic distributed generation (PV-DG) on power distribution systems, *CEATI 2010 Distribution Planning Workshop*, Toronto, Canada, June 2010, http://www.ceati.com/Meetings/DPW2010/program.html
7. Electricity Storage Association (ESA), Technology Comparison, http://www.electricitystorage.org/technology/storage_technologies/technology_comparison
8. Garrison, J. B. and Webber, M. E. An integrated energy storage scheme for a dispatchable solar and wind powered energy system. *Journal of Renewable Sustainable Energy*, 3, 2011, 043101.
9. Zalba, B., Marín, J. M., Cabeza, L. F., and Mehling, H. Review on thermal energy storage with phase change: Materials, heat transfer analysis and applications. *Applied Thermal Engineering*, 23(3), 2003, 251–283.
10. Bradley, T. H. and Frank, A. A. Design, demonstrations and sustainability impact assessments for plug-in hybrid electric vehicles. *Sustainable and Renewable Energy Reviews*, 13(1), 2009, 115–128.
11. Suppes, G. J. Plug-in hybrid with fuel cell battery charger. *International Journal of Hydrogen Energy*, 30(2), 2005, 113–121.
12. Suppes, G. J., Lopes, S., and Chiu, C. W. Plug-in fuel cell hybrids as transition technology to hydrogen infrastructure. *International Journal of Hydrogen Energy*, January 2004, 369–374.
13. SAE Electric Vehicle and Plug in Hybrid Electric Vehicle Conductive Charge Coupler, http://standards.sae.org/j1772_201001/
14. Range Anxiety: Fact or Fiction, http://news.nationalgeographic.com/news/energy/2011/03/110310-electric-car-range-anxiety/
15. U.S. DOE Alternative Fuel Vehicles (AFVs) and Hybrid Electric Vehicles (HEVs), http://www.afdc.energy.gov/afdc/data/vehicles.html#afv_hev
16. Wikipedia, Hybrid Vehicle Drivetrain, http://en.wikipedia.org/wiki/Hybrid_vehicle_drivetrains
17. Tomic, J. and Kempton, W. Using fleets of electric-drive vehicles for grid support. *Journal of Power Sources*, 168, 2007, 459–468.
18. Kempton, W. and Udo, V. A test of vehicle-to-grid (V2G) for energy storage and frequency regulation in the PJM system, http://www.magicconsortium.org, 2008.
19. Kirby, B. and Hirst, E. in ORNL/CON—474, Oak Ridge National Laboratory Report (ORNL), Oak Ridge, TN, 2000.
20. Kempton, W. and Tomic, J. *Journal of Power Sources*, 144, 2005, 280–294.
21. Quinn, C., Zimmerle, D., and Bradley, T. H. The effect of communication architecture on the availability, reliability, and economics of plug-in hybrid electric vehicle-to-grid ancillary services. *Journal of Power Sources*, 195(5), 2010, 1500–1509.
22. Gage, T. B. Development and evaluation of a plug-in HEV with vehicle-to-grid power flow, AC Propulsion Inc., San Dimas, CA, ICAT 01-2, 2003.
23. Kempton, W. and Tomic, J. Vehicle-to-grid power fundamentals: Calculating capacity and net revenue. *Journal of Power Sources*, 144, 2005, 268–279.
24. Kirby, B. and Hirst, E. Computer-specific metrics for the regulation and load following ancillary services. Oak Ridge National Laboratory Report (ORNL), Oak Ridge, TN, ORNL/CON—474, 2000.
25. Parks, K., Denholm, P., and Markel, T., Costs and emissions associated with plug-in hybrid electric vehicle charging in the Xcel energy Colorado service territory, National Renewable Energy Laboratory, NREL/TP-640-41410, Golden, CO, 2007.
26. Kintner-Meyer, M., Schneider, K., and Pratt, R. Impacts assessment of plug-in hybrid vehicles on electric utilities and regional U.S. power grids, Part 1: Technical analysis. Pacific Northwest National Laboratory, Richland, WA, 2006.
27. Davis, B. M., Grid integration of plug-in vehicles, *Plug-in Conference*, Raleigh, NC, July 18–21, 2011.
28. The network grid, November 4, 2009.
29. IEEE 1547-2003: Standard for Interconnecting Distributed Resources with Electric Power Systems.
30. FERC Large Generator Interconnection Requirements (Orders 661), http://www.ferc.gov/industries/electric/indus-act/gi/wind/appendix-G-lgia.doc?
31. EPRI, Standard language protocols for PV and storage grid integration: Developing a common method for communicating with inverter-based systems, May 2010.
32. Enslin, J. Network impacts of high penetration of photovoltaic solar power systems, Invited Panel Session, Paper PESGM2010-001626, *IEEE PES General Meeting*, Minneapolis, MN, July 25–29, 2010.
33. Short, W. and Denholm, P. Preliminary assessment of plug-in hybrid electric vehicles on wind energy markets, National Renewable Energy Lab (NREL), NREL/TP-620-39729, Golden, CO, 2006.

34. Kintner-Meyer, M. C. and Schneider, K. P. Impacts assessment of plug-in hybrid vehicles on electric utilities and regional U.S. power grids: Part 1: Technical analysis, Pacific Northwest National Laboratory, PNNL-SA-61669, Richland, WA, 2007.
35. Hadley, S. W. and Tsvetkova, A. A. in ORNL/TM-2007/150, Oak Ridge National Laboratory (ORNL), Oak Ridge, TN, 2007.
36. Denholm, P. and Short, W. in NREL/TP-620-40293, National Renewable Energy Laboratory (NREL), Golden, CO, 2006.
37. Rowand, M. *Plug-in 2009 Conference and Exposition*, Long Beach, CA, 2009.
38. Taylor, J., Maitra, A., Alexander, M., Brooks, D., and Duvall, M. Evaluation of the impact of PEV loading on distribution system operations, IEEE Power Engineering Society, Calgary, Canada, July 2009.
39. Brooks, A. in A Report prepared by AC Propulsion for the California Air Resources Board and the California Environmental Protection Agency, http://www.udel.edu/V2G, 2002.
40. Markel, T. Communication and control of electric vehicles supporting renewables, National Renewable Energy Laboratory, Golden, CO, 2009.
41. Morrow, K. and Karner, D. in INL/EXT-08-15058, Idaho National Laboratory (INL), Idaho Falls, ID, 2008.
42. SAE Charging Configurations and Ratings Terminology, http://www.sae.org/smartgrid/chargingspeeds.pdf
43. Xu, L., Marshall, M., and Dow, L. A framework for assessing the impact of plug-in electric vehicle to distribution systems, *2011 IEEE PSCE*, Phoenix, AZ, March 2011.
44. Dow, L., Marshall, M., Xu, L., Agüero, J. R., and Willis, L. A novel approach for evaluating the impact of electric vehicles on the power distribution system, *2010 IEEE PES General Meeting*, Minneapolis, MN, July 2010.
45. Agüero, J. R. Tools for success. *IEEE Power and Energy Magazine*, September/October 2011, 82–93.
46. Agüero, J. R. and Dow, L. Impact studies of electric vehicles, Quanta Technology, Raleigh, NC, 2011.
47. Richard Factor, The PriUPS Concept, http://www.priups.com/misc/intro.htm, 2010.
48. Wilson, W, http://hiwaay.net/~bzwilson/prius/priups.html

3.3 Smart Substations

1. http://ge.ecomagination.com/smartgrid

3.4 Transmission Systems

EMS

1. M.R. Endsley and D.J. Garland, *Designing for Situation Awareness: An Approach to User-Centered Design*, CRC Press, ISBN-10: 0805821338, ISBN-13: 978-0805821338, 2000.
2. Overbye, Wide-area power system visualization with geographic data views, *Power and Energy Society General Meeting—Conversion and Delivery of Electrical Energy in the 21st Century, 2008 IEEE*, pp. 1–3, July 20–24, 2008.
3. Parashar, Manu, Jianzhong Mo: Real time dynamics monitoring system (RTDMS): Phasor applications for the control room, *42nd Hawaii International Conference on System Sciences*, Waikoloa, Big Island, Hawaii, January 05–08, 2009.
4. Miller, The magical number seven, plus or minus two: Some limits on our capacity to process information, *Psychology Review*, 63, 81–96, 1956.
5. M. Gilger, Addressing information display weaknesses for situational awareness, *Military Communications Conference, 2006. MILCOM 2006. IEEE*, pp. 1–7, October 23–25, 2006.
6. Critical Infrastructure Protection Standards 002-3–009-3, North American Electric Reliability Corporation. December 16, 2009. Available online at http://www.nerc.com//page.php?cid=2|20. Accessed June 18, 2012.
7. International Electrotechnical Commission, Technical Specification IEC/TS 62351-1: Power Systems Management and Associated Information Exchange—Data and Commucations Security—Part1: Communication Network and System Security—Introduction to Security Issues, First Edition ed. Geneva, Switzerland: IEC, 2007.
8. Gordon Fyodor Lyon, Nmap Network Scanning: Official Nmap Project Guide to Network Discovery and Security Scanning. Insecure.Com, LLC, Sunnyvale, CA, 2008.
9. E. Tews and M. Beck, Practical attacks against WEP and WPA, *Proceedings of the Second ACM Conference on Wireless Network Security*, WiSec'09, Zurich, Switzerland, ACM, New York, pp. 79–86, March 16–19, 2009.

10. R.W. McGrew and R.B. Vaughn, Discovering vulnerabilities in control system human machine interface software, *Journal of Systems and Software*, 82(4), 583–589, April 2009.
11. D. Peterson, Quickdraw: Generating security log events for legacy SCADA and control system devices, *Conference For Homeland Security, 2009. CATCH'09. Cybersecurity Applications & Technology*, Wasington, DC, pp. 227–229, March 3–4, 2009, doi: 10.1109/CATCH.2009.33.
12. P. Oman and M. Phillips, Intrusion Detection and Event Monitoring in SCADA Networks, in *Critical Infrastructure Protection, Eric Goetz and Sujeet Shenoi*, Goetz, E. and Shenoi, S. (eds.), Springer, Boston, MA, pp. 161–173, 2007.
13. M. Hadley and K. Huston, *Secure SCADA Communication Protocol Performance Test Results*, Pacific Northwest National Laboratory, Richland, WA.
14. P.P. Tsang and S.W. Smith, YASIR: A low-latency, high-integrity security retrofit for legacy SCADA systems. In S. Jajodia, P. Samarati, and S. Cimato, eds. *SEC*, vol. 278 of IFIP, Springer, Heidelberg, Germany, pp. 445–459, 2008.

FACTS and HVDC

1. D. Povh*, D. Retzmann*, J. Kreusel**, Integrated AC/DC transmission systems—Benefits of power electronics for security and sustainability of power supply. *PSCC 2008*, Glasgow, U.K., July 14–17, 2008. Survey Paper 2, * part 1 and ** part 2.
2. European Technology Platform, *SmartGrids—Vision and Strategy for Europe's Electricity Networks of the Future*, European Commission, Luxembourg, Europe, 2006.
3. DENA Study Part 1, Energiewirtschaftliche Planung für die Netzintegration von Windenergie in Deutschland an Land und Offshore bis zum Jahr 2020, Cologne, Germany, February 24, 2005.
4. M. Luther and U. Radtke, Betrieb und Planung von Netzen mit hoher Windenergieeinspeisung, *ETG Kongress*, Nuremberg, Germany, October 23–24, 2001.
5. G. Beck, D. Povh, D. Retzmann, and E. Teltsch, Global blackouts—Lessons learned, *POWER-GEN Europe*, Milan, Italy, June 28–30, 2005.
6. W. Breuer, D. Povh, D. Retzmann, and E. Teltsch, Trends for future HVDC Applications, *16th CEPSI*, Mumbai, India, November 6–10, 2006.
7. G. Beck, D. Povh, D. Retzmann, and E. Teltsch, Use of HVDC and FACTS for power system interconnection and grid enhancement, *POWER-GEN Middle East*, Abu Dhabi, United Arab Emirates, January 30–February 1, 2006.
8. FACTS Overview, *IEEE and CIGRE*, Catalog Nr. 95 TP 108, 1996.
9. C. Schauder, The unified power flow controller—A concept becomes reality, *IEE Colloquium on Flexible AC Transmission Systems—The FACTS* (Ref. No. 1998/500), London, U.K., pp. 7/1–7/6, 1998.
10. Economic Assessment of HVDC Links, CIGRE Brochure Nr.186 (Final Report of WG14-20), June 2001.
11. C.D. Barker, N.M. Kirby, N.M. MacLeod, and R.S. Whitehouse, Widening the bottleneck: Increasing the utilisation of long distance AC transmission corridors, *IEEE T&D Expo*, New Orleans, LA, April 2010.
12. N.M. MacLeod, C.D. Barker, and N.M. Kirby, Connection of renewable energy sources through grid constraint points using HVDC power transmission systems, *IEEE T&D Expo*, New Orleans, LA, April 2010.
13. J. Dorn, H. Huang, and D. Retzmann, Novel voltage-sourced converters for HVDC and FACTS applications, *Cigre Symposium*, Osaka, Japan, November 1–4, 2007.
14. Working Group B4-WG 37 CIGRE, VSC Transmission, May 2004.
15. MANGO—modal snalysis for grid operation: A method for damping improvement through operating point adjustment, Pacific Northwest National Laboratory report (PNNL-19890), October 2010. http://certs.lbl.gov/pdf/pnnl-19890-mango.pdf

3.5 Distribution Systems

1. Fan, J. and Borlase, S., Advanced distribution management systems for smart grids, *IEEE Power and Engineering*, March/April 2009, 7(2), 63–68.
2. Voltage ratings for electrical power systems and equipment, American National Standard ANSI C84.1-1989.
3. Wilson, T.L., Measurement and Verfication of Distribution Voltage Optimization Results, presented at IEEE/PES 2010 General Meeting, 2010.
4. Stoupis, J. et al., Restoring confidence: Control-center and field-based feeder restoration, *ABB Review*, Q3/2009, 17–22.

5. Northcote-Green, J. and Wilson, R., *Control and Automation of Electrical Power Distribution Systems*, CRC Taylor & Francis Group, Boca Raton, FL, 2007, pp. 149–163, 251–264.
6. Short, T.A., *Electric Power Distribution Handbook*, CRC Press LLC, Boca Raton, FL, 2004, pp. 441–476.

3.6 Communications Systems

1. http://www.grid-net.com/pr-2010-12-06
2. http://www.ausgrid.com.au/Common/About-us/Newsroom/Media-Releases/2010/November/Smart-Grid-to-use-LTE. aspx
3. BBC News, 9 July 2010, http://www.bbc.co.uk/news/10569081
4. International Electrotechnical Commission, Technical Committee 57, IEC 61850-9-2 ed1.0 (2004-04), Communication networks and systems in substations, Part 9-2: Specific Communication Service Mapping (SCSM), Sampled values over ISO/IEC 8802-3, April 2004.
5. International Electrotechnical Commission, Technical Committee 57, IEC 61850-8-1 ed1.0 (2004-05), Communication networks and systems in substations, Part 8-1: Specific Communication Service Mapping (SCSM), Mappings to MMS (ISO 9506-1 and ISO 9506-2) and to ISO/IEC 8802-3, May 2004.
6. International Electrotechnical Commission, Technical Committee 57, IEC 61850-90-1 ed1.0 (2010-03), Communication networks and systems for power utility automation, Part 90-1: Use of IEC 61850 for the communication between substations, March 2010.
7. International Electrotechnical Commission, Technical Committee 57, IEC 61968-9 ed1.0 (2009-9), Application integration at electric utilities, System interfaces for distribution management, Part 9: Interfaces for meter reading and control, September 2009.
8. DNP User Group, DNP Specification Document, February 2010.
9. International Electrotechnical Commission, Technical Committee 57, IEC 60870-5-1 ed1.0 (1990-02), Telecontrol equipment and systems. Part 5: Transmission protocols, Section one: Transmission frame formats, February 1990.
10. Institute of Electrical and Electronics Engineers, Power Engineering Society, IEEE Std C37.118-2005, IEEE Standard for Synchrophasors for Power Systems, March 2006.
11. American National Standard for Utility Industry, ANSI C12.19-2008, End Device Data Tables Secretariat, 2009.
12. American National Standard for Protocol Specification for ANSI Type 2 Optical Port, ANSI C12.18-2006, 2006.
13. American National Standard for Protocol Specification for Telephone Modem Communication, ANSI C12.21-2006, 2006.
14. American National Standard for Protocol Specification for Interfacing to Data Communication Networks, ANSI C12.22-2008, 2008.
15. International Electrotechnical Commission, Technical Committee 65C, IEC 62439-3 ed1.0 (2010-02), Industrial communication networks, High availability automation networks, Part 3: Parallel Redundancy Protocol (PRP) and High-availability Seamless Redundancy (HSR), February 2010.
16. Institute of Electrical and Electronics Engineers, Instrumentation and Measurement Society, IEEE Std 1588-2008, IEEE standard for a precision clock synchronization protocol for networked measurement and control systems, July 2008.
17. Internet Engineering Task Force, Network time protocol version 4, Protocol and algorithms specification (draft), April 2010.
18. International Electrotechnical Commission, ISO/IEC 8802-3 ed6.0 (2000–12), Information technology, Telecommunications and information exchange between systems, Local and metropolitan area networks, Specific requirements, Part 3: Carrier sense multiple access with collision detection (CSMA/CD) access method and physical layer specifications, December 2000.
19. Internet Engineering Task Force, RFC 791, Internet Protocol, DARPA Internet Program Protocol Specification, September 1981.
20. Internet Engineering Task Force, RFC 793, Transmission Control Protocol, DARPA Internet Program Protocol Specification, September 1981.
21. Internet Engineering Task Force, RFC 768, User Datagram Protocol, August 1980.
22. L. L. Peterson and B. S. Davie, *Computer Networks: A Systems Approach*, 4th edn., Morgan Kaufmann, San Francisco, CA, March 2007.
23. R. W. Stevens, *TCP/IP Illustrated, Volume 1: The Protocols*, Addison-Wesley Professional, Boston, MA, January 1994.

24. International Standard Organization, International Electrotechnical Commission, ISO/IEC 7498-1, Information technology, Open Systems Interconnection, Basic reference model: The basic model, November 1994.
25. The use of Ethernet technology in the power utility environment, CIGRE Technical Brochure TB460, Working Group D2-23, April 2011.
26. The production of technical specifications for a third-generation mobile system based on the evolved GSM core networks. www.3gpp.org
27. C. Gellings, *The Smart Grid: Enabling Energy Efficiency and Demand Response*, Fairmont Press, Lilburn, GA, 2009.
28. International Electrotechnical Commission, Technical Committee 57, IEC 61850-7-420 ed1.0 (2009-3), Communication networks and systems for power utility automation, Part 7-420: Basic communication structure, Distributed energy resources logical nodes, March 2009.

3.7 Monitoring and Diagnostics

1. H. Eren, *Wireless Sensors and Instruments: Networks, Design, and Applications*, CRC Press, Boca Raton, FL, 2006.
2. J. S. Wilson, *Sensor Technology Handbook*, Elsevier, Burlington, MA, 2005.
3. M. Mousavi, V. Donde, J. Stoupis, J. Mcgowan, and L. Tang, Information not data: Real-time automated distribution event detection and notification for grid control, *ABB Review Journal*, 3, 2009, 38–44.

3.10 Smart Meters and Advanced Metering Infrastructure

1. Guerry Waters, Prepay customers—Without prepay meters!, Public Utility's Reports, Inc. Fortnightly's Spark, Letter #57 magazine, September 2008.
2. Edison Institute for Electric Efficiency report, Utility scale smart meter deployments, plans & proposals, May 2012 (http://www.edisonfoundation.net/iee/Documents/IEE_SmartMeterRollouts_0512.pdf).
3. eMeter Strategic Consulting, available at www.emeter.com/smart-grid-watch
4. VaasaETT Global Energy Think Tank, The potential of smart meter enabled programs to increase energy and systems efficiency: A mass pilot comparison, October 2011.
5. NIST, NIST framework and roadmap for smart grid interoperability standards, Release 1.0, NIST Special Publication 1108, January 2010.

3.11 Consumer Demand Management

1. E. Woychik, Integrated demand side management cost effectiveness framework white paper, Black & Veatch, for the California IDSM Task Force, May 2011, available at http://www.calmac.org/publications/IDSM_Final_White_Paper_12May2011.pdf
2. K. Spees, Meeting electric peak on the demand side: Wholesale and retail market impacts of real-time pricing and peak load management policy, s.l.: Carnegie Mellon University, Pittsburgh, PA, 2008.
3. North American Electric Reliability Council, Historic capacity and demand, 2007, http://www.nerc.com/page.php?cid=4|38|41
4. ISO-NE, Hourly historical data, 2008, http://www.iso-ne.com/markets/hstdata/hourly/index.html
5. R. Wilson, *Allocation, Information, and Markets: The New Palgrave*, MacMillan, New York, 1989.
6. A. Faruqi, R. Hledik, S. Newell, H. Pfeifenberger, The Power of 5 Percent, *Electricity Journal* (Elsevier), 20(8), 68–77 October 2007.
7. ERCOT demand response program helps restore frequency following Tuesday evening grid event, February 27, 2008, http://www.ercot.com/news/press_releases/2008/nr02-27-08
8. T. Konrad, The Texas 'wind' emergency, the smart grid and the dumb grid, March 24, 2009, http://seekingalpha.com/article/127471-the-texas-wind-emergency-the-smart-grid-and-the-dumb-grid
9. Pacific Northwest National Lab, Department of Energy putting power in the hands of consumers through technology, *Top Story*, 2010, http://www.pnl.gov/topstory.asp?id=285
10. 2010 Assessment of Demand Response and Advanced Metering: Staff Report, Federal Energy Regulatory Commission, February 2011.
11. Intel, Intel Intelligent Home Energy Management Proof of Concept, Intel Corporation, Santa Clara, CA [Online] 2010, http://edc.intel.com/embedded/homeenergy/

12. R. T. Guttromson, D. P. Chassin, Optimizing retail contracts for electricity markets, in Proceedings of the Grid Interoperability Forum, PNNL-SA-57497, Richland, WA, 2007.
13. Consumer engagement: Facts, myths and motivations, an EPRI Issues Analysis, 2011, EPRI product number 1024566.

3.12 Convergence of Technologies and Enterprise-Level Integration

1. Halevy, A.Y. (Ed.), N. Ashishy, D. Bittonz, M. Careyx, D. Draper, J. Pollock, A. Rosenthal, and V. Sikka, Enterprise information integration: Successes, challenges and controversies, *SIGACM-SIGMOD'05*, Baltimore, MD, 2005.
2. UtilityAMI, AMI-Enterprise System Requirements Specification, Version: v1.0, Release Date: October 14, 2009, UCAIug OpenSG UtilityAMI AMI-ENT Task Force. (AMI-Enterprise (AMI-ENT) is a utility led initiative under UtilityAMI and Open Smart Grid (OpenSG) within the UCA International Users Group (UCAIug). The AMI-Enterprise Task Force defines systems requirements, policies, principles, best practices, and services required for information exchange and control between AMI related systems and utility enterprise front and back office systems.)
3. Robinson, G., Synergies achieved through integrating smart grid components—A system of systems, *IEEE PES Transmission and Distribution Optimization Panel*, New Orleans, LA, April 21, 2010.
4. Parekh, K., J. Zhou, K. McNair, and G. Robinson, Utility enterprise information management strategies, *Grid InterOp*, Albuquerque, NM, 2007.

3.13 High-Performance Computing for Advanced Smart Grid Applications

1. G. E. Moore, Cramming more components onto integrated circuits, *Electronics Magazine*, 38(8), April 19, 1965.
2. Y. Chen, Z. Huang, and N. Zhou, An advanced framework for improving situational awareness in electric power grid operation, in: *Proceedings of the 18th World Congress of the International Federation of Automatic Control (IFAC)*, Milano, Italy, August 28–September 2, 2011.
3. Z. Huang and J. Nieplocha, Transforming power grid operations via high-performance computing, in: *Proceedings of PES-GM2008—The IEEE Power and Energy Society General Meeting 2008*, Pittsburgh, PA, July 20–24, 2008.
4. U.S.–Canada Power System Outage Task Force, Final report on the August 14, 2003 blackout in the United States and Canada: Causes and recommendations, April 2004, available at https://reports.energy.gov/
5. North American Synchrophasor Initiative (NASPI), available at https://www.naspi.org/
6. North American Synchrophasor Initiative (NASPI), Current PMU locations—Map of networked PMUs, available at https://www.naspi.org/site/Module/Resource/Resource.aspx
7. Netezza Performance Server 10200 Evaluation for the Pacific Northwest National Laboratory, Netezza Corporation, Marlborough, MA.
8. M. R. Hestenes and E. L. Stiefel, Methods of conjugate gradient for solving linear systems, *Journal of Research of the National Bureau of Standards*, 49, 409–436, 1952.
9. J. Nieplocha, A. Marquez, V. Tipparaju, D. Chavarría-Miranda, R. Guttromson, and Z. Huang, Towards efficient power system state estimators on shared memory computers, in: *Proceedings of PES-GM2006—The IEEE Power Engineering Society General Meeting 2006*, Montreal, Canada, June 18–22, 2006.
10. High Performance Preconditioner (hypre) Library, http://acts.nersc.gov/hypre/, Lawrence Livermore National Laboratory.
11. D. Hysom and A. Pothen, Efficient parallel computation of ILU(k) preconditioners, SC99, published on CDROOM, ISBN 1-58113-091-0, ACM Order #415990, IEEE Computer Society Press Order #RS00197.
12. D. Hysom and A. Pothen, A scalable parallel algorithm for incomplete factor preconditioning, *SIAM Journal on Scientific Computing*, 22(6), 2194–2215, 2001.
13. NERC transmission system standards—Normal and emergency conditions, North American Electricity Reliability Corporation, available at www.nerc.com
14. Z. Huang, Y. Chen, and J. Nieplocha, Massive contingency analysis with high performance computing, in: *Proceedings of PES-GM2009—The IEEE Power and Energy Society General Meeting 2009*, Calgary, Canada, July 26–30, 2009.
15. Y. Chen, Z. Huang, and D. Chavarria, Performance evaluation of counter-based dynamic load balancing schemes for massive contingency analysis with different computing environments, in: *Proceedings of the IEEE Power and Energy Society General Meeting 2010*, Minneapolis, MN, July 25–29, 2010.

16. Z. Huang, K. Schneider, and J. Nieplocha, Feasibility studies of applying Kalman Filter techniques to power system dynamic state estimation, in: *Proceedings of the 8th International Power Engineering Conference*, Singapore, 3–6 December 2007.
17. Y. Li, Z. Huang, N. Zhou, B. Lee, R. Diao, and P. Du, Application of ensemble Kalman Filter in power system state tracking and sensitivity, in: *Proceedings of the 2012 IEEE Power and Energy Society Transmission and Distribution Conference and Exposition*, Orlando, FL, May 7–10, 2012.
18. G. Welch and G. Bishop, An introduction to the Kalman Filter, TR 95-041, Department of Computer Science, University of North Carolina at Chapel Hill, April 2004.
19. Singular perturbations, coherency and aggregation of dynamic systems, General Electric Company (GE) final report, 1981.
20. J. Nieplocha, R. J. Harrison, and R. J. Littlefield, Global arrays: A nonuniform memory access programming model for high-performance computers, *The Journal of Supercomputing*, 10, 197–220, 1996.
21. Y. Zhang, V. Tipparaju, J. Nieplocha, and S. Hariri, Parallelization of the NAS conjugate gradient benchmark using the global arrays shared memory programming model, in: *Proceedings of the IPDPS'05*, Washington, DC, 2005.
22. Western interconnection 2006 congestion assessment study, Prepared by the Western Congestion Analysis Task Force, May 08, 2006.

3.14 Cybersecurity

1. http://whatis.techtarget.com/definition/cybersecurity.html

3.15 Smart Grid Standardization Work

1. 2009 ANSI Essential Requirements, Section 3.1 [Online]. Available at www.ansi.org/essentialrequirements/
2. Purcell, The consequences of silence [Online]. Available at http://www.iec.ch/online_news/etech/arch_2009/etech_0309/industry_1.htm?mlref=etech
3. NIST, Report to NIST on the smart grid interoperability standards roadmap [Online], available at http://www.nist.gov/smartgrid/InterimSmartGridRoadmapNISTRestructure.pdf
4. http://en.wikipedia.org/wiki/osi
5. http://www.grid-interop.com/2008/
6. 49 U.S.C., Subtitle VI, Part C, Chapter 329, §32902, 32904.
7. GridWise Architecture Council. GridWise Architecture Council Interoperability Checklist. Available online at http://www.gridwiseac.org/pdfs/gwac_decisionmakerchecklist.pdf
8. EPRI 1013802, Chapter 5.
9. http://www.smartgridnews.com/pdf/Smart_Grid_Scorecard.pdf

BIBLIOGRAPHY

A Systems View of the Modern Grid, Appendix A7: Optimizes Assets and Operates Efficiently, DOE/NETL. Modern Grid Team, v2.0, January, 2007, pgs A7-4.
AMI Security Profile 2.0, AMI-SEC, UCAIug, AMI-SEC TF, http://osgug.ucaiug.org/utilisec/amisec/default.aspx
Babnik, T., U. Gabrijel, B. Mahkovec, M. Perko, and G. Sitar. Wide area measurement system in action, *Power Tech 2007*, Lausanne, Switzerland, pp. 232–237 [COBISS.SI-ID 28917509].
Breulmann, H., E. Grebe, M. Lösing et al. Analysis and damping of inter-area oscillations in the UCTE/CENTREL power system, *CIGRE Session 2000*, Paris, France, 2000.
Catalog of control systems security: Recommendations for standards developers, US-CERT: Control Systems, United States Department of Homeland Security, US-CERT, Web. http://www.us-cert.gov/control_systems/
Čerina, Z., I. Šturlić, R. Matica, and V. Skendžić, Synchrophasor applications in the Croatian power system, *Western Protective Relay Conference 2009*, Spokane, WA, October 20–22, 2009.
Chow, J., Power System Toolbox, Version 2.0, Load Flow Tutorial and Functions, Cherry Tree Scientific Software, 2000.
Common cyber security vulnerabilities observed in control system assessments by the INL NSTB program, Idaho National Laboratory—National SCADA Test Bed Program, Idaho National Laboratories, Idaho Falls, ID, Web. http://www.inl.gov/scada/index.shtml
Common weakness enumeration, The MITRE Corporation, McLean, VA, Web. http://cwe.mitre.org/

Cyber security procurement language for control systems, US-CERT: Control Systems, United States Department of Homeland Security, US-CERT, Web. http://www.us-cert.gov/control_systems/

Dagle, J., North American synchrophasor initiative—An update of progress, *Proceedings of the 42nd Hawaii International Conference on System Sciences (HICSS)*, Waikoloa, HI, January 5–8, 2009.

Decker, I.C., D. Dotta, M.N. Agostini, S.L. Zimath, and A.S. Silva, Performance of a synchronized phasor measurements system in the Brazilian power system, *IEEE Power Engineering Society General Meeting*, Atlanta, GA, 2006.

EnerNex Corporation, Smart grid scorecard, January, 2008 [Online], available at http://www.smartgridnews.com/pdf/Smart_Grid_Scorecard.pdf

EPRI Technical Report, Integration of advanced automation and enterprise information infrastructures: Harmonization of IEC 61850 and IEC 61970/61968 Models, EPRI, Palo Alto, CA 2006, Product ID 1013802.

Giri, J., G. Castelli, and M. Parashar, Visualization & wide-area situational awareness of the power grid, Invited paper, *PACWORLD Conference Proceedings*, Dublin, Ireland, June 21–24, 2010.

Grünbaum, R., M. Noroozian, and B. Thorvaldsson, FACTS—Powerful systems for flexible power transmission, *ABB Review* 5/1999, 4–17.

Hauer, J.F., Preliminary examination of the Alberta trip on August 4, 2000, *Working Note for the WECC Modeling & Validation Work Group*, June 20, 2002.

Hauer, J.F., Realistic testing of procedures to identify western system dynamics from wide area measurements, *WAMS Working Note for the DOE EPSCoR Program*, draft of August 24, 2004.

Hauer, J.F., C.J. Demeure, and L.L. Scharf, Initial results in prony analysis of power system response signals, *IEEE Transactions on Power Systems*, 5(1), 80–89, February 1990.

Hauer, J.F., H. Lee, J. Burns, and R. Baker, Preliminary analysis of western system oscillation event on June 4, 2003: BPA & Canada, *Working Note for the WECC Disturbance Monitoring Work Group*, July 7, 2003.

Huang, Z., D. Kosterev, R. Guttromson, and T. Nguyen, Model validation with hybrid dynamic simulation, *Proceedings of the IEEE Power Engineering Society General Meeting 2006*, Montreal, Quebec, Canada, June 18–22, 2006.

ICOEUR—Intelligent coordination of operation and emergency control of EU and Russian power grids, http://icoeur.eu/

IntelliGrid methodology for developing requirements for energy systems, IEC/PAS 62559, January 2008.

ISA99, International Society of Automation, http://isa99.isa.org

Jackson, J., The utility smart grid business case: Problems, pitfalls and ten real-world recommendations, white paper from Smart Grid Research Consortium (www.smartgridresearchconsortium.org) August 3, 2011.

Journalist's Guide to the Federal Courts Glossary [Online], available at http://www.uscourts.gov/journalistguide/glossary.html

Kosterev, D.N., C.W. Taylor, and W.A. Mittelstadt, Model validation for the August 10, 1996 WSCC system outage, *IEEE Transactions on Power Systems*, 14(3), 967–979, August 1999.

Kundur, P., *Power System Stability and Control*, McGraw-Hill, Inc., New York, 1994.

Miller, J., What is smart grid? *Smart Grid News*, April 17, 2009.

Moraes, R.M., H.A.R. Volskis, and Y. Hu, Deploying a large-scale PMU system for the Brazilian interconnected power system, *IEEE Third International Conference on Electric Utility Deregulation and Restructuring and Power Technologies*, Nanjing, China, April 2008.

Moraes, R.M., H.A.R. Volskis, Y. Hu, K. Martin, A.G. Phadke, V. Centeno, and G. Stenbakken, PMU performance certification test process for WAMPAC systems, *CIGRÉ SC-B5 Annual Meeting & Colloquium*, Jeju, Korea, October 2009.

NERC critical infrastructure protection, NERC reliability standards, North American Electric Reliability Corporation, Web. http://www.nerc.com/

NIST, Report to NIST on the smart grid interoperability standards roadmap [Online], available at http://www.nist.gov/smartgrid/InterimSmartGridRoadmapNISTRestructure.pdf

NIST IR 7628, NIST Publications, National Institute of Standards and Technology, http://csrc.nist.gov/publications/PubsNISTIRs.html#NIST-IR-7628

NIST framework and roadmap for smart grid interoperability standards, release 1.0, NIST Publications, National Institute of Standards and Technology, http://www.nist.gov/public_affairs/releases/upload/smartgrid_interoperability_final.pdf

North American Electric Reliability Corporation (NERC), Standard TOP-007-0—Reporting SOL and IROL Violations, April 1, 2005.

North American Electric Reliability Corporation (NERC), Standard TOP-008-0—Response to Transmission Limit Violations, April 1, 2005.

North American Electric Reliability Corporation (NERC), Standard TOP-008-1—Response to Transmission Limit Violations, January 1, 2007.

North American Electric Reliability Corporation (NERC), Standard TOP-004-1—Transmission Operations, October 1, 2007.

North American Electric Reliability Corporation (NERC), Standard IRO-006-4—Reliability Coordination—Transmission Loading Relief, October 23, 2007.

North American Electric Reliability Corporation (NERC), Standard TOP-004-3—Real-time Transmission Operations (Draft), March 26, 2009.

Nuqui, R., *Electric Power Grid Monitoring with Synchronized Phasor Measurements*, VDM Verlag, Saarbrücken, Germany, 2009.

Pai, M.A., D.P. Sen Gupta, and K.R. Padiyar, *Small Signal Analysis of Power Systems*, Narosa Publishing House, New Delhi, India, 2004.

Pan, J., R. Nuqui, B. Berggren, S. Thorburn, and B. Jacobson, The balance of power—Advanced transmission grids are embedding HVDC Light®, *ABB Review* 3/2009, 28–32.

Penny McLean-Conner, *Customer Service Utility Style*, Pennwell Corp., Tulsa, OK, 2006.

Phadke, A.G., Synchronized phasor measurements in power systems, *Computer Applications in Power*, April 1993.

Phadke, A.G. and J. Thorp, *Computer Relaying for Power Systems*, Research Studies Press Ltd., Hertfordshire, U.K., 1988.

Phadke, A.G. and J. Thorp, *Synchronized Phasor Measurements and Their Applications (Power Electronics and Power Systems)*, Springer, New York, 2008.

Philipson, L. and H. L. Willis, *Understanding Electric Utilities and De-Regulation*, 2nd edn., CRC Press, Boca Raton, FL, 2006.

Portland General Electric website http://www.portlandgeneral.com

Quanta Technology LLC, Phasor Gateway Technical Specification for North American Synchro-Phasor Initiative Network (NASPInet), May 29, 2009, http://www.naspi.org/

Reinhardt, P., C. Carnal, and W. Sattinger, Reconnecting Europe, *Power Engineering International*, 23–25, January 2005.

Sattinger, W., WAMS initiatives in continental Europe, *IEEE Power & Energy Magazine*, 58–59, September/October 2008.

Sattinger, W., Awareness system based on synchronized phasor measurements, *IEEE General Meeting*, Calgary, Alberta, Canada, 2009.

Sattinger, W., R. Baumann, and Ph. Rothermann, A new dimension in grid monitoring, in *Transmission & Distribution World*, 2, 54–60, 2007.

Sattinger, W., P. Reinhard, and J. Bertsch, Operational experience with wide area monitoring systems, *CIGRE 2006 Session*, B5-216, Paris, France, 2006.

Toward a smarter grid, www.abb.com

Trotignon, M., F. Counan, J.F. Lesigne et al., Defence plan against major disturbances on the French EHV power system: Present realisation and prospects of evolution, *CIGRE*, Paper 39-306, Paris, France, 1992.

Trudnowski, D., J. Pierre, N. Zhou, J. Hauer, and M. Parashar, Performance of three mode-meter block-processing algorithms for automated dynamic stability assessment, *IEEE Transactions on Power Systems*, 23(2), 680–690, May 2008.

Turner, S., Simple techniques for fault location, *2002 Georgia Tech Protective Relaying Conference*, Atlanta, GA.

Western Electricity Coordinating Council (WECC), WECC Standard IRO-STD-006-0—Qualified Path Unscheduled Flow Relief, March, 2007.

Western Electricity Coordinating Council (WECC), (December 2006) E-Tags, Accessed November 15, 2009. Available: http://www.wecc.biz/committees/StandingCommittees/OC/ISAS/Shared%20Documents/Forms/AllItems.aspx

Western Electricity Coordinating Council (WECC), WECC Standard TOP-007-WECC-1—System Operating Limits, March 12, 2008.

Willis, H. L., *Power Distribution Planning Reference Book*, 2nd edn., CRC Press, Boca Raton, FL, 2004.

Wireless procurement language in support of advanced metering infrastructure security, Idaho National Laboratory—National SCADA Test Bed Program, Idaho National Laboratories, Idaho Falls, ID, Web. http://www.inl.gov/scada/index.shtml

Zhou, N., J. Pierre, D. Trudnowski, and R. Guttromson, Robust RLS methods for on-line estimation of power system electromechanical modes, *IEEE Transactions on Power Systems*, 22(3), 1240–1249, August 2007.

Zuo, J., R. Carroll, P. Trachian, J. Dong, S. Affare, B. Rogers, L. Beard, and Y. Liu, Development of TVA SuperPDC: Phasor applications, tools, and event replay, *2008 IEEE Power and Energy Society General Meeting—Conversion and Delivery of Electrical Energy in the 21st Century*, pp. 1–8, Pittsburgh, PA, July 20–24, 2008.

4 Smart Grid Barriers and Critical Success Factors

Stuart Borlase, Steven Bossart, Keith Dodrill, Erich Gunther, Gerald T. Heydt, Miriam Horn, Mladen Kezunovic, Joe Miller, Marita Mirzatuny, Mica Odom, Steve Pullins, Bruce A. Renz, and David M. Velazquez

CONTENTS

4.1 Utility Organizational and Business Process Transformation 500
4.2 Convergence of Operations Technology and Information Technology 502
4.3 Integrated System Approach 503
4.4 Cybersecurity 504
4.5 Data Privacy 505
    4.5.1 Grid Security in the United States 506
4.6 Benefits Realization 506
4.7 Performance Goals and Progress Metrics 511
4.8 Technology Investment and Innovation 511
4.9 Consumer Engagement and Empowerment 512
4.10 Vendor Partnerships 514
4.11 Standards Development, Coordination, and Acceleration 514
4.12 Policy and Regulation 517
    4.12.1 Examples of Policy Measures 521
        4.12.1.1 Climate and Energy Package 20-20-20 (EU) 521
        4.12.1.2 EISA, ACES (United States) 521
        4.12.1.3 Regulatory Review (United Kingdom) 521
4.13 Industry Expertise and Skills 522
    4.13.1 Indicators of Declining Workforce Supply in the United States 523
4.14 Knowledge and Future Education 524
    4.14.1 Forms and Goals of Future Learning 527
    4.14.2 Example of an Initiative to Build Knowledge: U.K. Power Academy 528
References 529

Our electricity infrastructure is on the cusp of game-changing investments: embedding sensors, communications, and controls—a brain and nervous system—throughout the entire system, from power plant all the way through to consumer devices. Those investments have the potential to unleash entrepreneurial innovation and open up the biggest of all global industries—and the single largest source of global warming pollution on the planet—to a vast new diversity of clean energy resources, including thousands we have not yet dreamed of. This smart grid—at its core—will be an enabler: facilitating deployment at scale of clean, low-carbon energy and transportation, and reducing our fossil fuel burden. By making demand flexible and responsive to available supply and incorporating electricity storage and grid awareness technologies, it will enable high penetration of variable renewable generation and plug-in electric vehicles without compromising grid stability. It will transform the historically passive electricity consumer to an aware and active participant in electricity markets, able to produce and sell electricity and ancillary services. It will deliver accurate, real-time price signals, influencing the load curve to realize a far more efficient system overall. In short, the smart grid will facilitate a new paradigm: structured around renewable and distributed resources, integrated with the transport sector through mass deployment of clean electric vehicles, and inclusive of many more suppliers of energy and services—adding up to a system that is vastly more efficient, flexible, reliable, cost-effective, and clean [1].

Adherence to a set of core principles will maximize the return on the enormous investments countries around the world will make over the next two decades in electric infrastructure. The fundamental question that each market will face is how to provide incentives for electricity companies to invest in and implement the right level of smart technology. Electricity companies, in this case, should be viewed in the broadest sense. They include both traditional utility network companies that will be responsible for the provision of the underlying electricity network infrastructure and a wide range of non-utility companies providing diverse technologies, solutions, applications, and services to deliver the full value from smart grid deployment (e.g., communications companies behind home-area networks, companies providing microgeneration and devices to support advanced end-user services, electric vehicle and battery manufacturers, and companies that will provide the associated e-vehicle charging and billing infrastructure). In market terms, a smart grid supports a whole new range of product offerings, services, and opportunities that create value for users, electricity companies, and the host governments.

Although smart grids provide an essential supporting infrastructure for energy efficiency and environmental measures (e.g., intermittent renewable generation), by themselves they create benefits outside of network operational efficiencies. Wider societal value scales up (e.g., through real-time consumer propositions and carbon reduction) only when all of these electricity companies interact to provide the range of commercial services that a smart grid supports. To be effective and efficient, any market stimulation must be designed to overcome barriers to the development of the supporting smart grid infrastructure and commercially viable associated products, offerings, and services. Different countries have different drivers for, and expectations of, a smart grid, and contain one or many different market and regulatory structures that will need to support their development.

Smart grid implementation brings many opportunities as well as extraordinary "change management" challenges. In fact, "challenge management" may be the more operative term. Smart grid implementation is complex and the impact both on the utility organization and electricity consumers is potentially formidable. This fact contrasts with the countless articles that have focused on the many purely technological challenges to which utilities must rise. One could suggest that even more critical may be the challenge of managing change along several dimensions and integrating multiple components into a seamless, real-time system.

Barriers and success for smart grid involves several factors, not only just from the utility standpoint, but also from the consumer, regulatory, and utility industry perspective. Another important aspect is the need for the education and skills to support the evolution of smart grid. A successful smart grid implementation is a multifunctional effort:

- Regulatory matters need to be addressed, including designing dynamic rates and obtaining key approvals to proceed with a reasonable cost recovery mechanism.
- Adequate funds need to be raised in the capital market.
- Vendor partners need to be selected.
- An extensive construction effort needs to be managed.
- Systems need to be implemented and integrated.
- Processes need to be redesigned.
- A robust customer communications campaign needs to be conducted.
- New customer-facing services need to be developed.

While these tasks appear to be functionally specific, maximum efficiency and efficacy can only be achieved if they are managed across organizational boundaries. This includes the most critical task of all: the realization of customer benefits.

The stakes are high when dealing with a service as vital to consumers, business, and industry as is electricity and are that much more so when one considers how stimulus funding and other factors have raised public awareness of the smart grid. Simply put, there is a need to "upgrade the airplane while it is in the air"—and do so seamlessly. With that in mind, consider the inherent challenges in the "to do" list provided earlier:

- Business requirements will need to be flexible enough to accommodate changes while being specific enough to enable construction.
- Vendor selection process will be complicated by a lack of extensive track records.
- While system implementation will be challenging enough, interoperability between these cutting edge systems and the utility's legacy systems (including the billing engine) must be achieved.
- Rate structures must be developed that properly and adequately balance customer risk and reward.
- The technology to be deployed, in the case of advanced metering infrastructure (AMI), touches every customer.
- Some of the benefits to be realized are under the control of the utility while others are dependent on changes in customer behavior.

As one utility Vice President of Business Transformation puts it, "If you look at this as just reading meters remotely, you've missed the boat." An important corollary to this is, "If the Smart Grid is being viewed as belonging to any one department within your utility, your effort is on the road to failure."

There are many reasons the smart grid is not emerging more quickly. Fundamentally, no single business owns or operates the grid. Individual players have little incentive to risk major change. With so many players in the grid system, finding a common interest in or vision for change is difficult but imperative. The benefits are so broad and far reaching that perhaps only government can account for the cumulative societal value. Longer-term financial incentives are needed to enable the larger infrastructure investments needed for grid modernization. Many barriers exist today and more will arise before the vision is realized. These challenges are daunting, but they can be overcome. With a clear vision, we can generate the alignment needed to overcome the barriers as well as those yet to be identified. Some options include the following:

1. Regulatory and legislative barriers—change statutes, policy, and regulations to eliminate those that inhibit progress and create those that encourage progress and create a "win-win" scenario for all stakeholders. For example,
    a. Capture and include the full set of societal benefits when addressing the costs of grid modernization.
    b. Provide regulatory framework to present rate cases for recovery of smart grid investments.
    c. Establish grid modernization goals, metrics, and coordination mechanisms to better manage the transition to the smart grid.
    d. Provide enhanced returns on smart grid investments.
    e. Modify the model that links utility profit to sales volumes.
2. Culture and communication barriers—increase the understanding and awareness of stakeholders on the value of the smart grid and encourage them to embrace the needed changes within their organizational cultures.

3. Industrial barriers—define the case for change, the "burning platform," and provide the necessary incentives to engage industry on grid modernization. Industry will respond when it understands there is a profitable market for grid modernization technologies and services.
4. Technical barriers—increase the speed of research, development, and deployment.
5. Increase funding to support research, development, and deployment for those technologies that are needed for grid modernization.
6. Work more closely with academia to develop the human resources with skills needed for the modernized grid.
7. Apply more priority and resources to the development of needed standards.
8. Clarify the pathway to the smart grid by developing a transition plan that shows the intermediate milestones for achieving its vision.

4.1 UTILITY ORGANIZATIONAL AND BUSINESS PROCESS TRANSFORMATION

A broad consensus for the smart grid vision throughout the power industry is gaining momentum but has not yet been institutionalized. A greater understanding of the advantages and benefits of the smart grid is needed. Moreover, once consensus is reached, a transition plan defining the pathway for transforming today's grid into the smart grid is needed.

Electric utility executives do not see a burning platform that would motivate them to change. Most say that their customers are happy, their reliability is good, and their customers want lower rates not higher ones. They are hesitant to make major investments in their systems. In fact, the financial markets are driving them to minimize investments and there is no force on the horizon to make them do otherwise. However, the consequences of "doing nothing" should be considered:

- Increasing number of major blackouts
- More local interruptions and power quality events
- Continued vulnerability to attack
- Less efficient wholesale markets
- Higher electricity prices
- Limited customer choice
- Rising product prices
- Greater environmental impact

More cooperation and the free exchange of information among the approximately 3000 diverse utilities is needed to successfully achieve the smart grid vision. Some industry observers believe that, as a result of deregulation, the industry's corporate culture has moved from cooperation and coordination to competition and confrontation. Relationships among utilities need to shift to a more collaborative model to foster sharing metrics that capture best practices and lessons learned in order to reduce time and investment in the smart grid transition.

Industry executives are reluctant to change processes and technologies. Some utility cultures are resistant to change and operate in "silos" organizationally. As a result, processes and technologies that are based on long standing practices and policies are difficult to change. Additionally, senior managers today may be more focused on marketing and legal issues, rather than the technical aspects of power systems. The result may be an over reliance on markets to address grid modernization issues rather than proactive investment in new processes and technologies. Integration of change management techniques into utility organizations might stimulate change in their culture.

Industry technical staffs are reluctant to change planning and design traditions and standards. Utility planning and design traditions and standards generally focus on the traditional model of the electric grid—centralized generation, legacy technologies, and little reliance on the consumer as an active resource. Smart grid principles have generally not yet been incorporated into technical policies and standards that limit the deployment of new processes and technologies that exist today. A significant change management effort is needed to encourage technical staffs to modify their current approach. Resources at many utilities (both human and financial) are limited and stressed. The amount of resources available to look beyond day-to-day operations is limited. While it may seem that slow progress is being made in the area of grid modernization from the project deployment perspective, there has been significant progress in aligning utilities around the core smart grid concepts that will ultimately build strategic plans. Early adopters are forging the way for followers that will benefit from an easier logical transition to modernize their grids.

None of the aforementioned can be done without multiple perspectives at the table, working with a common definition of success and common guiding principles, and commitment to collaborate to achieve the best outcome for the organization as a whole. Since smart grid is a company-wide challenge, not a technology deployment, there will necessarily be some new organizational components to consider. These could include, if they are not in place already, some notions that are relatively new to the utility industry, such as a senior business transformation executive, an enterprise architecture function, a design authority to which technology issues and opportunities are directed, a company-wide smart grid steering committee to ensure alignment across all of the activities described earlier, and a commitment to a change management discipline and process. Among other things, this change management process should include a standard approach for measuring performance and providing feedback across the stakeholder community. Openly sharing successes and unsuccessful efforts is at odds with the current utility culture. However, doing so would ultimately break down many of the barriers that would cause untimely starts and stops and potentially reduce the overall investment by eliminating rework (Figure 4.1).

Take a moment to reflect back on the tasks and challenges noted at the outset.

- Business requirements that are both flexible and specific
- Vendor selection under uncertainty
- System interoperability, both new and legacy
- Innovative rates that are effective and acceptable
- High profile technology deployment that needs to be as transparent as possible
- Behavioral-driven benefits

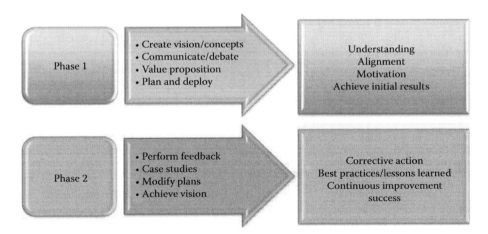

FIGURE 4.1 Components of managing change. (From Sharing smart grid experiences through performance feedback, National Energy Technology Laboratory, Morgantown, WV, http://www.netl.doe.gov/smartgrid/docs/PFP%20for%20the%20Smart%20Grid_Final_v1.0_031511.pdf)

These are not tasks and challenges that are purely technical in nature and these are not tasks and challenges that can be wrestled to the ground by any one group, or by a series of groups working independently. This effort requires subject matter expertise, certainly, but more importantly, it requires a cohesive application of that expertise across organizational boundaries to achieve the full range of operational, informational, and behavioral benefits made possible by the smart grid.

4.2 CONVERGENCE OF OPERATIONS TECHNOLOGY AND INFORMATION TECHNOLOGY

The smart grid changes many things for the customer and countless others for those who serve the customer. A relatively easy way to think about this issue is in its primary component parts: operational efficiencies and informational efficiencies.

A smart grid business case is driven in no small part by operational efficiencies, such as (1) cost reductions through automating or eliminating manual tasks and (2) improving outage duration metrics through the deployment of greater computing power in the field. As critical as it will be to apply sound program and project management skills to ensure that these multiple, interdependent efforts are aligned and effective, it will be just as important to leverage the informational capabilities of the technology being deployed.

While the concept of operational efficiencies is relatively obvious, the notion of informational efficiencies is not as apparent. Smart grid devices, from AMI technology to distribution automation components, are essentially various forms of grid sensors that will generate an enormous amount of data. The utilities that can develop the analytical infrastructure necessary to transform these data into information will be able to better manage their assets—which will translate into meaningful process improvements such as better repair/replace decisions or highly targeted preventive maintenance programs. As utilities further mine these data, optimization of distribution network performance based on near real-time (as opposed to historical) information becomes possible.

As with other disruptive technologies—such as those that revolutionized correspondence, the recording industry, and publishing, to name just three—the potential benefits are high for those that properly "manage the challenge."

The convergence of operations technology and information technology (IT) is not just happening within technology hardware and software, but also within the company's functional organizations. These two groups and sets of activities have been converging for some time, but smart grid greatly accelerates that convergence and forces some organizational decisions. For a utility to be successful, it would not be enough for IT to simply manage the back office integration of business systems (the typical purview of most such groups). The technology being deployed in the field via smart grid in many ways bears a greater resemblance to the technology that IT groups have been supporting than it does to what operations technology groups have been supporting.

The most successful utilities, if they have not done so already, will find a way to integrate the best of both. For example,

- Adopting a smart grid patch management process that leverages a tried and true IT process for devices, only "in the field" rather than a data center
- Leveraging the capabilities of those responsible for the corporation's data network and bringing that skill base to bear on smart grid communications infrastructure challenges
- Building a network monitoring process that establishes common visibility to all mission critical systems and networks, be they in the data center, system operations, a substation, a remote facility, or any other grid-attached location

Investments in security upgrades are difficult to justify. A standard approach is lacking for conducting security assessments, understanding consequences, and valuing security upgrades. Additionally,

Smart Grid Barriers and Critical Success Factors

limited access to government-held threat information makes the case for security investments even more difficult to justify. When examined independently, the costs and benefits of security investments can seem unjustifiable. It is difficult to place a value on preventing a cyber or physical attack through implementation of security measures.

4.3 INTEGRATED SYSTEM APPROACH

The smart grid approach to the design and operation of generation, transmission, and distribution systems requires an integrative strategy. That is, several technologies must be brought to bear on the design and operation philosophy as shown in Figure 4.2.

Demonstration and pilot projects are essential for the smart grid transition, so that the whole chain of technologies can be tested together. This allows weaker areas to be identified and refined, and deployment and commercial models to be tested. Although many pilot and demonstration projects exist globally, there are limitations on their effectiveness. Few are at a scale large enough to provide a thorough understanding of how they will operate in full-scale deployment or to make them economically and functionally viable. Limited, although not coordinated, learning and knowledge sharing is emerging from these projects.

Integration of smart grid solutions requires a versatile and flexible communication infrastructure. Communications requirements of the smart grid are much more demanding than of the legacy grid. The real-time requirements for exchange of data and information require low

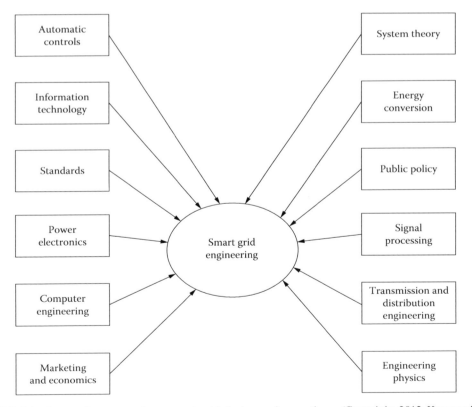

FIGURE 4.2 Integrative approach to smart grid design and operations. (Copyright 2012 Kezunovic, M. All rights reserved.)

latency and redundancy in communication paths. The back office data processing and storage requires communication support for distributed databases and processing facilities. The communication infrastructure also has to enable the system integrity protection schemes critical to reliable system operation and control. The increased interest and emphasis on communications design arises from

- Need for security in all systems and subsystems
- Conformity of protocols to accelerate the use of components from diverse vendors
- Making a set of sensory signals available for operation and control of all systems and subsystems

A smart grid can reduce the need for new central station power plants and transmission lines, but only if the planning processes are linked. As the California Public Utilities Commission (PUC) found, "the Smart Grid can decrease the need for other infrastructure investments and these benefits should be taken into account when planning infrastructure." Conventional supply-side solutions should become just one among many options considered for meeting demand growth, with costs and benefits of competing strategies comprehensively assessed. Utilities and regulators will need, for example, to consider whether wider reliance on demand response (DR) will offset peak demand at a lower financial and environmental cost than a new peaking plant and transmission to serve it. More broadly, they will need to determine which path best meets their constituent's goals: which, for instance, will ensure the lowest-cost achievement of state and federal energy and environmental policies and which foundational investments will best support future innovation, multiple functionalities, and economies of scale. Integrated solutions will inevitably deliver far more than piecemeal investments: AMI combined with smart appliances and home energy management systems, for instance, will deliver far more than AMI alone.

Grid operators must be able to reliably integrate programs for energy efficiency and DR along with distributed energy resources, some of which are inherently more intermittent, and various energy storage options as confidently as they rely on traditional centralized resources. Having the necessary two-way communications and protocols to see and value these resources—including smaller scale (200 kW–1 MW) resources—is key to enabling the sale of demand-side resources into wholesale energy markets on equal footing with traditional generation resources [3]. It will also allow regulators to consider these resources when modeling options to address the long-term demands on the system, including avoiding congestion, meeting environmental targets, and integrating new sources of generation and consumption.

4.4 CYBERSECURITY

Various cybersecurity intrusion studies have demonstrated the vulnerability of communication, automation, and control systems to unauthorized access. Many real-world cases of intrusion into critical infrastructures have occurred, including illegal access into electric power systems for transmission, distribution, and generation, as well as systems for water, oil and gas, chemicals, paper, and agricultural businesses. Confirmed damage from cyber intrusions include intentionally opened breaker switches and the shutdown of industrial facilities. Very few of the incidents have been publicly reported, and initiatives aimed at creating an open repository of industrial security incidents encounter resistance. Threats come from hackers, employees, insiders, contractors, competitors, traders, foreign governments, organized crime, and extremist groups. These potential attackers have a wide range of capabilities, resources, organizational support, and motives.

The possible vulnerability of the utility's system, business and customer operations, and consumer premises represent serious security risks; therefore, security must be approached and managed with an extreme level of care. Apart from active, malicious threats, accidental cyber threats are increasing as the complexities of modern data and control systems increase. Security risks are growing in diverse areas, including the following:

- Risk of accidental, unauthorized logical access to system components and devices and the associated risk of accidental operation
- Risk of individual component failure (including software and networks)
- Number of failure modes, both directly due to the increased number of components and indirectly due to increased (and often unknown) interdependencies among components, devices, and equipment
- Risk of accidentally misconfiguring components
- Failure to implement appropriate maintenance activities (e.g., patch management, system housekeeping)

Worldwide, initial security gaps have been highlighted by security companies and were discovered within pilot projects, which are not designed to resist sustained cyber attack. While such systems are now broadly secure against elementary hacking techniques, situations where an insider, who knows the system, can exploit the vulnerabilities are of particular concern to smart grid technology stakeholders. All parties involved in managing network operations centers or the relevant IT systems have to be trained and alert to tamper from the inside. Specially trained security officers need to be implemented in all potentially vulnerable areas.

Open communication and operating systems may be vulnerable to security issues. Although open systems are more flexible and improve system performance, they are not as secure as proprietary systems. The increasing use of open systems must be met with industry approved and adopted standards and protocols that ensure system security.

A utility needs to define its own selection of security controls for system automation, control systems, and smart devices, based on normative sources and as appropriate for the utility's regulatory regime and assessment of business risks. The security controls need to be defined within each security domain, and the information flows between the domains need to be based on agreed risk assessments, established corporate security policies, and possible legal requirements imposed by the government. In addition, limitations related to the existing legacy systems must be accommodated in a manner that does not hamper organizational security. Emerging smart grid systems and solutions have to be thoroughly tested by qualified laboratories to ensure that new digital communications and controls necessary for the smart power grid do not open up new opportunities for malicious attack. The responsibility for this security rests with all market participants—both industry and governments.

The idea of extending an internet protocol (IP)-based network to the meter level does open up the potential for both internal and external hacking. To protect against those threats, the structure of the system architecture should be considered carefully. By having a distributed intelligence in the grid we mitigate a single point of failure.

Every company thinking about providing equipment and services for smart grid technology enterprises should be cognizant of security and standards, with thought given to security certification for hardware and software providers.

4.5 DATA PRIVACY

The massive amount of potentially sensitive data collected in a smart grid, particularly with the implementation of consumer technologies, offerings, and services (e.g., advanced metering infrastructure [AMI] and demand side management [DSM]), inherently creates data privacy and security risks. Consumer involvement applications and solutions (e.g., AMI and DSM) put privacy interests at risk because information is collected on energy usage by a particular household or business.

Consumer-based smart grid technologies put privacy interests at risk because a core purpose is to collect information related to a particular household or business. With granularity to a minute, meters already collect a unique meter identifier, timestamp, usage data, and time synchronization every 15–60 min. Soon, they will also collect outage, voltage, phase, and frequency data, and detailed status and diagnostic information from networked sensors and smart appliances. Interpreted correctly, such data can convey precisely whether people were present in the home, when they were present, and what they were doing. Utilities implementing consumer technologies, offerings, and services within a smart grid environment that fails to address these issues will encounter consumer and political opposition, restricting their ability to realize the economic promise of smart grid technologies. They may face angry regulators and customers as well as liability issues.

In the consumer context, the right to privacy means the consumer's ability to set a boundary between permissible and impermissible uses of information about themselves. What is impermissible is a matter of culture, as expressed in law, markets, and what individuals freely accept without objection (i.e., consensus values). If customers believe a utility is misusing personally identifiable data or is generally enabling the use of personal information beyond what they deem acceptable (whether or not legal), then they are likely to resist the implementation of vital smart grid functionality related to consumer offerings and services. Consumers may refuse to consent (where required), hide their data, or awaken political opposition. Utilities may face customer liability claims or regulatory fines if inadequate privacy or security practices enable eavesdroppers, adversaries, or bad actors to acquire and use collected data to a customer's detriment. Utilities must take privacy and security concerns into account when designing consumer technologies, offerings, and services, and must persuade consumers, regulators, and politicians that privacy interests are adequately protected.

What constitutes permissible uses of personally identifiable information varies from culture to culture and over time; yet, what goes on inside a residence is generally an area of special privacy concern. The collected data reveal more about what goes on inside a residence than would otherwise be known to outsiders, and the collection and use of such data would reduce the scope of private information. Although privacy is generally considered a personal right, businesses typically have analogous rights.

Once a utility establishes the permissible uses of consumer data, it is in its best interest to assure that unauthorized uses do not occur. For example, if an electricity service provider is allowed to sell appliance-related data to a manufacturer or retailer, the utility will want to protect its economic interest by preventing access or use by others who might become competitive data brokers. Every utility will want to avoid regulatory sanctions for violating express or implied privacy policies, as well as damages claims based on compromised customer data or facilities.

Concerns about data privacy in smart gird environments and AMI, in particular, are now being widely discussed. In the Netherlands, for example, the formerly compulsory AMI roll out was subsequently made voluntary.

4.5.1 Grid Security in the United States

In August 2009, the U.S. President Barack Obama cited smart grid security as one reason for creating a new White House cybersecurity position. The announcement came after a string of press reports emerged about intrusions into power system security. These accounts included an anonymously sourced *Wall Street Journal* article claiming that foreign spies had infiltrated a grid system, and claims by cybersecurity firm IOActive that it had proven it could hack into smart meters, potentially cutting the power to millions of homes at once and causing the grid to fail.

4.6 BENEFITS REALIZATION

Business cases for investing in smart grid processes and technologies are often incomplete and therefore not compelling. It is often easier to demonstrate the value of the end point than it is to make a sound business case for the intermediate steps to get there. Societal benefits, often necessary to make investments in smart grid principles compelling, are normally not included in utility business cases. Additionally, lack of protection from inherent investment risks such as stranded investments further impacts the ability of these investments to pass financial hurdles. Meanwhile, the increased number of players and extent of new regulation has complicated decision making. Credit for societal benefits in terms of incentives and methods for reducing investment risks might stimulate the deployment of smart grid processes and technologies.

Smart grid cost–benefit analyses should take into consideration the full range of benefits of deployment, including the reduced use of high-polluting peak power plants; reduced land and wildlife impacts (through avoided construction of power plants and transmission lines); and the lowest-cost achievement of state and federal energy and environmental policies through efficiency and generation options made possible by smart grid investments [3,4].

Various components come into play when considering the impact of smart grid technologies. Utilities and customers can benefit in several ways. Rate increases are inevitable, but smart grids can offer the prospect of increased utility earnings, together with reduced rate increases (plus improved quality of service). Viewing smart grid programs in the context of, for example, a "green" program for customer choice, or a cost reduction program to moderate customer rate increases, can help define utility drivers and shape the smart grid roadmap. A smart grid program should have a robust business case where numerous groups in the utility have discussed and agreed upon the expected benefits and costs of smart grid candidate technologies and a realistic implementation plan. In some cases, the benefits are modestly incremental, but a smart grid plan should minimize the lag in realized benefits that typically occur after a step change in technology. A smart grid deployment is also intended to allow smoother and lower-cost migrations to new technologies and avoid the need to incur "forklift" costs. A good smart grid plan should move away from the "pilot" mentality and depend on wisely implemented field trials or "phased deployments" that provide the much-needed feedback of cost, benefit, and customer acceptance that can be used to update and verify the business case.

Some smart grid benefits are under the control of the utility while others are dependent on changes in customer behavior. This is a critical point. For most utilities:

- Operational benefits will be driven by activities such as the elimination of manual meter reading, the implementation of remote connect/disconnect capability, improvement in billing activities, reduction of off-cycle meter reading, optimization of assets, and reduction in maintenance costs, all of which require regulatory support for changes in business processes
- DR benefits will be driven in part by direct load control efforts, which in most jurisdictions will require customer adoption through voluntary enrollment
- DR benefits will be driven primarily by a form of dynamic pricing, which requires regulatory approval, establishment of a customer "opt in" or "opt out" approach, customer awareness, and changes in customer behavior
- Reliability benefits will depend on various factors, such as the expected efficiency of automated sectionalizing and reclosing (ASR) scheme operations, but as with all system performance measures, actual benefits will vary due to non-controllable factors, such as weather

The approach to meeting these challenges successfully will necessarily be transformational, evolutionary, and multidimensional. Overall, it is an effort with significant implications for how work gets done, how information is used, and how customers are served. Fundamentally, it is about moving beyond the traditional, "transactional" customer relationship and becoming a "trusted energy advisor."

At many utilities, the smart grid has been advanced first by the engineering group. Obviously, the role of engineering cannot be overstated. An ASR scheme, for example, is not something that can be bought shrink-wrapped. However, to help underscore the importance of cross-organizational teamwork in meeting these challenges effectively, consider the role that many other functions will need to play in achieving smart grid success.

While the typical non-price regulated entity seeks to earn a return on its investment through profit-maximizing pricing, product and marketing strategies, a price-regulated entity such as a power delivery utility does not have that level of autonomy. Benefits maximization rather than profit maximization is the key goal. A portion of these benefits are in the control of the utility—such as the reliability improvements gained through effective distribution automation implementation or the operational benefits gained through automated meter reading. The bulk of the potential benefits, however, are driven by changes in customer behavior, specifically their consumption levels and patterns.

To help drive that behavior, customer education is critical—as is the transition of a utility's customer care function from a transactional "call taker" to a trusted energy advisor. Done properly, greater utilization of these new technologies and innovative rate structures will be driven.

Few regulatory bodies, if any, are receptive at this time to a transition to mandatory dynamic pricing for all customers. Therefore, dynamic pricing will most likely be offered on a voluntary, "opt in" basis, and the most likely rate structure will be a critical peak rebate rather than critical peak pricing (i.e., a rate structure predicated on a "carrot" rather than a "carrot and the accompanying stick"). Both of these factors greatly increase the importance of marketing the potential benefits to the customer base. This necessarily implies the need for a growing understanding of what messages (and message delivery vehicles) are most effective in reaching different customer segments—something at which regulated utilities are typically not expert. Until now, they have never needed to be.

An example of an approach to benefits realization that recognizes the need for collaboration and education would be as follows:

1. Prioritize customer-facing smart grid benefits and work toward "early delivery"—while effectively managing stakeholder expectations.
2. Establish stakeholder-working groups that provide opportunity for detailed discussions about dynamic pricing programs and their benefits.
3. Conduct public regulatory hearings that assess and verify the cost and benefits of programs.
4. Provide greater availability of information to customers through improved website capabilities (and ensure customer care access to the same information to facilitate "energy advisor" conversations).
5. Launch proactive customer programs that provide a clear, simple message about the utility's offerings and programs to manage customer expectations. Ideally, these programs would be informed by market research that focuses on (a) increasing enrollment and retention in dynamic pricing programs, (b) improving behavioral responses to pricing options and usage information, and (c) ensuring that benefits flow to all customers.

One of the greatest obstacles in smart grid initiatives is approval from public utility commissions when a rate case is required by utilities to fund smart grid programs. The rates that regulated utilities are allowed to charge are based on the cost of service and an allowed return on equity (ROE). Once base rates are established, the rates remain fixed until the utility files for a rate change. Throw in an environment where power generation has been deregulated and the business case for a wires company still under regulation is more challenging. An additional challenge is presenting a rate case where the total system load decreases with DR and energy efficiency programs.

Utilities are looking for that magic "easy" button for smart grid deployments, but smart grid plans may be "subject to regulatory approval." Therefore, it is important to not only have a solid business case internally, but also a business proposition around the view of regulatory approval. The focus on the business case should also show regulators

1. How smart grid technology maintains low customer bills. Benefits may include
   a. Reduced O&M through lower meter related and outage costs
   b. Reduced cost of energy through DSM and IVVC
   c. Reduced capital expenditures through M&D (Monitoring and Diagnostics), DSM, and IVVC
2. What smart grid does to secure the "green image" of the state or service territory. Benefits may include
   a. Lower carbon emissions through reduced energy consumption and field force drive time via DSM, IVVC, AMI, and FDIR (fault detection, isolation, and restoration)
   b. Renewable energy source integration, facilitated by DSM and DER (distributed energy resources) to help with renewable energy intermittency
   c. Distributed generation and plug in hybrids facilitated by AMI and DA (distribution automation)
3. How the smart grid improves poor reliability. Benefits may include
   a. Significant SAIFI (system average interruption frequency index) and SAIDI (system average interruption duration index) improvement through AMI, FDIR, integrated OMS (outage management system), and FFA (field force automation)
   b. Improved power quality for an increasingly digital economy
   c. Ongoing M&D will further improve reliability
   d. Improved customer service through billing accuracy and reduced outages

Customer choice, energy efficiency, and customer value are seen as key to a successful smart grid implementation platform and the likely acceptance by regulators. The opportunities lie in leveraging the foundation of AMI to support a more comprehensive smart grid program. In response, utilities will be looking to regulators to provide incentives for smart grid programs, such as accelerated depreciation and higher returns for rate cases.

Bottom line for regulators and consumers: "Look for Smart Grid initiatives that are likely to reduce long term bills as well as emissions and outages."

The business case is intended as an overall guide for utilities in developing a long-term strategy. The business case relies on system estimates, sometimes educated guesses on benefits and costs, and compares the net present value of phased implementations of smart grid candidate technologies. Some parallels can be drawn between utility service areas and served customers, but not all utilities are at the same level or phase of technology implementation and therefore expected benefits and costs will vary. Industry standards, such as reliability indices, can be used as an overall comparison of the operations of a utility, but the comparison of smart grid implementations and expected benefits can vary widely across utilities with a similar number of customers. Not all business cases are the same—"your mileage may vary." While current smart grid initiatives are driven by regulatory pressure and tend to focus more on the meters as a direct impact on consumers, we are likely to see more technology-rich initiatives after well-proven smart grid evaluations ("staged deployments").

A key smart grid market barrier is business case fragmentation, particularly in competitive-leaning fragmented markets. A network business operating separately from generation and supply companies, with different companies operating in each part of the value chain, is an indicator of a fragmented market. In contrast, a concentrated market has one or two vertically integrated companies.

The importance of the business case will vary from country to country. In some centralized markets, the development of a smart grid may be a matter of policy, driven primarily by security of supply, environmental, or research and development (R&D) aspirations. In competitive-leaning markets, an economic business case may be more important, with clearly defined internal rate-of-return hurdles to jump.

Creating a complex business case for smart grid technologies is difficult: all networks within a market, and circuits within networks, will have different levels of capability required, all driven by interdependent supply and demand characteristics, making cost estimation difficult. Benefit estimation is similarly complex as benefits will depend on the levels of capability in different network areas and will comprise direct and indirect benefits that are difficult to quantify (e.g., carbon and pollution reduction, improvement in security of supply).

In a fragmented market, creating a commercial model means allocating investment, reward, and risk among the stakeholders. This allocation will be driven by the extent to which each party captures benefits and best manages different risks. However, the number of different entities involved makes the business case and commercial model particularly difficult. For example, a smart grid project benefits power generation companies through avoided capital expenditure required for generation, or support for the introduction of intermittent energy supplies (e.g., from wind). For networks, benefits include improved operational efficiency and reduced losses, and for retail, it can support the introduction of innovative offerings and help trim load curves. A networks-only investment into smart grid technologies will therefore support huge opportunities for other parties.

For a vertically integrated market, most of these benefits accrue to the lead incumbent company, and therefore the immediate benefits (excluding societal), investment, and risk are borne by the same party. This makes business cases more straightforward—assuming that societal benefits can be adequately captured.

The EPRI Electricity Sector Framework for the Future estimates $1.8 trillion in annual additive revenue by 2020 with a substantially more efficient and reliable grid [5]. To elaborate, according to the Galvin Electricity Initiative, "Smart Grid technologies would reduce power disturbance costs to the U.S. economy by $49 billion/year. Smart Grids would also reduce the need for massive infrastructure investments by between $46 billion and $117 billion over the next 20 years [6].

"Widespread deployment of technology that allows consumers to easily control their power consumption could add $5 billion to $7 billion per year back into the U.S. economy by 2015, and $15 billion to $20 billion per year by 2020." Assuming a 10% penetration, distributed generation technologies and smart, interactive storage capacity for residential and small commercial applications could add another $10 billion/year by 2020.

The efficient technologies can dramatically reduce total fuel consumption—and thereby potentially reduce fuel prices for all consumers. Virtually, the nation's entire economy depends on reliable energy. The availability of high-quality power could help determine the future of the U.S. economy. Additionally, a smart grid creates new markets as private industry develops energy-efficient and intelligent appliances, smart meters, new sensing and communications capabilities, and passenger vehicles [7].

Around the globe, countries are pursuing or considering pursuit of greenhouse gas legislation suggesting that public awareness of issues stemming from greenhouse gases has never before been at such a high level. According to the National Renewable Energy Laboratory (NREL), "utilities are pressured on many fronts to adopt business practices that respond to global environmental concerns. According to the FY 2008 Budget Request, NREL stipulates that, if we do nothing, U.S. carbon emissions are expected to rise from 1700 million tons of carbon per year today to 2300 (million tons of carbon) by the year 2030. In that same study, it was demonstrated that utilities, through implementation of energy efficiency programs and use of renewable energy sources, could not only displace that growth, but actually have the opportunity to reduce the carbon output to below 1000 (million tons of carbon) by 2030" [8].

Implementing smart grid technologies could reduce carbon emissions by

- Leveraging DR/load management to minimize the use of costly peaking generation, which typically uses energy resources that are comparatively fuel inefficient
- Facilitating increased energy efficiency through consumer education, programs leveraging usage information, and time-variable pricing

- Facilitating mitigation of renewable generation variability of output—mitigation of this variability is one of the chief obstacles to integration of large percent of renewable energy capacity into the bulk power system
- Integrating plug-in hybrid electric vehicles (PHEVs), distributed wind and photovoltaic solar energy resources, and other forms of distributed generation

4.7 PERFORMANCE GOALS AND PROGRESS METRICS

Smart grids can and should accurately measure their own performance, including overall efficiency, use of renewable energy, DR, and energy storage, as well as impacts on air pollution, water consumption, and land. Frameworks are now available to aid regulators and utilities in evaluating smart grid deployment plans and benefits and power purchasers in assessing the quality of their electric supply [9,10]. Power companies and utilities should be held accountable for delivering promised benefits—with allowed rates of return on investment linked to performance.

To avoid the problem of stranded assets and to maximize the economic and social return on investments, important questions of timing will need to be resolved. Some strategies, like voltage regulating devices (Volt/VAr), can already reliably reduce the strain on the electric system and save substantial amounts of energy: efficiency standards for power lines or market incentives could accelerate the deployment of such technologies. More consumer-focused options, such as residential time of use pricing, have been shown in pilots across the country to shave peak demand and reduce overall consumption, but will require outreach to consumers and time to adapt. This suggests a phased approach—investing in off-the-shelf options that deliver benefits in the near term while building the capacity for longer-term strategies, making use of lessons learned to guide deployments going forward.

4.8 TECHNOLOGY INVESTMENT AND INNOVATION

Many of the technologies necessary for smarter grids are available today as discrete capability building blocks. However, the levels of maturity and commercial viability differ. R&D efforts continue to advance the development of these technologies, particularly those essential to the advanced capabilities of smart grid solutions: communications, embedded sensing, automation, and remote control. The speed of technology research, development, and deployment in the power industry has been slower than in other industries. Technology development and deployment needs to be accelerated.

Each of these technologies has differing requirements for R&D to reduce technology and deployment risk, lower costs, and secure confidence that they can be implemented at scale. The challenge is to develop all component technologies necessary for an integrated smart grid solution to a level of maturity sufficient to deploy them all at scale at the same time. For this to occur, R&D for some components may need to be accelerated. An emerging area for R&D is the integration of all component technologies to ensure interoperable, coordinated, secure, and reliable electric system operations. This focus area includes the integration of high-penetration renewable energy (e.g., wind, solar), distributed generation, and electric vehicles into the electric grid.

The level of R&D spending in the utility sector is amazingly low. Utilities are among the lowest of all industries in R&D as a percent of revenue (<1%) (Figure 4.3). Competitive high-tech industries are 5–10 times higher. Yet, the move to make electricity competitive has not spurred more industry R&D. R&D costs are often not explicitly stated as a line item in rate cases. As a result, these costs are often the first to be cut when less than favorable rate case decisions are made.

Technology development efforts lack coordinated R&D for both individual technology components and integrated smart grid projects. Smart grids are potentially a global solution, albeit, in different forms for different markets. However, R&D is not entirely coordinated, and there is a natural tendency for institutes and companies to choose to develop those technologies most closely aligned

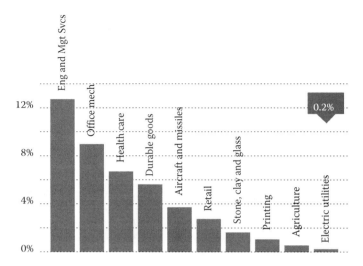

FIGURE 4.3 Industry R&D as a percentage of revenue. (From National Science Foundation.)

to their own capabilities and interests. This may leave some technologies with less focus than others. Given the high cost of R&D, technologies with less potential economic payback in their own right may well be left behind, leaving a maturity gap in the smart grid technology chain.

The integration of multiple key technologies has not yet occurred. The benefit realized from the integration of suites of technologies normally exceeds the sum of the benefits of the individual ones. For example, the deployment of integrated communication systems, including supercomputers, is needed to support the processing and analysis of the large data volumes that will be supplied by advanced technologies of the smart grid. Deployments of individual technologies often fail because they have not been adequately integrated with other needed technologies.

The price of many new technologies is currently not competitive with traditional alternatives and should be reduced to stimulate the level of deployment needed to achieve the smart grid. As the price ultimately is reduced, technical performance is proven and societal benefits recognized, the demand for these technologies will increase leading to further reductions in prices. Economies of scale and design innovation are needed to drive costs down. For example, our ability to store electrical energy remains limited. One of the most fundamental and unique limitations of electricity is that it cannot easily be stored for use at a later time. Although incremental progress is being made in energy storage research, the discovery of a transformative storage technology would greatly accelerate grid modernization.

4.9 CONSUMER ENGAGEMENT AND EMPOWERMENT

More work is needed to communicate the concepts and benefits of the smart grid to a wide variety of stakeholders and the general public and to encourage them to embrace the changes that will be needed to achieve the smart grid vision. Smart grid should be seen in the eyes of the customer, not just the utility—one from customer to consumerism.

Smart grid technologies demand behavioral changes in power consumption as DR involves consumer participation. Although consumers are becoming more aware of climate change and energy efficiency, the majority are not aware of the necessity to evolve electricity networks as a means of reducing emissions. The integration of renewable energy sources and DR will, in many cases, require making the existing network stronger and smarter and building new infrastructures. The public may negatively perceive changes in their electricity experience, particularly if it is accompanied by rising bills.

Stakeholders do not see a "burning platform" or a case for change. The societal consequences of inaction (i.e., not modernizing the grid) have not been clearly articulated to our diverse group of stakeholders. A lack of understanding of the fundamental value of a smart grid, and of the societal and economic costs associated with an antiquated one, has created the misperception that today's grid is good enough or at least not worth the sacrifices involved in improving it. Even the inconvenience and cost of infrequently occurring large-scale blackouts are quickly forgotten. More effort is needed to communicate and educate our citizens in the following areas:

- Today's grid is vulnerable to attack.
- An extended loss of the national grid would be catastrophic to our nation's security, economy, and quality of life.
- Today's grid will not address the security and economic challenges of the twenty-first century.
- A smart grid will be the platform, system, and network that will enable clean technologies, DSM and other options for addressing climate change to be deployed.
- A smart grid will help the United States become less dependent on foreign energy.
- A smart grid will be a more efficient and less costly grid.
- The performance of today's grid may lead to loss of jobs in the future as work is transferred to countries with more reliable and economic grids.

Effective consumer education is still lacking. The benefits of a smart grid have not been made clear to consumers. Some potential components of the consumers' value proposition include

- More effective monitoring and control of energy consumption to reduce overall electricity costs
- Participation in future electricity markets for DR, spinning reserve, energy, etc
- Enjoyment of future value added services that may be enabled by a smart grid

Public perception can create a key barrier to implementing policy and accelerating smart grid deployment. This is especially the case in open- and competitive-leaning markets that consult widely on policy implementation. Public pressure against a perceived societal disadvantage can force policy abandonment. For example, in the Netherlands, the rollout of smart meters was quashed by a small but vocal group concerned about the increased level of personal information that the meters would provide.

Utilities should educate customers before any technology deployment, building in costs for significant customer outreach and education. They should be ready to pass through AMI data, along with tools and incentives for customers to manage their onsite energy production, storage, and use—including the ability to safely share their data with third-party entrepreneurs. Customers should understand the real-time price of energy and services they consume, and deliver, to the grid. Ultimately, customers should pay—and be paid—that price (locational marginal pricing [LMP] or another agreed upon market signal). Pilots such as PowerCentsDC have shown consumer enthusiasm for time of use rates when they are carefully designed to provide choice and to help customers understand pricing options.

Consumer protections on disconnection and low-income assistance should be provided at the same or improved level and investment and technology risk should be shared by utilities and their customers. Where customers do pay upfront for these investments, with surcharges or other riders, utilities should be held accountable for delivering on the promised benefits. For instance, the California PUC included in its approval of a surcharge the requirement that utilities share projected operational savings—whether realized or not. That is, 8 months after the cost of the meter is included in the customer's bill, the Investor-Owned Utilities (IOUs) must credit customers $1.42/month in operational savings, even if the utility has not realized those savings. Cost recovery mechanisms that reward over-performance will incentivize utilities to seek out the most effective solutions.

Consumer involvement is a required ingredient for grid modernization and consumer education is the first step in gaining their involvement. Much remains to be done in the area of consumer education. The not in my backyard (NIMBY) philosophy must be resolved to reduce the excessive delays experienced today in deploying needed upgrades to the grid. Solutions are needed to reduce the concerns of citizens who object to the placement of new facilities near their homes and cities. New ideas are needed to make these new investments desirable rather than objectionable to nearby citizens. Communication of the smart grid vision with its goals of improving efficiency and environmental friendliness may help address this issue.

4.10 VENDOR PARTNERSHIPS

It is important to utilize technical talent, regardless of where it is located within an organization. It is also important to identify vendor partners that bring the necessary technical talent. A project of this breadth cannot be readily outsourced to even the most capable of vendors, but keeping them at arms' length will not be beneficial either.

Building a smart grid is not a single vendor task and utilities need to be prepared for a deliberate and complex vendor evaluation process to ensure that the right solutions, for both the short term and the long term, are selected. The vendor selection process itself needs to be very carefully and very rigorously done. If the sourcing process that a utility undergoes for smart grid is not painful, then it is probably not a sufficient process. A holistic view of smart grid technology is vital—not just the pieces of the puzzle, but how they fit together and what picture they form. If that complexity is not fully recognized and managed, the likelihood of making the right sourcing decisions declines significantly.

4.11 STANDARDS DEVELOPMENT, COORDINATION, AND ACCELERATION

Global standardization is essential for the deployment and successful operation of smart grids. While progress is being made, challenges remain due to fragmentation among stakeholders in the process of standards development, the lack of well-defined standards for smart grid interoperability, and intellectual property issues. At the same time, standards defined too early risk stifling innovative technological advances.

While smart grid technologies continue to progress, without well-defined and technology-neutral interoperability standards, further innovations and opportunities for deployment at scale are limited. Global cooperation for defining standards has not kept pace with technology innovation and development, which could impede large-scale development and rollout. Therefore, interoperability and scalability should be priorities, while taking care to avoid stifling innovation.

Since smart grid technologies encompass a diverse scope of technology sectors, including electricity infrastructure, telecommunication, and IT, misinterpretation and error may arise where there is a lack of interface standardization and related communication protocols. Therefore, even after standardization of the respective technologies, conformity testing and certification of interoperability may prove problematic for providers, since each technology must go through a conformity assessment specifically designed for the particular technology.

Existing international standards development organizations (SDOs) include the following:

- IEC—International Electrotechnical Commission (www.iec.ch)
- IEEE—Institute of Electrical and Electronics Engineers (www.ieee.org)
- ISO—International Organization for Standardization (www.iso.org)
- ITU—International Telecommunication Union (www.itu.int)

Smart Grid Barriers and Critical Success Factors

In addition to the ISOs, a large number of country or region-based standard associations influence the smart grid standards community. A key barrier is the lengthy process to develop and reach international consensus on a standard. For example, the average development time for IEC publications in 2008 was 30 months. Even after one of the SDOs has defined a standard, it still has to go through the harmonization process.

The smart grid is a large and complex marriage of the traditional electrical infrastructure and modern IT systems. This is truly a global effort involving thousands of utilities and vendors to implement and deploy the smart grid. In order to complete, a successful and cost effective deployment of the smart grid "international standards" will have to be followed by all who participate in its deployment. Why do we say this and why are standards so important to success? The following points characterize the importance of standards:

- Shareability—economies of scale, minimize duplication
- Ubiquity—readily utilize infrastructure, anywhere
- Integrity—high level of manageability and reliability
- Ease of use—logical and consistent rules to use infrastructure
- Cost effectiveness—value consistent with cost
- Interoperability—define how basic elements interrelate
- Openness—supports multiple uses and vendors, not proprietary
- Secure—systems must be protected
- Scalable—low- or high-density areas, phased implementation
- Quality—many entities testing and verifying

Across the United States, many utilities have already or are in the process of implementing smart technologies into their transmission, distribution, and customer systems based on several factors such as implementing legislative and regulatory policy, realizing operational efficiencies, and creating customer value. Smart grid value realization by utility customers and society at large is, in part, linked to the pace of technology implementation that enables a secure, smart and fully connected electric grid. Utilities agree that the development and adoption of open standards to ensure interoperability and security are essential for a smart grid. In many cases, utilities have defined open standards in the requirements for smart technology.

The smart grid is broad in its scope, so the potential standards landscape is also very large and complex. This is why "standards" adoption has become a challenge. However, the opportunity today is that utilities, vendors, and policy makers are actively engaged and there are mature standards that are applicable and much work on emerging standards and cybersecurity can be leveraged. Technology is not the primary barrier to adoption. The fundamental issue is organization and prioritization to focus on those first aspects that provide the greatest customer benefit toward the goal of achieving an interoperable and secure smart grid.

Although we are starting with a broad suite of standards that can help the country implement the smart grid in a timely and efficient manner, there are still a large number of standards that have to be developed on standards already in progress, which have to be completed in a timely manner. It is critical that we find a process that will accelerate the adoption of new smart grid standards. First, consider the challenges the industry has to overcome to accelerate the smart grid standards adoptions:

1. There are a large number of standards bodies and industry committees working and industry committees working in parallel with many duplicate and conflicting efforts. The industry must come together to focus concerted effort to accelerate the adoption of the stands which they are focus on.
2. The number of stakeholders, range of considerations, and applicable standards are very large and complex, which require a formal governance structure at a national level involving

both government and industry, with associated formal processes to prioritize and oversee the highest value tasks.
3. The smart grid implementation has already started and will be implemented as an "evolution" of successive projects over a decade or more. Standards adoption must consider the current state of deployment, development in progress, and vendor product development life cycles.
4. Interoperability is generally being discussed too broadly and should be considered in two basic ways, with a focus placed on prioritization and acceleration of the adoption of "intersystem" standards.

How are these challenges to be overcome and quickly?

1. A single national standards body should be established to take over formal control of smart grid standards adoption/NIST should immediately establish a more formal governance structure and related processes to prioritize smart grid standards selection and ensure an open unbiased process including key stakeholders. NIST shall coordinate the efforts of all existing standards bodies and industry committees to focus their attentions on a common goal and remove duplication of efforts.
2. Develop a smart grid "road map" that outlines a path and direction of deploying existing and future standards giving the industry clean direction forward.
3. Identify focus areas are as follows:
    a. Common information model
    b. Cybersecurity
    c. Interoperability base on open protocol
    d. Application interface standards
    e. Messaging, etc.
4. Governance principle definition includes the following:
    a. Openness
    b. Integrity
    c. Separation of duties and responsibility
    d. Compliance
5. Clearly defined test and verification methodologies and certification bodies shall be established to certify compliance to standards.

The grid will become "smarter" and more capable over time and the supporting standards must also evolve to support higher degrees of interoperability enabling more advanced capabilities over time. The implication of the smart grid evolution for standards adoption is that at any point in time the industry will be characterized by a mix of no/old technology, last generation smart technology, current generation smart technology, and "greenfield" technology opportunities. Smart grid implementation is an evolutionary process involving long project development life cycles from regulatory approvals through engineering and deployment. Given that technology life cycles are much shorter than the regulatory-to-deployment cycle, it is very likely that the grid will continuously evolve in the degree to which intelligence is both incorporated and leveraged.

The issue of evolution is particularly important because investments are a continuum based on policy imperatives, system reliability, and creating customer value. Policy makers and utilities must balance these considerations regarding certain smart grid investments before a complete set of standards has been adopted and customer benefit dictates moving forward. In a number of instances across the nation, utilities and regulators have given much thought to balancing accelerating customer benefits, project cost-effectiveness, and managing emerging technology risks. While there is no single "silver standards bullet" for legacy and projects currently in development, projects that are in the customers' and public policy interest should proceed. However, not having clear standards

Smart Grid Barriers and Critical Success Factors

going forward compounds the technology obsolescence risk—concrete action is needed in 2009 to standardize a few key aspects.

There is no technical reason to attempt to standardize all aspects of the smart grid today, if engineered and designed correctly. Nor is it likely possible given the lack of clear definition of all the elements and uses of the smart grid and complexity given the number of systems involved. Smart grid systems architected appropriately should be able to accept updated and new standards as they progress assuming the following standards evolution principles are recognized:

- Interoperability must be adopted as a design goal, regardless of the current state of standards.
- Interoperability through standards must be viewed as a continuum.
- Successive product generations must incorporate standards to realize interoperability value.
- Smart grid technology roadmaps must consider each product's role in the overall system and select standards compliant commercial products accordingly.
- Standards compliance testing to ensure common interpretation of standards is required.

These principles are being followed by many utilities implementing smart grid systems today by requiring capabilities such as remote device upgradability, support for robust system-wide security, and identifying key boundaries of interoperability to preserve the ability of smart grid investments to evolve to satisfy increasingly advanced capabilities.

Accelerating smart grid standards adoption can be achieved by focusing industry efforts on the right tasks in the right order. A systems engineering approach provides a formal, requirements-based method to decompose a complex "System of Systems," such as the smart grid, from a high intersystems view through a very structured process to a lower intrasystems view. Applying systems engineering to smart grid capabilities and supporting standards reveals that it is more important to create a unifying design for the entire system operationally, than to focus on implementing individual elements at the risk of future systems operations. This means that it is not necessary to first resolve interoperability of "intrasystem" interfaces within the utility's smart grid implementations before projects can proceed. This is true, as long as the important "intersystem" boundaries are well understood and the following interoperability design concepts are preserved.

4.12 POLICY AND REGULATION

According to some vocal experts, the biggest impediment to the smart electric grid transition is neither technical nor economic. Instead, the transition is limited today by obsolete regulatory barriers and disincentives that echo from an earlier era [12]. Public policy is commonly defined as a plan of action designed to guide decisions for achieving a targeted outcome. In the case of smart grids, new policies are needed if smart grids are actually to become a reality. This statement may sound dire, given the recent signing into law of the 2007 Energy Independence and Security Act (EISA) in the United States. And, in fact, work is underway in several countries to encourage smart grids and smart grid components such as smart metering. However, the risk still exists that unless stronger policies are enacted, grid modernization investments may fail to leverage the newer and better technologies now emerging, and smart grid efforts will never move beyond demonstration projects. This would be an unfortunate result when one considers the many benefits of a true smart grid: cost savings for the utility, reduced bills for customers, improved reliability, and better environmental stewardship. The U.S. Energy Policy Act of 2005 was a good first step in addressing barriers to grid modernization, but more is needed. Legislators and regulators have not yet taken a strong leadership role in support of grid modernization. A clear and consistent vision for the smart grid has not been adopted by legislators or regulators. Much has been said about individual technologies such as renewables and about specific energy issues such as environmental impact, but little has been said about the overall vision for a modernized grid—a vision that integrates the appropriate technologies, solves the various grid related issues, and provides the desired benefits to stakeholders and society.

Unclear policy increases market uncertainty with regard to how the overall market structure and rules will develop, which technologies merit investment, and the levels of capability required of the network. Market uncertainty varies depending on the market structure. Competitive-leaning markets place the emphasis on market participants, motivated by profitability and/or growth, to most efficiently allocate capital and select which technologies to apply. Centralized markets are subject to a more directed approach by the government or regulatory body acting within powers set by the government.

Governments and regulators in competitive-leaning markets will avoid the charge of picking winners and will attempt to let the market decide on the best structure and technology, often through lengthy consultations. However, where there is a lack of clarity about future market structure, roles, and rules, competitive markets tend to lock up. Companies then base investment decisions on the status quo and use tried-and-tested technologies to avoid the risk of picking the losing market technology. However, once the new market structure is defined and technologies are proven, rapid adoption of technology and associated innovation can be expected.

Network companies within centralized markets are more likely to be subject to government or regulatory directives, whereby direct mandates are given to the market participant(s) to invest in certain areas. Although this approach will gain quick results in the short term, a centralized market will, by its nature, provide political/regulatory risk, and not necessarily offer the rewards that could encourage the innovative new markets and services needed to realize the full potential value of a smart grid. This implies the need for a trade-off between (1) being sufficiently directive to provide clarity to companies on the future shape and rules for the market and (2) providing sufficient incentive for companies to invest in innovative technologies and services.

While regulation can help in implementing smart grid technologies, regulatory structure and other factors can create revenue uncertainties. If a company is required to invest in smart grid technologies, the revenue model must align with the benefits expected and provide assurance of return—at least for the payback period of investment.

Perhaps, the most glaring, and often quoted, disparity between current revenue drivers and smart grid drivers in many markets is the link between revenue and throughput. If smart grid technologies are successful, energy efficiency measures will be supported that will reduce throughput. The network company in this unreconstructed market would be investing to reduce its own revenue.

Network companies are rewarded for their success in delivering (approved) capital expenditure programs, providing the capacity to avoid network congestion—many by having revenue directly tied to the value of their asset base. Smart grid investment will reduce the need for network reinforcement capital expenditure. Networks could operate more efficiently, with less headroom than currently required, if peaks in demand are smoothed out through DR offerings enabled by smart grid technologies. On the face of it, investment in smart grid technologies in many regulatory environments would be undermining a utility's growth and revenue model by placing downward pressure on the requirement to build additional capacity.

To restructure the regulatory model to address issues such as revenue assurance, utilities and policymakers need a broad understanding of the primary role that smart grid technologies can play in meeting energy and environmental policy. This understanding will help them define a suitable regulatory regime that can align utilities' rewards with the benefits that their investments bring.

A number of initiatives exist to address these areas, in some cases setting out legislative and market frameworks to provide the right market environment for smart grid development, in others cases implementing full-scale projects that develop business cases and test financing arrangements. This section highlights a few examples of current initiatives for discussion and is not intended to be a comprehensive review.

Market uncertainty—can be largely overcome by defining a clear roadmap or strategy for smart grid development, together with supporting legislation that designates targets and incentives for those responsible for delivering the required infrastructure. The European Climate and Energy Package establishes a compelling rationale for the development of smart grids to support the introduction of renewable energy and energy efficiency measures that will help meet its targets. However, until this package is backed up by a stable and transparent carbon market that places a value on those targets, other interests will likely drive smart grid development (e.g., security of supply and reliability issues).

In the United States, the American Recovery and Reinvestment Act (ARRA) allocated a total of U.S.$4.5 billion to help subsidize smart grid modernization efforts and an additional U.S.$7.25 billion in loans for transmission infrastructure projects, coinciding with the U.S. Energy Independence and Security Act of 2007 [25] and the American Clean Energy and Security Act of 2009 [26]. Over the past year, such clear government support has generated a great deal of movement in smart grids and DR initiatives in the United States.

The recently proposed ACES 2009 includes a mandate for load-serving entities (distribution and retail companies) or state entities to publish peak demand reduction goals. Reduced capacity by DR is already being traded in a U.S. wholesale power market (e.g., the Pennsylvania-New Jersey Maryland [PJM] Regional Transmission Organization [RTO]).

In most cases, a clear mandate, together with associated incentives from the government, is required to quickly drive an industry forward. For example, in Italy ENEL originally developed its own business case for its smart meter program, originally recovering its investment through a significant cost reduction and increase in efficiency. Later, recognizing the benefits for the entire electricity system, the regulator decided to compensate this initiative through the tariff. This led to the rapid rollout of a program to replace 30 million electromechanical power meters with smart meters and the preparation of the supporting system hardware and software architecture.

Where a full-scale direct mandate does not match the market philosophy of a country, other measures are required. In the United Kingdom, a regulatory review by the Office of Gas and Electricity Markets (Ofgem) will set a long-term regulatory environment for the development of a low-carbon energy market. However, this is supplemented in the shorter term by a £500 million fund to promote the development of networks, including smart grid projects aimed at initiating smart grid infrastructure spending and innovation [27].

Global, regional, and national economic recovery and growth will serve as the cornerstones of increasing investments in development of smart electric power infrastructure and increased reliance on integrated communications and IT. Drivers will include national and state government issuance of clear policy directives and incentives concerning energy futures and development of smart infrastructure.

Current rate designs limit progress in grid modernization. Real-time rates that reflect actual wholesale market conditions are not yet widely implemented, preventing the level of demand side involvement needed in the smart grid. Net metering policies that provide consumers full retail credit for energy generated by them are also not widely deployed, reducing the incentive for consumers to install DER. Also, policies and regulations to encourage investment in power quality (PQ) programs, including those that provide pricing related to grades of power are not in place.

Some regulatory policies penalize utilities for supporting and investing in smart grid technologies. For example, from a strictly financial perspective, utilities are motivated to address system peak issues by investing in new generating facilities—which increases their revenue requirement—rather than supporting consumer-side DR opportunities—which reduces their revenues. New methods are needed to provide the incentive for marketers and utilities to invest in technologies that benefit society and are consistent with grid modernization even when those investments negatively impact their revenue stream. New regulatory models that decouple profit from revenue may be a solution to this conflict.

Incentives to stimulate smart grid investments that provide societal benefits are lacking. Regulatory policies often do not give credit to utilities for investments that provide substantial societal benefits (e.g., improvements in reliability and national security, reduction in our dependency on foreign oil, reductions in environmental impacts). Regulators play a vital role in ensuring that customers' interests are reflected in the decision making of the service provider. As such, regulators are a critically important gatekeeper in a smart grid project life cycle. This is particularly important for AMI, which (1) represents a wholesale replacement of not-yet-obsolete assets, (2) is a technology that will fundamentally transform the utility-customer relationship, and (3) offers potential benefits that cannot be realized without changes in customer behavior. To the latter point, the most obvious examples are the innovative rate structures such as critical peak pricing, which can leverage AMI technology to drive beneficial changes in customer usage patterns. To maximize their value, smart meters require smart rates, and smart rate design requires detailed dynamic pricing discussions among utilities, regulators, and customer advocates. Effective collaboration among these groups will result in programs and pricing tailored appropriately to the customer segments being served. Financial incentives at both the federal and state levels would enable such projects to pass financial hurdles that they otherwise could not—enabling the projects to proceed.

Deployment of smart grid technologies is costly, and without such incentives, utilities and energy providers are reluctant to invest in these needed technologies when they would bear all of the costs but where many of the benefits would accrue to other parties or to society as a whole. And, the absence of regulatory certainty inhibits technology deployments, as utilities and energy providers are left to weigh the risks of advanced technology investments with little assurance that the investment will be recoverable.

In addition to regulatory changes, changes in the tax law should be considered to make smart grid investments tax preferred. Tax incentives have been in place for many years for other preferred areas such as energy efficiency and renewable energy. Investment tax credits and incentives tied to more efficient operation of the grid and/or reduction in electricity costs might be helpful.

Regulations that support integrated electricity markets are needed. Federal and state regulations should support and not interfere with the development of large integrated wholesale electricity markets, which meet the needs of consumers and system operators. In the transition to fully enabled markets, there may be individual dissatisfaction along the way, but consumers, and society as a whole, will win. Regulations should support the ability of the smart grid to enable markets where they are appropriate.

Inconsistent policies among the states and with federal regulators prevent effective collaboration across a national footprint. Differing regulations among states that are electrically interconnected present challenges to the operation of a larger interconnected network and to the development of one that is even more integrated and dynamic. The optimal model for the electric industry has not been found and the lack of a consistent solution is a barrier to grid modernization. Coordination among local, state, and federal agencies on this issue is needed.

Alignment of regulatory policies to support grid modernization is generally weak. In general, regulatory policies were not designed with grid modernization in mind. Consequently, various specific policies need to be reviewed and updated. Some examples include the following:

- State legislatures and regulatory commissions currently focus on protecting consumers from the risks of consumer choice. Regulatory policy does not currently support the transition of consumers from passive protected users to proactive informed users as has occurred in other industries such as telecommunications and transportation.
- A significant reduction in R&D expenditures by utilities is an unintended drawback of deregulation and should be addressed to support grid modernization.
- Some existing utility assets are technologically obsolete and are incompatible with new smart grid technologies. Regulatory policies are needed for addressing the replacement or modification of these assets so that smart grid technologies can be integrated with them. Recovery of remaining book value of retired obsolete equipment is needed.
- Current policies often penalize consumers rather than utility shareholders for ineffective management decisions.

Smart Grid Barriers and Critical Success Factors 521

- The consequences of such actions as renewable portfolio standards (RPS), carbon tax, cap and trade, and carbon capture and sequestration will impact how the grid evolves and performs. Climate change legislation and regulations should be developed in the broader context of grid modernization so that both objectives can be effectively and efficiently met.

Regulatory structures should recognize and capture the full range of benefits of smart grid technology and provide assurance of appropriate cost recovery (particularly important given that the regulatory approval process varies by state, and may result in vastly different outcomes for a utility operating across state lines). The regulatory view should focus on the consumer and discussions with regulators should

- Encourage regulators to shift discussions toward "where's the value" away from "must comply"
- Include reliability directives and green initiatives
- Subtly emphasize the value of long-term, integrated smart grid components and not just short-term, individual projects to realize the synergies of satisfying multiple objectives; the role of the PUC in providing support for technology evolution and obsolescence is also critical and the recent precedent established by the U.S. Congress in October 2008, which allows utilities to depreciate smart grid investments over a 10 year period instead of 20 years, provides indication of such support required
- Reinforce the message of scarce options for new generation and the importance of energy efficiency, demand management, and load shifting programs in order to add urgency to the need for smart grid initiatives

4.12.1 Examples of Policy Measures

4.12.1.1 Climate and Energy Package 20-20-20 (EU)

The EU Climate and Energy Package 20-20-20, passed by European Parliament in 2008, resulted in specific carbon reduction targets on countries, their industry, and utilities. Voted for by more than 550 members and voted against by fewer than 100 members, this package focuses on three major policy areas: greenhouse gas emissions reduction, renewable energy, and energy efficiency. An important instrument to achieve the goal set by Europe is the Strategic Energy Technology Plan (SET Plan) including seven priority energy technologies to be deployed, one of which being the smart grids.

4.12.1.2 EISA, ACES (United States)

The United States enacted the Energy Independence and Security Act of 2007 (EISA 2007) to decrease their dependence on imported energy. It required standards around renewable fuels, vehicle efficiency, and electric appliance efficiency, and outlined a general federal policy on electric grid modernization.

In the American Clean Energy and Security Act of 2009 (ACES 2009), more specific details are set out to promote clean and efficient energy, to facilitate the deployment of a smart grid with DR applications, and to require electric utility providers to integrate electric vehicles into current grid infrastructure.

4.12.1.3 Regulatory Review (United Kingdom)

The U.K. regulator, Ofgem, is undertaking a full-scale regulatory review (RPI-X@20) to ensure that companies are rewarded for performance that aligns with the government roadmap toward delivery of a low-carbon energy sector.

4.13 INDUSTRY EXPERTISE AND SKILLS

A declining infusion of new thought is occurring. The technical experience base of utilities is graying. The talent pool is shrinking due to retirements and a shortage of new university graduates in the power engineering field. Additionally, fundamental knowledge and understanding of power system engineering principles is being lost as more and more of the technical analysis is done by computers rather than by human resources.

It is common knowledge that baby boomers in the United States are beginning to retire and leave the workforce. The electric power and energy industry is already beginning to experience shortages caused by these retirements. Over the next five years, roughly one-half of the utility industry engineers may retire or leave for other reasons. These experienced engineers provided the expertise needed to design, build, and maintain a safe and reliable electric power system. Over the years, they have planned for and expanded the system to serve a growing population, developed needed operating and maintenance practices, and brought about innovations to make improvements.

The departure of this engineering expertise is being met by hiring new engineers and by using supplementary methods, such as knowledge retention systems. The future engineering workforce will supplement traditional power system knowledge with new skills, such as in communication and IT. Traditional and new skills will still be necessary to successfully deploy advanced technologies while maintaining the aging infrastructure.

Meeting the functional needs of a smart grid will require consideration not only of the end state when a smart grid vision is realized, but the evolutionary period to that state during which the legacy infrastructure will be used side-by-side with new technologies. In order to integrate engineering elements in design and operation, the engineer must have a sufficient depth of understanding to put aside preconceived legacy notions. These legacy notions admittedly comprise the majority of power system engineering, but in order to realize new paradigms, a more holistic approach is required. For example, the use of time varying wind power, or solar power available in an uncertain schedule, the engineer needs to consider (1) at the design stage, control error tolerances, timing of controls, electronic designs of inverters needed to incorporate the alternative energy sources, and other basic system configurations and (2) in power system operation, the operating strategies of generation control, system control, and managing multiobjectives.

The integrative requirements of smart grid philosophies require that the depth of comprehension of engineers extend to the several areas illustrated in Figure 4.2. It appears that the legacy power engineering educational programs, while valuable for the installation of legacy systems, and maintenance of those systems, are not sufficient to accommodate the main elements of the smart grid. To insure that our society has the well-qualified power and energy engineers it needs, the following objectives must be sought [13]:

1. Develop and communicate an image of a power engineer based on a realistic vision of how engineers will be solving challenges facing companies, regions, the nation, and the world, thereby improving the quality of life. Youth want to choose jobs that make a difference in the world and make their life more meaningful.
2. Motivate interest in power and energy engineering careers and prepare students for a post-high-school engineering education in power and energy engineering. Students should be exposed to engineering even before high school. Teachers, counselors, and parents must be the target of information as well as the students.
3. Make the higher education experience relevant, stimulating, and effective in training high quality and professional power and energy engineers. Establish and maintain a direct link between power engineering and the solution of major challenges facing the United States and the world.
4. Increase university research funding to find innovative solutions for pressing challenges and to enhance student education.

Smart Grid Barriers and Critical Success Factors

Expertise and skill development are facilitated by government policies, such as the U.S. Green Jobs Act and Workforce Investment Act, which formalize investment in next-generation skills development. There are also international efforts like CCiNet (Climate Change Information Network) of the UNFCCC (United Nations Framework Convention on Climate Change), which includes education, training, and public participation programs. Currently, major initiatives specifically dedicated to develop smart grid skills are few in existence, with a noted exception being the workforce development in the United States for the electric power sector to implement a national clean-energy smart grid. This U.S.$100 million initiative—as part of the ARRA U.S.$4.5 billion investment to grid modernization—targets new curricula and training activities for the current and next-generation workforce, including cross-disciplinary training programs spanning the breadth of science, engineering, social science, and economics.

4.13.1 Indicators of Declining Workforce Supply in the United States

For electric and gas utility employees, the results of a survey by the Center for Energy Workforce Development (CEWD) in 2008 show that approximately 50% of all employees will be eligible for retirement within 10 years [14]. Figure 4.4 shows the electric and gas utility employee age distribution from this survey. The survey was comprised of 55 electric and gas utilities nationwide, as well as all electric cooperative organizations. As of 2010, indications are that nearly 45% of the eligible retirement age employees may need to be replaced by as early as 2013 [15].

The facts indicate there are workforce and education system problems summarized as follows [13]:

- Over the next 5 years, approximately 45% of engineers in electric utilities will be eligible for retirement or could leave engineering field for other reasons. If they are replaced, then there would be a need for over 7000 power engineers by electric utilities alone: two or three times more power engineers may be needed to satisfy needs of the entire economy.
- About 40% of the key power engineering faculty at U.S. universities will be eligible for retirement in 5 years with about 27% anticipated to actually retire. In other words, of the 170 engineering faculty working full time in power engineering education and research, some 50 senior faculty members will be retiring. This does not account for senior faculty who are already working less than full time in the area. Finally, even more faculty will be needed to increase the number of power engineering students to meet the demand for new engineers in the workplace.

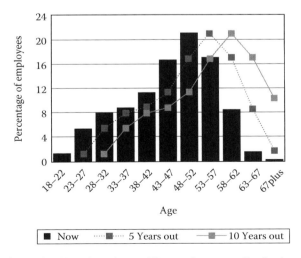

FIGURE 4.4 (See color insert.) Electric and gas utility employee age distribution.

- The pipeline of students entering into engineering is not strong enough to support the coming need, with surveys showing (1) that most high school students do not know much about engineering and do not feel confident enough in their math and science skills and (2) that few parents encourage their children, particularly girls, to consider an engineering career. Furthermore, often career counselors and teachers know little about engineering as a career. Workforce diversity is also a concern. Women constitute only 18% of the engineering enrollments and 12% of the electrical engineering students. Enrollment of underrepresented student populations should be higher.
- Enrollment by university students in power and energy engineering courses is increasing (perhaps fueled by interest in renewable energy systems and green technologies); however, the overall number of students interested in electrical engineering is declining. A shrinking pool of electrical engineering students limits the future supply of new power engineers.
- The hiring rate of new power engineering faculty is beginning to grow after years of insufficient hiring to replace retiring faculty; however, as time has passed, many historically strong university power engineering programs have ended or significantly declined.
- There are less than five very strong university power engineering programs in the United States. A very strong program has (1) four or more full-time power engineering faculty, (2) research funding per faculty member that supports a large but workable number of graduate students, (3) a broad set of undergraduate and graduate course offerings in electric power systems, power electronics, and electric machines, and (4) sizable class enrollments of undergraduates and graduate students in those courses. The general lack of research funding opportunities has made it difficult for faculty in existing programs and new emerging programs to meet university research expectations and for engineering deans to justify adding new faculty.

4.14 KNOWLEDGE AND FUTURE EDUCATION

Since implementing the smart grid initiative will take engineering professional resources of broad expertise and different profile than previously available, one may naturally ask the question as to where the new generation of electrical and electronic engineers shall come from with the specialized integrated skills needed by smart grid engineers.

Traditional power engineering skills include

- Power system dynamics and stability
- Electric power quality and concomitant signal analysis
- Transmission and distribution system operations
- Economic analysis, energy market, and planning
- Reliability and risk assessment

The traditional power engineering educational programs, while valuable for the installation of legacy systems, and maintenance of those systems, are not sufficient to accommodate the main elements of the smart grid. This is the case since simple replicative engineering is not sufficient to formulate new designs and new paradigms. The innovation extends to power system operation as well. The solution to this quandary appears to be in the integration of new technologies into the power engineering curriculum programs, and extending the depth of those programs through a master's level experience. It is desirable that the master's level experience be industry oriented in the sense that the challenges of the smart grid be presented to the student at the master's level.

Traditional power engineering education, the source of engineers for the future grid, need to be educated on a number of topics that are not traditionally included in a power engineering program. Among these are

- The design of wind energy systems
- The design of photovoltaic solar energy systems
- The design of solar thermal (concentrated solar energy) systems
- The calculation of reserve margin requirements for power systems with high penetration of renewable resources
- The modeling of uncertainty/variability in renewable energy systems
- Inclusion of cost-to-benefit calculations in generation expansion studies
- Conceptualization, design, and operation of energy storage systems, including bulk energy storage systems
- Discussion of the socio-political issues of renewable energy development

The desired elements of the cross cutting energy engineering skills for the next generation of "smart grid" power engineers appears to include all or most of the following elements. The exposure to these subjects is not recommended to be a casual, low-level exposure; rather, the exposure is recommended to be at a depth that analysis is possible in a classroom environment. Moreover, it is recommended that research be performed by the student so that synthesis can be accomplished. Some of the elements identified are discussed in the following.

Direct digital control: The importance of direct digital control is important in realizing most of the smart grid objectives. Direct digital control needs to be examined not only in terms of classical automatic control principles (including, if not emphasizing discrete control), but also how digital control relies on communication channels, how these controls need to be coordinated in terms of safety and operator permissive strategies, the impact of latency (e.g., Ref. [16]), new instrumentation, and how that instrumentation will impact the power system design and operation.

Identification of new roles of system operators: Components of the system that need to be fully automated versus components that are "operator permissive" controlled need to be identified. This must be presented to the students in a way that integrates computer engineering and power engineering. As an example, visualization of power systems is an especially important subject area (e.g., Ref. [17]).

Power system dynamics and stability: Power system stability is a classical subject. However, the new issues of this field relate to how maximal power marketing can occur and yet still insure operationally acceptable system operation and stability. The subject appears best taught as an in-depth semester course that includes modeling and practical examples. The examples should be examined by the students in a project format.

Electric power quality and concomitant signal analysis: With the advent of electronic switching as a means of energy control, electric power quality has taken on a new importance in power engineering education. Again, we find that simply a casual discussion of this topic is insufficient to achieve the analytical stage: rather it is recommended that a semester's course, complete with project work and mathematical rigor is needed as instruction. Power quality is discussed as an educational opportunity in Ref. [18].

Transmission and distribution hardware and the migration to middleware: New materials are revolutionizing transmission designs. Transmission expansion needs to be discussed in an in-depth fashion that includes elements of high voltage engineering and engineering physics, new solid-state transformer designs, and solid-state circuit breakers (e.g., Ref. [19]). Classical power engineering seems to leave a gap between software and hardware, and it is recommended that hardware-oriented courses at the master's level include issues of middleware applications. The use of intelligent electronic devices (IEDs) is deemed important. This development is especially important in the area of substation automation and synchronized phasor measurement systems [20].

New concepts in power system protection: With increased loading of power systems and dynamic behavior due to accommodating deregulated electricity markets and interfacing renewable resources, designing protective relaying solutions that are both dependable and secure has become a challenge. Introduction of microprocessor-based relays, high-speed communications, and synchronized phasor measurement systems made opportunities for adaptive and systemwide relaying. Learning how the relaying field evolves from traditional approaches designed for handling $N-1$ contingencies to new schemes for handling $N-m$ contingencies becomes an integral part of a modern power systems curriculum. The use of modern modeling and simulation tools is required [21].

Environmental and policy issues: Exposure to environmental and policy issues need to be included in the master's level in power engineering education. This exposure needs to go beyond "soft science" and it needs to appeal to the students' capability in mathematics and problem solving. The main issues are discussed in Ref. [22].

Reliability and risk assessment: There is little doubt that the importance of reliability of the power grid is widely recognized. However, when transformative changes are planned and implemented, the traditional tried and tested rules to ensure reliability cannot be relied on. Such changes need to be modeled and analyzed for reliability assessment based on sound mathematical foundations. Fortunately, now a large body of knowledge exists for modeling and analysis of power system reliability and risk assessment. The students at the master's and doctoral level should be provided this knowledge so that they can effectively use it in the integration and transformative process.

Economic analysis, energy markets, and planning: Planning can no longer be done incrementally, motivated largely to satisfy the next violation of planning reliability criteria. Investment strategies must be identified beyond the standard 5–20 year period at an interregional if not national level, to identify cost-effective ways to reach environmental goals, increase operational resiliency to large-scale disturbances, and facilitate energy market efficiency. Engineers capable of organizing and directing such planning processes require skills in electric grid operation and design, mathematics, optimization, economics, statistics, and computing, typically inherent only in PhD graduates [23]. Engineers from BS and MS levels will be needed to participate in these processes, and these engineers will require similar skills at the analysis level or above.

The smart grid approach combines advances in IT with the innovations in power system management to create a significantly more efficient power system for electrical energy. Modern society is migrating to an Internet-based business and societal model. As an example, it is common to pay bills, order equipment, make reservations, and perform many of the day-to-day tasks of living via the Internet. In power engineering, one needs only to examine such tools as the Open Access Same-Time Information System (OASIS) to realize that the same Internet model applies to power transmission scheduling [24]. The identical model appears in many power engineering venues including setting protective relays, transcommuting of engineering personnel, managing assets and inventory, scheduling maintenance, and enforcing certain security procedures. Cloud data storage and virtual networks may be a key to solve operational issues associated with concerns on the distribution grids such as localized peak loads caused by concentrated areas of charging electric vehicles. While the open Internet has security issues, similar models in an intranet or virtual private network may be used to enhance security. As this general model progresses, in many cases, one may wonder why certain procedures, whether in power engineering or elsewhere, have not been automated.

Automation is at the heart of the smart grid. That is, various decisions in operation may no longer be relegated to operators' action. Instead, operating decisions considering a wide range of multiobjectives might be "calculated" digitally and implemented automatically and directly. While safety, redundancy, and reliability considerations are clearly issues as this high level of direct digital control is implemented, it is believed to be possible to realize the objectives of the smart grid. To this end, the analogy between the needs of Internet opportunities and the needs of smart grid translates into a new philosophy in power engineering education: develop the cognitive and cyber skills while focusing on domains of specific expertise. This often translates to instruction tools that are highly interactive and having strong modeling and simulation background. Interestingly, the very same Internet philosophy may be applicable to the identification of where engineering expertise will be obtained—and how the complex issues of power engineering, public policy, and IT can be presented to students in undergraduate and graduate programs.

To tackle the smart grid research issues a variety of engineering and non-engineering disciplines need to be brought together. Almost every engineering discipline has its role in this development: electrical and computer engineering (grid generation, transmission, and distribution enhancements), petroleum engineering (alternative fuels for electricity generation), nuclear engineering (sustainable electricity production), chemical engineering (alternative and renewable electricity production), aerospace engineering (wind energy infrastructure), mechanical engineering (design of generators and energy-efficient buildings), civil engineering (environmental impacts), etc. In addition, a number of non-engineering disciplines are needed to resolve associated economic, societal, and environmental and policy issues: economics, sociology, architecture, chemistry, agriculture, economics, public policy, etc. The fact that some of the disciplines are allocated to different colleges should not be underestimated since bringing those resources together will require a concerted university-wide effort.

4.14.1 Forms and Goals of Future Learning

University education: The overall education model will include a combination of in-residence and distance education programs offered by universities, community colleges, and government and industry providers. In addition, the model will include certificate programs and professional development programs. Universities can hire non-tenured staff, such as adjunct professors, relatively quickly to supplement the available instruction time of university faculty. This will allow universities to expand educational opportunities to address the rising shortage of well-trained power engineers. However, actions must also be quickly taken by industry and government to build and sustain university power engineering programs through increased research support for faculty. Strong university power programs are needed to meet the needs for innovation, for future engineers, and for future educators. The following are recommendations for the university education:

- Work toward doubling the supply of power and energy engineering students.
- Continue enhancing education curricula and teaching techniques to insure an adequate supply of well qualified job candidates that can be successful in the energy jobs of the future.
- Increase research in areas that can contribute to meeting national objectives.
- Get involved in state and regional consortia to address workforce issues.
- Conduct seminars and encourage industry to provide information sessions to develop university student interest in power and energy engineering careers.
- Build communications and collaborations with industry, particularly between industry executives, department chairs, and college deans.
- Communicate with industry about education needs that may require innovative approaches to education.
- Insure that adequate educational opportunities exist for retraining engineers with education and experience in fields other than power engineering.
- Use college or university student recruiting programs to also spread the word about opportunities in power and energy engineering.

Career and technical education

- Identify and communicate needs and ideas on education materials, lesson plans, and computer-based learning related to energy and engineering.
- Encourage students to consider engineering as a career.

Continuing education

- Inform students about engineering career opportunities.
- Provide course opportunities that prepare students for an engineering education at a university.
- Work with universities to establish credit transfer programs so that students can continue education at a university after graduating from a community college.

Certification and professional licensing

- Provide education opportunities for trainees to obtain the certification or license for engineering career.
- Build tools and relationships to recruit and train people leaving the military and from underrepresented populations.

Training of non-engineering workforce segment

- Partner with professional societies in areas of career awareness, workforce development and education, and workforce planning.
- Provide aides in education planning and a career awareness video for engineers in cooperation with professional societies.
- Publish promotional material and presentations that target potential power and energy engineers and transitioning military personnel; adjust messaging to appeal to underrepresented groups.
- Develop industry-wide and regional solutions that maximize the efficiency of electric utility workforce development activities.
- Perform annual electric utility surveys to identify high priority energy industry engineering workforce needs.

Role of professional societies

- Take advantage of delays in retirements due to the economic downturn to more fully develop collaborations to implement wide-scale training and marketing programs.
- Keep the organization and its members knowledgeable of engineering workforce issues and mobilize the membership, where individuals, chapters, or regions as a whole, to get involved in responding.
- Develop training plans targeted toward life-long learning. The development needs to consider the adjustment of skills arising out of technological change and new fields.
- Explore ways to support retraining of engineers whose education and experience is in fields other than power and energy engineering.
- Provide opportunities to bridge promising student talent and industry.

4.14.2 Example of an Initiative to Build Knowledge: U.K. Power Academy

A decreasing trend of acceptance for study at U.K. power engineering courses ranging around 28% in the period 2002–2007 was observed. To cope with the progressively decreasing number of power engineers, a group of seven leading U.K. universities and 17 companies (including utilities, manufacturers, service providers, and transmission and distribution system owners and operators) joined forces founding the Power Academy. The aim of the Power Academy is to address the shortfall in engineering expertise in the electricity power industry by attracting new talent into the industry primarily at the undergraduate level leading to graduate employment.

Students supported by the Power Academy receive a bursary, a book allowance, a financial contribution toward university fees, the participation in a summer seminar, 8 weeks of training experience under the supervision of a company mentor and the free membership to the U.K. Institution of Engineering and Technology. This initiative has seen more than 240 applications in 2008 and the 3 leading companies have supported over 40 scholars each in the past 4 years.

It has been argued that, since its foundation, the Power Academy has been a success with increasing numbers of scholars and a significant majority of positive responses from an annual survey of their experiences. However, the level of influence it has on the degree choices by school and college graduates is yet to be determined and remains a challenging area, not only for the Power Academy partners, but also for the U.K. Government in respect to its overall skills agenda. Another benefit of the Power Academy partnership is that these collective efforts provide the best chance of addressing the challenge of shortfall of power engineering experience, and of influencing the government in its future course of action.

REFERENCES

1. National Institute of Standards and Technology (NIST), U.S. Department of Commerce, Roadmap for smart grid interoperability standards 26, 2010, http://www.nist.gov/public_affairs/releases/smartgrid_interoperability_final.pdf
2. Sharing smart grid experiences through performance feedback, US DOE Office of Electricity Delivery and Energy Reliability, National Energy Technology Laboratory, Morgantown, WV, March 2011, DOE/NETL-DE-FE0004001
3. Environmental Defense Fund, Opening comments to the California Public Utilities Commission in R.08-12-009, March 9, 2010, p. 12, http://docs.cpuc.ca.gov/EFILE/CM/114701.htm
4. Environmental Defense Fund, Comments to the New York Public Service Commission in Case 10-E-0285, Proceedings on Motion of the Commission to consider Regulatory Policies Regarding Smart Grid Systems and the Modernization of the Electric Grid, September 10, 2010.
5. Electric Power Research Institute, *Electricity Sector Framework for the Future Volume I: Achieving the 21st Century Transformation*, Electric Power Research Institute, Palo Alto, CA, 2003.
6. Galvin Electricity Initiative, The case for transformation, 2011, http://www.galvinpower.org
7. The Electricity Advisory Committee, Smart grid: Enabler of the new energy economy, December 2008, http://www.oe.energy.gov/final-smart-grid-report.pdf
8. National Renewable Energy Laboratory, Projected benefits of federal energy efficiency and renewable energy programs—FY 2008 Budget Request, 2007.
9. Galvin Electricity Initiative, Perfect power seal of approval, (2001) http://www.galvinpower.org; Electric Power Research Institute, A methodological approach to estimating the benefits and costs of smart grid demonstration projects, January 2010, http://my.epri.com/portal/server.pt?Abstract_id=000000000001020342
10. Evaluation Framework for Smart Grid Deployment Plans, Karen Herter, Ph.D., Herter Energy Research Solutions, Inc. in collaboration with Timothy O'Connor, Environmental Defense Fund and Lauren Navarro, Environmental Defense Fund.
11. Barriers to Achieving the Modern Grid, US DOE Office of Electricity Delivery and Energy Reliability, National Energy Technology Laboratory, Morgantown, WV, July 2007.
12. Yeager, K.E., Facilitating the transition to a smart electric grid, Galvin Electricity Initiative, 2007, http://www.galvinpower.org/files/Congressional_Testimony_5_3_07.pdf
13. Bose, A., A. Fluek, M. Lauby, D. Niebur, A. Randazzo, D. Ray, W. Reder, G. Reed, P. Sauer, and F. Wayno, *Preparing the U.S. Foundation for Future Electric Energy System: A Strong Power and Energy Engineering Workforce*, IEEE Power and Energy Society, New York, April 2009.
14. Center for Energy Workforce Development, Gaps in the energy workforce pipeline—2008 survey conducted by Chris Messer, Programming Plus++, October, 2008.

15. Reed, G.F. and W.E. Stanchina, The power and energy initiative at the University of Pittsburgh: Addressing the aging workforce issue through innovative education, collaborative research, and industry partnerships, *IEEE PES T&D Conference and Exposition*, New Orleans, LA, April 2010.
16. Browne, T.J., V. Vittal, G.T. Heydt, and A.R. Messina, A comparative assessment of two techniques for modal identification from power system measurements, *IEEE Transactions on Power Systems*, 23(3), 2008, 1408–1415.
17. Overbye, T., Visualization enhancements for power system situational assessment, *Proceedings of IEEE Power and Energy Society General Meeting*, July 2008, pp. 1–4. http://ieeexplore.ieee.org/xpl/login.jsp?tp=&arnumber=4596284&url=http%3A%2F%2Fieeexplore.ieee.org%2Fxpls%2Fabs_all.jsp%3Farnumber%3D4596284
18. Browne, T.J. and G.T. Heydt, Power quality as an educational opportunity, *IEEE Transactions on Power Systems*, 23(2), 2008, 814–815.
19. Yang, L., T. Zhao, J. Wang, and A.Q. Huang, Design and analysis of a 270 kW five-level DC/DC converter for solid state transformer using 10 kV SiC power devices, *Proceedings IEEE Power Electronics Specialist Conference*, June 2007, pp. 245–251. http://ieeexplore.ieee.org/xpl/login.jsp?tp=&arnumber=4341996&url=http%3A%2F%2Fieeexplore.ieee.org%2Fxpls%2Fabs_all.jsp%3Farnumber%3D4341996
20. Kezunovic, M., G.T. Heydt, C. DeMarco, and T. Mount, Is teamwork the smart solution? *IEEE Power and Energy Magazine*, 7(2), 2009, 69–78.
21. Kezunovic, M., User-friendly, open-system software for teaching protective relaying application and design concepts, *IEEE Transactions on Power Systems*, 18(3), 2003, 986–992.
22. Overbye, T., J. Cardell, I. Dobson, M. Kezunovic, P.K. Sen, and D. Tylavsky, The electric power industry and climate change, Power Systems Engineering Research Center, Report 07-16, Tempe, AZ, June 2007.
23. Power Systems Engineering Research Center, U.S. Energy Infrastructure Investment: Long-term strategic planning to inform policy development, Publication 09-02, March 2009, http://www.pserc.org/ecow/get/publicatio/2009public
24. DeMarco, C.L., Grand challenges: Opportunities and perils in ubiquitous data availability for the open access power systems environment, *Proceedings of IEEE Power Engineering Society Summer Meeting*, Vol. 3, July 2002, pp. 1693–1694. http://ieeexplore.ieee.org/xpl/login.jsp?tp=&arnumber=985258&url=http%3A%2F%2Fieeexplore.ieee.org%2Fiel5%2F7733%2F21229%2F00985258.pdf%3Farnumber%3D985258
25. U.S. Energy Independence and Security Act of 2007 (EISA). 2007. http://www.gpo.gov/fdsys/pkg/BILLS-110hr6enr/pdf/BILLS-110hr6enr.pdf
26. American Clean Energy and Security Act of 2009 (ACES). 2009. http://www.govtrack.us/congress/bills/111/hr2454
27. Office of the Gas and Electricity Markets (Ofgem). 2009. http://www.ofgem.gov.uk/Pages/OfgemHome.aspx

5 Global Smart Grid Initiatives

Matt Wakefield and Bartosz Wojszczyk

CONTENTS

- 5.1 Smart Grid Investments 532
- 5.2 Australia 534
  - 5.2.1 Lead Organizations 534
  - 5.2.2 Project Examples 534
  - 5.2.3 Country-Specific Drivers and Benefits 534
  - 5.2.4 Scale 534
- 5.3 Canada 536
  - 5.3.1 Lead Organizations 536
  - 5.3.2 Project Category/Technology 536
  - 5.3.3 Project Examples 536
  - 5.3.4 Regional Drivers and Benefits 537
  - 5.3.5 Country-Specific Drivers and Benefits 537
  - 5.3.6 Scale 537
- 5.4 China 538
  - 5.4.1 Lead Organizations 538
  - 5.4.2 Project Category/Technology 538
  - 5.4.3 Project Examples 538
  - 5.4.4 Regional Drivers and Benefits 538
  - 5.4.5 Country-Specific Drivers and Benefits 538
  - 5.4.6 Scale 539
- 5.5 Europe 539
  - 5.5.1 Lead Organizations 539
  - 5.5.2 Project Category/Technology 539
  - 5.5.3 Project Examples 539
  - 5.5.4 Regional Drivers and Benefits 539
  - 5.5.5 Country-Specific Drivers and Benefits 539
  - 5.5.6 Scale 539
- 5.6 India 540
  - 5.6.1 Lead Organizations 540
  - 5.6.2 Project Examples 540
  - 5.6.3 Country-Specific Drivers and Benefits 540
  - 5.6.4 Scale 540
- 5.7 Japan 540
  - 5.7.1 Lead Organizations 540
  - 5.7.2 Project Examples 541
  - 5.7.3 Country-Specific Drivers and Benefits 541
  - 5.7.4 Scale 541

5.8 Korea..541
    5.8.1 Lead Organizations...541
    5.8.2 Project Examples...541
    5.8.3 Country-Specific Drivers and Benefits 541
    5.8.4 Scale...541
5.9 Latin America..541
    5.9.1 Lead Organizations...541
    5.9.2 Project Category/Technology 542
    5.9.3 Project Examples...542
    5.9.4 Regional Drivers and Benefits 542
    5.9.5 Country-Specific Drivers and Benefits 542
    5.9.6 Scale...542
5.10 United States..542
    5.10.1 Lead Organizations...542
    5.10.2 Project Category/Technology 543
    5.10.3 Project Examples...543
    5.10.4 Regional Drivers and Benefits 543
    5.10.5 Country-Specific Drivers and Benefits 544
    5.10.6 Scale...544
5.11 Other Countries 544
    5.11.1 Lead Organizations...544
    5.11.2 Project Category/Technology 544
    5.11.3 Project Examples...545
    5.11.4 Regional Drivers and Benefits 545
    5.11.5 Country-Specific Drivers and Benefits 545
    5.11.6 Scale...545
5.12 Global Collaborative Efforts 546
    5.12.1 United States, Canada, Ireland, France, Australia, and Japan 546
        5.12.1.1 Lead Organizations...546
        5.12.1.2 Project Category/Technology...546
        5.12.1.3 Project Examples..546
        5.12.1.4 Regional Drivers and Benefits 546
        5.12.1.5 Country-Specific Drivers and Benefits 546
        5.12.1.6 Scale..546
5.13 Summary 546
References..547
Bibliography 548

5.1 SMART GRID INVESTMENTS

Why do electric utilities continue to invest in pilot and demonstration projects, and why do the words "pilot" and "demo" give the connotation of one-off projects with very little useful outcome? Electric utilities are very diverse. They have different regulatory environments, customer bases, grid characteristics (underground, overhead), geography, availability of renewable resources, and different generations of legacy enterprise systems and field equipment. Therefore, one utility's pilot or demonstration project is not likely to be aligned with another utility's needs. In addition, the benefits and synergies of emerging technologies, especially in advanced fields of applications, are not always easily conveyed from one utility to the next. However, the risks are too high for a utility to forego a pilot or demonstration project and to deploy a multimillion-dollar full-scale smart grid project based on the results and lessons of another utility whose environment is different than their own. That being said, there are significant benefits to be gained by sharing lessons learned from other utility pilot and demonstration projects to avoid making common mistakes and to identify technology gaps to focus research on the highest priority needs.

The well-known adage, the whole is greater than the sum of its parts, acknowledges the power of collaboration. The key for smart grid demonstration and pilot projects is not only for utilities, research organizations, regulatory organizations, vendors, and standards bodies to collaborate and freely exchange information, but also for the demonstration and pilot projects to focus on the goals of an integrated smart grid architecture today and a roadmap to embrace advances in technology in the future. By orchestrating and collaborating among utilities in smart grid projects using common methodologies and broadly sharing lessons learned, the industry will see the advancement of proven technologies performing smart grid applications where benefits are proven, monetized, and accepted by the industry.

According to forecasts from Pike Research [12], smart grids will attract around $200 billion in worldwide investment by the end of 2015. Many industry reports define a wide range of smart grid initiatives that are driven by regional, country, or utility-specific objectives and requirements. These technologies can be broadly captured under the following areas:

- *Low carbon*: For example, large-scale renewable generation, distributed energy resources (DER), electric vehicles (EVs), and carbon capture and sequestration (CCS)
- *Grid performance*: For example, advanced distribution and substation automation (self-healing), wide-area adaptive protection schemes (special protection schemes), wide-area monitoring and control systems (PMU-based situational awareness), asset performance optimization and conditioning (CBM), dynamic rating, advanced power electronics (e.g., FACTS, intelligent inverters, etc.), high temperature superconducting (HTS), and many others
- *Grid enhanced applications*: For example, distribution management systems (DMS); energy management systems (EMS); outage management systems (OMS); demand response (DR); advanced applications to enable active voltage and reactive power management (IVVC, CVVC); advanced analytics to support operational, nonoperational, and BI decision making; distributed energy resource management; microgrid and Virtual Power Plant (VPP); work force management; geospatial asset management (GIS); KPI dashboards and advanced visualization; and many others
- *Customer*: For example, advanced metering infrastructure (AMI), home/building automation (HAN), EMS and display portals, EV charging stations, smart appliances, and many others
- *Cybersecurity and data privacy*
- *Communication and integration infrastructure*

Smart grids are evolving, yet a marked acceleration is happening. Many smart grid technologies are based on existing mature solutions, but much more effort is needed to deploy them on a system-wide scale. More importantly, smart grid deployments must deliver the integrated architecture of applications that is the core concept of the smart grid vision, while incorporating business models supported by regulatory and market frameworks. Many countries are investing in smart grid technology pilot, demonstration, and deployment projects. The growth of these projects has been accelerated by federal stimulus investments—see Figure 5.1 and Table 5.1 for summaries of the top 10 countries receiving smart grid federal stimulus investments in 2010 [1].

The following sections provide an overview of global smart grid initiatives. Smart grid investments have resulted in countless projects, entirely or partly focused on smart grid technologies. The intention is not to list every smart grid related project around the world, but to summarize the activities, trends, and smart grid drivers in each major world region.

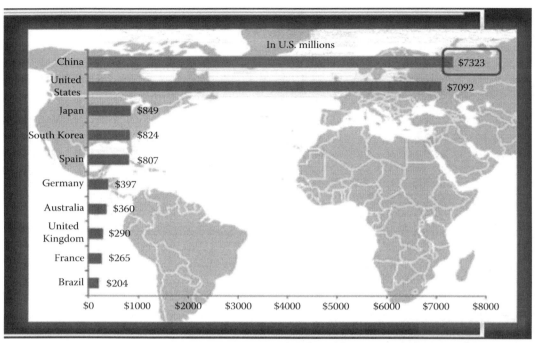

FIGURE 5.1 Ranking of top 10 countries receiving smart grid federal stimulus investments in 2010. (From Zpryme Research and Consulting, 2010.)

5.2 AUSTRALIA

5.2.1 Lead Organizations

Victorian power utility companies, including Jemena, UED, Powercor, SP AusNet, and CitiPower. The participants in Australia's Smart Grid/Smart City (SGSC) project include consortium partners AGL Energy, IBM Australia, GE Energy Australia, Sydney Water, Hunter Water Australia, and the Newcastle City Council.

5.2.2 Project Examples

Meters in Victoria State; Smart Grid/Smart City Project: The aim is to provide a smart grid environment for nearly 10,000 households by 2013 in Newcastle, Newington, the Sydney CBD, Ku-ring-gai, and the rural township of Scone.

5.2.3 Country-Specific Drivers and Benefits

Electricity distribution systems are seeking improvements in network reliability and ways to defer network capacity upgrades.

5.2.4 Scale

In Australia, the central government has established the mandated rollout of smart meters by 2013 with 100% completion. The Victorian government has mandated that smart meters and AMI be rolled out to all customers consuming less than 160 MWh of electricity per annum between 2009 and 2013. The state of New South Wales will complete its smart meter deployments by 2017.

TABLE 5.1
Top 10 Country Smart Grid Federal Stimulus Investments in 2010

Country	Key Stimulus Investments
China	The Chinese government has developed a large, long-term stimulus plan to invest in water systems, rural infrastructures and power grids, including a substantial investment in smart grids. Smart technologies are seen as a way to reduce energy consumption, increase the efficiency of the electricity network, and manage electricity generation from renewable technologies. China's State Grid Corporation has outlined plans for a pilot program by 2010 and deployment by 2030. China has engaged in several collaboration agreements on the deployment of renewable energy sources.
United States	Extensive smart grid funding available as part of the American Recovery and Reinvestment Act (ARRA). USD 4.5 billion allocated to grid modernization. USD 3.3 billion for the integration of proven technologies into existing electric grid infrastructure. USD 7.25 billion in loans for transmission infrastructure projects. USD 100 million for regional smart technologies demonstrations. USD 515 million for energy storage and demonstrations.
Japan	The Federation of Electric Power Companies of Japan plans to develop a smart grid by 2020 that incorporates solar power generation with investment support of USD 104 million from the Japanese government. The Japanese government has announced a national smart metering initiative and major utilities have announced smart grid programs.
South Korea	The Korean Government has launched a USD 65 million pilot program on Jeju Island with major players in the industry. The program consists of a fully integrated smart grid system for 6000 households, wind farms and four distribution lines. Korea plans to reduce electricity consumption by 10% and 41 million tons of CO_2 by 2030. Korea also plans to implement smart grids nationwide by 2030.
Spain	The utility Iberdrola is coordinating the PRIME smart metering project, which aims to develop a public, open, and standard automated metering infrastructure. The utility Endesa aims to deploy automated meter management to more than 13 million customers on the low voltage network from 2010 to 2015, building on past efforts by the Italian utility ENEL.
Germany	The German government's E-Energy funding program has several projects focusing on specific smart grid technologies.
Australia	The Australian government announced the AUD 100 million "Smart Grid, Smart City" initiative in 2009 to deliver a commercial scale smart grid demonstration project. Additional efforts in the area of renewable energy deployments are resulting in further study on smart grids.
United Kingdom	The U.K. energy regulator OFGEM has outlined an incentive initiative called the Registered Power Zone that will encourage distributors to develop and implement innovative solutions to connect distributed generators to the network.
France	The electricity distribution operator ERDF is deploying 300,000 smart meters in a pilot project based on an advanced communication protocol named Linky. If the pilot is deemed a success, ERDF will replace all of its 35 million meters with Linky smart meters from 2012 to 2016.
Brazil	APTEL, a utility association, has been working with the Brazilian government on narrowband power line carrier trials with a social and educational focus. Ampla, a power distributor in Rio de Janeiro State owned by the Spanish utility Endesa, has been deploying smart meters and secure networks to reduce losses from illegal connections. AES Eletropaulo, a distributor in São Paulo State, has developed a smart grid business plan using the existing fiber-optic backbone. The utility CEMIG has started a smart grid project based on system architecture developed by the IntelliGrid Consortium, an initiative of the United States-based Electric Power Research Institute (EPRI).

Source: Major Economies Forum (MEF) on Energy and Climate, Technology Action Plan (TAP): Smart Grids. Prepared by Italy and Korea in consultation with MEF Partners, December 2009. http://www.majoreconomiesforum.org/ (Accessed on March, 2011).

5.3 CANADA

5.3.1 LEAD ORGANIZATIONS

Activities associated with the development of the smart grid in Ontario are well known in North America. The Independent Electricity System Operator (IESO), the Ontario Power Authority (OPA), and the Ontario Energy Board (OEB), along with transmitters and distributors, are coordinating their efforts to enable renewable energy to be brought into service more quickly and efficiently. There are over 80 electricity distribution utilities in Ontario and one major Crown corporation, Hydro One. Hydro One chose Trilliant for its AMI deployment over its large territory [8]. While Ontario will achieve full deployment earlier than either California or Texas, the two states leading the smart meter charge in America, IDC Energy Insights forecasts that smart meter deployments in the rest of Canada will trail deployment in the United States. IDC believes that British Columbia, lead by the full deployment of BC Hydro (slated to begin mid-2011) will lead the next wave of smart meter deployment in Canada, followed by Quebec and Saskatchewan. Outside of BC Hydro, relatively few major utilities have advanced AMI implementation plans [9].

5.3.2 PROJECT CATEGORY/TECHNOLOGY

Demand response: These capabilities have been rolled out in several Canadian jurisdictions to date.

DG: In Ontario, the Energy Board has directed that it is the responsibility of the generator to mitigate any negative effects that connected supply may have on the distribution grid in terms of voltage variances and power quality. The optimal solution set to accomplish this, however, is still being examined.

ENMAX, the local electricity distributor in Calgary, deployed a major distribution automation project to improve the reliability of the distribution system. More than 175 new automated switches made by S&C Electric have been rolled out. Automated Meter Reading (AMR) is now in place to collect meter data. Recently, ENMAX announced a major partnership with Cisco Systems Inc. in managing data from building energy management systems.

A maritime consortium consisting of New Brunswick Power, Nova Scotia Power, Maritime Electric, and Saint John Energy received a grant from the CEF for a 4-year, C$38M demonstration project. The PowerShift Atlantic project aims to balance wind energy with dispatchable loads and thermal storage. The project will involve the installation of monitoring and control systems in 2000 buildings in PEI, NB, and NS.

5.3.3 PROJECT EXAMPLES

Hydro One Networks: Initial meter deployments began in late 2006 and, in December 2008, meter deployment surpassed the notable milestone of 700,000 meters installed

The Alberta Utilities Commission has been directed to review how smart grid technology, such as advanced metering or smart metering infrastructure, can be used to modernize the electricity system in Alberta.

At Fortis Alberta, the utility serving the southern part of Alberta province, a major AMR project was concluded in 2010. Over 400,000 meters were deployed. BC Hydro launched the Smart Metering Program and the Smart Grid Program in line with the new British Columbia Clean Energy Act. The deployment of smart meters was announced in January 2011. The development of a distribution management system (DMS) with Telvent Canada Ltd., based in Calgary, will usher in modern grid applications such as volt and VAr and network reconfiguration.

As of July 2010, the federal government has committed to a 4-year investment of $32 million toward a New Brunswick Power Smart Grid research project. Almost half of the funding is drawn from the Clean Energy Fund, with partners including Nova Scotia Power and the University of New Brunswick.

Global Smart Grid Initiatives

5.3.4 Regional Drivers and Benefits

In comparison to the United States, IDC has observed a greater tendency amongst Canadian policy-makers and utilities to favor caution over the desire to become a first mover. Some utilities, such as Newfoundland Power, have explicitly expressed a desire to await the maturation of the smart meter industry, noting that standardization of communications protocols and increased manufacturing capacity will drive down unit costs and improve the cost effectiveness of smart meter technology. Some drivers are

1. Enabling higher levels of efficiency and demand response (and better EM&V), distributed and renewable resources
2. Deferral of costly new power plants and power lines
3. Getting ahead of mass use of PHEVs—automate off-peak charging and V2G (on-peak discharging)
4. Giving customers more control over energy bills and letting them participate in electricity market
5. End-to-end system integration and system efficiencies
6. Calls for higher reliability
7. Stimulus funding [8]

5.3.5 Country-Specific Drivers and Benefits

Each province has its own priority while building their clean energy strategy and defining the role of a smart grid. Several utilities are proactive in increasing their grid reliability, while others are leading the way in reducing demand or increasing the use of renewable energy. Provinces with substantial wind, hydro or fossil fuel resources have so far tailored their electrical industry to their natural resources.

With greater smart grid activity across Canada, there is an increasing need for building awareness, sharing of knowledge and progress, and collaboration on both research and deployments. SmartGrid Canada has recently been formed as a not-for-profit alliance to foster that advancement. There is a goal of reducing greenhouse gas emissions to 17% by 2020. Heightened expectations ("SG as a panacea"). The electricity industry in Canada is expected to invest $11 billion in infrastructure replacement in each of the next 20 years just to replace existing assets.

Industry is also providing leadership through its participation in new initiatives, such as the Canadian Smart Microgrid Research Network and the Smart Grid Standards Task Force. Canada has a great role to play in smart grid development in North America and around the world, by enhancing the innovative culture and the technical expertise that is already evident in many areas across the country.

5.3.6 Scale

The installed Smart Meter base in Canada surged past 5 million in 2010 largely on the strength of Ontario's push toward full deployment. IDC Energy Insights estimates that more than 95% of the nearly 4.7 million eligible endpoints in Ontario had been outfitted with smart meters by the end of 2010. Alberta, British Columbia, Manitoba, New Brunswick, Nova Scotia, and Quebec have all flirted with smart metering, though, other than Fortis' PLC deployment in Alberta, installations outside of Ontario have been modest. Between now and 2025, it's estimated that Ontario must build an almost entirely new electricity system—including replacing approximately 80% of current generating facilities [10].

One important initiative is the Clean Energy Fund (CEF), which was announced by Natural Resources Canada on May 19, 2009 and included a demonstration component for Renewable and Clean Energy Technologies. Nineteen demonstration projects were announced in six technology areas, including energy storage and smart grids. These projects accounted for $146 million in CEF contributions [11].

5.4 CHINA

5.4.1 Lead Organizations

Of the 31 Chinese provinces, State Grid Corporation of China (SGCC) has been operating utilities in all except five southern provinces managed by China Southern Power Grid. Under the governance of SGCC, each province has its own power utility company [3].

5.4.2 Project Category/Technology

Projects are mainly focused on the following points:

1. Efficiency for power facilities
   a. Substation automation, WAMPAC, SCADA, and EMS
2. Efficiency in energy usage
   a. AMI, demand response and dynamic pricing, energy storage, smart appliances, HEMS/BEMS, EV management, and V2G
   b. Efficiency in Utility ICT Systems
   c. Cybersecurity, IP networking management, and infrastructure
3. Efficiency in derivative smart grid applications
   a. OMS, asset management and monitoring, MDMS, and distributed power generation

5.4.3 Project Examples

The "Strong Smart Grid" concept (including the Ring project to upgrade transmission lines) indicates that an UHV/UHVDC-based grid is the backbone of the Chinese smart grid.

5.4.4 Regional Drivers and Benefits

The following are key drivers for current smart grid deployments over the next few years:

1. Growing need for electricity under supply constraints
   a. According to the International Energy Agency (IEA), energy demand will continuously grow by 55% between 2005 and 2030
   b. The share of electricity in the energy mix is increasing
   c. Aging network equipment and limited power line capacity extension
2. Push for a low-carbon energy mix
   a. CO_2 emissions grow faster than energy demand
   b. Twenty percent of renewable energy and 20% reduction in CO_2 by 2020
3. New consumption and production modes for end customer
   a. Smart meter and smart appliances
   b. Home energy management systems and services
   c. Electrified vehicles: Plug-in hybrid and pure electric vehicles

5.4.5 Country-Specific Drivers and Benefits

Smart grid adoption driving force:

1. Diversify its energy economy
2. Reduce environmental problems
3. Stave off massive increases in energy imports

5.4.6 Scale

Currently, China operates a 640 km pilot mega-volt transmission line. By 2012, it intends to install an additional 17,600 km ("Strong Smart Grid" concept).

5.5 EUROPE

5.5.1 Lead Organizations

The leading organizations of the projects are split into the following categories:

1. Energy companies (e.g., EDF)
2. Distribution system operators (e.g., Enel Distribution)
3. Transmission system operators
4. Service providers (manufacturers, aggregators, retailers, IT companies, etc.)
5. Universities, research centers, and public organizations [2]

5.5.2 Project Category/Technology

About 27% of the projects fall in the smart meters category; these projects involve the installation of about 40 million devices for a total investment of around €3 billion. In two other countries, France and Finland, the great majority of the budget is also attributable to smart meter projects.

In France, the demonstration project "Pilot Linky" accounts for about 75% of the total spending, while in Finland the smart meters roll-out project by Fortum accounts for over 80% of the whole budget.

5.5.3 Project Examples

Integrated System projects represent about 34% of the projects and about 15% of the total budget. Most of the technologies are known, but their integration is the new challenge. This result highlights the need to consider the smart grid as a system rather than simply a collection of different technologies and applications.

5.5.4 Regional Drivers and Benefits

The reduction of CO_2 emissions is one of the drivers, even though only few of projects have tried to quantify the impacts of the deployed solutions over the business as usual scenario. Another major driver is the integration of large amounts of DERs, including renewables and storage, into the grid. The large-scale deployment of these technologies entails a high potential for emissions reductions and, at the same time, it can have a positive impact on the diversification of the energy mix, and therefore on energy security.

5.5.5 Country-Specific Drivers and Benefits

Projects are not uniformly distributed across Europe. The majority of them are located in the EU_{15} Member States, while most of the EU_{12} still lag behind. Most of the projects are concentrated in a few countries: Denmark, Germany, Spain and the United Kingdom together account for about half of the total number of projects. Denmark stands out in terms of the number of R&D and demonstration projects. This is partly explained by the fact that Denmark has already achieved a very high penetration of renewables and distributed generation and therefore needs to update its electricity system.

5.5.6 Scale

A few countries stand out in terms of spending. With a budget of over €2 billion, Italy accounts for almost half of the total spending. The great majority of this budget is however attributable to only one project, the Telegestore project, which consisted of the national roll-out of smart meters in Italy. The different pace at which smart grids are deployed across Europe could make trade and cooperation across national borders more difficult and jeopardize the achievement of the EU energy policy goals. The majority of projects are concentrated in the R&D and demonstration phases. Only 7% of the projects are in the deployment phase. R&D and demonstration projects account for a much smaller portion of the total budget. Most of these projects are small to medium size, with an average budget of €4.4 million for R&D projects and about €12 million for demonstration projects.

5.6 INDIA

5.6.1 Lead Organizations

BESCOM's (Bangalore utility) $100 million smart grid pilot project. Bangalore Electricity Supply Company (BESCOM) is preparing to launch the first pilot project related to the smart grid in India. It has already established the India Smart Grid Forum and Smart Grid Task Force to develop a framework and national policy:

1. IIT's (IIT Madras and IIT Kharagpur) collaboration with IBM for smart grid research on phasor measurement units (PMUs)
2. MDI's smart grid educational program
3. NDPL's collaboration with GE for smart grid development in efficiency for power distribution

5.6.2 Project Examples

The Restructured Accelerated Power Development and Reform Programme (R-APDRP):

1. *Phase 1*: Upgrade transmission and distribution networks and secure robust and transparent accountability in power services
2. *Phase 2*: Enhance power grid operational efficiency and automated control
3. *Phase 3*: Final phase of the R-APDRP roadmap and smart grid deployment in India

5.6.3 Country-Specific Drivers and Benefits

India suffers from electricity deficiency and a lack of national infrastructure caused by electricity theft and aging equipment.

5.6.4 Scale

Electricity Act (2003) implies:

1. Full provision of electricity to Indian households in 5 years
2. Support for consumption per capita of 1000 kWh
3. Mandatory installment of metering systems for Indian customers BESCOM's (Bangalore utility) $100 million smart grid pilot project

5.7 JAPAN

5.7.1 Lead Organizations

Toyota, Japan Wind Development, Panasonic Electric, and Hitachi. Key participants of smart network projects are NTT DOCOMO, NEC Corp., Sekisui House Ltd., and NAMCO BANDAI Games Inc.; Mitsubishi and Japan NEDO—for EV-based SG project.

5.7.2 PROJECT EXAMPLES

The "Smart Network Project" aims to build a sustainable low-carbon society through the development of smart grids, networked home appliances, and electric vehicles connected via common communications standards; EV-Smart Grid Project.

5.7.3 COUNTRY-SPECIFIC DRIVERS AND BENEFITS

Japan's vision is focused on the development of a low-carbon society—Ample Solar Power, Promotion of Energy Savings in general consumers.

5.7.4 SCALE

The potential capacity from solar in Japan is equivalent to the total power generation from 50 nuclear power plants.

5.8 KOREA

5.8.1 LEAD ORGANIZATIONS

KEPCO's investment of $4.2 billion annually toward all elements of power utility systems; Hyundai, SK Energy, POSCO, LG Electronics, SK Electronic Corp.

5.8.2 PROJECT EXAMPLES

Jeju Smart Grid Demonstration: Smart Transportation (electrified vehicles, charging solutions, and EV batteries)

1. Smart Place (AMI, smart meters, and HEMS)
2. Smart Renewables (renewable energy sources and microgrids)
3. Smart Power Grid (the advancement of existing transmission and distribution systems in power grids)
4. Smart Electricity Service (energy marketplace and electricity retail service)

5.8.3 COUNTRY-SPECIFIC DRIVERS AND BENEFITS

Korea faces an imminent problem related to energy security and maintaining steady growth in its economy. Due to the fast-changing global trends, surrounding energy and the environment, the Korean government established a "Green Growth" policy and smart grid roadmap as a part of its national policy.

5.8.4 SCALE

The Korean government has swiftly launched organizations, legislation, and support for its vision: nationwide smart grid completion by 2030. Currently, over 160 Korean enterprises are participating in the Jeju Smart Grid Demonstration Project, which has a 42-month timeline—from December 2009 to May 2013—and funding of $103 million ($55 million awarded by the Korean government).

5.9 LATIN AMERICA

5.9.1 LEAD ORGANIZATIONS

Brazil is leading the way. Echelon Corporation and ELO Sistemas Electronicos S.A. (ELO) announced a strategic alliance to bring smart grid solutions to the Latin American energy market.

ELO, the number one supplier of electronic meters and systems in the Brazilian electricity supply market, has become a value-added reseller of Echelon's market-leading Networked Energy Services (NES) System, an advanced metering infrastructure solution.

5.9.2 Project Category/Technology

The smart meter market alone will reach 104.5 million meters and $25.1 billion by 2020. Other smart grid technologies—including distribution automation—will also grow quickly in the region [8].

5.9.3 Project Examples

Brazil's energy regulator, ANEEL, announced tentative plans for a nationwide rollout of smart metering, expecting to replace approximately 63 million electricity meters in the country with smart meters by 2021.

5.9.4 Regional Drivers and Benefits

The Central America Electrical Interconnection System (SIEPAC), developed under the Plan Puebla Panama (PPP): this plan would bring $590 M in new transmission investments, with the goal of attracting more foreign investment in generation assets for the area. The Initiative for Regional Infrastructure Integration in South America (IIRSA), which spans beyond electricity integration to every area of major infrastructure. IIRSA would also harmonize regulatory frameworks and introduce new financing models.

5.9.5 Country-Specific Drivers and Benefits

While Brazil has a growing renewable energy sector, the aging infrastructure has made incorporating renewables difficult. The grid was initially designed for one-way power flow, from a central generating facility to the consumer. A Smarter Grid will be needed to help integrate this renewable power into the grid.

5.9.6 Scale

Brazil will be first to begin large-scale deployments. Chile and Argentina will be next to follow, with other countries set to launch deployments in the latter half of the decade.

5.10 UNITED STATES

5.10.1 Lead Organizations

A new GTM Research report identifies 10 states that are emerging as key smart grid policy leaders. The 10 states identified in the report are California, Colorado, Florida, Massachusetts, New Jersey, New York, North Carolina, Ohio, Pennsylvania, and Texas.

Major players are PG&E, Xcel Energy, Duke Energy, SDG&E, Southern California Edison Co., and American Electric Power [4].

California and Texas are the two states with the highest number of smart grid projects of around 20 in each of these states.

Utilities are also forming alliances and working closely with companies outside of their industry as part of the transformation of the power grid [5].

5.10.2 Project Category/Technology

The current smart grid activity in the United States mostly reflects activity undertaken to implement AMI. It is a generally accepted concept that AMI is often a precursor or foundational element to smart grid or that the activity of smart grid efforts would incorporate levels of AMI [6].

Most U.S. smart grid activity is limited to smart metering initiatives in pilot projects; no near term investable themes for utility investors. However, the industry mindset is shifting toward the importance of the communications layer to support distribution automation and other advanced applications [4].

ON World's survey of U.S. smart grid projects [13] found that about a third plan to enable two-way communications with their customers and another third are developing their demand response strategies. Over 1 million in-home devices are funded by the U.S. smart grid awards including energy displays, thermostats, and load controllers. Nationwide, utilities now have more than 200 smart grid projects underway.

5.10.3 Project Examples [4]

Pacific Gas & Electric plans full deployment of 10 million meters by mid-2012. By the end of 2009, PG&E had installed 4.5 million meters. The company expects to spend approximately USD 535 million and USD 165 million in 2010 and 2011, respectively, on this programme.

Baltimore Gas & Electric (BG&E) is planning to install smart meters for 1.2 million customers throughout its service area. The project will cost the utility approximately USD 800 million, and USD 200 million Federal stimulus grant will lower the cost to the residential customer by about 80%. The company expects at least USD 2.5 billion in savings for BG&E customers over the life of the project. Rollout will start in late 2011 with full deployment expected by mid-2014.

CenterPoint Energy (Houston, Texas)—to complete the installation of 2.2 million smart meters and further strengthen the reliability and self-healing properties of the grid by installing more than 550 sensors and automated switches that will help protect against system disturbances like natural disasters (total project value—USD 638 million).

Salt River Project, which has more than 480,000 customers served by smart meters. Almost half, or 222,000 have elected the utility's Time-of-Day price plans, making it the nation's third largest program of time-of-use pricing.

Southern California Edison Co., which began installing the first of a planned 5 million meters in late 2009 under its 5 year USD 1.6 billion SmartConnect program, estimated that the program will realize cost savings of USD 4.6 billion, assuming all meters are installed by 2013.

5.10.4 Regional Drivers and Benefits

Improve supply reliability, quality, grid resilience, and peak load reductions. Diversify energy dependencies and secure energy supply. Integrate with other initiatives—smart city, electric vehicles, renewables, empower, and inform customers. Reduction in peak demand is overwhelmingly the leading benefit that utilities expect.

Challenges: Some examples of poor execution in early smart metering pilots have increased consumer and regulator sensitivity to smart meters. [7].

The American Recovery and Reinvestment Act (ARRA) has placed an unprecedented funding resource in the hands of DOE, resulting in the Smart Grid Investment Grant (SGIG) program and the Smart Grid Demonstration (SGD) program ("Smart Grid Programs"). As part of the Smart Grid Programs, DOE will award approximately $4 billion to utilities, equipment suppliers, regional transmission organizations, states, and research organizations to jump start smart grid deployment and demonstration on a massive scale.

5.10.5 Country-Specific Drivers and Benefits

Today, a typical U.S. utility plans a deployment period of 5 or more years for AMI or smart grid. Factors that limit the potential program acceleration are primarily attributed to capital-spending constraints, workforce reduction issues, and supply-chain barriers. While the customers that are first to receive advanced meters or be served by automated distribution circuits start to see direct benefits immediately, the broader social benefits and cost savings to the utility normally are not fully achieved until the deployment reaches a critical mass [6].

The American Recovery and Reinvestment Act of 2009 (stimulus package) earmarked approximately USD 4 billion of Federal funds for smart grid projects and grid upgrades. Utilities have access to these funds to assist in the rollout of smart meters. A large percentage of North American utilities have formal smart grid strategies in place.

5.10.6 Scale

The $3.4 billion in grant awards are part of the American Reinvestment and Recovery Act.

1. Private industry is adding an additional $4.7 billion in matching funds

In total, the grants will result in the installation of more than 200,000 smart transformers and more than 850 phasor measurement units—sensors that can assist grid operators in monitoring grid efficiency.

During the next 4 years, KEMA's projection anticipates that a potential disbursement of $16 billion in smart grid incentives would act as a catalyst in driving associated smart grid1 projects that are worth $64 billion [6]. KEMA's report, "U.S. Smart Grid: Finding New Ways to Cut Carbon and Create Jobs," identifies 334 U.S. locations in 39 states that are already developing or manufacturing products for a smart grid. The region with the largest number of sites is the southeast, with California having the most sites of any one state.

The Federal Energy Regulatory Commission (FERC) has a target of 80 million meters to be installed by 2019. To date, California has made the most progress in implementation because State law requires the deployment of meters.

Texas is also deploying meters and Baltimore is about to begin a major initiative.

5.11 OTHER COUNTRIES

5.11.1 Lead Organizations

Taiwan: Taipower is the major player.

Malaysia: Tenaga Nasional Berhad, the largest electric utility company in Malaysia.

Singapore: Global players such as Accenture, IBM, Logica, and Siemens have been participating in a myriad of subprojects within the IES pilot projects, including smart metering, EV tests, renewable energy, and the smart grid.

The Philippines: Global players such as Schneider Electric and Siemens are participating in the untapped smart grid market in the Philippines. Manila Electric Co. (Meralco), the country's biggest power distributor; Philippine Long Distance Telephone Co. (PLDT).

5.11.2 Project Category/Technology

Singapore: In November 2009, the Energy Market Authority (EMA) launched the IES projects to test a range of smart grid technologies to enhance the capabilities of Singapore's power grid infrastructure.

5.11.3 PROJECT EXAMPLES

Malaysia: Tenaga Nasional Berhad, the largest electric utility company in Malaysia with over 7 million customers and the largest power company in Southeast Asia, worked in tandem with Itron to deploy meter data management software in 2007.

Indonesia: Itron has been working with Indonesian state-owned electricity utility PT Perusahaan Listrik Negara (PLN) to supply 800,000 keypad smart payment residential meters.

5.11.4 REGIONAL DRIVERS AND BENEFITS

These nations face common challenges, including a lack of electricity sources and aged and inefficient grid infrastructure. The overall investment in and expansion of power services to consumers are top priorities in power and utility services in these emerging countries. They believe that changes in architecture T&D networks are necessary to develop a smart grid or an advanced architecture with a controlled grid environment.

5.11.5 COUNTRY-SPECIFIC DRIVERS AND BENEFITS

Singapore: Singaporean market players aim to lure global high-tech and cleantech companies with smart grid capabilities by offering a positive R&D environment and an advanced power grid structure. Long-lasting market perceptions about benefits related to global R&D and APAC headquarters in Singapore contributes to economic growth in the country.

Indonesia: This strong economy is driving demand for energy measured by over 19% by 2011. As such, the Indonesian government has pledged to increase power generation, transmission, and distribution throughout the country over the next 5 years and provide "Electricity for All" by 2020.

The Philippines: Electric Power Industry Reform Act of 2002:

1. The National Transmission Corporation (TransCo) was created to assume the power transmission functions from the National Power Corporation (NPC).
2. NPC agreed to focus on the development and operation of small island grids and manage the generation assets and IPP contracts.
3. Wholesale Electricity Spot Market (WESM) for trading of energy among generators/suppliers and customers started in June 2006 + demand factors.

5.11.6 SCALE

Taiwan: Taipower has announced the following plans:

1. Smart meter trial tests for industrial customers with 23,000 units and for general consumers with 10,000 units by 2012
2. Smart meter deployment in 1 million households by 2015
3. Long-term goal of reaching 6 million households (50% of total households in Taiwan)

Malaysia: The software supports the utility's remote meter reading technology for 90,000 low-voltage C&I meters in order to automate meter data collection, analysis, and billing process.

Singapore: the EMA has independently called an open tender to procure about 4500 smart meters to be used in the IES pilot project. Implementation will occur in residential areas and C&I locations to test and evaluate workable solutions.

5.12 GLOBAL COLLABORATIVE EFFORTS

5.12.1 United States, Canada, Ireland, France, Australia, and Japan

5.12.1.1 Lead Organizations

Electric Power Research Institute—a 7 year international Smart Grid Demonstration Collaboration with 23 utilities.

5.12.1.2 Project Category/Technology

Integration of DER into a virtual power plant. Includes end-to-end integration of all types of resources from demand response, renewables, storage, and traditional distributed generation.

5.12.1.3 Project Examples

AEP—Virtual Power Plant Simulator; Electricité de France (EDF) Smart Grid Demonstration Project—PREMIO: Distributed Energy Resources Aggregation and Management; Consolidated Edison Smart Grid Demonstration Project—Interoperability of Demand Response Resources; KCP&L Smart Grid Demonstration Project—The Green Impact Zone.

5.12.1.4 Regional Drivers and Benefits

Integration of DER shows promise to have a significant impact on emissions reduction and addresses smart grid challenges such as advancing data standards, communications, and advanced integration techniques associated with Smart Grid Integration. A strong emphasis on cost–benefit analysis to determine practicality of smart grid deployments in numerous environments.

5.12.1.5 Country-Specific Drivers and Benefits

A significant research effort associated with this project is cost–benefit assessment and developing an industry framework for continued benefit assessment and leveraging lessons learned from one project to the next from an international perspective.

5.12.1.6 Scale

Twenty-three utilities around the world with 15 large scale demonstrations. The smart grid projects represent over $1.5 billion in smart grid investments with over $30 million directly focused on smart grid research.

5.13 SUMMARY

Most current smart grid projects focus on AMI, demand-side management, and distributed generation in order to utilize existing networks better. Demonstration projects have so far been undertaken on a limited scale and have been hindered by limited customer participation and a lack of a credible aggregator business model. The number of data (and security) challenges will greatly increase as planned large-scale projects are rolled out. Non-network solutions, such as information and communication technology, are being used in a growing number of smart grid projects, bringing a greater dependence on IT and data management systems to enable network operation [14].

Although significant effort and financial resources are being invested in smart technologies, the scale of demonstration, deployment, and coordination needs to be increased. Acknowledgement of this need has spawned additional efforts to increase collaboration on global piloting, demonstration, and deployment of smart technologies, especially in the areas of

- Policy, standards, and regulation
- Finance and business models

- Technology and systems development
- User and consumer engagement (including behavioral and social studies)
- Workforce skills and knowledge

An example of a smart grid collaborative effort to leverage the lessons learned across the industry is the U.S. Electric Power Research Institute (EPRI) 7 year International Smart Grid Demonstration project. This project has 23 utilities as members including utilities in the United States, Canada, Ireland, France, and Australia and a resident researcher from Tokyo and Japan. A strong emphasis on this effort is related to developing and applying a comprehensive cost–benefit analysis framework and research plan to ensure a strong collaborative foundation for the multitude of projects around the world.

Several concepts are emerging that extend the reach of the smart grids projects to broader energy and societal contexts. One of these is the smart community or smart city. A smart community integrates several energy supply and use systems within a given region in an attempt to optimize operation and allow for maximum integration of renewable energy resources—from large-scale wind farm deployments to microscale rooftop photovoltaics and residential EMS. This concept includes existing infrastructure systems, such as electricity, water, transportation, gas, waste, and heat, as well as future systems like hydrogen and EV charging. The goals of such integration through the use of information and communication technology (ICT) include increased sustainability, security, and reliability, as well as societal benefits such as job creation and better services. Smart communities are a logical extension of smart grids, which are ultimately expected to evolve in this direction.

REFERENCES

1. Zpryme. (2010). Smart Grid Snapshot: China Tops Stimulus Funding. Retrieved July 2011 from http://www.zpryme.com/SmartGridInsights/2010_Top_Ten_Smart_Grid_Stimulus_Countries_China_Spotlight_Zpryme_Smart_Grid_Insights.pdf
2. JRC Reference Report—Smart Grid projects in Europe: Lessons learned and current developments, European Commission Joint Research Centre Reference Report (JRC65215), Publications Office of the European Union, 2011.
3. Pike Research Report, Smart Grid in Asia Pacific, 2Q 2011.
4. GTM Research, The 2010 North American Utility Smart Grid Deployment Survey, David Leeds, GTM Research, February 2010.
5. Smart Grid Growing Plan, PricewaterhouseCoopers, 2010.
6. The U.S. Smart Grid Revolution KEMA's Perspectives for Job Creation, KEMA, Inc., January 2009.
7. The World Economic Forum's September 2010 report, Accelerating Successful Smart Grid Pilots, World Economic Forum in association with Accenture, September 2010. Accessed via http://www.weforum.org/reports/accelerating-successful-smart-grid-pilots
8. Lisa Schwartz, RAP, "Tour of Smart Grid Projects and State Policies," Sept. 9, 2009, at http://raponline.org/docs/RAP_Schwartz_SmartGridProjectsandPoliciesORwks_2009_09_09.pdf
9. Canadian Electricity Association, www.electricity.ca/
10. North America Utility Industry 2011 Top 10 Predictions, report by IDC Energy Insights, January 2011, Document #E1226479.
11. Smart Grid in Canada - Overview of the Industry in 2010, report by CanmetENERGY, 8 February 2011, Report 2011-027 (RP-TEC) 411-SGPLAN.
12. Smart Grid Technologies, Pike Research Report, December 2010, http://www.pikeresearch.com/research/smart-grid-technologies (Accessed in March, 2011).
13. ON World Inc, San Diego, CA, November 4, 2009, http://www.onworld.com/html/newssmartgrid.htm
14. Boots, M., Thielens, D., and Verheij, F. 2010. *International example developments in Smart Grids-Possibilities for application in the Netherlands (Confidential report for the Dutch Government),* KEMA Nederland B.V., Arnhem.

BIBLIOGRAPHY

1. Electric Power Research Institute, www.smartgrid.epri.com
2. KEMA on Smart Grid http://www.kema.com/services/ges/smart-grid/Default.aspx
3. Northeast Group, LLC, Smart grid market research reports at http://www.northeast-group.com/research.html
4. Smart Grid: Enabler of the New Energy Economy, a report by the Electricity Advisory Committee, December 2008, available at http://www.oe.energy.gov/eac.htm
5. Smart Grid Information Clearinghouse. www.sgiclearinghouse.org
6. U.S. Department of Energy, *The Smart Grid: An Introduction* (Washington, DC: U.S. Department of Energy, 2008), http://www.oe.energy.gov/1165.htm

6 Smart Grid: Where Do We Go from Here?

Stuart Borlase, Tim Heidel, Charles W. Newton, Bartosz Wojszczyk, and Eric Woychik

CONTENTS

6.1 The Next Few Years 549
6.2 Market Drivers and Enablers 550
    6.2.1 Technical Innovation 552
    6.2.2 Policy and Regulatory Priorities and the Role of Electricity Markets 554
    6.2.3 Economic Growth and Changes in the Global Electric Power Market 556
6.3 Fad, Failure, or Fame? 559
References 559

From where we are today the smart grid has promised many glories of electric grid change, promises of utility heroes, incredible technology breakthroughs, dreams of consumer enamourment, and the promise of additional jobs. There have certainly been some changes in the industry since intelligent grids and smart grids (and other related spin-offs) became the excitement of marketing people. Even during the writing of this book, there has been significant change in the thinking around smart grids and their benefits. The fundamental questions always remain, "what exactly is smart grid?," "how do we get there?," and the favorite "is there anything real about smart grid?." Utilities and vendors alike have been clamoring to be the first in smart grid, many ideas of smart grid have been thrown out, there has been week after week of some kind of smart grid meeting or conference, and there have been some serious efforts at making sense of it all through concerted efforts and initiatives worldwide.

It is clear that in the years and decades ahead the smart grid will increasingly take shape from the merging of two primary infrastructures: the electrical power system and the communications infrastructure. Smart grid programs will incorporate advanced applications to harness the full set of demand-side options (distributed energy resources, demand–response, energy efficiency, and storage), communications options, information management, advanced metering infrastructure (AMI), and automated control technologies. These resources will modernize, optimize, and transform electric power and gas infrastructures. Smart grid visions seek to bring together these technologies to make each grid element, from the customer's circuit to the bulk power grid, self-healing, more reliable, safer, and more efficient. At the same time, the smart grid promises to empower customers to use electricity more efficiently and contribute to a sustainable future with improvements to national security, economic growth, and climate change.

6.1 THE NEXT FEW YEARS

Newton-Evans Research Company, Inc. recently developed a smart grid-related market outlook through 2015 based on insights gathered from dozens of recently completed industry and utility surveys and secondary research activities. Equipment manufacturers, systems integration specialists, consultants, and software developers, together with other representatives of the power industry's "supply side" were surveyed or interviewed for their outlooks and opinions on where the "smart grid" market and each smart grid market segment is headed in the near term (2011–2012) and the mid-term (2013–2015).

At the global level, 2012 is likely to be the year in which the "basket" of smart-grid related expenditures will approach $10 billion for the first time. The expenditure estimates in Table 6.1 exclude internal utility spending for staff services to support smart grid procurements. The value of such services can account for as much as one-third or more of a total smart grid project cost. For control systems in particular, there are significant staff resources applied in both the developmental and operational stages of an implementation. There are other costs not reported here such as the cost to construct and furnish an operations center, or the ongoing (O&M) costs of operating a utility private telecommunications network.

Newton-Evans Research Company, Inc. indicated that some smart grid business opportunities would continue unabated in the years ahead, especially for the following:

- Systems operations (visualization and situational awareness)
- Physical and cybersecurity advances
- Data analytics
- Advanced metering (in states/provinces/countries with regulatory foresight and where the business case and customer education requirements have been made)
- Advanced communications networks with greater bandwidth and lower latency—serving as enablers for several smart grid components (*including phasor measurement implementations at transmission substations; distribution network automation; and metering data acquisition*)
- Utility and enterprise systems integration activities

Figure 6.1 gives one view of global smart grid and information technology (IT) expenditure in 2020. Note the inclusion of "overlap" portions of administrative IT investments and certain operational IT investments as well as "pure" smart grid-related spending in the formulation of what may well grow to become a $21 billion portion of the electric power industry.

6.2 MARKET DRIVERS AND ENABLERS

Each electricity market has its own set of ever evolving political, regulatory, and commercial drivers that will influence the capabilities required of the electricity network over time. Each electricity network will require a different level of technological development to deliver the desired capabilities. Levels of smart grid capabilities, barriers to development and deployment, and solutions will also differ significantly among countries, markets, and individual networks. No single smart grid solution will be appropriate to all electricity markets or networks. Creating the smart grid will require a significant effort by a wide range of stakeholders—federal, state, and local government, utilities, vendors, policy and regulatory agencies, advocacy groups, consumers, academia, and others. Many of these stakeholders can be expected to have different, and sometimes conflicting, priorities and goals.

The growth and evolution of the electric power industry over the past century has been, in part, a story of continuous technological innovation. The National Academies of Engineering recently named the construction of electric grids the "supreme engineering achievement of the twentieth century [1]." However, as described in Chapter 1, just as importantly, the pace and direction of change in the industry has always been heavily influenced by ever shifting regulatory designs and policy priorities and the pace of global economic growth. These three factors: (1) technical innovation, (2) shifting regulatory and policy priorities, and (3) global economic growth will have a strong impact on the evolution of smart grids over the next several decades. Finally, change within the

TABLE 6.1
Global Spending for Major Components of Smart Grid (in Millions of U.S. Dollars)

	2008	2009	2010	2011	2012	2013	2014	2015	Total
Control systems	$665.0	$630.0	$740.0	$830.0	$885.0	$945.0	$1,010.0	$1,050.0	$6,755.0
Systems + integration	$525.0	$495.0	$580.0	$650.0	$690.0	$740.0	$775.0	$800.0	$5,255.0
Third-party Sw/Svcs	$140.0	$135.0	$160.0	$180.0	$195.0	$205.0	$235.0	$250.0	$1,500.0
Outage management	$90.0	$90.0	$98.0	$120.0	$145.0	$175.0	$190.0	$225.0	$1,133.0
Substation automation	$1,200.0	$1,165.0	$1,265.0	$1,370.0	$1,435.0	$1,535.0	$1,640.0	$2,035.0	$11,645.0
SA programs	$300.0	$275.0	$290.0	$325.0	$355.0	$380.0	$405.0	$635.0	$2,965.0
RTUs	$650.0	$665.0	$700.0	$735.0	$755.0	$800.0	$850.0	$975.0	$6,130.0
Other IEDs	$250.0	$225.0	$275.0	$310.0	$325.0	$355.0	$385.0	$425.0	$2,550.0
AMI-AMR	$2,158.4	$2,359.2	$2,620.2	$3,112.1	$3,363.1	$3,588.9	$3,814.8	$4,040.7	$25,057.3
Protection and control	$1,750.0	$1,775.0	$1,900.0	$2,020.0	$2,065.0	$2,100.0	$2,125.0	$2,155.0	$15,890.0
Utility telecoms	$510.0	$525.0	$545.0	$565.0	$590.0	$625.0	$660.0	$695.0	$4,715.0
Telecoms for control sys.	$350.0	$345.0	$375.0	$380.0	$385.0	$400.0	$415.0	$430.0	$3,080.0
Telecoms for AMI	$160.0	$180.0	$170.0	$185.0	$205.0	$225.0	$245.0	$265.0	$1,635.0
Distribution automation	$1,440.0	$1,600.0	$1,735.0	$1,940.0	$2,085.0	$2,365.0	$2,540.0	$2,765.0	$16,470.0
DA software/platforms	$125.0	$130.0	$145.0	$170.0	$195.0	$225.0	$260.0	$300.0	$1,550.0
Smart distribution equip.	$1,125.0	$1,250.0	$1,325.0	$1,475.0	$1,550.0	$1,775.0	$1,900.0	$2,060.0	$12,460.0
Telecoms for DA	$190.0	$220.0	$265.0	$295.0	$340.0	$365.0	$380.0	$405.0	$2,460.0
Total	$7,813.4	$8,144.2	$8,903.2	$9,957.1	$10,568.1	$11,333.9	$11,979.8	$12,965.7	$81,665.3

Source: © Copyright 2012 Newton-Evans Research. All rights reserved.

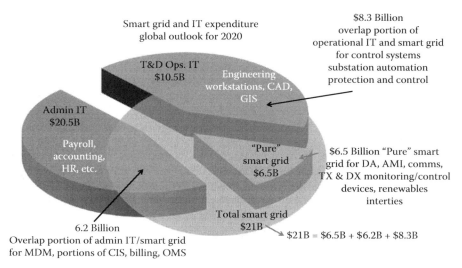

FIGURE 6.1 Smart grid and IT expenditure global outlook for 2020. (Copyright 2012 Newton-Evans Research. All rights reserved.)

electric power industry has also historically been prompted by unforeseen, rare emergency events (e.g., 1965 and 2003 blackouts, the 2000–2001 California energy market crisis). While it is impossible to predict the likelihood of disruptive events in the future, they could also play a critical role in determining the degree to which smart grid visions are realized.

6.2.1 Technical Innovation

The wide deployment of smart grid will yield an abundance of new operational and non-operational devices and technologies connected to the electric grid including smart meters, advanced monitoring, protection, control and automation equipment, electric vehicle (EV) chargers, dispatchable and non-dispatchable distributed generation resources, energy storage, etc. Effective and real-time management and support of these devices will introduce enormous challenges for grid operations and maintenance. The integration of relatively large-scale new generation and active load technologies into the electric grid introduces real-time system control and operational challenges around reliability, security of the power supply, and customer resource options.

These challenges include

- Managing an increasing number of operating contingencies that differ from "system as design" expectations (e.g., in response to wind and solar variability)
- Facilitating the introduction of intermittent renewable and distributed energy resources with limited controllability and dispatchability
- Mitigating power quality issues (voltage and frequency variations) that cannot be readily addressed by conventional solutions
- Integrating highly distributed, advanced control, and operations logic into system operations
- Developing sufficiently fast response capabilities for quickly developing disturbances
- Improving system reliability despite increasing volatility of generation and demand patterns under increasing wholesale market demand elasticity
- Increasing the adaptability of advanced protection schemes to rapidly changing operational behavior (due to the intermittent nature of renewable and distributed generation [DG] resources)

These challenges, if not addressed properly, will result in degradation of service, diminished asset service life, and unexpected grid failures, which will impact the financial performance of the utility's business operations and public relationship image. Meeting these challenges effectively, though, will allow the utility to maximize return on investments in advanced technologies, and represent a strong financial opportunity.

Regional and national demonstrations will help utilities overcome many of the individual challenges introduced above and bring the needed exposure of smart grid principles to energy company executives. Through such projects, the substantial value of integrating suites of technologies rather than deploying multiple, often redundant ones in isolation can be demonstrated. This would create interest, excitement, and the societal, political, and economic stimuli needed to accelerate the deployment of the successful integrated suite of technologies demonstrated. Regional and national demonstrations would also provide the data needed by regulators to assist them in crafting policies and regulations.

Engineering, designing, and operating the smart grid with a comprehensive, integrated vision in mind will enable overall system stability and integrity. An integrated smart grid solution, from field devices to the utility's control room, utilizing intelligent sensors and monitoring, advanced grid analytical and operational and non-operational applications, and sophisticated visualization tools will enable wide-area and real-time operational anomaly detection and system "health" predictability. However, given the current economic climate and regulatory uncertainty, it is unrealistic to expect most utilities to be able to finance an end-to-end smart grid deployment all at once. Instead, utilities should formulate a comprehensive long-term vision for these systems and then carefully consider the most efficient staged deployment that will help them get there.

Many basic smart grid components are likely to be included in the early phases of smart grid deployment plans. Advanced control systems, outage management systems, substation and distribution automation programs, AMI and automated meter reading systems, and new protection and control systems all appear poised for significant growth in the years ahead.

Utility telecommunications systems will also likely see significant early investment in many smart grid plans. Cybersecurity investments will be part and parcel of any intelligently developed IT plan for modern organizations. In today's smart grid marketplace, cybersecurity offerings for operational control systems are available from specialist companies including Industrial Defender, Tripwire, Tec-Sec, Securicon, and others, as well as from control systems providers. Smart grid (read control systems) cybersecurity expenditures are significant, but mostly "internal" with the commercial portion of utility shipments hovering at about $50–75 million per year, inclusive of cybersecurity software bundled with control systems.

Beyond the basic infrastructure components, utilities are also likely to make substantial investments to specialized applications including demand–response systems, home energy management systems (HEMS), and systems that facilitate the introduction of distributed generation, energy storage, and electric vehicles. However, while these areas will contribute additional hundreds of millions of dollars to the overall smart grid "basket" over the next decade, their contribution is less certain than many of the basic components indicated above.

There is an established market for demand–response. In North America, market leaders such as Enernoc and Comverge have been mainstays in the industry for a decade now. The market is comprised of end-user premises-installed devices/switches for heating, cooling, lighting together with the requisite control software to activate programs when demand levels are reached.

Grid scale energy storage is one of the key industry research and development (R&D) areas being studied and developed around the world. Today, the industry is in its infancy, with a handful of commercial contracts signed over the past year, worth perhaps as much as $50 million worldwide.

The HEMS industry is also in its infancy today. There is still a battle over communications methodologies and protocols and for agreeing on one or more of the actual design alternatives available. It is easy to forecast a multi-billion dollar potential in this potential market, but in the real world of today, the market for such devices is currently limited, with most such devices consisting of "smart/programmable thermostats." This area will probably remain an option for developed nations for years to come. Commercial and industrial facility/building energy management systems represent another large, developed market, not considered in this study.

Finally, there are currently fewer than 200,000 electric vehicles operating in North America at mid-year 2011 and another few hundred thousand units elsewhere around the world. It is the EV charging stations that form the smart grid component. In 2011, the total value of worldwide shipments of electric vehicle charging stations is likely to be well under $100 million. The U.S. Department of Energy has provided more than $100 million (as part of the ARRA multi-year initiative from 2009) to push the industry forward. As of July 2011, China has reported disappointing EV sales to date and forecasts cut by 50% to a level of only 250,000 EVs sold by 2015. (*Source*: Lux Populi's weekly newsletter update—July 24, 2011).

As the smart grid takes shape, incompatible equipment will have to be replaced if it cannot be retrofitted to be compatible with smart grid technologies. Unlike many of today's other technologies, which are upgraded frequently such as the personal computer and the cell phone, grid technologies are rarely upgraded—advanced meters are an example. Early retirement of grid assets to incorporate new technologies needed by the smart grid may become an issue with regulators since keeping equipment beyond its depreciated life minimizes the capital cost to consumers. Recovering the cost of retired obsolete equipment that has not been fully depreciated must also be addressed. Utilities may have to modify their smart grid visions to be compatible with legacy equipment.

6.2.2 Policy and Regulatory Priorities and the Role of Electricity Markets

The pace of smart grid deployment will also depend on the design of regulatory systems and federal, state, and local policy priorities. At the federal level, the United States has developed market-related policies to support demand–response and the smart grid. In 2007, the U.S. Congress passed Title XIII of the Energy Independence and Security Act to require the Federal Energy Regulatory Commission (FERC) to focus more on issues associated with the smart grid. As a result, the smart grid and the role of demand–response are top priorities of the FERC, including the basic policy to provide comparability of demand-side and supply-side resources—so that demand–response to be treated more like to supply-side resources at the wholesale level. When market policies allow demand–response and other smart grid services open access and comparability in value monetary value can be defined and accrued. Wholesale trading of power and trading of demand-side resources are then considered to be *comparable*.

FERC's policies seek to link smart grid technologies with the wholesale grid, real-time coordination of information, and further use of demand–response and DG. The FERC is also required to adopt interoperability standards and protocols to enable smart grid functionality and interoperability in regional and wholesale electricity markets. More specifically, Title XIII of the Energy Independence and Security Act of 2007 calls for a number of smart grid characteristics as follows:

1. Increased use of digital information and control technology to improve reliability, security, and efficiency of the electric grid
2. Dynamic optimization of grid operations and resources, with full cybersecurity
3. Deployment and integration of distributed resources and generation, including renewable resources
4. Development and incorporation of demand–response, demand-side resources, and energy efficiency resources
5. Deployment of "smart" technologies (real-time, automated, interactive technologies that optimize the physical operation of appliances and consumer devices) for metering, communications concerning grid operations and status, and distribution automation

6. Integration of "smart" appliances and consumer devices
7. Deployment and integration of advanced electricity storage and peak-shaving technologies, including plug-in electric and hybrid electric vehicles, and thermal storage air conditioning
8. Provision to consumers of timely information and control options
9. Development of standards for communication and interoperability of appliances and equipment connected to the electric grid, including the infrastructure serving the grid
10. Identification and lowering of unreasonable or unnecessary barriers to adoption of smart grid technologies, practices, and services

A number of policy initiatives have been initiated in response, most notably the FERC's smart grid policy [2]. In that statement, five major challenges are highlighted: existing cybersecurity issues, issues associated with changes to the nation's generation mix such as increasing reliance on variable renewable generation resources, issues that could arise with increased and more variable electricity loads associated with transportation technology, and the overarching need for standardization of communication and coordination across intersystem interfaces.

FERC, which regulates wholesale interstate power transactions, has placed very high priorities on non-discriminatory open access to high-voltage transmission grids, and both demand–response and the smart grid. There are now 10 separate electricity markets in North America that are organized as either independent system operators or regional transmission (ISO/RTO) organizations.* The general view is that the growth and expansion of ISOs/RTOs has lead to further unbundling of electricity grid services, which facilitates a number of smart grid services and the monetization of electricity services. ISOs/RTOs provide for bulk, high-voltage power grid transactions, but as a result are limited, as they do not reach into lower-voltage retail utility and distribution services. Similar electric market structures have been adopted in a host of other countries and regions, including Australia, Brazil, Canada (Alberta, Manitoba, New Brunswick, Ontario), Chile, Czech Republic, Northern Europe, Ireland, New Zealand, Russia, Singapore, and the United Kingdom, though in many of these regions and countries more advanced policies and market mechanisms are needed to further enable electricity markets and smart grid services to develop.

The question is whether wholesale electricity markets organized as ISOs/RTOs will further develop policies and market drivers to encourage the smart grid. Some ISOs/RTOs seem to have been more successful in providing policies and market mechanisms that enable smart grid services to participate and directly monetize market benefits. The Pennsylvania–Jersey–Maryland (PJM) market stands out for its success in these regards but still faces major challenges. While the progress of ISOs/RTOs to date in the United States has been significant in these markets, these efforts do not directly impact regional electric and gas utilities that are not subject to FERC's jurisdiction. As much of smart grid technology is located at levels within retail and distribution utilities, FERC policies on wholesale power transactions and the smart grid do not directly apply. Moreover, efforts to create open access for demand–response providers has met with mixed success, as some utilities seek to avoid the reach of FERC's policies. Still, FERC's smart grid policies and open transmission access both enable extensive technology integration and critically important market value to monetize smart grid benefits. Still, these are very encouraging developments, some of which are being replicated in certain other ISO/RTO markets and regions such as in Northern Europe, California, Texas, and Maryland.

---

* These ISOs/RTOs are as follows: Alberta Energy System Operator, California Independent System Operator, Independent System Operator (of Ontario), ISO New England, Midwest ISO, New Brunswick System Operator, New York ISO, Pennsylvania–Jersey–Maryland (RTO), Southwest Power Pool. See the ISO/RTO Council web site at ttp://www.isorto.org./site/c.jhKQIZPBImE/b.2603295/k.BEAD/Home.htm

6.2.3 Economic Growth and Changes in the Global Electric Power Market

The pace of economic growth throughout the world will also impact the pace of smart grid realization. We need to gain an appreciation for, and understanding of, economic fundamentals and for the way in which the world's economic activities and developments affect spending for electric power generation, transmission and distribution (T&D) infrastructure, and smart grid development. The continuing difficult economic outlook will no doubt bring delays in procurement and implementation of many large-scale smart grid initiatives and programs, especially in the developed nations of the world. With a negative outlook for further government stimulus legislation in the near term both in the United States and in other developed economies, this may place additional limits on utility investments. The regulatory agencies remain the drivers for much of potential smart grid investment.

It is clear that there are now, and will continue to be, pockets of opportunities for suppliers and integrators in each smart grid segment. Some world regions and some countries in particular are growing at a faster clip than the global average growth in gross domestic output (of goods and services). World total output growth is expected to be in the 4.0% range for 2011 and for 2012. Western nations are expected to have lower rates of growth (below 3%), while emerging and developing economies are expected to grow at about 6.4% on average over 2011 and about 6.1% for 2012. China, projected to grow at about 9%, and India, anticipating growth in the 7.5% range outpace other ASEAN nations. Brazil and Mexico apparently will grow at more moderate levels, between 4% and 4.7%. Sub-Saharan African growth (5.2%–5.8%) will likely outpace that of the Middle East (about 4.0%).

It is also important to note that access to capital markets and availability of human resources together with the outlook for electricity demand in each country will be vital in any estimates of likely growth of smart grid activities.

The U.S. Department of Energy's Energy Information Administration publishes an international energy outlook each year. The 2010 outlook includes the following highlights:

- Over the span of 2007–2035, non-OECD* countries are likely to account for 86% of the increase in global energy use.
- Renewables are the fastest growing energy source (but from a relatively small base).
- Coal will continue to fuel the largest share of the world's electricity, with growth in coal consumption pegged at 56%. China and India will account for the great majority of this increase.
- Economic activity and population drive the increases in energy use; energy intensity improvements will moderate this trend.
- Nuclear power generation may increase as much as 74% over the span of 2010–2035.
- Total renewable energy use may grow by 111%.
- Ninety percent of all transmission substations are remotely monitored and controlled by RTUs and/or PLCs.
- Forty-nine percent of all the distribution substations are remotely monitored and controlled by RTUs and/or PLCs.
- About 43% of installed RTUs or PLCs are estimated to be more than 10 years old.
- Capital investment in electric power infrastructure and in "smart grid" development is lagging due to the global recession of 2008–2010. However, the outlook for spending in 2010 is still positive, thanks in large part to various government stimulus programs around the world.
- Once utility planners foresee improvement in commercial and industrial outlooks, spurring increased use of electricity, smart grid related budgets would grow at a smart pace.
- The developing nations of the world will continue to rely on nongovernmental organization funding to fuel their expansion of electricity to more citizens, as provided by the United Nations Development Programme (UNDP), World Bank, Islamic Development Bank, and similar sources of capital.

---

* Organization for Economic Cooperation and Development.

Smart Grid: Where Do We Go from Here?

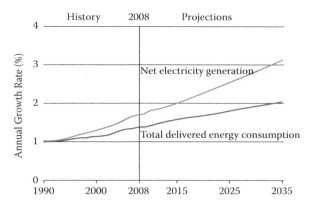

FIGURE 6.2 Forecasted growth in world electric power generation and consumption to 2035. (From U.S. Department of Energy, Energy Information Administration, International Energy Outlook 2009, http://www.eia.gov/forecasts/ieo/electricity.cfm)

Internationally, the U.S. DoE projects that electric power generation and consumption will grow at an average annual rate of 2.3% through 2035 (Figure 6.2). Electricity generation increases by 87% from 2007 to 2035 in the DoE's International Energy Outlook 2010 Reference case. Non-OECD countries account for 61% of the world's electricity use in 2035. We think this robust growth will be led by China, India, Brazil, and the GCC. In the developing nations of the world, the demand for electricity will continue to climb rapidly, as millions more people each year join the ranks of the middle classes. How to meet this increasing demand for power will absolutely be a key and recurring dilemma for utilities, governments, and industry.

In general, growth in the OECD countries,* where electricity markets are well established and consuming patterns are mature, is slower than in the non-OECD countries, where a large amount of demand goes unmet at present. The International Energy Agency estimates that nearly 32% of the population in the developing non-OECD countries (excluding non-OECD Europe and Eurasia) did not have access to electricity in 2005—a total of about 1.6 billion people. Regionally, sub-Saharan Africa fares the worst: more than 75% of the population remains without access to power. High projected economic growth rates support strong increases in demand for electricity among the developing regions of the world.

The OECD countries will continue to generate more than one-half of the global electricity production overall, at least through 2010, but by 2015 non-OECD countries will be producing about 52% of the world's electricity. Non-OECD countries already have more coal-fired production than do the OECD countries. By 2025, the non-OECD countries will generate twice as much electricity using coal-fired resources than will the OECD nations. By 2030, the non-OECD countries will generate 57% of all the electricity produced from renewable energy sources. Note that the non-OECD countries are expected to increase their use of electricity at triple the OECD rate over the forecast period (Table 6.2).

The period of 2011–2030 will very likely result in increased use of renewables; increased reliance on smart meters and energy efficiencies; and more demand–response, more extra high voltage (EHV) transmission, and higher medium voltage (MV) rated lines for distribution. Increased demand for electricity in the Western nations will be largely offset by increases in energy efficiencies and demand–response programs and a relatively slow moderate pace of economic growth.

---

* OECD member countries include Australia, Austria, Belgium, Canada, Chile, Czech Republic, Denmark, Finland, France, Germany, Greece, Hungary, Iceland, Ireland, Italy, Japan, Korea, Luxembourg, Mexico, the Netherlands, New Zealand, Norway, Poland, Portugal, Slovak Republic, Spain, Sweden, Switzerland, Turkey, United Kingdom, and the United States.

TABLE 6.2
OECD and Non-OECD Net Electricity Generation: 2006–2030
(Trillion Kilowatt Hours)

Region	2006	2010	2015	2020	2025	2030	Average Annual % Change
OECD	9.9	10.6	11.3	11.9	12.6	13.2	1.2
Non-OECD	8.0	10.0	12.0	14.1	16.3	18.6	3.5
World total	17.9	20.6	23.2	26.0	28.9	31.8	2.4

Source: Table prepared by Newton-Evans Research Company, Inc. based on U.S. Department of Energy, Energy Information Administration, International Energy Outlook 2009, http://www.eia.gov/forecasts/ieo/electricity.cfm

Renewables have been a factor in power generation for more than 100 years, but it is the newer generation of renewables, principally wind and solar, that must somehow become integrated to the grid that is really part and parcel of smart grid architecture. Overall expenditures for renewable grid integration, excluding the high voltage and medium voltage switchgear, but including controls and related substations and control systems installed at or near renewable energy production farms today, may account for as much as $75–$125 million globally. These estimates were excerpted from the Newton-Evans Research study entitled "The Worldwide Smart Grid in 2011: A Reality Check and Five Years Outlook Through 2015". However, major shares of some renewable energy resources (private wind and solar farms, for example) are owned by nonutilities and their portion of expenses for grid integration activities may not be included in our study.

The chart from the U.S. Department of Energy, energy information administration (EIA) depicts the U.S. outlook for fuel mix between 2010 and 2035. Note the change in fuel mix, with much of the relative growth coming from renewables (Figure 6.3).

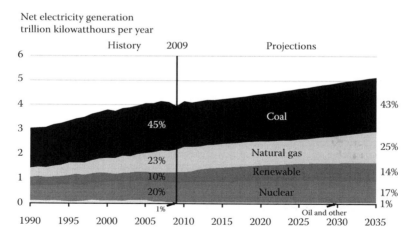

FIGURE 6.3 Projected fuel mix for U.S. electricity generation to 2035. (From U.S. Department of Energy, Energy Information Administration, International Energy Outlook 2009, http://www.eia.gov/forecasts/ieo/electricity.cfm)

6.3 FAD, FAILURE, OR FAME?

It is without a doubt that the smart grid hype has gained much interest and backing worldwide, driven by regulatory initiatives in advanced metering and from funding opportunities, not only in the United States, but also in other regions of the world. Smart grid has helped bring more focus on cleaner, more efficient, and greener solutions to electric energy usage. The discussion about smart grid in combination with the political awareness regarding carbon emissions and the finiteness of fossil fuels has helped heighten the interest, awareness, and commitment from consumers in energy reduction and savings. While advanced technologies have been applied to the electric utility industry for many years, the smart grid brings promises of synergies and advancements in integrating technologies and focusing on both the consumer and the utility.

Overall, one must ask whether there is really any new spending for "smart grid" that would not have occurred if there had been no special name given to the inclusion of modern, digital technology to help monitor and control grid-related operations and activities and to measure electricity usage. The term "smart grid" is a relatively recent development, coined only 5 years ago. The development of multiple systems, subsystems, equipment, and devices to effectively monitor, measure, control, and protect the electricity grids around the world has been underway for the past half century or longer. Each generation of electrical equipment manufactured for electric power utilities and industrial companies has been more powerful, with smaller footprints, and has been designed to be increasingly intelligent. It seems to Newton-Evans Research Company, Inc. that what is new here is the attempt to better integrate so many disparate operations and engineering activities and to provide a more intelligent and visual approach to information management, telecommunications planning and operations, as well as to the inclusion of physical and cybersecurity concerns that are also directly related to the foundation of an intelligent electricity grid system. In essence, smart grid is the set of enabling technology that will more easily and accurately inform management and staff by providing improved situational awareness through better visualization of real-time conditions.

However, the pace and ultimate magnitude of change that will be experienced with smart grid remains unclear today. Will the smart grid prove to be a passing fad filled with dreams of significant utility transformation? Will it slowly fade with diminishing incentives and funding and fall short of that enormous promise, providing marginal gains but leaving intact our increasingly expensive and unsustainable fossil-fuel goliath?... Or will smart grid eventually triumph?

Right now smart grid seems to be dormant, in some kind of pupal stage waiting to emerge in a glorious fanfare of system-wide technology implementations and final realization of significant benefits to the utility, consumer, society, and environment. Changes in the utility industry have historically been very slow and while there has been a tremendous momentum in the smart grid evaluation and foundation work of standards, interoperability, and security, it will take a few more years to determine if the efforts are just a dot.com fad that will pass, or if the support and excitement will drive the industry into a new era. Hopefully, in a few years Thomas Edison would have a hard time recognizing the electric grid if he were alive today.

REFERENCES

1. G. Constable and B. Somerville, *A Century of Innovation: Twenty Engineering Achievements that Transformed Our Lives* (Joseph Henry Press, Washington, DC, 2003).
2. http://www.ferc.gov/whats-new/comm-meet/2009/071609/E-3.pdf

Index

A

Advanced asset management
    asset health, 351, 353
    asset managers and smart grid committees, 355
    benefits, asset management, 351, 354
    business intelligence solution, 351, 352
    customers, shareholders, employees and regulators, 355
    distribution asset performance, dashboard, 351, 353
    International Electrotechnical Commission (IEC), 356
    network management (NM), 354
    reliability, 355–356
    transmission and distribution (T&D), 355
    workforce management, 354
Advanced distribution management systems
    analytics and visualization, 220
    dashboard metrics, reporting and historical data, 220
    databases and data exchange, 220
    DER and DRM, 218
    description, 218
    enterprise integration and enhanced security, 220
    FDIR, IVVC and LM/LE, 219
    integration, interfaces, standards and open systems, 219
    microgrids, distributed and customer generation, 219–220
    monitoring, control and data acquisition, 219
    new technologies and applications, 220
    TP, DPF, ONR, CA, SCA and RPC, 219
Advanced metering infrastructure (AMI)
    ANSI C.12.18 and C.12.19, 362
    benefits, building, 220
    CIM, 364
    communications, *see* AMI communications
    communications systems, 356
    description, 356
    drivers and benefits, 361–362
    electric meter, *see* Electric meter
    802.16e-2005 standard, 364
    functions, 356
    IEC, *see* International Electrotechnical Commission (IEC)
    meter reading, 359–361
    monitoring, control and data acquisition, 219
    NIST and NERC, 363
    security, *see* AMI security (AMI-SEC)
    and sensor technologies, 259–260
    SEP, 364
    smart grid technology framework, 75
    VVO, 245
Advanced mobile phone system (AMPS), 274–275
Advanced visualization framework (AVF), 158, 160
Air pollution, 26
AMDT, *see* Amorphous metal core distribution transformers (AMDT)

American National Standards Institute (ANSI)
    ANSI C.12.18 and C.12.19, 362
    ANSI C12 guidelines, 481
American Recovery and Reinvestment Act of 2009 (ARRA)
    objectives, 33
    projects and programs, funding, 34
    as Stimulus/Recovery Act, 33
AMI, *see* Advanced metering infrastructure (AMI)
AMI communications
    demand management deployments, 267
    developments, 266
    3G networks, 276
    new technology standards, 267–268
    RF mesh, 281
    scope, communications, 268
    technological advances, 267
AMI security (AMI-SEC)
    in California and Italy, 369
    customer interface, 367–368
    enterprise applications, 368
    firmware upgrades, 367
    functions, 367
    integration environments, 366
    internal device management, 367
    local connectivity and meter data reads, 367
    remote configuration, 367
    requirements, 364–366
    smart grid, 366
    strategy, 364
    testing and diagnostics, 367
    threats, 366
    time synchronization, 367
Amorphous metal core distribution transformers (AMDT), 264–265
AMPS, *see* Advanced mobile phone system (AMPS)
ANSI, *see* American National Standards Institute (ANSI)
ARRA, *see* American Recovery and Reinvestment Act of 2009 (ARRA)
Asset management
    advanced, *see* Advanced asset management
    centralized, data-driven asset management
        data collection, 347–348
        decision making process, 348
        description, 347
        integration and analysis, 348
        work execution, 349
    consumers
        energy demand and consumption, 346
        on-site generation, 345–346
    description, 327–328
    drivers
        financial and regulatory, 329
        reliability, 328–329
        safety, 328
    electric utility, *see* Utility asset management

geospatial integration, 349–350
implementation, 331–332
optimizing asset utilization, 329–331
utility executives, 327
auto-DR, 378
Automation, 526–527
AVF, see Advanced visualization framework (AVF)

B

Baltimore Gas & Electric (BG&E), 543
Bangalore Electricity Supply Company (BESCOM), 540
Base-load/CHP generator, 85
Battery electric vehicle (BEV)
 description, 97
 PHEV, 99
Battery energy storage (BES), 93
Benefits realization
 ASR scheme, 507–508
 clean and renewable energy, 26–27
 components, 20–21
 consumer, 22
 cost–benefit analyses, 507
 demand response, 22–24
 description, 20
 environmental, 25–26
 EPRI Electricity Sector Framework for the Future, 510
 EVs, 27–28
 market barrier, 509
 price-regulated entity, 508
 programs, 21
 real-time information and pricing, 22
 reduction, carbon emissions, 510–511
 regulators, 508–509
 synergies, 28
 utilities, 21, 507
BEV, see Battery electric vehicle (BEV)
"Big-O" notation, 430, 439
Biomass and biogas, 84
Biomass energy, 31
BIPS, see Brazilian Interconnected Power System (BIPS)
Brazilian Interconnected Power System (BIPS)
 description, 200
 operation, 202
Brazil's energy regulator, ANEEL, 542
Brazil, WAMPAC
 BIPS, 200–201
 deployment plan, components, 202–203
 hydroelectric, 201
 implementation, 202
 seasonal rainfall and water flow, 201–202
 transmission grid synchronized measurements, 202

C

CAES, see Compressed air energy storage (CAES)
Capacitor coupled voltage transformer (CCVT), 344
CCVT, see Capacitor coupled voltage transformer (CCVT)
Cellular network
 AMPS, 274–275
 description, 274, 275
 2G and 3G networks, 275–276
 4G networks, 276–277
 integrated stratums, 280–281
 rapid expandability, 280
 strengths and weaknesses
  bill, materials cost, 278
  commercial and industrial customers, 277
  counterparty risk, 279
  "Dropped calls", 278
  impact, natural disasters, 279–280
  obsolescence, 279
  regulatory commissions, 279
  sensors and meters, 277–278
  technical issue, 278
  widespread global use, 277
 transparent commonality, 281
CenterPoint Energy, United States, 543
Centre for Sustainable Electricity and Distributed Generation, 50
Characteristics, smart grid
 active participation, consumers, 70
 asset utilization and operate efficiently, 71
 description, 72
 developments, programs, 70
 DOE elements, 72–73
 generation and storage options, 70–71
 goals, 70
 against physical and cyber attack, 71–72
 power quality (PQ), 71
 products, services, and markets, 71
 self-healing, 71
CHP, see Combined heat and power (CHP)
CIM, see Common information model (CIM)
CIP, see Critical infrastructure protection (CIP)
CISs, see Customer information systems (CISs)
Clean and renewable energy, 26–27
Combined heat and power (CHP), 85
Common information model (CIM)
 application security, 481
 common misconceptions and concerns, 418–419
 definition, 364
 design framework, 480–481
 electrical utility industry, issues, 477
 61850 harmonization, 481
 IEC 61968-9, 363
Communications
 AMI developments, 266
 challenges, see Communications challenges
 development, wireless solutions, 266
 integration roadmap, smart grid, 294–296
 regulatory and consumer interest, 266
 requirements, see Communications requirements
 standards and protocols, see Communications standards and protocols
Communications challenges
 cybersecurity, wireless networks, see Cybersecurity
 harnessing technology complexity, 287
 legacy integration, migration and technology life cycle, 287–288
 management and organization, 294
 service planning and evolution trends
  Ethernet and SONET/SDH, 288
  IP networks, 288–289
  WDM, 289
  wireless technologies and telecommunications, 289–290

Index

Communications requirements
  AMI, 267–268
  HANs, 273–274
  smart grid operations
    automation, 268, 270
    condition monitoring and asset management, 271
    description, 268, 269
    energy management and control center, 270
    energy metering, 271
    mobile workforce, 271
    operational communications domains, 268, 270
    predictability, 272
    protection relays, 268
    robustness, 272–273
    security, surveillance and safety, 271
    substation nonoperational data, 271
    video monitoring, 271
    WAMPAC, 270–271
Communications standards and protocols
  ANSI C12.19, ANSI C12.18, ANSI C12.21 and ANSI C12.22, 285–286
  classification, 282
  DNP3 and IEC 60870-5, 283–285
  exchange of information, 282
  IEC 61850, 282, 283
  IEC 61968-9 and MultiSpeak, 285
  IEEE C37.118, 285
  ISO OSI 7, 283, 284
  PRP and HSR, 283, 286
  time synchronization, 286–287
Communication systems
  communications, *see* Communications
  wireless network solutions
    cellular, *see* Cellular network
    RF mesh, 281–282
Companion specification for energy metering (COSEM), 363
Compressed air energy storage (CAES), 92, 95
Computational complexity, HPC
  "Big-O" notation, 430
  dynamic simulation and contingency analysis, 431
  dynamic state estimation, 432
  power flow solution and contingency analysis, 431
  power grid modeling and analysis, 430–431
  requirements, power grid, 432
  small signal stability, 432
  state estimation, 431
Conservation voltage reduction (CVR), 221, 226, 240–241, 243
Consumer Demand Management, 31
Consumer engagement and empowerment, 512–514
Control centers, EMS
  applications, 156, 157
  automation, 161
  continual development, 162
  grid operator visualization, 157
  modern-day EMS control center environment, 157, 158
  operator decisions, 159
  PDCs, 191
  PMUs, 161
Control system cybersecurity considerations, EMS
  availability, 166–167
  confidentiality, 168–169
  IEC 61850, 162
  IEEE C37.118 protocol, 162
  integrity, 167–168
  isolation, control system network, *see* Corporate networks
  NERC CIP Standards, 165–166
  network penetration threats, 162–163
  SCADA systems, 162
  synchrophasor systems, installations, 169
Control systems, microgrids, 110
Corporate networks
  access control, 164
  description, 163
  e-mail, 164
  external networks, 163
  IDS, 165
  network arrangements, 164
  PKI, 164–165
  wireless LANs, 163
  WWW connections, 163
COSEM, *see* Companion specification for energy metering (COSEM)
Country-specific drivers and benefits
  Australia, 534
  Canada, 537
  China, 538
  Europe, 539
  global collaborative efforts, 546
  India, 540
  Indonesia, 545
  Japan, 541
  Korea, 541
  Latin America, 542
  The Philippines, 545
  Singapore, 545
  United States, 544
Critical infrastructure protection (CIP)
  NERC standards, 471–472
  security requirements, 363
Critical/nondiscretionary loads, 377
Cross cutting energy engineering skills
  cloud data storage and virtual networks, 526
  elements, 525–526
Customer information systems (CISs)
  IT applications, 399
  outage management, 406
CVR, *see* Conservation voltage reduction (CVR)
Cybersecurity, 504–505
  authenticity, 446
  availability, 445–446
  communications model, 447–448
  components, 364
  confidentiality, 444–445
  control, 446
  definition, 444
  functions
    auditing, 450–451
    authentication, 449–450
    authorization, 450
    key management, 451
    layered security model, 448–449
    message integrity, 451–452
    network integrity, 452
    system integrity, 452
  integrity, 445

smart grid
  authentication and authorization services, 458–459
  certificate services, 459
  network security services, 459–460
  NIST, 458
threats
  people, 453–454
  process, 454
  technology, 454–457
traditional approach, 444
usability, 446–447
wireless networks
  application domains, 292–294
  circulative and aggressive security, 290–291
  encompassing security, 290
  functional domains, 291–292
  NERC CIP standards, 290

D

Data complexity, HPC, 427–429
Data privacy, 505–506
Decision support systems (DSS)
  components, 160
  implementation, 160, 161
  look-ahead analytical tool, 160
  predictive decision making, 160
  preventive, 160
  reactive, 159
Demand management
  conserved vs. deferred energy, 374–376
  consumer loads, 377–378
  empowering consumers, see Empowering consumers
  enablers, smart grid
    aggregation/disaggregation, 393
    area, regional and national coordination regimes, 382
    consumer behavior, 389–392
    consumer response estimation, 393
    delivery T&D infrastructure, 383
    demand dispatch, 393
    DER technology, 382–383
    digital technology, 382
    DRMS, 392–393
    information networks and finance, 383
    load modeling and forecasting, 385–387
    load responses, 387–388
    measurement and verification, 393
    peak shifting, 383–385
    price signals, 388–389
    purchasing power and consuming energy, 383
    smarter consumers, 383
    system engineers, 383
    technical areas, 382
  energy consumption, 370
  load patterns and behavior, 371–374
  mechanisms, 370–371
  power generation
    baseload units, 376
    intermediate/responsive and load units, 376
    peaking units, 376–377
  supply electricity, 369–370
  utility–customer interaction, 378–379

value of
  bolster grid reliability, 380
  calculation, benefits, 381
  in California, 379
  DR technologies, 379–380
  economic benefits, 380
  installation, interval metering, 381–382
  long-standing goal, 380
  PNNL Olympic Peninsula Project, 382
  programs, 379
  Texas grid, 380
  wholesale market, DR and energy storage, 380–381
  wind energy, 381
Demand response
  CO_2 reductions and emissions, 24
  dynamic pricing, 23
  environmental gains, 23–24
  peak shaving, 23
  PJM, 24
  programs, 22–23
  2003 Synapse model, 24
Demand response management system (DRMS)
  aggregation/disaggregation, 393
  consumer behavior, 389–392
  consumer response estimation, 393
  demand dispatch, 393
  HEM, 392–393
  impact, DMS, 218
  load modeling and forecasting, 385–387
  load responses, 387–388
  measurement and verification, 393
  OT applications, 399
  price signals, see Price signals
DERs, see Distributed energy resources (DERs)
Design and run times, ESM architecture
  application layer, 417
  business modeling and design layer, 417
  business process and intelligence layer, 418
  integration layer, 418
Device language message specification (DLMS), 363
Dispatcher training simulator (DTS), 218
Distributed energy resources (DERs)
  devices and demand response programs, 260–261
  distribution network, 218
  smart grid standards, 481
  technology, 382–383
  VVO, 245
Distribution management systems (DMSs)
  advanced, see Advanced distribution management systems
  advanced network applications, 402
  "chair rolls", 216
  contingency analysis (CA), 217, 218
  EMS monitors, 156
  FDIR, 216–217
  grid-enhanced applications, 75
  IVVC, DPF and LM/LE, 217
  OCP/OVP and DTS, 218
  and OMS, 110
  ONR, SOM, SCA and RPC, 217
  operational/real-time system, 399–400
  OT applications, 399
  primary functions, 130

Index

SCADA, *see* Supervisory control and data acquisition (SCADA)
SCADA and OMS, 402–403
smart grid advances, 218
topology processor (TP), 217, 218
Distribution systems
  DMSs, *see* Distribution management systems
  FDIR, 245–261
  high-efficiency distribution transformers
    AMDT, 264–265
    calculation, total ownership cost (TOC), 264
    energy saving programs and efficiency requirements, 264
    RGO and AM, 265, 266
    types, losses, 264
  OMS, 261–263
  VVC, 221–245
DLMS, *see* Device language message specification (DLMS)
DMSs, *see* Distribution management systems (DMSs)
DRMS, *see* Demand response management system (DRMS)
DSS, *see* Decision support systems (DSS)
DTS, *see* Dispatcher training simulator (DTS)

E

EAC, *see* Electricity Advisory Committee (EAC)
Electric grid systems, EVs
  backup generators, 125
  charging configurations and ratings terminology, 121, 122
  equipment
    conventional approaches, 124
    loading, 121–122
    maintenance, and life cycle, 121
    vs. PEV market penetration, 123–124
    PEVs charging scenarios, 122–123
  household electrical services, 126
  PEVs penetrations, 120
  power quality, 125
  "smart" charging strategies, 120–121
  standards, 121
  transportation sector, 120
  voltage regulation and feeder losses, 125
Electricity Advisory Committee (EAC), 35–36
Electricity market
  capabilities, 550
  economic growth and changes, *see* Global economic growth, electricity market
  electric power industry, 550, 552
  FERC policy and priorities, 554–555
  ISO/RTO, 555
  technical innovation, 552–554
Electricity Networks Strategy Group (ENSG), 49
Electricity system, smart grid
  area, regional and national coordination regimes, 382
  delivery T&D infrastructure, 383
  DER technology, 382–383
  information networks and finance, 383
Electric meter
  categories, 357
  "central wallet", 359
  cumulative energy usage, 357
  electromechanical, *see* Electromechanical meter
  kilowatt hour (kWh), 356–357
  standard business model, 359
Electric Power Research Institute (EPRI)
  California-based, 44
  IntelliGrid™ methodology, 44–45
  NIST, 36, 464
  smart grid
    definition, 18
    demonstration, 45–47
  utility and premise, 477
Electric utility industries
  Arab oil embargo of 1973–1974, 5
  Asia–Pacific region, 11
  deregulation, 5–6
  electrification and regulation, 3–4
  elements, 2
  FERC, *see* Federal Energy Regulatory Commission (FERC)
  heterogeneity, 2–3
  Latin America, 10
  Middle East and Africa, 10–11
  National Energy Act of 1978, PURPA, 5
  Northeast Blackout of 1965, 4
  Northeast Blackout of 2003, 9
  operating structure, 1–2
  power stations, 9
  regulatory structures, 2
  Western and Eastern Europe, 10
  Western Energy crisis of 2000–2001, 6, 8–9
Electric vehicles (EVs)
  BEV, 97
  energy buffering, 102–103
  environmental benefits, 27–28
  "flexible load", 28
  grid support, 99–102
  grid systems, *see* Electric grid systems, EVs
  HEV, 97–98
  integration, 27
  PHEVs, 99
  regulatory and market forces, 96–97
Electromechanical meter
  digital smart meter, 358
  drawbacks
    inaccurate and inefficient, 357
    no consumer visibility, energy usage, 358
    no remote monitoring/control functionality, 358
    tamper prone, 358
  electronic content, 358
  smart meters, 358
Empowering consumers
  education and participation, 396–397
  energy management, 396
  real-time information, 393–395
  smart loads and appliances, 395–396
EMS, *see* Energy management systems (EMS)
Energy buffering, EVs
  dispatchable generation, 102
  energy consumption, PEVs, 102
  time-of-use (TOU) rate, 103
Energy Independence and Security Act of 2007 (EISA), Title XIII
  characteristics, 33
  description, 32
  provisions, 32–33

Energy management systems (EMS)
    applications, control center, 156, 157
    control centers, 161–162
    control system cybersecurity considerations, *see* Control system cybersecurity considerations, EMS
    DSS, 159–161
    grid operator visualization, 157–159
    hardware components, 129
    modern-day EMS control center environment, 157, 158
    monitors and manages flows, 156
    "normal synchronous operation", 156
    operational/real-time system, 399–400
    OT applications, 399
    primary functions, 130
    service territories, 156–157
Energy storage
    BES, 93
    CAES, 95
    centralized energy storage applications, 91
    comparisons, technologies, 91, 92
    coordinated implementation, 91
    description, 86, 93
    division, methods, 91–92
    FES, 94
    installed grid-connected energy storage, 89, 90
    mitigation, PV-DG and PEV, 91, 94
    pumped hydro, 95
    regulatory and market forces, 86–89
    SMES, 93–94
    thermal energy storage, 95–96
    ultracapacitors, 95
ENMAX, Canada, 536
ENSG, *see* Electricity Networks Strategy Group (ENSG)
Enterprise information management (EIM)
    applications, 411–412
    benefits, 419–420
    developing and implementing, 419
    ESM, *see* Enterprise semantic model (ESM)
    framework, 412–414
    processes and tools, 411
Enterprise resource planning (ERP)
    asset and work management, 406
    IT applications, 399
Enterprise semantic model (ESM)
    architecture, 416–418
    goals, 413–414
    information sources, 418–419
    role, 414–416
Enterprise SOA, 411
EPRI, *see* Electric Power Research Institute (EPRI)
EPRI's smart grid demonstration initiative
    collaborating utilities, 45–46
    elements, 46
    large-scale, design, 46
    Resource Center, 46–47
    utility members, 46
    "virtual power plant", 45
ERP, *see* Enterprise resource planning (ERP)
802.16e-2005 standard, 364
European Renewable Energy Directive, 49
EVs, *see* Electric vehicles (EVs)
EV-Smart Grid project, Japan, 540, 541

F

FACTSs, *see* Flexible AC transmission systems (FACTSs)
Fault detection, isolation and service restoration (FDIR)
    drivers, objectives and benefits, 247
    equipment, *see* FDIR equipment
    faults, distribution systems, 246
    implementation, *see* FDIR implementation
    level, optimization, 219
    location, electrical fault, 216
    power outages, 245–246
    remote measurements, 216
    RSA application, 216–217
FDIR, *see* Fault detection, isolation and service restoration (FDIR)
FDIR equipment
    automatic recloser, 249–250
    automatic sectionalizer, 249
    high-performance fault testing, 251–253
    load-break switch, 248–249
    manual switch, 248
    sensors, 250–251
    single-phase dropout recloser, 251
    source transfer gear, 250, 251
    substation circuit breaker, 247–248
FDIR implementation
    AMI and sensor technologies, 259–260
    closed-loop fault clearing systems, 258
    deployment considerations, 258, 260
    large-scale energy storage, 256
    loop restoration, pulseclosers, 255
    peer-to-peer communications, 255, 256
    reclosers, fault hunting loop schemes, 254–255
    reliability needs, smarter grid, 260–261
    SCADA/DMS and remotely controlled switching devices
        automatic switching, 259
        manual switching, 258–259
    source transfer applications, 256
    substation breaker
        automatic sectionalizers, 254
        fault-detecting switches, 253
        midpoint recloser, 253–254
    substation computer-based schemes, 255–257
Federal Energy Regulatory Commission (FERC)
    definition, 39
    Energy Policy Act of 2005, 9
    interest and responsibilities, 40
    investor-owned utilities, 12
    ISO/RTO, 6, 7
    policies, 554–555
    policy, 39–40
    priorities, 40, 555
    Western Energy crisis, 8
Federal smart grid, U.S.
    ARRA, 33–34
    EISA, Title XIII, 32–33
    FERC, *see* Federal Energy Regulatory Commission (FERC)
    NIST, 36–39
    U.S. DOE, *see* U.S. Department of Energy (U.S. DOE)
FERC, *see* Federal Energy Regulatory Commission (FERC)

Index

FES, *see* Flywheel energy storage (FES)
Fixed series compensation (FSC), 172, 174
Flexible AC transmission systems (FACTSs)
   developments
      series compensation, 172, 173
      transmission solution, 172, 173
      voltage profile, 172, 174
   IPFC, 179
   reactive power compensation, 171–172
   series compensation
      benefits, 175
      FSC, 172, 174
      installation, 172, 175
      TCSC, 175, 176
      transient stability, 173
      voltage stability, 173, 175
   shunt compensation
      STATCOM, 177–178
      SVC, 175–177
   SSSC, 178, 179
   stepwise interconnection, 211
   synchronous condenser, 180
   transmission efficiency, 210
   UPFC, 178–179
   VFT, 179–180
Flywheel energy storage (FES), 92, 94
Forms and goals, learning
   career and technical education, 527–528
   certification and professional licensing, 528
   continuing education, 528
   role, professional societies, 528
   training, non-engineering workforce segment, 528
   university education, 527
FSC, *see* Fixed series compensation (FSC)

G

Generic Object Oriented System Event (GOOSE)
   application, 139
   messages, 141, 149, 150
   peer-to-peer communication, 136–137
Geographic information system (GIS)
   development, model-based VVO, 244
   geospatial technologies, *see* Geospatial technologies
   IT applications, 399
   IT/back office, operational/real-time system, 400
   MWFM, 335–337
   OMS
      customer connectivity, 262
      data extraction, 262
      system connectivity, 261
   outage management, 405–406
   view, asset management, 349, 350
Geospatial smart grid
   core spatial functionality
      geovisualization, 311–312
      queries and reporting, 312
      system of truth, 311
   description, 310
   engaging consumers, 320
   mobile geospatial technologies, *see* Mobile geospatial technologies
   operating and maintaining
      network analysis, 315
      outage restoration, 315–316
   planning and designing
      communications network design, 314–315
      grid design, 313–314
      system planning, 313
Geospatial technologies
   application, geospatial tools, 326
   architecture, 326
   changing grid, 309–310
   cloud, 326–327
   geospatial smart grid, *see* Geospatial smart grid
   neo-geo, 327
   smart grid impact
      coping, scale, 321
      description, 320
      distributed users, 322
      moving to realtime, 321
      security and privacy, 325
      sensors and data sources, 324–325
      standards and data quality, 323–324
      usability, 322–323
      visualization, 323
   technology roadmap, 304–309
Geothermal energy, 31
Geothermal power, 84
GIS, *see* Geographic information system (GIS)
Global economic growth, electricity market
   average growth, 556
   description, 556
   energy outlook, 2010, 556
   OECD, 557, 558
   power generation and consumption, 557
   renewables, 557–558
GOOSE, *see* Generic Object Oriented System Event (GOOSE)
Grid operator visualization
   ALSTOM technology, 158–159
   AVF, 158
   real-time grid conditions, 157–158
   situational awareness (SA)
      capabilities, 159
      definitions, 158
Grid support, EVs
   aggregative architecture, 100–102
   ancillary service market, 99
   communications network, 101
   costs, estimation, 99
   deterministic architecture, 99–100
   G2V and V2G, 99
Grid-to-vehicle charging (G2V), 99
GridWise
   architecture group, 465
   definition, 382
GridWise Alliance, 44, 47
GridWise Architecture Council (GWAC), 47, 465, 472
GWAC, *see* GridWise Architecture Council (GWAC)

H

HANs, *see* Home area networks (HANs)
HEMS, *see* Home energy management systems (HEMS)

HEV, *see* Hybrid electric vehicles (HEV)
High-availability seamless redundancy (HSR), 282, 283, 286
High-performance computing (HPC)
    computational complexity
        "Big-O" notation, 430
        dynamic simulation and contingency analysis, 431
        dynamic state estimation, 432
        power flow solution and contingency analysis, 431
        power grid modeling and analysis, 430–431
        requirements, power grid, 432
        small signal stability, 432
        state estimation, 431
    computer processor hardware, 426
    data complexity, 427–429
    enable new functions
        dynamic state estimation, 438–440
        modeling assumption, 437
        predetermined limits, 438
        quasi-steady-state assumption, 438
        real-time path rating, 440–443
    existing functions
        parallel contingency analysis, 433–437
        parallelized state estimation, 433
        power grid, 433
    grid computational paradigm, 426, 427
    handle, data, 427
    modeling complexity, 429–430
    power grid, 425
    real-time power grid operations, 426, 427
    smart grid, 425–426, 443–444
    techniques and computational capabilities, 426
High-voltage direct current (HVDC)
    benefits, 181
    description, 181
    developments
        AC cable transmission, 183
        configurations and technologies, 183, 184
        VSC, 183, 184
    functionality, 183
    intelligent subsystems, 183
    interconnections, 181, 182
    post-fault response, AC networks, 181, 182
    sources, 181
    stepwise interconnection, 211
    thyristor-based "conventional" HVDC
        circuit diagram, 184, 185
        description, 184–185
        installation, 185
        valve hall, 185–186
    VSC-based HVDC
        circuit diagram, 186, 187
        converter station, 187
        description, 186
        feature, 187–188
        phase reactor, 186–187
Home area networks (HANs)
    alliances, utility industry, 462
    AMI/communications network, 360–361
    demand management systems, 273
    description, 396
    electric meter, 273
    home energy management, 273–274
    HomePlug GreenPhy, 273
    smart grid, issues, 424
    standards, 476–477
    UCAIug, 462, 463, 465–466
    wireless 4G networks, 274
    ZigBee Alliance, 273
Home energy management systems (HEMS), 553–554
HSR, *see* High-availability seamless redundancy (HSR)
HVDC, *see* High-voltage direct current (HVDC)
Hybrid electric vehicles (HEV)
    description, 97–98
    PHEV, *see* Plug-in hybrid electric vehicles (PHEVs)
Hydro One Networks, Canada, 536
Hydropower plants, 84
Hydro resources, 30

I

IDS, *see* Intrusion detection system (IDS)
IDSM, *see* Integrated DSM (IDSM)
IEC, *see* International Electrotechnical Commission (IEC)
IEC 61850
    advantages, 142
    application and communications, 138, 139
    bay level, 141
    configuration benefits, 142, 143
    description, 132, 136, 283
    division, 138
    engineering approach, 138
    Ethernet, 140
    free configuration, 137
    GOOSE, 139
    identification, 136–137
    implementation guideline, 141
    interfaces, 137, 138
    interoperability, 137
    logical nodes (LNs), 140
    long-term stability, 137
    process-bus-based applications, 142
    process level technology, 141
    SCSM, 140
    serial communication and digital systems, 137
    station level, 141
    substation design, *see* IEC 61850-based substation design
    transmission, SVs, 139–140
IEC 61850-based substation design
    Communication Networks and Systems for Utility Automation, 142–143
    functional integration and flexibility, 146
    integrated protection and control system, 146
    logical interfaces, 147
    merging unit (MU), 147–148
    paradigm shift
        circuits, 145
        conventional substation design, 145–146
        primary equipment, 144
        secondary equipment, 144–145
    physical device, 146
    station and process bus architecture
        advantages, 152
        alternative substation design, 151–152
        communications architecture, 151
        functional architecture, 150

Index

station-bus-based architectures
　advantages, 148
　　functional architecture, 148, 149
　　GOOSE message repetition mechanism, 149–150
　substation architectures, 148
IEC System for Conformity Testing and Certification of Electrotechnical Equipment and Components (IECEE), 464
IEDs, *see* Intelligent electronic devices (IEDs)
IEEE, *see* Institute of Electrical and Electronics Engineers (IEEE)
Independent system operators/regional transmission organizations (ISO/RTO), 6, 7, 555
Industry expertise and skills
　declining workforce supply, United States, 523–524
　energy engineers and power, 522
　engineering expertise, 522
　holistic approach, 522
Information technology (IT)
　applications, 399
　and operations, 478
　and OT, *see* OT and IT systems
Initiative to build knowledge (U.K. Power Academy), 528–529
Institute of Electrical and Electronics Engineers (IEEE), 469
Integrated DSM (IDSM), 370
Integrated system approach, 503–504
Integrated System projects, Europe, 539
Integrated volt/VAr control (IVVC)
　approaches, *see* volt/VAr control implementation
　description, 217, 221
　load reduction, CVR, 225, 226
　operational and asset improvements, 219
　switched capacitor banks, 232–233
Intelligent electronic devices (IEDs)
　advantages, 127
　application features, 128
　communications capabilities, 132–133
　description, 127
　IEC 61850, 136, 141, 142
　protection relays and meters, 133
　server-based SCADA architecture, 133
Interline power flow controller (IPFC), 179
International Electrotechnical Commission (IEC)
　families, 460
　IEC 61850, *see* IEC 61850
　IEC 60870-5 and DNP3, 283–285
　IEC 61968-9 and MultiSpeak, 285
　IEC 61968-9 common information model, 363
　IEC 62056 DLMS-COSEM standard, 363
　IECEE CB, 464
　IEC 61850 standards and outside substation, 462, 481
　SG3, 465
　smart grids, 356
　Strategic Group (SG 3), 49
　Technical Committee (TC) 57, 47–48
　transmission standards and technology, 469
International treaties
　2009 Copenhagen Accord, 40
　MEF, 41
　national and regional action
　　Africa/Latin America, 43–44
　　Asia/Pacific, 42–43
　　Europe, 43

　　negotiations and collaborations, 42
　　and NGOs, 44
　"2011 Technology Roadmap on Smart Grids", 40
　WEF, 41
Intrusion detection system (IDS)
　availability, 166, 167
　control system network, 164
　description, 165
　signature-based, 165
　statistical, 165
IPFC, *see* Interline power flow controller (IPFC)
ISO/RTO, *see* Independent system operators/regional transmission organizations (ISO/RTO)
IVVC, *see* Integrated volt/VAr control (IVVC)

J

Jeju Smart Grid Demonstration, Korea, 541

L

LCNF, *see* Low-Carbon Networks Fund (LCNF)
Lead organizations
　Australia, 534
　Canada, 536
　China, 538
　Europe, 539
　global collaborative efforts, 546
　India, 540
　Japan, 540
　Korea, 541
　Latin America, 541–542
　Malaysia, 544
　The Philippines, 544
　Singapore, 544
　Taiwan, 544
　United States, 542
LM/LE, *see* Load modeling/load estimation (LM/LE)
Load modeling/load estimation (LM/LE), 217, 219
Low-Carbon Networks Fund (LCNF), 49

M

Major Economies Forum on Energy and Climate (MEF), 41
MANGO, *see* Modal Analysis for Grid Operations (MANGO)
Master stations, SCADA systems
　components, 128–129
　description, 128
　modern EMS architecture, 129
　primary functions
　　DA system, 130
　　DMS, 130
　　EMS, 130
　　SCADA/AGC system, 129
　types, 129
MDMS, *see* Meter data management system (MDMS)
MEF, *see* Major Economies Forum on Energy and Climate (MEF)
Message integrity
　encryption, 451–452
　nonrepudiation, 451
　signing, 451

Meter data management system (MDMS), 361
Microgrid pilot projects
   Europe, 107
   Japan, 107
   North America, 107
   San Diego Gas and Electric, 108
Microgrids
   benefits, 105–106
   characteristics, 103–104
   control systems, 110, 111
   definition, 103
   DMS/OMS and AMI system, 110
   drivers, 104
   functions, 110, 112
   physical systems, 109–110
   pilot projects, 107–108
   technical challenges, 106–107
   types, 108–109
Mobile geospatial technologies
   data collection and update, 319–320
   inspections, 318, 319
   map viewing, 316–317
   routing and navigation, 318–319
   workforce management, 317–318
"Mobile WiMAX", 364
Mobile workforce management (MWFM)
   aging assets and infrastructure, 336
   customer service, inspection and maintenance, 336–337
   description, 335, 354
   divisions, 336
   drivers, 336
   end-to-end and top-to-bottom benefits, 341, 342
   reporting, 341
   scheduling algorithms, 340
   transmission and distribution (T&D) systems, 355
Modal Analysis for Grid Operations (MANGO)
   description, 208
   vs. modulation stability control, 208–209
   proposed MANGO framework, 209
   signal stability, 209
Model-based VVO approach
   control variables, 243
   description, 242–243
   development, 244
   optimum settings, 243
   power engineering calculations and analysis, 243–244
Monitoring and diagnostics
   applications, 303
   architectures, smart sensors, *see* Smart sensors
   data, intelligent algorithms and communications, 296–297
   future trends, 304
   intelligent machine algorithms, 303–304
   measurement and instrumentation, 297
   predictive maintenance, 303
   sensors and application areas, 302–303
   transmission and distribution assets, 301–302
   wireless sensor networks, 301
MWFM, *see* Mobile workforce management (MWFM)

N

National Energy Act of 1978, PURPA, 5
National Energy Technology Laboratory (NETL), 70

National Institute of Standards and Technology (NIST)
   AMI, 363
   definition, 36
   model, smart grid, 36, 37
   roadmap, 36–37
   SGFAC, 39
   SGIP, 37–39
   smart grid interoperability standards
      "action plans", 464
      framework and roadmap, 458
National Renewable Energy Laboratory (NREL), 510
NETL, *see* National Energy Technology Laboratory (NETL)
Network integrity
   firewalls, 452
   intrusion detection and prevention, 452
Newton-Evans Research Company, Inc.
   business opportunities, 550
   components, smart grid, 550, 551
   electricity generation, OECD and non-OECD, 557, 558
   and information technology (IT) expenditure, 550, 552
   market outlook, 549–550
NGOs, *see* Nongovernmental organizations (NGOs)
NIST, *see* National Institute of Standards and Technology (NIST)
Nongovernmental organizations (NGOs), 44
North American Electric Reliability Corporation (NERC)
   AMI, 363
   CIP standards, 471–472
The North American Electric Reliability Corporation (NERC) Critical Infrastructure Protection (CIP) Standards
   electronic security perimeter, 165–166
   electronic security perimeters, 169
   NERC CIP 007-3, Cyber Security, 164
   responsible entities, 162, 167
NREL, *see* National Renewable Energy Laboratory (NREL)
Nuclear, 29–30

O

OASIS, *see* Open Access Same Time Information System (OASIS)
OCP/OVP, *see* Optimal capacitor placement/optimal voltage regulator placement (OCP/OVP)
OECD, *see* Organization for Economic Co-operation and Development (OECD)
Office of Gas and Electricity Markets (Ofgem)
   ENSG, 49
   implementation, 43
   LCNF, 49
Ofgem, *see* Office of Gas and Electricity Markets (Ofgem)
OMS, *see* Outage management systems (OMS)
ONR, *see* Optimal network reconfiguration (ONR)
Open Access Same Time Information System (OASIS), 526
Open systems interconnection (OSI), 447, 467
Operational/real-time system
   IT/back office, 400
   SCADA, 399–400
Operations technology (OT)
   applications, 399
   and IT, *see* OT and IT systems

Index

Optimal capacitor placement/optimal voltage regulator placement (OCP/OVP), 218
Optimal network reconfiguration (ONR), 217, 219
Organization for Economic Co-operation and Development (OECD)
    energy outlook, 2010, 556
    and non-OECD, 557, 558
OSI, *see* Open systems interconnection (OSI)
OT, *see* Operations technology (OT)
OT and IT systems, 502–503
    "back-office" systems, 425
    benefits
        best practice and high productivity, 422
        changes, business structure, 423
        consistency, 422
        customer expectations, 422–423
        media attention, 423
        operating expenses, 421
        quality, service, 423
        regulatory performance, 423
        reliability and operating revenue, 422
        vision fosters teamwork, 422
        working capital expenses, 421
    control and monitoring, 420
    data, 424–425
    dichotomy, utility operations, 420, 421
    obstacles
        business process, 423
        change management, 424
        data acquisition, volume and rate, 424
        installation, complexity, 423
        performance and availability, 423
        standards, adoption, 423
    operations group, aim, 420
    procurement process, 424
    role, 421
    smart grid, issues, 424
    utility, 421
Outage management systems (OMS)
    advanced DMS network applications, 402
    AMI synergies, 404
    applications, 241, 243
    connectivity maps, 261
    customer connectivity, 262
    data extraction, 262
    and DMS, 110
    DMS fault-location algorithm, 263
    functions, 261
    grid-enhanced applications, 75
    IT/back office, 400
    and mobile workforce management, 403
    modern computer-based, 261–263
    operational/real-time system, 399
    outage engine algorithms, 262
    RSA application, 263
    SCADA and DMS, 402–403, 408
    smart grid implementations, 263

P

Pacific Gas & Electric (PG&E) plans, 543
Parallel redundancy protocol (PRP), 282, 283, 286
PCTs, *see* Programmable, controllable thermostats (PCTs)
PDC, *see* Phasor data concentrator (PDC)

Performance goals and progress metrics, 511
PEVs, *see* Plug-in electric vehicles (PEVs)
Phasor data concentrator (PDC)
    bulk electric transmission system, 166
    description, 190
    functions, 190–191
    IEEE C37.118 protocol, 162
    levels, 190, 191
    PMUs, 190
    Super-PDC, *see* Super-phasor data concentrator (Super-PDC)
Phasor measurement units (PMUs)
    angle information and differences, 189
    connection, transmission substation, 189, 190
    description, 189
    large-scale deployment, 194
    modern power systems, 162
    North American Power Systems, 427–428
    PDC, 190–191
    and PDC certification test process, 203
    Super-PDC, 196
PHEVs, *see* Plug-in hybrid electric vehicles (PHEVs)
Physical systems, microgrids
    description, 109
    energy storage, 109–110
    generators, 110
    metering, 110
    power electronics, 109
    protection equipment, 110
    sensors, 109
    switches, 109
Pilot and demonstration projects, 532–533
PKI, *see* Public key infrastructure (PKI)
Plug-in electric vehicles (PEVs)
    aggregation, 101
    charging, 120–123
    electrical demand, 102
    electric transit, 96
    electrification, 96
    energy consumption, 102
    grid support, 99–100
    HEVs, 96
    integration, 91
    penetration levels, 125
Plug-in hybrid electric vehicles (PHEVs)
    advantage, 99
    customer interface challenge, 368
    description, 99
    G2V, 99
    HEVs, 96
    OT and IT systems, 421, 422
    smart grid standards, 482
    smart grid standards landscape, 482
    time-of-use (TOU) rate, 103
PMUs, *see* Phasor measurement units (PMUs)
Policy and regulation
    alignment, 520–521
    AMI, 520
    ARRA, 519
    climate and energy package 20-20-20 (EU), 521
    consumer and discussions, regulators, 521
    EISA, ACES (United States), 521
    governments and regulators, 518
    market uncertainty, 518–519

network companies, 518
public, 517
regulatory changes, 520
regulatory review (United Kingdom), 521
supporting and investing, 519
Power generation, electric loads
baseload units, 376
intermediate/responsive and load units, 376
peaking units, 376–377
Power system developments
AC transmission, 170
FACTS and HVDC, functions, 171, 172
synchronous AC interconnections, 171
Price signals
consumer DR
critical-peak pricing (CPP), 388
"decoupled rates", 389
dynamic pricing (DP)/real-time pricing (RTP), 388–389
peak-time rebate (PTR), 388
time of use (TOU), 388
and utilities, 378
mechanism, 389
Programmable, controllable thermostats (PCTs), 480
Project category/technology
Canada, 536
China, 538
Europe, 539
global collaborative efforts, 546
Latin America, 542
Singapore, 544
United States, 543
PRP, *see* Parallel redundancy protocol (PRP)
PT Perusahaan Listrik Negara (PLN), Indonesia, 545
Public key infrastructure (PKI), 164–165, 451
Public Utility Holding Company Act of 1935 (PUHCA), 3–4
Public Utility Regulatory Policies Act (PURPA), 5
PUHCA, *see* Public Utility Holding Company Act of 1935 (PUHCA)
Pumped hydro, 95
PURPA, *see* Public Utility Regulatory Policies Act (PURPA)

R

Radio-frequency (RF) mesh, 267, 281–282
Regional drivers and benefits
Canada, 537
China, 538
Europe, 539
global collaborative efforts, 546
Latin America, 542
United States, 543
Regulatory and market forces
energy storage
ability, 87
frequency regulation, 87–88
Internal Revenue Code, 87
peak shaving/load shifting, 88–89
PEVs, 87
renewable integration, 89
spinning reserve, 88
EVs
benefits, 96–97
degree of electrification, 96
HEVs, 96
implementation, 96
PEV, 97
transit, 96
renewable generation
entities, 79
RPS program, 79
wind and solar generation, 79–81
Relay protection coordination (RPC), 217–219
Remote terminal unit (RTU)
communication, 131
description, 130
internal software modules, 130
SCADA system data flow, 130, 131
software, 130, 131
Renewable energy generation
biomass, 31
description, 30
geothermal, 31
hydro, 30
percentages, 30
RPS-type mechanisms, 30
solar, 31
wind, 30
Renewable energy resources
biomass and biogas, 84
CHP, 85
description, 82
fuel cells, 84
geothermal power, 84
hydropower, 84
solar PV, 83
STE, 83
tidal power, 84–85
wave power, 84
wind power, 83
Renewable generation
centralized, 81
DER, 81–82
description, 81, 113
dispatchability and control, 119
distributed generation (DG), 82
equipment loading, maintenance, and life cycle, 116
frequency control, 119
intentional and unintentional islanding, 117
intermittency, 114–115
levels, intermittent resources, 86
penetration, intermittent resources, 113–114
planning and operational solutions, 86
power quality, 119
primary and secondary lines, 115
protection systems, 116–117
regulatory and market forces, 79–81
short-circuit current levels, 115
substation, 115
system losses and reactive power flow, 116
technologies, *see* Renewable energy resources
voltage regulation and control, 117–119
Renewable portfolio standards (RPS), 30, 79
Restoration switching analysis (RSA) application, 216–217, 259, 263
Restructured Accelerated Power Development and Reform Programme (R-APDRP), 540

Index

RF mesh, *see* Radio-frequency (RF) mesh
RPC, *see* Relay protection coordination (RPC)
RPS, *see* Renewable portfolio standards (RPS)
RSA application, *see* Restoration switching analysis (RSA) application
RTU, *see* Remote terminal unit (RTU)

S

Salt River Project, United States, 543
San Diego Gas and Electric microgrid project, 108
SCA, *see* Short-circuit analysis (SCA)
SCADA, *see* Supervisory control and data acquisition (SCADA)
Scale
    Australia, 534
    Canada, 537
    China, 539
    Europe, 539–540
    global collaborative efforts, 546
    India, 540
    Japan, 541
    Korea, 541
    Latin America, 542
    Malaysia, 545
    Singapore, 545
    Taiwan, 545
    United States, 544
SCSM, *see* Specific communication service mapping (SCSM)
SDOs, *see* Standards development organizations (SDOs)
SEA, *see* Smart Energy Alliance (SEA)
Security functions
    auditing, 450–451
    authentication, 449–450
    authorization, 450
    key management, 451
    layered security model, 448–449
    message integrity, 451–452
    network integrity, 452
    system integrity, 452
Security threats
    people
        accidental impact, 453–454
        intentional impact, 453
        prevention, 454
    process, 454
    technology
        communications, 456
        denial of service (DoS), 456–457
        hardware, 454
        information, 456
        input validation, 455
        message interception, 457
        message tampering, 457
        operational maintenance, 455–456
        session management, 455
        software, 455
Sensors, 127–128
SEP, *see* Smart Energy Profile (SEP)
SGDP, *see* Smart Grid Demonstration Projects (SGDP)
SGFAC, *see* Smart Grid Federal Advisory Committee (SGFAC)
SGIC, *see* Smart Grid Information Clearinghouse (SGIC)
SGIGs, *see* Smart Grid Investment Grants (SGIGs)
SGIP, *see* Smart Grid Interoperability Panel (SGIP)
Short-circuit analysis (SCA), 217–219
Smart Energy Alliance (SEA), 47
Smart Energy Profile (SEP), 364
Smart energy resources
    electric distribution system, 112
    energy storage, *see* Energy storage
    EVs, *see* Electric vehicles (EVs)
    integration standards, 112–113
    large-scale DERs, 112
    microgrids, *see* Microgrids
    renewable generation, *see* Renewable generation
Smart grid
    business opportunities, 550
    carbon emissions and fossil fuels, 559
    constituents, 17
    definitions, 18
    "dumb" grid, 17
    efforts, 16–17
    electrical and communications infrastructures, 17–18
    expenditure, 550, 551
    and information technology (IT) expenditure, 550, 552
    "intelligent grid", 16
    market drivers and enablers, *see* Electricity market
    Newton-Evans Research Company, Inc., 549–550, 558
    programs, 549
    utility transformation, 559
Smart Grid Demonstration Projects (SGDP), 34–35
Smart grid drivers
    characteristics, 19–20
    customer empowerment, 19
    description, 18–19
    economic competitiveness, 19
    energy reliability and security, 19
    environmental sustainability, 19
    indicator, industry momentum, 20
    policy and legislative, 19
Smart Grid Federal Advisory Committee (SGFAC), 39
Smart grid industry initiatives
    Centre for Sustainable Electricity and Distributed Generation, 50
    ENSG, 49
    EPRI, *see* Electric Power Research Institute (EPRI)
    European Renewable Energy Directive, 49
    GridWise Alliance, 47
    GWAC, 47
    IEC, *see* International Electrotechnical Commission (IEC)
    LCNF, Ofgem, 49
    SEA, 47
    SGIC, 50
    U.K. Low Carbon Transition Plan, 49
Smart Grid Information Clearinghouse (SGIC), 50
Smart Grid Interoperability Panel (SGIP)
    definition, 37
    membership, 38–39
    structure, 37–38
Smart Grid Investment Grants (SGIGs), 34–35
Smart grid investments
    collaborative effort, 547
    effort and financial resources, 546–547
    federal stimulus investments, 533–535
    pilot and demonstration projects, 532–533
    utility-specific objectives and requirements, 533

Smart grid market
    and IT expenditure, 53–55
    market drivers
        CAPEX budgeting, 51–52
        economic forecasting services, 52
        electricity consumption, 50, 51
        electric power usage, 51, 52
        primary electricity end-user segments, 50, 51
    products, 50
    T&D services, 52–53
Smart grid security requirements
    authentication and authorization, 365
    cryptographic key management, 365–366
    firmware management, 366
    interoperability standards, 366
    logging and auditing, 365
    password and system account management, 365
    system, information integrity and confidentiality, 365
    time synchronization, 365
    user, resource and role management, 365
Smart Grid/Smart City Project, Australia, 534
Smart grid technology framework
    AMI, 75
    areas, 75
    components, 74
    description, 73–74
    functionalities and capabilities, 76–79
    interoperability, 74
    selection and rollout, factors, 75–76
    stakeholder-driven definition, 73
Smart sensors
    centralized control room level, 299–300
    digital data, 297
    embedded intelligence and communications, 297–298
    functions, 298
    local level, 298, 299
    station/feeder level
        analysis and interpretation, 298, 299
        mesh topology, 299, 300
        peer-to-peer communications, 299
Smart substations
    centralized functions and decisions, 152
    challenges, 152–153
    communications capabilities, 131–132
    data flow, 133, 134
    description, 126
    engineering and design, 153–154
    enterprise integration, 155
    functions, 126–127
    IEC 61850, *see* IEC 61850
    information infrastructure, 154
    integration architecture, 134, 136
    migration path, 134, 135
    network-based architecture, 132–133
    operational and nonoperational data, 133–134
    operations and maintenance, 154–155
    protection, monitoring, and control devices, *see* Intelligent electronic devices (IEDs)
    real-time data, 127
    SCADA, 128–131
    sensors, 127–128
    smart feeder applications, 134–135
    smart meters, 153
    testing and commissioning, 155

SMES, *see* Superconducting magnetic energy storage (SMES)
Solar energy, 31
Solar photovoltaic (PV) generation, 83
Solar thermal energy (STE), 83
SOM, *see* Switch order management (SOM)
Southern California Edison Co., 543
Specific communication service mapping (SCSM), 140
SSSC, *see* Static synchronous series compensator (SSSC)
Standardization work
    advancing smart grid standards, 480–483
    alliances, 465
    AMI communications technologies, 470
    applications, 468
    best practices, 479
    CAFE, 473
    communications domain decomposition, 468
    consumer, 471
    distribution, 470
    enterprise integration, 471
    gap analysis, 469
    generation, 469
    interoperability tests, 472
    issues
        CIM, 477
        under development, 473
        enterprise application integration, 474–475
        HAN standards, 476–477
        high initial investment, 478
        holistic security, 479
        interoperability weak spots, 474
        legacy transmission and distribution automation, 477
        market power, smaller utilities, 474
        merging organizations, 478–479
        utility and premise, 477
        wireless mesh and BPL, 475–476
    legislation and regulations, 480
    NERC CIP standards, 471–472
    SDO, *see* Standards development organizations (SDOs)
    smart grid standards assessment, 465–467
    successful standardization efforts, 472
    and technology
        agreements, requirements and specifications, 462
        alliances, 462
        anticompetitive behavior, 463
        continuum, 461
        de facto and *de jure* "standards", 463
        elements, continuum, 463, 464
        home area network (HAN) market space, 462
        IEC, 462
        Microsoft Windows SDK, 462
        UCAIug, 462, 463
        utility industry, North America, 463
    transmission, 469
    user groups, 465–466
Standards development, coordination and acceleration, 514–517
Standards development organizations (SDOs)
    anticompetitive behavior, 463
    balloting and candidate voters, 463–464
    NIST, 464
    smart grid standards, 464–465
    utility industry, North America, 463

Index

STATCOM, *see* Static synchronous compensator (STATCOM)
Static synchronous compensator (STATCOM)
    advantages, 180
    BESS application, 89, 90
    circuit diagram, 177
    description, 177
    energy storage, 178
    FACTS systems, 171
    functions, 177–178
    UPFC configuration, 178–179
Static synchronous series compensator (SSSC), 178, 179
Static VAr compensator (SVC)
    benefits, 176
    description, 175
    installation, 176, 177
    and STATCOM technology, 177–178
    technology, 175–176
    thyristor control, 175
STE, *see* Solar thermal energy (STE)
Superconducting magnetic energy storage (SMES), 93–94
Super-PDC, *see* Super-phasor data concentrator (Super-PDC)
Super-phasor data concentrator (Super-PDC)
    architecture, 196, 197
    description, 196
    generation II Super-PDC system, 198
    pilot projects, 198
    TVA, 196, 198
Supervisory control and data acquisition (SCADA)
    automatic switching, 259
    communications, 270
    components, 212–213
    and control systems
        faults, 215
        IEC 61850, 215
        lockout alarm, 215
        monitoring and measuring capability, 214
    description, 128
    DMS and OMS, 402–403
    EAM, 408
    equipment types, 213–214
    infrastructure, 401–402
    load-break switch, 248–249
    manual switching, 258–259
    master stations, *see* Master stations, SCADA systems
    model-based VVO, 243, 244
    operational/real-time system, 399–400
    OT applications, 399
    and protection, 478
    protocol architecture, 132
    protocols, 284
    RSA application, 263
    RTU, 130–131
    sensors, 332
    server-based SCADA architecture, 133
    transmission and distribution, 214
SVC, *see* Static VAr compensator (SVC)
Switch order management (SOM), 217, 218
Synchronous condenser, 180
System integrity
    configuration management, 452
    malware protection, 452
    testing and validation, 452

T

TCSC, *see* Thyristor-controlled series compensation (TCSC)
T&D, *see* Transmission and distribution (T&D)
Technical innovation, electricity market
    demand–response systems, 553
    devices, operational and non-operational, 552
    electric vehicles (EV) chargers, 554
    grid scale energy storage, 553
    HEMS, 553–554
    integrated smart grid solution, 553
    operational challenges, 552–553
    utility telecommunications systems, 553
Technologies and enterprise level integration
    AMI and OMS synergies
        communications infrastructure and meter, 405
        data, 405
        meters and communications, 404
        restoration notifications, 405
        voltage violation alarms, 405
    data *vs.* application, 410, 411
    distribution operation applications
        architecture, 401
        business intelligence, distribution organizations, 402
        electric utilities, 400–401
        OMS, *see* Outage management systems (OMS)
        operator graphical user interface, 402
        SCADA, 401–402
        single dynamic distribution network model, 402
    EIM, *see* Enterprise information management (EIM)
    enterprise service bus, 409–410
    multiple smart grid functions, outage management, 405–406
    OT and IT systems, 420–425
    service-oriented architecture, 411
    synergies, 398–400
    workforce, asset and network management systems, 406–409
Technology drivers
    characteristics, 70–73
    framework, 73–79
    transformation, 67–70
Technology investment and innovation, 511–512
Technology roadmap
    age of paper, 305–306
    description, 304–305
    developing world, 309
    emergence, digital maps, 306–307
    enterprise
        development, 309
        extending maps, field, 309
        interoperability, 307–308
        mobile applications, 309
    GIS, 307
Tenaga Nasional Berhad, Malaysia, 545
Tennessee Valley Authority (TVA), 196, 198
Thermal energy storage, 95–96
Thyristor-controlled series compensation (TCSC), 175, 176
Tidal power, 84–85
Traditional power engineering skills, 524
Transformation, smart grid
    decentralized information technology, 69–70

description, 69
distribution network, 69
generation facilities, 68
power exchanges and trading, 68
requirements, 67–68
substations, 68–69
Transmission and distribution (T&D) infrastructure, 383
Transmission systems
availability, electric power, 210–211
current flow, 169
description, 155–156
devices, 169
EMS, *see* Energy management systems (EMS)
environmental constraints, 169–170
FACTSs, *see* Flexible AC transmission systems (FACTSs)
HVDC, *see* High-voltage direct current (HVDC)
hybrid system interconnections, 211
large synchronous power grids, 210
power system developments, 170–171
prospects, 211, 212
renewable energy generation, 210
WAMPAC, *see* Wide area monitoring, protection and control (WAMPAC)
TVA, *see* Tennessee Valley Authority (TVA)

U

UHV/UHVDC-based grid, Chinese smart grid, 538
U.K. Low Carbon Transition Plan, 49
Ultracapacitors, 95
Unified power flow controller (UPFC), 178–179
Universities/Research & Development and Demonstration Projects, 199
UPFC, *see* Unified power flow controller (UPFC)
U.S. Department of Energy (U.S. DOE)
ARRA funding, 34
definitions, 18, 34
EAC, 35–36
EISA, 32
elements, smart grid characteristics, 72–73
EVs, 27–28
goals, NETL, 70
SGDP and SGIGs, 34–35
Smart Grid Task Force, 35
U.S. DOE, *see* U.S. Department of Energy (U.S. DOE)
U.S. electric utility industry
challenges, 28–29
coal, 29
Consumer Demand Management, 31
demands, 31
energy storage, 31
natural gas, 29
nuclear, 29–30
oil, 30
regulatory challenges, 32
renewable generation, 30–31
technical challenges, 32
transmission expansion, 31
Utility asset management
asset condition monitoring
electrical losses, 334
investment planning, 335
load factors, 332–334
outage frequency and duration, 334
percentage loading, 332
CCVT, 344
circuit breaker monitoring, 344
description, 332
fault location, 344–345
just-in-time maintenance, 345
through-fault monitoring, 344
TLM system, 343
transformer monitoring systems, 343
workforce management, *see* Workforce management
Utility Communications Architecture International Users Group (UCAIug), 462, 465–466
Utility organizational and business process transformation, 500–502
Utility regulatory systems
investor-owned utilities, 12
natural monopoly, 11–12
pricing controls, 12
profitability, 12–13

V

Validation, estimation and editing (VEE), 357
Variable frequency transformer (VFT), 179–180
VEE, *see* Validation, estimation and editing (VEE)
Vehicle-to-grid (V2G)
aggregative architecture, 100, 101
community energy storage (CES), 91
and G2V, 99
Vendor partnerships, 514
VFT, *see* Variable frequency transformer (VFT)
V2G, *see* Vehicle-to-grid (V2G)
Voltage-sourced converter (VSC)
IPFC, 179
STATCOM technology, 177–178
VSC-HVDC, 183, 186–188
volt/VAr control (VVC)
CVR, IVVC and VVO, 221
drivers, objectives and benefits
CO_2 emissions, 226
load reduction, CVR, 225, 226
losses reduction, 224
purchase energy, 224
security and reliability, 225
voltage regulators and capacitor banks, 225–226
ways, electricity generation, 224–225
equipment, *see* VVC equipment
implementation, *see* volt/VAr control implementation
inefficiency, power delivery system, 221–222
load, voltage effect, 223–225
VAr control, VAr compensation and power factor correction, 221
voltage fluctuations, 222–223
VVO, 239–245
volt/VAr control implementation
centralized approach, 237–238
decentralized approach, 236–237
hybrid approach, 239, 240
local intelligence approach, 235–236
local intelligence, centralized approach, 238–239

Index

VAr control, 234–235
voltage control, 234
volt/VAr optimization (VVO)
   centralized approach, 242
   decentralized approach, 241–242
   description, 221
   model-based approach
      control variables, 243
      description, 242–243
      development, 244
      optimum settings, 243
      power engineering calculations and analysis, 243–244
   model hierarchy, 241
   objectives, 239–241
   optimization *vs.* Control, 239
   and smart grid, 245
VSC, *see* Voltage-sourced converter (VSC)
VVC, *see* volt/VAr control (VVC)
VVC equipment
   distribution feeders
      description, 231
      pole-top capacitor banks, 232–234
      single-phase line regulators, 232, 233
   substations
      bus regulation, 228, 229
      capacitor banks, 230–231
      power transformers, 226–228
      single-phase voltage regulators, 228–230
VVO, *see* volt/VAr optimization (VVO)

W

WAC, *see* Wide area control (WAC)
WAMPAC, *see* Wide area monitoring, protection and control (WAMPAC)
WAMS, *see* Wide area monitoring systems (WAMS)
WAN, *see* Wide area network (WAN)
WAP, *see* Wide area protection (WAP)
Wavelength-division multiplexing (WDM), 289
Wave power, 84
WDM, *see* Wavelength-division multiplexing (WDM)
WECC, *see* Western Electricity Coordinating Council (WECC)
WEF, *see* World Economic Forum (WEF)
Western Electricity Coordinating Council (WECC), 430, 433, 434, 439–440, 441, 442, 443
Western energy crisis of 2000–2001
   deregulation, electric market, 6
   factors, 8
   industry regulation, 8–9
   prices, spot market, 6, 8
   stranded assets, 6
Wide area control (WAC), 206–207
Wide area monitoring, protection and control (WAMPAC)
   Brazil, 200–203
   characteristics, 192
   communications requirements, 270–271
   drivers and benefits, 191–192
   Europe
      industrial applications, 199–200
      Universities/R&D and Demonstration Projects, 199
      WAMS, 199–200
   interoperability, 203–204
   levels, renewable energy, 192
   management, intermittent generation, 193, 209
   PDC, 190–191
   PMUs, 189
   power quality, 193–194, 209
   reliability, stability, and security, 193
   synchronized sampling, power system signals, 188
   time synchronization, 189–190
   transmission efficiency, 194, 210
   United States
      NASPI architecture, 194, 195
      phasor gateway and functions, 194–196
      Super-PDC, 196–198
   WAC, 206–207
   WAM, *see* Wide area monitoring systems (WAMS)
   WAP, 206
   wide area stability, 207–209
Wide area monitoring systems (WAMS)
   deployment plan, Brazil, 202–203
   Europe, 198–200
   measurement-based techniques, 204–205
   phase angle separation information, 205
   steady-state network analysis applications, 206
   Super-PDC, 191
   and WACS applications, 206
Wide area network (WAN), 466–468, 477
Wide area protection (WAP), 206
Wide area stability
   electromechanical oscillations, 207
   MANGO, 208–209
   ModeMeter, 208
   undamped oscillations, 207, 208
Wind energy, 30
Wind power, 83
Workforce management
   aging infrastructure and aging workforce call
      abilities and benefits, integrated approach, 339
      description, 337–338
      hyperlinks and attachments, 339
      inspections and maintenance procedures, 338
      mobile data terminal (MDT), 339
      silo approach, 338–339
      tool and capability, 339–340
   description, 335
   dispatch, 340
   end-to-end and top-to-bottom benefits, 341–343
   forecasting, 340
   mobile and reporting, 341
   MWFM, 335–337
   scheduling, 340
World Economic Forum (WEF), 41, 43

X

XML firewalls, 455